COMPLEX VARIABLES WITH APPLICATIONS

SECOND EDITION

A. DAVID WUNSCH

University of Massachusetts Lowell

ADDISON-WESLEY PUBLISHING COMPANY

Reading, Massachusetts • Menlo Park, California • New York
Don Mills, Ontario • Wokingham, England • Amsterdam • Bonn
Sydney • Singapore • Tokyo • Madrid • San Juan • Milan • Paris

Sponsoring Editor: Laurie Rosatone
Production Supervisor: Peggy McMahon
Manufacturing Manager: Roy Logan
Cover Design: Meredith Lightbown
Composition Services: P&R Typesetters, Ltd
Technical Art Consultant: Loretta Bailey

Library of Congress Cataloging-in-Publication Data

Wunsch, A. David.
Complex variables with applications / A. David Wunsch.—2nd ed.
p. cm.
Includes bibliographical references and index.
ISBN 0-201-12299-5
1. Functions of complex variables. I. Title.
QA331.7.W86 1994 93-22875
515′.9—dc20 CIP

3 4 5 6 7 8 9 10 MA 969594

To my family and good friends
who encouraged me with this work

Introduction

I hope that this second edition of my book preserves the spirit of the first. My intention has been to create an inviting applied complex variables text for engineering and science students who have had no mathematics beyond the usual obligatory courses in elementary calculus and differential equations. As a teacher I am only too aware of how imperfectly these subjects are learned and remembered. Thus the reader will find turning up here review topics in such troublesome areas of real calculus as infinite series, line integration, limits, and continuity.

As in the first edition, I have included numerous worked examples. I harbor the naive hope that I have presented the subject matter clearly and simply enough so that a student can learn most of the material directly from the book without recourse to the classroom. I say naive since this doubtless is the intention of most textbook authors—and yet one does not see masses of teachers made redundant by lucid writing. Since, realistically, this text will generally be used in conjunction with a college course, I have broken each chapter into sections each of which can be covered, more or less, in a 50-minute lecture hour. The previous edition did sometimes put a strain on those 50 minutes, and so in certain places I have now distributed material over two sections where one had been used.

I imagine my reader as someone who will be using complex variable theory in science or engineering or a teacher of such people. Thus I have presented applications of the mathematics early in the text—the practical uses of the material occur as a leitmotif from Chapter 2 on to the very end. As in the first edition, I have shown some slight favoritism toward my own field—electrical engineering—with the rationale that electrical engineers are probably the biggest users of complex variables. To strengthen and give variety to the book I have added to the applications included in the first edition. The **z** transform, which is much used in the field of signal processing, now makes an appearance. The chapter on Laplace transforms has been bolstered to include generalized functions (delta functions, etc.). Line sources of heat, fluid, and electric flux now appear in Chapter 8. I have included a section on fractals in Chapter 6. The topic is new and fashionable and appears naturally in any discussion of limits of sequences of complex numbers. Most students are intrigued by the accompanying graphical patterns.

For the mathematically curious, I have added some more sophisticated topics to the first edition. There is a new section on integration around infinity. Its usefulness in evaluating real integrals is demonstrated. This section (6.9) is perhaps the most difficult in the book and can be skipped by those who find it daunting. The proof of the Fundamental Theorem of Algebra now occurs early in the text. Analytic continuation and a theorem on the zeros of analytic functions now appear in the chapter on series. The use of complex variables to sum numerical series is also investigated as is a method for summing Fourier series. The quantity of harder exercise problems has been increased, but the amount of easy problems has not been reduced. Altogether there are significantly more problems in this book than in its progenitor, and their quantity is sufficient to allow a teacher to assign different sets from year to year.

Students with desk top microcomputers generally enjoy using them in a course, particularly when they can generate attractive graphics. Thus I have called attention at various places to available inexpensive computer software that can serve as an adjunct to the text. Indeed some of the figures in the book have been taken from the screen of the author's own Macintosh computer. A particularly handy piece of software called **f(z)** that can be used for checking homework assignments or mainly for just having fun is mentioned in Chapters 7 and 8 in connection with mapping.

As in the first edition my notation remains a little idiosyncratic. I prefer to use i for the imaginary operator rather than the electrical engineer's j. In doing so I am in conformity with most complex variables books, and I can acknowledge our debt to Leonhard Euler who invented the i symbol over 200 years ago. However, I could not resist employing in places the engineer's $\angle$ to indicate the argument of a complex number. This widely used operator has the advantage of suggesting an angle, as it should. I also use the symbol cis, since it is more traditional and has the same meaning. Students should become used to both notations. I employ the old fashioned term log to mean a natural logarithm; there doesn't seem much to be gained by switching to the newer ln.

Although the book is designed to be used in a one semester course of three credit hours there is more material here than can be covered in this time. I suggest that instructors try to teach most of the first five chapters and then pick and choose parts of the remaining three according to their interests. A manual containing detailed solutions to all of the exercise problems is available to bona fide college faculty from the publisher. I welcome hearing from anyone who finds mistakes in this book or in the accompanying manual. My address is given below.

ACKNOWLEDGMENTS

Mr. Michael F. Brown has been an invaluable assistant to me in preparing this second edition. He has read the manuscript carefully and offered many suggestions for its improvement. In addition, he has worked out all the exercises in the book and in a number of cases caught errors in the draft of the solutions manual. He is a careful reader and a fine mathematician.

Professor Francesco Bacchialoni has taught complex variables with me at the University of Massachusetts Lowell for many years. He has offered a number of useful criticisms of the first edition of my book—some of which I have incorporated into this new volume. Professor Stephen Spurk of Lowell has been helpful in spotting errors in the first edition. A number of my students have also assisted me in this way—I hope that they will forgive me for not naming them all here. Among them are Sophie Ting and her husband Shen Ting, who have been of special help. My colleague Professor Roger Baumann proposed my writing the first edition. I still feel indebted to him.

I began writing the second edition while on sabbatical in the Division of Applied Sciences at Harvard University. I would like to thank Prof. Ronold W. P. King of that institution for arranging my stay there and providing me with the congenial intellectual environment for which he is well known.

My editor at Addison-Wesley, Mr. Michael Payne has given me wise counsel and held me to a reasonable writing schedule. I thank him as well as his able assistant Ms. Laurie Rosatone. Addison-Wesley has provided me with a number of consultants to whom I am grateful for reading my manuscript and offering good suggestions. They are the following: Professors Newman H. Fisher of San Francisco State University, Mark Elmer of SUNY—Oswego, Edward Kolesar of the Air Force Institute of Technology, and Peter Colwell of Iowa State University. The word processing for this project has been ably performed by Stuart Stephens and Jeff Knight.

Peggy McMahon of Addison-Wesley has supervised the production of this book. She has held me to a pace that has been both brisk and humane. I have enjoyed working with her. My skillful copy editor Andrew Schwartz has impressed me with both his knowledge of English and mathematics.

A. David Wunsch
Department of Electrical Engineering
University of Massachusetts Lowell
Lowell, Massachusetts 01854

August 1993

Contents

CHAPTER 1
Complex Numbers

Look if you like, but you will have to leap.
— W.H. Auden

1.1 INTRODUCTION

In order to prepare ourselves for a discussion of complex numbers, complex variables, and, ultimately, functions of a complex variable, let us review a little of the previous mathematical education of a hypothetical reader.

A young child learns early about those whole numbers that we with more sophistication call the positive integers. Zero, another integer, is also a concept that the child soon grasps.

Adding and multiplying two integers, the result of which is always a positive integer or zero, is learned in elementary school. Subtraction is studied, but the problems are carefully chosen; 5 minus 2 might, for instance, be asked but not 2 minus 5. The answers are always positive integers or zero.

Perhaps several years later this student is asked to ponder 2 minus 5 and similar questions. Negative integers, a seemingly logical extension of the system containing the positive integers and zero, are now required. Nevertheless, to avoid some inconsistencies one rule must be accepted that does not appeal directly to intuition, namely, $(-1)(-1) = 1$. The reader has probably forgotten how artificial this equation at first seems.

With the set of integers (the positive and negative whole numbers and zero) any feat of addition, subtraction, or multiplication can be performed by the student, and the answer will still be an integer. Some simple algebraic equations such as $m + x = n$ (m and n are any integers) can be solved for x, and the answer will be an integer. However, other algebraic equations—the solutions of which involve division—present difficulties. Given the equation $mx = n$, the student sometimes obtains an integer x as a solution. Otherwise he or she must employ a kind of number called a fraction, which is specified by writing a pair of integers in a particular order; the fraction n/m is the solution of the equation just given if $m \neq 0$.

The collection of all the numbers that can be written as n/m, where n and m are any integers (excluding $m = 0$), is called the rational number system since it is based on the *ratio* of whole numbers. The rationals include both fractions and whole numbers. Knowing this more sophisticated system, our hypothetical student can solve any linear algebraic equation. The result is a rational number.

Later, perhaps in our student's early teens, irrational numbers are learned. They come from two sources: algebraic equations with exponents, the quadratic $x^2 = 2$, for example; and, geometry, the ratio of the circumference to the diameter of a circle, π, for example.

For $x^2 = 2$ the unknown x is neither a whole number nor a fraction. The student learns that x can be written as a decimal expression 1.41421356... requiring an infinite number of places for its complete specification. The digits do not display a repetitive pattern. The number π also requires an infinite number of nonrepeating digits when written as a decimal.†

Thus for the third time the student's repertoire of numbers must be expanded. The rationals are now supplemented by the irrationals, namely, all the numbers that must be represented by infinite nonrepeating decimals. The totality of these two kinds of numbers is known as the *real number system.*

The difficulties have not ended, however. Our student, given the equation $x^2 = 2$, obtains the solution $x = \pm 1.414\ldots$, but given $x^2 = -2$ or $x^2 = -1$, he or she faces a new complication since no real number times itself will yield a negative real number. To cope with this dilemma, a larger system of numbers—the *complex system*—is usually presented in high school. This system will yield solutions not only to equations like $x^2 = -1$ but also to complicated polynomial equations of the form

$$a_n z^n + a_{n-1} z^{n-1} + \cdots + a_0 = 0,$$

where $a_0, a_1, \ldots, a_n$ are complex numbers, n is a positive integer, and z is an unknown.

The following discussion, presented partly for the sake of completeness, should overlap much of what the reader probably already knows about complex numbers.

A complex number, let us call it z, is a number that is written in the form

$$z = a + ib \quad \text{or, equivalently,} \quad z = a + bi.$$

The letters a and b represent real numbers, and the significance of i will soon become clear.‡

---

† For a proof that $\sqrt{2}$ is irrational see Exercise 10 at the end of this section. A number such as 4.32432432... in which the digits repeat is rational. However, a number like .101001000100001... , which displays a pattern, but where the digits do not repeat, is irrational. For further discussion see C. B. Boyer, *A History of Mathematics* (Princeton: Princeton University Press, 1985), Chapter 25.

‡ Most electrical engineering tests use j instead of i, since i is reserved to mean current. However, mathematics books invariably use i.

We say that a is the real part of z and that b is the imaginary part. This is frequently written as

$$a = \text{Re}(z), \qquad b = \text{Im}(z).$$

Note that *both* the real part *and* the imaginary part of the complex number are real numbers. The complex number $-2 + 3i$ has a real part -2 and an imaginary part of 3.

> Two complex numbers are said to be equal if, and only if, the real part of one equals the real part of the other and the imaginary part of one equals the imaginary part of the other.

That is, if

$$z = a + ib, \qquad w = c + id, \tag{1.1–1}$$

and

$$z = w,$$

then

$$a = c, \qquad b = d.$$

We do not establish a hierarchy of size for complex numbers; if we did, the familiar inequalities used with real numbers would not apply. Using real numbers we can say, for example, that $5 > 3$, but it makes no sense to assert that either $(1 + i) > (2 + 3i)$ or $(2 + 3i) > (1 + i)$. An inequality like $a > b$ will always imply that both a and b are real numbers.

> The words *positive* and *negative* are never applied to complex numbers, and the use of these words implies that a real number is under discussion.

We add and subtract the two complex numbers in Eq. (1.1–1) as follows:

$$z + w = (a + ib) + (c + id) = (a + c) + i(b + d), \tag{1.1–2}$$

$$z - w = (a + ib) - (c + id) = (a - c) + i(b - d). \tag{1.1–3}$$

Their product is defined by

$$zw = (a + ib)(c + id) = (ac - bd) + i(ad + bc). \tag{1.1–4}$$

The results in Eqs. (1.1–2) through (1.1–4) are obtainable through the use of the ordinary rules of algebra and one additional crucial fact: When doing the multiplication $(a + ib)(c + id)$ we must take

$$i \cdot i = i^2 = -1. \tag{1.1–5}$$

Real numbers obey the commutative, associative, and distributive laws. We readily find, with the use of the definitions shown in Eqs. (1.1–2) and (1.1–4), that complex numbers do also. Thus if w, z, and q are three complex numbers, we have

commutative law:

$$\begin{aligned} w + z &= z + w \quad &&\text{(for addition)},\\ wz &= zw \quad &&\text{(for multiplication)}, \end{aligned} \tag{1.1–6}$$

associative law:

$$w + (z + q) = (w + z) + q \qquad \text{(for addition)},$$
$$w(zq) = (wz)q \qquad \text{(for multiplication)}, \tag{1.1–7}$$

distributive law:

$$w(z + q) = wz + wq. \tag{1.1–8}$$

Now, consider two complex numbers, z and w, whose imaginary parts are zero. Let $z = a + i0$ and $w = c + i0$. The sum of these numbers is

$$z + w = (a + c) + i0,$$

and for their product we find

$$(a + i0)(c + i0) = ac + i0.$$

These results show that those complex numbers whose imaginary parts are zero behave mathematically like real numbers. We can think of the complex number $a + i0$ as the real number a in different notation. The complex number system therefore *contains* the real number system.

We speak of complex numbers of the form $a + i0$ as "purely real" and, for historical reasons, those of the form $0 + ib$ as "purely imaginary." The term containing the zero is usually deleted in each case so that $0 + i$ is, for example, written i.

A multiplication (or addition) involving a real number and a complex number is treated, by definition, as if the real number were complex but with zero imaginary part. For example, if k is real,

$$(k)(a + ib) = (k + i0)(a + ib) = ka + ikb. \tag{1.1–9}$$

The complex number system has quantities equivalent to the zero and unity of the real number system. The expression $0 + i0$ plays the role of zero since it leaves unchanged any complex number to which it is added. Similarly, $1 + i0$ functions as unity since a number multiplied by it is unchanged. Thus $(a + ib)(1 + i0) = a + ib$.

Expressions such as $z^2, z^3, \ldots,$ imply successive self-multiplication by z and can be calculated algebraically with the help of Eq. (1.1–5). Thus to cite some examples:

$$i^3 = i^2 \cdot i = -i, \qquad i^4 = i^3 \cdot i = -i \cdot i = 1, \qquad i^5 = i^4 \cdot i = i,$$
$$(1 + i)^3 = (1 + i)^2(1 + i) = (1 + 2i - 1)(1 + i) = 2i(1 + i) = -2 + 2i.$$

We still have not explained why the somewhat cumbersome complex numbers can yield solutions to problems unsolvable with the real numbers. Consider, however, the quadratic equation $z^2 + 1 = 0$, or $z^2 = -1$. As mentioned earlier, no real number provides a solution. Let us rewrite the problem in complex notation:

$$z^2 = -1 + i0. \tag{1.1–10}$$

We know that $(0 + i)^2 = i^2 = -1 + i0$. Thus $z = 0 + i$ (or $z = i$) is a solution of Eq. (1.1–10). Similarly, one verifies that $z = 0 - i$ (or $z = -i$) is also. We can say that $z^2 = -1$ has solutions $\pm i$. Thus we assert that in the complex system -1 has two square roots: i and $-i$, and that i is *one* of these square roots.

In the case of the equation $z^2 = -N$, where N is a nonnegative real number, we can proceed in a similar fashion and find that $z = \pm i\sqrt{N}$.† Hence, the complex system is capable of yielding two square roots for any negative real number. Both roots are purely imaginary.

For the quadratic equation

$$az^2 + bz + c = 0 \qquad (a \neq 0) \quad \text{and} \quad a, b, c \text{ are real numbers,} \tag{1.1–11}$$

we are initially taught the solution

$$z = \frac{-b \pm \sqrt{b^2 - 4ac}}{2a} \tag{1.1–12}$$

provided that $b^2 \geq 4ac$. With our complex system this restriction is no longer necessary.

Using the method of "completing the square" in Eq. (1.1–11), and taking $i^2 = -1$, we have, when $b^2 \leq 4ac$,

$$z^2 + \frac{b}{a}z + \frac{c}{a} = 0,$$

$$\left(z + \frac{b}{2a}\right)^2 = \frac{b^2}{4a^2} - \frac{c}{a} = i^2\left(\frac{c}{a} - \frac{b^2}{4a^2}\right),$$

$$\left(z + \frac{b}{2a}\right) = \pm i\sqrt{\frac{c}{a} - \frac{b^2}{4a^2}} = \pm i\frac{\sqrt{4ac - b^2}}{2a},$$

and finally

$$z = \frac{-b \pm i\sqrt{(4ac - b^2)}}{2a}.$$

We will soon see that a, b, and c in Eq. (1.1–11) can themselves be complex, and we can still solve Eq. (1.1–11) in the complex system.

Having enlarged our number system so that we now use complex numbers, with real numbers treated as a special case, we will find that there is no algebraic equation whose solution requires an invention of any new numbers. In particular, we will show in Chapter 4 that the equation

$$a_n z^n + a_{n-1} z^{n-1} + \cdots + a_0 = 0$$

where a_n, a_{n-1}, etc., can be complex, z is an unknown, and $n > 0$ is an integer, has a solution in the complex number system. This is the Fundamental Theorem of Algebra.

The story presented earlier of a hypothetical student's growing mathematical sophistication in some ways parallels the actual expansion of the number system by

---

† The expression $\sqrt{N}$, where N is a positive real number, will mean the *positive* square root of N, and $\sqrt[n]{N}$ will mean the positive nth root of N.

mathematicians over the ages.[†] Complex numbers were "discovered" by people trying to solve certain algebraic equations. For example, in 1545, Girolamo Cardan (1501–1576), an Italian mathematician, attempted to find two numbers whose sum is 10 and whose product is 40. He concluded by writing $40 = (5 + \sqrt{-15})(5 - \sqrt{-15})$, a result he considered meaningless.

Later, the term "imaginary" was applied to expressions like $a + \sqrt{-b}$ (where a is real and b is a positive real) by Rene Descartes (1596–1650), the French philosopher and mathematician of the Age of Reason. This terminology, with its aura of the fictional, is perhaps unfortunate and is still used today in lieu of the word "complex." We shall often speak of the "imaginary part" of a complex number—a usage that harkens back to Descartes.

Although still uncomfortable with the concept of imaginary numbers, mathematicians had, by the end of the eighteenth century, made rather heavy use of them in both physical and abstract problems. The Swiss mathematician Leonhard Euler (1707–1783) invented in 1779 the i notation, which we still use today. By 1799 Karl Friedrich Gauss (1777–1855), a German mathematician, had used complex numbers in his proof of the Fundamental Theorem of Algebra. Finally, an Irishman, Sir William Rowan Hamilton (1805–1865) presented in 1835 the modern rigorous theory of complex numbers, which dispenses entirely with the symbols i and $\sqrt{-1}$. We will briefly look at this method in the next section.

EXERCISES

Consider the hierarchy of increasingly sophisticated number systems:

integers

rational numbers

real numbers

complex numbers

For each of the following equations what is the most elementary number system, of the four listed above, in which a solution for x is obtainable?

1. $4x + 2 = 0$

2. $x^2 + 2x + 1 = 0$

3. $x^2 + 2x = 0$

4. $x^2 + x + 2 = 0$

5. $x^2 - 2 = 0$

6. $x^2 + 2 = 0$

7. $4x^2 - 1 = 0$

[†] An excellent brief history of complex numbers can be found in the article "Thinking the Unthinkable: The Story of Complex Numbers (with a Moral)" by Israel Kleiner in *The Mathematics Teacher* 81:7 (October 1988): pp. 583–592.

8. An infinite decimal such as $e = 2.718281\ldots$ is an irrational number since there is no repetitive pattern in the successive digits. However, an infinite decimal such as $12.1212121\ldots$ is a rational number. Because the digits do repeat in a specific manner, we can write this number as the ratio of two integers, as the following steps will show.

First we rewrite the number as $12[1.01010101\ldots]$ or $12[1 + 10^{-2} + 10^{-4} + 10^{-6} + \cdots]$.

a) Recall from your knowledge of infinite geometric series that $1/(1 - r) = 1 + r + r^2 \ldots$, where r is a real number such that $-1 < r < 1$. Sum the series $[1 + 10^{-2} + 10^{-4} + 10^{-6} + \cdots]$.

b) Use the result of part (a) to show that $12.1212\ldots$ equals $1200/99$. Verify this answer by division on a pocket calculator.

c) Using the same technique, express $143.143\ldots$ as the ratio of integers.

9. Using the technique of Exercise 8, express $3.040404\ldots$ as the ratio of integers.

10. a) Show that if an integer is a perfect square and even, then its square root must be even.

b) Assume that $\sqrt{2}$ is a rational number. Then it must be expressible in the form $\sqrt{2} = m/n$, where m and n are integers and m/n is an irreducible fraction (m and n have no common factors). From this equation we have $m^2 = 2n^2$. Explain why this shows m is an even number.

c) Rearranging the last equation, we have $n^2 = m^2/2$. Why does this show n is even?

d) What contradiction has been caused by our assuming that $\sqrt{2}$ is rational? Although it is easy to show that $\sqrt{2}$ is irrational, it is not always so simple to prove that other numbers are irrational. For example, a proof that $2^{\sqrt{2}}$ is irrational was not given until the twentieth century. For a discussion of this subject, see R. Courant and H. Robbins, *What is Mathematics?* (Oxford, England: Oxford University Press, 1941), p. 107.

In the following exercises, perform the operations and express the results in the form $a + ib$.

11. $(3 + 4i) + (1 + 2i)$

12. $(3 + 4i)(1 - 2i)$

13. $(1 + 2i)(1 - 2i)(3 + 4i)$

14. $\text{Im}[(3 + 4i)(1 - 2i)]$

15. $(3 - 4i)(3 - 4i)(3 + 4i)(3 + 4i)$

16. $(f + ig)^n(f - ig)^n$, where $n \geq 0$ is an integer, and f and g are real

Let z be any complex number. In the following exercises, show that:

17. $\text{Re}(iz) = -\text{Im}\, z$

18. $\text{Im}(iz) = \text{Re}\, z$

19. $\text{Re}(z^2) = (\text{Re}\, z)^2 - (\text{Im}\, z)^2$

20. $\text{Im}(z^2) = 2(\text{Re}\, z)(\text{Im}\, z)$

Which of the following statements are true in general for arbitrary complex numbers z_1 and z_2?

21. $\text{Re}(z_1 + z_2) = \text{Re}\, z_1 + \text{Re}\, z_2$

22. $\text{Re}(z_1 z_2) = \text{Re}\, z_1\, \text{Re}\, z_2$

23. $\mathrm{Re}(\beta z_1) = \beta \,\mathrm{Re}\, z_1$, where β is real

24. $\mathrm{Im}(z_1 - z_2) = -\mathrm{Im}(z_2 - z_1)$

25. $\mathrm{Im}[(z_1 - z_2)^2] = -\mathrm{Im}[(z_2 - z_1)^2]$

26. If $n \geq 0$ is any integer, what are the four possible values of i^n? Show that $i^{n+4} = i^n$.

27. Use the preceding result to find $i^{12,735}$ and $(1 + i)^{3074}$. Express your result as $a + ib$.

Hint: Find $(1 + i)^2$ first.

For the following equations, x and y are real numbers. Solve for x and y. Begin by equating real and imaginary parts, thus obtaining two real equations.

28. $i(x + iy) = x + 1 + i2y$

29. $x^2 - y^2 + i2xy = -ix + y$

30. $(x + iy)^2 = 0 + i$

31. $\sqrt{x^2 + y^2} = 1 - 2x + iy$

32. $\sin(e^x) + i \cos x = 1 + i \sin y$

1.2 MORE PROPERTIES OF COMPLEX NUMBERS

A pair of complex numbers are said to be *conjugates* of each other if they have identical real parts and imaginary parts that are identical except for being opposite in sign.

If $z = a + ib$, then the conjugate of z, written $\bar{z}$ or z^*, is $a - ib$. Thus $\overline{(-2 + i4)}$ is $-2 - i4$. Note that $\overline{(\bar{z})} = z$; if we take the conjugate of a complex number twice, the number emerges unaltered.

Other important identities for complex numbers $z = a + ib$ and $\bar{z} = a - ib$ are

$$z + \bar{z} = 2a + i0 = 2 \,\mathrm{Re}\, z = 2 \,\mathrm{Re}\, \bar{z}, \tag{1.2–1}$$

$$z - \bar{z} = 0 + 2ib = 2i \,\mathrm{Im}\, z. \tag{1.2–2}$$

Therefore the sum of a complex number and its conjugate is twice the real part of the original number. If from a complex number we subtract its conjugate, we obtain a quantity that has a real part of zero and an imaginary part twice that of the imaginary part of the original number.

The product of a complex number and its conjugate is a real number. Thus if $z = a + ib$ and $\bar{z} = a - ib$, we have

$$z\bar{z} = (a + ib)(a - ib) = a^2 + b^2 + i0 = a^2 + b^2. \tag{1.2–3}$$

This fact is particularly useful when we seek to derive the quotient of a pair of complex numbers. Suppose, for the three complex numbers α, z, and w,

$$\alpha z = w, \qquad z \neq 0, \tag{1.2–4}$$

where $z = a + ib$, $w = c + id$. Quite naturally, we call α the quotient of w and z, and write $\alpha = w/z$. To determine the value of α, we multiply both sides of Eq. (1.2–4) by

$\bar{z}$. We have

$$\alpha(z\bar{z}) = w\bar{z}. \tag{1.2–5}$$

Now $z\bar{z}$ is a real number. We can remove it from the left in Eq. (1.2–5) by multiplying the entire equation by another real number $1/(z\bar{z})$. Thus $\alpha = w\bar{z}/(z\bar{z})$, or

$$\frac{w}{z} = \frac{w\bar{z}}{z\bar{z}}. \tag{1.2–6}$$

This formula says that to compute $w/z = (c + id)/(a + ib)$ we should multiply the numerator and the denominator by the conjugate of the denominator, that is,

$$\frac{c + id}{a + ib} = \frac{(c + id)(a - ib)}{(a + ib)(a - ib)} = \frac{(ac + bd) + i(ad - bc)}{a^2 + b^2},$$

or

$$\frac{c + id}{a + ib} = \frac{ac + bd}{a^2 + b^2} + i\frac{(ad - bc)}{a^2 + b^2}. \tag{1.2–7}$$

Using Eq. (1.2–7) with $c = 1$ and $d = 0$, we can obtain a useful formula for the reciprocal of $a + ib$; that is,

$$\frac{1}{a + ib} = \frac{a}{a^2 + b^2} - \frac{ib}{a^2 + b^2}. \tag{1.2–8}$$

Note in particular that with $a = 0$, and $b = 1$, we find $1/i = -i$. This result is easily checked since we know that $1 = (-i)(i)$.

Since all the preceding expressions can be derived by application of the conventional rules of algebra, and the identity $i^2 = -1$, to complex numbers, it follows that other rules of ordinary algebra, such as the following, can be applied to complex numbers:

$$\frac{z_1}{z_2} = z_1\left(\frac{1}{z_2}\right), \qquad \frac{1}{z_1 z_2} = \left(\frac{1}{z_1}\right)\left(\frac{1}{z_2}\right), \qquad \frac{z_1 z_2}{z_3 z_4} = \left(\frac{z_1}{z_3}\right)\left(\frac{z_2}{z_4}\right). \tag{1.2–9}$$

There are a few other properties of the conjugate operation that we should know about.

The conjugate of the sum of two complex numbers is the sum of their conjugates. Thus if $z_1 = x_1 + iy_1$ and $z_2 = x_2 + iy_2$, then

$$\overline{(z_1 + z_2)} = (x_1 + x_2) - i(y_1 + y_2) = (x_1 - iy_1) + (x_2 - iy_2) = \bar{z}_1 + \bar{z}_2.$$

A similar statement applies to the difference of two complex numbers and also to products and quotients, as will be proved in the exercises. In summary:

$$\overline{(z_1 + z_2)} = \bar{z}_1 + \bar{z}_2, \tag{1.2–10a}$$

$$\overline{(z_1 - z_2)} = \bar{z}_1 - \bar{z}_2, \tag{1.2–10b}$$

$$\overline{z_1 z_2} = \bar{z}_1 \bar{z}_2, \tag{1.2–10c}$$

$$\overline{\left(\frac{z_1}{z_2}\right)} = \frac{\bar{z}_1}{\bar{z}_2}. \tag{1.2–10d}$$

Formulas such as these can sometimes save us some labor. For example, consider

$$\frac{1+i}{3-4i} + \frac{1-i}{3+4i} = x + iy.$$

There can be a good deal of work involved in finding x and y. Note, however, from Eq. (1.2–10d) that the second fraction is the conjugate of the first. Thus from Eq. (1.2–1) we see that $y = 0$, whereas $x = 2\,\text{Re}((1+i)/(3-4i))$. The real part of $(1+i)/(3-4i)$ is found from Eq. (1.2–7) to be $(3-4)/25 = -1/25$. Thus the required answer is $x = -(2/25)$ and $y = 0$.

Equations (1.2–10a–d) can be extended to more than two complex numbers, for example, $\overline{z_1 z_2 z_3} = \overline{z_1 z_2}\bar{z}_3 = \bar{z}_1\bar{z}_2\bar{z}_3$, or in general $\overline{z_1 z_2 \cdots z_n} = \bar{z}_1\bar{z}_2 \cdots \bar{z}_n$. Similarly $\overline{z_1 + z_2 + \cdots + z_n} = \bar{z}_1 + \bar{z}_2 + \cdots + \bar{z}_n$.

Euler's notation, $z = a + ib$, has never been entirely palatable to pure mathematicians. One reason is the presence of the plus sign. It is not clear whether this type of addition is the same as occurs in conventional algebra. Also, one might ask whether or not the implied multiplication between i and b is identical to ordinary multiplication.

A formulation of complex number theory, which dispenses with Euler's notation and the i operator, was presented in the mid-nineteenth century by the Irish mathematician William Rowan Hamilton. Any student who has done FORTRAN computer programming with complex numbers knows that neither the program nor the machine uses i. The approach used is identical to Hamilton's.

In his method a complex number is defined as a pair of real numbers expressed in a particular order. If this seems artificial, recall that a fraction is also expressed as a pair of numbers stated in a certain order. Hamilton's complex number z is written (a, b), where a and b are real numbers. The order is important (as it is for fractions), and such a number, in general, is not the same as (b, a). In the ordered pair (a, b), we call the first number, a, the real part of the complex number and the second, b, the imaginary part. This kind of expression is often called a *couple*. Two such complex numbers (a, b) and (c, d) are said to be equal: $(a, b) = (c, d)$ if and only if $a = c$ and $b = d$. The sum of these two complex numbers is defined by

$$(a, b) + (c, d) = (a + c, b + d), \tag{1.2–11}$$

and their product is computed as

$$(a, b)\cdot(c, d) = (ac - bd, ad + bc). \tag{1.2–12}$$

The product of a real number k and the complex number (a, b) is defined by $k(a, b) = (ka, kb)$.

Consider now all the couples whose second number in the pair is zero. Such couples handle mathematically like the ordinary real numbers. For instance, we have $(a, 0) + (c, 0) = (a + c, 0)$. Also, $(a, 0)\cdot(c, 0) = (ac, 0)$. Therefore we will say that the real numbers are really those couples, in Hamilton's notation, for which the second element is zero.

Another important identity involving ordered pairs, which is easily proved from Eq. (1.2–12), is

$$(0, 1)\cdot(0, 1) = (-1, 0). \tag{1.2–13}$$

It implies that $z^2 + 1 = 0$ when written with couples has a solution. Thus

$$z^2 + (1, 0) = (0, 0) \tag{1.2–14}$$

is satisfied by $z = (0, 1)$ since

$$(0, 1)(0, 1) + (1, 0) = (-1, 0) + (1, 0) = (0, 0).$$

The student should readily see the analogy between the $a + ib$ and the (a, b) notation. The former terminology will more often be used in these pages.

EXERCISES

Compute the numerical values of the following expressions. Give the answers in the form $a + ib$, where a and b are real numbers.

1. $\dfrac{1}{1-i}$

2. $\dfrac{2+i}{1-i}$

3. $\dfrac{2+i}{1-i} + \dfrac{2-i}{1+i} + i$

4. $\left[\dfrac{2+i}{1-i} + \dfrac{2-i}{1+i} + i\right]^2$

5. $\left[\dfrac{1+i}{1-i}\right]^3$

6. $\left[\dfrac{1+i}{1-i}\right]^{1041}$

7. $\left[(1+i) + \dfrac{1}{1+i}\right]^2$

8. $\left[\dfrac{2+i}{3+4i} - \dfrac{2i}{3-4i}\right]^2$

Let $z_1 = x_1 + iy_1$ and $z_2 = x_2 + iy_2$. Without using Eqs. (1.2–10a–d), show that:

9. $\overline{(z_1 - z_2)} = \bar{z}_1 - \bar{z}_2$

10. $\overline{(z_1 z_2)} = \bar{z}_1 \bar{z}_2$

11. $\overline{\left(\dfrac{1}{z_1}\right)} = \dfrac{1}{\bar{z}_1}$

12. $\overline{\left(\dfrac{z_1}{z_2}\right)} = \dfrac{\bar{z}_1}{\bar{z}_2}$

13. $\operatorname{Re}(z_1 z_2) = \operatorname{Re}(\bar{z}_1 \bar{z}_2)$

14. $\operatorname{Im}(z_1 z_2) = -\operatorname{Im}(\bar{z}_1 \bar{z}_2)$

Let z_1, z_2, and z_3 be three arbitrary complex numbers. Which of the following statements are true in general? You may use the results in (1.2–10) and their generalizations.

15. $\overline{z_1 z_2 \bar{z}_3} = \bar{z}_1 z_2 z_3$

16. $\overline{z_1(z_2 + z_3)} = \bar{z}_1 \bar{z}_2 + \bar{z}_1 \bar{z}_3$

17. $\dfrac{1}{z_3}\overline{\left(\dfrac{z_1}{z_2}\right)} = \dfrac{\bar{z}_1}{\bar{z}_2 \bar{z}_3}$

18. $z_1^2 + z_2^2 = \overline{\bar{z}_1^2 + \bar{z}_2^2}$

What restrictions, if any, must be placed on z_1 and z_2 so that the following equations are satisfied? Let $z_1 = x_1 + iy_1$ and $z_2 = x_2 + iy_2$.

19. $\operatorname{Re}(z_1 z_2) = \operatorname{Re}(z_1)\operatorname{Re}(z_2)$

20. $z_1 \bar{z}_2 = \bar{z}_1 z_2$

21. $\text{Re}(z_1/z_2) = \text{Re}(z_1)/\text{Re}(z_2)$

22. a) Let k, l, m, and n be integers. Show that there must exist integers p and q satisfying $(k^2 + l^2)(m^2 + n^2) = p^2 + q^2$ and find explicit formulas for p and q in terms of k, l, m, n.

Hint: $[(k + il)(m + in)][(k - il)(m - in)] = (p + iq)(p - iq)$.

b) If $(100 + 225)(9 + 400) = p^2 + q^2$, find positive integer values for p and q. Check your result with a pocket calculator.

23. Suppose, following Hamilton, we regard a complex number as a pair of ordered real numbers. We want the appropriate definition for the quotient $(c, d)/(a, b)$. Let us put $(c, d)/(a, b) = (e, f)$, where e and f are real numbers to be determined. Assume $(a, b) \neq (0, 0)$. If our definition is to be plausible, then

$$(c, d) = (a, b) \cdot (e, f).$$

a) Perform the indicated multiplication by using the product rule for couples.

b) Equate corresponding members (real numbers and imaginary numbers) on both sides of the equation resulting from part (a).

c) In part (b) a pair of simultaneous linear equations were obtained. Solve these equations for e and f in terms of a, b, c, and d. How does the result compare with that of Eq. (1.2–7)?

1.3 COMPLEX NUMBERS AND THE ARGAND PLANE

Modulus

The *magnitude* or *modulus* of a complex number is the positive square root of the sums of the squares of its real and imaginary parts.

If the complex number is z, then its modulus is written $|z|$. If $z = x + iy$, we have, from the definition,

$$|z| = \sqrt{x^2 + y^2}. \tag{1.3–1}$$

The modulus of a complex number is a nonnegative real number. Although we cannot say one complex number is greater (or less) than another, we can say the modulus of one number exceeds that of another; for example, $|4 + i| > |2 + 3i|$ since

$$|4 + i| = \sqrt{16 + 1} = \sqrt{17} > |2 + 3i| = \sqrt{4 + 9} = \sqrt{13}.$$

A complex number has the same modulus as its conjugate because

$$|\bar{z}| = \sqrt{x^2 + (-y)^2} = \sqrt{x^2 + y^2} = |z|.$$

The product of a complex number and its conjugate is the squared modulus of the complex number. To see this note that $z\bar{z} = x^2 + y^2$. Thus, from Eq. (1.3–1),

$$z\bar{z} = |z|^2. \tag{1.3–2}$$

The square root of this expression is also useful:

$$|z| = \sqrt{z\bar{z}}. \tag{1.3–3}$$

We will now prove that

> The modulus of the product of two complex numbers is equal to the product of their moduli.

Let the numbers be z_1 and z_2. Their product is $z_1 z_2$. Let $z = z_1 z_2$ in Eq. (1.3–3). We then have

$$|z_1 z_2| = \sqrt{z_1 z_2 (\overline{z_1 z_2})} = \sqrt{z_1 z_2 \bar{z}_1 \bar{z}_2} = \sqrt{z_1 \bar{z}_1} \sqrt{z_2 \bar{z}_2}.$$

Using Eq. (1.3–3) to rewrite the two radicals on the far right we have, finally,

$$|z_1 z_2| = |z_1||z_2|. \tag{1.3–4}$$

Similarly, $|z_1 z_2 z_3| = |z_1 z_2||z_3| = |z_1||z_2||z_3|$, and in general,

> The modulus of a product of numbers is the product of the moduli of each factor, regardless of how many factors are present.

It is left as an exercise to show

$$\left|\frac{z_1}{z_2}\right| = \frac{|z_1|}{|z_2|}. \tag{1.3–5}$$

> The modulus of the quotient of two complex numbers is the quotient of their moduli.

After reading a few more pages the reader will see that the modulus of the sum of two complex numbers is not in general the same as the sum of their moduli.

EXAMPLE 1

What is the modulus of $-i + ((3 + i)/(1 - i))$?

Solution

We first simplify the fraction

$$\frac{3 + i}{1 - i} = \frac{(3 + i)(1 + i)}{(1 - i)(1 + i)} = 1 + 2i.$$

Thus

$$-i + \frac{3 + i}{1 - i} = 1 + i \quad \text{and} \quad |1 + i| = \sqrt{2}.$$ ◀

EXAMPLE 2

Find $|(3 + 4i)^5/(1 + i\sqrt{3})|$.

Solution

From Eq. (1.3–5) we have

$$\left|\frac{(3 + 4i)^5}{1 + i\sqrt{3}}\right| = \frac{|(3 + 4i)^5|}{|1 + i\sqrt{3}|}.$$

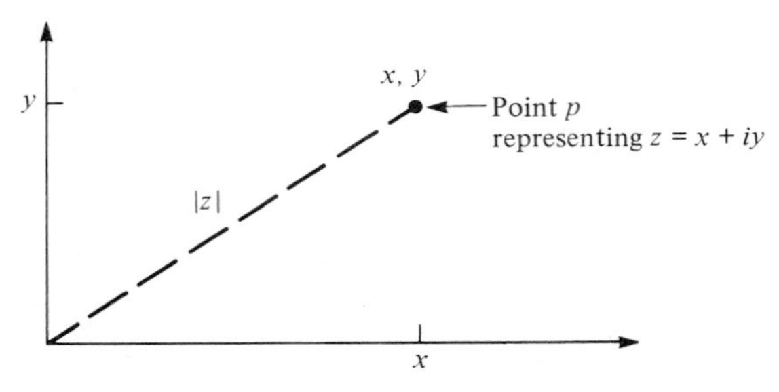

Figure 1.3–1

Now, $|(3+4i)^5| = |3+4i|^5 = (\sqrt{3^2+4^2})^5$. Also, $|1+i\sqrt{3}| = \sqrt{1^2+(\sqrt{3})^2}$. Thus

$$\left|\frac{(3+4i)^5}{1+i\sqrt{3}}\right| = \frac{(\sqrt{3^2+4^2})^5}{\sqrt{1^2+(\sqrt{3})^2}} = \frac{5^5}{2}.$$ ◀

Complex or Argand Plane

If the complex number $z = x + iy$ were written as a couple $z = (x, y)$, we would perhaps be reminded of the notation for the coordinates of a point in the xy-plane. The expression $|z| = \sqrt{x^2 + y^2}$ also recalls the Pythagorean expression for the distance of that point from the origin.

It should come as no surprise to learn that the xy-plane (Cartesian plane) is frequently used to represent complex numbers. When used for this purpose, it is called the Argand plane,† the z-plane, or the complex plane. Under these circumstances, the x- or horizontal axis is called the axis of real numbers, whereas the y- or vertical axis is called the axis of imaginary numbers.

In Fig. 1.3–1 the point p, whose coordinates are x, y, is said to represent the complex number $z = x + iy$. The modulus of z, that is $|z|$, is the distance of x, y from the origin.

Another possible representation of z in this same plane is as a vector. We display $z = x + iy$ as a directed line that begins at the origin and terminates at the point x, y, as shown in Fig. 1.3–2. The length of the vector is $|z|$.

Thus a complex number can be represented by either a point or a vector in the xy-plane. We will use both methods. Often we will refer to the point or vector as if it were the complex number itself rather than merely its representation.

Since the length of either leg of a right triangle cannot exceed the length of the hypotenuse, Fig. 1.3–2 reveals the following:

$$|\text{Re}\, z| = |x| \le |z|, \tag{1.3–6a}$$

$$|\text{Im}\, z| = |y| \le |z|. \tag{1.3–6b}$$

† The plane is named for Jean Argand (1768–1822), a Swiss mathematician who proposed this representation of complex numbers in 1806.

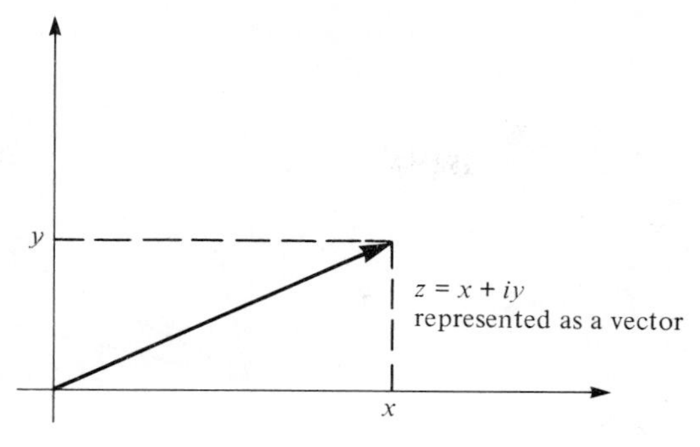

Figure 1.3–2

The | | signs have been placed around Re z and Im z since we are concerned here with physical length (which cannot be negative). Even though we have drawn Re z and Im z positive in Fig. 1.3–2, we could have easily used a figure in which one or both were negative, and Eq. (1.3–6) would still hold.

When we represent a complex number as a vector, we will regard it as a sliding vector, that is, one whose starting point is irrelevant. Thus the line directed from the origin to $x = 3$, $y = 4$ is the complex number $3 + 4i$, and so is the directed line from $x = 1$, $y = 2$ to $x = 4$, $y = 6$ (see Fig. 1.3–3). Both vectors have length 5 and point in the same direction. Each has projections of 3 and 4 on the x- and y-axes, respectively.

There are an unlimited number of directed line segments that we can draw in order to represent a complex number. All have the same magnitude and point in the same direction.

There are simple geometrical relationships between the vectors for $z = x + iy$, $-z = -x - iy$, and $\bar{z} = x - iy$, as can be seen in Fig. 1.3–4. The vector for $-z$ is the vector for z reflected through the origin, whereas $\bar{z}$ is the vector z reflected about the real axis.

The process of adding the complex number $z_1 = x_1 + iy_1$ to the number $z_2 = x_2 + iy_2$ has a simple interpretation in terms of their vectors. Their sum, $z_1 + z_2 = x_1 + x_2 + i(y_1 + y_2)$, is shown vectorially in Fig. 1.3–5. We see that the

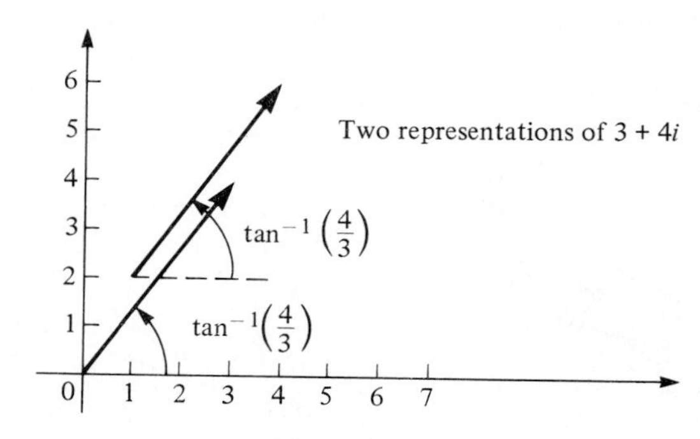

Figure 1.3–3

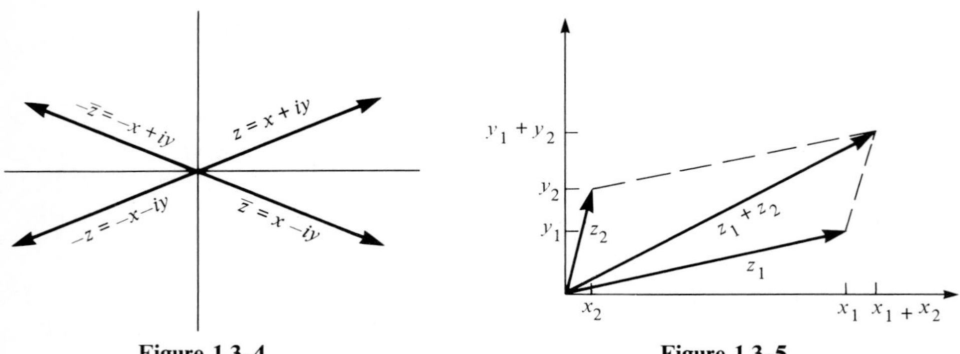

Figure 1.3–4

Figure 1.3–5

vector representing the sum of the complex numbers z_1 and z_2 is obtained by adding vectorially the vector for z_1 and the vector for z_2. The familiar parallelogram rule, which is used in adding vectors such as force, velocity, or electric field, is employed in Fig. 1.3–5 to perform the summation. We can also use a "tip-to-tail" addition, as shown in Fig. 1.3–6.

The "triangle inequalities" are derivable from this geometric picture. The length of any leg of a triangle is less than or equal to the sums of the lengths of the legs of the other two sides (see Fig. 1.3–7). The length of the vector for $z_1 + z_2$ is $|z_1 + z_2|$, which must be less than or equal to the combined length $|z_1| + |z_2|$. Thus

$$|z_1 + z_2| \leq |z_1| + |z_2|. \tag{1.3–7}$$

This triangle inequality is also derivable from purely algebraic manipulations (see Exercise 34).

Two other useful triangle inequalities are derived in Exercises 29 and 30. They are

$$|z_1 - z_2| \leq |z_1| + |z_2|$$

and

$$|z_1 + z_2| \geq ||z_1| - |z_2||.$$

Equation (1.3–7) shows as promised that the modulus of the sum of two complex numbers need not equal the sum of their moduli. By adding three complex numbers

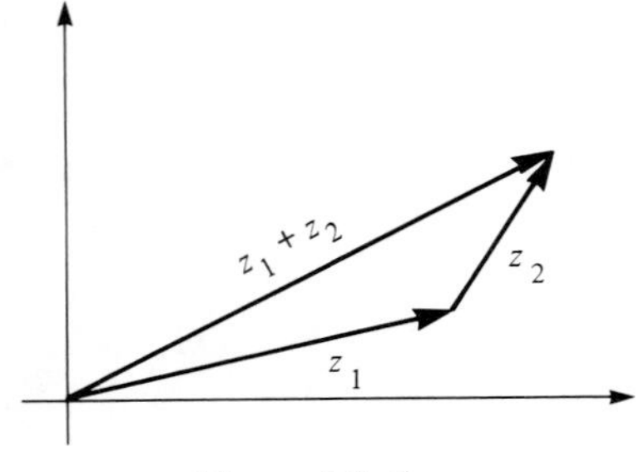

Figure 1.3–6

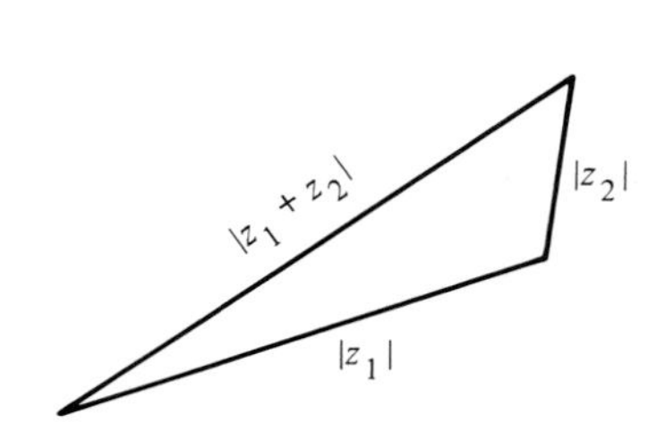

Figure 1.3–7

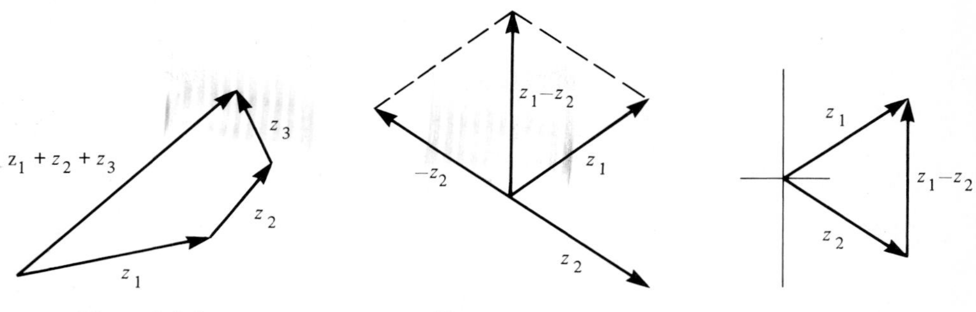

Figure 1.3–8 Figure 1.3–9 Figure 1.3–10

vectorially, as shown in Fig. 1.3–8, we see that

$$|z_1 + z_2 + z_3| \le |z_1| + |z_2| + |z_3|.$$

This obviously can be extended to a sum having any number of elements:

$$|z_1 + z_2 + \cdots + z_n| \le |z_1| + |z_2| + \cdots + |z_n|. \tag{1.3–8}$$

The subtraction of two complex numbers also has a counterpart in vector subtraction. Thus $z_1 - z_2$ is treated by adding together the vectors for z_1 and $-z_2$, as shown in Fig. 1.3–9. Another familiar means of vector subtraction is shown in Fig. 1.3–10.

Polar Representation

Often, points in the complex plane, which represent complex numbers, are defined by means of polar coordinates (see Fig. 1.3–11). The complex number $z = x + iy$ is represented by the point p whose Cartesian coordinates are x, y, or whose polar coordinates are r, θ. We see that r is identical to the modulus of z, that is, the distance of p from the origin; and θ is the angle that the ray joining the point p to the origin makes with the positive x-axis. We call θ the *argument* of z and write $\theta = \arg z$. Occasionally, θ is referred to as the angle of z. Unless otherwise stated, θ will be expressed in radians. When the $^\circ$ symbol is used, θ will be given in degrees.

The angle θ is regarded as positive when measured in the counterclockwise direction and negative when measured clockwise. The distance r is never negative. For a point at the origin, r becomes zero. Here θ is undefined since a ray like that shown in Fig. 1.3–11 cannot be constructed.

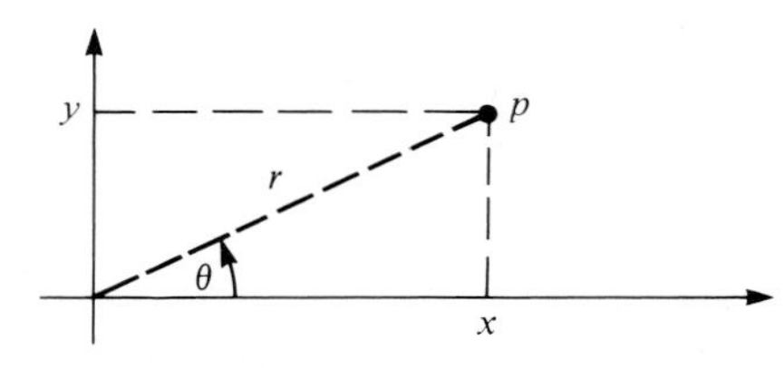

Figure 1.3–11

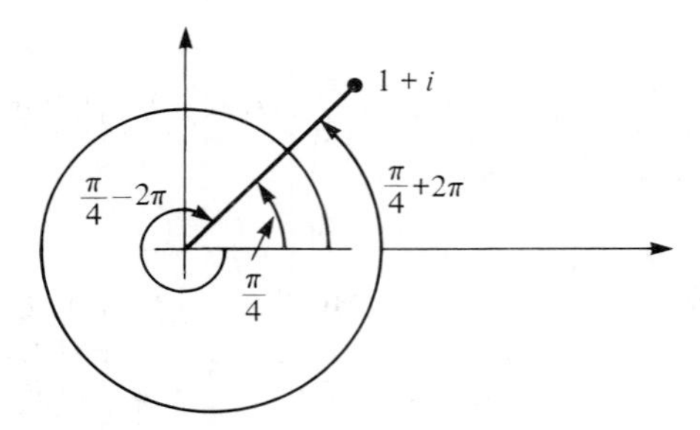

Figure 1.3–12

Since $r = \sqrt{x^2 + y^2}$, we have

$$r = |z|, \tag{1.3–9a}$$

and a glance at Fig. 1.3–11 shows

$$\tan \theta = y/x. \tag{1.3–9b}$$

An important feature of θ is that it is multivalued. Suppose that for some complex number we have found a correct value of θ in radians. Then we can add to this value any positive or negative integer multiple of 2π radians and again obtain a valid value for θ. If θ is in degrees we can add integer multiples of $360°$. For example, suppose $z = 1 + i$. Let us find the polar coordinates of the point that represents this complex number. Now, $r = |z| = \sqrt{1^2 + 1^2} = \sqrt{2}$, and from Fig. 1.3–12 we see that $\theta = \pi/4$ radians, or $\pi/4 + 2\pi$ radians, or $\pi/4 + 4\pi$, or $\pi/4 - 2\pi$, etc. Thus in this case $\theta = \pi/4 + k2\pi$, where $k = 0, \pm 1, \pm 2, \ldots$.

In general, all the values of θ are contained in the expression

$$\theta = \theta_0 + k2\pi, \qquad k = 0, \pm 1, \pm 2, \ldots, \tag{1.3–10}$$

where θ_0 is some particular value of arg z. If we work in degrees, $\theta = \theta_0 + k\ 360°$ describes all values of θ.

The *principal value of the argument* (or *principal argument*) of a complex number z is that value of arg z that is greater than $-\pi$ and less than or equal to π.

Thus the principal value of θ satisfies

$$-\pi < \theta \leq \pi.^{\dagger} \tag{1.3–11}$$

The reader can restate this in degrees. Note that the principal value of the argument when z is a negative real number is π (or $180°$).

† The definition presented here for the principal argument is the most common one. However, some texts use other definitions, for example, $0 \leq \theta < 2\pi$.

EXAMPLE 3

Using the principal argument, find the polar coordinates of the point that represents the complex number $-1-i$.

Solution

The polar distance r for $-1-i$ is $\sqrt{2}$, as we can see from Fig. 1.3–13. The principal value of θ is $-3\pi/4$ radians. It is *not* $5\pi/4$ since this number exceeds π. In computing *principal* values we should never do what was done with the dashed line in Fig. 1.3–13, namely, cross the negative real axis. From Eq. (1.3–10) we see that all the values of $\arg(-1-i)$ are contained in the expression

$$\theta = \frac{-3\pi}{4} + 2k\pi, \qquad k = 0, \pm 1, \pm 2, \ldots.$$

Note that by using $k = 1$ in the above, we obtain the nonprincipal value $5\pi/4$. ◀

The inverse of Eq. (1.3–9b),

$$\theta = \tan^{-1}(y/x),$$

which might be used to find θ, especially when one uses a pocket calculator, requires a comment. From our knowledge of elementary trigonometry, we know that if y/x is established, this equation does not contain enough information to define the set of values of θ. With y/x known, the sign of x or y must also be given if the appropriate set is to be determined.

For example, if $y/x = 1/\sqrt{3}$ we can have $\theta = \pi/6 + 2k\pi$ or $\theta = -5\pi/6 + 2k\pi$, $k = 0, \pm 1, \pm 2, \ldots$. Now a positive value of y (which puts z in the first quadrant) dictates choosing the first set. A negative value of y (which puts z in the third quadrant) requires choosing the second set.

If $y/x = 0$ (because $y = 0$) or if y/x is undefined (because $x = 0$) we must know respectively the sign of x or y to find the set of values of θ.

Let the complex number $z = x + iy$ be represented by the vector shown in Fig. 1.3–14. The point at which the vector terminates has polar coordinates r and θ. The angle θ need not be a principal value. From Fig. 1.3–14 we have $x = r\cos\theta$ and

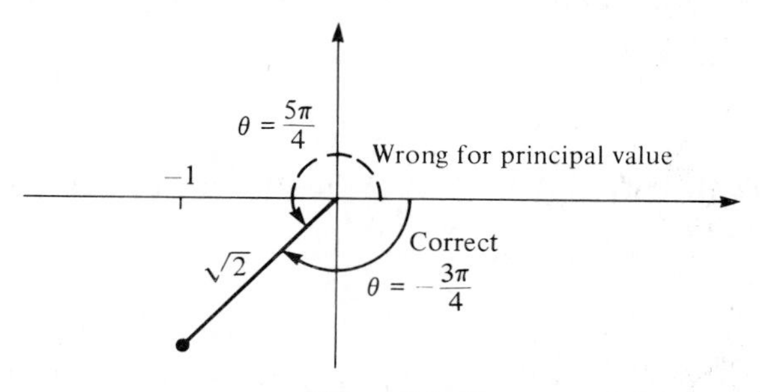

Figure 1.3–13

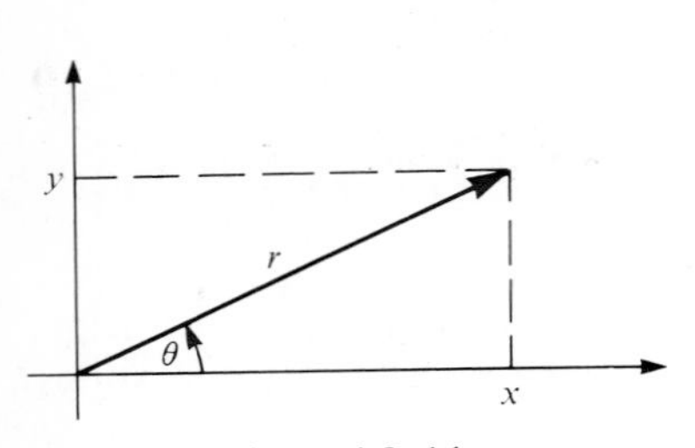

Figure 1.3–14

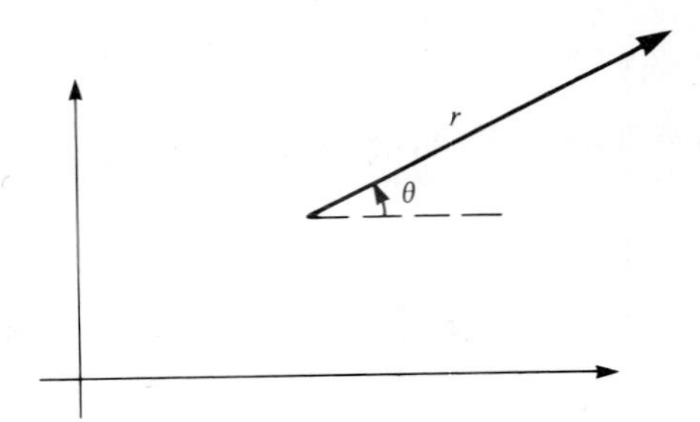

Figure 1.3–15

$y = r \sin \theta$. Thus $z = r \cos \theta + ir \sin \theta$, or

$$z = r(\cos \theta + i \sin \theta). \tag{1.3–12}$$

We call this the *polar form* of a complex number as opposed to the rectangular (Cartesian) form $x + iy$. The expression $\cos \theta + i \sin \theta$ is often abbreviated cis θ. We will often use $\underline{/\theta}$ to mean cis θ. Our complex number $x + iy$ becomes $r\underline{/\theta}$. This is a useful notation because it tells not only the length of the corresponding vector but also the angle made with the real axis. Note that $i = 1\underline{/\pi/2}$ and $-i = 1\underline{/-\pi/2}$.

A vector such as the one in Fig. 1.3–15 can be translated so that it emanates from the origin. It too represents a complex number $r\underline{/\theta}$.

The complex numbers $r\underline{/\theta}$ and $r\underline{/-\theta}$ are conjugates of each other, as can be seen in Fig. 1.3–16. Equivalently, we find that the conjugate of $r(\cos \theta + i \sin \theta)$ is $r(\cos(-\theta) + i \sin(-\theta)) = r(\cos \theta - i \sin \theta)$.

The polar description is particularly useful in the multiplication of complex numbers. Consider $z_1 = r_1$ cis θ_1 and $z_2 = r_2$ cis θ_2. Multiplying z_1 by z_2 we have

$$z_1 z_2 = r_1(\cos \theta_1 + i \sin \theta_1) r_2(\cos \theta_2 + i \sin \theta_2). \tag{1.3–13}$$

With some additional multiplication we obtain

$$z_1 z_2 = r_1 r_2[(\cos \theta_1 \cos \theta_2 - \sin \theta_1 \sin \theta_2) + i(\sin \theta_1 \cos \theta_2 + \cos \theta_1 \sin \theta_2)]. \tag{1.3–14}$$

The reader should recall the identities

$$\cos(\theta_1 + \theta_2) = \cos \theta_1 \cos \theta_2 - \sin \theta_1 \sin \theta_2,$$
$$\sin(\theta_1 + \theta_2) = \sin \theta_1 \cos \theta_2 + \cos \theta_1 \sin \theta_2,$$

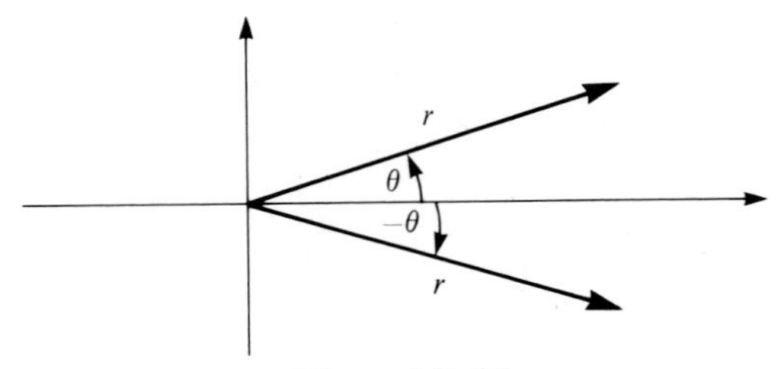

Figure 1.3–16

which we now install in Eq. (1.3–14), to get

$$z_1z_2 = r_1r_2[\cos(\theta_1 + \theta_2) + i\sin(\theta_1 + \theta_2)]. \tag{1.3–15}$$

In other notation Eq. (1.3–15) becomes

$$z_1z_2 = r_1r_2\underline{/\theta_1 + \theta_2}. \tag{1.3–16}$$

The two preceding equations contain the following important fact.

When two complex numbers are multiplied together, the resulting product has a modulus equal to the product of the moduli of the two factors and an argument equal to the sum of the arguments of the two factors.

To multiply three complex numbers we readily extend this method. Thus

$$z_1z_2z_3 = (z_1z_2)(z_3) = r_1r_2\underline{/\theta_1 + \theta_3}r_3\underline{/\theta_3} = r_1r_2r_3\underline{/\theta_1 + \theta_2 + \theta_3}.$$

Any number of complex numbers can be multiplied in this fashion.

The modulus of the entire product is the product of the moduli of the factors, and the argument of the product is the sum of the arguments of the factors.

EXAMPLE 4

Verify Eq. (1.3–16) by considering the product $(1 + i)(\sqrt{3} + i)$.

Solution

Multiplying in the usual way (see Eq. 1.1–4), we obtain

$$(1 + i)(\sqrt{3} + i) = (\sqrt{3} - 1) + i(\sqrt{3} + 1).$$

The modulus of the preceding product is

$$\sqrt{(\sqrt{3} - 1)^2 + (\sqrt{3} + 1)^2} = \sqrt{8},$$

whereas the product of the moduli of each factor is

$$\sqrt{1^2 + 1^2}\sqrt{(\sqrt{3})^2 + 1^2} = \sqrt{2}\sqrt{4} = \sqrt{8}.$$

The factor $(1 + i)$ has an argument of $\pi/4$ radians and $\sqrt{3} + i$ has an argument of $\pi/6$ (see Fig. 1.3–17).

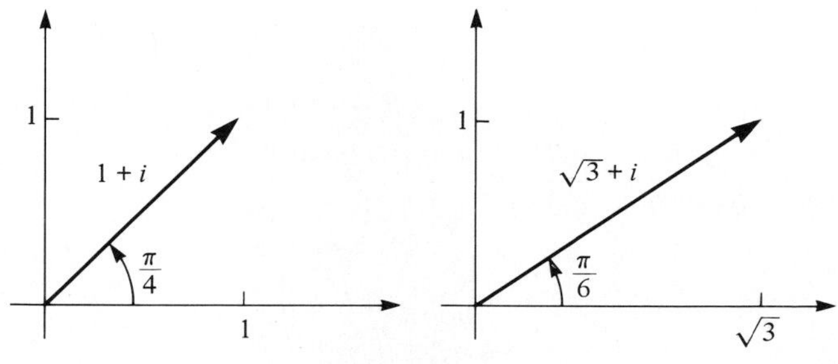

Figure 1.3–17

The complex number $(\sqrt{3} - 1) + i(\sqrt{3} + 1)$ has an angle in the first quadrant equal to

$$\tan^{-1}\frac{\sqrt{3}+1}{\sqrt{3}-1} = \frac{5\pi}{12}.$$

But $5\pi/12 = \pi/4 + \pi/6$, that is, the sums of the arguments of the two factors. ◀

The argument of a complex number has, of course, an infinity of possible values. When we add together the arguments of two factors in order to arrive at the argument of a product, we obtain only one of the possible values for the argument of that product. Thus in the preceding example $(\sqrt{3} - 1) + i(\sqrt{3} + 1)$ has arguments $5\pi/12 + 2\pi$, $5\pi/12 + 4\pi$, and so forth, none of which were obtained through our procedure. However, any of these results can be derived by our adding some whole multiple of 2π on to the number $5\pi/12$ actually obtained.

Equation (1.3–16) is particularly useful when we multiply complex numbers that are given to us in polar rather than in rectangular form. For example, the product of $2\angle\pi/2$ and $3\angle 3\pi/4$ is $6\angle 5\pi/4$. We can convert the result to rectangular form:

$$6\angle\frac{5\pi}{4} = 6\cos\left(\frac{5\pi}{4}\right) + i6\sin\left(\frac{5\pi}{4}\right) = \frac{-6}{\sqrt{2}} - i\frac{6}{\sqrt{2}}.$$

The two factors in this example were written with their principal arguments $\pi/2$ and $3\pi/4$. However, when we added these angles and got $5\pi/4$, we obtained a nonprincipal argument. In fact, the principal argument of $6\angle 5\pi/4$ is $-3\pi/4$.

When principal arguments are added together in a multiplication problem, the resulting argument need not be a principal value. Conversely, when nonprincipal arguments are combined, a principal argument may result.

When we multiply z_1 by z_2 to obtain the product $z_1 z_2$, the operation performed with the corresponding vectors is neither scalar multiplication (dot product) nor vector multiplication (cross product), which are perhaps familiar to us from elementary vector analysis. Similarly, we can divide two complex numbers as well as, in a sense, their vectors. This too has no counterpart in any previously familiar vector operation.

It is convenient to use polar coordinates to find the reciprocal of a complex number. With $z = r\angle\theta$ we have

$$\frac{1}{z} = \frac{1}{r(\cos\theta + i\sin\theta)} = \frac{\cos\theta - i\sin\theta}{r(\cos\theta + i\sin\theta)(\cos\theta - i\sin\theta)}$$
$$= \frac{\cos\theta - i\sin\theta}{r(\cos^2\theta + \sin^2\theta)} = \frac{\cos\theta - i\sin\theta}{r} = \frac{1}{r}\angle-\theta.$$

Hence,

$$\frac{1}{z} = \frac{1}{r\angle\theta} = \frac{1}{r}\angle-\theta. \tag{1.3–17}$$

Thus the modulus of the reciprocal of a complex number is the reciprocal of the modulus of that number, and the argument of the reciprocal of a complex number is the negative of the argument of that number.

Consider now the complex numbers $z_1 = r_1\underline{/\theta_1}$ and $z_2 = r_2\underline{/\theta_2}$. To divide z_1 by z_2 we multiply z_1 by $1/z_2 = 1/r_2\underline{/-\theta_2}$. Thus

$$\frac{z_1}{z_2} = r_1\underline{/\theta_1}\,\frac{1}{r_2}\underline{/-\theta_2} = \frac{r_1}{r_2}\underline{/\theta_1 - \theta_2}. \tag{1.3–18}$$

The modulus of the quotient of two complex numbers is the quotient of their moduli, and the argument of the quotient is the argument of the numerator less the argument of the denominator.

EXAMPLE 5

Evaluate $(1 + i)/(\sqrt{3} + i)$ by using the polar form of complex numbers.

Solution

We have

$$\frac{1+i}{\sqrt{3}+i} = \frac{\sqrt{2}\underline{/\text{arc tan}(1/1)}}{2\underline{/\text{arc tan}(1/\sqrt{3})}} = \frac{\sqrt{2}\underline{/\pi/4}}{2\underline{/\pi/6}} = \frac{1}{\sqrt{2}}\underline{/\pi/4 - \pi/6} = \frac{1}{\sqrt{2}}\underline{/\pi/12}.$$

The above result is convertible to rectangular form:

$$\frac{1+i}{\sqrt{3}+i} = \frac{\cos(\pi/12)}{\sqrt{2}} + i\frac{\sin(\pi/12)}{\sqrt{2}}.$$

This problem could have been done entirely in rectangular notation with the aid of Eq. (1.2–7). There are computations, however, where switching to polar notation saves us some labor, as in the following example. ◀

EXAMPLE 6

Evaluate

$$\frac{(1+i)(3+i)(-2-i)}{(i)(3+4i)(5+i)} = a + ib = r\underline{/\theta},$$

where a and b are to be determined.

Solution

Let us initially seek the polar answer $r\underline{/\theta}$. First, r is obtained from the usual properties of the moduli of products and quotients:

$$r = \frac{|(1+i)(3+i)(-2-i)|}{|(i)(3+4i)(5+i)|} = \frac{|1+i||3+i||-2-i|}{|i||3+4i||5+i|} = \frac{\sqrt{2}\sqrt{10}\sqrt{5}}{1\sqrt{25}\sqrt{26}} = \frac{2}{\sqrt{26}}.$$

The argument θ is the argument of the numerator, $(1 + i)(3 + i)(-2 - i)$, less that of the denominator, $(i)(3 + 4i)(5 + i)$. Thus

$$\theta = \left(\arc\tan\frac{1}{1} + \arc\tan\frac{1}{3} + \arc\tan\frac{-1}{-2}\right) - \left(\arc\tan\frac{1}{0} + \arc\tan\frac{4}{3} + \arc\tan\frac{1}{5}\right),$$

$$\theta \doteq (0.785 + 0.322 + 3.605) - (1.571 + 0.927 + 0.197) \doteq 2.017.$$

Therefore,

$$r\underline{/\theta} \doteq \frac{2}{\sqrt{26}}\underline{/2.017},$$

and

$$a + ib \doteq \frac{2}{\sqrt{26}}[\cos 2.017 + i \sin 2.017] \doteq -0.169 + i0.354.$$

◀

EXERCISES

Find the modulus of each of the following complex expressions:

1. $3 + 4i$

2. $(3 + 4i)(1 + 2i)(i)$

3. $\dfrac{(3 + 4i)(1 + i)}{(3 - 4i)}$

4. $\left[\dfrac{(3 + 4i)(1 + i)}{3 - 4i}\right]^4$

5. $\left(\dfrac{x + iy}{x - iy}\right)^n$, $n \geq 0$ is an integer

6. $i + \dfrac{(3 + 4i)(1 + i)}{3 - 4i}$

The following vectors represent complex numbers. State these numbers in the form $a + ib$.

7. The vector directed from $(1, 2)$ to $(-3, 4)$.

8. The vector of length 10, originating at $(1, -2)$, and making an angle of $\pi/6$ radians in the positive sense with the positive x-axis.

9. The vector terminating at $(2, 2)$. It has length 5 and passes through $(1, 0)$.

10. The vector that begins at $(2, 3)$ and terminates on the line $y = -x$. The line and vector intersect at right angles.

11. The vector that begins at $(0, 0)$, has length 5, and terminates in the first quadrant on the parabola $y = x^2$.

Find the principal argument of these complex numbers. A pocket calculator can be helpful.

12. $2\,\text{cis}(3.14)$

13. $2\,\text{cis}(3.15)$

14. $2\,\text{cis}(-2.99\pi)$

15. $2\,\text{cis}(2001)$

16. $-2\,\text{cis}(2001)$

17. $2\,\text{cis}(2.01\pi \times 10^8)$

Find in the form $a + ib$ the complex numbers represented by the points having the following polar coordinates.

18. $r = 2,\ \theta = 3$

19. $r = 2,\ \theta = 3\frac{1}{3}\pi$

Convert the following expressions to the form $r \operatorname{cis} \theta$. State r and give all possible values of θ in radians. Indicate the principal value.

20. $\sqrt{3} + i$

21. $-\sqrt{3} - i$

22. $(3 + 4i)(3 + 4i)(1 + i)$

23. $(\sqrt{3} + i)^4(1 - i)^3$

Reduce the following expressions to the form $r \operatorname{cis} \theta$. Give, in radians, only the principal value of θ.

24. $\dfrac{(1 + i)(-1 - i\sqrt{3})}{3\sqrt{\pi/8}}$

25. $\dfrac{(1 + i)^2(2 + i)}{(3 + i)\, 2 \operatorname{cis}(3\pi/4)}$

26. $\left(\dfrac{1 + i}{1 - i}\right)^3 + 2i$

27. $\dfrac{[\operatorname{cis}(\pi/6)]^4}{[\operatorname{cis}(-\pi/6)]^4}$

28. Under what circumstances will the equality sign hold true in Eq. (1.3–7)?

29. a) Let z_1 and z_2 be complex numbers. By replacing z_2 with $-z_2$ in Eq. (1.3–7), show that

$$|z_1 - z_2| \le |z_1| + |z_2| \tag{1.3–19}$$

Interpret this result with the aid of a triangle.

b) What must be the relationship between z_1 and z_2 in order to have the equality hold in part (a)?

30. a) Let L, M, and N be the lengths of the legs of a triangle, with $M \ge N$. Convince yourself with the aid of a drawing of the triangle that $L \ge M - N \ge 0$.

b) Let z_1 and z_2 be complex numbers. Consider the vectors representing z_1, z_2, and $z_1 + z_2$. Using the result of part (a) show that

$$|z_1 + z_2| \ge |z_1| - |z_2| \ge 0 \qquad \text{if } |z_1| \ge |z_2|,$$

$$|z_1 + z_2| \ge |z_2| - |z_1| \ge 0 \qquad \text{if } |z_2| \ge |z_1|.$$

Explain why both formulas can be reduced to the single expression

$$|z_1 + z_2| \ge \big||z_1| - |z_2|\big|. \tag{1.3–20}$$

31. a) By considering the expression $(p - q)^2$, where p and q are nonnegative real numbers, show that

$$p + q \le \sqrt{2}\sqrt{p^2 + q^2}.$$

b) Use the preceding result to show that for any complex number z we have $|\operatorname{Re} z| + |\operatorname{Im} z| \le \sqrt{2}\,|z|$.

c) Verify the preceding result for $z = 1 - i\sqrt{3}$.

d) Find a value for z such that the equality sign holds in (b).

32. a) By considering the product of $1 + ia$ and $1 + ib$, and the argument of each factor, show that

$$\arctan(a) + \arctan(b) = \arctan\left(\frac{a+b}{1-ab}\right),$$

where a and b are real numbers.

b) Use the preceding formula to prove that

$$\pi = 4[\arctan(\tfrac{1}{2}) + \arctan(\tfrac{1}{3})].$$

Check this result with a pocket calculator.

c) Extend the technique used in (a) to find a formula for $\arctan(a) + \arctan(b) + \arctan(c)$.

33. a) Consider vectors representing complex numbers f and g. Show that these vectors are perpendicular if and only if $|f - g|^2 = |f|^2 + |g|^2$.

Hint: Draw a right triangle that employs these vectors.

b) Show that the preceding equation is equivalent to the requirement $\text{Re}(f\bar{g}) = 0$.

34. a) Consider the inequality $|z_1 + z_2|^2 \le |z_1|^2 + |z_2|^2 + 2|z_1||z_2|$. Prove this expression by algebraic means (no triangles).

Hint: Note that $|z_1 + z_2|^2 = (z_1 + z_2)(\overline{z_1 + z_2}) = (z_1 + z_2)(\bar{z}_1 + \bar{z}_2)$. Multiply out $(z_1 + z_2)(\bar{z}_1 + \bar{z}_2)$, and use the facts that for a complex number, say w,

$$w + \bar{w} = 2\,\text{Re}\,w, \quad \text{and} \quad |\text{Re}\,w| \le |w|.$$

b) Observe that $|z_1|^2 + |z_2|^2 + 2|z_1||z_2| = (|z_1| + |z_2|)^2$. Show that the inequality proved in part (a) leads to the triangle inequality $|z_1 + z_2| \le |z_1| + |z_2|$.

35. a) Beginning with the product $(z_1 - z_2)(\overline{z_1 - z_2})$, show that

$$|z_1 - z_2|^2 = |z_1|^2 + |z_2|^2 - 2\,\text{Re}(z_1\bar{z}_2).$$

b) Recall the law of cosines from elementary trigonometry: $a^2 = b^2 + c^2 - 2bc\cos\alpha$. Show that this law can be obtained from the formula you derived in part (a).

Hint: Identify lengths b and c with $|z_1|$ and $|z_2|$. Take z_2 as being positive real.

1.4 INTEGER AND FRACTIONAL POWERS OF A COMPLEX NUMBER

Integer Powers

In the previous section we learned to multiply any number of complex quantities together by means of polar notation. Thus with n complex numbers $z_1, z_2, \ldots, z_n$ we have

$$z_1 z_2 z_3 \cdots z_n = r_1 r_2 r_3 \cdots r_n \underline{/\theta_1 + \theta_2 + \theta_3 + \cdots + \theta_n}, \tag{1.4–1}$$

where $r_j = |z_j|$ and $\theta_j = \arg z_j$.

If all the values, z_1, z_2, and so on, are identical so that $z_j = z$ and $z = r\underline{/\theta}$, then Eq. (1.4–1) simplifies to

$$z^n = r^n\underline{/n\theta} = r^n \operatorname{cis}(n\theta) = r^n[\cos(n\theta) + i\sin(n\theta)] \tag{1.4–2}$$

The modulus of z^n is the modulus of z raised to the nth power, whereas the argument of z^n is n times the argument of z.

The preceding was proved valid when n is a positive integer. If we define $z^0 = 1$ (as for real numbers), Eq. (1.4–2) applies when $n = 0$ as well. The expression 0^0 remains undefined.

With the aid of a suitable definition, we will now prove that Eq. (1.4–2) also is applicable for negative n.

Let m be a positive integer. Then, from Eq. (1.4–2) we have $z^m = r^m[\cos(m\theta) + i\sin(m\theta)]$. We define z^{-m} as being identical to $1/z^m$. Thus

$$z^{-m} = \frac{1}{r^m[\cos(m\theta) + i\sin(m\theta)]}. \tag{1.4–3}$$

If on the right in Eq. (1.4–3) we multiply numerator and denominator by the expression $\cos m\theta - i\sin m\theta$, we have

$$z^{-m} = \frac{1}{r^m}\frac{\cos m\theta - i\sin m\theta}{\cos^2 m\theta + \sin^2 m\theta} = r^{-m}[\cos m\theta - i\sin m\theta].$$

Now, since $\cos(m\theta) = \cos(-m\theta)$ and $-\sin m\theta = \sin(-m\theta)$, we obtain

$$z^{-m} = r^{-m}[\cos(-m\theta) + i\sin(-m\theta)], \qquad m = 1, 2, 3, \ldots. \tag{1.4–4}$$

If we let $-m = n$ in the preceding equation, it becomes

$$z^n = r^n[\cos(n\theta) + i\sin(n\theta)], \qquad n = -1, -2, -3, \ldots. \tag{1.4–5}$$

We can incorporate this result into Eq. (1.4–2) by allowing n to be any integer in that expression.

Equation (1.4–2) allows us to raise complex numbers to integer powers when the use of Cartesian coordinates and successive self-multiplication would be very tedious. For example, consider $(1 + i\sqrt{3})^{11} = a + ib$. We want a and b. We could begin with Eq. (1.1–4), square $(1 + i\sqrt{3})$, multiply the result by $(1 + i\sqrt{3})$, and so forth. Or, if we remember the binomial theorem, we could apply it to $(1 + i\sqrt{3})^{11}$, and then combine the twelve resulting terms. Instead, we observe that $(1 + i\sqrt{3}) = 2\underline{/\pi/3}$, and

$$\left(2\underline{/\frac{\pi}{3}}\right)^{11} = 2^{11}\underline{/\frac{11\pi}{3}} = 2^{11}\left[\cos\left(\frac{11\pi}{3}\right) + i\sin\left(\frac{11\pi}{3}\right)\right] = 2^{10} - i2^{10}\sqrt{3}.$$

Equation (1.4–2) can yield an important identity. First, we put $z = r(\cos\theta + i\sin\theta)$ so that

$$[r(\cos\theta + i\sin\theta)]^n = r^n(\cos n\theta + i\sin n\theta).$$

Taking $r = 1$ in this expression, we then have

$$(\cos\theta + i\sin\theta)^n = \cos n\theta + i\sin n\theta, \qquad n = 0, \pm 1, \pm 2, \ldots, \tag{1.4–6}$$

which is known as DeMoivre's theorem.[†]

This formula can yield some familiar trigonometric identities. For example, with $n = 2$,

$$(\cos\theta + i\sin\theta)^2 = \cos 2\theta + i\sin 2\theta.$$

Expanding the left side of the preceding expression, we arrive at

$$\cos^2\theta + 2i\sin\theta\cos\theta - \sin^2\theta = \cos 2\theta + i\sin 2\theta.$$

Equating corresponding parts (real and imaginary), we obtain the pair of identities $\cos^2\theta - \sin^2\theta = \cos 2\theta$ and $2\sin\theta\cos\theta = \sin 2\theta$.

Fractional Powers

Let us try to raise z to a fractional power, that is, we want $z^{1/m}$, where m is a positive integer. We define $z^{1/m}$ so that $(z^{1/m})^m = z$. Suppose

$$z^{1/m} = \rho\underline{/\phi}. \tag{1.4–7}$$

Raising both sides to the mth power we have

$$z = (\rho\underline{/\phi})^m = z = \rho^m\underline{/m\phi} = \rho^m[\cos(m\phi) + i\sin(m\phi)]. \tag{1.4–8}$$

Using $z = r\underline{/\theta} = r(\cos\theta + i\sin\theta)$ on the left side of the above, we obtain

$$r(\cos\theta + i\sin\theta) = \rho^m[\cos(m\phi) + i\sin(m\phi)]. \tag{1.4–9}$$

For this equation to hold the moduli on each side must agree. Thus

$$r = \rho^m \quad \text{or} \quad \rho = r^{1/m}.$$

Since ρ is a positive real number, we must use the positive root of $r^{1/m}$. Hence

$$\rho = \sqrt[m]{r}. \tag{1.4–10}$$

The angle θ in Eq. (1.4–9) need not equal $m\phi$. The best that we can do is to conclude that these two quantities differ by an integral multiple of 2π, that is, $m\phi - \theta = 2k\pi$, which means

$$\phi = \frac{1}{m}[\theta + 2k\pi], \qquad k = 0, \pm 1, \pm 2, \ldots. \tag{1.4–11}$$

Thus from Eqs. (1.4–7), (1.4–10), and (1.4–11),

$$z^{1/m} = \rho\underline{/\phi} = \sqrt[m]{r}\left[\cos\left(\frac{\theta}{m} + \frac{2k\pi}{m}\right) + i\sin\left(\frac{\theta}{m} + \frac{2k\pi}{m}\right)\right].$$

[†] This novel and useful formula was discovered by a French born Huguenot, Abraham DeMoivre (1667–1754), who lived much of his life in England. He was a disciple of Sir Isaac Newton.

The number k on the right side of this equation can assume any integer value. Suppose we begin with $k = 0$ and allow k to increase in unit steps. With $k = 0$, we are taking the sine and cosine of θ/m; with $k = 1$, the sine and cosine of $\theta/m + 2\pi/m$, and so on. Finally, with $k = m$, we take the sine and cosine of $\theta/m + 2\pi$. But $\sin(\theta/m + 2\pi)$ and $\cos(\theta/m + 2\pi)$ are numerically equal to the sine and cosine of θ/m.

If $k = m + 1$, $m + 2$, etc., we merely repeat the numerical values for the cosine and sine obtained when $k = 1, 2$, etc. Thus all the *numerically distinct* values of $z^{1/m}$ can be obtained by our allowing k to range from 0 to $m - 1$ in the preceding equation. Hence, $z^{1/m}$ has m different values, and they are given by the equation

$$z^{1/m} = \sqrt[m]{r}\,\underline{\left/\frac{\theta + 2\pi k}{m}\right.} = \sqrt[m]{r}\left[\cos\left(\frac{\theta}{m} + \frac{2k\pi}{m}\right) + i\sin\left(\frac{\theta}{m} + \frac{2k\pi}{m}\right)\right],$$

$$k = 0, 1, 2, \ldots, m - 1; \quad m \geq 1. \tag{1.4–12}$$

Actually, we can let k range over any m successive integers (e.g., $k = 2 \to m + 1$) and still generate all the values of $z^{1/m}$.

From our previous mathematics we know that a positive real number, say 9, has two different square roots, in this case ± 3. Equation (1.4–12) tells us that any complex number also has two square roots (we put $m = 2$, $k = 0, 1$) and three cube roots ($m = 3$, $k = 0, 1, 2$) and so on.

The geometrical interpretation of Eq. (1.4–12) is important since it can quickly permit the plotting of those points in the complex plane that represent the roots of a number. The moduli of all the roots are identical and equal, $\sqrt[m]{r}$ (or $\sqrt[m]{|z|}$). Hence, the roots are representable by points on a circle having radius $\sqrt[m]{r}$. Each of the values obtained from Eq. (1.4–12) has a different argument. While k increases as indicated in Eq. (1.4–12), the arguments grow from θ/m to $\theta/m + 2(m - 1)\pi/m$ by increasing in increments of $2\pi/m$. The points representing the various values of $z^{1/m}$, which we plot on the circle of radius $\sqrt[m]{r}$, are thus spaced uniformly at an angular separation of $2\pi/m$. One of the points ($k = 0$) makes an angle of θ/m with the positive x-axis. We thus have enough information to plot all the points (or all the corresponding vectors).

Equation (1.4–12) was derived under the assumption that m is a positive integer. If m is a negative integer the equation is still valid, *except* we now generate all roots by allowing k to range over $|m|$ successive values (e.g., $k = 0, 1, 2, \ldots, |m| - 1$). The $|m|$ roots are uniformly spaced around a circle of radius $\sqrt[m]{r}$, and one root makes an angle θ/m with the positive x-axis. Each value of $z^{1/m}$ will satisfy $(z^{1/m})^m = z$ for this negative m.

EXAMPLE 1

Find all values of $(-1)^{1/2}$ by means of Eq. (1.4–12).

Solution

Here $r = |-1| = 1$ and $m = 2$ in Eq. (1.4–12). For θ we can use *any* valid argument of -1. We will use π. Thus

$$(-1)^{1/2} = \sqrt{1}\left[\cos\left(\frac{\pi}{2} + k\pi\right) + i\sin\left(\frac{\pi}{2} + k\pi\right)\right], \qquad k = 0, 1.$$

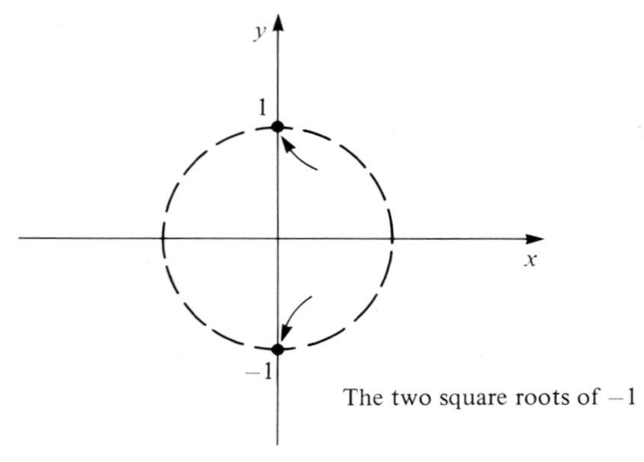

Figure 1.4–1

With $k = 0$ in this formula we obtain $(-1)^{1/2} = i$, and $k = 1$ yields $(-1)^{1/2} = -i$. Points representing the two roots are plotted in Fig. 1.4–1. Their angular separation is $2\pi/m = 2\pi/2 = \pi$ radians. ◀

EXAMPLE 2

Find all values of $1^{1/m}$ where m is a positive integer.

Solution

Taking $1 = r\,\text{cis}(\theta)$ where $r = 1$ and $\theta = 0$ and applying Eq. (1.4–12), we have

$$1^{1/m} = \sqrt[m]{1}[\cos(2k\pi/m) + i\sin(2k\pi/m)] = \text{cis}(2k\pi/m), \qquad k = 0, 1, 2, \ldots, m-1.$$

These m values of $1^{1/m}$ all have modulus 1. When displayed as points on the unit circle they are uniformly spaced and have an angular separation of $2\pi/m$ radians. Note that one value of $1^{1/m}$ is necessarily unity. ◀

EXAMPLE 3

Find all values of $(1 + i\sqrt{3})^{1/5}$.

Solution

We anticipate five roots. We use Eq. (1.4–12) with $m = 5$, $r = |1 + i\sqrt{3}| = 2$, and $\theta = \tan^{-1}\sqrt{3} = \pi/3$. Our result is

$$(1 + i\sqrt{3})^{1/5} = \sqrt[5]{2}\left[\cos\left(\frac{\pi}{15} + \frac{2\pi}{5}k\right) + i\sin\left(\frac{\pi}{15} + \frac{2\pi k}{5}\right)\right], \qquad k = 0, 1, 2, 3, 4.$$

Expressed as decimals, these answers become approximately

$$\begin{aligned} 1.123 + i0.241, &\qquad k = 0,\\ 0.120 + i1.142, &\qquad k = 1,\\ -1.049 + i0.467, &\qquad k = 2,\\ -0.769 - i0.854, &\qquad k = 3,\\ 0.574 - i0.995, &\qquad k = 4. \end{aligned}$$

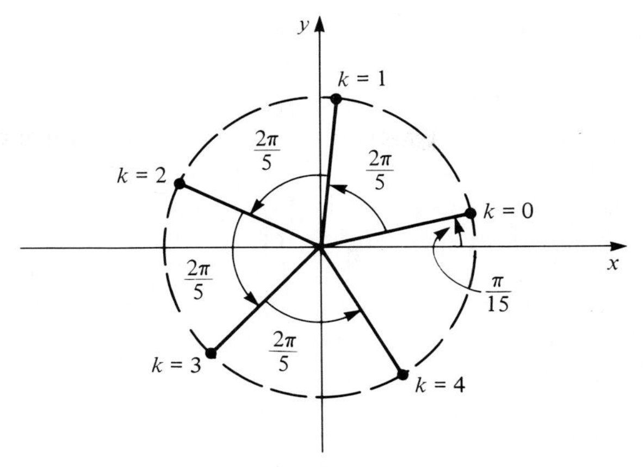

Figure 1.4–2

Vectors representing the roots are plotted in Fig. 1.4–2. They are spaced $2\pi/5$ radians or 72° apart. The vector for the case $k = 0$ makes an angle of $\pi/15$ radians, or 12°, with the x-axis. Any of these results, when raised to the fifth power, must produce $1 + i\sqrt{3}$. For example, let us use the root for which $k = 1$. We have, with the aid of Eq. (1.4–2),

$$\left(\sqrt[5]{2}\left[\cos\left(\frac{\pi}{15} + \frac{2\pi}{5}\right) + i\sin\left(\frac{\pi}{15} + \frac{2\pi}{5}\right)\right]\right)^5 = 2\left[\cos\left(\frac{\pi}{3} + 2\pi\right) + i\sin\left(\frac{\pi}{3} + 2\pi\right)\right]$$

$$= 2\left[\frac{1}{2} + \frac{i\sqrt{3}}{2}\right] = 1 + i\sqrt{3}. \qquad ◀$$

Our original motivation for expanding our number system from real to complex numbers was to permit our solving equations whose solution involved the square roots of negative numbers. It might have worried us that trying to find $(-1)^{1/3}$, $(-1)^{1/4}$, and so forth could lead us to successively more complicated number systems. From the work just presented we see that this is not the case. The complex system, together with Eq. (1.4–12), is sufficient to yield any such root.

From our discussion of the fractional powers of z, we can now formulate a consistent definition of z raised to any rational power, (e.g., $z^{4/7}$, $z^{-2/3}$) and, as a result, solve such equations as $z^{4/3} + 1 = 0$. We use the definition

$$z^{n/m} = (z^{1/m})^n.$$

With Eq. (1.4–12) we perform the inner operation and with Eq. (1.4–2) the outer one. Thus

$$z^{n/m} = (\sqrt[m]{r})^n\left[\cos\left(\frac{n}{m}\theta + \frac{2kn\pi}{m}\right) + i\sin\left(\frac{n}{m}\theta + \frac{2kn\pi}{m}\right)\right], \qquad k = 0, 1, 2, \ldots, m - 1, \tag{1.4–13}$$

where, as before, $\theta = \arg z$ and $|z| = r$. It is not difficult to show that the expressions $z^{-n/m}$ and $z^{n/-m}$ yield the same set of values. Thus, for example, we could compute $z^{4/-7}$ by using (1.4–13) with $m = 7$, $n = -4$.

If n/m is an irreducible fraction, then our letting k range from 0 to $m-1$ in Eq. (1.4–13) results in m numerically distinct roots. In the complex plane these roots are arranged uniformly around a circle of radius $(\sqrt[m]{r})^n$.

However, if n/m is reducible (i.e., n and m contain common integral factors) then, when k varies from 0 to $m-1$, some of the values obtained from Eq. (1.4–13) will be numerically identical. This is because the expression $2kn\pi/m$ will assume at least two values that differ by an integer multiple of 2π. In the extreme case where n is exactly divisible by m, all the values obtained from Eq. (1.4–13) are identical. This confirms the familiar fact that z raised to an integer power has but one value. The fraction n/m should be reduced as far as possible, *before* being used in Eq. (1.4–13), if we don't wish to waste time generating identical roots.

Assuming n/m is an irreducible fraction and z, a given complex number, let us consider the m possible values of $z^{n/m}$. Choosing any one of these values and raising it to the m/n power, we obtain n numbers. Only one of these is z. Thus the equation $(z^{n/m})^{m/n} = z$ must be interpreted with some care.

EXAMPLE 4

Solve the following equation for w:

$$w^{4/3} + 2i = 0. \qquad (1.4\text{–}14)$$

Solution

We have $w^{4/3} = -2i$, which means $w = (-2i)^{3/4}$. We now use Eq. (1.4–13) with $n = 3$, $m = 4$, $r = |z| = |-2i| = 2$, and $\theta = \arg(-2i) = -\pi/2$. Thus

$$w = (-2i)^{3/4} = (\sqrt[4]{2})^3 \angle \frac{3}{4}\left(\frac{-\pi}{2}\right) + 2k\frac{3}{4}\pi, \qquad k = 0, 1, 2, 3;$$

$$w \doteq 1.68 \angle \frac{-3\pi}{8}, \qquad k = 0;$$

$$w \doteq 1.68 \angle \frac{9\pi}{8}, \qquad k = 1;$$

$$w \doteq 1.68 \angle \frac{21\pi}{8}, \qquad k = 2;$$

$$w \doteq 1.68 \angle \frac{33\pi}{8}, \qquad k = 3.$$

The four results are plotted in the complex plane shown in Fig. 1.4–3. They are uniformly distributed on the circle of radius $(\sqrt[4]{2})^3$. Each of these results is a solution of Eq. (1.4–14) provided we use the appropriate 4/3 power. A check of the answer is performed in Exercise 16. ◀

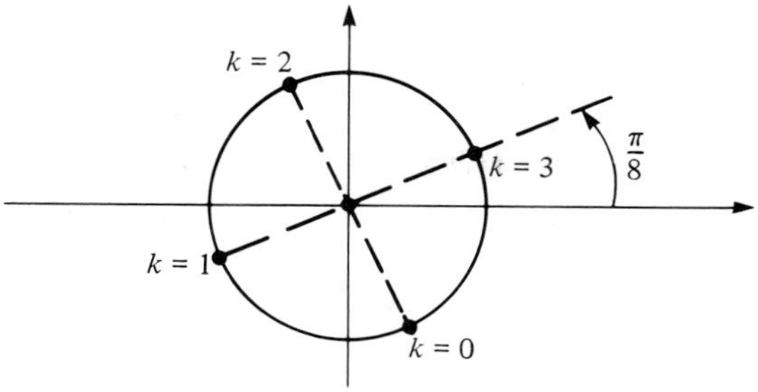

Figure 1.4–3

EXERCISES

Express each of the following in the form $a + ib$ and also in the polar form $r\underline{/\theta}$. Give θ as a principal value.

1. $(-\sqrt{3} + i)^9$ **2.** $(-\sqrt{3} - i)^{-5}$ **3.** $(3 + 4i)^{12}(1 + i)^{-12}$

4. $[2 \operatorname{cis}(-\pi/4)]^5$ **5.** $(\sqrt{3} - i)^4[\operatorname{cis}(\pi/10)]^2$ **6.** $(1 + 2i)^{19}$

7. With the aid of DeMoivre's theorem, express $\cos 3\theta$ as a real sum of terms containing only functions like $\cos^m \theta \sin^n \theta$, where m and n are nonnegative integers.

8. Repeat Exercise 7 to find a similar expression for $\sin 5\theta$.

Express the following in the form $a + ib$. Give all values. Make a polar plot of the points that represent your results.

9. $i^{1/2}$ **10.** $(1 - i)^{1/2}$ **11.** $(1 - i)^{-1/3}$

12. $(-1 - i\sqrt{3})^{-1/3}$ **13.** $1^{1/2}1^{-1/2}$ **14.** $(16i)^{1/4}1^{1/4}$

15. Consider the quadratic equation $az^2 + bz + c = 0$, where $a \neq 0$, and a, b, and c are complex numbers. Use the method of completing the square to show that $z = (-b + (b^2 - 4ac)^{1/2})/(2a)$. How many solutions does this equation have in general?

In high school you learned that if a, b, and c are real numbers, then the roots of the quadratic equation are either a pair of real numbers or a pair of complex numbers whose values are complex conjugates of each other. If a, b, and c are not restricted to being real does the preceding still apply?

16. Consider one of the solutions to Example 4, for example, the case $k = 2$. Raise this solution to the 4/3 power, state all the resulting values, and show that just one satisfies (1.4–14).

Find all solutions of each of the following equations. Give the answer as $a + ib$. Use the result of Exercise 15 where needed.

17. $w^2 - i = -1$ **18.** $w^3 - i = -\sqrt{3}$

19. $4w^2 + 4w + i = 0$ **20.** $w^4 - 2\sqrt{3}w^2 + 4 = 0$

21. $w^6 - 2w^3 + 2 = 0$ **22.** $w^5 + 16w - w^4 - 16 = 0$

23. a) Show that $z^n - 1 = (z - 1)(z^{n-1} + z^{n-2} + \cdots + 1)$, where n is an integer ≥ 1.

b) Use the preceding result to find and plot all solutions of $z^4 + z^3 + z^2 + z + 1 = 0$.

24. Using the technique suggested in Exercise 23, find and plot all solutions of $z^5 + z^4 + z^3 + z^2 + z + 1 = 0$.

25. Let $m \neq 0$ be an integer. We know that $z^{1/m}$ has m values and that $z^{-1/m}$ does also. For a given z and m we select at random a value of $z^{1/m}$ and one of $z^{-1/m}$.

a) Is their product necessarily one?

b) Is it always possible to find a value for $z^{-1/m}$ so that, for a given $z^{1/m}$, we will have $z^{1/m}z^{-1/m} = 1$?

26. Let $m \neq 0$ be a positive integer and $z = r \operatorname{cis} \theta$. Show that the m possible values of $z^{1/m}$ are given by $\sqrt[m]{r} \operatorname{cis}(\theta/m)\, 1^{1/m}$, where $1^{1/m} = \operatorname{cis}(2k\pi/m)$, $k = 0, 1, 2, \ldots, m-1$, are the m values of the mth root of 1.

Give all values of the following in the form $a + ib$ and draw their vector representations in the complex plane.

27. $i^{2/3}$ **28.** $(1 + i)^{7/2}$ **29.** $(1 + i)^{2/3}$

30. $(1 + i)^{4/6}$ **31.** $(8^{2/3})(8^{-2/3})$ **32.** $(1 + i)^{9/3}$

33. a) Consider the multivalued real expression $|1^{1/m} - i^{1/m}|$, where $m \geq 1$ is an integer. Show that its minimum possible value is $2 \sin(\pi/4m)$.

b) Find a comparable formula giving the maximum possible value of $|1^{1/m} + i^{1/m}|$.

34. Prove that

$$\left(\frac{1 + i \tan \theta}{1 - i \tan \theta}\right)^n = \frac{1 + i \tan n\theta}{1 - i \tan n\theta}$$

35. a) Show that if m and n are positive integers with $m \neq 0$ and if n/m is an irreducible fraction, then the set of values of $z^{n/m}$, defined by (1.4–13) as $(z^{1/m})^n$, is identical to the set of values of $(z^n)^{1/m}$.

b) If n/m is reducible (m and n contain common integer factors), then $(z^{1/m})^n$ and $(z^n)^{1/m}$ do not produce identical sets of values. Compare all the values of $(1^{1/4})^2$ with all the values of $(1^2)^{1/4}$ to see that this is so.

36. It is possible to extract the square root of the complex number $z = x + iy$ without resorting to polar coordinates. Let $a + ib = (x + iy)^{1/2}$, where x and y are known real numbers, and a and b are unknown real numbers.

a) Square both sides of this equation and show that this implies

$$1)\ \ x = a^2 - b^2,$$

$$2)\ \ y = 2ab.$$

Now assume $y \neq 0$.

b) Use equation (2) above to eliminate b from equation (1), and show that equation (1) now leads to a quadratic equation in a^2. Prove that

$$3)\ \ a^2 = \frac{x \pm \sqrt{x^2 + y^2}}{2}.$$

Explain why we must reject the minus sign in equation (3). Recall our postulate about the number a.

c) Show that

$$4)\quad b^2 = \frac{-x + \sqrt{x^2 + y^2}}{2}.$$

d) From the square roots of equations (3) and (4) we obtain

$$5)\quad a = \frac{\pm\sqrt{x + \sqrt{x^2 + y^2}}}{\sqrt{2}},$$

$$6)\quad b = \frac{\pm\sqrt{-x + \sqrt{x^2 + y^2}}}{\sqrt{2}}.$$

Assume y is positive. What does equation (2) say about the signs of a and b? Hence, if a, obtained from equation (5), is positive, then b, obtained from equation (6), is also. Show that if a is negative then so is b. Thus with $y > 0$ there are two possible values for $z^{1/2} = a + ib$.

e) Assume y is negative. Again show that $a + ib$ has two values and that a is positive and b is negative for one value and vice versa for the other.

f) Assume that $y = 0$. Show that $(x + iy)^{1/2}$ has a pair of values that are real, zero, or imaginary according to whether x is positive, zero, or negative. Use (1) and (2). Give a and b in terms of x.

g) Use equations (5) and (6) to obtain both values of $i^{1/2}$. Check your results by obtaining the same values by means of Eq. (1.4–12).

37. Recall from elementary algebra the formula for the sum of a finite geometric series:

$$1 + p + p^2 + \cdots + p^n = \frac{1 - p^{n+1}}{1 - p}, \qquad n \text{ is a nonnegative integer, } p \neq 1.$$

This formula is also valid when p is a complex number since the same derivation applies. Show by using this formula and DeMoivre's theorem that the sum of all n values of $z^{1/n}$ is zero when $n \geq 2$. Give a vector interpretation of your result.

38. Use the formula for the sum of a geometric series in Exercise 37 and DeMoivre's theorem to derive the following formulas for $0 < \theta < 2\pi$:

$$1 + \cos\theta + \cos 2\theta + \cdots + \cos n\theta = \frac{\cos(n\theta/2)\sin[(n+1)\theta/2]}{\sin(\theta/2)},$$

$$\sin\theta + \sin 2\theta + \sin 3\theta + \cdots + \sin n\theta = \frac{\sin(n\theta/2)\sin[(n+1)\theta/2]}{\sin(\theta/2)}.$$

39. If n is an integer greater than or equal to 2, prove that

$$\cos\left(\frac{2\pi}{n}\right) + \cos\left(\frac{4\pi}{n}\right) + \cdots + \cos\left[\frac{2(n-1)\pi}{n}\right] = -1,$$

and that

$$\sin\left(\frac{2\pi}{n}\right) + \sin\left(\frac{4\pi}{n}\right) + \cdots + \sin\left[\frac{2(n-1)\pi}{n}\right] = 0.$$

Hint: Use the result of Exercise 37 and take $z = 1$.

1.5 LOCI, POINTS, SETS, AND REGIONS IN THE COMPLEX PLANE

In the previous section we saw that there is a specific point in the z-plane that represents any complex number z. Similarly, as we will see, curves and areas in the z-plane can represent equations or inequalities in the variable z.

Consider the equation $\text{Re}(z) = 1$. If this is rewritten in terms of x and y, we have $\text{Re}(x + iy) = 1$, or $x = 1$. In the complex plane the locus of all points satisfying $x = 1$ is the infinite vertical line shown in Fig. 1.5–1. Now consider the inequality $\text{Re}\, z < 1$, which is equivalent to $x < 1$. All points that satisfy this inequality must lie in the region† to the left of the vertical line in Fig. 1.5–1. We show this in Fig. 1.5–2.

Similarly, the double inequality $-2 \leq \text{Re}\, z \leq 1$, which is identical to $-2 \leq x \leq 1$, is satisfied by all points lying between and on the vertical lines $x = -2$ and $x = 1$. Thus $-2 \leq \text{Re}\, z \leq 1$ defines the infinite strip shown in Fig. 1.5–3.

More complicated regions can be likewise described. For example, consider $\text{Re}\, z \leq \text{Im}\, z$. This implies $x \leq y$. The equality holds when $x = y$, that is, for all the points on the infinite line shown in Fig. 1.5–4. The inequality $\text{Re}\, z < \text{Im}\, z$ describes those points that satisfy $x < y$, that is, they lie to the left of the 45° line in Fig. 1.5–4. Thus $\text{Re}\, z \leq \text{Im}\, z$ represents the shaded region shown in the figure and includes the boundary line $x = y$.

The description of circles and their interiors is particularly important and easily accomplished. The locus of all points representing $|z| = 1$ is obviously the same as those for which $\sqrt{x^2 + y^2} = 1$, that is, the circumference of a circle, centered at the origin, of unit radius. The inequality $|z| < 1$ describes the points inside the circle (their modulus is less than unity), whereas $|z| \leq 1$ represents the inside and the circumference.

We need not restrict ourselves to circles centered at the origin. Let $z_0 = x_0 + iy_0$ be a complex constant. Then the points in the z-plane representing solutions of $|z - z_0| = r$, where $r > 0$, form the circumference of a circle, of radius r, centered at x_0, y_0. This statement can be proved by algebraic or geometric means. The latter course is followed in Fig. 1.5–5, where we use the vector representation of complex numbers. A vector for z_0 is drawn from the origin to the fixed point x_0, y_0 while another, for z, goes to the variable point whose coordinates are x, y. The vector difference $z - z_0$ is also shown. If this quantity is kept constant in magnitude, then

---

† A precise definition of "region" is given further in the text. For the moment we will regard the word as meaning some portion of the total z-plane.

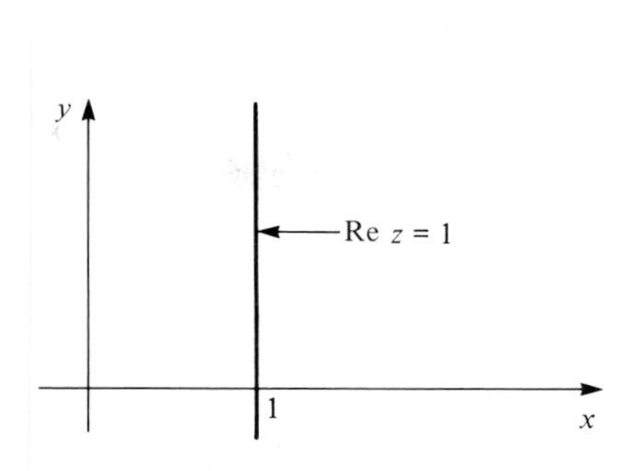

Figure 1.5–1

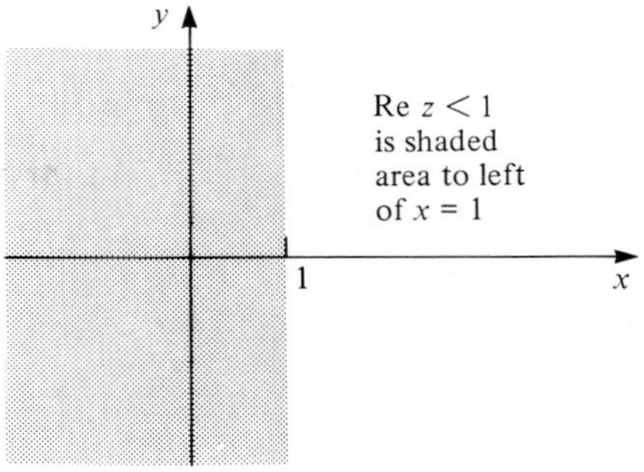

Figure 1.5–2

z is obviously confined to the perimeter of the circle indicated. The points representing solutions of $|z - z_0| < r$ lie inside the circle, whereas those for which $|z - z_0| > r$ lie outside.

Finally, let r_1 and r_2 be a pair of nonnegative real numbers such that $r_1 < r_2$. Then the double inequality $r_1 < |z - z_0| < r_2$ is of interest. The first part, $r_1 < |z - z_0|$, specifies those points in the z-plane that lie outside a circle of radius r_1 centered at x_0, y_0, whereas the second part $|z - z_0| < r_2$ refers to those points inside a circle of radius r_2 centered at x_0, y_0. Points that simultaneously satisfy both inequalities must lie in the *annulus* (a disc with a hole in the center) of inner radius r_1, outer radius r_2, and center z_0.

EXAMPLE 1

What region is described by the inequality $1 < |z + 1 - i| < 2$?

Solution

We can write this as $r_1 < |z - z_0| < r_2$, where $r_1 = 1, r_2 = 2, z_0 = -1 + i$. The region described is the shaded area *between*, but not including, the circles shown in Fig. 1.5–6. ◀

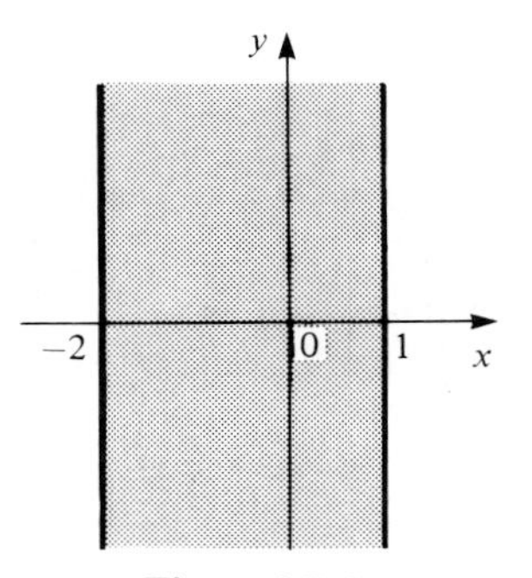

Figure 1.5–3

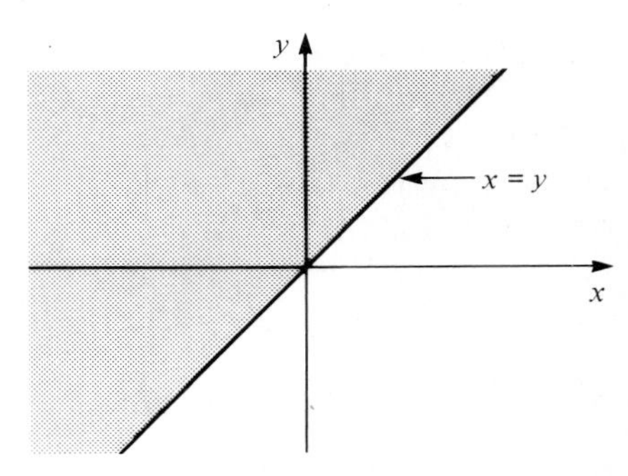

Figure 1.5–4

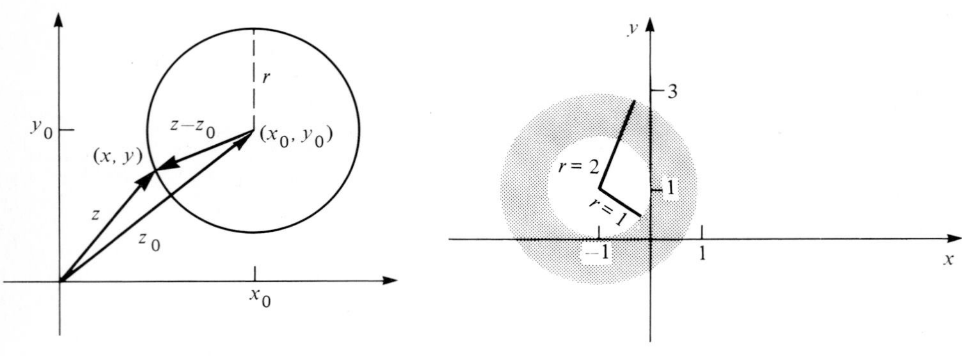

Figure 1.5–5

Figure 1.5–6

Points and Sets

We need to have a small vocabulary with which to describe various points and collections of points (called *sets*) in the complex plane. The following terms are worth studying and memorizing since most of the language will reappear in subsequent chapters.

The points belonging to a set are called its *members* or *elements*.

A *neighborhood* of radius r of a point z_0 is the collection of all the points inside a circle, of radius r, centered at z_0. These are the points satisfying $|z - z_0| < r$. A given point can have various neighborhoods since circles of different radii can be constructed around the point.

A *deleted neighborhood* of z_0 consists of the points inside a circle centered at z_0 but excludes the point z_0 itself. These points satisfy $0 < |z - z_0| < r$. Such a set is sometimes called a *punctured disc* of radius r centered at z_0.

An *open set* is one in which every member of the set has some neighborhood, all of whose points lie entirely within that set. For example, the set $|z| < 1$ is open. This inequality describes all the points inside a unit circle centered at the origin. As shown in Fig. 1.5–7, it is possible to enclose every such point with a circle C_0 (perhaps very tiny) so that all the points inside C_0 lie within the unit circle. The set $|z| \le 1$ is not open. Points on and inside the circle $|z| = 1$ belong to this set. But every

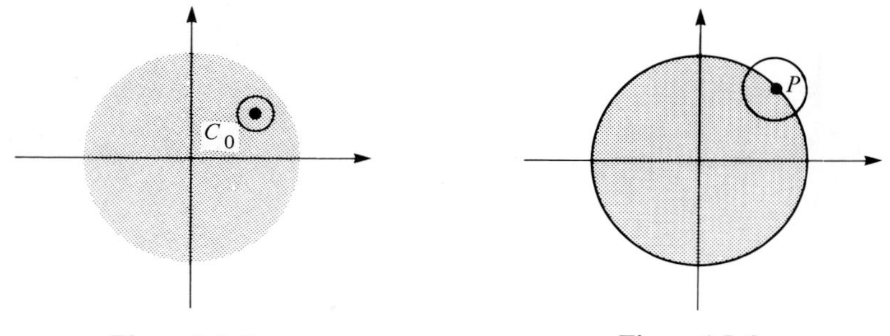

Figure 1.5–7

Figure 1.5–8

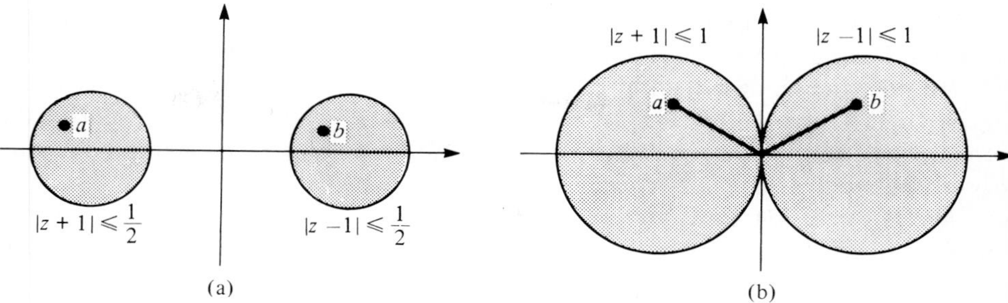

Figure 1.5–9

neighborhood, no matter how tiny, of a point such as P that lies on $|z| = 1$ (see Fig. 1.5–8) contains points outside the given set.

A *connected set* is one in which any two points of the set can be joined by some path of straight line segments, all of whose points belong to the set. Thus the set of points shown shaded in Fig. 1.5–9(a) is not connected since we cannot join a and b by a path within the set. However, the set of points in Fig. 1.5–9(b) is connected.

A *domain* is an open connected set. For example $\text{Re } z < 3$ describes a domain. However, $\text{Re } z \leq 3$ does not describe a domain since the set defined is not open.

We will often speak of *simply* and *multiply connected domains.* Loosely speaking, a simply connected domain contains no holes, but a multiply connected domain has one or more holes. An example of the former is $|z| < 2$, and an example of the latter is $1 < |z| < 2$, which contains a circular hole. More precisely, when any closed curve is constructed in a simply connected domain, every point inside the curve lies in the domain. On the other hand, it is always possible to construct some closed curve inside a multiply connected domain in such a way that one or more points inside the curve do not belong to the domain (see Fig. 1.5–10). A doubly connected domain has one hole, a triply connected domain two holes, etc.

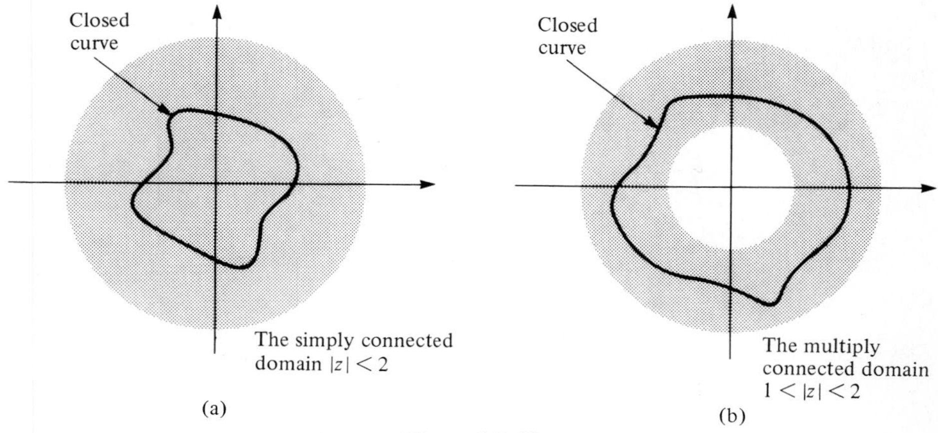

Figure 1.5–10

A *boundary point* of a set is a point whose every neighborhood contains at least one point belonging to the set and one point not belonging to the set.

Consider the set described by $|z - 1| \leq 1$, which consists of the points inside and on the circle shown in Fig. 1.5–11. The point $z = 1 + i$ is a boundary point of the set since, inside every circle such as C_0, there are points belonging to and not belonging to the given set. *Although in this case the boundary point is a member of the set in question, this need not always be so.* For example, $z = 1 + i$ is a boundary point but not a member of the set $|z - 1| < 1$. One should realize that an open set cannot contain any of its boundary points.

An *interior point* of a set is a point having some neighborhood, all of whose elements belong to the set. Thus $z = 1 + i/2$ is an interior point of the set $|z - 1| \leq 1$ (see Fig. 1.5–11).

An *exterior point* of a set is a point having a neighborhood all of whose elements do not belong to the set. Thus, $1 + 2i$ is an exterior point of the set $|z - 1| \leq 1$.

An *accumulation point*, let us call it P, of a set S is a point whose every neighborhood contains at least one member of S besides P. Note that P need not belong to S. The term *limit point* is also used to mean accumulation point.

The set of points $z = (1 + i)/n$, where n assumes the value of all finite positive integers, has $z = 0$ as an accumulation point. This is because as n becomes increasingly positive, elements of the set are generated that lie increasingly close to $z = 0$. Every circle centered at the origin will contain members of the set. In this example, the accumulation point does not belong to the set.

A *region* is a domain plus possibly some, none, or all the boundary points of the domain. Thus every domain is a region, but not every region is a domain. The set defined by $2 < \text{Re}\, z \leq 3$ is a region. It contains some of its boundary points (on $\text{Re}\, z = 3$) but not others (on $\text{Re}\, z = 2$) (see Fig. 1.5–12). This particular region is not a domain.

A *closed region* consists of a domain plus all the boundary points of the domain.

A *bounded set* is one whose points can be enveloped by a circle of some finite radius. For example, the set occupying the square $0 \leq \text{Re}\, z \leq 1$, $0 \leq \text{Im}\, z \leq 1$ is bounded since we can put a circle around it (see Fig. 1.5–13). A set that cannot be encompassed by a circle is called *unbounded*. An example is the infinite strip in Fig. 1.5–12.

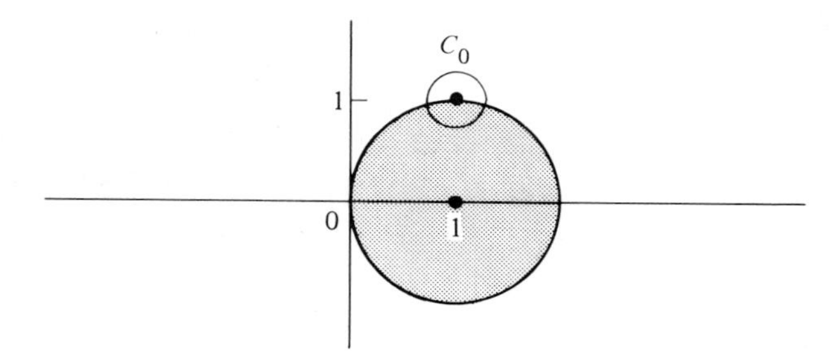

Figure 1.5–11

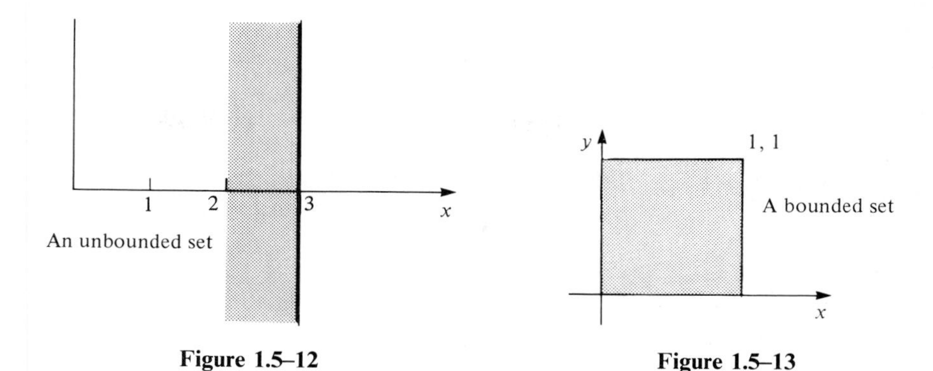

Figure 1.5–12

Figure 1.5–13

The Complex Number Infinity and the Point at Infinity

When dealing with real numbers, we frequently use the concept of infinity and speak of "plus infinity" and "minus infinity." For example, the sequence 1, 10, 100, 1000,... diverges to plus infinity, and the sequence $-1, -2, -4, -8, \ldots$ diverges to minus infinity.

In dealing with complex numbers we also speak of infinity, which we call "the complex number infinity." It is designated by the usual symbol ∞. We do not give a sign to the complex infinity nor do we define its argument. Its modulus, however, is larger than any preassigned real number.

We can imagine that the complex number infinity is represented graphically by a point in the Argand plane—a point, unfortunately, that we can never draw in this plane. The point can be reached by proceeding along any path in which $|z|$ grows without bound, as, for instance, is shown in Fig. 1.5–14.

In order to make the notion of a *point at infinity* more tangible, we use an artifice called the stereographic projection illustrated in Fig. 1.5–15.

Consider the z-plane, with a third orthogonal axis, the ζ-axis,† added on. A sphere of radius $1/2$ is placed with center at $x = 0$, $y = 0$, $\zeta = 1/2$. The north pole, N, lies at $x = 0$, $y = 0$, $\zeta = 1$ while the south pole, S, is at $x = 0$, $y = 0$, $\zeta = 0$. This is called the *Riemann number sphere.*

Let us draw a straight line from N to the point in the xy-plane that represents a complex number z. This line intersects the sphere at exactly one point, which we label z'. We say that z' is the projection on the sphere of z. In this way *every* point in the complex plane can be projected on to a corresponding unique point on the sphere. Points far from the origin in the xy-plane are projected close to the top of the sphere, and, as we move farther from the origin in the plane, the corresponding projections on the sphere cluster more closely around N. Thus we conclude that N on the sphere corresponds to the point at infinity, although we are not able to draw $z = \infty$ in the complex plane.

† Obviously, we don't want to call this the z-axis.

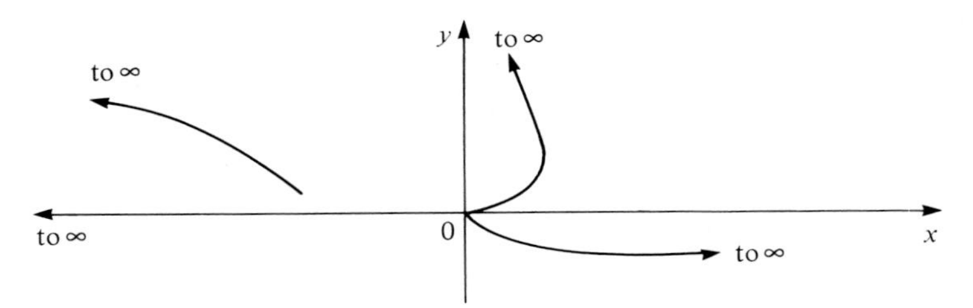

Figure 1.5–14

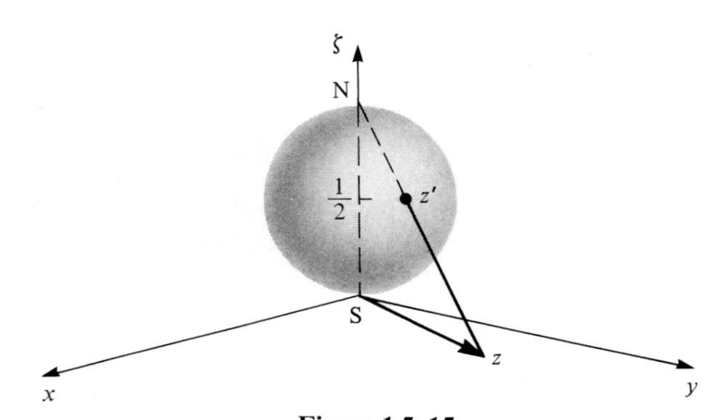

Figure 1.5–15

When we regard the z-plane as containing the point at infinity, we call it the "extended z-plane." When ∞ is not included, we merely say "the z-plane" or "the finite z-plane."

Except when explicitly stated, we will not regard infinity as a number in any sense in this text. However, when we employ the extended complex plane, we will treat infinity as a number satisfying these rules:

$$\frac{z}{\infty} = 0; \qquad z \pm \infty = \infty, \quad (z \neq \infty); \qquad \frac{z}{0} = \infty, \quad (z \neq 0);$$

$$z \cdot \infty = \infty, \quad (z \neq 0); \qquad \frac{\infty}{z} = \infty, \quad (z \neq \infty).$$

We do not define

$$\infty + \infty, \qquad \infty - \infty, \qquad \text{or} \qquad \frac{\infty}{\infty}.$$

EXERCISES

Describe with words or a sketch the portion of the complex plane corresponding to the following equations or inequalities. State which problems have no solutions.

1. $\operatorname{Re} z = -2$ **2.** $\operatorname{Re} z = 3 \operatorname{Im} z$

3. $\operatorname{Im} z \geq \operatorname{Re} z$

4. $\operatorname{Im} z \geq |z|$

5. $|z + 3 - 4i| > 5$

6. $|z + 3 - 4i| \leq 5$

7. $|z + 3 - 4i| \leq 0$

8. $\operatorname{Im}(z + 2i) = \operatorname{Re}(z - 3)$

9. $z\bar{z} \geq 2 \operatorname{Re} z$

10. $\operatorname{Im} z \geq \operatorname{Re}(z^2)$

11. $|z| \leq e^{-|z|}$ Do this numerically with a pocket calculator.

12. $\sin |z| \geq 1/\sqrt{2}$

13. Find the points on the circle $|z - 1 - i| = 1$ that have the nearest and furthest linear distance to the point $z = -1 + i0$. In addition, state what these two distances are.

    Hint: Consider a vector beginning at $-1 + i0$ and passing through the center of the circle. Where can this vector intersect this circle?

Represent the following regions by means of equations or inequalities in the variable z.

14. All the points lying on and exterior to a circle of unit radius centered at $-1 - i$.

15. All the points occupying an annular region centered at $3 + i$. The inner radius is 2; the outer is 4. Exclude points on the inner boundary but include those on the outer one.

16. All the points, except the center, on and within a circle of radius 2, centered at $3 + 4i$.

17. Consider the open set described by $|z| < 1$. Find a neighborhood of the point $0.99i$ that lies entirely within the set. Represent the neighborhood with an inequality involving z. In the same way specify a deleted neighborhood of $0.99i$ within the set.

The *union* of two sets A and B, written $A \cup B$, means the set of all points belonging to A *or* B, while the *intersection* of A and B, written $A \cap B$, means the set of all points belonging to *both* A and B. Which of the following are connected sets? Which of the connected sets are domains?

18. The set $A \cup B$, where A consists of the points in $|z| < 1$ while B consists of the points in $|z - 1| < 1$. Sketch $A \cup B$.

19. The set $A \cap B$, where A and B are as given in Exercise 18. Sketch $A \cap B$.

20. The set $A \cup B$, where A consists of the points in $|z| < 1$ and B consists of the points in $\operatorname{Re} z \geq 1$.

21. The set described in Exercise 20, except that B now consists of the points in $\operatorname{Re} z > 1$.

What are the boundary points of the sets defined below? State which boundary points belong to the given set.

22. $0 < |z - 3| \leq 2$

23. $3 < |z + 1 + i| < 4$

24. $\sin |z| < 1$

25. $\operatorname{cis}(1/n)$, where $n < \infty$ assumes all positive integer values.

A set is said to be *closed* if it contains all its boundary points. Which of the following are closed sets?

26. $-1 < \operatorname{Re} z < 1$

27. $-1 < \operatorname{Re} z \leq 1$

28. $-1 \leq \operatorname{Re} z \leq 1$

29. $0 < |z| \leq 2$

30. $|z| \leq 2$

31. The union of the sets in Exercises 28 and 29.

32. What are the accumulation points of the sets defined in Exercises 26 and 29? Which of the accumulation points do not belong to the given set?

33. Is a boundary point of a set necessarily an accumulation point of that set? Explain.

34. According to the Bolzano–Weierstrass Theorem,† a bounded set having an infinite number of points must have at least one accumulation point. Consider the set consisting of the solutions of $y = 0$ and $\sin(\pi/x) = 0$ lying in the domain $0 < |z| < 1$. What is the accumulation point for this set? Prove your result by showing mathematically that every neighborhood of this point contains at least one member of the given set.

35. a) When all the points on the unit circle $|z| = 1$ are projected stereographically on to the sphere of Fig. 1.5–15, where do they lie?

b) Where are all the points inside the unit circle projected?

c) Where are all the points outside the unit circle projected?

36. Use stereographic projection to justify the statement that the two semiinfinite lines $y = x$, $x \geq 0$ and $y = -x$, $x \leq 0$ intersect twice, once at the origin and once at infinity. What is the projection of each of these lines on the Riemann number sphere?

37. a) If $z = x_1 + iy_1$ in Fig. 1.5–15 and if z' (the projection of z onto the number sphere) has coordinates x', y', ζ', show algebraically that:

$$\zeta' = \frac{x_1^2 + y_1^2}{x_1^2 + y_1^2 + 1}, \qquad x' = x_1\left(1 - \frac{x_1^2 + y_1^2}{x_1^2 + y_1^2 + 1}\right), \qquad y' = y_1\left(1 - \frac{x_1^2 + y_1^2}{x_1^2 + y_1^2 + 1}\right)$$

b) Consider a circle of radius r lying in the xy-plane of Fig. 1.5–15. The circle is centered at the origin. The stereographic projection of this circle onto the number sphere is another circle. Find the radius of this circle by using the equations derived in (a). Check your result by finding the answer geometrically.

---

† See, for example, Tom Apostol, *Mathematical Analysis*, 2nd ed. (Reading, MA: Addison-Wesley, 1974), p. 54.

This theorem is one of the most important in the mathematics of infinite processes (analysis). It was first proved by the Czech priest Bernhard Bolzano (1781–1848) and was later used and publicized by the German mathematician Karl Weierstrass (1815–1897).

CHAPTER 2

The Complex Function and Its Derivative

2.1 INTRODUCTION

In studying elementary calculus the reader doubtless received ample exposure to the concept of a real function of a real variable. To review briefly: When y is a function of x, or $y = f(x)$, we mean that when a value is assigned to x there is at our disposal a method for determining a corresponding value of y. We term x the independent variable and y the dependent variable in the relationship. Often y will be specified only for certain values of x and left undetermined for others. If the quantity of values involved is relatively small, we might express the relationship between x and y by presenting a list showing a numerical value for y for each x.

Of course, there are other ways to express a functional relationship besides using such a table. The most common method involves a mathematical formula as in the expression $y = e^x$, $-\infty < x < \infty$, which, in this case, yields a value of y for any value of x. Occasionally, we require several formulas, as in the following: $y = e^x$, $x > 0$; $y = \sin x$, $x < 0$. Taken together, these expressions determine y for any value of x except zero, that is, y is undefined at $x = 0$.

The term *multivalued function* is used in mathematics and will occur at various places in this book. To see where this phrase might be used, consider the expression $y = x^{1/2}$. Assigning a positive value to x we find two possible values for y; they differ only in their sign. Because we obtain two, not one, values for y, the statement $y = x^{1/2}$ does not by itself give a function of x. However, because there is a set of possible values of y (two in fact) for each $x > 0$ we speak of $y = x^{1/2}$ as describing a multivalued function of x, for positive x. A multivalued function is not really a function. In general, if we are given an expression in which two or more values of the dependent variable are obtained for some set of values of the independent variable, we say that we are given a multivalued function. In this book the word "function" by itself is applied in the strict sense unless we use the adjective "multivalued."

The easiest way to visualize most functional relationships is by means of a graphical plot, and the reader doubtless spent time in high school drawing y versus x, in the Cartesian plane, for various functions.

Some, but not all, of these concepts carry over directly into the study of functions of a complex variable. Here we use an independent variable, usually z, that can assume complex values. We will be concerned with functions typically defined in a domain or region of the complex z-plane. To each value of z in the region there will correspond a value of a dependent variable, let us say w, and we will say that w is a function of z, or $w = f(z)$, in this region. Often the region will be the entire z-plane.† We must assume that w, like z, is capable of assuming values that are complex, real, or purely imaginary. Some examples follow.

	$w = f(z)$	Region in which w is defined
a)	$w = 2z$	all z
b)	$w = e^{\lvert z\rvert}$	all z
c)	$w = 2i\lvert z\rvert^2$	all z
d)	$w = (z + 3i)/(z^2 + 9)$	all z except $\pm 3i$

Example (a) is quite straightforward. If z assumes a complex value, say $3 + i$, then $w = 6 + 2i$. If z happens to be real, w is also.

In example (b), w assumes only real values irrespective of whether z is real, complex, or purely imaginary; for example, if $z = 3 + i$, $w = e^{\sqrt{10}} \doteq 23.6$.

Conversely, in example (c), w is purely imaginary for all z; for example, if $z = 3 + i$, $w = 2i|3 + i|^2 = 20i$.

Finally, in example (d), $(z + 3i)/(z^2 + 9)$ cannot define a function of z when $z = 3i$, since the denominator vanishes there. If $z = -3i$ both the numerator and denominator vanish, an indeterminate form $0/0$ results, and the function is again undefined.

The function $w(z)$ is sometimes expressed in terms of the variables x and y rather than directly in z. For example, $w(z) = 2x^2 + iy$ is a function of the variable z since, with z known, x and y are determined. Thus if $z = 3 + 4i$, then $w(3 + 4i) = 2 \cdot 3^2 + 4i = 18 + 4i$. Often an expression for w, given in terms of x and y, can be rewritten rather

† The term "domain of definition" (of a function) is often used to describe the set of values of the independent variable for which the function is defined. A domain in this sense may or may not be a domain in the sense in which we use it (i.e., as defined in Section 1.5).

simply in terms of z; in other cases the z-notation is rather cumbersome. In any case, the identities

$$x = \frac{z + \bar{z}}{2}, \qquad y = \frac{1}{i}\frac{(z - \bar{z})}{2} \tag{2.1–1}$$

are useful if we wish to convert from the xy-variables to z. An example follows.

EXAMPLE 1

Express w directly in terms of z if

$$w(z) = 2x + iy + \frac{x - iy}{x^2 + y^2}.$$

Solution

Using Eq. (2.1–1), we rewrite this as

$$w(z) = (z + \bar{z}) + \frac{i(z - \bar{z})}{i2} + \frac{\bar{z}}{z\bar{z}} = \frac{3z}{2} + \frac{\bar{z}}{2} + \frac{1}{z}. \qquad \blacktriangleleft$$

In general, $w(z)$ possesses both real and imaginary parts, and we write this function in the form $w(z) = u(z) + iv(z)$, or

$$w(z) = u(x, y) + iv(x, y). \tag{2.1–2}$$

where u and v are real functions of the variables x and y. In Example 1 we have

$$u = 2x + \frac{x}{x^2 + y^2} \quad \text{and} \quad v = y - \frac{y}{x^2 + y^2}.$$

A difference between a function of a complex variable $u + iv = f(z)$ and a real function of a real variable $y = f(x)$ is that while we can usually plot the relationship $y = f(x)$ in the Cartesian plane, graphing is not so easily done with the complex function. Two numbers x and y are required to specify any z, and another pair of numbers is required to state the resulting values of u and v. Thus, in general, a four-dimensional space is required to plot $w = f(z)$ with two dimensions reserved for the independent variable z and the other two used for the dependent variable w.

For obvious reasons four-dimensional graphs are not a convenient means for studying a function. Instead, other techniques are employed to visualize $w = f(z)$. This matter is discussed at length in Chapter 8, and some readers may wish to skip to Sections 8.1–8.3 after finishing this one. A small glimpse of one useful technique is in order here, however.

Two coordinate planes, the z-plane with x- and y-axes and the w-plane with u- and v-axes, are drawn side-by-side. Now consider a complex number A, which lies in the z-plane within a region for which $f(z)$ is defined. The value of w that corresponds to A is $f(A)$. We denote $f(A)$ by A'. The pair of numbers A and A' are now plotted

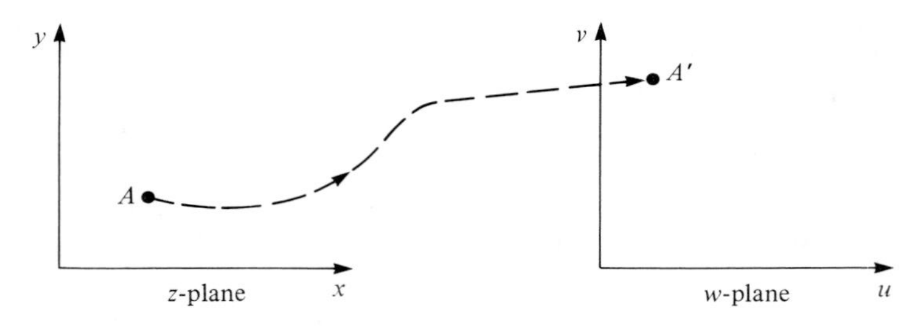

Figure 2.1–1

in the z- and w-planes, respectively (see Fig. 2.1–1). We say that the complex number A' is the *image* of A under the mapping $w = f(z)$ and that the points A and A' are images of each other.

In order to study a particular function $f(z)$ we can plot some points in the z-plane and also their corresponding images in the uv-plane. In the following table and in Fig. 2.1–2 we have investigated a few points in the case of $w = f(z) = z^2 + z$.

z	$w = z^2 + z$
$A = 0$	$0 = A'$
$B = 1$	$2 = B'$
$C = 1 + i$	$1 + 3i = C'$
$D = i$	$-1 + i = D'$

After determining the image of a substantial number of points we may develop some feeling for the behavior of $w = f(z)$. We have not yet discussed which points to choose in this endeavor. A systematic method that involves finding the images of points comprising entire curves in the z-plane is discussed in Sections 8.1–8.3.

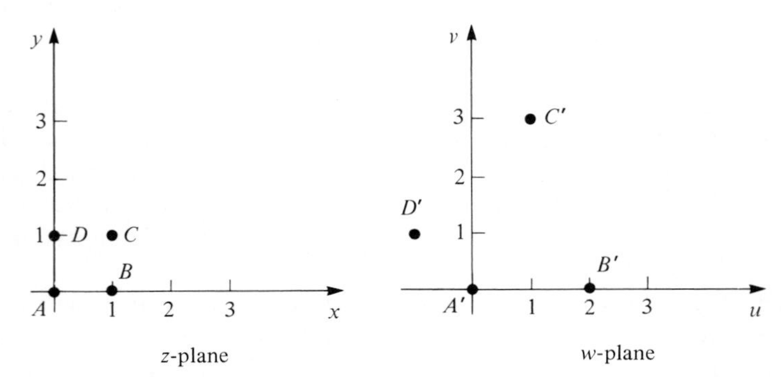

Figure 2.1–2

EXERCISES

Suppose $f(z) = (z - i)(z - 1)/[(z)(z^2 + 1)]$. State the point or points where $f(z)$ fails to be defined in each of the following domains.

1. $|z| < 3/4$

2. $0 < |z| < 0.99$

3. $|z| < 2$

4. $|z - 1| < \sqrt{2}$

5. $|z - i| < 2$

For each of the following functions find $f(1 + 2i)$.

6. $\dfrac{z}{z - \bar{z}}$

7. $\dfrac{1}{|z|}$

8. $\dfrac{1}{\sin x + i \cos y}$

9. $\dfrac{x - iy}{1 + z}$

10. $(x^2 + y^2) \sin x + i \cos y$

Write the following functions of z in the form $u(x, y) + iv(x, y)$, where u and v are explicit real functions of x and y.

11. $(z - i)^2$

12. $|z|^2 + i$

13. $z^{-1} + i$

14. $(\bar{z})^{-2} + i$

15. z^3

Rewrite the following complex functions entirely in terms of the complex variable z, its conjugate, and constants.

16. $-2xy + i(x^2 - y^2)$

17. $-2xy + i(x^2 + y^2)$

18. $x^2 + iy^2$

19. $x^2 + y^2$

For each of these functions, tabulate the value of the function for these values of z: $1 + i0$, $1 + i$, $0 + i$, $-1 + i$, -1. Indicate graphically the correspondence between values of w and values of z by means of a diagram like Fig. 2.1–2.

20. $w = 1/z$

21. $w = iz$

22. $w = \arg z$ (principal value)

23. $w = \text{Log}|z| + i \arg z$ (natural log and principal argument)

24. $w = e^x \cos y + ie^x \sin y$

Let $f(z) = z^2 + 1$.

25. Find $f(f(i))$.

26. Find $f\left(f\left(\dfrac{1}{1 + i}\right)\right)$.

27. Find $f(f(z))$ in the form $u(x, y) + iv(x, y)$.

2.2 LIMITS AND CONTINUITY

In elementary calculus the reader learned the notion of the limit of a function as well as the definition of continuity as applied to real variables. These concepts apply with some modification to functions of a complex variable. Let us first briefly review the real case.

The function $f(x)$ has a limit f_0 as x tends to x_0 (written $\lim_{x\to x_0} f(x) = f_0$) if the difference between $f(x)$ and f_0 can be made as small as we wish provided we choose x sufficiently close to x_0. In mathematical terms, given any positive number ε, we have

$$|f(x) - f_0| < \varepsilon \tag{2.2–1}$$

if x satisfies

$$0 < |x - x_0| < \delta, \tag{2.2–2}$$

where δ is a positive number typically dependent upon ε. Note that x never precisely equals x_0 in Eq. (2.2–2) and that $f(x_0)$ need not be defined for the limit to exist.

An obvious example of a limit is $\lim_{x\to 1}(1 + 2x) = 3$. To demonstrate this rigorously note that Eq. (2.2–1) requires $|1 + 2x - 3| < \varepsilon$, which is equivalent to

$$|x - 1| < \varepsilon/2. \tag{2.2–3}$$

Since $x_0 = 1$, Eq. (2.2–2) becomes

$$0 < |x - 1| < \delta. \tag{2.2–4}$$

Thus Eq. (2.2–3) can be satisfied if we choose $\delta = \varepsilon/2$ in Eq. (2.2–4).

A more subtle example proved in elementary calculus is

$$\lim_{x\to 0} \frac{\sin x}{x} = 1.$$

An intuitive verification can be had from a plot of $\sin x/x$ as a function of x and the use of $\sin x \approx x$ for $|x| \ll 1$.

Let us consider two functions that fail to possess limits at certain points. The function $f(x) = 1/(x - 1)^2$ fails to possess a limit at $x = 1$ because this function becomes unbounded as x approaches 1. The expression $|f - f_0|$ in Eq. (2.2–1) is unbounded for x satisfying Eq. (2.2–2) regardless of what value is assigned to f_0.

Now consider $f(x) = u(x)$, where $u(x)$ is the unit step function (see Fig. 2.2–1) defined by

$$u(x) = 0, \quad x < 0, \qquad u(x) = 1, \quad x \geq 0.$$

We investigate the limit of $f(x)$ at $x = 0$. With $x_0 = 0$, Eq. (2.2–2) becomes $0 < |x| < \delta$. Notice x can lie to the right or to the left of 0. The left side of Eq. (2.2–1) is now either $|1 - f_0|$ or $|f_0|$ according to whether x is positive or negative. If $\varepsilon < 1/2$, it is impossible to simultaneously satisfy the inequalities $|1 - f_0| < \varepsilon$ and $|f_0| < \varepsilon$, irrespective of the value of f_0. We see that a function having a "jump" at x_0 cannot have a limit at x_0.

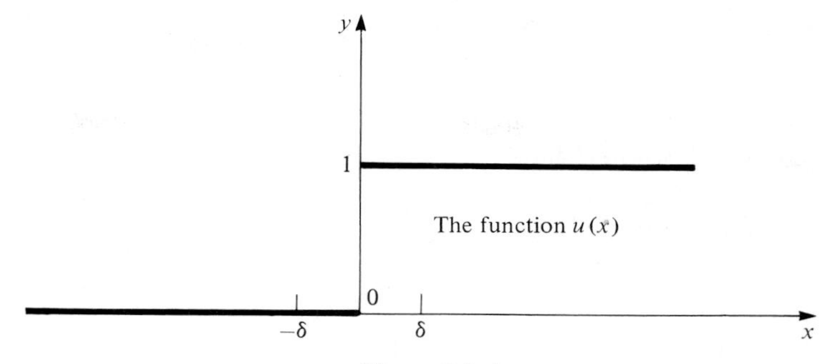

Figure 2.2–1

For $f(x)$ to be *continuous* at a point x_0, $f(x_0)$ must be defined and $\lim_{x\to x_0} f(x)$ must exist. Furthermore, these two quantities must agree, that is,

$$\lim_{x\to x_0} f(x) = f(x_0). \tag{2.2–5}$$

A function that fails to be continuous at x_0 is said to be *discontinuous* at x_0.

The functions $1/(x-1)^2$ and $u(x)$ fail to be continuous at $x = 1$ and $x = 0$ respectively because they do not possess limits at these points. The function

$$f(x) = \begin{cases} \dfrac{\sin x}{x}, & x \neq 0, \\ 2, & x = 0, \end{cases}$$

is discontinuous at $x = 0$. We have that $\lim_{x\to 0} f(x) = 1$, and $f(0) = 2$; thus Eq. (2.2–5) is not satisfied. However, one can show that $f(x)$ is continuous for all $x \neq 0$.

The concept of a limit can be extended to complex functions of a complex variable according to the following definition.

DEFINITION Limit

Let $f(x)$ be a complex function of the complex variable z, and let f_0 be a complex constant. If for every real number $\varepsilon > 0$ there exists a real number $\delta > 0$ such that

$$|f(z) - f_0| < \varepsilon \tag{2.2–6}$$

for all z satisfying

$$0 < |z - z_0| < \delta, \tag{2.2–7}$$

then we say that

$$\lim_{z\to z_0} f(z) = f_0;$$

that is, $f(z)$ has a limit f_0 as z tends to z_0. □

The definition asserts that ε, the upper bound on the magnitude of the difference between $f(z)$ and its limit f_0, can be made arbitrarily small, provided that we confine z to a deleted neighborhood of z_0. The radius, δ, of this deleted neighborhood typically depends on ε and becomes smaller with decreasing ε.

To employ the preceding definition we require that $f(z)$ be defined in a deleted neighborhood of z_0. The definition does not use $f(z_0)$. Indeed we might well have a function that is undefined at z_0 but has a limit at z_0.

EXAMPLE 1

As a simple example of this definition, show that

$$\lim_{z \to i}(z + i) = 2i.$$

Solution

We have $f(z) = z + i$, $f_0 = 2i$, $z_0 = i$. From (2.2–6) we need

$$|z + i - 2i| < \varepsilon,$$

or equivalently

$$|z - i| < \varepsilon, \tag{2.2–8}$$

which according to (2.2–7) must hold for

$$0 < |z - i| < \delta. \tag{2.2–9}$$

Taking δ as, say, ε (this is not the only possible choice; e.g., $\varepsilon/2$ will work), we see that (2.2–8) will be satisfied as long as z lies in the deleted neighborhood of i described in (2.2–9). ◀

When we investigated the limit as $x \to x_0$ of the real function $f(x)$ we were concerned with values of x lying to the right and left of x_0. If the limit f_0 exists, then as x approaches x_0 from either the right or left $f(x)$ must become increasingly close to f_0. In the case of the step function $u(x)$ previously considered, $\lim_{x \to 0} f(x)$ fails to exist because as x shrinks toward zero from the right (positive x), $f(x)$ remains at 1; but if x shrinks toward zero from the left (negative x), $f(x)$ remains at zero.

In the complex plane the concept of limit is more complicated because there are infinitely many *paths*, not just two directions, along which we can approach z_0. Four such paths are shown in Fig. 2.2–2. If $\lim_{z \to z_0} f(z)$ exists, $f(z)$ must tend toward the same complex value no matter which of the infinite number of paths of approach to z_0 is selected. Fortunately, in Example 1 the precise nature of the path used did not figure in our calculation. This is not always the case, as for the following two functions that fail to have limits at certain points. We demonstrate this by considering particular paths.

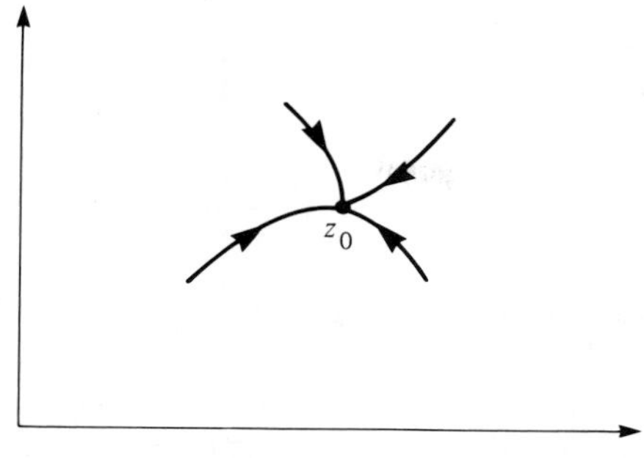

Figure 2.2–2

EXAMPLE 2

Let $f(z) = \arg z$ (principal value). Show that $f(z)$ fails to possess a limit on the negative real axis.

Solution

Consider a point z_0 on the negative real axis. Refer to Fig. 2.2–3. Every neighborhood of such a point contains values of $f(z)$ (in the second quadrant) that are arbitrarily near to π and values of $f(z)$ (in the third quadrant) that are arbitrarily near to $-\pi$. Approaching z_0 on two different paths such as C_1 and C_2, we see that $\arg z$ tends to two different values. Therefore $\arg z$ fails to possess a limit at z_0. ◀

EXAMPLE 3

Let

$$f(z) = \frac{x^2 + x}{x + y} + \frac{i(y^2 + y)}{x + y}.$$

This function is undefined at $z = 0$. Show that $\lim_{z \to 0} f(z)$ fails to exist.

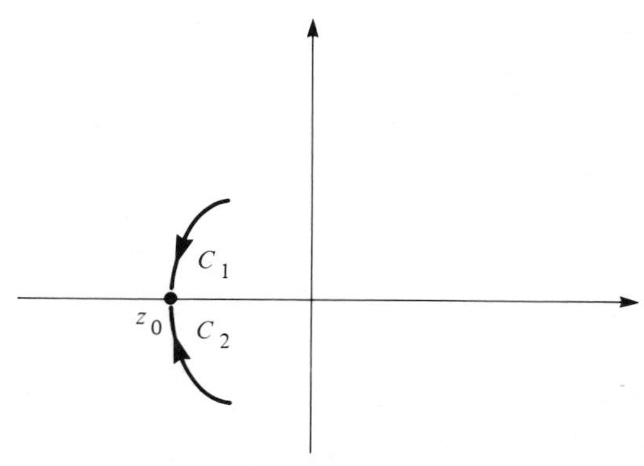

Figure 2.2–3

Solution

Let us move toward the origin along the y-axis. With $x = 0$ in $f(z)$ we have

$$f(z) = \frac{i(y^2 + y)}{y} = i(y + 1).$$

As the origin is approached, this expression becomes arbitrarily close to i.

Next we move toward the origin along the x-axis. With $y = 0$ we have $f(z) = x + 1$. As the origin is approached this expression becomes arbitrarily close to 1. Because our two results disagree, $\lim_{z \to 0} f(z)$ fails to exist. ◀

Sometimes we will be concerned with the limit of a function $f(z)$ as z tends to infinity. If this limit exists and has the value f_0, we write $\lim_{z \to \infty} f(z) = f_0$. This means that given any $\varepsilon > 0$, there exists a number r such that $|f(z) - f_0| < \varepsilon$ for all $|z| > r$. Thus the magnitude of the difference between $f(z)$ and f_0 can be made smaller than any preassigned positive number ε provided the point representing z lies more than the distance r from the origin. Usually r depends on ε. Some typical limits that can be established with this definition are $\lim_{z \to \infty}(1/z^2) = 0$ and $\lim_{z \to \infty}(1 + z^{-1}) = 1$. Exercise 12 deals with the rigorous treatment of a limit at ∞.

Formulas pertaining to limits that the reader studied in elementary calculus have counterparts for functions of a complex variable. These counterparts, stated here without proof, can be established from the definition of a limit.

THEOREM 1 Let $f(z)$ have limit f_0 as $z \to z_0$ and $g(z)$ have limit g_0 as $z \to z_0$. Then

$$\lim_{z \to z_0} (f(z) + g(z)) = f_0 + g_0, \tag{2.2–10a}$$

$$\lim_{z \to z_0} (f(z)g(z)) = f_0 g_0, \tag{2.2–10b}$$

$$\lim_{z \to z_0} [f(z)/g(z)] = f_0/g_0, \qquad \text{if } g_0 \neq 0. \quad \square \tag{2.2–10c}$$

The definition of continuity for complex functions of a complex variable is analogous to that for real functions of a real variable.

DEFINITION Continuity

A function $w = f(z)$ is continuous at $z = z_0$ provided the following conditions are both satisfied:

a) $f(z_0)$ is defined;

b) $\lim_{z \to z_0} f(z)$ exists, and

$$\lim_{z \to z_0} f(z) = f(z_0). \quad \square \tag{2.2–11}$$

Generally we will be dealing with functions that fail to be continuous only at certain points or along some locus in the z-plane. We can usually recognize points of discontinuity as places where a function becomes infinite or undefined or exhibits an abrupt change in value.

If a function is continuous at all points in a region, we say that it is continuous in the region.

The principal value of $\arg z$ is discontinuous at all points on the negative real axis because it fails to have a limit at every such point. Moreover, $\arg z$ is undefined at $z = 0$, which means that $\arg z$ is discontinuous there as well.

EXAMPLE 4

Investigate the continuity at $z = i$ of the function

$$f(z) = \begin{cases} \dfrac{z^2 + 1}{z - i}, & z \neq i, \\ 3i, & z = i. \end{cases}$$

Solution

Because $f(i)$ is defined, part (a) in our definition of continuity is satisfied. To investigate part (b) we must first determine $\lim_{z \to i} f(z)$. Since the value of this limit does not depend on $f(i)$, we first study $f(z)$ for $z \neq i$. We factor the numerator in the above quotient as follows:

$$f(z) = \frac{z^2 + 1}{z - i} = \frac{(z - i)(z + i)}{z - i}, \qquad z \neq i,$$

and cancel $z - i$ common to both numerator and denominator. (Since $z \neq i$ we are not dividing by zero.) Thus $f(z) = z + i$ for $z \neq i$. From this we might conclude that $\lim_{z \to i} f(z) = 2i$. This was in fact rigorously done in Example 1, to which the reader should now refer.

Because $f(i) = 3i$ while $\lim_{z \to i} f(z) = 2i$, and these two results are not in agreement, condition (b) in our definition of continuity is not satisfied at $z = i$. Thus, $f(z)$ is discontinuous at $z = i$. It is not hard to show that $f(z)$ is continuous for all $z \neq i$. Notice too that a function identical to our given $f(z)$ but satisfying $f(i) = 2i$ is continuous for all z. ◀

There are a number of important properties of continuous functions that we will be using. Although the truth of the following theorem may seem self-evident, in certain cases the proofs are not easy, and the reader is referred to a more advanced text for them.†

---

† See for example, R. V. Churchill and J. W. Brown, *Complex Variables and Applications*, 5th ed. (New York: McGraw-Hill, 1990, Section 14).

THEOREM 2

a) *Sums, differences,* and *products* of continuous functions are themselves continuous functions. The *quotient* of a pair of continuous functions is continuous except where the denominator equals zero.

b) A continuous function of a continuous function is a continuous function.

c) Let $f(z) = u(x, y) + iv(x, y)$. The functions $u(x, y)$ and $v(x, y)$ are continuous† at any point where $f(z)$ is continuous. Conversely, at any point where u and v are continuous, $f(z)$ is also.

d) If $f(z)$ is continuous in some region R, then $|f(z)|$ is also continuous in R. If R is bounded and closed there exists a positive real number, say M, such that $|f(z)| \leq M$ for all z in R. M can be chosen so that the equality holds for at least one value of z in R. □

We can use part (a) of the theorem to investigate the continuity of the quotient $(z^2 + z + 1)/(z^2 - 2z + 1)$. Since $f(z) = z$ is obviously a continuous function of z (this is proved rigorously in Exercise 1), so is the product $z \cdot z = z^2$. Any constant is a continuous function. Thus the sum $z^2 + z + 1$ is continuous for all z and by similar reasoning so is $z^2 - 2z + 1$. The quotient of these two polynomials is therefore continuous except where $z^2 - 2z + 1 = (z - 1)^2$ is zero. This occurs only at $z = 1$.

A similar procedure applies to any rational function $P(z)/Q(z)$, where P and Q are polynomials of any degree in z. Such an expression is continuous except for values of z satisfying $Q(z) = 0$.

The usefulness of part (b) of the theorem will be more apparent in the next chapter, where we will study various transcendental functions of z. We will learn what is meant by $f(z) = e^z$, where z is complex,‡ and we will find that this function is continuous for all z. Now, $g(z) = 1/z^2$ is continuous for all $z \neq 0$. Thus $f(g(z)) = \exp(1/z^2)$ is also continuous for $z \neq 0$.

As an illustration of parts (c) and (d), consider $f(z) = e^x \cos y + ie^x \sin y$ in the disc-shaped region R given by $|z| \leq 1$. Since $u = e^x \cos y$ and $v = e^x \sin y$ are continuous in R, $f(z)$ is also. Thus $|f(z)|$ must be continuous in R. Now, $|f(z)| = \sqrt{\exp(2x)[\cos^2 y + \sin^2 y]} = e^x$, which is indeed a continuous function. The maximum value achieved by $|f(z)|$ in R will occur when e^x is maximum, that is, at $x = 1$. Thus $|f(z)| \leq e$ in R, and the constant M in part (d) of the theorem here equals e.

EXERCISES

1. a) Let $f(z) = z$. Show by using an argument like that presented in Example 1 that $\lim_{z \to z_0} f(z) = z_0$, where z_0 is an arbitrary complex number.

---

† Continuity for $u(x, y)$, a *real* function of two real variables, is defined in a way analogous to continuity for $f(z)$. For continuity at (x_0, y_0) the difference $|u(x, y) - u(x_0, y_0)|$ can be made smaller than any positive ε for all (x, y) lying inside a circle of radius δ centered at (x_0, y_0).

‡ e^z is also written $\exp(z)$.

b) Using the definition of continuity, explain why $f(z)$ is continuous for any z_0.

2. Let $f(z) = c$ where c is an arbitrary constant. Using the definitions of limit and continuity, prove that $f(z)$ is continuous for all z.

Assuming the continuity of the functions $f(z) = z$ and $f(z) = c$, where c is any constant (proved in Exercises 1 and 2), use various parts of Theorem 2 to prove the continuity of the following functions in the domain indicated. Take $z = x + iy$.

3. $f(z) = z^3 + z + 1$ all z

4. $f(z) = 1/(z^2 + 1)$ all $z \neq \pm i$

5. $f(z) = |z|$ all z

6. $f(z) = |z| + x$ all z

7. $f(z) = 1/(x^2 - y^2 + z)$ all z except 0 and -1

8. $z/\bar{z}$ all $z \neq 0$

9. The function $(\sin x + i \sin y)/(x - iy)$ is obviously undefined at $z = 0$. Show that it fails to have a limit as $z \to 0$ by comparing the values assumed by this function as the origin is approached along the following three line segments: $y = 0$, $x > 0$; $x = 0$, $y > 0$; $x = y$, $x > 0$.

10. Is the following function continuous at $z = 3i$? Give an explanation like the one provided in Example 4.

$$f(z) = \begin{cases} (z^2 + 9)/(z - 3i), & z \neq 3i, \\ 6i, & z = 3i. \end{cases}$$

11. The function $f(z) = z(z^2 - 16)/(z^2 - 4z)$ is defined and continuous for all z except $z = 0$ and $z = 4$. How should we define $f(0)$ and $f(4)$ so that $f(z)$ is continuous throughout the z-plane?

12. Show that $\lim_{z\to\infty}(1 + z^{-2}) = 1$.

Hint: We require $|f(z) - f_0| < \varepsilon$ for $|z| > r$. Show that we can take $r = 1/\sqrt{\varepsilon}$.

13. Consider $f(z) = z^2$.

a) In the region R described by $|z| \leq 2$, we have $|f(z)| \leq M$. Find M assuming $|f(z)| = M$ for some z in R.

b) Repeat part (a) but take R as the region $|z - 1| \leq 2$.

c) Repeat part (a) but take $f(z) = 1/z$ and R as $|z - 1 - i| \leq 1$.

14. a) Consider $f(z) = xy + 1$. Explain why this function is continuous everywhere in the z-plane. Use Theorem 2.

b) Consider the square shaped region: $|x| \leq 1$, $|y| \leq 1$. In this region $|f(z)| \leq M$. Find M assuming $|f(z)| = M$ somewhere in the region.

c) Where in the region is the equality satisfied in (b)?

2.3 THE COMPLEX DERIVATIVE

Review

Before discussing the derivative of a function of a complex variable, let us briefly review some facts concerning the derivative of a function of a real variable $f(x)$. The derivative of $f(x)$ at x_0, which is written $f'(x_0)$, is given by

$$f'(x_0) = \lim_{\Delta x \to 0} \frac{f(x_0 + \Delta x) - f(x_0)}{\Delta x}. \tag{2.3–1}$$

If the limit in this expression fails to exist, $f'(x_0)$ is undefined and $f(x)$ has no derivative (is not differentiable) at x_0.

If $f(x)$ is not continuous at x_0, $f'(x_0)$ does not exist. However, a function can be continuous and *still not* have a derivative. In Eq. (2.3–1), Δx is a small increment, shrinking progressively to zero, in the argument of $f(x)$. The increment can be either a positive or a negative number. If $f'(x_0)$ is to exist, *identical* finite results must be obtained from the right side of Eq. (2.3–1) for both the positive and negative choices. If two different numbers are obtained, $f'(x_0)$ does not exist.

As an example of how this can occur, consider $f(x) = 2|x|$ plotted in Fig. 2.3–1. It is not hard to show that $f(x)$ is continuous for all x. Let us try to compute $f'(0)$ by means of Eq. (2.3–1). With $x_0 = 0$, $f(x_0) = 0$, and $f(x_0 + \Delta x) = 2|\Delta x|$, we have

$$f'(0) = \lim_{\Delta x \to 0} \frac{2|\Delta x|}{\Delta x}. \tag{2.3–2}$$

Unfortunately, if Δx is positive, $2|\Delta x|/\Delta x$ has the value 2, whereas if Δx is negative, it has the value -2. The limit in Eq. (2.3–2) cannot exist, and neither does $f'(0)$. Of course, the values 2 and -2 are the slopes of the curve to the right and left of $x = 0$.

Computing the derivative of the above $f(x)$ at any point $x_0 \neq 0$, we find that the limit on the right in Eq. (2.3–1) exists. It is independent of the sign of Δx; that

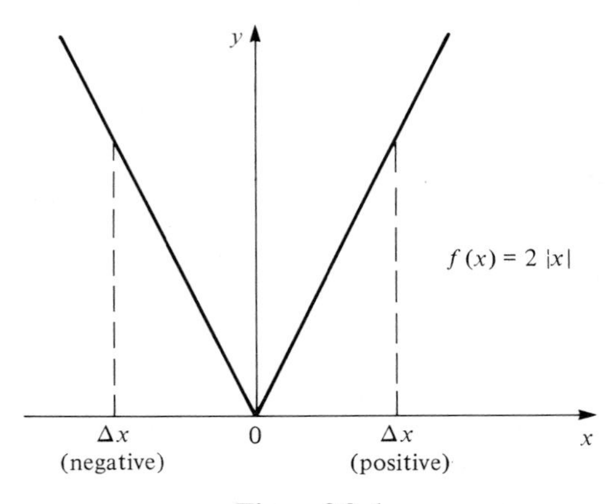

Figure 2.3–1

is, the same value is obtained irrespective of whether we approach x_0 from the right ($\Delta x > 0$) or from the left ($\Delta x < 0$). The reader should do this simple exercise.

Complex Case

Given a function of a complex variable $f(z)$, its derivative at z_0, $f'(z_0)$, or $(df/dz)_{z_0}$, is defined as stated below provided the limit shown exists.

DEFINITION Derivative

$$f'(z_0) = \lim_{\Delta z \to 0} \frac{f(z_0 + \Delta z) - f(z_0)}{\Delta z}. \quad \square \tag{2.3–3}$$

This definition is identical in form to Eq. (2.3–1), the corresponding expression for real variables. Like the derivatives for the function of a real variable, a function of a complex variable must be continuous at a point to possess a derivative there (see Exercise 20), but continuity by itself does not guarantee the existence of a derivative.

Despite being like (2.3–1) in form, Eq. (2.3–3) is in fact more subtle. We saw that in Eq. (2.3–1) there were two directions from which we could approach x_0. As Fig. 2.3–2 suggests, there are an infinite number of different directions along which $z_0 + \Delta z$ can approach z_0 in Eq. (2.3–3). Moreover, we need not approach z_0 along a straight line but can choose some sort of arc or spiral. If the limit in Eq. (2.3–3) exists, that is, if $f'(z_0)$ exists, then the quotient in Eq. (2.3–3) must approach the same value irrespective of the direction or locus along which Δz shrinks to zero.

In the case of the function $f(z) = z^n$ ($n = 0, 1, 2, \ldots$), it is easy to verify the existence of the derivative and to obtain its value. Now $f(z_0) = z_0^n$, and $f(z_0 + \Delta z) = (z_0 + \Delta z)^n$. This last expression can be expanded with the binomial theorem:

$$(z_0 + \Delta z)^n = z_0^n + nz_0^{n-1}(\Delta z) + \frac{n(n-1)}{2}(z_0)^{n-2}(\Delta z)^2 + \text{higher powers of } \Delta z.$$

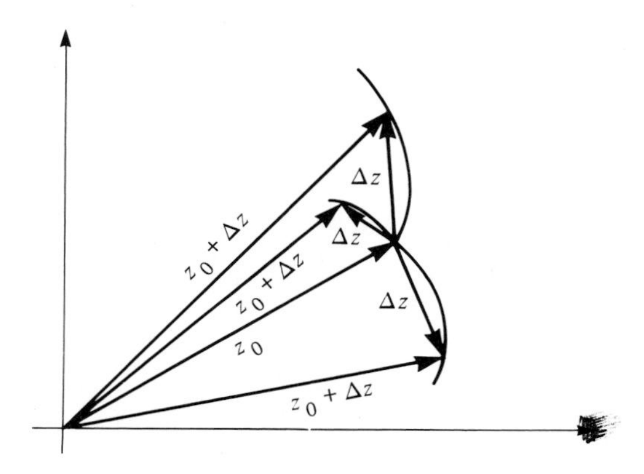

Figure 2.3–2

Thus

$$f'(z_0) = \lim_{\Delta z \to 0} \frac{f(z_0 + \Delta z) - f(z_0)}{\Delta z}$$

$$= \lim_{\Delta z \to 0} \left[\frac{z_0^n + nz_0^{n-1}\Delta z + \frac{n(n-1)}{2}(z_0)^{n-2}(\Delta z)^2 + \cdots - z_0^n}{\Delta z} \right]$$

$$= \lim_{\Delta z \to 0} \left[nz_0^{n-1}\frac{\Delta z}{\Delta z} + \frac{n(n-1)}{2}(z_0)^{n-2}\frac{(\Delta z)^2}{\Delta z} + \cdots + \right] = n(z_0)^{n-1}.$$

We do not need to know the path along which Δz shrinks to zero in order to obtain this result. The result is independent of the way in which $z_0 + \Delta z$ approaches z_0. Dropping the subscript zero, we have

$$\frac{d}{dz} z^n = nz^{n-1}. \tag{2.3–4}$$

Thus if n is a nonnegative integer, the derivative of z^n exists for all z. When n is a negative integer, a similar derivation can be used to show that Eq. (2.3–4) holds for all $z \neq 0$. With n negative z^n is undefined at $z = 0$, and this value of z must be avoided.

A more difficult problem occurs if we are given a function of z in the form $f(z) = u(x, y) + iv(x, y)$ and we wish to know whether its derivative exists. If the variables x and y change by incremental amounts Δx and Δy, the corresponding incremental change in z, called Δz is $\Delta x + i\Delta y$ (see Fig. 2.3–3a). Suppose, however, that Δz is constrained to lie along the horizontal line passing through z_0 depicted in Fig. 2.3–3(b). Then, y is constant and $\Delta z = \Delta x$. Now with $z_0 = x_0 + iy_0$, $f(z) = u(x, y) + iv(x, y)$, and $f(z_0) = u(x_0, y_0) + iv(x_0, y_0)$ we will assume that $f'(z_0)$

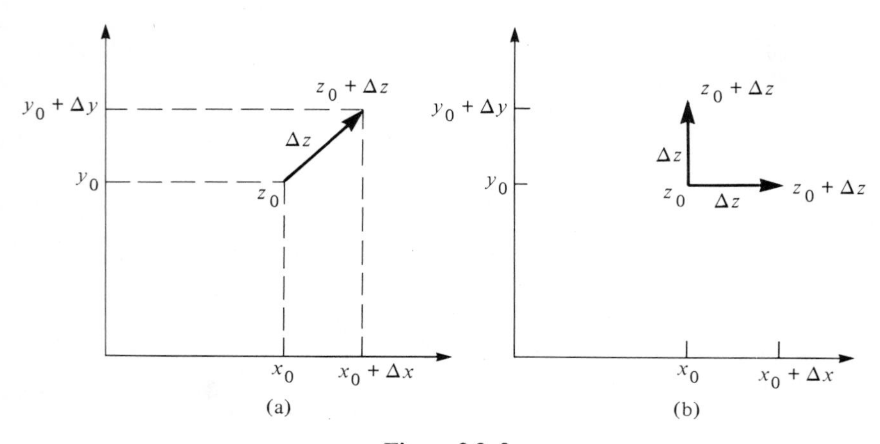

Figure 2.3–3

exists and apply Eq. (2.3–3):

$$f'(z_0) = \lim_{\Delta x \to 0} \frac{f(z_0 + \Delta x) - f(z_0)}{\Delta x}$$

$$= \lim_{\Delta x \to 0} \frac{u(x_0 + \Delta x, y_0) + iv(x_0 + \Delta x, y_0) - u(x_0, y_0) - iv(x_0, y_0)}{\Delta x}.$$

We can rearrange the preceding expression to read

$$f'(z_0) = \lim_{\Delta x \to 0} \left[\frac{u(x_0 + \Delta x, y_0) - u(x_0, y_0)}{\Delta x} + i \frac{v(x_0 + \Delta x, y_0) - v(x_0, y_0)}{\Delta x} \right]. \tag{2.3–5}$$

We should recognize the definition of two partial derivatives in Eq. (2.3–5). Passing to the limit we have

$$f'(z_0) = \left(\frac{\partial u}{\partial x} + i \frac{\partial v}{\partial x} \right)_{x_0, y_0}. \tag{2.3–6}$$

Instead of having $z_0 + \Delta z$ approach z_0 from the right, as was just done, we can allow $z_0 + \Delta z$ to approach z_0 from above. If Δz is constrained to lie along the vertical line passing through z_0 in Fig. 2.3–3(b), $\Delta x = 0$ and $\Delta z = i\Delta y$. Thus proceeding much as before, we have

$$f'(z_0) = \lim_{\Delta y \to 0} \frac{f(z_0 + i\Delta y) - f(z_0)}{i\Delta y}$$

$$= \lim_{\Delta y \to 0} \left[\frac{u(x_0, y_0 + \Delta y) + iv(x_0, y_0 + \Delta y) - u(x_0, y_0) - iv(x_0, y_0)}{i\Delta y} \right]. \tag{2.3–7}$$

Passing to the limit and putting $1/i = -i$ we get

$$f'(z_0) = \left(-i \frac{\partial u}{\partial y} + \frac{\partial v}{\partial y} \right)_{x_0, y_0}. \tag{2.3–8}$$

Assuming $f'(z_0)$ exists, Eqs. (2.3–6) and (2.3–8) provide us with two methods for its computation. Equating these expressions, we obtain the result

$$\left(\frac{\partial u}{\partial x} + i \frac{\partial v}{\partial x} \right) = \left(-i \frac{\partial u}{\partial y} + \frac{\partial v}{\partial y} \right). \tag{2.3–9}$$

The real part on the left side of Eq. (2.3–9) must equal the real part on the right. A similar statement applies to the imaginaries. Thus we require at *any* point where $f(z)$ exists the set of relationships shown below.

CAUCHY–RIEMANN EQUATIONS

$$\frac{\partial u}{\partial x} = \frac{\partial v}{\partial y}, \tag{2.3–10a}$$

$$\frac{\partial v}{\partial x} = -\frac{\partial u}{\partial y}. \tag{2.3–10b}$$

This set of important relationships is known as the Cauchy–Riemann (or C–R) equations. They are named after the French mathematician Augustin Cauchy (1789–1857) who was widely thought to be their discoverer and for the German mathematician George Friedrich Bernhard Riemann (1826–1866) who found early and important application for them in his work on functions of a complex variable. It is now known that another Frenchman, Jean D'Alembert (1717–1783), had arrived at these equations by 1752, ahead of Cauchy. We shall encounter Cauchy's name again. He is one of the giants of nineteenth century mathematics and has made more fundamental contributions to complex variables than anyone.

If the equations fail to be satisfied for some value of z, say z_0, we know that $f'(z_0)$ cannot exist since our allowing Δz to shrink to zero along *two* different paths (Fig. 2.3–3b) leads to two contradictory limiting values for the quotient in Eq. (2.3–3). Thus we have shown that satisfaction of the C–R equations at a point is a *necessary* condition for the existence of the derivative at that point. The mere fact that a function satisfies these equations does not guarantee that *all* paths along which $z_0 + \Delta z$ approaches z_0 will yield identical limiting values for the quotient in Eq. (2.3–3). In more advanced texts[†] the following theorem is proved for $f(z) = u + iv$.

THEOREM 3

If u, v and their first partial derivatives $(\partial u/\partial x, \partial v/\partial x, \partial u/\partial y, \partial v/\partial y)$ are continuous throughout some neighborhood of z_0, then satisfaction of the Cauchy–Riemann equations at z_0 is both a *necessary* and *sufficient* condition for the existence of $f'(z_0)$. □

With the conditions of this theorem fulfilled, the limit on the right in Eq. (2.3–3) exists; that is, all paths by which $z_0 + \Delta z$ approaches z_0 yield the same finite result in this expression.

EXAMPLE 1

Investigate the differentiability of $f(z) = z^2 = (x + iy)^2 = x^2 - y^2 + i2xy$.

Solution

We already know (see Eq. 2.3–4) that $f'(z)$ exists, but let us verify this result by means of the C–R equations. Here, $u = x^2 - y^2$, $v = 2xy$, $\partial u/\partial x = 2x = \partial v/\partial y$, and $\partial v/\partial x = 2y = -\partial u/\partial y$. Thus Eqs. (2.3–10) are satisfied for all z. Also, because u, v, $\partial u/\partial x$, $\partial v/\partial y$, etc., are continuous in the z-plane, $f'(z)$ exists for all z. ◀

EXAMPLE 2

Investigate the differentiability of $f(z) = z\bar{z} = |z|^2$.

[†] See J. Bak and D. Newman, *Complex Analysis* (New York: Springer-Verlag, 1982), pp. 30–35.

Solution

Here the C–R equations are helpful. We have $u + iv = |z|^2 = x^2 + y^2$. Hence, $u = x^2 + y^2$ and $v = 0$, therefore, $\partial u/\partial x = 2x$, $\partial v/\partial y = 0$, $\partial u/\partial y = 2y$, and $\partial v/\partial x = 0$. When these expressions are substituted into Eqs. (2.3–10), we obtain $2x = 0$ and $2y = 0$. These equations are simultaneously satisfied only where $x = 0$ and $y = 0$, that is, at the origin of the z-plane. Thus this function of z possesses a derivative only for $z = 0$. ◀

Let us consider why the derivative of $|z|^2$ fails to exist except at one point.

Consider the definition in Eq. (2.3–3) and refer to Fig. 2.3–4. At an arbitrary point, $z_0 = x_0 + iy_0$, we have $f(z_0) = |z_0|^2 = |x_0 + iy_0|^2 = x_0^2 + y_0^2$. With $\Delta z = \Delta x + i\Delta y$ then, $f(z_0 + \Delta z) = |z_0 + \Delta z|^2 = |(x_0 + \Delta x) + i(y_0 + \Delta y)|^2 = x_0^2 + 2x_0\Delta x + (\Delta x)^2 + y_0^2 + 2y_0\Delta y + (\Delta y)^2$. Thus

$$\lim_{\Delta z \to 0} \frac{f(z_0 + \Delta z) - f(z_0)}{\Delta z} = \lim_{\substack{\Delta x \to 0 \\ \Delta y \to 0}} \frac{2x_0\Delta x + 2y_0\Delta y + (\Delta x)^2 + (\Delta y)^2}{\Delta x + i\Delta y}. \tag{2.3–11}$$

Now, suppose we allow Δz to shrink to zero along a straight line passing through z_0 with slope m. This means $\Delta y = m\Delta x$. With this relationship in Eq. (2.3–11) we have

$$\lim_{\Delta x \to 0} \frac{2x_0\Delta x + 2y_0 m\Delta x + (\Delta x)^2 + m^2(\Delta x)^2}{\Delta x(1 + im)}$$

$$= \lim_{\Delta x \to 0} \left[\frac{2x_0 + 2y_0 m}{1 + im} + \frac{\Delta x}{1 + im} + \frac{m^2\Delta x}{1 + im} \right] = \frac{2x_0 + 2y_0 m}{1 + im}.$$

Unless $x_0 = 0$ and $y_0 = 0$, this result is certainly a function of the slope m, that is, of the direction of approach to z_0. For example, if we approach z_0 along a line parallel to the x-axis, we put $m = 0$ and find that the result is $2x_0$. However, if we approach z_0 along a line making a 45° angle with the horizontal, we have $\Delta y = \Delta x$, or $m = 1$. The expression becomes $(2x_0 + 2y_0)/(1 + i)$.

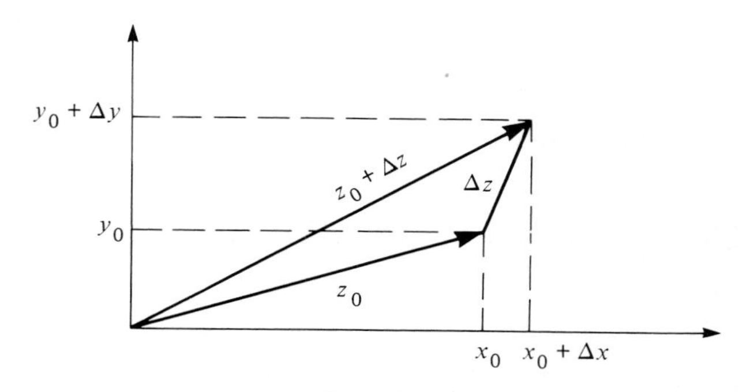

Figure 2.3–4

EXERCISES

Sketch the following real functions $f(x)$ over the indicated interval. In each case find the one value of x where the derivative with respect to x does not exist. State whether or not the function is continuous at this location. A formal proof is not required.

1. $f(x) = |\sin x|, \frac{\pi}{2} < x < \frac{3\pi}{2}$

2. $f(x) = (1 - x)^{1/3}$, $0 < x < 9$, where the real root is used

For what values of the complex variable z do the following functions have derivatives?

3. $\bar{z}$

4. c (a constant)

5. $x^2 - y^2 - i2xy$

6. z^{10}

7. z^{-5}

8. x

9. $x^2 - y^2 - 2xy + i(2xy + x^2 - y^2)$

10. $e^x \cos y - ie^x \sin y$

11. $e^{(|z-1|^2)}$

12. $\cos x + i \sin y$

13. $z^5 + \bar{z}$

14. $\arg z$ (principal value)

15. $f(z) = 2$, $|z| > 3$, but $f(z) = 1$, $|z| \leq 3$

16. $f(z) = 2$, $|z| > 3$, but $f(z) = 2$, $|z| < 3$

17. $f(z) = |z|^2$, $|z| \leq 1$, but $f(z) = z^2$, $|z| > 1$

18. $f(z) = z$, $|z| \geq 1$, but $f(z) = 1$, $|z| < 1$

19. Let $f(z) = u(x, y) + iv(x, y)$. Assume that the second derivative $f''(z)$ exists. Show that

$$f''(z) = \frac{\partial^2 u}{\partial x^2} + i\frac{\partial^2 v}{\partial x^2} \quad \text{and} \quad f''(z) = -\frac{\partial^2 u}{\partial y^2} - i\frac{\partial^2 v}{\partial y^2}$$

Hint: See the derivation of Eqs. (2.3–6) and (2.3–8).

20. Show that if $f'(z_0)$ exists, then $f(z)$ must be continuous at z_0.

Hint: Let $z = z_0 + \Delta z$. Consider

$$\lim_{\Delta z \to 0}\left[\frac{f(z_0 + \Delta z) - f(z_0)}{\Delta z}\right] \lim_{\Delta z \to 0} \Delta z$$

Refer to Eq. (2.2–10b) of Theorem 1.

2.4 THE DERIVATIVE AND ANALYTICITY

Finding the Derivative

If we can establish that the derivative of $f(z) = u + iv$ exists for some z, it is a straightforward matter to find $f'(z)$. We can work directly with the definition shown in Eq. (2.3–3).

In addition, either Eq. (2.3–6), $f'(z) = \partial u/\partial x + i\partial v/\partial x$, or Eq. (2.3–8), $f'(z) = \partial v/\partial y - i\partial u/\partial y$, can be used. For example, the function $f(z) = x^2 - y^2 - y + i(2xy + x)$ is found, from the C–R equations, to have a derivative for all z. With $u = x^2 - y^2 - y$ and with $v = 2xy + x$ we have, from Eq. (2.3–6), $f'(z) = 2x + i(2y + 1)$. An identical result comes from Eq. (2.3–8).

In Section 2.3 we observed that $dz^n/dz = nz^{n-1}$, where n is any integer. This formula is identical in form to the corresponding expression in real variable calculus, $dx^n/dx = nx^{n-1}$. Thus to differentiate such expressions as z^2, $1/z^3$, etc. the usual method applies and these derivatives are respectively, $2z$ and $-3z^{-4}$.

The reason that the procedure used in differentiating x^n and z^n is identical lies in the similarity of the expressions

$$\lim_{\Delta z\to 0} \frac{f(z + \Delta z) - f(z)}{\Delta z} \quad \text{and} \quad \lim_{\Delta x\to 0} \frac{f(x + \Delta x) - f(x)}{\Delta x},$$

which define the derivatives of functions of complex and real variables.

All the identities of real differential calculus that are obtained through direct manipulation of the definition of the derivative can be carried over to functions of a complex variable.

Specifically, if $f(z)$ and $g(z)$ are differentiable for some z, then

THEOREM 4

$$\frac{d}{dz}(f(z) \pm g(z)) = f'(z) \pm g'(z); \tag{2.4–1a}$$

$$\frac{d}{dz}(f(z)g(z)) = f'(z)g(z) + f(z)g'(z); \tag{2.4–1b}$$

$$\frac{d}{dz}\left(\frac{f(z)}{g(z)}\right) = \frac{f'(z)g(z) - f(z)g'(z)}{[g(z)]^2}, \quad \text{provided } g(z) \neq 0; \tag{2.4–1c}$$

$$\frac{d}{dz}f(g(z)) = \frac{df}{dg}g'(z). \quad \square \tag{2.4–1d}$$

Thus a function formed by the addition, subtraction, multiplication, or division of differentiable functions is itself differentiable. Equations (2.4–1a–c) provide a means for finding its derivative. Another useful formula is the "chain rule" (2.4–1d) for finding the derivative of a function of a function. It is applied in the ways familiar to us from elementary calculus, for example,

$$\begin{aligned}\frac{d}{dz}(z^3 + z^2 + 1)^{10} &= 10(z^3 + z^2 + 1)^9 \frac{d}{dz}(z^3 + z^2 + 1) \\ &= 10(z^3 + z^2 + 1)^9(3z^2 + 2z).\end{aligned}$$

The equations contained in Eq. (2.4–1) are of no use in establishing the differentiability or in determining the derivative of any expression involving $|z|$ or $\bar{z}$. We can first rewrite such expressions in the form $u(x, y) + iv(x, y)$ and then apply the

C–R equations to investigate differentiability. If the derivative exists, it can then be found from Eqs. (2.3–6) or (2.3–8). Alternatively, we might choose to investigate differentiability and determine the derivative by means of the definition given in Eq. (2.3–3).

Using the definition of the derivative, we can obtain L'Hôpital's Rule for functions of a complex variable. The rule will be used at numerous places in later chapters. It states the following:

L'HÔPITAL'S RULE

If $g(z_0) = 0$ and $h(z_0) = 0$, and if $g(z)$ and $h(z)$ are differentiable at z_0 with $h'(z_0) \neq 0$, then

$$\lim_{z \to z_0} \frac{g(z)}{h(z)} = \frac{g'(z_0)}{h'(z_0)}. \quad \square \tag{2.4–2}$$

The preceding rule is formally the same as that used in elementary calculus for evaluating indeterminate forms involving functions of a real variable.

To prove (2.4–2) we observe that since $g(z_0) = 0$, $h(z_0) = 0$, then

$$\frac{g(z)}{h(z)} = \frac{g(z) - g(z_0)}{z - z_0} \Bigg/ \frac{h(z) - h(z_0)}{z - z_0}, \qquad z \neq z_0. \tag{2.4–3}$$

Putting $z = z_0 + \Delta z$ in the preceding, we have that

$$\frac{g(z)}{h(z)} = \frac{g(z_0 + \Delta z) - g(z_0)}{\Delta z} \Bigg/ \frac{h(z_0 + \Delta z) - h(z_0)}{\Delta z}. \tag{2.4–4}$$

Passing to the limit $z \to z_0$ in (2.4–3) is equivalent to taking the limit $\Delta z \to 0$ in (2.4–4). Now, recall from (2.2–10c) that the limit of the quotient of two functions is the quotient of their limits (provided the denominator is nonzero). Applying this fact, letting $\Delta z \to 0$ in (2.4–4), and using the definition of the derivative we have the desired result:

$$\lim_{z \to z_0} \frac{g(z)}{h(z)} = \lim_{\Delta z \to 0} \frac{g(z + \Delta z) - g(z)}{\Delta z} \Bigg/ \lim_{\Delta z \to 0} \frac{h(z + \Delta z) - h(z)}{\Delta z} = \frac{g'(z_0)}{h'(z_0)}.$$

EXAMPLE 1

Find

$$\lim_{z \to 2i} \frac{z - 2i}{z^4 - 16}.$$

Solution

Taking $g(z) = z - 2i$, $h(z) = z^4 - 16$, $z_0 = 2i$, we see that $g(z_0) = 0$, $h(z_0) = 0$, $g'(z_0) = 1$, and $h'(z_0) = 4(2i)^3 = -32i$. L'Hôpital's Rule can be applied since $g(z_0) = 0$, $h(z_0) = 0$ while $h'(z_0) \neq 0$. The desired limit is $1/(-32i)$. ◄

Comment: If $g(z_0) = 0 = h(z_0)$, $h'(z_0) = 0$ while $g'(z_0) \neq 0$, then L'Hôpital's Rule does not apply. In fact, one can show that $\lim_{z \to z_0}(g(z)/h(x))$ does not exist, and the magnitude of this quotient grows without bound as $z \to z_0$.

On the other hand, if $g(z_0)$, $h(z_0)$, $g'(z_0)$, and $h'(z_0)$ are all zero, the limit may exist, but Eq. (2.4–2) cannot yield its value. An extension of L'Hôpital's Rule can be applied to these situations.† It asserts that if $g(z)$, $h(z)$, and their first n derivatives vanish at z_0 and $h^{(n+1)}(z_0) \neq 0$, then

$$\lim_{z \to z_0} \frac{g(z)}{h(z)} = \frac{g^{(n+1)}(z_0)}{h^{(n+1)}(z_0)}.$$

Analytic Functions

The concept of an *analytic function*, although seemingly simple, is at the very core of complex variable theory, and a grasp of its meaning is essential.

DEFINITION Analyticity

A function $f(z)$ is analytic at z_0 if $f'(z)$ exists not only at z_0 but at every point belonging to some neighborhood of z_0. □

Thus for a function to be analytic at a point it must not only have a derivative at that point but must have a derivative everywhere within some circle of nonzero radius centered at the point.

DEFINITION Analyticity in a domain

If a function is analytic at every point belonging to some domain, we say that the function is *analytic in that domain.* □

It is quite possible for a function to possess a derivative at some point yet fail to be analytic at that point. In Example 2 of Section 2.3 we considered $f(z) = |z|^2$ and found it to have a derivative only for $z = 0$. Every circle that we might draw about the point $z = 0$ will contain points at which $f'(z)$ fails to exist. Hence, $f(z)$ is not analytic at $z = 0$ (or anywhere else).

EXAMPLE 2

For what values of z is the function $f(z) = x^2 + iy^2$ analytic?

Solution

From the C–R equations, with $u = x^2$, $v = y^2$, we have

$$\frac{\partial u}{\partial x} = 2x = 2y = \frac{\partial v}{\partial y} \quad \text{and} \quad \frac{\partial v}{\partial x} = 0 = -\frac{\partial u}{\partial y}.$$

---

† This is proved in Exercise 9 of Section 6.2.

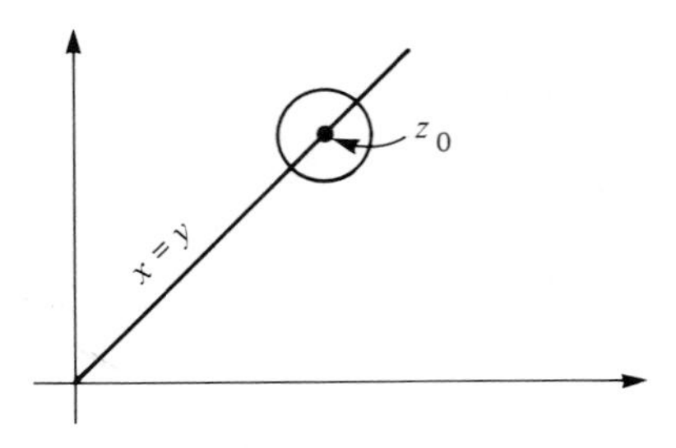

Figure 2.4–1

Thus $f(z)$ is differentiable only for values of z that lie along the straight line $x = y$. If z_0 lies on this line, any circle centered at z_0 will contain points for which $f'(z)$ does not exist (see Fig. 2.4–1). Thus $f(z)$ is nowhere analytic. ◀

Equations (2.4–1a–d), which yield the derivatives of sums, products, and so forth, can be extended to give the following theorem on analyticity:

THEOREM 5 If two functions are analytic in some domain, the *sum*, *difference*, and *product* of these functions are also analytic in the domain. The *quotient* of these functions is analytic in the domain except where the denominator equals zero. An analytic function of an analytic function is analytic. □

EXAMPLE 3

Using Theorem 5 prove that if $f(z)$ is analytic at z_0 and $g(z)$ is not analytic at z_0, then $h(z) = f(z) + g(z)$ is not analytic at z_0.

Solution

Assume $h(z)$ is analytic at z_0. Then $h(z) - f(z)$ is the difference of two analytic functions and, by Theorem 5, must be analytic. But we contradict ourselves since $h(z) - f(z)$ is $g(z)$, which is not analytic. Thus our assumption that $h(z)$ is analytic is false. ◀

DEFINITION Entire function

A function that is analytic throughout the finite z-plane is called an *entire function.* □

Any constant is entire. Its derivative exists for all z and is zero. The function $f(z) = z^n$ is entire if n is a nonnegative integer (see Eq. 2.3–4). Now $a_n z^n$, where a_n is any constant, is the product of entire functions and is also entire. A polynomial expression $a_n z^n + a_{n-1} z^{n-1} + \cdots + a_0$ is entire since it is a sum of entire functions.

DEFINITION Singularity

If a function is not analytic at z_0 but is analytic for at least one point in every neighborhood of z_0, then z_0 is called a *singularity* (or *singular point*) of that function. □

A rational function

$$f(z) = \frac{a_n z^n + a_{n-1} z^{n-1} + \cdots + a_0}{b_m z^m + b_{m-1} z^{m-1} + \cdots + b_0},$$

where m and n are nonnegative integers and a_n, b_m, etc. are constants, is the quotient of two polynomials and thus the quotient of two entire functions. The function $f(z)$ is analytic except for values of z satisfying $b_m z^m + b_{m-1} z^{m-1} + \cdots + b_0 = 0$. The solutions of this equation are singular points of $f(z)$.

EXAMPLE 4

For what values of z does

$$f(z) = \frac{z^3 + 2}{z^2 + 1}$$

fail to be analytic?

Solution

For z satisfying $z^2 + 1 = 0$ or $z = \pm i$. Thus $f(z)$ has singularities at $+i$ and $-i$. ◀

In Chapter 3 we will be studying some transcendental functions of z. Here the portion of our theorem that deals with analytic functions of analytic functions will prove useful. For example, we will define $\sin z$ and learn that this is an entire function. Now $1/z^2$ is analytic for all $z \neq 0$. Hence, $\sin(1/z^2)$ is analytic for all $z \neq 0$.

Analyticity of Functions Expressed in Polar Variables

Sometimes instead of expressing a function of z in the form $f(z) = u(x, y) + iv(x, y)$ it is convenient to change to the polar system r, θ so that $x = r\cos\theta$, $y = r\sin\theta$. Thus $f(z) = u(r, \theta) + iv(r, \theta)$. In Exercise 22 we show that the appropriate form of the Cauchy–Riemann equations is now

$$\frac{\partial u}{\partial r} = \frac{1}{r}\frac{\partial v}{\partial \theta}, \tag{2.4–5a}$$

$$\frac{\partial v}{\partial r} = -\frac{1}{r}\frac{\partial u}{\partial \theta}. \tag{2.4–5b}$$

The equations can be used for all r, θ *except* where $r = 0$. In the same problem we show that if the derivative of $f(z)$ exists it can be found from either of the following:

$$f'(z) = \left(\frac{\partial u}{\partial r} + i\frac{\partial v}{\partial r}\right)(\cos\theta - i\sin\theta), \tag{2.4–6}$$

$$f'(z) = \left(\frac{\partial u}{\partial \theta} + i\frac{\partial v}{\partial \theta}\right)\left(\frac{-i}{r}\right)(\cos\theta - i\sin\theta). \tag{2.4–7}$$

Note that Theorem 5 applies equally well to analytic functions expressed in polar or Cartesian form.

EXAMPLE 5

Investigate the analyticity of

$$f(z) = r^2 \cos^2\theta + ir^2 \sin^2\theta$$

for $z \neq 0$.

Solution

Letting $u = r^2\cos^2\theta$, $v = r^2\sin^2\theta$, we find that (2.4–5a) and (2.4–5b) become, respectively,

$$2r\cos^2\theta = 2r\sin\theta\cos\theta,$$
$$2r\sin^2\theta = 2r\sin\theta\cos\theta.$$

If $\cos\theta = 0$, the first equation will be satisfied, but the second cannot since it reduces to $\sin^2\theta = 0$, which is not satisfied when $\cos\theta = 0$. Thus $\cos\theta \neq 0$. Similarly although $\sin\theta = 0$ solves the second equation it will not solve the first. Thus $\sin\theta \neq 0$. Canceling $r \neq 0$ from both sides of the above equations, dividing the first by $\cos\theta$ and the second by $\sin\theta$, we find that both equations reduce to $\sin\theta = \cos\theta$, or $\tan\theta = 1$. Thus $\theta = \pi/4$ and $5\pi/4$, while r may have any value except zero. Sketching the rays $\theta = \pi/4$ and $\theta = 5\pi/4$, we find that $f(z)$ has derivatives only on the line $x = y$. Our polar Cauchy–Riemann equations cannot establish whether the derivative exists at the origin. Because there is no domain throughout which $f(z)$ has a derivative, $f(z)$ is nowhere analytic. Note that the present example is really the same as Example 2 but $f(z)$ has been recast in polar form. The conclusion is the same. ◀

EXERCISES

Where in the complex plane are the following functions analytic? State which are entire functions. If the derivative exists in a domain, find an expression for $f'(z)$ in terms of z or x and y. If $f'(1+i)$ exists, state its numerical value.

1. $2z^2 + 3$

2. $z + z^{-1}$

3. $1/(z^4 + 2z^2 + 1)$

4. $-xy + \frac{i}{2}(x^2 - y^2)$

5. $(y+1)^2 + i(x+1)^2$

6. $y + (x-1)^2 + i[(y-1)^3 - x]$

7. $x^3 - 3xy^2 + i(3x^2y - y^3)$

8. $e^{x^2-y^2}[\cos(2xy) + i\sin(2xy)]$

9. $\dfrac{z^2}{e^x \cos y + ie^x \sin y}$

10. $f(z) = z$, $|z| \leq 1$ but $f(z) = 1/z$, $|z| > 1$

11. $\dfrac{1}{(y - ix + 1 + i + z)^4}$

Use L'Hôpital's Rule to evaluate these limits:

12. $\dfrac{(z-1) + (z^2 - 1)}{z^{16} - 1}$ as $z \to 1$

13. $\dfrac{z^9 + z - 2i}{z^{15} + i}$ as $z \to i$

Where in the complex plane are the following functions analytic? The origin need not be considered. Use the polar form of the Cauchy–Riemann equations.

14. $r \cos \theta + ir$

15. $r^4 \sin 4\theta - ir^4 \cos 4\theta$

16. $\text{Log}\, r^2 + i2\theta$, $-\pi < \theta \leq \pi$ (natural log)

17. a) Prove that if an analytic function is purely real in some domain, then it must be constant in that domain.

    b) Repeat the question in part (a) with the word "imaginary" substituted for "real."

18. Suppose $f(z) = u + iv$ is analytic. Under what circumstances will $g(z) = u - iv$ be analytic?

    Hint: Consider the functions $f(z) + g(z)$ and $f(z) - g(z)$. Then refer to Exercise 17. You may also use Theorem 5.

19. Consider an analytic function $f(z) = u + iv$ whose modulus $|f(z)|$ is equal to a constant k throughout some domain. Show that this can occur only if $f(z)$ is constant throughout the domain.

    Hint: The case $k = 0$ is trivial. Assuming $k \neq 0$, we have $u^2 + v^2 = k^2$ or $k^2/(u + iv) = u - iv$. Now refer to Exercise 18.

20. a) Assume that both $f(z)$ and $f(\bar{z})$ are defined in a domain D and that $f(z)$ is analytic in D. Assume that $f(\bar{z}) = \overline{f(z)}$ in D. Show that $f(\bar{z})$ cannot be analytic in D unless $f(z)$ is a constant.

    Hint: $f(z) + f(\bar{z})$ is real. Why? Now use the result of Exercise 17.

    b) Use the preceding result to argue in a few lines that $(\bar{z})^3 + \bar{z}$ is nowhere analytic.

21. a) Assume that $f(z)$ is analytic and nonzero at z_0 and assume that $g(z)$ is not analytic at z_0. Prove, using Theorem 5, that $h(z) = f(z)g(z)$ is not analytic at z_0.

    Hint: Assume $h(z)$ is analytic. Then consider $h(z)/f(z)$ and apply Theorem 5. Explain why a contradiction is obtained.

    b) Use the preceding to explain why $z^4\bar{z}$ is nowhere analytic. Note that the point $z = 0$ requires some special attention.

22. Polar form of the C–R equations.

a) Suppose, for the analytic function $f(z) = u(x, y) + iv(x, y)$, that we express x and y in terms of the polar variables r and θ, where $x = r\cos\theta$ and $y = r\sin\theta$ ($r = \sqrt{x^2 + y^2}$, $\theta = \tan^{-1}(y/x)$). Then, $f(z) = u(r, \theta) + iv(r, \theta)$. We want to rewrite the C–R equations entirely in the polar variables. From the chain rule for partial differentiation we have

$$\frac{\partial u}{\partial x} = \left(\frac{\partial u}{\partial r}\right)_\theta \left(\frac{\partial r}{\partial x}\right)_y + \left(\frac{\partial u}{\partial \theta}\right)_r \left(\frac{\partial \theta}{\partial x}\right)_y.$$

Give the corresponding expressions for $\partial u/\partial y$, $\partial v/\partial x$, $\partial v/\partial y$.

b) Show that

$$\left(\frac{\partial r}{\partial x}\right)_y = \cos\theta,$$

$$\left(\frac{\partial \theta}{\partial x}\right)_y = \frac{-\sin\theta}{r},$$

and find corresponding expressions for $(\partial r/\partial y)_x$ and $(\partial\theta/\partial y)_x$. Use these four expressions in the equations for $\partial u/\partial x$, $\partial u/\partial y$, $\partial v/\partial x$, and $\partial v/\partial y$ found in part (a). Show that u and v satisfy the equations

$$\frac{\partial h}{\partial x} = \frac{\partial h}{\partial r}\cos\theta - \frac{1}{r}\frac{\partial h}{\partial \theta}\sin\theta,$$

$$\frac{\partial h}{\partial y} = \frac{\partial h}{\partial r}\sin\theta + \frac{1}{r}\frac{\partial h}{\partial \theta}\cos\theta,$$

where h can equal u or v.

c) Rewrite the C–R equations (2.3–10a, b) using the two equations from part (b) of this exercise. Multiply the first C–R equation by $\cos\theta$, the second by $\sin\theta$, and add to show that

$$\frac{\partial u}{\partial r} = \frac{1}{r}\frac{\partial v}{\partial \theta}. \tag{2.4–5a}$$

Now multiply the first C–R equation by $-\sin\theta$, the second by $\cos\theta$, and add to show that

$$\frac{\partial v}{\partial r} = \frac{-1}{r}\frac{\partial u}{\partial \theta}. \tag{2.4–5b}$$

The relationships of Eqs. (2.4–5a, b) are the *polar form of the C–R equations.* If the first partial derivatives of u and v are continuous at some point whose polar coordinates are r, θ ($r \neq 0$), then Eqs. (2.4–5a, b) provide a necessary and sufficient condition for the existence of the derivative at this point.

d) Use Eq. (2.3–6) and the C–R equations in polar form to show that if the derivative of $f(r, \theta)$ exists it can be found from

$$f'(z) = \left[\frac{\partial u}{\partial r} + i\frac{\partial v}{\partial r}\right][\cos\theta - i\sin\theta] \tag{2.4–6}$$

or from

$$f'(z) = \left[\frac{\partial u}{\partial \theta} + i\frac{\partial v}{\partial \theta}\right]\left(\frac{-i}{r}\right)[\cos\theta - i\sin\theta]. \tag{2.4–7}$$

2.5 HARMONIC FUNCTIONS

Given a real function of x and y, say $\phi(x, y)$, we wish to determine if there exists an analytic function $f(z)$ of either of the forms $f(z) = \phi(x, y) + iv(x, y)$ or $f(z) = u(x, y) + i\phi(x, y)$. In other words, can ϕ be regarded as either the real or the imaginary part of an analytic function? It is relatively easy to answer this question.

Consider an analytic function $f(z) = u + iv$. Then, the Cauchy–Riemann equations

$$\frac{\partial u}{\partial x} = \frac{\partial v}{\partial y}, \tag{2.5–1a}$$

$$\frac{\partial u}{\partial y} = -\frac{\partial v}{\partial x} \tag{2.5–1b}$$

are satisfied by u and v. Now let us assume that we can differentiate Eq. (2.5–1a) with respect to x and Eq. (2.5–1b) with respect to y. We obtain

$$\frac{\partial^2 u}{\partial x^2} = \frac{\partial}{\partial x}\frac{\partial v}{\partial y}, \tag{2.5–2a}$$

$$\frac{\partial^2 u}{\partial y^2} = -\frac{\partial}{\partial y}\frac{\partial v}{\partial x}. \tag{2.5–2b}$$

It can be shown† that if the second partial derivatives of a function are continuous, then the order of differentiation in the cross partial derivatives is immaterial. Thus $\partial^2 v/\partial x\,\partial y = \partial^2 v/\partial y\,\partial x$. With this assumption we add Eqs. (2.5–2a) and (2.5–2b). The right-hand sides cancel, leaving

$$\frac{\partial^2 u}{\partial x^2} + \frac{\partial^2 u}{\partial y^2} = 0. \tag{2.5–3}$$

Alternatively, we might have differentiated Eq. (2.5–1a) with respect to y and Eq. (2.5–1b) with respect to x. Assuming that the second partial derivatives of u are continuous, we add the resulting equations and obtain

$$\frac{\partial^2 v}{\partial x^2} + \frac{\partial^2 v}{\partial y^2} = 0. \tag{2.5–4}$$

---

† See W. Kaplan, *Advanced Calculus* (Reading, MA: Addison-Wesley, 1991), Section 2.15.

Thus both the real and imaginary parts of an analytic function must satisfy a differential equation of the form shown below.

LAPLACE'S EQUATION[†]

$$\frac{\partial^2 \phi}{\partial x^2} + \frac{\partial^2 \phi}{\partial y^2} = 0 \qquad (2.5\text{–}5)$$

Laplace's equation for functions of the polar variables r and θ is derived in Exercise 25 of the present section.

DEFINITION Harmonic function

Functions satisfying Laplace's equation in a domain are said to be *harmonic* in that domain. □

An example of a harmonic function is $\phi(x, y) = x^2 - y^2$ since $\partial^2\phi/\partial x^2 = 2$, $\partial^2\phi/\partial y^2 = -2$ and Laplace's equation is satisfied throughout the z-plane. A function satisfying Laplace's equation only for some set of points that does not constitute a domain is not harmonic. An example of this is presented in Exercise 1 of this section.

Equations (2.5–3) and (2.5–4) can be summarized as follows:

THEOREM 6

If a function is analytic in some domain, its real and imaginary parts are harmonic in that domain. □

A converse to the preceding theorem can be established, provided we limit ourselves to simply connected domains.[‡]

THEOREM 7 Given a real function $\phi(x, y)$, which is harmonic in a simply connected domain D, there exists in D an analytic function whose *real part* equals $\phi(x, y)$. There also exists in D an analytic function whose *imaginary part* is $\phi(x, y)$. □

Given a harmonic function $\phi(x, y)$, we may wish to find the corresponding harmonic function $v(x, y)$ such that $\phi(x, y) + iv(x, y)$ is analytic. Or, given $\phi(x, y)$, we might seek $u(x, y)$ such that $u(x, y) + i\phi(x, y)$ is analytic. In either case we can determine the unknown function up to an additive constant. The method is best illustrated with an example.

EXAMPLE 1

Show that $\phi = x^3 - 3xy^2 + 2y$ can be the real part of an analytic function. Find the imaginary part of the analytic function.

[†] Laplace's equation was derived by the French mathematician Pierre Simon Laplace (1749–1827), who discovered it while studying gravitation and its relation to planetary motion.

[‡] J. Bak and D. Newman, *Complex Analysis* (New York: Springer-Verlag, 1982), Section 16.1.

Solution

We have

$$\frac{\partial^2 \phi}{\partial x^2} = 6x \quad \text{and} \quad \frac{\partial^2 \phi}{\partial y^2} = -6x,$$

which sums to zero throughout the z-plane. Thus ϕ is harmonic. To find $v(x, y)$ we use the C–R equations and take $u(x, y) = \phi(x, y)$:

$$\frac{\partial u}{\partial x} = 3x^2 - 3y^2 = \frac{\partial v}{\partial y}, \tag{2.5–6}$$

$$-\frac{\partial u}{\partial y} = 6xy - 2 = \frac{\partial v}{\partial x}. \tag{2.5–7}$$

Let us solve Eq. (2.5–6) for v by integrating on y:

$$v = \int (3x^2 - 3y^2)dy \quad \text{or} \quad v = 3x^2y - y^3 + C(x). \tag{2.5–8}$$

It is important to recognize that the "constant" C, although independent of y, can depend on the variable x. The reader can verify this by substituting v from Eq. (2.5–8) into Eq. (2.5–6).

To evaluate $C(x)$ we substitute v from Eq. (2.5–8) into Eq. (2.5–7) and get $6xy - 2 = 6xy + dC/dx$. Obviously, $dC/dx = -2$. We integrate and obtain $C = -2x + D$, where D is a true constant, independent of x and y. Putting this value of C into Eq. (2.5–8) we finally have

$$v = 3x^2y - y^3 - 2x + D. \tag{2.5–9}$$

Since v is a real function, D must be a real constant. Its value cannot be determined if we are given only u. However, if we are told the value of v at some point in the complex plane, the value of D can be found. For example, given $v = -2$ at $x = -1$, $y = 1$, we substitute these quantities into Eq. (2.5–9) and find that $D = -6$. ◀

DEFINITION Harmonic conjugate

Given a harmonic function $u(x, y)$, we call $v(x, y)$ the *harmonic conjugate* of $u(x, y)$ if $u(x, y) + iv(x, y)$ is analytic. □

This definition is not related to the notion of the conjugate of a complex number. In Example 1, just given, $3x^2y - y^3 - 2x + D$ is the harmonic conjugate of $x^3 - 3xy^2 + 2y$ since $f(z) = x^3 - 3xy^2 + 2y + i(3x^2y - y^2 - 2x + D)$ is analytic. However, $x^3 - 3xy^2 + 2y$ *is not* the harmonic conjugate of $3x^2y - y^3 - 2x + D$ since $f(z) = 3x^2y - y^3 - 2x + D + i(x^3 - 3xy^2 + 2y)$ is not analytic. This matter is explored more fully in Exercise 15 of this section where we investigate the circumstances under which $u + iv$ and $v + iu$ can both be analytic.

Conjugate functions have an interesting geometrical property. Given a harmonic function $u(x, y)$ and a real constant C_1, we find that the equation $u(x, y) = C_1$ is satisfied along some locus, typically a curve, in the xy-plane. Given a collection of

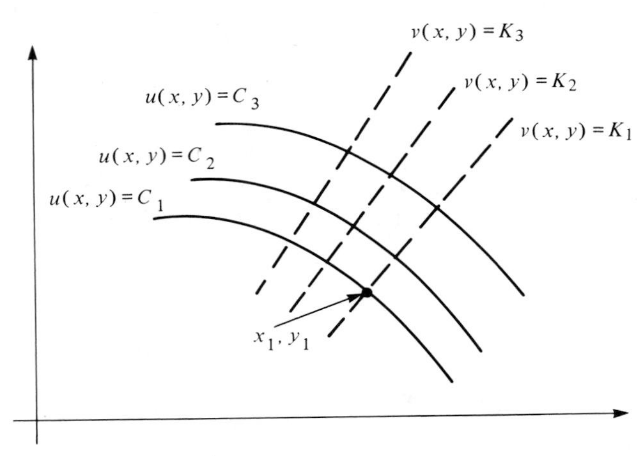

Figure 2.5–1

such constants, $C_1, C_2, C_3, \ldots$, we can plot a family of curves by using the equations $u(x, y) = C_1$, $u(x, y) = C_2$, etc. A typical family for some $u(x, y)$ is shown in solid lines in Fig. 2.5–1.

Suppose $v(x, y)$ is the harmonic conjugate of $u(x, y)$, and $K_1, K_2, K_3, \ldots$, are real constants. We can plot on the same figure the family of curves given by $v(x, y) = K_1$, $v(x, y) = K_2$, etc., which are indicated by dashed lines. We will now prove the following important theorem pertaining to the two families of curves.

> ***THEOREM 8*** Let $f(z) = u(x, y) + iv(x, y)$ be an analytic function and C_1, $C_2, C_3, \ldots$ and $K_1, K_2, K_3, \ldots$ be real constants. Then the family of curves in the xy-plane along which $u = C_1$, $u = C_2$, etc., is orthogonal to the family given by $v = K_1$, $v = K_2, \ldots$; that is, the intersection of a member of one family with that of another takes place at a 90° angle, except possibly at any point where $f'(z) = 0$. □

For a proof let us consider the intersection of the curve $u = C_1$ with the curve $v = K_1$ in Fig. 2.5–1. Recall the total differential

$$du = \frac{\partial u}{\partial x} dx + \frac{\partial u}{\partial y} dy.$$

On the curve $u = C_1$, u is constant, so that $du = 0$. Thus

$$0 = \frac{\partial u}{\partial x} dx + \frac{\partial u}{\partial y} dy. \tag{2.5–10}$$

Let us use Eq. (2.5–10) to determine dy/dx at the point of intersection x_1, y_1. We then have

$$\left.\frac{dy}{dx}\right|_{x_1, y_1} = \left(-\frac{\partial u/\partial x}{\partial u/\partial y}\right)_{x_1, y_1}. \tag{2.5–11}$$

This is merely the slope of the curve $u = C_1$ at the point being considered. Similarly, the slope of $v = K_1$ at x_1, y_1 is

$$\left.\frac{dy}{dx}\right|_{x_1, y_1} = \left(-\frac{\partial v/\partial x}{\partial v/\partial y}\right)_{x_1, y_1}. \qquad (2.5\text{–}12)$$

With the C–R equations

$$-\frac{\partial v}{\partial x} = \frac{\partial u}{\partial y} \quad \text{and} \quad \frac{\partial v}{\partial y} = \frac{\partial u}{\partial x}$$

we can rewrite Eq. (2.5–12) as

$$\frac{dy}{dx} = \left(\frac{\partial u/\partial y}{\partial u/\partial x}\right)_{x_1, y_1}. \qquad (2.5\text{–}13)$$

Comparing Eqs. (2.5–11) and (2.5–13), we observe that the slopes of the curves $u = C_1$ and $v = K_1$ at the point of intersection x_1, y_1 are negative reciprocals of one another. Hence, the intersection takes place at a 90° angle. An identical procedure applies at any other intersection involving the families of curves. Notice that if $f'(z) = 0$ at some point, then according to Eqs. (2.3–6) and (2.3–8) the first partial derivatives of u and v vanish. The slope of the curves cannot now be found from Eqs. (2.5–11) and (2.5–12). The proof breaks down at such a point.

EXAMPLE 2

Consider the function

$$f(z) = \tfrac{1}{2}\,\text{Log}(x^2 + y^2) + i \arg z \qquad \text{(natural log)},$$

where we use the principal value of $\arg z$. Thus, $-\pi < \arg z \leq \pi$. Show that this function satisfies the Cauchy–Riemann equations in any domain not containing the origin and/or points on the negative real axis (where $\arg z$ is discontinuous), and that Theorem 8 holds for this function.

Solution

We let $u = 1/2\,\text{Log}(x^2 + y^2)$ and $v = \arg z$. To apply the C–R equations we need v in terms of x and y. We may use either $v = \arg z = \tan^{-1}(y/x)$ or $v = \arg z = \cot^{-1}(x/y)$. The multivalued functions $\tan^{-1}(y/x)$ and $\cot^{-1}(x/y)$ are evaluated so that v will be the principal value of $\arg z$.

At most points we can use either the $\tan^{-1}$ or $\cot^{-1}$ expressions in the C–R equations. However, when $x = 0$ we employ the $\cot^{-1}$ formula, while when $y = 0$ we employ $\tan^{-1}$. In this way we avoid having to differentiate functions whose arguments are infinite. Differentiating both u and v, we observe that the C–R equations are satisfied. The reader should verify that the formulas for the derivatives of v produced by the expressions employing $\tan^{-1}$ and $\cot^{-1}$ are identical. Thus

$$\frac{\partial u}{\partial x} = \frac{x}{x^2 + y^2} = \frac{\partial v}{\partial y} \quad \text{and} \quad \frac{\partial u}{\partial y} = \frac{y}{x^2 + y^2} = \frac{-\partial v}{\partial x}.$$

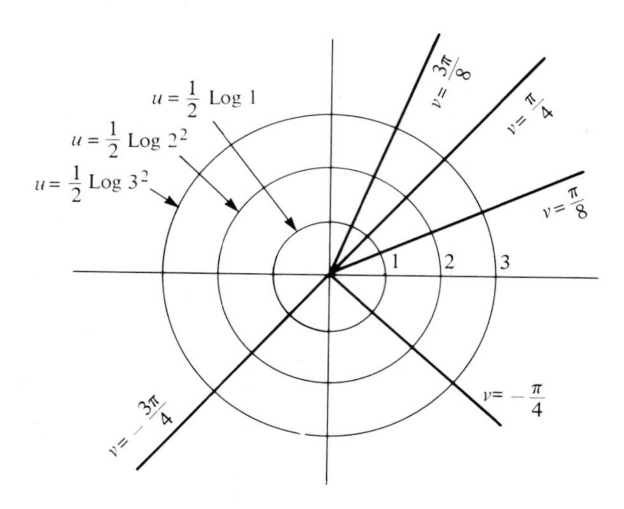

Figure 2.5–2

Loci along which u is constant are merely the circles along which $x^2 + y^2$ assumes constant values, for example, $x^2 + y^2 = 1$, $x^2 + y^2 = 2^2$, $x^2 + y^2 = 3^2$, etc.

Since $v = \arg z$, the curves along which v assumes constant values are merely rays extending outward from the origin of the z-plane. Families of curves of the form $u = C$ and $v = K$ are shown in Fig. 2.5–2. The orthogonality of the intersections should be apparent. ◀

EXERCISES

1. Find the values of z for which $x^4 - y^3$ satisfies Laplace's equation. Why is this function not harmonic?
2. Find the value of the integer n if $x^n - y^n$ is harmonic.

Which of the following functions are harmonic? In what domain?

3. $x + y$
4. xy
5. $\dfrac{y}{x^2 + y^2}$
6. $e^{x^2 - y^2}$
7. $\sin x \cosh y$
8. Find two values of k such that $\cos x[e^y + e^{ky}]$ is harmonic.
9. If $g(x)[e^{2y} - e^{-2y}]$ is harmonic, $g(0) = 0$, $g'(0) = 1$, find $g(x)$.

Show by direct calculation with Laplace's equation that:

10. $\text{Re}(1/z)$ is harmonic throughout any domain not containing $z = 0$.

11. $\text{Im}(z^3)$ is harmonic in any domain.

Let $\phi(x, y) = 6x^2y^2 - x^4 - y^4 + y - x + 1$.

12. Show that $\phi(x, y)$ could be the real or imaginary part of an analytic function.

13. If $\phi(x, y)$ is the real part of an analytic function, find the imaginary part.

14. If $\phi(x, y)$ is the imaginary part of an analytic function, find the real part. Discuss whether the answer here is the same as for Exercise 13.

15. Suppose that $f(z) = u + iv$ is analytic and that $g(z) = v + iu$ is also. Show that v and u must both be constants.

Hint: $-if(z) = v - iu$ is analytic (the product of analytic functions). Thus $g(z) \pm if(z)$ is analytic and must satisfy the C–R equations. Now refer to Exercise 17 of Section 2.4.

16. Find the harmonic conjugate of $e^x \cos y + e^y \cos x + xy$.

17. Find the harmonic conjugate of $\tan^{-1}(x/y)$ where $-\pi < \tan^{-1}(x/y) \leq \pi$.

18. Show, if $u(x, y)$ and $v(x, y)$ are harmonic functions, that $u + v$ must be a harmonic function but that uv need not be a harmonic function.

If $v(x, y)$ is the harmonic conjugate of $u(x, y)$, show that the following are harmonic functions of the variables x and y.

19. uv **20.** $e^u \cos v$ **21.** $\sin u \cosh v$

22. Consider $f(z) = z^2 = u + iv$.

a) Find the equation describing the curve along which $u = 1$ in the xy-plane. Repeat for $v = 2$.

b) Find the point of intersection, in the first quadrant, of the two curves found in part (a).

c) Find the numerical value of the slope of each curve at the point of intersection, which was found in part (b), and verify that the slopes are negative reciprocals.

23. a) Show that $f(z) = e^x \cos y + ie^x \sin y = u + iv$ is entire.

b) Consider the curve along which $u = 1$ and the curve along which $v = 2$. Plot these loci, in the xy-plane, in the domain $0 < y < \pi/2$. Use a pocket calculator with log and trigonometric functions for the numerical data.

c) Find mathematically the point of intersection of the curves in part (b).

d) Take derivatives to find the slopes of the curves at their intersection, and verify that they are negative reciprocals.

24. Consider $f(z) = z^3 = u + iv$.

a) Find the equation describing the curve along which $u = 1$ in the xy-plane. Repeat for $v = 1$. In each case, sketch the curves in the first quadrant.

b) Find mathematically the point of intersection (x_0, y_0) in the first quadrant of the two curves. This is most easily done if you let $z = r \text{ cis } \theta$. First find the intersection in polar coordinates.

c) Find the slope of each curve at the point of intersection. Verify that these are negative reciprocals of each other.

25. a) Let $x = r\cos\theta$ and $y = r\sin\theta$, where r and θ are the usual polar coordinate variables. Let $f(z) = u(r, \theta) + iv(r, \theta)$ be a function that is analytic in some domain that does not include $z = 0$. Use Eqs. (2.4–5a, b) and an assumed continuity of second partial derivatives to show that in this domain u and v satisfy the differential equation

$$\frac{\partial^2 \phi}{\partial r^2} + \frac{1}{r^2}\frac{\partial^2 \phi}{\partial \theta^2} + \frac{1}{r}\frac{\partial \phi}{\partial r} = 0. \qquad (2.5\text{–}14)$$

This is Laplace's equation in the polar variables r and θ.

b) Show that $u(r, \theta) = r^2 \cos 2\theta$ is a harmonic function.

c) Find $v(r, \theta)$, the harmonic conjugate of $u(r, \theta)$, and show that it too satisfies Laplace's equation everywhere.

2.6 SOME PHYSICAL APPLICATIONS OF HARMONIC FUNCTIONS

A number of interesting cases of natural phenomena that are described to a high degree of accuracy by harmonic functions will be discussed in this section.

Steady-state Heat Conduction[†]

Heat is said to move through a material by *conduction* when energy is transferred by collisions involving adjacent molecules and electrons. For conduction the time rate of flow of heat energy at each point within the material can be specified by means of a vector. Typically this vector will vary in both magnitude and direction throughout the material. In general, a variation with time must also be considered. However, we shall limit ourselves to steady-state problems where this will be unnecessary. Thus the intensity of heat conduction within a material is given by a vector function of spatial coordinates. Such a function is often known as a *vector field*. In the present case the vector field is called the *heat flux density* and is given the symbol Q.

Because of their close connection with complex variable theory, we will consider here only two-dimensional heat flow problems. The flow of heat takes place within a plate that we will regard as being parallel to the complex plane. The broad faces of the plate are assumed perfectly insulated. No heat can be absorbed or emitted by the insulation.

As shown in Fig. 2.6–1, some of the edge surfaces of the plate are connected to heat sources (which send out thermal energy) or heat sinks (which absorb thermal energy). The remaining edge surfaces are insulated. Heat energy cannot flow into or out of any perfectly insulated surface. Thus the heat flux density vector Q will be assumed tangent to any insulated boundary. Since the properties of the heat sources

---

[†] For a detailed discussion of this subject, see F. Kreith and W. Block, *Basic Heat Transfer* (New York: Harper and Row, 1980).

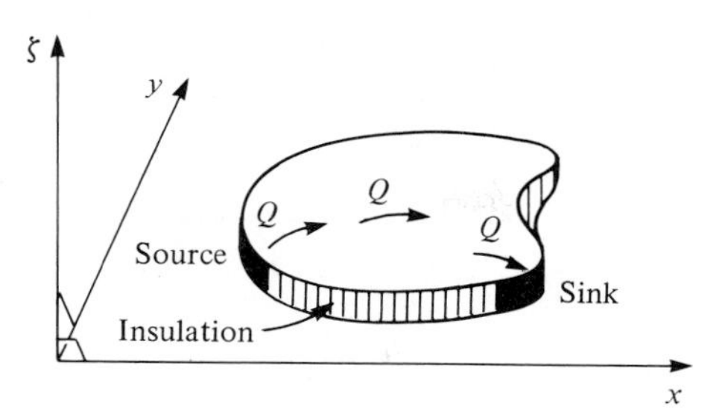

Figure 2.6–1

and sinks are assumed independent of the coordinate ζ, which lies perpendicular to the xy-plane, the vector field Q within the plate depends on only the two variables x and y. The insulation on the broad faces of the plate ensures that Q has components along the x- and y-axes only; that is, Q has components $Q_x(x, y)$ and $Q_y(x, y)$. In conventional vector analysis we would write

$$Q = Q_x(x, y)\mathbf{a}_x + Q_y(x, y)\mathbf{a}_y, \tag{2.6–1}$$

where $\mathbf{a}_x$ and $\mathbf{a}_y$ are unit vectors along the x- and y-axes.

One must be careful not to confuse the vector that locates a particular point in a two-dimensional configuration with the vector representing Q at that point. For instance, if $Q_x = y + 1, Q_y = x$, then at the point $x = 1, y = 1$, we have $Q_x = 2, Q_y = 1$.

The direction of Q, at a particular point, is the direction in which thermal energy is being transported most rapidly.

Now consider a flat differential surface of area dS (see Fig. 2.6–2).[†] The heat flux f through any surface is the flow of thermal energy through that surface per unit time. For dS, a differential flux of heat df passes through the surface given by

$$df = Q_n\, dS, \tag{2.6–2}$$

where Q_n is that component of Q normal to dS. The component Q_t, which is parallel to dS, carries no heat through the surface.

To obtain the heat flux f crossing a surface that is not flat and not of differential size, we must integrate the normal component of Q over the surface. The heat flux entering a volume is the total heat flux traversing inward through the surface bounding the volume.

Under steady-state conditions, the temperature in a conducting material is independent of time. The net flux of heat into any volume of the conductor is zero; otherwise, the volume would get hotter or colder depending on whether the entering flux was positive or negative. By requiring that the net flux entering a differential

[†] dS will be used as both the name of the differential surface as well as the size of its differential area.

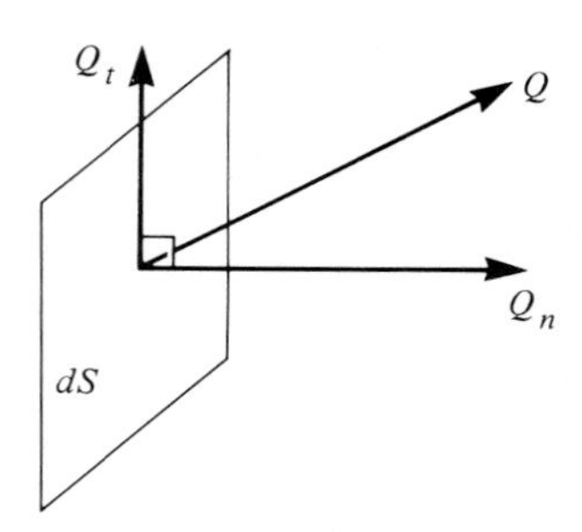

Figure 2.6–2

volume centered at x, y be zero, one can show that the components of Q satisfy

$$\frac{\partial Q_x}{\partial x} + \frac{\partial Q_y}{\partial y} = 0. \tag{2.6–3}$$

The equation is satisfied in the steady state by the two-dimensional heat flux density vector at any point where there are no heat sources or sinks. Equation (2.6–3) is the local or point form of the law of conservation of heat. The reader may recognize $\partial Q_x/\partial x + \partial Q_y/\partial y$ as the divergence of Q.

It is a familiar fact that the rate at which heat energy is conducted through a material is related to the temperature differences occurring within the material and also to the distances over which these differences occur, that is, to the rate of change of temperature with distance. Let us continue to assume two-dimensional heat flow, where the heat flow vector $Q(x, y)$ has components Q_x and Q_y. Let $\phi(x, y)$ be the temperature in the heat conducting medium. Then it can be shown that the components of the vector Q are related to $\phi(x, y)$ by

$$Q_x = -k \frac{\partial \phi}{\partial x}(x, y), \tag{2.6–4a}$$

$$Q_y = -k \frac{\partial \phi}{\partial y}(x, y). \tag{2.6–4b}$$

Here k is a constant called *thermal conductivity*. Its value depends on the material being considered. The reader may recognize Eqs. (2.6–4a, b) as being equivalent to the statement that Q is "minus k times the gradient of the temperature ϕ." The temperature ϕ serves as a "potential function" from which the heat flux density vector can be calculated by means of Eqs. (2.6–4a, b). With the aid of these equations we can rewrite Eq. (2.6–3) in terms of temperature:

$$-k \frac{\partial^2 \phi}{\partial x^2} - k \frac{\partial^2 \phi}{\partial y^2} = 0 \quad \text{or} \quad \frac{\partial^2 \phi}{\partial x^2} + \frac{\partial^2 \phi}{\partial y^2} = 0. \tag{2.6–5}$$

Thus under steady-state conditions, and where there are no sources or sinks, the temperature inside a conductor is a harmonic function.

Because the temperature $\phi(x, y)$ is a harmonic function, it can be regarded as the real part of a function that is analytic within a domain of the xy-plane corresponding to the interior of the conducting plate. This analytic function, which we call $\Phi(x, y)$ is known as the *complex temperature*. We then have

$$\Phi(x, y) = \phi(x, y) + i\psi(x, y). \tag{2.6–6}$$

Thus the real part of the complex temperature $\Phi(x, y)$ is the actual temperature $\phi(x, y)$. The imaginary part of the complex temperature, namely $\psi(x, y)$, we will call the *stream function* because of its analogy with a function describing streams along which particles flow in a fluid.

The curves along which $\phi(x, y)$ assumes constant values are called *isotherms* or *equipotentials*. These curves are just the edges of the surfaces along which the temperature is equal to a specific value. Several examples are represented by solid lines in Fig. 2.6–3.

From Theorem 8 we realize that the family of curves along which $\psi(x, y)$ is constant must be perpendicular to the isotherms. The ψ = constant curves are called *streamlines*. Several are depicted by dashed lines in Fig. 2.6–3.

The slope of a curve along which $\psi(x, y)$ is constant is of interest. If we refer back to the derivation of Eq. (2.5–13) and replace u with ϕ and v with ψ, we find that the slope of a streamline passing through x, y is given by

$$\frac{dy}{dx} = \left(\frac{\partial\phi/\partial y}{\partial\phi/\partial x}\right). \tag{2.6–7}$$

Now suppose we draw, at the same point, the local value of the heat flux density vector Q. From Eqs. (2.6–4a, b) we see that the slope of this vector is

$$\frac{Q_y}{Q_x} = \frac{\partial\phi/\partial y}{\partial\phi/\partial x}. \tag{2.6–8}$$

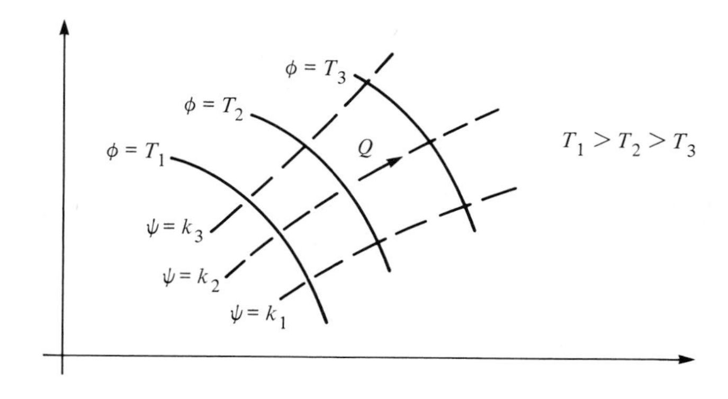

Figure 2.6–3

Comparing Eqs. (2.6–7) and (2.6–8) and noting the identical slopes, we can now conclude the following theorem.

> ***THEOREM 9*** The heat flux density vector at a given point within a heat conducting medium is tangent to the streamline passing through that point. □

We have illustrated this theorem by drawing Q at one point in Fig. 2.6–3. Note that the streamline establishes the slope of Q but not the actual direction of the vector. The direction is established by our realizing that the direction of heat flow is from a warmer to a colder isotherm.

A diagram of the family of curves along which $\psi(x, y)$ assumes constant values provides us with a picture of the paths along which heat is flowing. Moreover, since these streamlines are orthogonal to the isotherms, we conclude that:

> The heat flux density vector, calculated at some point, is perpendicular to the isotherm passing through that point.

It is often convenient to introduce a function called the *complex heat flux density*, which is defined by

$$\mathbf{q}(z) = Q_x(x, y) + iQ_y(x, y). \tag{2.6–9}$$

Since

$$\mathrm{Re}(\mathbf{q}) = Q_x \quad \text{and} \quad \mathrm{Im}(\mathbf{q}) = Q_y, \tag{2.6–10}$$

the vector associated with this function at any point x, y is precisely the heat flux density vector Q at that point. With the aid of Eqs. (2.6–4a, b) we rewrite $\mathbf{q}$ in terms of the temperature:

$$\mathbf{q} = -k\left(\frac{\partial \phi}{\partial x} + i\frac{\partial \phi}{\partial y}\right). \tag{2.6–11}$$

The complex temperature $\Phi(z) = \phi(x, y) + i\psi(x, y)$ is an analytic function. If we want to know its derivative with respect to z, there are two convenient formulas, shown in Eqs. (2.3–6) and (2.3–8), at our disposal. Choosing the former (with $\phi = u$, $\psi = v$), we have

$$\frac{d\Phi}{dz} = \frac{\partial \phi}{\partial x} + i\frac{\partial \psi}{\partial x}. \tag{2.6–12}$$

Now ϕ and ψ satisfy the C–R equations

$$\frac{\partial \phi}{\partial x} = \frac{\partial \psi}{\partial y} \quad \text{and} \quad \frac{\partial \psi}{\partial x} = -\frac{\partial \phi}{\partial y}.$$

With the second of these equations used in the imaginary part of Eq. (2.6–12) we obtain

$$\frac{d\Phi}{dz} = \frac{\partial \phi}{\partial x} - i\frac{\partial \phi}{\partial y}.$$

Now, note that

$$\overline{\left(\frac{d\Phi}{dz}\right)} = \frac{\partial\phi}{\partial x} + i\,\frac{\partial\phi}{\partial y}. \tag{2.6–13}$$

Comparing Eqs. (2.6–13) and (2.6–11), we obtain the following convenient formula for the complex heat flux density:

$$\mathbf{q} = -k\overline{\left(\frac{d\Phi}{dz}\right)}. \tag{2.6–14}$$

The real and imaginary parts of this expression then yield Q_x and Q_y. Of course, Q_x and Q_y are also obtainable if we find the temperature $\phi = \operatorname{Re}\Phi$ and then apply Eqs. (2.6–4a, b).

Fluid Flow

Because many of the concepts applying to heat conduction carry over directly to fluid mechanics, we can be a bit briefer about this topic.

Let us assume we are dealing with an "ideal fluid," that is, one that's incompressible (its mass density does not alter) and nonviscous, (there are no losses due to internal friction). We assume a steady state, which means that the velocity of flow at any point of the fluid is independent of time. Like heat flow, fluid flow originates in sources and terminates in sinks.

If a rigid impermeable obstruction is placed in the moving fluid, the fluid will move tangent to the surface of the object much as heat flows parallel to an insulated boundary.

Earlier, we restricted ourselves to two-dimensional heat flow configurations. Heat conduction was parallel to the xy-plane and depended on only the variables x and y. Here we restrict ourselves to two-dimensional fluid flow parallel to the xy-plane. The *fluid velocity* V will be a vector field dependent in general on the coordinates x and y. It is analogous to the heat flux density Q. The components of V along the coordinate axes are V_x and V_y. The velocity V is the vector associated with the *complex velocity* defined by

$$\mathbf{v} = V_x(x, y) + iV_y(x, y). \tag{2.6–15}$$

This expression is analogous to the complex heat flux density $\mathbf{q} = Q_x(x, y) + iQ_y(x, y)$.

Under certain conditions there is a fluid mechanical analogue of Eqs. (2.6–4a, b). There exists a real function of x and y, the *velocity potential* $\phi(x, y)$, such that

$$V_x = \frac{\partial\phi}{\partial x}, \tag{2.6–16a}$$

$$V_y = \frac{\partial\phi}{\partial y}. \tag{2.6–16b}$$

This condition is described by saying that the velocity is the *gradient* of the velocity potential. For V_x and V_y to be derivable from ϕ, as stated in Eq. (2.6–16), it is necessary that the fluid flow be what is called *irrotational.* Irrotational flow is approximated in many physical problems.† It is characterized by the absence of vortices (whirlpools).

Under steady-state conditions the total mass of fluid contained within any volume of space remains constant in time. For any volume not containing sources or sinks as much fluid flows in during any time interval as flows out. This should remind us of the steady-state conservation of heat. In fact, for an incompressible fluid *the velocity components V_x and V_y satisfy the same conservation equation (2.6–3) as do the corresponding components Q_x and Q_y of the heat flux density vector.* Using Eq. (2.6–16) to eliminate V_x and V_y from the conservation equation satisfied by the velocity vector, we have

$$\frac{\partial^2 \phi}{\partial x^2} + \frac{\partial^2 \phi}{\partial y^2} = 0. \tag{2.6–17}$$

Thus the velocity potential is a harmonic function.

An analytic function $\Phi(z)$ that has real part $\phi(x, y)$ can now be defined. Its imaginary part $\psi(x, y)$ is called the *stream function.* Thus

$$\Phi(z) = \phi(x, y) + i\psi(x, y). \tag{2.6–18}$$

We call Φ the *complex velocity potential*, or simply the complex potential. The curves along which $\phi(x, y)$ assumes constant values are called equipotentials, and, as before, the curves of constant ψ are called streamlines. The two families of curves are orthogonal.

> Like the heat flux density vector the fluid velocity vector at every point is tangent to the streamline passing through that point.

If we follow the progress of some specific moving droplet of fluid, we find that its path is a streamline. The fluid velocity vector at a given point is perpendicular to the equipotential passing through that point.

There is a simple relationship between the complex potential and the fluid velocity. From Eqs. (2.6–13) and (2.6–16) we have

$$\overline{\left(\frac{d\Phi}{dz}\right)} = V_x + iV_y = \mathbf{v}. \tag{2.6–19}$$

Electrostatics‡

In the theory of electrostatics it is stationary (nonmoving) electric charge that plays the role of the sources and sinks we mentioned when discussing heat conduction and fluid flow.

Acccrding to the theory there are two kinds of charge: positive and negative. Charge is often measured in *coulombs.* Positive charge acts as a source of electric

---

† See, e.g., R. Sabersky, A. Acosta, and E. Hauptman, *Fluid Flow, A First Course in Fluid Mechanics*, 2nd ed. (New York: Macmillan, 1971), Sections 3.5, 6.2–6.4.

‡ A clearly written reference on this subject is W. H. Hayt's *Engineering Electromagnetics*, 4th ed. (New York: McGraw-Hill, 1981).

flux, negative charge acts as a sink. In other words, electric flux emanates outward from positive charge and is absorbed into negative charge.

The concentration of electric flux at a point in space is described by the *electric flux density vector* D. Although the notion of electric flux is something of a mathematical abstraction, we can make a physical measurement to determine D at any location. This is accomplished by our putting a point-sized test charge of q_0 coulombs at the spot in question. The test charge will experience a force because of its interaction with the source and (and sink) charges.† The vector force F is given by

$$F = q_0 \frac{D}{\varepsilon}. \tag{2.6–20}$$

Here, ε is a positive constant, called the *permittivity*. Its numerical value depends on the medium in which the test charge is embedded.

Often, instead of using the vector D directly, we employ the electric field vector E, which is defined by

$$\varepsilon E = D. \tag{2.6–21}$$

Thus Eq. (2.6–20) becomes

$$F = q_0 E \quad \text{or} \quad F/q_0 = E.$$

We see that the vector force on a test charge divided by the size of the charge yields the electric field E. Then Eq. (2.6–21) tells us the vector flux density D at the test charge.

We will be considering two-dimensional problems in electrostatics. This requires some explanation. All the electric charges involved in the creation of the electric flux are assumed to exist along lines, or cylinders, of infinite extent that lie perpendicular to the xy-plane. Let ζ be the coordinate perpendicular to the xy-plane. We assume that the distribution of charge along these sources or sinks of flux is independent of ζ. Any obstructions (for example, metallic conductors) placed within the electric flux must also be of infinite length, and perpendicular to the xy-plane. For this sort of configuration the electric flux density vector D is parallel to the xy-plane. Its components D_x and D_y depend, in general, on the variables x and y but are independent of ζ. Maxwell's equations show that the electric flux density vector created by static charges can be derived from a scalar potential. This electrostatic potential ϕ, usually measured in volts, bears much the same relation to the electric flux density as does the temperature to the heat flux density or the velocity potential to the fluid velocity.

The components of the electric flux density vector are obtained from $\phi(x, y)$ as follows:

$$D_x = -\varepsilon \frac{\partial \phi}{\partial x}, \qquad D_y = -\varepsilon \frac{\partial \phi}{\partial y}. \tag{2.6–22}$$

---

† This is an example of a field force that acts at a distance from its source, even through vacuum. Gravity is another example of such a force.

These equations are analogous to Eqs. (2.6–4a, b) for the heat flux density. If E_x and E_y are the components of the electric field, then, from Eq. (2.6–21), we have $D_x = \varepsilon E_x$, and $D_y = \varepsilon E_y$.

A glance at Eq. (2.6–22) then shows that for the electric field,

$$E_x = -\frac{\partial \phi}{\partial x}, \qquad E_y = -\frac{\partial \phi}{\partial y}. \tag{2.6–23}$$

One can define the electric flux crossing a surface in much the same way as one defines the heat flux crossing that surface. The amount of electric flux df that passes through a flat surface dS is obtained from Eq. (2.6–2) with D_n, the normal component of the electric flux density vector, substituted for Q_n. Thus $df = D_n\,dS$. The flux crossing a nondifferential surface is obtained by an integration of the normal component of D over that surface.

According to Maxwell's first equation the total electric flux entering any volume that contains no net electric charge is zero. This reminds us of an identical condition obeyed by the heat flux for a source-free volume. In fact, *at any point in space where there is no electric charge, the electric flux density vector satisfies the same conservation equation (2.6–3) as does the heat flux density vector.* If the components of D are eliminated from this conservation equation by means of Eq. (2.6–22), a familiar result is found:

$$\frac{\partial^2 \phi}{\partial x^2} + \frac{\partial^2 \phi}{\partial y^2} = 0.$$

Hence, the electrostatic potential is a harmonic function in any charge-free region.

As expected, we define an analytic function, the complex electrostatic potential $\Phi = \phi + i\psi$, whose real part is the actual electrostatic potential. As before, the imaginary part is called the *stream function.*

The electric flux density vector is tangent to the streamlines generated from ψ.

The electric flux density D and electric field vector E are the vectors corresponding to the following complex functions:

$$\mathbf{d}(z) = D_x(x, y) + iD_y(x, y),$$
$$\mathbf{e}(z) = E_x(x, y) + iE_y(x, y).$$

These are called the *complex electric flux density* and the *complex electric field,* respectively, and they satisfy

$$\mathbf{d} = -\varepsilon\left(\overline{\frac{d\Phi}{dz}}\right) \quad \text{and} \quad \mathbf{e} = -\left(\overline{\frac{d\Phi}{dz}}\right).$$

Our discussion of heat, fluids, and electrostatics is summarized in Table 1. There are other physical situations, for example, material diffusion, magnetostatics, and gravitation, where harmonic functions also prove useful.

Table 1

	Heat Conduction	Fluid Flow	Electrostatics
Flux density vector	Q = heat flux density	V = velocity	D = electric flux density
Complex flux function	$\mathbf{q} = Q_x + iQ_y$	$\mathbf{v} = V_x + iV_y$	$\mathbf{d} = D_x + iD_y$
Harmonic potential function ϕ	temperature	velocity potential	electrostatic potential
Flux density components	$Q_x = -k\dfrac{\partial\phi}{\partial x}$ $Q_y = -k\dfrac{\partial\phi}{\partial y}$	$V_x = \dfrac{\partial\phi}{\partial x}$ $V_y = \dfrac{\partial\phi}{\partial y}$	$D_x = -\varepsilon\dfrac{\partial\phi}{\partial x}$ $D_y = -\varepsilon\dfrac{\partial\phi}{\partial y}$
Complex flux density from complex potential $\Phi = \phi + i\psi$	$\mathbf{q} = -k\left(\overline{\dfrac{d\Phi}{dz}}\right)$	$\mathbf{v} = \left(\overline{\dfrac{d\Phi}{dz}}\right)$	$\mathbf{d} = -\varepsilon\left(\overline{\dfrac{d\Phi}{dz}}\right)$

EXAMPLE 1

A complex potential is of the form

$$\Phi(z) = Az + B, \qquad \text{where } A \text{ and } B \text{ are real numbers.} \tag{2.6–24}$$

Discuss its associated equipotentials, streamlines, and flux density in terms of electrostatics, heat conduction, and fluid flow.

Solution

The potential function is

$$\phi(x, y) = \text{Re}(Az + B) = Ax + B, \tag{2.6–25}$$

and the stream function is

$$\psi(x, y) = \text{Im}(Az + B) = Ay. \tag{2.6–26}$$

The equipotentials (or isotherms) are the surfaces on which $\phi(x, y)$ assumes fixed values. From Eq. (2.6–25) we see that these appear in the z-plane as lines on which x is constant. Some of these lines are drawn in Fig. 2.6–4. The streamlines on which ψ assumes fixed values are, according to Eq. (2.6–26), lines along which y is constant. These are indicated by dashes in the figure.

The reader who has studied electrostatics will recognize the potential distribution in the Fig. 2.6–4 as that existing between the plates of a parallel plate capacitor whose plates are perpendicular to the x-axis. The complex electric flux density for this

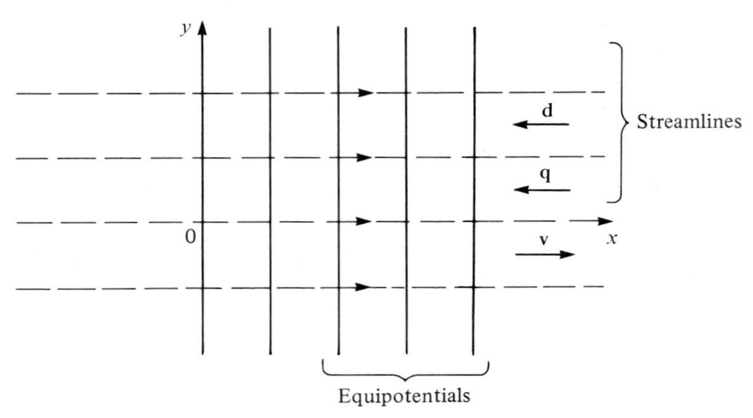

Figure 2.6–4

configuration is, from the bottom line in Table 1,

$$\mathbf{d} = -\varepsilon \overline{\frac{d}{dz}(Az + B)} = -\varepsilon A = D_x + iD_y,$$

which implies $D_x = -\varepsilon A$, $D_y = 0$. The electric flux density vector is parallel to the x-axis. If $A > 0$, it points toward the left in Fig. 2.6–4.

If $\Phi(z)$ is the complex temperature, then the isotherms are the equipotentials in Fig. 2.6–4. The complex heat flux density is

$$\mathbf{q} = -k \overline{\frac{d}{dz}(Az + B)} = -kA = Q_x + iQ_y$$

(see Table 1), which implies that $Q_x = -kA$, $Q_y = 0$. The heat flow is uniform and, if $A > 0$, in the negative x-direction.

Finally, if $\Phi(z)$ describes fluid flow, the fluid velocity is

$$V_x + iV_y = \overline{\frac{d}{dz}(Az + B)} = A,$$

so that $V_x = A$, $V_y = 0$. The fluid flow is thus uniform and in the direction of the streamlines in Fig. 2.6–4. For $A > 0$, flow is to the right. ◀

EXERCISES

1. Suppose that everywhere within a medium the components of the heat flux density vector Q are $Q_x = 3$, $Q_y = -4$ calories per square centimeter per second.

a) Beginning with Eqs. (2.6–4a, b) find the temperature $\phi(x, y)$ in degrees. Assume $\phi(0, 0) = 0$ and that the conductivity k of the medium equals 0.1 calorie per centimeter degree second.

b) Find the stream function $\psi(x, y)$. Assume $\psi(0, 0) = 0$.

c) Sketch the equipotentials on which ϕ equals 0, 40, -40.

d) Sketch the streamlines $\psi = 0, 40, -40$. Verify that the lines are parallel to Q.

2. Suppose the complex potential describing a certain fluid flow is given by $\Phi(z) = 1/z$ (meter^2/sec) for $z \neq 0$.

a) Find the complex fluid velocity at $x = 1$, $y = 1$ (meter) by differentiating the complex potential. State V_x and V_y.

b) Find the components V_x and V_y, at the same point, by finding and using the velocity potential $\phi(x, y)$.

c) Show that the equation of the equipotential that passes through $x = 1$, $y = 1$ is $(x - 1)^2 + y^2 = 1$. Plot this curve.

d) Find the equation of the streamline passing through $x = 1$, $y = 1$. Plot this curve.

3. Suppose that $\Phi(z) = e^x \cos y + ie^x \sin y$ represents the complex potential, in volts, for some electrostatic configuration.

a) Use the complex potential to find the complex electric field at $x = 1$, $y = 1/2$.

b) Obtain the complex electric field at the same point by first finding and using the electrostatic potential $\phi(x, y)$.

c) Assuming the configuration lies within a vacuum, find the components D_x and D_y of the electric flux density vector at $x = 1$, $y = 1/2$. In m.k.s. units, $\varepsilon = 8.85 \times 10^{-12}$ for vacuum.

d) What is the value of ϕ at $x = 1$, $y = 1/2$? Plot the equipotential surface passing through this point.

e) What is the value of ψ at $x = 1$, $y = 1/2$? Plot the streamline passing through this point.

4. a) Explain why $\mathbf{d}(x, y) = y + ix$ can be the complex electric flux density in a charge-free region, but $\mathbf{d}(x, y) = x + iy$ cannot.

b) Assume that the complex electric flux density $y + ix$ exists in a medium for which $\varepsilon = 9 \times 10^{-12}$. Find the electrostatic potential $\phi(x, y)$. Assume $\phi(0, 0) = 0$. Sketch the equipotentials $\phi(x, y) = 0$, $\phi(x, y) = 1/\varepsilon$.

c) Find the stream function $\psi(x, y)$. Assume $\psi(0, 0) = 0$.

d) Find the complex potential Φ and express it explicitly in terms of z.

e) Find the components of the electric field at $x = 1$, $y = 1$ by three different methods: from $\mathbf{d}$, from $\Phi(z)$, and from $\phi(x, y)$. Show with a sketch the vector for this field and the equipotential passing through $x = 1$, $y = 1$.

5. a) Fluid flow is described by the complex potential $\Phi(z) = (\cos \alpha - i \sin \alpha)z$. $\alpha > 0$. Sketch the associated equipotentials and give their equations.

b) Sketch the streamlines and give their equations.

c) Find the components V_x and V_y of the velocity vector at (x, y). What angle does the velocity vector make with the positive x-axis?

CHAPTER 3

The Basic Transcendental Functions

We are well acquainted with numerous functions that exist in the mathematics of real variables. We know the algebraic functions (sums, products, quotients, and powers of x^m and $x^{m/n}$), and we know transcendental functions such as e^x, $\log x$, $\sinh x$, and so forth. However, in the complex plane we have seen only algebraic functions of z. In this chapter we will enlarge our collection of functions of a complex variable. We will extend our definitions of some elementary transcendental functions of a real variable (e^x, $\sin x$, etc.) so as to yield functions of the complex variable z. These new functions reduce to our previously known real functions if z happens to be real.

3.1 THE EXPONENTIAL FUNCTION

We would like this function to have the following properties:

a) e^z reduces to our known e^x if z happens to assume real values.†

b) e^z is an analytic function of z.

The function $e^x \cos y + ie^x \sin y$ will be our definition of e^z (or $\exp z$). Thus

$$e^z = e^{x+iy} = e^x[\cos y + i \sin y]. \tag{3.1–1}$$

† Recall that e^x can be defined by $e^x = \lim_{n\to\infty}(1 + x/n)^n$.

Equation (3.1–1) clearly satisfies condition (a). (Put $y = 0$.) Observe that as in the real case, $e^0 = 1$. To verify condition (b) we have

$$u + iv = e^x \cos y + ie^x \sin y, \tag{3.1–2}$$

where

$$u = \text{Re } e^z = e^x \cos y, \qquad v = \text{Im } e^z = e^x \sin y. \tag{3.1–3}$$

The pair of functions u and v have first partial derivatives:

$$\frac{\partial u}{\partial x} = e^x \cos y, \qquad \frac{\partial v}{\partial y} = e^x \cos y, \qquad \text{and so on,}$$

which are continuous everywhere in the xy-plane. Furthermore, u and v satisfy the Cauchy–Riemann equations,

$$\frac{\partial u}{\partial x} = \frac{\partial v}{\partial y}, \qquad \frac{\partial v}{\partial x} = -\frac{\partial u}{\partial y},$$

everywhere in this plane. *Thus e^z is analytic for all z and is therefore an entire function.* Condition (b) is clearly satisfied.

The derivative $d(e^z)/dz$ is easily found from Eqs. (2.3–6) and (3.1–3). Thus

$$\frac{d}{dz} e^z = \frac{\partial}{\partial x} e^x \cos y + i \frac{\partial}{\partial x} e^x \sin y = e^x \cos y + ie^x \sin y,$$

or

$$\frac{d}{dz} e^z = e^z.$$

This is a reassuring result since we already knew that e^x satisfies

$$\frac{d}{dx} e^x = e^x.$$

Note that if $g(z)$ is an analytic function, then, by the chain rule of differentiation, (see Eq. (2.4–1d)), we have

$$\frac{d}{dz} e^{g(z)} = e^{g(z)} g'(z).$$

The function e^z shares another property with e^x. We know that if x_1 and x_2 are real, then $e^{x_1} e^{x_2} = e^{(x_1 + x_2)}$. We can show that if $z_1 = x_1 + iy_1$ and $z_2 = x_2 + iy_2$ are a pair of complex numbers, then $e^{z_1} e^{z_2} = e^{z_1 + z_2}$. Observe that

$$e^{z_1} = e^{x_1}[\cos y_1 + i \sin y_1] \quad \text{and} \quad e^{z_2} = e^{x_2}[\cos y_2 + i \sin y_2],$$

so that

$$\begin{aligned} e^{z_1} e^{z_2} &= e^{x_1} e^{x_2}[\cos y_1 + i \sin y_1][\cos y_2 + i \sin y_2] \\ &= e^{x_1 + x_2}[(\cos y_1 \cos y_2 - \sin y_1 \sin y_2) + i(\sin y_1 \cos y_2 + \cos y_1 \sin y_2)]. \end{aligned}$$

The real part of the expression in the brackets is, from elementary trigonometry, $\cos(y_1 + y_2)$. Similarly, the imaginary part is $\sin(y_1 + y_2)$. Hence,

$$e^{z_1}e^{z_2} = e^{x_1+x_2}[\cos(y_1 + y_2) + i\sin(y_1 + y_2)]. \tag{3.1–4}$$

Now with the help of Eq. (3.1–1), we have

$$\begin{aligned} e^{z_1+z_2} &= e^{(x_1+x_2)+i(y_1+y_2)} \\ &= e^{(x_1+x_2)}[\cos(y_1 + y_2) + i\sin(y_1 + y_2)]. \end{aligned}$$

Since the right side of the preceding equation is identical to the right side of Eq. (3.1–4), it is obvious that

$$e^{z_1}e^{z_2} = e^{z_1+z_2}. \tag{3.1–5}$$

Letting $z_1 = z_2$ in the preceding relationship, we have $(e^z)^2 = e^{2z}$, which is easily generalized to

$$(e^z)^m = e^{mz}, \tag{3.1–6}$$

where $m \geq 0$ is an integer. Just as $e^{x_1}/e^{x_2} = e^{x_1-x_2}$ we can show (see Exercise 14) that

$$e^{z_1}/e^{z_2} = e^{z_1-z_2}. \tag{3.1–7}$$

If $z_1 = 0$ and $z_2 = z$ in the preceding equation we obtain $1/e^z = e^{-z}$. We can use this result to show that Eq. (3.1–6) holds when m is a negative integer.

Now observe that Eq. (3.1–1) is equivalent to the polar form

$$e^z = e^x \operatorname{cis} y = e^x \angle y. \tag{3.1–8}$$

Taking the magnitude of both sides of the preceding, we have

$$|e^z| = e^x. \tag{3.1–9}$$

Thus, *the magnitude of e^z is determined entirely by the real part of z.*

Equation (3.1–8) also says that one value for $\arg(e^z)$ is y. Because of the multivaluedness of the polar angle we have, in general,

$$\arg(e^z) = y + 2k\pi, \qquad k = 0, \pm 1, \pm 2, \tag{3.1–10}$$

which shows that the *argument of e^z is determined up to an additive constant $2k\pi$ by the imaginary part of z.*

An important consequence of Eq. (3.1–9) (see Exercise 12) is that the equation $e^z = 0$ has no solution in the complex plane.

Although e^x is not a periodic function of x, e^z varies periodically as we move in the z-plane along any straight line parallel to the y-axis. Consider e^{z_0} and $e^{z_0+i2\pi}$, where $z_0 = x_0 + iy_0$. The points z_0 and $z_0 + i2\pi$ are separated a distance 2π on the line $\operatorname{Re} z = x_0$. From our definition (see Eq. 3.1–1) we have

$$\begin{aligned} e^{z_0} &= e^{x_0}(\cos y_0 + i\sin y_0), \\ e^{z_0+i2\pi} &= e^{x_0+i(y_0+2\pi)} = e^{x_0}[\cos(y_0 + 2\pi) + i\sin(y_0 + 2\pi)]. \end{aligned}$$

Since $\cos y_0 = \cos(y_0 + 2\pi)$ (and similarly for the sine), then $e^{z_0} = e^{z_0 + i2\pi}$.

Thus e^z is periodic with imaginary period $2\pi i$.

Of particular interest is the behavior of $e^{i\theta}$ when θ is a real variable. In Eq. (3.1–8) we put $x = 0$, $y = \theta$ to get

$$e^{i\theta} = 1\underline{/\theta} = \cos\theta + i\sin\theta. \tag{3.1–11}$$

Thus if θ is real, $e^{i\theta}$ is a complex number of modulus 1, which lies at an angle θ with respect to the positive real axis (see Fig. 3.1–1). As θ increases, the complex number $1\underline{/\theta}$ progresses counterclockwise around the unit circle. Observe, in particular, that

$$e^{i0} = 1, \qquad e^{i\pi/2} = i, \qquad e^{i\pi} = -1, \qquad e^{i3\pi/2} = -i = e^{-i\pi/2}.$$

The relationship $e^{i\theta} = \cos\theta + i\sin\theta$ is known as *Euler's identity* and is named after the 18th century Swiss mathematician mentioned in Chapter 1 as the inventor of the i notation. He is also credited with the popularization of e to mean the base of natural logarithms.

With $\theta = \pi$ in Euler's identity, and a slight rearrangement of the equation, we obtain the legendary formula

$$e^{i\pi} + 1 = 0,$$

which neatly and unexpectedly relates the five most important numbers $(0, 1, i, e, \pi)$ in mathematics.

Recalling DeMoivre's Theorem $(\cos\theta + i\sin\theta)^n = \cos n\theta + i\sin n\theta$, we see that with the aid of Euler's identity this can be equivalently expressed as

$$(e^{i\theta})^n = e^{in\theta}.$$

This is a special case of Eq. (3.1–6).

The exponential notation $e^{i\theta}$ is often used in lieu of cis θ or $1\underline{/\theta}$ when we represent the variable z in polar coordinates. Thus $z = r\operatorname{cis}\theta$ becomes $z = re^{i\theta}$. Using this

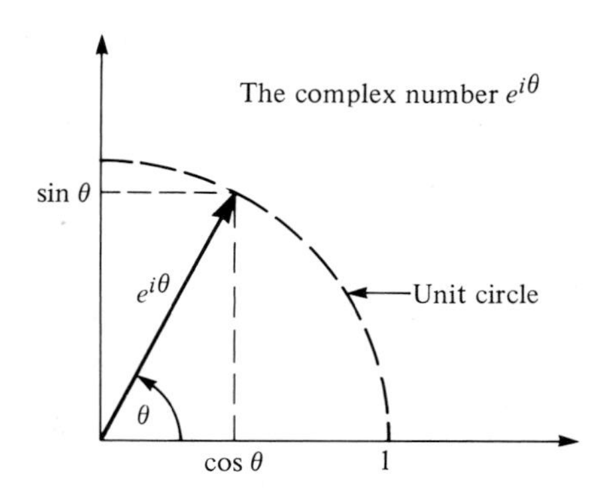

Figure 3.1–1

representation, we do the following multiplication as an example. Consider

$$1 + i\sqrt{3} = 2\underline{/\frac{\pi}{3}} = 2e^{i\pi/3} \quad \text{and} \quad 1 + i = \sqrt{2}\underline{/\frac{\pi}{4}} = \sqrt{2}e^{i(\pi/4)},$$

so that

$$(1 + i\sqrt{3})(1 + i) = 2\sqrt{2}e^{i(\pi/3 + \pi/4)}.$$

With the aid of Eq. (3.1–11) the right side of this equation can be rewritten as

$$2\sqrt{2}\left[\cos\left(\frac{\pi}{3} + \frac{\pi}{4}\right) + i\sin\left(\frac{\pi}{3} + \frac{\pi}{4}\right)\right].$$

Some important physical applications of the complex exponential in electrical engineering and mechanics can be found in the appendix to this chapter.

EXERCISES

Express each of the following in the form $a + ib$, where a and b are real numbers.

1. e^{3+4i} **2.** e^{3-4i} **3.** $e^{-3}e^{4i}$ **4.** e^{i}

5. $e^{[1/(1-i)]}$ **6.** $e^{i \arccos 1}$ **7.** $e^{e^{i}}$ **8.** $e^{-e^{-i}}$

9. $e^{i^{1/2}}$ (all values) **10.** $(e^{i})^{1/2}$ (all values)

11. If $f(z) = e^{z}$, show that $f(z + in\pi) = (-1)^{n}f(z)$, where n is any integer.

12. Prove that $e^{z} = 0$ has no solution in the z-plane.

13. Find all solutions of $e^{z} = 1 + i0$ by equating corresponding parts (reals and imaginaries) on both sides of this equation.

14. Use the definition of e^{z} given in Eq. (3.1–1) to show that

$$\frac{e^{z_1}}{e^{z_2}} = e^{z_1 - z_2},$$

where $z_1 = x_1 + iy_1$ and $z_2 = x_2 + iy_2$.

Express each of the following functions in the form $u(x, y) + iv(x, y)$, where u and v are real functions.

15. e^{z^2} **16.** $e^{1/\bar{z}}$ **17.** $e^{e^{z}}$ **18.** $e^{(z+z^{-1})}$

Find $f'(1 + i)$ if $f(z)$ is given by

19. $e^{e^{z}}$ **20.** $e^{(z+z^{-1})}$

Where in the complex plane are the following equations satisfied?

21. $|e^{z}| = 1$ **22.** $|e^{z^2}| = 1$ **23.** $|e^{z-1/z}| = 1$

24. Let $\beta \neq 0$ be a complex number. Use L'Hôpital's Rule to find, if it exists,

$$\lim_{z\to 0} \frac{e^{\beta e^{z}} - e^{\beta}}{e^{iz} - 1}.$$

In the following regions, what are the maximum and minimum values assumed by $|e^z|$? State where in the regions these values are achieved.

25. $|z - (1 + i)| \le 1$ **26.** $|z + i| \le 3$

27. The absolute magnitude of the expression

$$P = 1 + e^{j\psi} + e^{j2\psi} + \cdots + e^{j(N-1)\psi} = \sum_{n=0}^{N-1} e^{jn\psi}$$

is of interest in many problems involving radiation from N identical physical elements (e.g., antennas, loudspeakers). Here ψ is a real quantity that depends on the separation of the elements and the position of an observer of the radiation. $|P|$ can tell us the strength of the radiation observed.

a) Using the formula for the sum of a finite geometric series (see Exercise 37, Section 1.4), show that

$$|P(\psi)| = \left|\frac{\sin N\psi/2}{\sin \psi/2}\right|.$$

b) Find $\lim_{\psi\to 0} |P(\psi)|$.

c) Use a pocket calculator or a simple computer program to plot $|P(\psi)|$ for $0 \le \psi \le 2\pi$ when $N = 3$.

28. Let $z = re^{i\theta}$, where r and θ are the usual polar variables.

a) Show that $\operatorname{Re}[(1 + z)/(1 - z)] = (1 - r^2)/(1 + r^2 - 2r\cos\theta)$. Why must this function satisfy Eq. (2.5–14) throughout any domain not containing $z = 1$?

b) Find $\operatorname{Im}[(1 + z)/(1 - z)]$ in a form similar to the result given in (a).

29. Fluid flow is described by the complex potential $\Phi(z) = \phi(x, y) + i\psi(x, y) = e^z$.

a) Find the velocity potential $\phi(x, y)$ and the stream function $\psi(x, y)$ explicitly in terms of x and y.

b) In the strip $|\operatorname{Im} z| \le \pi/2$ sketch the equipotentials $\phi = 0, +1/2, +1, +2$ and the streamlines $\psi = 0, \pm 1/2, \pm 1, \pm 2$.

c) What is the fluid velocity vector at $x = 1$, $y = \pi/4$?

3.2 TRIGONOMETRIC FUNCTIONS

From the Euler identity (Eq. 3.1–11) we know that when θ is a real number, we have

$$e^{i\theta} = \cos\theta + i\sin\theta. \tag{3.2–1}$$

Now, if we alter the sign preceding θ in the above, we have

$$e^{-i\theta} = \cos(-\theta) + i\sin(-\theta) = \cos\theta - i\sin\theta. \tag{3.2–2}$$

The addition of Eq. (3.2–2) to Eq. (3.2–1) results in the purely real expression

$$e^{i\theta} + e^{-i\theta} = 2\cos\theta,$$

or, finally,

$$\cos\theta = \frac{e^{i\theta} + e^{-i\theta}}{2}. \tag{3.2–3}$$

If, instead, we had subtracted Eq. (3.2–2) from Eq. (3.2–1), we would have obtained

$$e^{i\theta} - e^{-i\theta} = 2i\sin\theta,$$

or

$$\sin\theta = \frac{e^{i\theta} - e^{-i\theta}}{2i}. \tag{3.2–4}$$

Equations (3.2–3) and (3.2–4) serve to define the sine and cosine of *real* numbers in terms of complex exponentials.

It is natural to define $\sin z$ and $\cos z$, where z is complex, as follows:

$$\sin z = \frac{e^{iz} - e^{-iz}}{2i}, \tag{3.2–5}$$

$$\cos z = \frac{e^{iz} + e^{-iz}}{2}. \tag{3.2–6}$$

These definitions make sense for several reasons:

a) When z is a real number, the definitions shown in Eqs. (3.2–5) and (3.2–6) reduce to the conventional definitions shown in Eqs. (3.2–3) and (3.2–4) for the sine and cosine of real arguments.

b) e^{iz} and e^{-iz} are analytic throughout the z-plane. Therefore, $\sin z$ and $\cos z$, which are defined by the sums and differences of these functions, are also.

c) $d\sin z/dz = i[e^{iz} + e^{-iz}]/2i = \cos z$. Also, $d\cos z/dz = -\sin z$.

It is easy to show that $\sin^2 z + \cos^2 z = 1$ and that the identities satisfied by the sine and cosine of real arguments apply here too, for example,

$$\sin(z_1 \pm z_2) = \sin z_1 \cos z_2 \pm \cos z_1 \sin z_2,$$
$$\cos(z_1 \pm z_2) = \cos z_1 \cos z_2 \mp \sin z_1 \sin z_2,$$

and so on.

Using Eqs. (3.2–5) and (3.2–6), we can compute the numerical value of the sine and cosine of any complex number we please. A somewhat more convenient procedure exists, however. Recall the hyperbolic sine and cosine of real argument θ illustrated in Fig. 3.2–1 and defined by the following equations:

$$\sinh\theta = \frac{e^{\theta} - e^{-\theta}}{2}, \tag{3.2–7}$$

$$\cosh\theta = \frac{e^{\theta} + e^{-\theta}}{2}. \tag{3.2–8}$$

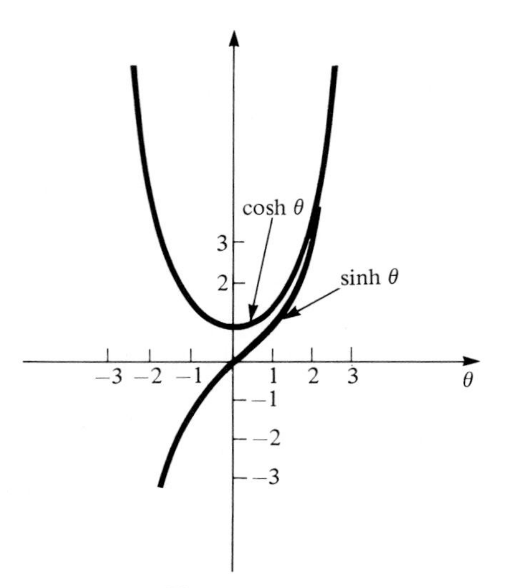

Figure 3.2–1

Consider now the expression in Eq. (3.2–5):

$$\sin z = \frac{e^{i(x+iy)} - e^{-i(x+iy)}}{2i} = \frac{e^{ix-y} - e^{-ix+y}}{2i} = \frac{e^{-y}e^{ix} - e^{y}e^{-ix}}{2i}.$$

The Euler identity (Eq. 3.2–1 or 3.2–2) can now be used to rewrite e^{ix} and e^{-ix} in the preceding equation. We then have

$$\begin{aligned}\sin z &= \frac{e^{-y}(\cos x + i \sin x)}{2i} - \frac{e^{y}(\cos x - i \sin x)}{2i}\\ &= \sin x\frac{(e^{y} + e^{-y})}{2} + i \cos x\frac{(e^{y} - e^{-y})}{2}.\end{aligned}$$

When the expressions involving y in the last equation are compared with the hyperbolic functions in Eqs. (3.2–7) and (3.2–8), we see that

$$\sin z = \sin x \cosh y + i \cos x \sinh y. \tag{3.2–9}$$

Since most pocket calculators are equipped with the trigonometric and hyperbolic functions of real arguments, the evaluation of $\sin z$ becomes a simple matter. A similar expression can be found for $\cos z$:

$$\cos z = \cos x \cosh y - i \sin x \sinh y. \tag{3.2–10}$$

The other trigonometric functions of complex argument are easily defined by analogy with real argument functions, that is,

$$\tan z = \frac{\sin z}{\cos z} = \frac{1}{\cot z}, \qquad \sec z = \frac{1}{\cos z}, \qquad \operatorname{cosec} z = \frac{1}{\sin z}.$$

The derivatives of these functions are as follows:

$$\frac{d}{dz}\tan z = \sec^2 z,$$

$$\frac{d}{dz}\sec z = \tan z \sec z,$$

$$\frac{d}{dz}\operatorname{cosec} z = -\cot z \operatorname{cosec} z.$$

Expressions for $\tan(x + iy)$ and $\cot(x + iy)$ in terms of trigonometric and hyperbolic functions of real arguments are available. They are derived in the exercises and given by Eqs. (3.2–13) and (3.2–14) in Exercises 28 and 29.

The sine or cosine of a real number is a real number whose magnitude is less than or equal to 1. Not only is the sine or cosine of a complex number, in general, a complex number, *but the magnitude of a sine or cosine of a complex number can exceed 1* (see Exercise 1 in this section).

EXAMPLE 1

Express $\sin(i\theta)$, where θ is a real number, in the form $a + ib$, and express $\cos(i\theta)$ in a similar form.

Solution

We use Eq. (3.2–9) with $x = 0$ and $y = \theta$. The result is

$$\sin(i\theta) = \sin 0 \cosh \theta + i \cos 0 \sinh \theta,$$

or

$$\sin(i\theta) = i \sinh \theta. \tag{3.2–11}$$

Similarly, with the aid of Eq. (3.2–10) we have

$$\cos(i\theta) = \cos 0 \cosh \theta - i \sin 0 \sinh \theta,$$

or

$$\cos(i\theta) = \cosh \theta. \quad ◀ \tag{3.2–12}$$

The cosine of a pure imaginary number is always a real number while the sine of a pure imaginary is always pure imaginary.

EXAMPLE 2

Show that all zeros of $\cos z$ in the z-plane lie along the x-axis.

Solution

Consider the equation $\cos z = 0$, where $z = x + iy$. From Eq. (3.2–10) this becomes $\cos x \cosh y - i \sin x \sinh y = 0$. Both the real and imaginary part of the left-hand side of this equation must equal zero. We thus have

$$\cos x \cosh y = 0 \quad \text{and} \quad \sin x \sinh y = 0.$$

Consider the first equation. Since $\cosh y$ is never zero for a real number y, (see Fig. 3.2–1) evidently $\cos x = 0$. This means $x = \pm\pi/2, \pm 3\pi/2, \ldots$, etc.; in other words, $x = \pm(2n + 1)\pi/2$, where $n = 0, 1, 2, \ldots$. Now consider the second equation. The first equation dictated that x be an odd multiple of $\pm\pi/2$; therefore, $\sin x$ in the second equation is ± 1. Thus $\sin x \sinh y = 0$ is only satisfied if $\sinh y = 0$. A glance at Fig. 3.2–1 shows this to be possible only when $y = 0$.

We see that $\cos z = 0$ only at those points that simultaneously satisfy $y = 0$, $x = \pm(2n + 1)\pi/2$. The first condition places all these points on the x-axis while the second spaces them at intervals of π. The values of z that solve $\cos z = 0$ are precisely the same as the solutions of $\cos x = 0$. A similar statement applies to $\sin x$ and $\sin z$ and is derived in one of the exercises below. ◀

EXERCISES

Find the numerical values of the following in the form $a + ib$, where a and b are real numbers.

1. $\sin(1 - 2i)$ **2.** $\cos(2 + i)$ **3.** $\tan(2 - i)$

4. $\sin(i^{1/3})$ (all values) **5.** $\cos(e^{1+i})$ **6.** $e^{\cos(1+i)}$

7. $\sin(\arg(1 + i))$ **8.** $e^{i \sin i}$ **9.** $\sin(i \sin i)$

10. $\cos(i \arg(1 + i))$ (all values) **11.** $\tan e^i$

Use the definitions shown in Eqs. (3.2–5) and (3.2–6) to establish each of the following:

12. $\sin^2 z + \cos^2 z = 1$,

13. $\dfrac{d}{dz} \sin z = \cos z$,

14. $\sin^2 z = \frac{1}{2} - \frac{1}{2}\cos 2z$,

15. $\sin(z_1 + z_2) = \sin z_1 \cos z_2 + \cos z_1 \sin z_2$,

16. $\sin(z + 2k\pi) = \sin z$, $\cos(z + 2k\pi) = \cos z$, where $k = 0, \pm 1, \pm 2, \ldots$.

17. When we studied Euler's identity in Section 3.1, $e^{i\theta} = \cos\theta + i\sin\theta$, we restricted θ to being real. Show, however, that θ can be complex, i.e., $e^{iz} = \cos z + i \sin z$ for any complex z.

18. Show that the equation $\sin z = 0$ has solutions in the complex z-plane only where $z = n\pi$ and $n = 0, \pm 1, \pm 2, \ldots$. Thus like $\cos z$, $\sin z$ has zeros only on the real axis.

19. Show that $\sin z - \cos z = 0$ has solutions only for real values of z. What are the solutions?

Where in the complex plane do each of the following functions fail to be analytic?

20. $\tan z$ **21.** $\dfrac{1}{\cos(iz)}$ **22.** $\dfrac{1}{\sin z \sin[(1 + i)z]}$ **23.** $\dfrac{1}{\sqrt{3}\sin z - \cos z}$

24. Let $f(z) = \cos(e^z)$. Express $f(z)$ and $f'(z)$ in the form $u(x, y) + iv(x, y)$. Where is $f(z)$ analytic?

25. Show that $|\cos z| = \sqrt{\sinh^2 y + \cos^2 x}$.

Hint: Recall that $\cosh^2 \theta - \sinh^2 \theta = 1$.

26. Show that $|\sin z| = \sqrt{\sinh^2 y + \sin^2 x}$.

27. Show that $|\sin z|^2 + |\cos z|^2 = \sinh^2 y + \cosh^2 y$.

28. Show that

$$\tan z = \frac{\sin(2x) + i\sinh(2y)}{\cos(2x) + \cosh(2y)}. \tag{3.2–13}$$

29. Show that

$$\cot z = \frac{\sin(2x) - i\sinh(2y)}{\cosh(2y) - \cos(2x)}. \tag{3.2–14}$$

3.3 HYPERBOLIC FUNCTIONS

In the previous section we used definitions of $\sin z$ and $\cos z$ that we constructed by studying the definitions of the sine and cosine functions of real arguments. A similar procedure will work in the case of $\sinh z$ and $\cosh z$, the hyperbolic functions of complex argument.

Equations (3.2–7) and (3.2–8), which define $\sinh \theta$ and $\cosh \theta$ for a real number θ, suggest the following definitions for complex z:

$$\sinh z = \frac{e^z - e^{-z}}{2}, \tag{3.3–1}$$

$$\cosh z = \frac{e^z + e^{-z}}{2}. \tag{3.3–2}$$

If z is a real number, these definitions reduce to those we know for the hyperbolic functions of real arguments. We see that $\sinh z$ and $\cosh z$ are composed of sums or differences of the functions e^z and e^{-z}, which are analytic in the z-plane. Thus $\sinh z$ and $\cosh z$ are analytic for all z. It is easy to verify that $d(\sinh z)/dz = \cosh z$ and that $d(\cosh z)/dz = \sinh z$. From Eqs. (3.2–5), (3.2–6), (3.3–1), and (3.3–2) one easily verifies that

$$\sinh(iz) = i \sin z$$

and

$$\cosh(iz) = \cos z.$$

All the identities that pertain to the hyperbolic functions of real variables carry over to these functions, for example, we may prove with Eqs. (3.3–1) and (3.3–2) that

$$\cosh^2 z - \sinh^2 z = 1, \tag{3.3–3}$$

$$\cosh(z_1 \pm z_2) = \cosh z_1 \cosh z_2 \pm \sinh z_1 \sinh z_2, \tag{3.3–4}$$

$$\sinh(z_1 \pm z_2) = \sinh z_1 \cosh z_2 \pm \cosh z_1 \sinh z_2. \tag{3.3–5}$$

Expressions for $\sinh z$ and $\cosh z$, involving real functions of real variables and analogous to Eqs. (3.2–9) and (3.2–10), are easily derived. They are

$$\sinh z = \sinh x \cos y + i \cosh x \sin y, \tag{3.3–6}$$

$$\cosh z = \cosh x \cos y + i \sinh x \sin y. \tag{3.3–7}$$

Other hyperbolic functions are readily defined in terms of the hyperbolic sine and cosine:

$$\tanh z = \frac{\sinh z}{\cosh z}, \qquad \operatorname{sech} z = \frac{1}{\cosh z}, \qquad \operatorname{cosech} z = \frac{1}{\sinh z}, \qquad \coth z = \frac{1}{\tanh z}.$$

The hyperbolic functions differ significantly from their trigonometric counterparts. Although all the roots of $\sin z = 0$ and $\cos z = 0$ lie along the real axis in the z-plane, it is shown in the exercises that all the roots of $\sinh z = 0$ and $\cosh z = 0$ lie along the imaginary axis. The trigonometric functions are periodic, with period 2π (see Exercise 16, Section 3.2) but the hyperbolic functions have period $2\pi i$, that is, $\sinh(z + 2\pi i) = \sinh z$, $\cosh(z + 2\pi i) = \cosh z$.

EXERCISES

Use Eqs. (3.3–1) and (3.3–2) to prove the following.

1. $\sinh z = \sinh x \cos y + i \cosh x \sin y$

2. $\cosh z = \cosh x \cos y + i \sinh x \sin y$

3. $\cosh^2 z - \sinh^2 z = 1$

4. $\sinh(z + 2\pi i) = \sinh z$ and $\cosh(z + 2\pi i) = \cosh z$

5. $\sinh(i\theta) = i \sin \theta$ and $\cosh(i\theta) = \cos \theta$. Thus the hyperbolic sine of a pure imaginary number is a pure imaginary number while the hyperbolic cosine of a pure imaginary number is a real number.

Express each of the following in the form $a + ib$, where a and b are real numbers.

6. $\sinh(1/(1 + i))$

7. $\cosh(1 - i\sqrt{3})$

8. $\tanh[(1 + i\sqrt{3})^{1/2}]$ (all values)

9. Find the numerical value of the derivative of $\cosh(\sinh(e^{z^2}))$ at $z = i$.

10. Consider the equation $\sinh(x + iy) = 0$. Use Eq. (3.3–6) to equate the real and imaginary parts of $\sinh z$ to zero. Show that this pair of equations can be satisfied if and only if $z = in\pi$, where $n = 0, \pm 1, \pm 2, \ldots$. Thus the zeros of $\sinh z$ all lie along the imaginary axis in the z-plane.

11. a) Carry out an argument similar to that in the previous problem to show that the zeros of $\cosh z$ must satisfy $z = \pm(2n + 1)\pi i/2$, where $n = 0, 1, 2, 3, \ldots$.

b) Where in the z-plane is $\tanh z$ analytic?

Where do the following functions fail to be analytic?

12. $\dfrac{1}{\sinh z - \cosh z}$ **13.** $\dfrac{1}{\sinh[(1+i)z]}$ **14.** $\dfrac{1}{\sinh(z^2)}$

15. Show that the equation $\sinh z - \sin z = 0$ has no solution on the line $x = 1$.

Prove the following.

16. $|\sinh z|^2 = \sinh^2 x + \sin^2 y$ **17.** $|\cosh z|^2 = \sinh^2 x + \cos^2 y = \cosh^2 x - \sin^2 y$

3.4 THE LOGARITHMIC FUNCTION

If x is a positive real number, then, as the reader knows, $e^{\log x} = x$. The logarithm† of x, which is a real number, is easily found from a calculator or numerical tables. Recall, however, that $\log 0$ is undefined.

In this section we will learn how to obtain the logarithm of a complex number z. We must anticipate that the logarithm of z may itself be a complex number. Our $\log z$ will have the property

$$e^{\log z} = z. \tag{3.4–1}$$

We shall see that $\log z$ is multivalued, that is, for a given value of z there is more than one value of $\log z$ capable of satisfying Eq. (3.4–1).

We will presently show that the following definition of $\log z$ will satisfy Eq. (3.4–1):

$$\log z = \log|z| + i \arg z, \qquad z \neq 0 \tag{3.4–2}$$

The logarithm of zero will remain undefined.

If z is expressed in polar variables $z = re^{i\theta}$, we know that $r = |z|$ and $\theta = \arg z$. Hence, Eq. (3.4–2) becomes

$$\log z = \log r + i\theta, \qquad r \neq 0. \tag{3.4–3}$$

Notice that the real part of the above expression, the logarithm of the modulus of z, is the logarithm of a positive real number. This quantity is evaluated by familiar means. The imaginary part θ is the argument (or angle) of z expressed in radians.

As promised, we see that $e^{\log z}$ is z because

$$\begin{aligned} e^{\log z} &= e^{\log r + i\theta} = e^{\log r} e^{i\theta} = e^{\log r}[\cos\theta + i\sin\theta] \\ &= r[\cos\theta + i\sin\theta] = x + iy = z. \end{aligned} \tag{3.4–4}$$

Here we have used Euler's identity (Eq. 3.2–1) to rewrite $e^{i\theta}$ and have replaced $e^{\log r}$ by r, which is obviously valid for $r > 0$.

The chief difficulty with Eq. (3.4–2) or (3.4–3) is that $\theta = \arg z$ is not uniquely defined. We know that if θ_1 is some valid value for θ, then so is $\theta_1 + 2k\pi$, where $k = 0, \pm 1, \pm 2, \ldots$. Thus the numerical value of the imaginary part of $\log z$ is directly

---

† All logarithms in this book are base e (natural) logarithms.

affected by our particular choice of the argument of z. We thus say that the logarithm of z, as defined by Eq. (3.4–2) or (3.4–3), is a multivalued function of z. Each value of $\log z$ satisfies $e^{\log z} = z$.

Even the logarithm of a positive real number, which we have been thinking is uniquely defined, has, according to Eq. (3.4–3), more than one value. However, when we consider all the possible logarithms of a positive real number, there is only one that is real; the others are complex. It is the real value that we find in numerical tables.

The *principal value* of the logarithm of z, denoted by $\text{Log}\, z$, is obtained when we use the *principal argument* of z in Eqs. (3.4–2) and (3.4–3). Recall (see Section 1.3) that the principal argument of z, which we will designate θ_p, is the argument of z satisfying $-\pi < \theta_p \le \pi$. Thus we have

$$\text{Log}\, z = \text{Log}\, r + i\theta_p, \qquad r = |z| > 0, \qquad \theta_p = \arg z, \qquad -\pi < \theta_p \le \pi. \quad (3.4\text{–}5)$$

Note that we put $\text{Log}\, r$ (instead of $\log r$) in the above equation since the natural logarithms of positive real numbers, obtained from tables or calculators, are principal values.

Any value of $\arg z$ can be obtained from the principal value θ_p by means of the formula $\arg z = \theta = \theta_p + 2k\pi$, where k has a suitable integer value. Thus all values of $\log z$ are obtainable from the expression

$$\log z = \text{Log}\, r + i(\theta_p + 2k\pi), \qquad k = 0, \pm 1, \pm 2, \ldots. \quad (3.4\text{–}6)$$

With $k = 0$ in this expression we obtain the principal value, $\text{Log}\, z$.

Observe that if we choose to use some nonprincipal value of $\arg z$, instead of θ_p, in Eq. (3.4–6), then Eq. (3.4–6) would still yield all possible values of $\log z$, although the principal value would not be generated by our putting $k = 0$.

The fact that we can find the logarithm of any number in the complex plane except $0 + i0$, and that we are not restricted to finding the logarithm of positive reals, was first described by Euler in the mid-nineteenth century. We have mentioned him before in connection with complex exponentials. He also was the first to assert that the logarithm of any number is multivalued.

The various notations used in specifying logarithms may cause some confusion to the reader. The function key labeled ln (or LN) on most basic pocket calculators yields the real value of the natural (base e) logarithm of positive real numbers. Thus, $\ln r$ of the calculator is identical to our $\text{Log}\, r$ when r is a positive real. However, calculators typically also have a function key designated log (or LOG). Use of this key yields base 10 logarithms of positive reals; its use does not give the logarithms employed here, and it will be of no help in solving the exercises of this book.

EXAMPLE 1

Find $\text{Log}(-1 - i)$ and find all values of $\log(-1 - i)$.

Solution

The complex number $-1 - i$ is illustrated graphically in Fig. 1.3–13. The principal argument of this number θ_p is $-3\pi/4$, while r, the absolute magnitude, is $\sqrt{2}$. From

Eq. (3.4–5),

$$\operatorname{Log}(-1-i) = \operatorname{Log}\sqrt{2} + i\left(\frac{-3\pi}{4}\right) \doteq 0.34657 - i\frac{3\pi}{4}.$$

All values of $\log(-1-i)$ are easily written down with the aid of Eq. (3.4–6):

$$\log(-1-i) = \operatorname{Log}\sqrt{2} + i\left(2k\pi - \frac{3\pi}{4}\right)$$

$$\doteq 0.34657 + i\left(2k\pi - \frac{3\pi}{4}\right), \qquad k = 0, \pm 1, \pm 2, \ldots .$$ ◄

EXAMPLE 2

Find $\operatorname{Log}(-10)$ and all values of $\log(-10)$.

Solution

The principal argument of any negative real number is π. Hence, from Eq. (3.4–5) $\operatorname{Log}(-10) = \operatorname{Log} 10 + i\pi \doteq 2.303 + i\pi$. From Eq. (3.4–6) we have

$$\log(-10) = \operatorname{Log} 10 + i(\pi + 2k\pi).$$

We can check this result as follows:

$$e^{\log(-10)} = e^{\operatorname{Log} 10 + i(\pi + 2k\pi)} = e^{\operatorname{Log} 10}[\cos(\pi + 2k\pi) + i\sin(\pi + 2k\pi)] = -10.$$ ◄

In elementary calculus one learns the identity $\log(x_1 x_2) = \log x_1 + \log x_2$, where x_1 and x_2 are positive real numbers. The statement

$$\log(z_1 z_2) = \log z_1 + \log z_2, \tag{3.4–7}$$

where z_1 and z_2 are complex numbers and where we allow for the multiple values of the logarithms, requires some interpretation. The expressions $\log z_1$ and $\log z_2$ are multivalued. So is their sum, $\log z_1 + \log z_2$. If we choose particular values of each of these logarithms and add them, we will obtain *one of the possible values* of $\log(z_1 z_2)$. To establish this let $z_1 = r_1 e^{i\theta_1}$ and $z_2 = r_2 e^{i\theta_2}$. Thus $\log z_1 = \operatorname{Log} r_1 + i(\theta_1 + 2m\pi)$ and $\log z_2 = \operatorname{Log} r_2 + i(\theta_2 + 2n\pi)$. Specific integer values are assigned to m and n. By adding the logarithms we obtain

$$\log z_1 + \log z_2 = \operatorname{Log} r_1 + \operatorname{Log} r_2 + i(\theta_1 + \theta_2 + 2\pi(m+n)). \tag{3.4–8}$$

Now $\operatorname{Log} r_1 + \operatorname{Log} r_2 = \operatorname{Log}(r_1 r_2)$ since r_1 and r_2 are positive real numbers. Thus Eq. (3.4–8) becomes

$$\log z_1 + \log z_2 = \operatorname{Log}(r_1 r_2) + i(\theta_1 + \theta_2 + 2\pi(m+n)). \tag{3.4–9}$$

Notice that $r_1 r_2 = |z_1 z_2|$ while $\theta_1 + \theta_2 + 2\pi(m+n)$ is one of the values of $\arg(z_1 z_2)$. Thus Eq. (3.4–9) is *one* of the possible values of $\log(z_1 z_2)$.

Suppose $z_1 = i$ and $z_2 = -1$. Then if we take $\log(z_1) = i\pi/2$ and $\log z_2 = i\pi$ (principal values), we have $\log z_1 + \log z_2 = i3\pi/2$. Now $z_1 z_2 = -i$, and, if we use the principal value, $\log(z_1 z_2) = -i\pi/2$. Note that here $\log z_1 + \log z_2 \neq \log(z_1 z_2)$.

However, $\log z_1 + \log z_2 = i3\pi/2$ is a valid value of $\log(-i)$. It just happens not to be the value we first computed. The statement $\log(z_1/z_2) = \log z_1 - \log z_2$ must also be interpreted in a manner similar to that of Eq. (3.4–7).

Putting $z = z_1 = z_2$ in (3.4–7), we have $\log z^2 = 2\log z$, which like Eq. (3.4–7) can be satisfied for appropriate choices of the logarithms on each side of the equation. An extension,

$$\log z^n = n \log z, \qquad n \text{ any integer}, \tag{3.4–10}$$

will also be valid for certain values of the logarithms. The same statement applies to $\log z^{n/m} = \frac{n}{m}\log z$, where n and m are any integers, except we exclude $m = 0$.

We are, of course, familiar with an identity from elementary calculus, $\log e^x = x$, where x is a positive real number. The corresponding complex statement is

$$\log e^z = z + i2k\pi, \qquad k = 0, \pm 1, \pm 2, \ldots, \tag{3.4–11}$$

and requires a comment. The expression e^z is, in general, a complex number. Its logarithm is multivalued. One of these values will correspond to z, and the others will not. We should not think that the principal value of the logarithm of e^z must equal z. There is nothing sacred about the principal value.

EXAMPLE 3

Let $z = 1 + 3\pi i$. Find all values of $\log e^z$ and state which one is the same as z.

Solution

We have $e^z = e^{1+3\pi i} = e[\cos 3\pi + i \sin 3\pi] = -e$. Thus

$$\log e^{1+3\pi i} = \log(-e) = \text{Log}|-e| + i(\arg(-e)).$$

Now $\text{Log}|-e| = \text{Log}\, e = 1$ while the principal argument of $-e$ (a negative real number) is π. Thus $\arg(-e) = \pi + 2k\pi$. Therefore

$$\log e^{1+3\pi i} = 1 + i(\pi + 2k\pi), \qquad k = 0, \pm 1, \pm 2, \ldots.$$

The choice $k = 1$ will yield $\log e^{1+3\pi i} = 1 + 3\pi i$. However, the principal value of $\log e^{1+3\pi i}$ is obtained with $k = 0$ and yields $\text{Log}\, e^{1+3\pi i} = 1 + \pi i$. ◀

EXERCISES

Find all values of the logarithm of each of the following numbers and state the principal value. Put answers in the form $a + ib$.

1. 1 **2.** $10i$ **3.** $e^{i\pi/3}$

4. e^{ie} **5.** $-1 + i\sqrt{3}$ **6.** $\text{Log}\, i$

7. $\text{Log}(1 + i\sqrt{3})$ **8.** $\sinh(1 + i)$ **9.** e^{ie^i}

10. $e^{\log(1+i\sqrt{3})}$ **11.** $e^{i^{1/2}}$ (give two principal values)

Find numerical values of each of the following. If the result is multivalued, state all the values.

12. $\sinh(\log i)$ **13.** $\log(\sinh i)$ **14.** $e^{\text{Log}(\text{Log}(\text{Log}\, i))}$

Use logarithms to find all the solutions of the following equations.

15. $e^z = e^{2+i}$ **16.** $e^z = e^{2z}$

17. $(e^z - 1)^2 = e^{2z}$ **18.** $(e^z - 1)^2 = e^z$

19. $(e^z - 1)^3 = 1$ **20.** $e^{4z} + e^{2z} + 1 = 0$ **21.** $e^{e^z} = 1$

Prove that if θ is real, then

22. $\text{Re}[\log(1 + e^{i\theta})] = \text{Log}\left|2\cos\left(\frac{\theta}{2}\right)\right|$ if $e^{i\theta} \neq -1$

23. $\text{Re}[\log(re^{i\theta} - 1)] = \frac{1}{2}\text{Log}(1 - 2r\cos\theta + r^2)$ if $r \geq 0$ and $re^{i\theta} \neq 1$

24. a) Consider the identity $\log z_1 + \log z_2 = \log(z_1 z_2)$. If $z_1 = -ie$ and $z_2 = -2$, find specific values for $\log z_1$, $\log z_2$ and $\log(z_1 z_2)$ that satisfy the identity.

b) For z_1 and z_2 given in part (a) find specific values of $\log z_1$, $\log z_2$ and $\log(z_1/z_2)$ so that the identity $\log(z_1/z_2) = \log z_1 - \log z_2$ is satisfied.

25. Consider the identity $\log z^n = n \log z$, where n is an integer, which is valid for appropriate choices of the logarithms on each side of the equation. Let $z = 1 + i$ and $n = 5$.

a) Find values of $\log z^n$ and $\log z$ that satisfy $n \log z = \log z^n$.

b) For the given z and n is $n\,\text{Log}\,z = \text{Log}\,z^n$ satisfied?

c) Suppose $n = 2$ and z is unchanged. Is $n\,\text{Log}\,z = \text{Log}\,z^n$ then satisfied?

26. What is wrong with this? $1^2 = (-1)^2$ and so $\log 1^2 = \log(-1)^2$. Thus $2\log 1 = 2\log(-1)$, and so $\log 1 = \log(-1)$. Since a possible value of $\log 1$ is zero we conclude that $\log(-1) = 0$. Describe the first invalid step.

27. This problem considers the relationship between $\log(8i)^{1/3}$ and $1/3 \log(8i)$.

a) Show that $(1/3)\log(8i) = \text{Log}\,2 + i(\pi/6 + (2/3)k\pi)$, where $k = 0, \pm 1, \pm 2, \ldots$.

b) Show that $\log(8i)^{1/3} = \text{Log}\,2 + i(\pi/6 + (2/3)m\pi + 2n\pi)$, where $m = 0, 1, 2$ and $n = 0, \pm 1, \pm 2, \ldots$. Thus there are three distinct sets (corresponding to $m = 0, 1, 2$) of values of $\log(8i)^{1/3}$. Each set has an infinity of members.

c) Show that the set of possible values of $\log(8i)^{1/3}$ is identical to the set of possible values of $(1/3)\log(8i)$. This discussion can be generalized to apply to $(1/p)\log z$ (where p is an integer) and $\log(z^{1/p})$.

3.5 ANALYTICITY OF THE LOGARITHMIC FUNCTION

To investigate the analyticity of $\log z$, let us first study the analyticity of the single-valued function $\text{Log}\,z$, that is, the function created from the principal values.†
We have

$$\text{Log}\,z = \text{Log}\,r + i\theta, \qquad r > 0, \qquad -\pi < \theta \leq \pi. \tag{3.5–1}$$

† The modifier "single-valued" in the phrase "single-valued function" is, strictly speaking, redundant since, by definition, a function is single valued.

Obviously, this function is not continuous at $z = 0$ since it is not defined there; it is also not continuous along the negative real axis because θ does not possess a limit at any point along this axis (see Fig. 3.5–1).

Observe that any point on the negative real axis has an argument $\theta = \pi$. On the other hand, points in the third quadrant, which are taken arbitrarily close to the negative axis, have an argument tending toward $-\pi$. The principal argument of z goes through a "jump" of 2π as we cross the negative real axis.

However, Log z is single valued and continuous in the domain D, which consists of the z-plane with the points on the negative real axis and origin "cut out." The troublesome points of discontinuity have been removed. Using the polar system with $z = re^{i\theta}$, we could describe D with the inequalities $r > 0$, $-\pi < \theta < \pi$.

Continuity is a prerequisite for analyticity. Having discovered a domain D in which Log z is continuous, we can now ask whether this function is analytic in that domain. This has already been answered affirmatively in Example 2 of Section 2.5. The function investigated there is precisely that in Eq. (3.5–1) since $\operatorname{Log} r = \operatorname{Log}\sqrt{x^2 + y^2} = (1/2)\operatorname{Log}(x^2 + y^2)$, and $\theta = \arg z$. However, it is convenient in this section to repeat the same discussion in polar coordinates.

Let us write Log z as $u(r, \theta) + iv(r, \theta)$. From Eq. (3.5–1) we find

$$u = \operatorname{Log} r, \qquad v = \theta. \tag{3.5–2}$$

These functions are both defined and continuous in D. From Eq. (2.4–5) we obtain the Cauchy–Riemann equations in polar form:

$$\frac{\partial u}{\partial r} = \frac{1}{r}\frac{\partial v}{\partial \theta}, \qquad \frac{\partial v}{\partial r} = -\frac{1}{r}\frac{\partial u}{\partial \theta}.$$

For u and v defined in Eq. (3.5–2) we have

$$\frac{\partial u}{\partial r} = \frac{1}{r}, \qquad \frac{1}{r}\frac{\partial v}{\partial \theta} = \frac{1}{r}, \qquad \frac{\partial v}{\partial r} = 0, \qquad -\frac{1}{r}\frac{\partial u}{\partial \theta} = 0.$$

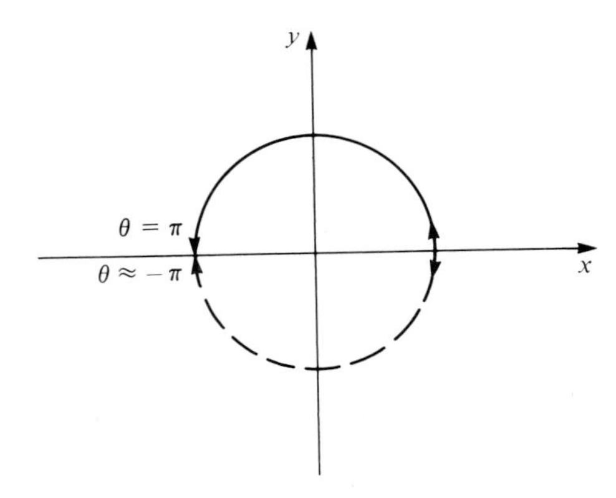

Figure 3.5–1

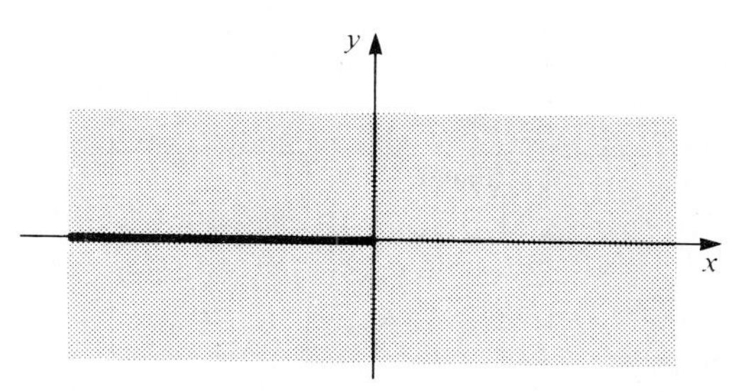

Domain of analyticity of Log z (shaded)

Figure 3.5–2

Obviously, u and v do satisfy the Cauchy–Riemann equations. Moreover, the partial derivatives $\partial u/\partial r$, $\partial v/\partial\theta$, etc. are continuous in domain D. Thus the derivative of Log z must exist everywhere in this domain, and Log z is analytic there. The situation is illustrated in Fig. 3.5–2. We can readily find the derivative of Log z within this domain of analyticity.

If $f(z(r,\theta)) = u(r,\theta) + iv(r,\theta)$ is analytic, then Eq. (2.4–6) provides us with a formula for $f'(z)$. Using this equation with the substitution $e^{-i\theta} = \cos\theta - i\sin\theta$ we have

$$f'(z) = e^{-i\theta}\left(\frac{\partial u}{\partial r} + i\frac{\partial v}{\partial r}\right). \tag{3.5–3}$$

Thus with u and v defined by Eq. (3.5–2) we obtain

$$\frac{d}{dz}\operatorname{Log} z = \frac{e^{-i\theta}}{r} = \frac{1}{re^{i\theta}} = \frac{1}{z} \tag{3.5–4}$$

in the domain D. An alternative derivation involving $u(x, y)$ and $v(x, y)$ is given in Exercise 1 of this section.

Equation (3.5–4) reminds us that $d(\log x)/dx = 1/x$ in real variable calculus. Note that Eq. (3.5–4) is inapplicable along the negative real axis and the origin, which are outside D.

The single-valued function $w(z) = \operatorname{Log} z$ for which z is restricted to the domain D is said to be a branch of $\log z$.

DEFINITION Branch

A *branch* of a multivalued function is a single-valued function *analytic* in some domain. At every point of the domain the single-valued function must assume exactly one of the various possible values that the multivalued function can assume. □

Thus, to specify a branch of a multivalued function we must have at our disposal a means for selecting one of the possible values of this function *and* we must also state the domain of analyticity of the resulting single-valued function.

We have used the notation Log z to mean the principal value of $\log z$. The principal value is defined for all z except $z = 0$. We will also use Log z to mean the principal branch of the logarithmic function. This function is defined for all z except $z = 0$ and values of z on the negative real axis. We have called its domain of analyticity D. Whether Log z refers to the principal value or the principal branch should be clear from the context.

Both the principal value and the principal branch yield the same values, except if z is a negative real number. Then the principal branch cannot be evaluated, but the principal value can.

There are other branches of log z that are analytic in the domain D of Fig. 3.5–2. If we put $k = 1$ in Eq. (3.4–6), we obtain $f(z) = \text{Log}\, r + i\theta$ where $\pi < \theta \le 3\pi$. If z is allowed to assume any value in the complex plane, we find that this function is discontinuous at the origin and at all points on the negative real axis. However, none of these points is present in D. When z is confined to D we have $r > 0$ and $\pi < \theta < 3\pi$, and as before, $df/dz = 1/z$.

The domain D was created by removing the semiinfinite line $y = 0$, $x \le 0$ from the xy-plane. This line is an example of a branch cut.

DEFINITION Branch cut

A line used to create a domain of analyticity is called a *branch line* or *branch cut.* □

It is possible to create other branches of log z that are analytic in domains other than D. Consider

$$f(z) = \text{Log}\, r + i\theta, \qquad \text{where } -3\pi/2 < \theta \le \pi/2.$$

Like the principal value Log z, this function is defined throughout the complex plane except at the origin. It is discontinuous at the origin and at all points on the positive imaginary axis. As it stands, it is not a branch of log z. However, when z is restricted to the domain D_1 shown in Fig. 3.5–3 it becomes a branch. D_1 is created by removing the origin and the positive imaginary axis from the complex plane. When z is restricted to D_1 we require that $r > 0$ and $-3\pi/2 < \theta < \pi/2$. As in the discussion of the principal branch, we can show that the derivative of this branch exists everywhere in D_1 and equals $1/z$.

The reader can readily verify that the logarithmic functions

$$f(z) = \text{Log}\, r + i\theta, \qquad -\frac{3\pi}{2} + 2k\pi < \theta \le \frac{\pi}{2} + 2k\pi, \quad k = 0, \pm 1, \pm 2, \ldots,$$

are, for each k, analytic branches, provided z is confined to the domain D_1.

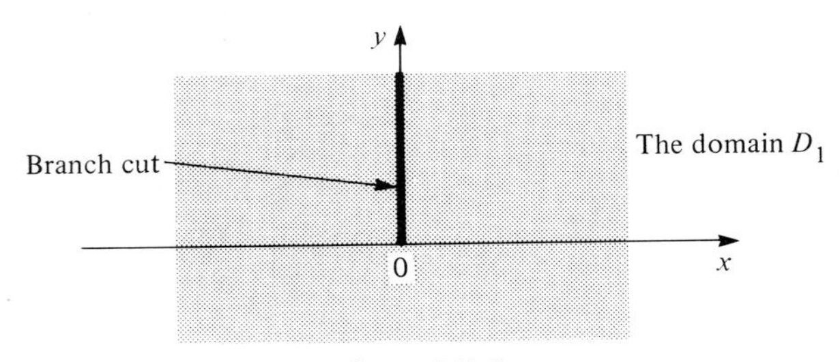

Figure 3.5–3

The domains D and D_1 are just two of the infinite number of possible domains in which we can find branches of $\log z$. Both of these domains were created by using a branch cut (or branch line) in the xy-plane. When all the possible branch lines that we can draw, whenever we are trying to establish branches of a multivalued function, share a common point, we call that point a *branch point* of the multivalued function. In the case of $\log z$ the origin is a branch point. Indeed, the two branch cuts that we investigated for this function passed through $z = 0$. Procedures for finding branch points of other functions are given in Section 3.8.

EXAMPLE 1

Consider the logarithmic function

$$\log z = \operatorname{Log} r + i\theta, \qquad -\frac{\pi}{2} < \theta \le \frac{3\pi}{2}.$$

a) What is the largest domain in the complex plane in which this function defines an analytic branch of the logarithmic function?

b) With this choice of branch what is the numerical value of $\log(-1 - i)$?

Solution

Part (a): The function obviously fails to be continuous at the origin since $\log r$ is undefined there. In addition, θ fails to be continuous along the negative imaginary axis. For points on this axis $\theta = 3\pi/2$, while points in the fourth quadrant, which are taken arbitrarily close to this axis, have a value of θ near to $-\pi/2$, as Fig. 3.5–4 indicates. A branch cut in the xy-plane, extending from the origin outward along the negative imaginary axis, will eliminate all the singular points of the given function. Thus, the given function will yield an analytic branch of the logarithmic function in a domain consisting of the xy-plane with the origin and negative imaginary axis removed.

Part (b): An analytic function varies continuously within its domain of analyticity. Thus to reach $-1 - i$ from a point on the positive x-axis, we must use the counterclockwise path shown in Fig. 3.5–5. The argument θ at a point on the positive

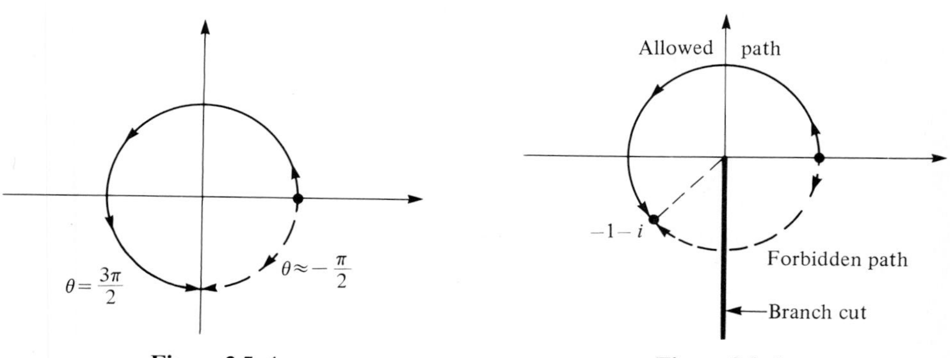

Figure 3.5–4

Figure 3.5–5

x-axis must be $2k\pi$ (where k is an integer). Since in the domain of analyticity we have $-\pi/2 < \theta < 3\pi/2$, evidently $k = 0$. Thus the argument θ begins at 0 radians and reaches the value $5\pi/4$ at $-1 - i$. It is not possible to use the broken clockwise path shown in Fig. 3.5–5 to reach the same point. In so doing we would strike the branch cut and thereby leave the domain of analyticity. Thus to answer the question,

$$\log(-1 - i) = \text{Log}|-1 - i| + i\frac{5\pi}{4} = \text{Log}\sqrt{2} + i\frac{5\pi}{4} \doteq 0.3466 + i\frac{5\pi}{4}. \quad ◀$$

EXAMPLE 2

a) Find the largest domain of analyticity of $f(z) = \text{Log}[z - (3 + 4i)]$.

b) Find the numerical value of $f(0)$.

Solution

Part (a): The function Log w is analytic in the domain consisting of the entire w-plane with the semiinfinite line Im $w = 0$, Re $w \leq 0$ removed. If $w = z - (3 + 4i)$, we ensure analyticity in the z-plane by removing the points that simultaneously satisfy $\text{Im}(z - (3 + 4i)) = 0$ and $\text{Re}(z - (3 + 4i)) \leq 0$. These two conditions can be rewritten

$$\text{Im}((x + iy) - (3 + 4i)) = 0 \quad \text{or} \quad y = 4,$$
$$\text{Re}((x + iy) - (3 + 4i)) \leq 0 \quad \text{or} \quad x \leq 3.$$

The full domain of analyticity is shown in Fig. 3.5–6.

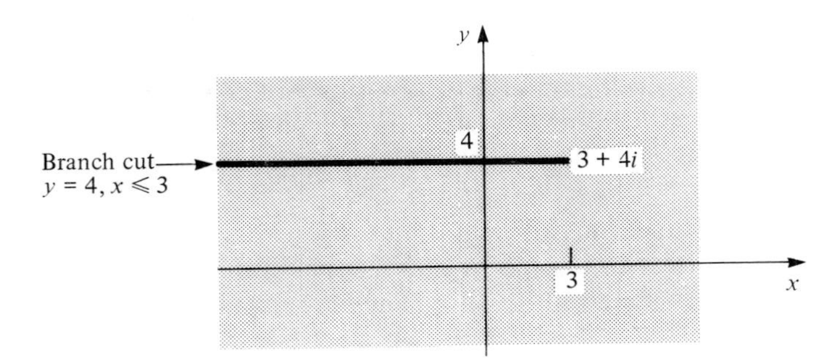

Figure 3.5–6

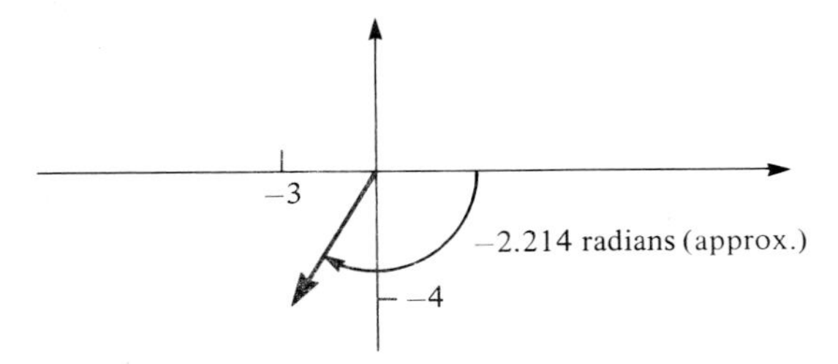

Figure 3.5–7

Part (b): $f(0) = \text{Log}(-3 - 4i) = \text{Log } 5 + i \arg(-3 - 4i)$. Since we are dealing with the principal branch, we require in the domain of analyticity that $-\pi < \arg(-3 - 4i) < \pi$. From Fig. 3.5–7 we find this value of $\arg(-3 - 4i)$ to be approximately -2.214. Thus $f(0) \doteq \text{Log } 5 - i2.214$. ◀

EXERCISES

1. Use $\text{Log } z = 1/2 \text{ Log}(x^2 + y^2) + i \arg z$, where $\arg z = \tan^{-1}(y/x)$ or, where appropriate $(x = 0)$, $\arg z = \pi/2 - \tan^{-1}(x/y)$, and Eq. (2.3–6) or (2.3–8) to show that $d(\text{Log } z)/dz = 1/z$ in the domain of Fig. 3.5–2. The inverse functions are here evaluated so that $\arg z$ is the principal value.

2. Suppose

$$f(z) = \log z = \text{Log } r + i\theta, \qquad 0 \le \theta < 2\pi.$$

   a) Find the largest domain of analyticity of this function.

   b) Find the numerical value of $f(-e^2)$.

   c) Explain why we cannot determine $f(e^2)$ within the domain of analyticity.

Consider a branch of $\log z$ analytic in the domain created with the branch cut $x = 0$, $y \le 0$. If for this branch $\log(-1) = i\pi$, find

3. $\log 1$ 4. $\log(ie)$ 5. $\log(-e - ie)$
6. $\log(-\sqrt{3} - i)$ 7. $\log(\text{cis}(-3\pi/4))$

Consider a branch of $\log z$ analytic in the domain created with the branch cut $x = y$, $y \ge 0$. If for this branch $\log 1 = i2\pi$, find

8. $\log i$ 9. $\log(\sqrt{3} + i)$ 10. $\log(-\sqrt{3} + i)$

11. Consider the function $f(z) = \text{Log}(z - i)$.

   a) Describe the branch cut that must be used to create the largest domain of analyticity for this function.

   b) Find the numerical value of $f(-i)$.

   c) Explain why $g(z) = [\text{Log}(z - i)]/(z - 2i)$ has a singularity in the domain found in part (a), but $h(z) = [\text{Log}(z - i)]/(z + 2 - i)$ is analytic throughout the domain.

12. a) Show that $-\text{Log } z = \text{Log}(1/z)$ is valid throughout the domain of analyticity of $\text{Log } z$.

   b) Find a nonprincipal branch of $\log z$ such that $-\log z = \log(1/z)$ is not satisfied somewhere in your domain of analyticity of $\log z$. Prove your result.

13. Show that $\text{Log}[(z - 1)/z]$ is analytic throughout the domain consisting of the z-plane with the line $y = 0, 0 \le x \le 1$ removed. Thus, a branch cut is not always infinitely long.

14. Show that $f(z) = \text{Log}(z^2 + 1)$ is analytic in the domain shown in Fig. 3.5–8.

   Hint: Points satisfying $\text{Re}(z^2 + 1) \le 0$ and $\text{Im}(z^2 + 1) = 0$ must not appear in the domain of analyticity. This requires a branch cut (or cuts) described by $\text{Re}((x + iy)^2 + 1) \le 0$, $\text{Im}((x + iy)^2 + 1) = 0$. Find the locus that satisfies both these equations.

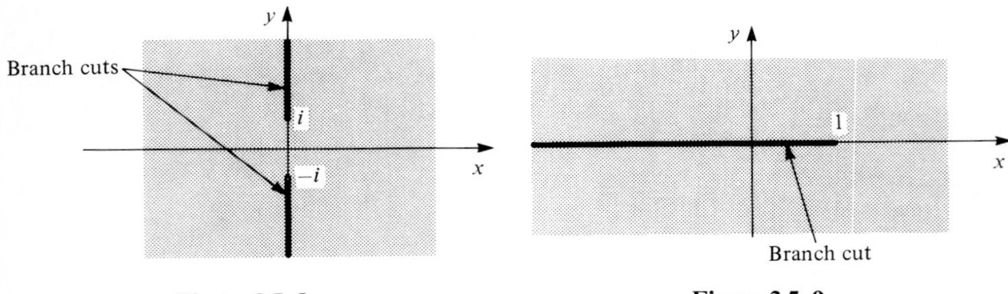

Figure 3.5–8

Figure 3.5–9

15. a) Show that Log(Log z) is analytic in the domain consisting of the z-plane with a branch cut along the line $y = 0$, $x \leq 1$ (see Fig. 3.5–9).

Hint: Where will the inner function, Log z, be analytic? What restrictions must be placed on Log z to render the outer logarithm an analytic function?

b) Find $d(\text{Log}(\text{Log}\, z))/dz$ within the domain of analyticity found in part (a).

c) What branch cut should be used to create the maximum domain of analyticity for Log(Log(Log z))?

16. The complex electrostatic potential $\Phi(x, y) = \phi + i\psi = \text{Log}(1/z)$, where $z \neq 0$, can be created by an electric line charge located at $z = 0$ and lying perpendicular to the xy-plane.

a) Sketch the streamlines for this potential.

b) Sketch the equipotentials for $\phi = -1, 0, 1$, and 2.

c) Find the components of the electric field at an arbitrary point x, y.

3.6 COMPLEX EXPONENTIALS

If we have the task of computing the numerical value of $7^{1.43}$, we can use a book of logarithms (and antilogs) and proceed as follows: We write 7 as $e^{\text{Log}\, 7}$ so that we have $7^{1.43} = (e^{\text{Log}\, 7})^{1.43}$. The latter expression is evaluated as $e^{(1.43\, \text{Log}\, 7)}$. This procedure should suggest a definition for general expressions of the form z^c, where z and c are complex numbers. We use, for $z \neq 0$,

$$z^c = e^{c(\log z)}. \tag{3.6–1}$$

We evaluate $e^{c(\log z)}$ by means of Eq. (3.1–1). As we well know, the logarithm of z is multivalued. For this reason, depending on the value of c, z^c may have more than one numerical value. The matter is fully explored in Exercise 14 of this section. It is not hard to show that if c is a rational number n/m, then Eq. (3.6–1) yields numerical values identical to those obtained for $z^{n/m}$ in Eq. (1.4–13).

EXAMPLE 1

Compute $9^{1/2}$ by means of Eq. (3.6–1).

Solution

We, of course, already know the two possible numerical values of $9^{1/2}$. Pretending otherwise, from Eq. (3.6–1) we have

$$9^{1/2} = e^{1/2 \log 9} = e^{1/2[\text{Log } 9 + i(2k\pi)]} = e^{1/2 \text{ Log } 9 + ik\pi}$$

$$= e^{1/2 \text{ Log } 9}[\cos(k\pi) + i \sin(k\pi)], \qquad k = 0, \pm 1, \pm 2, \ldots .$$

As k ranges over all integers, the expression in the brackets yields only two possible numbers, $+1$ and -1. The term $e^{1/2 \text{ Log } 9}$ equals $e^{\text{Log } 3} = 3$. Thus $9^{1/2} = \pm 3$, and the familiar results are obtained. ◀

If in computing the value of z^c by means of Eq. (3.6–1) we employ the principal value of the logarithm, we obtain what is called the principal value of z^c. In the preceding example, we can put $k = 0$ and see that the principal value of $9^{1/2}$ is 3.

EXAMPLE 2

Compute 9^{π} by means of Eq. (3.6–1). Here we are raising a number to a real, but irrational, power. Although we know there is exactly one distinct value of z^n, where n is an integer, we have not yet determined the number of values that will occur if n is irrational.

Solution

From Eq. (3.6–1) we have

$$9^{\pi} = e^{\pi \log 9} = e^{\pi[\text{Log } 9 + i2k\pi]} = e^{\pi \text{ Log } 9 + i2k\pi^2}$$

$$= e^{\pi \text{ Log } 9}[\cos(2k\pi^2) + i \sin(2k\pi^2)]$$

$$\doteq e^{6.903}[\cos(2k\pi^2) + i \sin(2k\pi^2)]$$

$$\doteq 995.04[\cos(2k\pi^2) + i \sin(2k\pi^2)], \qquad k = 0, \pm 1, \pm 2, \ldots .$$

With $k = 0$ we have that the principal value of 9^{π} is approximately 995.04. By allowing k to vary we generate other, complex, values of 9^{π}. All these values are numerically distinct; that is, there are no repetitions as k assumes new values. To see that this must be so assume that integers k_1 and k_2 yield identical values of 9^{π}. Then we would require

$$\cos(2k_1\pi^2) + i \sin(2k_1\pi^2) = \cos(2k_2\pi^2) + i \sin(2k_2\pi^2).$$

This equality can only hold if

$$2k_1\pi^2 - 2k_2\pi^2 = m\pi, \qquad \text{where } m \text{ is an even integer}$$

or if

$$\pi = \frac{m}{2k_1 - 2k_2}.$$

Since π is irrational, it cannot be expressed as the ratio of integers. Thus our assumption that there are two identical roots must be false. ◀

In our previous example we generate an infinite set of numerically distinct values of 9^{π} when we allow k to range over all the integers. This result is generalized in Exercise 14 of this section and shows that:

If c is any irrational number, then z^c possesses an infinite set of different values.

Let us now consider examples in which c is complex.

EXAMPLE 3

Find all values of i^i, and show that they are all real.

Solution

Taking $z = i$ and $c = i$ in (3.6–1), we obtain

$$i^i = e^{i \log i} = e^{i[i(\pi/2 + 2k\pi)]} = e^{-[\pi/2 + 2k\pi]}, \qquad k = 0, \pm 1, \pm 2, \ldots.$$

What is curious is that by beginning with two purely imaginary numbers we have obtained an infinite set of purely real numbers. This result, which is counterintuitive, seems first to have been derived by Euler around 1746. Note that with $k = 0$ we obtain the principal value $i^i = e^{-\pi/2}$. ◀

EXAMPLE 4

Find $(1 + i)^{3+4i}$.

Solution

Our formula in Eq. (3.6–1) yields

$$\begin{aligned}(1 + i)^{3+4i} &= e^{(3+4i)(\log(1+i))} \\ &= e^{(3+4i)[\operatorname{Log}\sqrt{2}+i(\pi/4+2k\pi)]} \\ &= e^{3\operatorname{Log}\sqrt{2}-\pi-8k\pi+i(4\operatorname{Log}\sqrt{2}+3\pi/4+6k\pi)}.\end{aligned}$$

With the aid of Eq. (3.1–1) this becomes

$$\begin{aligned}e^{3\operatorname{Log}\sqrt{2}-\pi-8k\pi}[\cos(4\operatorname{Log}\sqrt{2} + 3\pi/4 + 6k\pi) + i\sin(4\operatorname{Log}\sqrt{2} + 3\pi/4 + 6k\pi)] \\ = (1 + i)^{3+4i}, \qquad k = 0, \pm 1, \pm 2, \ldots.\end{aligned}$$

Note that $6k\pi$ can be deleted in the preceding equation. As k ranges over the integers, an infinite number of complex, numerically distinct values are obtained for $(1 + i)^{3+4i}$. The principal value, with $k = 0$, is

$$e^{3\operatorname{Log}\sqrt{2}-\pi}[\cos(4\operatorname{Log}\sqrt{2} + 3\pi/4) + i\sin(4\operatorname{Log}\sqrt{2} + 3\pi/4)]. \qquad ◀$$

Examples 1–4 are specific demonstrations of the following general statement for $z \neq 0$:

z^c has an infinite set of possible values except if c is a rational number.

There is one case in which the rule does not apply: If $z = e$, then by definition, we compute e^c by means of Eq. (3.1–1) and obtain just one value. Otherwise we would no longer have, for example, such familiar results as $e^{i\pi} = -1$.

If z is regarded as a variable and c is not an integer then z^c is a multivalued function of z. This function possesses various branches whose derivatives can be found. The principal branch, for example, is obtained with the use of the principal branch of $\log z$ in Eq. 3.6–1. This branch of z^c is analytic in the same domain as $\text{Log}\, z$. We find the derivative of any branch as follows:

$$z^c = e^{c \log z},$$

$$\frac{d}{dz} z^c = \frac{d}{dz} e^{c \log z} = \frac{ce^{c \log z}}{z} = \frac{ce^{c \log z}}{e^{\log z}} = ce^{(c-1)\log z} = cz^{c-1},$$

which is a familiar-looking result from real calculus. We can rewrite this as

$$\frac{d}{dz} z^c = \frac{cz^c}{z}. \tag{3.6–2}$$

Care should be taken to employ the same branch of z^c on both sides of this equation.

EXAMPLE 5

Find $(d/dz)z^{2/3}$ at $z = -8i$ when the principal branch is used.

Solution

Using (3.6–2), with $c = 2/3$, we see that we must evaluate $(2/3)z^{2/3}/z$ at $-8i$. Using the principal branch, we have, from (3.6–1),

$$(2/3)\,\frac{e^{(2/3)\,\text{Log}(-8i)}}{-8i} = (2/3)\,\frac{e^{(2/3)[\text{Log}(8) + i(-\pi/2)]}}{-8i} = (1/3)\,\text{cis}(\pi/6). \quad ◀$$

The expression c^z, where c is a constant and z a variable, is equal to $e^{z \log c}$. Having chosen a valid value for $\log c$, we find that we now have a single-valued function of z analytic in the entire z-plane. The derivative of this expression is found as follows:

$$\frac{d}{dz} c^z = \frac{d}{dz} e^{z \log c} = e^{z \log c}(\log c) = c^z \log c. \tag{3.6–3}$$

EXAMPLE 6

Find $(d/dz)i^z$.

Solution

We will use Eq. (3.6–3) taking $\log i = \text{Log}\, i = i\pi/2$. Thus

$$\frac{d}{dz} i^z = i^z\left(\frac{i\pi}{2}\right). \quad ◀$$

The multivalued function $g(z)^{h(z)}$ is defined as $e^{h(z)\log(g(z))}$. Such functions are considered in Exercises 19–21 of this section. Their principal branch is obtained if we use the principal branch of the logarithm.

EXERCISES

Find all values of the following in the form $a + ib$, and state the principal value.

1. 1^i **2.** $(-i)^{-i}$ **3.** $(\sqrt{3} + i)^{4-3i}$

4. $i^{(e^i)}$ **5.** $2^{\sqrt{2}}$ **6.** $(e^{i\pi^2})^i$

7. $[\text{Log}(i)]^{\sin i}$ **8.** $10^{\cosh(1+i)}$ **9.** $(1 + i\tan 2)^i$

10. $(\tan 2 + i)^i$ **11.** $(\sin i)^{\text{Log}(\sin i)}$

Using Eq. (3.1–5) or Eq. (3.1–7) prove that for any complex values α, β, and z

12. The values of $1/z^{\beta}$ are identical to the values of $z^{-\beta}$.

13. The values of $z^{\alpha}z^{\beta}$ are identical to the values of $z^{\alpha+\beta}$.

14. Use Eq. (3.6–1) to show that

a) if n is an integer, then z^n has only one value and it is the same as the one given by Eq. (1.4–2);

b) if n and m are integers and n/m is an irreducible fraction, then $z^{n/m}$ has just m values and they are identical to those given by Eq. (1.4–13);

c) if c is an irrational number, then z^c has an infinity of different values;

d) if c is complex with $\text{Im}\, c \neq 0$, then z^c has an infinity of different values.

15. The following puzzle appeared without attribution in the Spring 1989 Newsletter of the Northeastern Section of the Mathematical Association of America. What is the flaw in this argument?

Euler's identity?

$$e^{i\theta} = (e^{i\theta})^{2\pi/2\pi} = (e^{2\pi i})^{\theta/2\pi} = (1)^{\theta/2\pi} = 1.$$

Using the principal branch of the function, evaluate

16. $f'(i)$ if $f(z) = z^{1/4}$

17. $f'(-64i)$ if $f(z) = z^{7/6}$

18. $f'(-9i)$ if $f(z) = z^{i/2}$

Let $f(z) = z^z$, where the principal branch is used. Evaluate

19. $f'(z)$ **20.** $f'(i)$

21. Let $f(z) = z^{\sin z}$ where the principal branch is used. Find $f'(i)$.

22. Find $(d/dz)2^{\cosh z}$ using principal values. Where in the complex z-plane is $2^{\cosh z}$ analytic?

23. Find $f'(i)$ if $f(z) = i^{(e^z)}$ and principal values are used.

24. Let $f(z) = 10^{(z^3)}$. This function is evaluated such that $f'(z)$ is real when $z = 1$. Find $f'(1 + i)$. Where in the complex plane is $f(z)$ analytic?

25. Let $f(z) = 10^{(e^z)}$. This function is evaluated such that $|f(i\pi/2)| = e^{-2\pi}$. Find $f'(z)$ and $f'(i\pi/2)$.

3.7 INVERSE TRIGONOMETRIC AND HYPERBOLIC FUNCTIONS

If we know the logarithm of a complex number w, we can find the number itself by means of the identity $e^{\log w} = w$. We have used the fact that the exponential function is the inverse of the logarithmic function.

Suppose we know the sine of a complex number w. Let us see whether we can find w and whether w is uniquely determined.

Let $z = \sin w$. The value of w is referred to as arc sin z, or $\sin^{-1} z$, that is, the complex number whose sine is z. To find w note that

$$z = \frac{e^{iw} - e^{-iw}}{2i}. \tag{3.7–1}$$

Now with $p = e^{iw}$ and $1/p = e^{-iw}$ in Eq. (3.7–1) we have

$$z = \frac{p - 1/p}{2i}.$$

Multiplying the above by $2ip$ and doing some rearranging, we find that

$$2izp = p^2 - 1, \quad \text{or} \quad p^2 - 2izp - 1 = 0.$$

With the quadratic formula we solve this equation for p:

$$p = zi + (1 - z^2)^{1/2}, \quad \text{or} \quad e^{iw} = zi + (1 - z^2)^{1/2}.$$

We now take the logarithm of both sides of this last equation and divide the result by i to obtain

$$w = \frac{1}{i}\log(zi + (1 - z^2)^{1/2})$$

and, since $w = \sin^{-1} z$,

$$\sin^{-1} z = -i\log(zi + (1 - z^2)^{1/2}). \tag{3.7–2}$$

We thus, apparently, have an explicit formula for the complex number whose sine equals any given number z. The matter is not quite so simple, however, since the result is multivalued. There are two equally valid choices for the square root in Eq. (3.7–2). Having selected one such value, there are then an infinite number of possible values of the logarithm of $zi + (1 - z^2)^{1/2}$. Altogether, we see that because of the square root and logarithm, there are two different sets of values for $\sin^{-1} z$, and each set has an infinity of members. An exception occurs when $z = \pm 1$; then, the two infinite sets become identical to each other and there is one infinite set. To assure ourselves of the validity of Eq. (3.7–2) let us use it to compute a familiar result.

EXAMPLE 1

Find $\sin^{-1}(1/2)$.

Solution

We see from Eq. (3.7–2) that

$$\sin^{-1}\left(\frac{1}{2}\right) = -i\log\left[\frac{i}{2} + \left(\frac{3}{4}\right)^{1/2}\right].$$

With the positive square root of 3/4 we have

$$\sin^{-1}\left(\frac{1}{2}\right) = -i\log\left[\frac{\sqrt{3}}{2} + \frac{i}{2}\right] = -i\log(1\angle\pi/6) = \frac{\pi}{6} + 2k\pi, \qquad k = 0, \pm 1, \pm 2, \ldots,$$

whereas with the negative square root of 3/4 we have

$$\sin^{-1}\left(\frac{1}{2}\right) = -i\log\left[\frac{-\sqrt{3}}{2} + \frac{i}{2}\right] = -i\log\left(1\angle\frac{5\pi}{6}\right) = \frac{5\pi}{6} + 2k\pi,$$
$$k = 0, \pm 1, \pm 2, \ldots.$$

To make these answers look more familiar let us convert them to degrees. The first result says that angles with the sines of 1/2 are 30°, 390°, 750°, etc., while the second states that they are 150°, 510°, 870°, etc. We, of course, knew these results already from elementary trigonometry. ◀

In a high school trigonometry class, where one uses only real numbers, the following example would not have a solution.

EXAMPLE 2

Find all the numbers whose sine is 2.

Solution

From Eq. (3.7–2) we have $\sin^{-1} 2 = -i\log(2i + (-3)^{1/2})$. The two values of $(-3)^{1/2}$ are $\pm i\sqrt{3}$. With the positive sign our results are

$$\sin^{-1} 2 = -i\log[2i + i\sqrt{3}] = -i\left[\operatorname{Log}(2 + \sqrt{3}) + i\left(\frac{\pi}{2} + 2k\pi\right)\right]$$
$$\doteq \left(\frac{\pi}{2} + 2k\pi\right) - i1.317, \qquad k = 0, \pm 1, \pm 2,$$

whereas with the negative sign we obtain

$$\sin^{-1} 2 = -i\log(2i - i\sqrt{3}) = -i\left[\operatorname{Log}(2 - \sqrt{3}) + i\left(\frac{\pi}{2} + 2k\pi\right)\right]$$
$$\doteq \left(\frac{\pi}{2} + 2k\pi\right) + i1.317.$$

To verify these two sets of results, we use Eq. (3.2–9) and have

$$\sin\left[\left(\frac{\pi}{2} + 2k\pi\right) \pm i1.317\right] = \sin\left(\frac{\pi}{2} + 2k\pi\right)\cosh(1.317) \pm i\cos\left(\frac{\pi}{2} + 2k\pi\right)\sinh(1.317)$$

$$= \cosh(1.317) = 2.000 \qquad \text{(to four figures).}$$ ◀

The equation $z = \cos w$ can be solved for w, which we call arc cos z or $\cos^{-1} z$. The procedure is similar to the one just given for $\sin^{-1} z$. Thus

$$\cos^{-1} z = -i\log(z + i(1 - z^2)^{1/2}). \tag{3.7–3}$$

Also, $z = \tan w$ can be solved for w (or for $\tan^{-1} z$) with the following result:

$$\tan^{-1} z = \frac{i}{2}\log\left(\frac{i + z}{i - z}\right). \tag{3.7–4}$$

It is not hard to show that the expressions for $\cos^{-1} z$ and $\sin^{-1} z$ just derived are purely real numbers if and only if z is a real number and $-1 \le z \le 1$. Thus $z = \sin w$ and $z = \cos w$ have real number solutions w only for z satisfying $-1 \le z \le 1$. Otherwise w is a complex number.

Using Eqs. (3.2–9) and (3.2–10) we readily verify that $\sin(w) = \cos(2k\pi + \pi/2 - w)$, where k is any integer. This is a generalization of a result learned in elementary trigonometry. Thus if $z = \sin w = \cos(2k\pi + \pi/2 - w)$, we can say that $\sin^{-1}(z) = w$ and $\cos^{-1}(z) = 2k\pi + \pi/2 - w$, from which we derive

$$\sin^{-1}(z) + \cos^{-1}(z) = 2k\pi + \pi/2.$$

Since $\sin^{-1}(z)$ and $\cos^{-1}(z)$ are multivalued functions, we can assert that there must exist values of these functions such that the previous equation will be satisfied. Not all values of the inverse trigonometric functions will satisfy the above equation, however. Suppose, for example, we take $z = 1/\sqrt{2}$, $\sin^{-1}(z) = \pi/4$, $\cos^{-1}(z) = -\pi/4$. The equation is not satisfied. Using instead $\cos^{-1}(z) = \pi/4$, we see that it is satisfied for $k = 0$. In Exercise 3 we verify that if identical branches of $(z^2 - 1)^{1/2}$ are used in Eqs. (3.7–2) and (3.7–3), then $\sin^{-1}(z) + \cos^{-1}(z) = 2k\pi + \pi/2$ is satisfied for all choices of the logarithm.

The functions appearing on the right in Eqs. (3.7–2), (3.7–3), and (3.7–4) are examples of inverse trigonometric functions. The inverse hyperbolic functions are similarly established. Thus

$$\sinh^{-1} z = \log(z + (z^2 + 1)^{1/2}), \tag{3.7–5}$$

$$\cosh^{-1} z = \log(z + (z^2 - 1)^{1/2}), \tag{3.7–6}$$

$$\tanh^{-1} z = \frac{1}{2}\log\left(\frac{1 + z}{1 - z}\right). \tag{3.7–7}$$

All the inverse functions we have derived in this section are multivalued. Analytic branches exist for all these functions. For example, a branch of $\sin^{-1} z$ can be obtained from Eq. (3.7–2) if we first specify a branch of $(1 - z^2)^{1/2}$ and then a branch of the logarithm. Having done this, we can differentiate our branch within its domain of analyticity. The subject of branches in general is given more attention in the next section. Differentiating Eq. (3.7–2), we have

$$\frac{d}{dz}\sin^{-1} z = \frac{d}{dz}(-i\log[zi + (1 - z^2)^{1/2}]) = \frac{1}{(1 - z^2)^{1/2}}. \tag{3.7–8}$$

For this identity to hold we must use the same branch of $(1 - z^2)^{1/2}$ in defining $\sin^{-1} z$ and in the expression for its derivative. Other formulas that are derived through the differentiation of branches are

$$\frac{d}{dz}\cos^{-1} z = \frac{-1}{(1 - z^2)^{1/2}}, \tag{3.7–9}$$

$$\frac{d}{dz}\tan^{-1} z = \frac{1}{(1 + z^2)}, \tag{3.7–10}$$

$$\frac{d}{dz}\sinh^{-1} z = \frac{1}{(1 + z^2)^{1/2}}, \tag{3.7–11}$$

$$\frac{d}{dz}\cosh^{-1} z = \frac{1}{(z^2 - 1)^{1/2}}, \tag{3.7–12}$$

$$\frac{d}{dz}\tanh^{-1} z = \frac{1}{(1 - z^2)}. \tag{3.7–13}$$

EXERCISES

1. a) Derive Eq. (3.7–3). b) Derive Eq. (3.7–4). c) Derive Eq. (3.7–5).

2. a) Show that if we differentiate a branch of arc cos z we obtain Eq. (3.7–9).

b) Obtain Eq. (3.7–8) by noting that

$$z^2 = \sin^2 w = (1 - \cos^2 w) = 1 - \left(\frac{dz}{dw}\right)^2$$

can be solved for dw/dz.

c) Obtain Eq. (3.7–11) directly from Eq. (3.7–5); obtain it also by a procedure similar to part (b) of this exercise.

3. Show that if we use identical branches of $(z^2 - 1)^{1/2}$ in Eqs. (3.7–2) and (3.7–3) we obtain $\sin^{-1}(z) + \cos^{-1}(z) = 2k\pi + \pi/2$ for all choices of the logarithm in those equations.

Find all possible solutions to the following equations.

4. $\cos w = 2$ **5.** $\cosh w = i$ **6.** $\sin w = 1 + i$

7. $\sinh w = i\sqrt{3}$ **8.** $\sinh^2 w = i$ **9.** $\tan z = \log i$

10. $\sin[\cos w] = 0$ **11.** $\sinh[\cos w] = 0$ **12.** $\cosh^{-1} w = 3 + 4i$

13. Explain whether or not the following two equations are true in general.

a) $\tan^{-1}(\tan z) = z$ b) $\tan(\tan^{-1} z) = z$

14. Explain why the values of $\sinh^{-1} x$ that are given by tables or a pocket calculator are the same as those given by $\mathrm{Log}(x + \sqrt{x^2 + 1})$. Note the branches used for the functions in Eq. (3.7–5).

15. a) Show that if z is real, i.e., $z = x$, then from (3.7–5) we have $\sinh^{-1} x \approx \mathrm{Log}(2x)$ if $x \gg 1$ and $\sinh^{-1} x \approx -\mathrm{Log}(2|x|)$ if $x \ll -1$. Begin with the result in Exercise 14.

b) Using a pocket calculator that has inverse hyperbolic functions and natural logs, compare $\sinh^{-1} x$ and $\log(2x)$ for $x = 1, 2, 3, 4$.

16. Show that $\tanh^{-1}(e^{i\theta}) = (1/2)\log(i\cot(\theta/2))$.

17. Find a formula similar to the one above for $\tan^{-1}(e^{i\theta})$.

18. Use Eq. (3.7–2) and the definition of the cosine to show that $\cos(\sin^{-1} z) = (1 - z^2)^{1/2}$.

19. Find a formula similar to the one above for $\sinh[\cosh^{-1} z]$.

3.8 MORE ON BRANCH POINTS AND BRANCH CUTS

Let us study the branches and domains of analyticity of functions of the form $(z - z_0)^c$, where z_0 and c are complex constants. If c is an integer, such functions are single valued (do not have different branches) and will not be considered here. However, if c is a noninteger, these functions are multivalued. From Eq. (3.6–1) we have

$$(z - z_0)^c = e^{c \log(z - z_0)}. \tag{3.8–1}$$

Since $\log(z - z_0)$ has a branch point at z_0, so does $(z - z_0)^c$. To study branches of $(z - z_0)^c$ and branches of algebraic combinations of such expressions, we introduce a set of polar coordinate variables measured from the branch points and examine the changes in value of our functions as their branch points are encircled.

Consider, for example, $f(z) = z^{1/2}$. Letting $r = |z|$, $\theta = \arg z$, we use Eq. (1.4–12) with $m = 2$ to find that

$$f(z) = \sqrt{r}\, e^{i(\theta/2 + k\pi)}, \qquad k = 0, 1. \tag{3.8–2}$$

The reader should verify that this result is obtained if we put $z = re^{i\theta}$, $z_0 = 0$, and $c = 1/2$ in Eq. (3.8–1). Suppose we set $k = 0$ in Eq. (3.8–2) and negotiate clockwise the circle of radius r shown in Fig. 3.8–1. Beginning at the point marked a and taking $\theta = \pi$, we have from Eq. (3.8–2) that $f(z) = \sqrt{r}\, e^{i\pi/2} = i\sqrt{r}$. Now, moving clockwise along the circle to b, where θ has fallen to $\pi/2$, Eq. (3.8–2) shows that $f(z) = \sqrt{r}\, e^{i\pi/4}$. Continuing on to c, where $\theta = 0$, we have $f(z) = \sqrt{r}$.

Moving clockwise to a and starting a second trip around the circle, we now have at a, $\theta = -\pi$ and $f(z) = \sqrt{r}\, e^{-i\pi/2} = -i\sqrt{r}$. Proceeding to b, where $\theta = -3\pi/2$, we find $f(z) = \sqrt{r}\, e^{-i3\pi/4}$. Advancing to c, where $\theta = -2\pi$, we have $f(z) = -\sqrt{r}$. Coming again to a and starting a third trip around the circle, we have $\theta = -3\pi$ and $f(z) = \sqrt{r}\, e^{-i3\pi/2} = i\sqrt{r}$. This was our original starting value at a.

A third trip around the circle yields values of $f(z)$ identical to those obtained on the first excursion. No new values of $f(z)$ are generated by subsequent journeys around the circle, as the reader should verify.

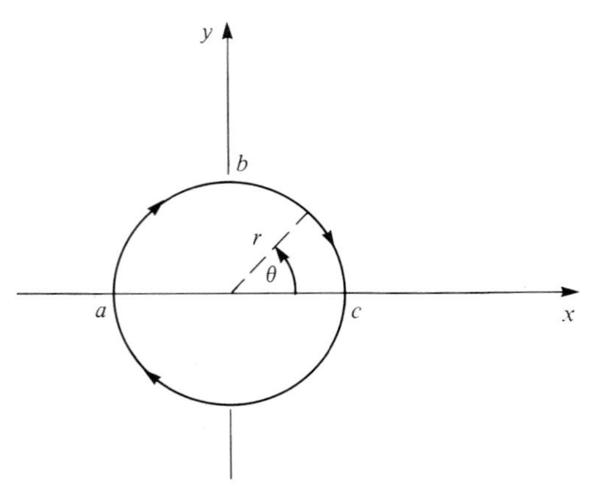

Figure 3.8–1

In encircling the branch point $z = 0$ twice we have encountered values of $z^{1/2}$ for two different branches of this function. In our first trip around the circle we found values of $z^{1/2} = e^{1/2 \operatorname{Log} z}$. This is the principal branch and is analytic in the same domain (see Fig. 3.5–2) as Log z. In our second trip we found values of the other branch of $z^{1/2}$, which is analytic throughout this domain. It is given by $z^{1/2} = e^{1/2[\operatorname{Log} z + i2\pi]} = -e^{1/2 \operatorname{Log} z}$.

What we have seen is true in general whenever we have a branch point.

Encirclement of a branch point causes us to move from one branch of a function to another.†

We need not employ circular paths (as was just done) for this to happen. Any closed path surrounding the branch point will do.

To prevent our proceeding from one branch of a function to another, when we move along a path, we can construct a branch cut in the z-plane and agree never to cross this cut.

The branch cut is regarded as a barrier to encirclement of the branch point.

Alternatively, we can create a domain consisting of the z-plane minus all the points on the branch cut. One finds branches of the function that are analytic throughout this "cut" plane. We can specify a particular branch by *giving its value at one point in the cut plane* (see Example 1 in this section). By using this branch on paths that are confined to the domain, we cannot pass from one branch of the function to another.

---

† For functions having more than one branch point (see Example 3 in this section) encirclement of just one branch point always causes progression to another branch; encirclement of two or more branch points does not necessarily cause such a transition, as we shall see.

EXAMPLE 1

Consider a branch of $z^{1/2}$ that is analytic in the domain consisting of the z-plane less the points on the branch cut $y = 0, x \le 0$. When $z = 4$, the multivalued function $z^{1/2}$ equals $+2$ or -2. Suppose for our branch $z^{1/2} = 2$ when $z = 4$. What value does this branch assume when

$$z = 9[-1/2 - i\sqrt{3}/2]?$$

Solution

With $|z| = r$ and $\theta = \arg z$ we have that

$$z^{1/2} = \sqrt{r}\, e^{i(\theta/2 + k\pi)}, \qquad k = 0, 1. \tag{3.8–3}$$

We will take $\theta = 0$ when $z = 4$. Then the condition $(4)^{1/2} = 2$ requires that $k = 0$ in Eq. (3.8–3). As we move along a path to $9[-1/2 - i\sqrt{3}/2]$ in Fig. 3.8–2 the argument θ in Eq. (3.8–3) changes continuously from 0 to $-2\pi/3$, and $r = |z|$ increases from 4 to 9. With $k = 0$ in Eq. (3.8–3) we have at $z = 9[-1/2 - i\sqrt{3}/2]$ that

$$z^{1/2} = \sqrt{9}\, e^{i(1/2)(-2\pi/3)} = 3[1/2 - i\sqrt{3}/2].$$

Notice that for our choice of branch we cannot reach $9[-1/2 - i\sqrt{3}/2]$ by way of the broken path in Fig. 3.8–2. This would take us out of the domain of analyticity of the branch. It would also involve crossing the branch cut.

For the branch under discussion, we might have taken $\theta = \arg z = 2\pi$ when $z = 4$. The condition $4^{1/2} = 2$ would require that we select $k = 1$ in (3.8–3). Following the allowed path in Fig. 3.8–2, which now goes from $\theta = 2\pi$ to $\theta = 2\pi - 2\pi/3$, we would again conclude that when $z = 9[-1/2 - i\sqrt{3}/2]$ we have $z^{1/2} = 3[1/2 - i\sqrt{3}/2]$.

The derivative of any branch of $z^{1/2}$ in its domain of analyticity is, from Eq. (3.6–2),

$$\frac{d}{dz} z^{1/2} = \frac{1}{(2z^{1/2})}.$$

The same branch of $z^{1/2}$ must be used on both sides of this equation. ◀

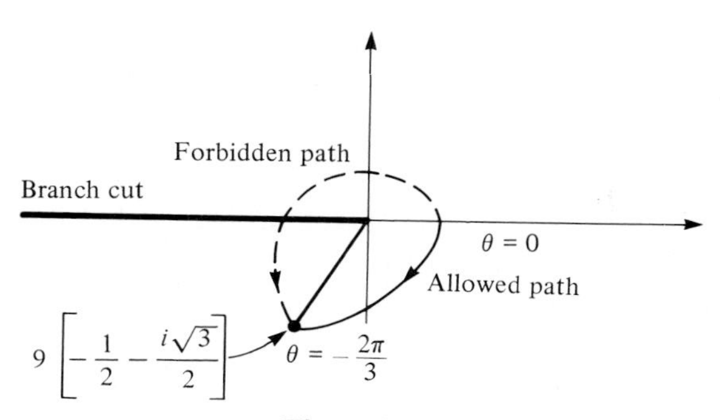

Figure 3.8–2

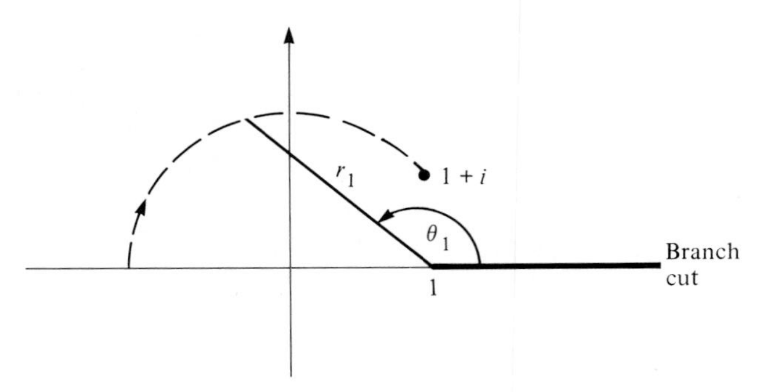

Figure 3.8–3

EXAMPLE 2

For $(z-1)^{1/3}$ let a branch cut be constructed along the line $y=0$, $x \geq 1$. If we select a branch whose value is a negative real number when $y=0$, $x<1$, what value does this branch assume when $z=1+i$?

Solution

Introduce the variables $r_1 = |z-1|$, $\theta_1 = \arg(z-1)$ (see Fig. 3.8–3). We have from Eq. (1.4–12) with $m=3$ that

$$(z-1)^{1/3} = \sqrt[3]{r_1}\, e^{i(\theta_1/3 + 2k\pi/3)}, \qquad k = 0, 1, 2. \tag{3.8–4}$$

Taking $\theta_1 = \pi$ on the line $y=0$, $x<1$ we have here that

$$(z-1)^{1/3} = \sqrt[3]{r_1}\, e^{i(\pi/3 + 2k\pi/3)}. \tag{3.8–5}$$

The left side of the equation can be a negative real number if we select $k=1$ in Eq. (3.8–5). Proceeding to $1+i$ from anywhere on the line $y=0$, $x<1$, we find that θ_1 has shrunk to $\pi/2$, and $r_1 = 1$. The path used in Fig. 3.8–3 for this purpose cannot cross the branch cut. With these values of r_1 and θ_1 in Eq. (3.8–4) (and $k=1$), at $1+i$ we have

$$(z-1)^{1/3} = i^{1/3} = \sqrt[3]{1}\, e^{i5\pi/6} = -\sqrt{3}/2 + i/2. \quad ◀$$

EXAMPLE 3

Consider the multivalued function $f(z) = z^{1/2}(z-1)^{1/2}$.

a) Where are the branch points of the function? Verify that these are branch points by encircling them and passing from one branch of $f(z)$ to another.
b) Show that if we encircle both branch points we do not pass to a new branch of $f(z)$.
c) Show some possible choices of branch cut that would prevent passage from one branch of $f(z)$ to another.

Solution

Part (a): The first factor $z^{1/2}$ has a branch point at $z = 0$, whereas the second $(z - 1)^{1/2}$ has a branch point at $z = 1$. Thus we suspect that the product has branch points at $z = 0$ and $z = 1$. We will verify that $z = 1$ is a branch point. The proof for $z = 0$ is quite similar and will not be presented.

We have (see Fig. 3.8–4a) that

$$z^{1/2} = \sqrt{r}\, e^{i(\theta/2 + k\pi)}, \tag{3.8–6}$$

where $\theta = \arg z$ and $r = |z|$; and

$$(z - 1)^{1/2} = \sqrt{r_1}\, e^{i(\theta_1/2 + m\pi)}, \tag{3.8–7}$$

where $|z - 1| = r_1$ and $\theta_1 = \arg(z - 1)$. Thus

$$f(z) = z^{1/2}(z - 1)^{1/2} = \sqrt{r}\,\underline{/\tfrac{1}{2}\theta + k\pi}\;\sqrt{r_1}\,\underline{/\tfrac{1}{2}\theta_1 + m\pi}, \tag{3.8–8}$$

where k and m are assigned integer values. Let us now encircle $z = 1$ using the path $|z - 1| = \delta$, where $\delta < 1$ (see Fig. 3.8–4b). Beginning at point a, we take $\theta_1 = 0$, $\theta = 0$, $r_1 = \delta$, $r = 1 + \delta$. With these values in Eq. (3.8–8) we have

$$f(z) = \sqrt{1 + \delta}\,\underline{/k\pi}\;\sqrt{\delta}\,\underline{/m\pi} = \sqrt{\delta + \delta^2}\,\underline{/(k + m)\pi}. \tag{3.8–9}$$

Moving counterclockwise once around the circle $|z - 1| = \delta$ and returning to point a we find now that $\theta_1 = 2\pi$, while θ, after some variation, has returned to zero. With these values in Eq. (3.8–8) we have

$$f(z) = \sqrt{1 + \delta}\,\underline{/k\pi}\;\sqrt{\delta}\,\underline{/\pi + m\pi} = -\sqrt{\delta + \delta^2}\,\underline{/(k + m)\pi}. \tag{3.8–10}$$

Because the value obtained for $f(z)$ at a is now not the value originally obtained (see Eq. 3.8–9), we have progressed to another branch of $f(z)$. The preceding discussion does not require the use of a circular path. Any closed path that encloses $z = 1$ and excludes $z = 0$ will lead to the same result.

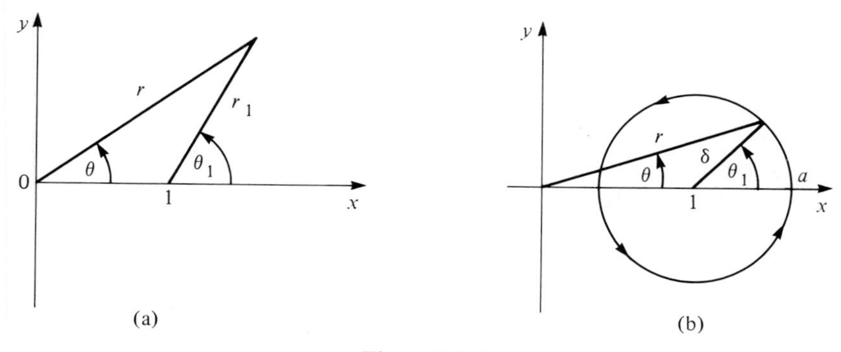

Figure 3.8–4

Part (b): An arbitrary closed path surrounds the branch points $z = 0$ and $z = 1$, as shown in Fig. 3.8–5. Let us evaluate $f(z)$ at the arbitrary point P lying on the path. Here $\arg z = \alpha$ and $\arg(z - 1) = \beta$. Substituting these values for θ and θ_1, respectively, in Eq. (3.8–8) and combining the arguments we have

$$f(z) = \sqrt{r}\sqrt{r_1} \underline{/ \tfrac{1}{2}(\alpha + \beta) + (k + m)\pi}. \tag{3.8–11}$$

Moving once around the path in Fig. 3.8–5 in the indicated direction and returning to P, we now have $\arg z = \theta = \alpha + 2\pi$ and $\arg(z - 1) = \theta_1 = \beta + 2\pi$. Using these values in Eq. (3.8–8), we obtain

$$f(z) = \sqrt{r}\sqrt{r_1} \underline{/ \tfrac{1}{2}(\alpha + \beta) + 2\pi + (k + m)\pi}. \tag{3.8–12}$$

If Eqs. (3.8–11) and (3.8–12) are converted to Cartesian form, identical numerical values of $f(z)$ are obtained since the difference of 2π in their arguments is of no consequence. Hence, by encircling both the branch points $z = 0$ and $z = 1$ we do not pass to a new branch of $f(z)$. There are functions, however, where encirclement of two or more branch points does cause passage to another branch. This matter is considered in Exercise 10.

Part (c): If we make a circuit around just one branch point of $z^{1/2}(z - 1)^{1/2}$, we move from one branch of this function to another. Some examples of branch cuts that prevent encirclement of one branch point are shown in Fig. 3.8–6(a) and (b). We have just seen that if we make a circuit along any path that encloses both branch points we do not pass to a new branch. In Fig. 3.8–6(c) we have constructed a branch cut that ensures that the encirclement of one branch point also requires the encirclement of the other.

Comment: The particular choice of branch cut is dictated by the desired domain of analyticity for our branch. For example, in Fig. 3.8–6(a) we can obtain a branch of $f(z)$ analytic throughout a domain consisting of the z-plane with all points on the lines $y = 0$, $x \leq 0$ and $y = 0$, $x \geq 1$ removed. If, however, we required a branch of

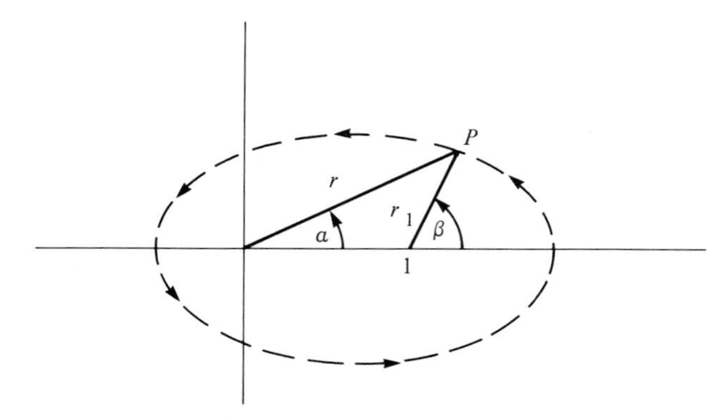

Figure 3.8–5

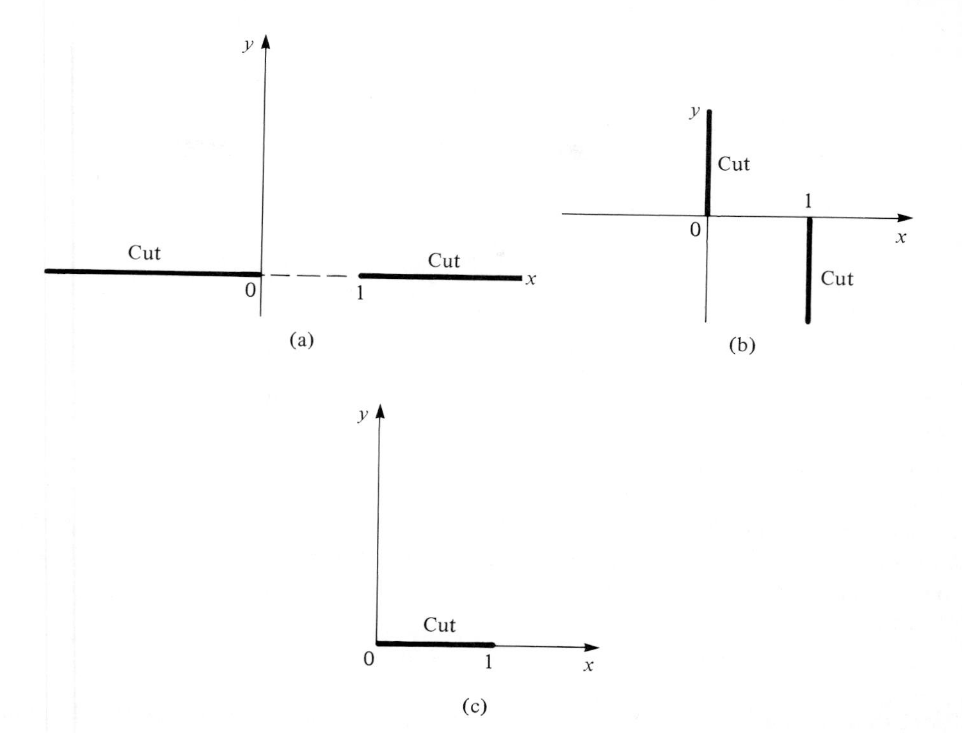

Figure 3.8–6

$f(z)$ analytic at say $y = 0$, $x = 2$, we might consider using the branch cuts shown in Fig. 3.8–6(b) or (c). ◀

EXAMPLE 4

Suppose we use a branch $f(z)$ of $z^{1/2}(z - 1)^{1/2}$ analytic throughout the cut plane shown in Fig. 3.8–6(b). If $f(1/2) = i/2$, what is $f(-1)$?

Solution

Using the notation of Example 3, we have from Eq. (3.8–8) that

$$f(z) = \sqrt{r}\underline{/\tfrac{1}{2}\theta + k\pi}\,\sqrt{r_1}\underline{/\tfrac{1}{2}\theta_1 + m\pi}. \tag{3.8–13}$$

At $z = 1/2$ we take $\theta = \arg z = 0$, $r = |z| = 1/2$, $\theta_1 = \arg(z - 1) = \pi$, $r_1 = |z - 1| = 1/2$ (see Fig. 3.8–7). Thus Eq. (3.8–13) becomes

$$f\left(\frac{1}{2}\right) = \frac{1}{2}\underline{/\frac{\pi}{2} + (k + m)\pi} = \frac{i}{2}\,e^{i(k+m)\pi}.$$

Taking $k + m = 0$ (or any other even integer), we obtain the condition $f(1/2) = i/2$ for our branch.

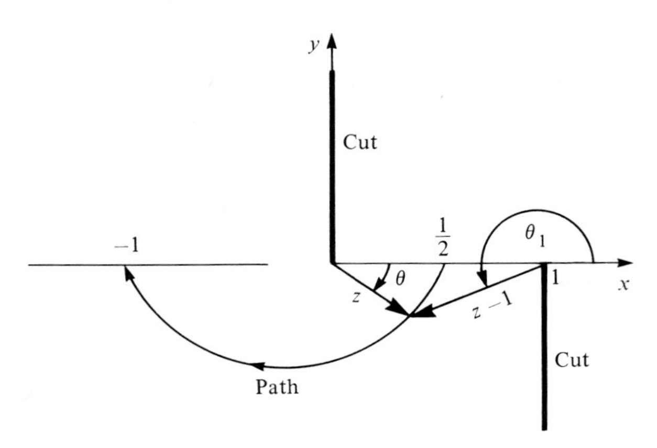

Figure 3.8–7

Now proceeding to $z = -1$ along the path indicated in Fig. 3.8–7, we find that $\theta = -\pi, r = |z| = 1, \theta_1$ is after some variation again π, $r_1 = |z - 1| = |-2| = 2$. Using these values in Eq. (3.8–13) together with $k + m = 0$, we obtain

$$f(-1) = \sqrt{1}\,\underline{/-\frac{\pi}{2} + k\pi}\,\sqrt{2}\,\underline{/\frac{\pi}{2} + m\pi} = \sqrt{2}\,e^{i(k+m)\pi} = \sqrt{2}.$$

Note that the path taken from 1/2 to -1 in Fig. 3.8–7 remains within the domain of analyticity. ◀

EXERCISES

A certain branch of $z^{1/2}$ is defined by means of the branch cut $x = 0$, $y \leq 0$. If this branch $f(z)$ has the value -2 when $z = 4$, what values does $f(z)$ assume at the following points? Also state the value of $f'(z)$ at each point.

1. 9 **2.** $9i$ **3.** $-1 - i$ **4.** $9 - 9i\sqrt{3}$

A branch of $(z - 1)^{2/3}$ is defined by means of the branch cut $x = 1$, $y \leq 0$. If this branch $f(z)$ equals 1 when $z = 0$ what is the value of $f(z)$ and $f'(z)$ at the following points?

5. $1 + 8i$ **6.** -1 **7.** $-i$ **8.** $1/2 - i/2$

9. A branch of $(z^2 - 1)^{1/2}$ is defined by means of a branch cut consisting of the line segment $-1 \leq x \leq 1$, $y = 0$.

a) Prove that this function has branch points at $z = \pm 1$.

b) Show that if we encircle these branch points by moving once around the ellipse $x^2/2 + y^2 = 1$, we do not pass to a new branch of the function. Present an argument like that in Example 3, part (b).

10. Consider the multivalued function $z^{1/3}(z - 1)^{1/3}$.

a) Show, by encircling each of them, that this function has branch points at $z = 0$ and $z = 1$.

b) Show that, unlike Example 3, the line segment $y = 0$, $0 \le x \le 1$, which connects the two branch points, cannot serve as a branch cut for defining a branch of this function.

c) State suitable branch cuts for defining a branch.

Suppose a branch of $(z^2 - 1)^{1/3}$ equals -1 when $z = 0$. There are branch cuts defined by $y = 0$, $|x| \ge 1$. What value does this branch assume at the following points?

11. i **12.** $-i$ **13.** $1 + i$

14. If two functions each have a branch point at z_0, does their product necessarily have a branch point at z_0? Illustrate with an example.

15. If $f(z)$ has a branch point at z_0, does $1/f(z)$ necessarily have a branch point at z_0? Explain.

Suppose a branch of $z^{-1/4}(z^2 + 1)$ assumes negative real values for $y = 0$, $x > 0$. There is a branch cut along the line $x = y$, $y \ge 0$. What value does this function assume at these points?

16. -1 **17.** $2i$ **18.** $-1 - i$

19. a) Consider the function $f(z) = \log(1 + z^{1/2})$, where a branch of $z^{1/2}$ will be defined with the aid of the branch cut $y = 0$, $x \le 0$. Show that if we choose $z^{1/2} > 0$ when z is positive real, then we can find a branch of the logarithm such that $f(z)$ is analytic throughout the cut plane defined by the preceding branch cut.

b) Suppose we use this same branch cut, but choose $z^{1/2} < 0$ when z is positive real. Explain why we cannot find a branch of the logarithm such that $f(z)$ is analytic throughout this cut plane.

c) For the branch of the logarithm that you chose in part (a), find $f'(i)$.

Consider $\sin^{-1} z = -i \log(zi + (1 - z^2)^{1/2})$. Suppose we use a branch of this function defined as follows: The principal branch of the log is employed, $(1 - z^2)^{1/2} = 1$ when $z = 0$, and the two branch cuts are given by $y \ge 0$, $x = \pm 1$. What is the value of this function and its derivative at each of the following points?

20. i **21.** 3 **22.** $1 - i$

Consider the branch of $z^{1/2}$ defined by a branch cut along $y = 0$, $x \le 0$. If for this branch $1^{1/2} = -1$, state whether the following equations have solutions within the domain of analyticity of the branch. Give the solution if there is one.

23. $z^{1/2} - 3 = 0$ **24.** $z^{1/2} + 3 = 0$ **25.** $z^{1/2} + 1 + i\sqrt{3} = 0$

26. $z^{1/2} - 1 - i\sqrt{3} = 0$

27. Find all solutions in the complex plane of $i^z + i^{-z} = 0$. Use principal values of all functions.

28. Consider the principal branch of z^i. Let $z^i = a + ib$, where a and b are real. If $z = r \operatorname{cis} \theta$, $-\pi < \theta < \pi$, show that

a) $r = e^{\cos^{-1}(ae^{\theta})}$ and $r = e^{\sin^{-1}(be^{\theta})}$

b) Draw in the complex z-plane a locus on which a assumes the value 1/2. Repeat for b, using the same value. A programmable calculator or desktop computer is helpful but not necessary. Also draw the branch cut for our branch of z^i.

APPENDIX TO CHAPTER 3

PHASORS

In the analysis of electrical circuits and many mechanical systems we must deal with functions that oscillate sinusoidally in time, grow or decay exponentially in time, or oscillate with an amplitude that grows or decays exponentially in time. Designating time as t, we find that many such functions $f(t)$ can be described by

$$f(t) = \text{Re}[Fe^{st}], \tag{A3–1}$$

where

$$s = \sigma + i\omega \tag{A3–2}$$

is called the complex frequency of oscillation of $f(t)$, and F is a complex number, independent of t, written

$$F = F_0 e^{i\theta}, \tag{A3–3}$$

where $F_0 = |F|$ and $\theta = \arg F$. We will always use real values for σ and ω.

The complex number F appearing in Eqs. (A3–1) and (A3–3) is called the *phasor* associated with $f(t)$.

DEFINITION Phasor

The phasor associated with a given function of time, $f(t)$, is a complex number F, independent of t, such that the real part of the product of F with a complex exponential e^{st} yields $f(t)$. □

We will usually use an uppercase letter to mean a phasor and the corresponding lowercase letter to represent the associated function of t. The one exception is that the letter I means a phasor electric current but the lowercase Greek iota, ι, will be the corresponding function of t. Thus the lowercase i retains its usual meaning. As we shall see, phasors are useful in the solution of the linear differential equations with constant coefficients used to describe many electrical and mechanical configurations.

The expression Fe^{st} appearing in Eq. (A3–1) is an example of a complex function of a real variable (since t remains real). Let us consider some specific cases in Eq. (A3–1). Suppose the phasor F in Eq. (A3–3) is positive real and the complex frequency in Eq. (A3–2) is real. Then with $s = \sigma$ and $F = F_0 > 0$ in Eq. (A3–1) we obtain

$$f(t) = \text{Re}[F_0 e^{\sigma t}] = F_0 e^{\sigma t}. \tag{A3–4}$$

Here, $f(t)$ grows or decays with increasing t according to whether σ is positive or negative. If $\sigma = 0$, then $f(t)$ is constant.

Assuming that both F in Eq. (A3–3) and s in Eq. (A3–2) are complex, we have, from Eq. (A3–1),

$$f(t) = \text{Re}[F_0 e^{i\theta} e^{(\sigma + i\omega)t}] = \text{Re}[F_0 e^{\sigma t} e^{i(\omega t + \theta)}].$$

Since (see Eq. 3.1–11) $\text{Re}\, e^{i(\omega t + \theta)} = \cos(\omega t + \theta)$, we have

$$f(t) = F_0 e^{\sigma t} \cos(\omega t + \theta). \tag{A3–5}$$

Equation (A3–5) describes an $f(t)$ that oscillates with radian frequency ω (usually taken as positive). The amplitude of the oscillations, $F_0e^{\sigma t}$, grows or decays with increasing t according to whether σ is positive or negative. If $\sigma = 0$, the amplitude of the oscillations remains unchanged. These three possible situations are illustrated in Fig. A3–1.

The function $f(t)$ described by Eq. (A3–5) displays a cosinusoidal variation with t. Taking $\theta = \phi - \pi/2$ in this equation, we have

$$f(t) = F_0e^{\sigma t}\cos(\omega t + \phi - \pi/2) = F_0e^{\sigma t}\sin(\omega t + \phi),$$

and a sinusoidal time variation is obtained. Some examples of functions of time, their complex frequencies, and their phasors, are given in the following table:

$f(t)$	F	$s = \sigma + i\omega$
$2\cos(10t + \pi/6)$	$2e^{i\pi/6}$	$10i$
$3\sin(5t + \pi/10)$	$3e^{i(-\pi/2+\pi/10)} = -3ie^{i\pi/10}$	$5i$
$3e^{-t}\sin(5t)$	$-3i$	$-1 + 5i$
$4e^{-3t}$	4	-3

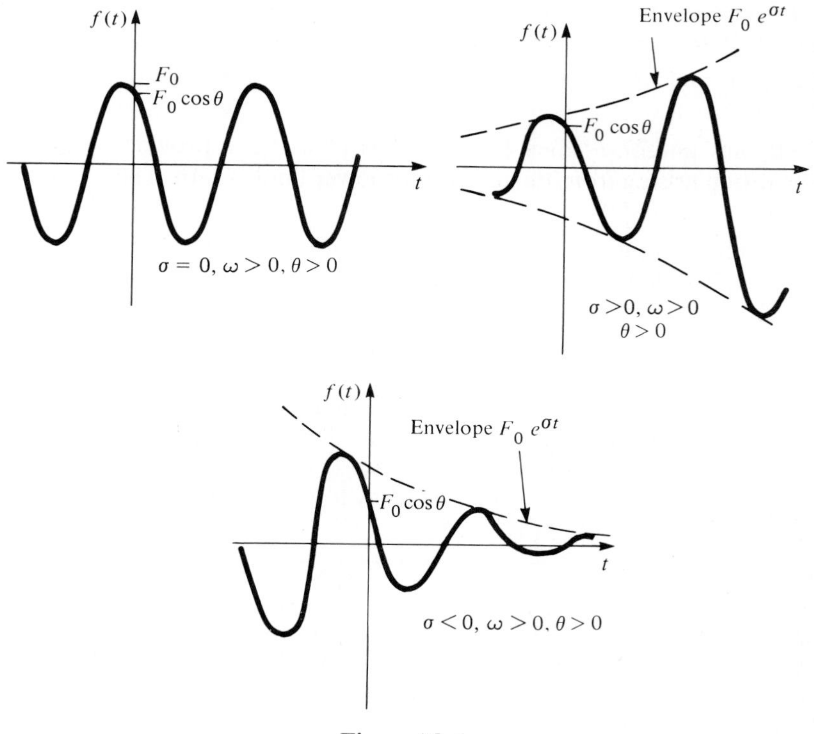

Figure A3–1

Many functions, for example, $f(t) = t \cos t$ or $\cos(t^2)$ or $\sin|t|$ or e^{t^2} are not representable in the form specified by Eq. (A3–1) and do not have phasors. Functions that are the sums of functions having different complex frequencies also are not representable in the form of Eq. (A3–1) and do not have phasors, for example,

$$f(t) = \cos(t) + \cos(2t), \qquad f(t) = e^{-t} + \sin t, \qquad f(t) = e^{-t}\cos t + \cos t.$$

The properties of phasors that make them particularly useful in the steady state solution of linear differential equations with real constant coefficients and real forcing functions having phasor representations are given below.† The proofs of these properties are, with one exception, left to the exercises.

1. a) If $f(t)$ is expressible in the form shown in Eq. (A3–1), then its phasor F is unique provided $\omega \neq 0$. There is no other phasor that can be substituted on the right in Eq. (A3–1) to yield $f(t)$. If $\omega = 0$, then only $\mathrm{Re}(F)$ is unique.‡
   b) If $f(t)$ and $g(t)$ are identically equal for all t, then their phasors are equal provided $\omega \neq 0$. If $\omega = 0$, then the real parts of their phasors must be equal.
2. For a given complex frequency $s = \sigma + i\omega$ there is only one function of t corresponding to a given phasor.
3. The phasor for the sum of two or more time functions having identical complex frequencies is the sum of the phasors for each. The phasor for $Mf(t)$, where M is a real number, is MF, where F is the phasor for $f(t)$.
4. For a given complex frequency the function of t corresponding to the sum of two or more phasors is the sum of the time functions for each.
5. If $f(t)$ has phasor F, then df/dt has phasor sF. By extension, d^nf/dt^n has phasor s^nF.
6. If $f(t)$ has phasor F, then $\int^t f(t')\,dt'$ has phasor F/s provided the constant of integration arising from the unspecified lower limit is zero. The relationship fails if $s = 0$.

To establish property 5, which is of major importance, we differentiate both sides of Eq. (A3–1) as follows:

$$\frac{df}{dt} = \frac{d}{dt}\mathrm{Re}[Fe^{st}]. \tag{A3–6}$$

The operations d/dt and Re can be interchanged since the time derivative is a real operator. For example, if $x(t)$ and $y(t)$ are real functions of t, then $(d/dt)\,\mathrm{Re}[x(t) + iy(t)] = dx/dt$. We have also that $\mathrm{Re}[(d/dt)[x(t) + iy(t)]] = dx/dt$. Exchanging operators on the right side of Eq. (A3–6), we have $df/dt = \mathrm{Re}[sFe^{st}]$. Thus (see Eq. A3–1) df/dt

---

† If the coefficients and/or the forcing function are complex, the methods described in this appendix can be slightly modified to produce a solution. See, e.g., W. Kaplan, *Operational Methods for Linear Systems* (Reading, MA: Addison-Wesley, 1962), Sections 1.9–1.11.

‡ When $\omega = 0$ this lack of uniqueness does not matter when we use phasors in the solution of differential equations. However, by convention, when $\omega = 0$, $\mathrm{Im}(F)$ is taken as zero.

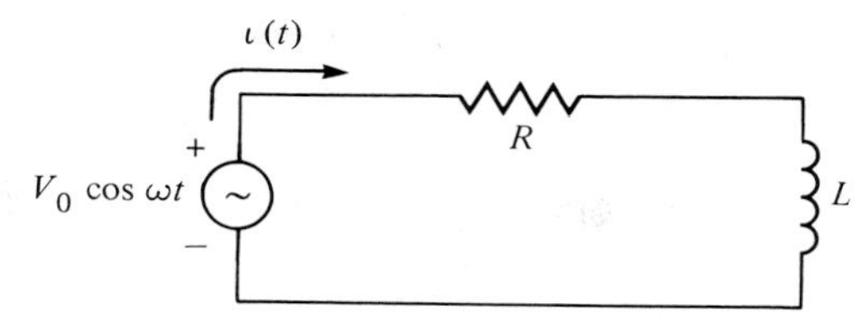

Figure A3–2

possesses a phasor sF. This discussion can be extended to yield the phasor of higher-order derivatives.

Phasors are applied to physical problems in which it is assumed that an electric circuit or mechanical configuration is excited by a real voltage, current, or mechanical force describable by Eq. (A3–1). The excitation, or forcing function, has been applied for a sufficiently long time so that all transients in the configuration have died out. Thus all voltages, currents, velocities, displacements, and so forth exhibit the same complex frequency s as the excitation.

Each quantity in the differential equation describing the physical problem is converted to its phasor. Property 5 is used to transform time derivatives in the differential equation into products of phasors and their complex frequency. The given differential equation is thus converted into an easily solved algebraic equation involving phasors. The required real function of time describing the physical problem can now be recovered from Eq. (A3–1). The uniqueness of the solution is guaranteed by the requirement that it exhibit the same complex frequency as the excitation. An example of the method follows.

EXAMPLE 1

In Fig. A3–2 a series electrical circuit containing an inductor of L henries and a resistor of R ohms is driven by a voltage source $v(t) = V_0 \cos(\omega t)$, where $V_0 > 0$. We want the unknown current $\imath(t)$. According to basic electric circuit theory† the voltage across the resistor is $R\imath(t)$ while that across the inductor is $L\,d\imath/dt$. According to Kirchhoff's voltage law the sum of these expressions must equal the source voltage. Thus

$$R\imath(t) + L\frac{d\imath}{dt} = V_0 \cos(\omega t). \tag{A3–7}$$

The phasor for the driving voltage $V_0 \cos(\omega t)$ is V_0, and the complex frequency is $s = i\omega$. The current also has this complex frequency.

If I is the phasor for $\imath(t)$, then by property 5 the phasor for $d\imath/dt$ must be $sI = i\omega I$. The phasor for the left side of Eq. (A3–7) is easily found from property 3 and equals $RI + i\omega LI = (R + i\omega L)I$. The phasor for the left side of this equation must equal

---

† See, for example, W. Hayt and J. Kemmerly, *Engineering Circuit Analysis*, 4th ed. (New York: McGraw-Hill, 1986).

the phasor for the right side (see property 1b). Thus

$$(R + i\omega L)I = V_0.$$

Solving for I, we have

$$I = \frac{V_0}{R + i\omega L} = \frac{V_0 e^{i\theta}}{\sqrt{R^2 + \omega^2 L^2}}, \tag{A3–8}$$

where

$$\theta = -\tan^{-1}\frac{\omega L}{R}. \tag{A3–9}$$

We can use Eq. (A3–1), taking $F = I$, $f = \imath$, and $s = i\omega$ to obtain $\imath(t)$. Thus using I from Eq. (A3–8) and θ from Eq. (A3–9) we have

$$\imath(t) = \mathrm{Re}\frac{V_0 e^{i\theta}}{\sqrt{R^2 + \omega^2 L^2}} e^{i\omega t}$$

$$= \frac{V_0 \cos(\omega t + \theta)}{\sqrt{R^2 + \omega^2 L^2}} = \frac{V_0 \cos\left(\omega t - \tan^{-1}\dfrac{\omega L}{R}\right)}{\sqrt{R^2 + \omega^2 L^2}}. \tag{A3–10}$$

The reader can verify that this result satisfies the differential equation (A3–7). ◀

There are problems where the linear differential equations (with constant coefficients) describing a physical configuration are not solvable with phasors. This occurs if the complex frequency of the generator or other excitation is equal to the "natural" or resonant frequency of the physical system. Then the solution is not of the form $e^{\sigma t}\cos(\omega t + \sigma)$, $e^{\sigma t}$, etc. and does not possess a phasor. The subject is discussed in many texts.[†]

Often integral equations or integrodifferential equations (containing integrals and derivatives of the unknown) can be solved with phasors. Here property 6 is helpful. Exercise 20 demonstrates its use.

EXERCISES

State the time function $v(t)$ corresponding to the following phasors V and complex frequency s.

	V	s		V	s
1.	3	$1 - i$	**5.**	$1 + ie^{i\pi/3}$	$e^{i\pi/6}$
2.	$-3i$	$-1 + i$	**6.**	i	i
3.	$3e^{-i\pi/4}$	$-1 - i$	**7.**	2	3
4.	$1 + i$	$1 + 2i$			

[†] See, for example, W. Kaplan, *Advanced Mathematics for Engineers* (Reading, MA: Addison-Wesley, 1981), Section 1.14.

Find the phasor corresponding to each of the following functions of time. In each case state the complex frequency. If the phasor does not exist, give the reason.

8. e^{2t}

9. $e^{-2t}\cos(3t)$

10. $6e^{-3t}\sin(2t)$

11. $2e^{4t}\sin(2t - \pi/6)$

12. $\sin t + 2\cos t$

13. $e^{-t}\sin t + 2\cos t$

14. $e^{-t}\sin t + 2e^{-t}\cos t$

15. $e^{-t}\sin(t + \pi/4) + 2e^{-t}\cos t$

16. Prove property 1(a) for phasors when $\omega \neq 0$.

17. Prove properties 3 and 4 for phasors.

18. Establish property 6 for phasors by integrating both sides of (A3–1). Justify the exchange of the order of any operations.

19. Consider an electric circuit identical to that shown in Fig. A3–2 except that the voltage source has been changed to $v(t) = V_0 e^{\sigma t}$. The differential equation describing the current $\imath(t)$ is now

$$R\imath(t) + L\frac{d\imath}{dt} = V_0 e^{\sigma t}.$$

Assume $\sigma \neq -R/L$. Find the phasor current I and use it to find the actual current $\imath(t)$.

20. In Fig. A3–3 a series circuit containing a resistor of R ohms and a capacitor of C farads is driven by the voltage generator $V_0 \sin \omega t$. The voltage across the capacitor is given by $(1/C)\int^t \imath(t')dt'$, where $\imath$ is the current in the circuit. According to Kirchhoff's voltage law this current satisfies the integral equation:

$$V_0 \sin(\omega t) = R\imath(t) + (1/C)\int^t \imath(t')dt'.$$

Obtain I, the phasor current, and use it to find $\imath(t)$. Assume $\omega > 0$.

21. A mass m is attached to the end of a spring and lies in a viscous fluid, as shown in Fig. A3–4. The coordinate $x(t)$ locating the mass also measures the elongation of the spring. Besides the spring force, the mass is subjected to a fluid damping force proportional

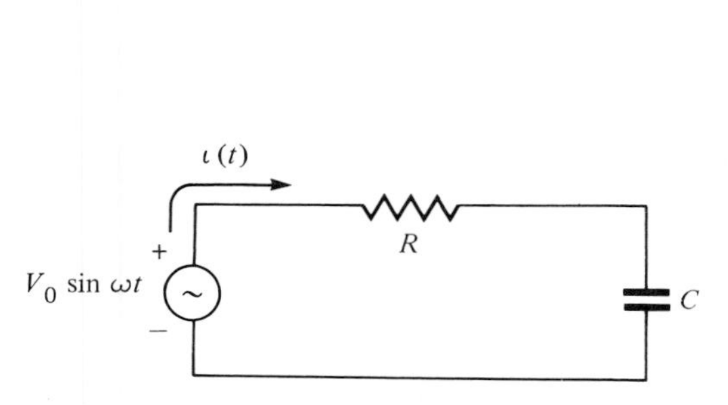

Figure A3–3

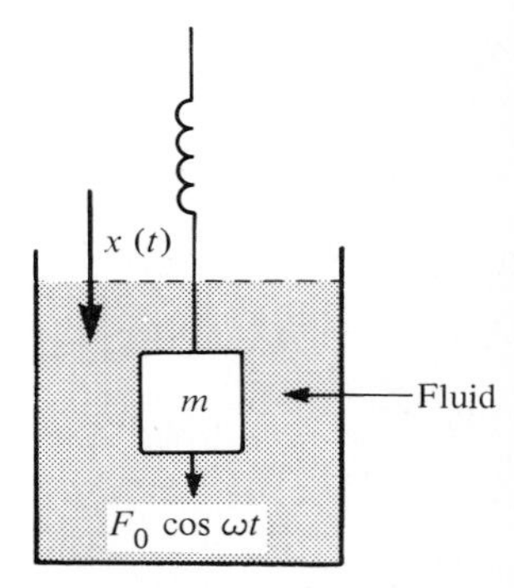

Figure A3–4

to the velocity of motion and also to an external mechanical force $F_0 \cos \omega t$. From Newton's second law of motion the differential equation governing $x(t)$ is

$$m \frac{d^2x}{dt^2} + \alpha \frac{dx}{dt} + kx = F_0 \cos \omega t, \qquad \omega > 0.$$

Here k is a constant determined by the stiffness of the spring and α is a damping constant determined by the fluid viscosity.

a) Find X, the phasor for $x(t)$. b) Use X to find $x(t)$.

CHAPTER 4

Integration in the Complex Plane

4.1 INTRODUCTION TO LINE INTEGRATION

When studying elementary calculus, the reader first learned to differentiate real functions of real variables and later to integrate such functions. Both indefinite and definite integrals were considered.

We have a similar agenda for complex variables. Having learned to differentiate in the complex plane and having studied the allied notion of analyticity, we turn our attention to integration. The indefinite integral, which (as for real variables) reverses the operation of differentiation, will not be considered first, however. Instead, we will initially look at a particular kind of definite integral called a line integral or *contour integral.*

Like the definite integral studied in elementary calculus, the line integral is a limit of a sum. However, the physical interpretation of this new integral is more elusive. We are used to interpreting the definite integrals of elementary calculus as the area under the curve described by the integrand. Generally, a line integral does not have such a simple interpretation.

Ordinarily, it cannot be considered as the area under a curve. Surprisingly, the study of line integrals will lead us to a useful theorem regarding the existence of derivatives of all orders of an analytic function and will provide us with further insight into the meaning of analyticity. Some practical physical problems solved with line integrals will be presented. In Chapter 6 we will show how evaluation of line integrals can often lead to the rapid integration of real functions; for example, an expression like $\int_{-\infty}^{+\infty} x^2/(x^4 + 1)\, dx$ is quickly evaluated if we first perform a fairly simple line integration in the complex plane.

In our discussion of integrals we require the concept of a *smooth arc* in the xy-plane. Loosely speaking, a smooth arc is a curve on which the tangent is defined everywhere and where the tangent changes its direction continuously as we move along the curve. One way to define a smooth arc is by means of a pair of equations dependent upon a real parameter, which we will call t. Thus

$$x = \psi(t), \tag{4.1–1a}$$

$$y = \phi(t), \tag{4.1–1b}$$

where $\psi(t)$ and $\phi(t)$ are real continuous functions with continuous derivatives $\psi'(t)$ and $\phi'(t)$ in the interval $t_a \le t \le t_b$. We also assume that there is no t in this interval for which both $\psi'(t)$ and $\phi'(t)$ are simultaneously zero. It is sometimes helpful to think that t represents time. As t advances from t_a to t_b, Eqs. (4.1–1a,b) define a locus that can be plotted in the xy-plane. This locus is a smooth arc.

An example of a smooth arc generated by such parametric equations is $x = t$, $y = 2t$, for $1 \le t \le 2$. Eliminating the parameter t, which connects the variables x and y, we find that the locus determined by the parametric equations lies along the line $y = 2x$. As t progresses from 1 to 2, we generate that portion of this line lying between (1, 2) and (2, 4) (see Fig. 4.1–1a).

Consider as another example the equations $x = \sqrt{t}$, $y = t$ for $1 \le t \le 4$. As t progresses from 1 to 4, the locus generated is the portion of the parabolic curve $y = x^2$ shown in Fig. 4.1–1(b). In Figs. 4.1–1(a,b) there are arrows that indicate the sense in which the arc is generated as t increases from t_a to t_b. For the right-hand arc the tangent has been constructed at some arbitrary point.

The slope of the tangent for any curve is dy/dx, which is identical to $(dy/dt)/(dx/dt) = \phi'(t)/\psi'(t)$ provided $\psi'(t) \ne 0$. If $\psi'(t) = 0$, the slope becomes infinite and the tangent is vertical. Since $\phi'(t)$ and $\psi'(t)$ are continuous, the tangent to the curve defined in Eqs. (4.1–1) alters its direction continuously as t advances through the interval $t_a \le t \le t_b$.

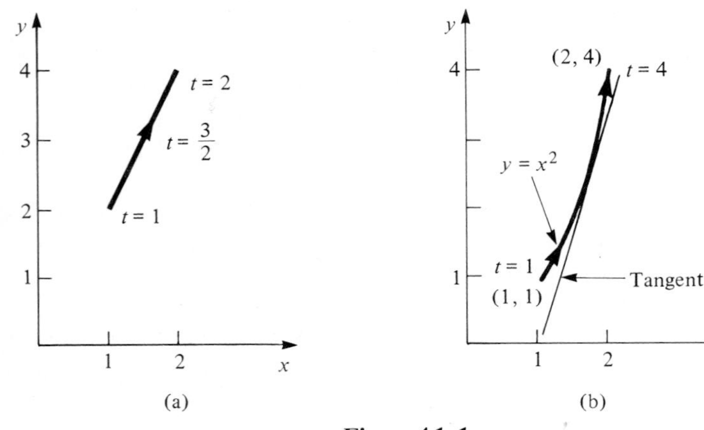

Figure 4.1–1

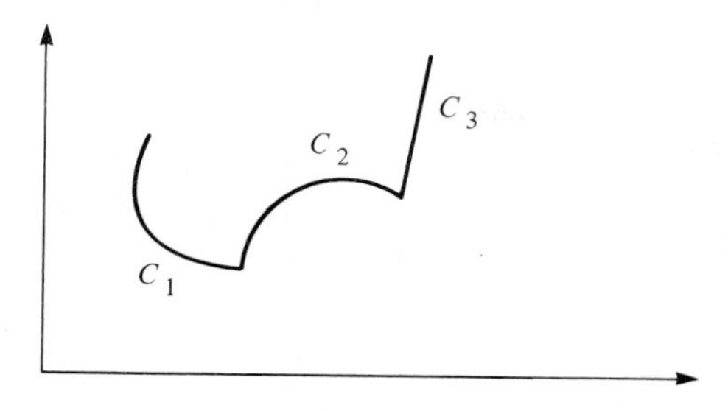

Figure 4.1–2

In our discussion of line integrals we must utilize the concept of a piecewise smooth curve, sometimes referred to as a contour.

DEFINITION Piecewise smooth curve (contour)

A *piecewise smooth curve* is a path made up of a finite number of smooth arcs connected end to end. □

Figure 4.1–2 shows three arcs C_1, C_2, C_3 joined to form a piecewise smooth curve.

Where two smooth arcs join, the tangent to a piecewise smooth curve can change discontinuously.

Real Line Integrals

We will begin our discussion of line integrals by using only real functions. An example of a real line integral—the integral for the length of a smooth arc†—is already known to the student. An approximation to the length of the arc is expressed as the sum of the lengths of chords inscribed on the arc. The actual length of the arc is obtained in the limit as the length of each chord in the sum becomes zero and the number of chords becomes infinite. In elementary calculus one learns to express this sum as an integral. The length of a piecewise smooth curve, such as is shown in Fig. 4.1–2, is obtained by adding together the lengths of the smooth arcs $C_1, C_2, \ldots$ that make up the curve.

Another type of real line integral involves not only a smooth arc C but also a function of x and y, say $F(x, y)$. It is important to realize that $F(x, y)$ is not the equation of C. Typically C is given by some equation, which, for the moment, we do not need to specify. Now, $\int_A^B F(x, y)\, ds$ integrated over C is defined as follows (refer to the arc C in Fig. 4.1–3).

We subdivide C, which goes from A to B, into n smaller arcs. The first arc goes from the point X_0, Y_0 to X_1, Y_1; the second arc goes from X_1, Y_1 to X_2, Y_2, etc.‡

† See G. Thomas and R. Finney, *Calculus and Analytic Geometry*, 8th ed. (Reading, MA: Addison-Wesley, 1992), Section 5.4.

‡ If the arc has been defined by a pair of parametric equations like those shown in Eqs. (4.1–1), where $t_0 \le t \le t_n$, we can generate the points X_0, Y_0; X_1, Y_1, etc. as follows: $X_0, Y_0 = \psi(t_0), \phi(t_0)$; $X_1, Y_1 = \psi(t_1)\phi(t_1)$; $X_n, Y_n = \psi(t_n), \phi(t_n)$, where $t_0 < t_1 < t_2 \cdots < t_n$.

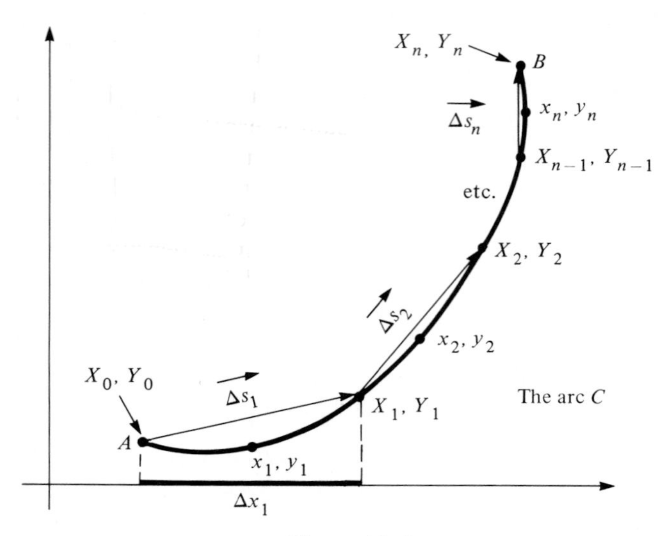

Figure 4.1–3

Corresponding to each of these arcs are the vector chords $\overrightarrow{\Delta s_1}, \overrightarrow{\Delta s_2}, \ldots, \overrightarrow{\Delta s_n}$. The first of these vectors is a directed line segment from X_0, Y_0 to X_1, Y_1; the second of these vectors is a directed line segment from X_1, Y_1 to X_2, Y_2, etc. These vectors, when summed, form a single vector going from A to B. The vectors form chords having lengths $\Delta s_1, \Delta s_2, \ldots$. The length of the vector $\overrightarrow{\Delta s_k}$ is thus Δs_k.

Let x_1, y_1 be a point at an arbitrary location on the first arc; x_2, y_2 a point somewhere on the second arc, etc. We now evaluate $F(x, y)$ at the n points (x_1, y_1), $(x_2, y_2), \ldots, (x_n, y_n)$. We define the line integral of $F(x, y)$ from A to B along C as follows:

DEFINITION $\int_A^B F(x, y)\, ds$

$$\int_A^B F(x, y)\, ds = \lim_{n\to\infty} \sum_{k=1}^{n} F(x_k, y_k)\, \Delta s_k, \tag{4.1–2}$$

where, as n, the number of subdivisions of C becomes infinite, the length Δs_k of each chord goes to zero. □

Of course, if the limit of the sum in this definition fails to exist, we say that the integral does not exist. It can be shown that if $F(x, y)$ is continuous on C then the integral will exist.†

Evaluation of the preceding type of integral is similar to the familiar problem of evaluating integrals for arc length. A typical procedure is outlined in Exercise 1 of this section.

---

† W. Kaplan, *Advanced Calculus*, 4th ed. (Reading, MA: Addison-Wesley, 1991), Sections 5.1–5.3.

If $F(x, y)$ in Eq. (4.1–2) happens to be unity everywhere along C, then the summation on the right simplifies to $\sum_{k=1}^{n} \Delta s_k$. This is the sum of the lengths of the chords lying along the arc C in Fig. 4.1–3. This summation yields approximately the length of C. As $n \to \infty$, the limit of this sum is exactly the arc length. In general, however, $F(x, y) \neq 1$ and the sum in Eq. (4.1–2) consists essentially of the sum of the lengths of the n straight line segments that approximate C, each of which is weighted by the value of the function $F(x, y)$ evaluated close to that segment. If the curve C is thought of as a cable and if $F(x, y)$ describes its mass density per unit length, then $F(x_k, y_k)\,\Delta s_k$ would be the approximate mass of the kth segment. When the summation is carried to the limit $n \to \infty$, it yields exactly the mass of the entire cable.

The line integral of a function taken over a piecewise smooth curve is obtained by adding together the line integrals over the smooth arcs in the curve. The integral of $F(x, y)$ along the contour of Fig. 4.1–2 is given by

$$\int F(x, y)\,ds = \int_{C_1} F(x, y)\,ds + \int_{C_2} F(x, y)\,ds + \int_{C_3} F(x, y)\,ds.$$

There is another type of line integral involving $F(x, y)$ and a smooth arc C that we can define. Refer to Fig. 4.1–3. Let Δx_1 be the projection of $\overrightarrow{\Delta s_1}$ on the x-axis, Δx_2 the projection of $\overrightarrow{\Delta s_2}$, etc. Note that although Δs_k is positive (because it is a length), Δx_k, which equals $X_k - X_{k-1}$, can be positive or negative depending on the direction of $\overrightarrow{\Delta s_k}$. We make the following definition:

DEFINITION $\int_A^B F(x, y)\,dx$

$$\int_A^B F(x, y)\,dx = \lim_{n\to\infty} \sum_{k=1}^{n} F(x_k, y_k)\,\Delta x_k, \tag{4.1–3}$$

where all $\Delta x_k \to 0$ as $n \to \infty$. □

A similar integral is definable when we instead use the projections of $\overrightarrow{\Delta s_k}$ on the y-axis. These projections are Δy_1, Δy_2, and so on, so that $\Delta y_k = Y_k - Y_{k-1}$. Hence:

DEFINITION $\int_A^B F(x, y)\,dy$

$$\int_A^B F(x, y)\,dy = \lim_{n\to\infty} \sum_{k=1}^{n} F(x_k, y_k)\,\Delta y_k, \tag{4.1–4}$$

where all $\Delta y_k \to 0$ as $n \to \infty$. □

The integrals in Eqs. (4.1–3) and (4.1–4) can be shown to exist† when $F(x, y)$ is continuous along the smooth arc C. Some procedures for the evaluation of this type of integral are discussed in Example 1 of this section. Integrals along piecewise smooth curves can be defined if we add together the integrals along the arcs that make up the curves. In general, the values of the integrals defined in Eqs. (4.1–2), (4.1–3), and

---

† Kaplan, ibid.

(4.1–4) depend not only on the function $F(x, y)$ in the integrand and the limits of integration but also on the path used to connect these limits.

What happens if we were to reverse the limits of integration in Eq. (4.1–3) or Eq. (4.1–4)? If we were to compute $\int_B^A F(x, y)\,dx$, we would go through a procedure identical to that used in computing $\int_A^B F(x, y)\,dx$, except that the vectors shown in Fig. 4.1–3 would all be reversed in direction; their sum would extend from B to A. The projections Δx_k would be reversed in sign from what they were before. Hence, along contour C,

$$\int_B^A F(x, y)\,dx = -\int_A^B F(x, y)\,dx. \tag{4.1–5}$$

A reversal in sign also occurs when we exchange A and B in the integral defined by Eq. (4.1–4). Note however that

$$\int_B^A F(x, y)\,ds = \int_A^B F(x, y)\,ds \tag{4.1–6}$$

because the Δs_k used in the definitions of these expressions involves lengths that are positive for both directions of integration.

Integrals of the type defined in Eqs. (4.1–2), (4.1–3), and (4.1–4) can be broken up into the sum of other integrals taken along portions of the contour of integration. Let A and B be the endpoints of a piecewise smooth curve C. Let Q be a point on C. Then, one can easily show that

$$\int_A^B F(x, y)\,dx = \int_A^Q F(x, y)\,dx + \int_Q^B F(x, y)\,dx. \tag{4.1–7}$$

Identical results hold for integrals of the form $\int_A^B F(x, y)\,dy$ and $\int_A^B F(x, y)\,ds$. Other identities that apply to all three of these kinds of integrals, but which will be written only for integration on the variable x, are

$$\int_A^B kF(x, y)\,dx = k\int_A^B F(x, y)\,dx, \qquad k \text{ is any constant}; \tag{4.1–8a}$$

$$\int_A^B [F(x, y) + G(x, y)]\,dx = \int_A^B F(x, y)\,dx + \int_A^B G(x, y)\,dx. \tag{4.1–8b}$$

EXAMPLE 1

Consider a contour consisting of that portion of the curve $y = 1 - x^2$ that goes from the point $A = (0, 1)$ to the point $B = (1, 0)$ (see Fig. 4.1–4). Let $F(x, y) = xy$. Evaluate

a) $\displaystyle\int_A^B F(x, y)\,dx$, b) $\displaystyle\int_A^B F(x, y)\,dy$.

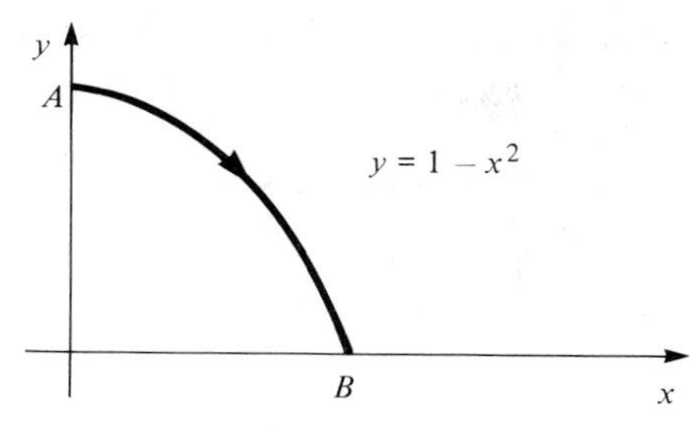

Figure 4.1–4

Solution

Part (a):

$$\int_A^B F(x, y)\,dx = \int_{0,1}^{1,0} xy\,dx.$$

Along the path of integration y changes with x. The equation of the contour of integration $y = 1 - x^2$ can be used to express y as a function of x in the preceding integrand. Thus

$$\int_A^B F(x, y)\,dx = \int_{0,1}^{1,0} (x)(1 - x^2)\,dx = \frac{x^2}{2} - \frac{x^4}{4}\bigg|_{x=0}^{x=1} = \frac{1}{4}.$$

We have converted our line integral to a conventional integral performed between constant limits; the evaluation is simple.

Part (b):

$$\int_A^B F(x, y)\,dy = \int_{0,1}^{1,0} xy\,dy.$$

To change this to a conventional integral we may regard x as a function of y along C. Since $y = 1 - x^2$ and $x \geq 0$ on the path of integration, we have $x = \sqrt{1 - y}$. Thus

$$\int_{0,1}^{1,0} xy\,dy = \int_1^0 y\sqrt{1 - y}\,dy = \frac{-4}{15}.$$

This result is negative because $F(x, y)$ is everywhere positive along the path of integration while the increments in y are everywhere negative as we proceed from A to B along the given contour.

An alternative method for performing the given integration is to notice that

$$\int F(x, y)\,dy = \int F(x, y)\,\frac{dy}{dx}\,dx.$$

On C, with $F(x, y) = xy$, $y = 1 - x^2$, and $dy/dx = -2x$ we obtain

$$\int_0^1 (xy)(-2x)\,dx = \int_0^1 x(1 - x^2)(-2x)\,dx = \int_0^1 (-2x^2 + 2x^4)\,dx = \frac{-4}{15}.$$

In part (a) we could have integrated on y instead of on x by a similar maneuver. Note that in Exercise 1 of this section the integral $\int xy\,ds$ along the same contour is evaluated. ◀

EXERCISES

1. Using the contour of Example 1, show that

$$\int_{0,1}^{1,0} xy\,ds = \int_0^1 x(1-x^2)\sqrt{1+4x^2}\,dx = \int_0^1 y\sqrt{5/4-y}\,dy.$$

Hint: Recall from elementary calculus that $ds = (\pm)\sqrt{1+(dy/dx)^2}\,dx = \pm\sqrt{1+(dx/dy)^2}\,dy$, and that $ds \geq 0$. Evaluate the contour integral by integrating either on x or y. One is slightly easier.

Let C be that portion of the curve $y = x^2$ lying between (0, 0) and (1, 1). Let $F(x, y) = x + y + 1$. Evaluate these integrals along C.

2. $\displaystyle\int_{0,0}^{1,1} F(x, y)\,dx$ 3. $\displaystyle\int_{0,0}^{1,1} F(x, y)\,dy$

Let C be that portion of the curve $x^2 + y^2 = 1$ lying in the first quadrant. Let $F(x, y) = x^2 y$. Evaluate these integrals along C.

4. $\displaystyle\int_{0,1}^{1,0} F(x, y)\,dx$ 5. $\displaystyle\int_{0,1}^{1,0} F(x, y)\,dy$ 6. $\displaystyle\int_{0,1}^{1,0} F(x, y)\,ds$

7. Show that $\int_{0,-1}^{0,1} y\,dx = -\pi/2$. The integration is along that portion of the circle $x^2 + y^2 = 1$ lying in the half plane $x \geq 0$. Be sure to consider signs in taking square roots.

8. Evaluate $\int_{3,0}^{0,-1} x\,dy$ along the portion of the ellipse $x^2 + 9y^2 = 9$ lying in the first, second, and third quadrants.

4.2 COMPLEX LINE INTEGRATION

We now study the kind of integral encountered most often with complex functions: the complex line integral. We will find that it is closely related to the real line integrals just discussed.

We begin, as before, with a smooth arc that connects the points A and B in the xy-plane. We now regard the xy-plane as being the complex z-plane. The arc is divided into n smaller arcs and, as shown in Fig. 4.2–1, successive endpoints of the subarcs have coordinates $(X_0, Y_0), (X_1, Y_1), \ldots, (X_n, Y_n)$. Alternatively, we could say that the endpoints of these smaller arcs are at $z_0 = X_0 + iY_0$, $z_1 = X_1 + iY_1$, etc. A series of vector chords are then constructed between these points. As in our discussion of real line integrals, the vectors progress from A to B when we are integrating from A to B along the contour. Let Δz_1 be the complex number corresponding to the vector going from (X_0, Y_0) to (X_1, Y_1), let Δz_2 be the complex number for the vector going from (X_1, Y_1) to (X_2, Y_2), etc. There are n such complex numbers. In general,

$$\Delta z_k = \Delta x_k + i\Delta y_k, \tag{4.2–1}$$

where Δx_k and Δy_k are the projections of the kth vector on the real and imaginary axes. Thus

$$\Delta z_k = (X_k - X_{k-1}) + i(Y_k - Y_{k-1}).$$

Let $z_k = x_k + iy_k$ be the complex number corresponding to a point lying, at an arbitrary position, on the kth arc. This arc is subtended by the vector chord Δz_k. Some study of Fig. 4.2–1 should make the notation clear.

Let us consider $f(z) = u(x, y) + iv(x, y)$, a continuous function of the complex variable z. We can evaluate this function at $z_1, z_2, \ldots, z_n$. We now define the line integral of $f(z)$ taken over the arc C.

DEFINITION Complex line integral

$$\int_A^B f(z)\,dz = \lim_{n\to\infty} \sum_{k=1}^{n} f(z_k)\,\Delta z_k, \tag{4.2–2}$$

where all $\Delta z_k \to 0$ as $n \to \infty$. □

As before, the integral only exists if the limit of the sum exists. If $f(z)$ is continuous in a domain containing the arc, it can be shown that this will be the case.†

In general, we must assume that the value of the integral depends not only on A and B, the location of the ends of the path of integration, but also on the specific path C used to connect these points. The reader is cautioned against interpreting the integral as the area under the curve in Fig. 4.2–1.

The line integral of a function over a piecewise smooth curve is computed by using Eq. (4.2–2) to determine the integral of the function over each of the arcs that make up the curve. The values of these integrals are then added together.

Let us try to develop some intuitive feeling for the sum on the right in Eq. (4.2–2). We can imagine that the arc in Fig. 4.2–1 is approximated, in shape, by the straight lines forming the n vectors. As n approaches infinity in the sum, there are more, and shorter, vectors involved in the sum. The broken line formed by these vectors more closely fits the curve C.

In the summation the complex numbers associated with each of these n vectors are added together after first having been multiplied by a complex weighting function $f(z)$ evaluated close to that vector. The function is evaluated on the nearby curve. If the weighting function were identically equal to 1, the sum in Eq. (4.2–2) would become $\sum_{k=1}^{n} \Delta z_k$. Graphically, this sum is represented by the vector addition of the n vectors shown in Fig. 4.2–1. Adding them, we obtain, for all n, a single vector extending from A to B.

---

† See E. T. Copson, *An Introduction to the Theory of Functions of a Complex Variable* (London: Oxford University Press, 1960), Section 4.13.

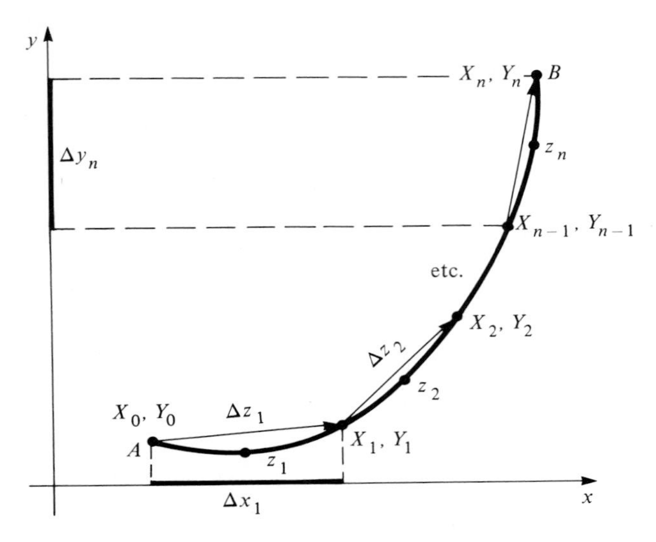

Figure 4.2–1

A summation up to a finite value of n in Eq. (4.2–2) can be used to approximate the integral on the left in this equation. Such a procedure is often used when one performs a complex line integration on the computer. Let us consider an example.

EXAMPLE 1

The function $f(z) = z + 1$ is to be integrated from $0 + i0$ to $1 + i$ along the arc $y = x^2$, as shown in Fig. 4.2–2. We will consider one-term series and two-term series approximations to the result.

a) One-term series: A single vector, associated with the complex number $\Delta z_1 = 1 + i$, goes from (0, 0) to (1, 1). The point z_1 can be chosen anywhere on the arc shown although our result will depend on the location we select. We arbitrarily choose it to lie so that its x-coordinate is in the middle of the projection of Δz_1 on the x-axis. Since z_1 is on the curve $y = x^2$, we have $\operatorname{Re} z_1 = 1/2$, $\operatorname{Im} z_1 = 1/4$. Now, because $f(z) = z + 1$, we find that $f(z_1) = (1/2 + i/4 + 1)$. Thus

$$\int_{0+i0}^{1+i} (z+1)\,dz \doteq f(z_1)\,\Delta z_1 = \left(\frac{1}{2} + \frac{i}{4} + 1\right)(1+i) = 1.25 + 1.75i.$$

b) Two-term series: Referring to Fig. 4.2–3, we see that

$$\Delta z_1 = \frac{1}{2} + \frac{i}{4}, \qquad \Delta z_2 = \frac{1}{2} + \frac{3i}{4}, \qquad z_1 = \frac{1}{4} + \frac{i}{16},$$

$$f(z_1) = \frac{1}{4} + \frac{i}{16} + 1, \qquad z_2 = \frac{3}{4} + i\frac{9}{16}, \qquad f(z_2) = \frac{3}{4} + i\frac{9}{16} + 1.$$

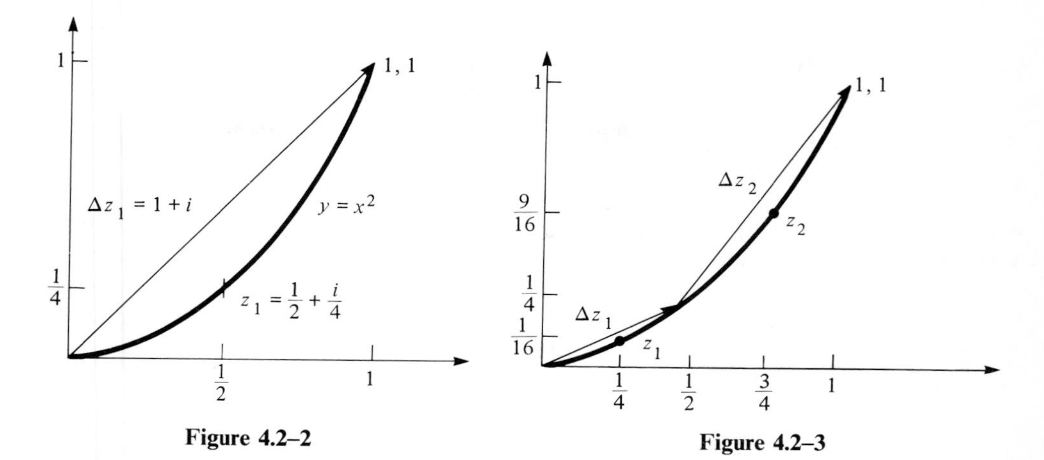

Figure 4.2–2

Figure 4.2–3

We have used the sum of two vectors to connect (0, 0) with (1, 1) and have chosen z_1 and z_2 according to the same criterion used in part (a). Thus

$$\int_{0+i0}^{1+i} (z + 1)\, dz \doteq f(z_1)\,\Delta z_1 + f(z_2)\,\Delta z_2$$

$$= \left(\frac{5}{4} + \frac{i}{16}\right)\left(\frac{1}{2} + \frac{i}{4}\right) + \left(\frac{7}{4} + \frac{i9}{16}\right)\left(\frac{1}{2} + \frac{3i}{4}\right) = 1.0625 + i1.9375.$$

In both parts (a) and (b) the approximation to the result $\int f(z)\, dz$ depends on the locations of z_1 and z_2, which we have chosen in an arbitrary fashion. However, in Eq. (4.2–2) the values chosen for z_k are immaterial to the result in the limit as $n \to \infty$ and $\Delta z_k \to 0$; i.e., $\int f(z)\, dz$ does not depend on the locations of $z_1, z_2, \ldots$.

We will see in Exercise 1 at the end of this section that the exact value of the given integral is $1 + 2i$, a result that is surprisingly well approximated by the two-term series. Note that this result is unrelated to the area under the curve $y = x^2$. ◀

When the function $f(z)$ is written in the form $u(x, y) + iv(x, y)$, line integrals involving $f(z)$ can be expressed in terms of real line integrals. Thus referring back to Eq. (4.2–2) and noting that $z_k = x_k + iy_k$ and that $\Delta z_k = \Delta x_k + i\Delta y_k$, we have

$$\int_A^B f(z)\, dz = \lim_{n\to\infty} \sum_{k=1}^{n} (u(x_k, y_k) + iv(x_k, y_k))(\Delta x_k + i\Delta y_k). \tag{4.2–3}$$

We now multiply the terms under the summation sign in Eq. (4.2–3) and separate the real and imaginary parts. Thus

$$\int_A^B f(z)\, dz = \lim_{n\to\infty} \left[\sum_{k=1}^{n} u(x_k, y_k)\,\Delta x_k - \sum_{k=1}^{n} v(x_k, y_k)\,\Delta y_k + i \sum_{k=1}^{n} v(x_k, y_k)\,\Delta x_k + i \sum_{k=1}^{n} u(x_k, y_k)\,\Delta y_k \right]. \tag{4.2–4}$$

Upon comparing the four summations in Eq. (4.2–4) with the definitions of the real line integrals $\int F(x, y)\,dx$, $\int F(x, y)\,dy$ (see Eqs. 4.1–3 and 4.1–4) we find that

$$\int_C f(z)\,dz = \int_C u(x, y)\,dx - \int_C v(x, y)\,dy + i\int_C v(x, y)\,dx + i\int_C u(x, y)\,dy. \quad (4.2\text{–}5)$$

The letter C signifies that all these integrals are to be taken, in a specific direction, along contour C. The continuity of u and v (or the continuity of $f(z)$) is sufficient to guarantee the existence of all the integrals in Eq. (4.2–5). The four real line integrals on the right are of a type that we have already studied; thus Eq. (4.2–5) provides us with a method for computing complex line integrals. Note that, as a useful mnemonic, Eq. (4.2–5) can be obtained from the following manipulation:

$$\int f(z)\,dz = \int (u + iv)(dx + i\,dy) = \int u\,dx - v\,dy + iv\,dx + iu\,dy.$$

We merely multiplied out the integrand $(u + iv)(dx + i\,dy)$.

When the path of integration for a complex line integral lies parallel to the real axis, we have $dy = 0$. There then remains

$$\int_C f(z)\,dz = \int_C [u(x, y) + iv(x, y)]\,dx, \qquad y \text{ is constant.}$$

This is the conventional type of integral encountered in elementary calculus, except that the integrand is complex if $v \neq 0$.

EXAMPLE 2

a) Compute $\int_{0+i}^{1+2i} \bar{z}^2\,dz$ taken along the contour $y = x^2 + 1$ (see Fig. 4.2–4a).
b) Perform an integration like that in part (a) using the same integrand and limits, but take as a contour the piecewise smooth curve C shown in Fig. 4.2–4b.

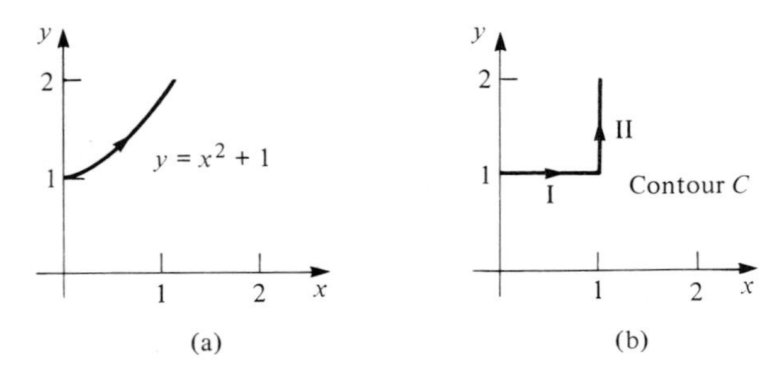

Figure 4.2–4

Solution

Part (a): To apply Eq. (4.2–5) we put $f(z) = (\bar{z})^2 = (x - iy)^2 = x^2 - y^2 - 2ixy = u + iv$. Thus with $u = x^2 - y^2$, $v = -2xy$, we have

$$\int_{0+i}^{1+2i} (\bar{z})^2\, dz = \int_{0,1}^{1,2} (x^2 - y^2)\, dx + \int_{0,1}^{1,2} 2xy\, dy + i \int_{0,1}^{1,2} -2xy\, dx + i \int_{0,1}^{1,2} (x^2 - y^2)\, dy. \tag{4.2–6}$$

In the first and third integrals on the right, $\int (x^2 - y^2)\, dx$ and $\int -2xy\, dx$, we substitute the relationship $y = x^2 + 1$ that holds along the contour. These line integrals become ordinary integrals whose limits are $x = 0$ and $x = 1$. After integration their values are found to be $-23/15$ and $-3/2$, respectively. The equation $y = x^2 + 1$ yields $x = \sqrt{y - 1}$ on the contour. This can be used to convert the second and fourth integrals on the right, $\int 2xy\, dy$ and $\int (x^2 - y^2)\, dy$, to ordinary integrals with limits $y = 1$ and $y = 2$. The integrals are found to have the numerical values $32/15$ and $-11/6$, respectively. Having evaluated the four line integrals on the right side of Eq. (4.2–6) we finally obtain

$$\int_{0+i}^{1+2i} \bar{z}^2\, dz = \frac{3}{5} - i\frac{10}{3}.$$

Part (b): Referring to Fig. 4.2–4(b), we break the path of integration into a part taken along path I and a part taken along path II.

Along I we have $y = 1$, so that $f(z) = (\bar{z})^2 = (\overline{x + i})^2 = x^2 - 1 - 2xi = u + iv$. Thus $u = x^2 - 1$, $v = -2x$. Since $y = 1$, $dy = 0$. The limits of integration along path I are (0, 1) and (1, 1). Using this information in Eq. (4.2–5), we obtain

$$\int_{\mathrm{I}} f(z)\, dz = \int_0^1 (x^2 - 1)\, dx + i \int_0^1 -2x\, dx = -\frac{2}{3} - i.$$

Along path II, $x = 1$, $dx = 0$, $f(z) = (\overline{1 + iy})^2 = 1 - y^2 - 2iy = u + iv$. The limits of integration are (1, 1) and (1, 2). Referring to Eq. (4.2–5), we have

$$\int_{\mathrm{II}} f(z)\, dz = \int_1^2 2y\, dy + i \int_1^2 (1 - y^2)\, dy = 3 - \frac{4}{3} i.$$

The value of the integral along C is obtained by summing the contributions from I and II. Thus

$$\int_C \bar{z}^2\, dz = -\frac{2}{3} - i + 3 - \frac{4}{3} i = \frac{7}{3} - \frac{7}{3} i.$$

This result is different from that of part (a) and illustrates how the value of a line integral between two points can depend on the contour used to connect them. ◀

Since Eq. (4.2–5) allows us to express a complex line integral in terms of real line integrals, the properties of real line integrals contained in Eqs. (4.1–5), (4.1–7), and (4.1–8) also apply to complex line integrals. Thus the following relationships are satisfied by integrals taken along a piecewise smooth curve C that connects points A and B:

$$\int_A^B f(z)\,dz = -\int_B^A f(z)\,dz; \tag{4.2–7a}$$

$$\int_A^B \Gamma f(z)\,dz = \Gamma \int_A^B f(z)\,dz, \qquad \text{where } \Gamma \text{ is any constant;} \tag{4.2–7b}$$

$$\int_A^B [f(z) + g(z)]\,dz = \int_A^B f(z)\,dz + \int_A^B g(z)\,dz; \tag{4.2–7c}$$

$$\int_A^B f(z)\,dz = \int_A^Q f(z)\,dz + \int_Q^B f(z)\,dz, \qquad \text{where } Q \text{ lies on } C. \tag{4.2–7d}$$

Sometimes it is easier to perform a line integration without using the variables x and y in Eq. (4.2–5). Instead, we integrate on a single real variable that is the parameter used in generating the contour of integration. Let a smooth arc C be generated by the pair of parametric equations in (4.1–1). Then

$$z(t) = x(t) + iy(t) = \psi(t) + i\phi(t) \tag{4.2–8}$$

is a complex function of the real variable t with derivative

$$\frac{dz}{dt} = \frac{d\psi}{dt} + i\frac{d\phi}{dt}. \tag{4.2–9}$$

As t advances from t_a to t_b in the interval $t_a \le t \le t_b$, the locus of $z(t)$ in the complex plane is the arc C connecting $z_a = \psi(t_a),\ \phi(t_a)$ with $z_b = \psi(t_b),\ \phi(t_b)$. To evaluate $\int_C f(z)\,dz$ we can make a change of variables as follows:

$$\int_C f(z)\,dz = \int_{t_a}^{t_b} f(z(t))\,\frac{dz}{dt}\,dt, \tag{4.2–10}$$

where the left-hand integration is performed along C from z_a to z_b. A rigorous justification for this equation is given in several texts.† Note that the integral on the right involves complex functions integrated on a real variable. This integration is performed with the familiar methods of elementary calculus. An application of Eq. (4.2–10) is given in the following example.

EXAMPLE 3

Evaluate $\int_C z^2\,dz$, where C is the parabolic arc $y = x^2$, $1 \le x \le 2$ shown in Fig. 4.1–1(b). The direction of integration is from (1, 1) to (2, 4).

---

† See, for example, E. T. Copson, *op. cit.*

Solution

We showed in discussing the parametric description of the curve in Fig. 4.1–1(b) that this arc can be generated by the equations $x = \sqrt{t}$, $y = t$, where $1 \le t \le 4$. Thus, from Eq. (4.2–8), the arc can be described as the locus of $z(t) = \sqrt{t} + it$ for $1 \le t \le 4$; notice that $dz/dt = 1/(2\sqrt{t}) + i$. The integrand is $f(z) = z^2 = (\sqrt{t} + it)^2 = t - t^2 + 2i(\sqrt{t})^3$. Using Eq. (4.2–10) with $t_a = 1$ and $t_b = 4$, we have

$$\int_C z^2\,dz = \int_1^4 [t - t^2 + 2i(\sqrt{t})^3]\left[\frac{1}{2\sqrt{t}} + i\right]dt$$

$$= \int_1^4 \left\{\left[\frac{\sqrt{t}}{2} - \frac{5}{2}(\sqrt{t})^3\right] + i(2t - t^2)\right\}dt = \frac{-86}{3} - 6i$$

Comment: A contour typically has more than one parametric representation. Another representation of C is used in Exercise 12. With this new parametrization $\int_C z^2\,dz$ is again evaluated with the same result. ◀

Bounds on Line Integrals; the "ML Inequality"

Given a line integral to evaluate $\int_C f(z)\,dz$, we can often, without going through the labor of performing the integration, obtain an upper bound on the absolute value of the answer. That is, we can find a positive number that is known to be greater than or equal to the magnitude of the still unknown integral.

We defined $\int_C f(z)\,dz$ by means of Eq. (4.2–2) and Fig. 4.2–1. A related integral will now be defined with the use of the smooth arc C of the same figure.

DEFINITION $\int_C |f(z)||dz|$

$$\int_C |f(z)||dz| = \lim_{n\to\infty} \sum_{k=1}^{n} |f(z_k)||\Delta z_k|, \qquad \text{all } |\Delta z_k| \to 0 \text{ as } n \to \infty. \quad \square \tag{4.2–11}$$

This integration results in a nonnegative real number. Since $|\Delta z_k| = \Delta s_k$ (refer to Figs. 4.1–3 and (4.2–1), we see from Eq. (4.1–2) that the preceding integral is identical to $\int_C |f(z)|\,ds$. Note that if $|f(z)| = 1$, then Eq. (4.2–11) simplifies to

$$\int_C |dz| = \lim_{n\to\infty} \sum_{k=1}^{n} |\Delta z_k| = \lim_{n\to\infty} \sum_{k=1}^{n} \Delta s_k = L, \tag{4.2–12}$$

where L, the length of C, is the sum of the chord lengths of Fig. 4.2–1 in the limit indicated. Let us compare the magnitude of the sum appearing on the right side of Eq. (4.2–2) with the sum on the right side of Eq. (4.2–11).

Recall that the magnitude of a sum of complex numbers is less than or equal to the sum of their magnitudes, and the magnitude of the product of two complex numbers equals the product of the magnitude of the numbers.

Using these two facts, it follows that

$$\left|\sum_{k=1}^{n} f(z_k)\,\Delta z_k\right| \le \sum_{k=1}^{n} |f(z_k)||\Delta z_k|. \tag{4.2–13}$$

The preceding inequality remains valid as $n \to \infty$ and $|\Delta z_k| \to 0$. Thus, combining Eqs. (4.2–2), (4.2–11), and (4.2–13), we have

$$\left|\int_C f(z)\,dz\right| \le \int_C |f(z)||dz|, \tag{4.2–14a}$$

which will occasionally prove useful. A special case of the preceding formula, applicable to complex functions of a real variable, is derived in Exercise 17. If $g(t)$ is such a function, we have, for $b > a$,

$$\left|\int_a^b g(t)\,dt\right| \le \int_a^b |g(t)|\,dt \tag{4.2–14b}$$

Now assume that M, a positive real number, is an upper bound for $|f(z)|$ on C. Thus $|f(z)| \le M$ for z on C. In particular, $|f(z_1)|$, $|f(z_2)|$, etc. on the right in Eq. (4.2–13) satisfy this inequality. Using this fact in Eq. (4.2–13), we obtain

$$\left|\sum_{k=1}^{n} f(z_k)\,\Delta z_k\right| \le \sum_{k=1}^{n} |f(z_k)||\Delta z_k| \le \sum_{k=1}^{n} M|\Delta z_k| = M\sum_{k=1}^{n} |\Delta z_k|. \tag{4.2–15}$$

Now observe that $\sum_{k=1}^{n} |\Delta z_k| \le L$ since the sum of the chord lengths, as in Fig. 4.2–1, cannot exceed the length L of the arc C. Combining this inequality with Eq. (4.2–15) we have

$$\left|\sum_{k=1}^{n} f(z_k)\,\Delta z_k\right| \le ML.$$

As $n \to \infty$, the preceding inequality still holds. Passing to this limit, with $\Delta z_k \to 0$, and referring to the definition of the line integral in Eq. (4.2–2), we have

ML INEQUALITY

$$\left|\int_C f(z)\,dz\right| \le ML. \tag{4.2–16}$$

The above is known as the *ML inequality*; stated in words, it reads:

> If there exists a constant M such that $|f(z)| \le M$ everywhere along a smooth arc C and if L is the length of C, then the magnitude of the integral of $f(z)$ along C cannot exceed ML.

EXAMPLE 4

Find an upper bound on the absolute value of $\int_{1+i0}^{0+i1} e^{1/z}\,dz$, where the integral is taken along the contour C, which is the quarter circle $|z| = 1$, $0 \le \arg z \le \pi/2$ (see Fig. 4.2–5).

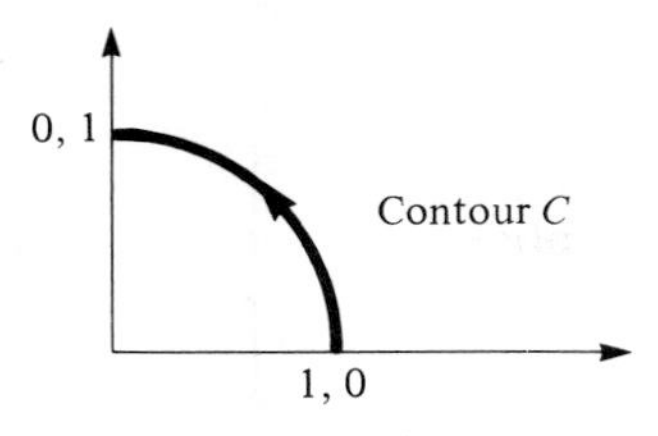

Figure 4.2–5

Solution

Let us first find M, an upper bound on $|e^{1/z}|$. We require that on C

$$|e^{1/z}| \leq M. \tag{4.2–17}$$

Now, notice that

$$e^{1/z} = e^{1/(x+iy)} = e^{x/(x^2+y^2)-iy/(x^2+y^2)} = e^{x/(x^2+y^2)}e^{i(-y)/(x^2+y^2)}.$$

Hence

$$|e^{1/z}| = |e^{x/(x^2+y^2)}||e^{-i(y)/(x^2+y^2)}| = |e^{x/(x^2+y^2)}|.$$

Since $e^{x/(x^2+y^2)}$ is always positive, we can drop the magnitude signs on the right side of the preceding equation. On contour C, $x^2 + y^2 = 1$. Thus

$$|e^{1/z}| = e^x \qquad \text{on } C.$$

The maximum value achieved by e^x on the given quarter circle occurs when x is maximum, that is, at $x = 1$, $y = 0$. On C, therefore, $e^x \leq e$. Thus

$$|e^{1/z}| \leq e$$

on the given contour. A glance at Eq. (4.2–17) now shows that we can take M as equal to e.

The length L of the path of integration is simply the circumference of the given quarter circle, namely, $\pi/2$. Thus, applying the ML inequality,

$$\left|\int_{1+i0}^{0+i1} e^{1/z}\,dz\right| \leq e\frac{\pi}{2}.$$ ◀

EXERCISES

1. In Example 1 we determined the approximate value of $\int_{0+i0}^{1+i}(z+1)\,dz$ taken along the contour $y = x^2$. Find the exact value of the integral and compare it with the approximate result.

2. Consider $\int_{0+i0}^{1+2i} z\,dz$ performed along the contour $y = 2x(2-x)$. Find the approximate value by means of the two-term series $f(z_1)\Delta z_1 + f(z_2)\Delta z_2$. Take z_1, z_2, Δz_1, Δz_2 as shown in Fig. 4.2–6. Now find the exact value of the integral and compare it with the approximate result.

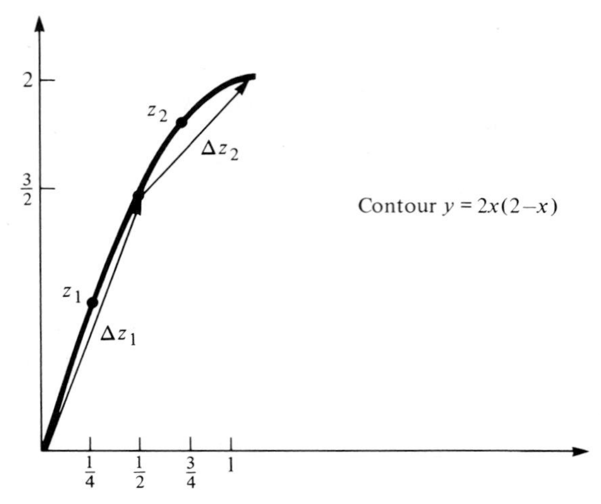

Figure 4.2–6

3. Consider $\int_{0+i0}^{1+2i} dz$ along the contour of Exercise 2. Evaluate this by using a two-term series approximation as is done in that problem. Explain why this result agrees perfectly with the exact value of the integral.

Evaluate $\int_i^1 \bar{z}\, dz$ along the contour C, where C is:

4. The straight line segment lying along $x + y = 1$.
5. The parabola $y = (1 - x)^2$.
6. The portion of the circle $x^2 + y^2 = 1$ in the first quadrant. Compare the answers to Exercises 4, 5, and 6.
7. Evaluate $\int e^z\, dz$
   a) from $z = 0$ to $z = 1$ along the line $y = 0$,
   b) from $z = 1$ to $z = 1 + i$ along the line $x = 1$,
   c) from $z = 1 + i$ to $z = 0$ along the line $y = x$. Verify that the sum of your three answers is zero. This is a specific example of a general result given in the next section.

The function $z(t) = e^{it} = \cos t + i \sin t$ can provide a useful parametric representation of circular arcs (see Fig. 3.1–1). If t ranges from 0 to 2π we have a representation of the whole unit circle, while if t goes from α to β we generate an arc extending from $e^{i\alpha}$ to $e^{i\beta}$ on the unit circle. Use this parametric technique to perform the following integrations.

8. $\displaystyle\int_1^{-1} \frac{1}{z}\, dz$ along $|z| = 1$, upper half plane
9. $\displaystyle\int_1^{-1} \frac{1}{z}\, dz$ along $|z| = 1$, lower half plane
10. $\displaystyle\int_1^{i} \bar{z}^4\, dz$ along $|z| = 1$, first quadrant

11. Show that $x = 2\cos t$, $y = \sin t$, where t ranges from 0 to 2π, yields a parametric representation of the ellipse $x^2/4 + y^2 = 1$. Use this representation to evaluate $\int_2^i \bar{z}\, dz$ along the portion of the ellipse in the first quadrant.

12. In Example 3 we evaluated $\int_{1+i}^{2+4i} z^2\, dz$ along the parabola $y = x^2$ by means of the parametric representation $x = \sqrt{t}$, $y = t$. Show that the representation $x = t$, $y = t^2$ can also be used, and perform the integration using this parametrization.

13. a) Find a parametric representation of the shorter of the two arcs lying along $(x - 1)^2 + (y - 1)^2 = 1$ that connects $z = 1$ with $z = i$.

Hint: See discussion preceding Exercises 8–10 above, where parametrization of a circle is discussed.

b) Find $\int_1^i \bar{z}\, dz$ along the arc of (a), using the parametrization you have found.

14. Consider $I = \int_{0+i0}^{2+i} e^{z^2}\, dz$ taken along the line $x = 2y$. Without actually doing the integration, show that $|I| \le \sqrt{5}\, e^3$.

15. Consider $I = \int_1^i (1/\bar{z}^4)\, dz$ taken along the line $x + y = 1$. Without actually doing the integration, show that $|I| \le 4\sqrt{2}$.

16. Consider $I = \int_i^1 e^{i \operatorname{Log} \bar{z}}\, dz$ taken along the parabola $y = 1 - x^2$. Without doing the integration, show that $|I| \le e^{\pi/2}$ (1.479).

17. a) Let $g(t)$ be a complex function of the real variable t.

Express $\int_a^b g(t)\, dt$ as the limit of a sum. Using an argument similar to the one used in deriving Eq. (4.2–14), show that for $b > a$ we have

$$\left| \int_a^b g(t)\, dt \right| \le \int_a^b |g(t)|\, dt. \tag{4.2–18}$$

b) Use Eq. (4.2–18) to prove that

$$\left| \int_0^1 \sqrt{t}\, e^{it}\, dt \right| \le \frac{2}{3}.$$

4.3 CONTOUR INTEGRATION AND GREEN'S THEOREM

In the preceding section we discussed piecewise smooth curves, called contours, that connect two points A and B. If these two points happen to coincide, the resulting curve is called a *closed contour.*

DEFINITION Simple closed contour

A *simple closed contour* is a contour that creates two domains, a bounded one and an unbounded one; each domain has the contour for its boundary. The bounded domain is said to be the *interior* of the contour. □

Examples of two different closed contours, one of which is simple, are shown in Fig. 4.3–1.

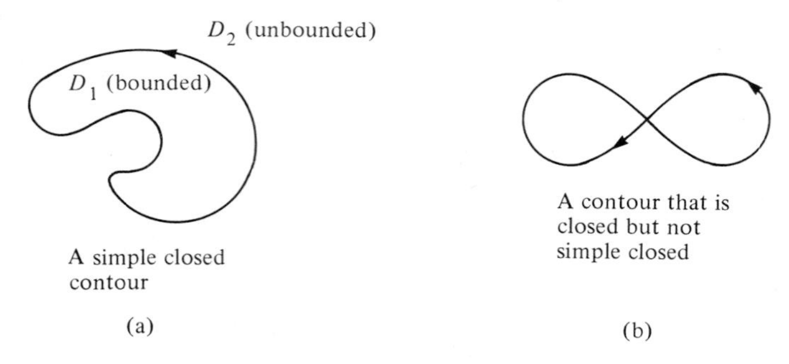

Figure 4.3–1

We will often be concerned with line integrals taken around a simple closed contour.

> The integration is said to be performed in the *positive sense* around the contour if the interior of the contour is on our left as we move along the contour in the direction of integration.

For the curve in Fig. 4.3–1(a) the positive direction of integration is indicated by the arrow.

When an integration around a simple closed contour is done in the positive direction, it will be indicated by the operator $\circlearrowleft\!\!\oint$ while an integration in the negative sense is denoted by $\circlearrowright\!\!\oint$. Note that

$$\circlearrowleft\!\!\oint f(z)\,dz = -\circlearrowright\!\!\oint f(z)\,dz,$$

$$\circlearrowleft\!\!\oint f(x, y)\,dx = -\circlearrowright\!\!\oint f(x, y)\,dx,$$

$$\circlearrowleft\!\!\oint f(x, y)\,dy = -\circlearrowright\!\!\oint f(x, y)\,dy.$$

The following important theorem, known as *Green's theorem* in the plane, applies to real functions.[†] It will, however, aid us in integrating complex analytic functions around closed contours.

THEOREM 1 Green's theorem in the plane

Let $P(x, y)$ and $Q(x, y)$ and their first partial derivatives be continuous functions throughout a region R consisting of the interior of a simple closed

---

[†] George Green (1793–1841), an Englishman, published this theorem in 1828 as part of an essay on electricity and magnetism.

contour C plus the points on C. Then

$$\oint_C P\,dx + Q\,dy = \iint_R \left(\frac{\partial Q}{\partial x} - \frac{\partial P}{\partial y}\right) dx\,dy. \quad \square \tag{4.3–1}$$

Thus Green's theorem converts a line integral around C into an integral over the area enclosed by C. A brief proof of the theorem is presented in the appendix to this chapter.

Complex line integrals can be expressed in terms of real line integrals (see Eq. 4.2–5) and it is here that Green's theorem proves useful. Consider a function $f(z) = u(x, y) + iv(x, y)$ that is not only analytic in the region R (of the preceding theorem) but whose first derivative is continuous in R. Since $f'(z) = \partial u/\partial x + i\partial v/\partial x = \partial v/\partial y - i\partial u/\partial y$, it follows that the first partial derivatives $\partial u/\partial x$, $\partial v/\partial x$, etc. are continuous in R also. Now refer to Eq. (4.2–5). We restate this equation and perform the integrations around the simple closed contour C:

$$\oint_C f(z)\,dz = \oint_C u\,dx - v\,dy + i\oint_C v\,dx + u\,dy. \tag{4.3–2}$$

We can rewrite the pair of integrals appearing on the right by means of Green's theorem. For the first integral we apply Eq. (4.3–1) with $P = u$ and $Q = -v$. For the second integral we use Eq. (4.3–1) with $P = v$ and $Q = u$. Hence

$$\oint_C u\,dx - v\,dy = \iint_R \left(-\frac{\partial v}{\partial x} - \frac{\partial u}{\partial y}\right) dx\,dy, \tag{4.3–3a}$$

$$\oint_C v\,dx + u\,dy = \iint_R \left(\frac{\partial u}{\partial x} - \frac{\partial v}{\partial y}\right) dx\,dy. \tag{4.3–3b}$$

Recalling the C–R equations $\partial u/\partial x = \partial v/\partial y$, $\partial v/\partial x = -\partial u/\partial y$, we see that both integrands on the right in Eq. (4.3–3) vanish. Thus both line integrals on the left in these equations are zero. Referring back to Eq. (4.3–2), we find that $\oint f(z)\,dz = 0$.

Our proof, which relied on Green's theorem, required that $f'(z)$ be continuous in R since otherwise Green's theorem is inapplicable. Cauchy was the first to derive our result in 1814. Green's Theorem, as such, had not yet been stated, but Cauchy used an equivalent formula. Thus he also demanded a continuous $f'(z)$.

There is a less restrictive proof, formulated in the late nineteenth century by Goursat,[†] that eliminates this requirement on $f'(z)$. The result contained in the previous equation, together with the less restrictive conditions of Goursat's derivation,

---

[†] See, for example, R. Remmert, *Theory of Complex Functions* (New York: Springer-Verlag, 1991), Section 7.1. This book also contains an interesting historical note on the Cauchy–Goursat Theorem.

are known as the Cauchy–Goursat theorem or sometimes just the Cauchy integral theorem.

THEOREM 2 Cauchy–Goursat

Let C be a simple closed contour and let $f(z)$ be a function that is analytic in the interior of C as well as on C itself. Then

$$\oint_C f(z)\,dz = 0. \quad \square \tag{4.3–4}$$

An alternative statement of the theorem is this:

Let $f(z)$ be analytic in a simply connected domain D. Then, for any simple closed contour C in D, we have $\oint_C f(z)\,dz = 0$.

The Cauchy–Goursat Theorem is one of the most important theorems in complex variable theory. One reason is that it is capable of saving us a great deal of labor when we seek to perform certain integrations. For example, such integrals as $\oint_C \sin z\,dz$, $\oint_C e^z\,dz$, $\oint_C \cosh z\,dz$ must be zero when C is any simple closed contour. The integrands in each case are entire functions.

Note that the direction of integration in Eq. (4.3–4) is immaterial since $-\oint_C f(z)\,dz = \oint_C f(z)\,dz$.

We can verify the truth of the Cauchy–Goursat theorem in some simple cases. Consider $f(z) = z^n$, where n is a nonnegative integer. Now, since z^n is an entire function, we have, according to the theorem,

$$\oint_C z^n\,dz = 0, \qquad n = 0, 1, 2, \ldots, \tag{4.3–5}$$

where C is any simple closed contour.

If n is a negative integer, then z^n fails to be analytic at $z = 0$. The theorem cannot be applied when the origin is inside C. However, if the origin lies outside C, the theorem can again be used and we have that $\oint_C z^n\,dz$ is again zero.

Let us verify Eq. (4.3–5) in a specific case. We will take C as a circle of radius r centered at the origin. Let us switch to polar notation and express C parametrically by using the polar angle θ. At any point on C, $z = re^{i\theta}$ (see Fig. 4.3–2). As θ advances from 0 to 2π or through any interval of 2π radians, the locus of z is the circle C generated in the counterclockwise sense. Note that $dz/d\theta = ire^{i\theta}$. Employing

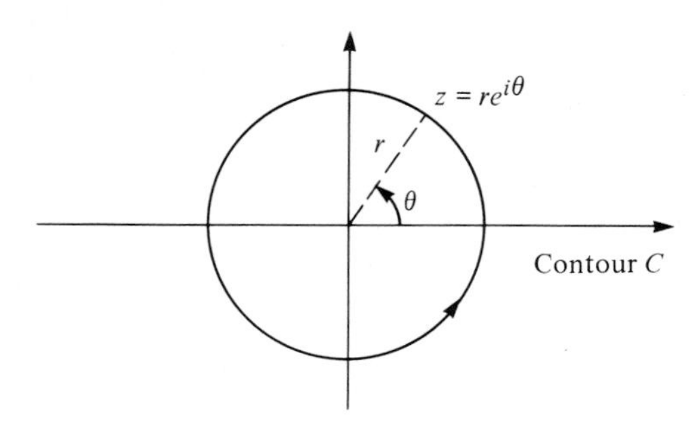

Figure 4.3–2

Eq. (4.2–10), with θ used instead of t, we have

$$\oint_{|z|=r} z^n\,dz = \int_0^{2\pi} (re^{i\theta})^n ire^{i\theta}\,d\theta = ir^{n+1}\int_0^{2\pi} e^{i(n+1)\theta}\,d\theta. \tag{4.3–6}$$

Proceeding on the assumption $n \geq 0$, we integrate Eq. (4.3–6) as follows:

$$ir^{n+1}\int_0^{2\pi} e^{i(n+1)\theta}\,d\theta = ir^{n+1}\int_0^{2\pi} \operatorname{cis}(n+1)\theta\,d\theta$$

$$= \frac{r^{n+1}}{n+1}[\cos(n+1)\theta + i\sin(n+1)\theta]_0^{2\pi} = 0. \tag{4.3–7}$$

This is precisely the result predicted by the Cauchy–Goursat theorem.

Suppose n is a negative integer and C, the contour of integration, is still the same circle. Because z^n is not analytic at $z = 0$ and $z = 0$ is enclosed by C, we *cannot* use the Cauchy–Goursat theorem. To find $\oint_C z^n\,dz$, we must evaluate the integral directly. Fortunately Eq. (4.3–6) is still valid if n is a negative integer. Moreover, Eq. (4.3–7) is still usable except, because of a vanishing denominator, when $n = -1$. Thus

$$\oint_{|z|=r} z^n\,dz = 0, \qquad n = -2, -3, \ldots. \tag{4.3–8}$$

Finally, to evaluate $\oint z^{-1}\,dz$ we employ Eq. (4.3–6) with $n = -1$ and obtain

$$\oint_{|z|=r} z^{-1}\,dz = i\int_0^{2\pi} e^0\,d\theta = 2\pi i.$$

In summary, if n is any integer,

$$\oint_{|z|=r} z^n\,dz = \begin{cases} 0, & n \neq -1, \\ 2\pi i, & n = -1. \end{cases} \tag{4.3–9}$$

An important generalization of this result is contained in Exercise 16 at the end of this section. With z_0 an arbitrary complex constant it is shown that

$$\oint_{|z-z_0|=r} (z-z_0)^n\,dz = \begin{cases} 0, & n \neq -1, \\ 2\pi i, & n = -1, \end{cases} \tag{4.3–10}$$

where the contour of integration is a circle centered at z_0.

EXAMPLE 1

The Cauchy–Goursat theorem asserts that $\oint_C z\,dz$ equals zero when C is the triangular contour shown in Fig. 4.3–3. Verify this result by direct computation.

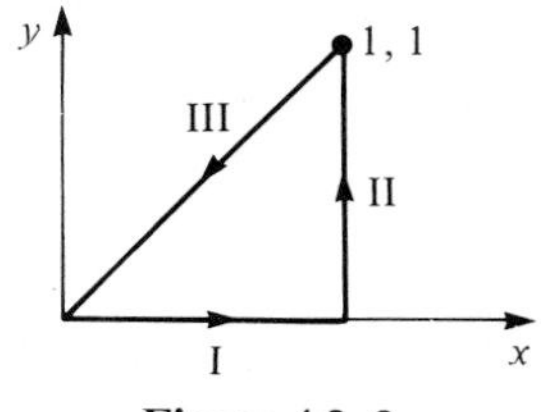

Figure 4.3–3

Solution

This kind of problem is familiar from the previous section, so we can be brief.

Along I, $y = 0$, $dz = dx$, $\int_{\text{I}} z\,dz = \int_0^1 x\,dx = 1/2$.

Along II, $x = 1$, $dz = i\,dy$, $\int_{\text{II}} z\,dz = \int_0^1 (1 + iy)i\,dy = -1/2 + i$.

Along III, $x = y$, $dz = dx + i\,dx$, $\int_{\text{III}} z\,dz = \int_1^0 (x + ix)(dx + i\,dx) = -i$.

The sum of these three integrals is zero.

Comment: In Exercise 7 of Section 4.2 we considered $\int e^z\,dz$ over paths I, II, and III of the present example. The sum of those three integrals should also be zero. ◀

There are situations in which an extension of the Cauchy–Goursat theorem establishes that two contour integrals are equal without necessarily telling us the value of either integral. The extension is as follows:

THEOREM 3 Deformation of contours

Consider two simple closed contours C_1 and C_2 such that all the points of C_2 lie interior to C_1. If a function $f(z)$ is analytic not only on C_1 and C_2 but at all points of the doubly connected domain D whose boundaries are C_1 and C_2, then

$$\oint_{C_1} f(z)\,dz = \oint_{C_2} f(z)\,dz. \quad \square$$

This theorem is easily proved. The contours C_1 and C_2 are displayed in solid line in Fig. 4.3–4(a). The domain D is shown shaded. We illustrate, using broken lines, a pair of straight line cuts, which connect C_1 and C_2. By means of these cuts we have created a pair of simple closed contours, C_U and C_L, which are drawn, slightly separated, in Fig. 4.3–4(b). The integral of $f(z)$ is now taken around C_U and also around C_L. In each case the Cauchy–Goursat theorem is applicable since $f(z)$ is analytic on and interior to both C_U and C_L. Thus

$$\oint_{C_U} f(z)\,dz = 0 \quad \text{and} \quad \oint_{C_L} f(z)\,dz = 0.$$

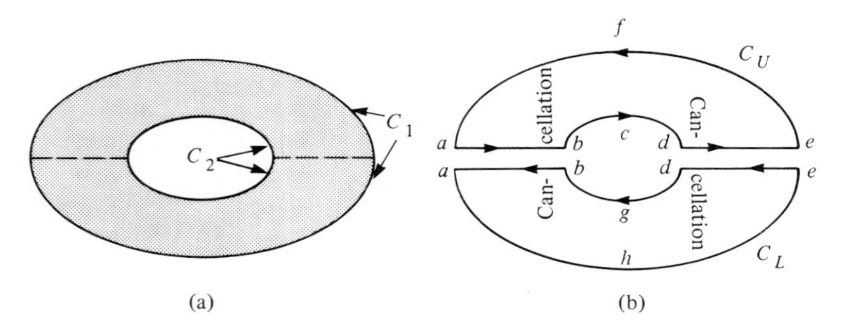

Figure 4.3–4

Adding these results yields

$$\oint_{C_U} f(z)\,dz + \oint_{C_L} f(z)\,dz = 0. \tag{4.3–11}$$

Now refer to Fig. 4.3–4(b) and observe that the integral along the straight line segment from a to b on C_U is the negative of the integral on the line from b to a on C_L. A similar statement applies to the integral from d to e on C_U and the integral from e to d on C_L.

If we write out the integrals in Eq. (4.3–11) in detail and combine those portions along the straight line segments that cancel, we are left only with integrations performed around C_1 and C_2 in Fig. 4.3–4(a). Thus (note the directions of integration)

$$\oint_{C_2} f(z)\,dz + \oint_{C_1} f(z)\,dz = 0,$$

or

$$\oint_{C_1} f(z)\,dz = -\oint_{C_2} f(z)\,dz = \oint_{C_2} f(z)\,dz.$$

We have eliminated the minus sign in the middle expression and obtained the right-hand expression by reversing the direction of integration. The preceding equation is the desired result.

Another, more general, way of stating the theorem just proved is this:

> ***THEOREM 4*** The line integrals of an analytic function $f(z)$ around each of two simple closed contours will be identical in value if one contour can be continuously deformed into the other without passing through any singularity of $f(z)$. □

In Fig. 4.3–4 we can regard C_2 as a deformed version of C_1 or vice versa. We call this approach *the principle of deformation of contours.*

Although in our derivation of Theorem 3 the contours C_1 and C_2 were assumed to be nonintersecting, Theorem 4 relaxes this restriction. Suppose in Fig. 4.3–5 $f(z)$ is analytic on and inside C_2 except possibly at points interior to C_0. Suppose also

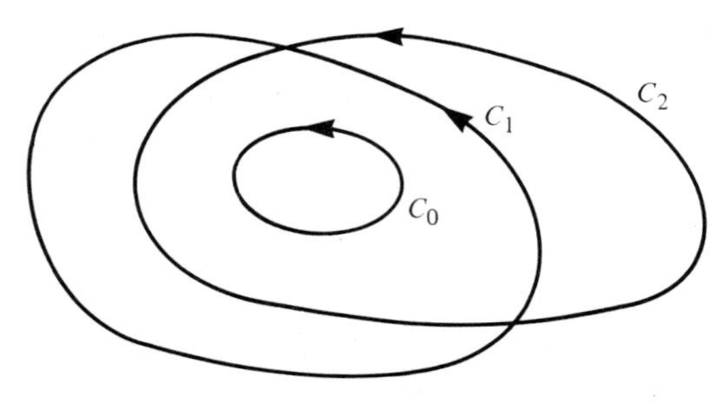

Figure 4.3–5

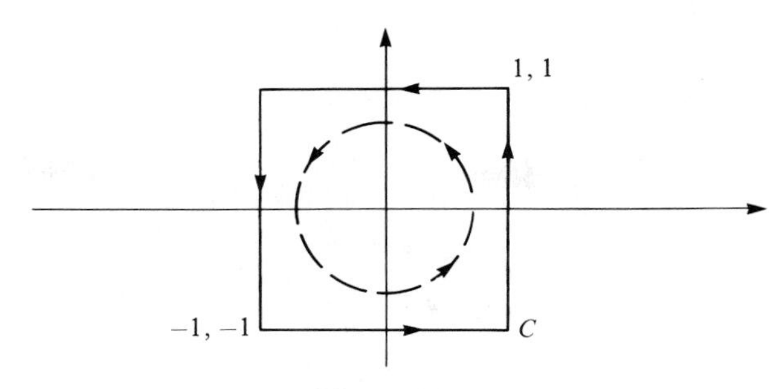

Figure 4.3–6

that $f(z)$ is analytic on and inside C_1 except possibly at points interior to C_0. Assume that C_0 lies inside C_1 and C_2 and does not intersect either contour. Note that C_2 and C_1 can intersect. By Theorem 3 we have $\oint_{C_1} f(z)\,dz = \oint_{C_0} f(z)\,dz$ and $\oint_{C_2} f(z)\,dz = \oint_{C_0} f(z)\,dz$. Thus

$$\oint_{C_2} f(z)\,dz = \oint_{C_1} f(z)\,dz.$$

EXAMPLE 2

What is the value of $\oint_C dz/z$, where the contour C is the square shown in Fig. 4.3–6?

Solution

If the integration is performed instead around the circle, drawn with broken lines, we obtain, using Eq. (4.3–9) (with $n = -1$), the value $2\pi i$. Since $1/z$ is analytic on this circle, on the given square, and at all points lying between these contours, the principle of deformation of contours applies. Thus

$$\oint_C \frac{dz}{z} = 2\pi i. \qquad ◀$$

EXAMPLE 3

Let $f(z) = \cos z/(z^2 + 1)$. The contours C_1, C_2, C_3, C_4 are illustrated in Fig. 4.3–7. Explain why the following equations are valid:

a) $$\oint_{C_1} f(z)\,dz = \oint_{C_2} f(z)\,dz;$$

b) $$\oint_{C_3} f(z)\,dz = \oint_{C_4} f(z)\,dz.$$

Solution

Except at points satisfying $z^2 + 1 = 0$, $f(z)$ is analytic. Hence, $f(z)$ is analytic in any domain not containing $z = \pm i$. If the contour C_1 is continuously deformed into the contour C_2, no singularity of $f(z)$ is crossed. Thus we establish equation (a). Similarly,

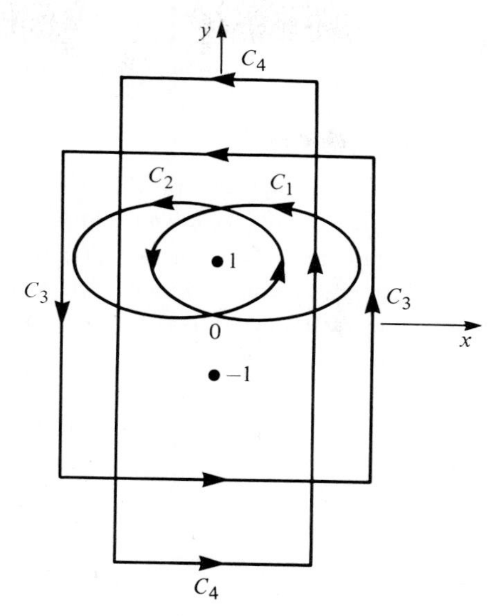

Figure 4.3–7

C_3 can be deformed into C_4 to establish (b). Note that we *cannot* conclude that $\oint_{C_2} f(z)\,dz$ equals $\oint_{C_3} f(z)\,dz$ since the domain bounded by C_2 and C_3 contains the singular point of $f(z)$ at $z = -i$.

EXERCISES

1. a) Let C be an arbitrary simple closed contour. Use Green's theorem to find a simple interpretation of the line integral $(1/2)\oint_C(-y\,dx + x\,dy)$.

b) Consider $\oint e^y\,dx - e^x\,dy$ performed around the square with corners at (0, 0), (0, 1), (1, 0), (1, 1). Evaluate this integral by doing an equivalent area integral inside the square.

c) Suppose you know the area enclosed by a simple closed contour C. Show with the aid of Green's Theorem that you can easily evaluate $\oint \bar{z}\,dz$ around C.

To which of the following integrals is the Cauchy–Goursat Theorem directly applicable?

2. $\displaystyle\oint_{|z|=1} \frac{\cos z}{z+2}\,dz$ **3.** $\displaystyle\oint_{|z+2|=2} \frac{\cos z}{z+2}\,dz$ **4.** $\displaystyle\oint_{|z-1|=4} \frac{\cos z}{z+2}\,dz$

5. $\displaystyle\oint_{|z+i|=1} \operatorname{Log} z\,dz$ **6.** $\displaystyle\oint_{|z-1-i|=1} \operatorname{Log} z\,dz$ **7.** $\displaystyle\oint_{|z|=\pi} \frac{dz}{1+e^z}$

8. $\displaystyle\oint_{|z|=3} \frac{dz}{1-e^z}$ **9.** $\displaystyle\oint_{|z|=b} \frac{dz}{z^2+bz+1},\ 0 < b < 1$

10. $\displaystyle\oint \frac{dz}{\sin z + e^z - 1}$ around the square with corners at $\pm(1 \pm i)$

11. In the discussion of Green's theorem in the appendix to this chapter it is shown that if $P(x, y)$ and $Q(x, y)$ are a pair of functions with continuous partial derivatives $\partial P/\partial y$ and $\partial Q/\partial x$ inside some simply connected domain D and if $\oint_C P\,dx + Q\,dy = 0$ for every simple closed contour in D, then $\partial Q/\partial x = \partial P/\partial y$ in D.

Let $f(z) = u(x, y) + iv(x, y)$ be a function such that the first partial derivatives of u and v are continuous in a simply connected domain D. Given that $\oint f(z)\,dz = 0$ for every simple closed contour in D, use the preceding result to show that $f(z)$ must be analytic in D.

This is a converse of the Cauchy–Goursat theorem. There is another derivation that eliminates the requirement that the partial derivatives be continuous in D. Only $u(x, y)$ and $v(x, y)$ are assumed continuous. The resulting converse of the Cauchy–Goursat theorem is known as *Morera's theorem.*

Prove the following results by means of the Cauchy–Goursat Theorem. Begin with $\oint e^z\,dz$ performed around $|z| = 1$. Use the parametric representation $z = e^{i\theta}$, $0 \le \theta \le 2\pi$. Separate your equation into real and imaginary parts.

12. $\displaystyle\int_0^{2\pi} e^{\cos\theta}[\cos(\sin\theta + \theta)]\,d\theta = 0,$

13. $\displaystyle\int_0^{2\pi} e^{\cos\theta}[\sin(\sin\theta + \theta)]\,d\theta = 0.$

Prove that the following identities hold for any integer $n \ge 0$.

Hint: Use the contour and technique of the preceding problem but a different integrand.

14. $\displaystyle\int_0^{2\pi} e^{\sin n\theta}\cos(\theta - \cos n\theta)\,d\theta = 0,$

15. $\displaystyle\int_0^{2\pi} e^{\sin n\theta}\sin(\theta - \cos n\theta)\,d\theta = 0.$

16. Let n be any integer, r a positive real number, and z_0 a complex constant. Show that, for $r > 0$,

$$\oint_{|z-z_0|=r} (z - z_0)^n\,dz = \begin{cases} 0, & n \ne -1, \\ 2\pi i, & n = -1. \end{cases}$$

Hint: Refer to the derivation of Eq. (4.3–9) and follow a similar procedure. Consider the change of variable $z = z_0 + re^{i\theta}$ indicated in Fig. 4.3–8.

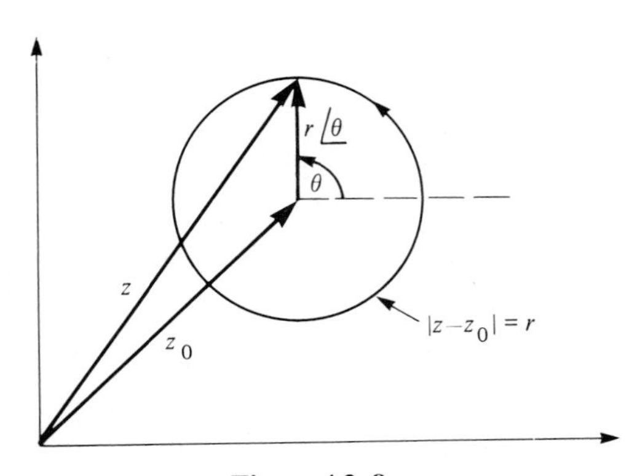

Figure 4.3–8

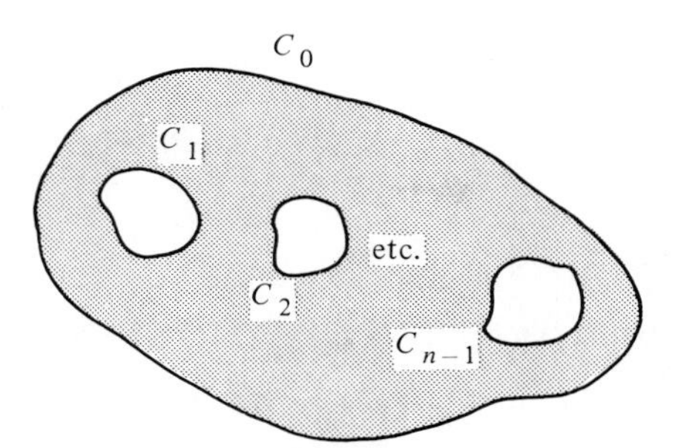

Figure 4.3–9

Evaluate the following integrals. The result contained in the preceding problem as well as the principle of deformation of contours may be useful. The contour is the square with corners at $z = 0, 3i, 3, 3 + 3i$.

17. $\oint \frac{dz}{z - 1 - i}$

18. $\oint \frac{dz}{z + 1 + i}$

19. $\oint \frac{dz}{(z - 1 - i)^5}$

20. $\oint \frac{dz}{(z + 1 + i)^5}$

21. $\oint \left(\frac{dz}{2z - 1 - i}\right)$

22. $\oint \left(\frac{1}{z - 1 - i} + \frac{1}{2 - i - z}\right) dz$

23. $\oint ((z - 1 - i)^{-1} + z - 1 - i)^2 \, dz$

24. $\oint ((z - 1 - i)^{-1} + z - 1 - i)^3 \, dz$

25. Show that

$$\oint_{|z-3|=2} \frac{\operatorname{Log} z}{(z + 1)(z - 3)} dz = \oint_{|z-3|=2} \frac{\operatorname{Log} z}{4(z - 3)} dz.$$

Hint: Write $1/[(z + 1)(z - 3)]$ as a sum of partial fractions.

26. Consider the n-tuply connected domain D whose nonoverlapping boundaries are the simple closed contours $C_0, C_1, \ldots, C_{n-1}$ as shown in Fig. 4.3–9.† Let $f(z)$ be a function that is analytic in D and on its boundaries. Show that

$$\oint_{C_0} f(z)\, dz = \oint_{C_1} f(z)\, dz + \oint_{C_2} f(z)\, dz + \cdots + \oint_{C_{n-1}} f(z)\, dz.$$

Hint: Consider the derivation of the principle of deformation of contours. Make a set of cuts similar to those made in Fig. 4.3–4 in order to link up the boundaries.

---

† An n-tuply connected domain has $n - 1$ holes. See Section 1.5.

27. Use the result derived in Exercise 26 to show

$$\oint_{|z|=2} \frac{\sin z}{(z^2-1)}\,dz = \oint_{|z-1|=1/2} \frac{\sin z}{(z^2-1)}\,dz + \oint_{|z+1|=1/2} \frac{\sin z}{(z^2-1)}\,dz.$$

4.4 PATH INDEPENDENCE AND INDEFINITE INTEGRALS

The Cauchy–Goursat theorem is a useful tool when we must integrate an analytic function around a closed contour. When the contour is not closed, there exist techniques, derivable from this theorem, that can assist us in evaluating the integral. For example, we can prove the following:

THEOREM 4 Principle of path independence

Let $f(z)$ be a function that is analytic throughout a simply connected domain D, and let z_1 and z_2 lie in D. Then, if we use contours lying in D, the value of $\int_{z_1}^{z_2} f(z)\,dz$ will not depend on the particular contour used to connect z_1 and z_2. □

The preceding theorem is sometimes known as the *principle of path independence.* It is really just a restatement of the Cauchy–Goursat theorem. To establish this principle we will consider two nonintersecting contours C_1 and C_2, each of which connects z_1 and z_2. Each contour is assumed to lie within the simply connected domain D in which $f(z)$ is analytic. We will show that

$$\underset{\text{along } C_1}{\int_{z_1}^{z_2} f(z)\,dz} = \underset{\text{along } C_2}{\int_{z_1}^{z_2} f(z)\,dz}. \tag{4.4–1}$$

We begin by reversing the sense of integration along C_1 in Eq. (4.4–1) and placing a minus sign in front of the integral to compensate. Thus

$$-\underset{\text{along } C_1}{\int_{z_2}^{z_1} f(z)\,dz} = \underset{\text{along } C_2}{\int_{z_1}^{z_2} f(z)\,dz},$$

or, with an obvious rearrangement,

$$0 = \underset{\text{along } C_2}{\int_{z_1}^{z_2} f(z)\,dz} + \underset{\text{along } C_1}{\int_{z_2}^{z_1} f(z)\,dz}.$$

The preceding merely states that the line integral of $f(z)$ taken around the closed loop formed by C_1 and C_2 (see Fig. 4.4–1) is zero. The correctness of this result follows directly from the Cauchy–Goursat theorem.

Although we have assumed that C_1 and C_2 in Eq. (4.4–1) do not intersect, a slightly different derivation dispenses with this restriction. Exercise 1 in this section treats the case of a single intersection.

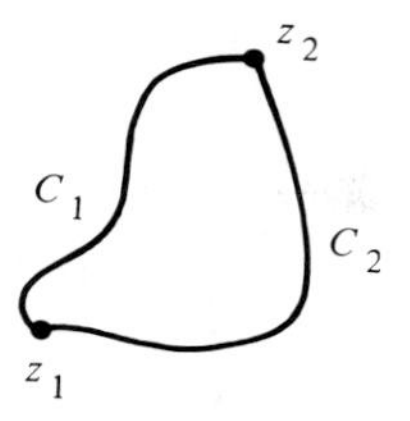

Figure 4.4–1

Since *any* two contours (in domain D) that connect z_1 and z_2 can be used in deriving Eq. (4.4–1), it follows that all such paths lying in D must yield the same result. Thus the value of $\int_{z_1}^{z_2} f(z)\,dz$ is independent of the path connecting z_1 and z_2 as long as that path lies within a simply connected domain in which $f(z)$ is analytic.

EXAMPLE 1

Compute $\int_1^i (1/z)\,dz$ where the integration is along the arc C_1, which is the portion of $x^4 + y^4 = 1$ lying in the first quadrant (see Fig. 4.4–2).

Solution

We could try to perform the integration using Eq. (4.2–5). However, the manipulations quickly become rather tedious. Since $1/z$ is analytic except at $z = 0$ we can switch to the quarter circle C_2 described by $|z| = 1$, $0 \le \arg z \le \pi/2$. This arc, shown in Fig. 4.4–2, also connects 1 with i. Note that the domain lying to the right of the line $y = -x$ is one containing C_1, C_2 and is a domain of analyticity of $1/z$. Hence we have $\int_{C_1} (1/z)\,dz = \int_{C_2} (1/z)\,dz$.

To integrate along C_2 we make the change of variable $z = e^{i\theta}$, and integrate on the parameter θ as it varies from 0 to $\pi/2$. We have used this technique before (see Fig. 4.4–2). Recall that $dz/d\theta = ie^{i\theta}$. We obtain

$$\int_{C_2} \frac{1}{z}\,dz = \int_0^{\pi/2} \frac{1}{e^{i\theta}} ie^{i\theta}\,d\theta = \frac{\pi}{2}\,i,$$

which is the answer to the given problem. ◀

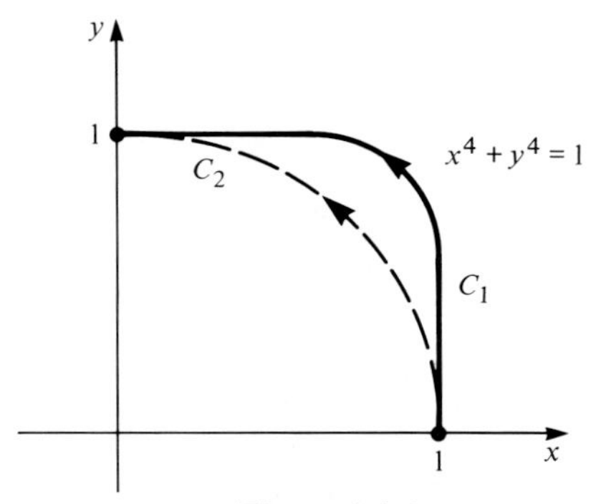

Figure 4.4–2

In elementary calculus integration is generally performed by our recognizing that the integrand $f(x)$ is the derivative of some particular function $F(x)$. Then $F(x)$ is evaluated at the limits of integration. Thus

$$\int_{x_1}^{x_2} f(x)\,dx = \int_{x_1}^{x_2} \left(\frac{dF}{dx}\right) dx = \int_{x_1}^{x_2} dF = F(x_2) - F(x_1).$$

To cite a specific example:

$$\int_1^4 x^2\,dx = \int_1^4 \frac{d}{dx}\left(\frac{x^3}{3}\right) dx = \int_1^4 d\left(\frac{x^3}{3}\right) = \frac{x^3}{3}\bigg|_1^4 = \frac{63}{3}.$$

Does a similar procedure work for complex line integrals? We shall see that with certain restrictions it does.

Let $F(z)$ be analytic in a domain D. Assume $dF/dz = f(z)$ in D. Consider $\int_{z_1}^{z_2} f(z)\,dz$ integrated along a smooth arc C lying in D and connecting z_1 with z_2. We will assume that C has a parametric representation of the form $z(t) = x(t) + iy(t)$, where $t_1 \le t \le t_2$, and we will assume that dz/dt exists in this same interval.

Observe that $z(t_1) = z_1$ and $z(t_2) = z_2$. Now from Eq. (4.2–10) we have

$$\int_{z_1}^{z_2} f(z)\,dz = \int_{t_1}^{t_2} f(z(t))\frac{dz}{dt}\,dt.$$

We can replace $f(z(t))$ on the right by dF/dz. Hence,

$$\int_{z_1}^{z_2} f(z)\,dz = \int_{t_1}^{t_2} \frac{dF}{dz}\frac{dz}{dt}\,dt. \tag{4.4–2}$$

The expression on the right, $(dF/dz)(dz/dt)$ is, from the chain rule of differentiation, merely dF/dt. Thus

$$\int_{z_1}^{z_2} f(z)\,dz = \int_{t_1}^{t_2} \frac{dF}{dt}\,dt.$$

The integrand on the right, dF/dt, is a complex function of the real variable t. We can perform this integration by the techniques familiar to us from real calculus, i.e.,

$$\int_{t_1}^{t_2} \frac{dF}{dt}\,dt = \int_{t_1}^{t_2} dF = F[z(t_2)] - F[z(t_1)] = F(z_2) - F(z_1).$$

Thus, we have

$$\int_{z_1}^{z_2} f(z)\,dz = F(z_2) - F(z_1), \tag{4.4–3}$$

If the contour C is not restricted to being a smooth arc but is permitted to be a piecewise smooth curve, the result in Eq. (4.4–3) is still valid. However, a slightly more elaborate proof, not presented here, is required since dz/dt may fail to exist at those points along C where smooth arcs are joined together.

The preceding discussion is summarized in the following theorem.

THEOREM 5 Integration of functions that are the derivatives of analytic functions

Let $F(z)$ be analytic in a domain D. Let $dF/dz = f(z)$ in D. Then, if z_1 and z_2 are in D,

$$\int_{z_1}^{z_2} f(z)\,dz = F(z_2) - F(z_1), \tag{4.4–4}$$

where the integration can be performed along any contour in D that connects z_1 and z_2. □

Thus within the constraints of the theorem the conventional rules of integration apply, and ordinary tables of integrals (which are based on such rules) can be employed. For example, the theorem justifies the following evaluation.

$$\int_{1+i}^{2+2i} z^2\,dz = \int_{1+i}^{2+2i} \frac{d}{dz}\frac{z^3}{3}\,dz = \frac{z^3}{3}\bigg|_{1+i}^{2+2i} = \frac{1}{3}[(2+2i)^3 - (1+i)^3]$$

Since $z^3/3$ is an entire function, the path of integration was not, and need not, be specified.

EXAMPLE 2

Evaluate $\int_{-i}^{+i} 1/z\,dz$ along the contour C shown in Fig. 4.4–3.

Solution

Recall from Chapter 3 that $(d/dz)\log z = 1/z$. The logarithm is a multivalued function. In order to specify a particular analytic branch of the log, let us call it $F(z)$, we must employ a branch cut. Any branch of the logarithm whose branch cut does not intersect C can be used to perform the given integration. A possible cut is shown by the solid bold line in the figure. Since the branch cut contains all the singular points of $F(z)$, the contour C lies in a domain of analyticity of $F(z)$.

Using our analytic branch of $\log z$, we have

$$\int_{-i}^{+i} \frac{1}{z}\,dz = \log z\bigg|_{-i}^{+i} = \text{Log}|z| + i\arg z\bigg|_{-i}^{+i}.$$

Note that $\text{Log}|i| = \text{Log}|-i| = 0$. Thus $\int_{-i}^{+i} 1/z\,dz = i(\arg i - \arg(-i))$. Along contour C the argument of z varies continuously. At $-i$ the argument of z is $-\pi/2 + 2k\pi$,

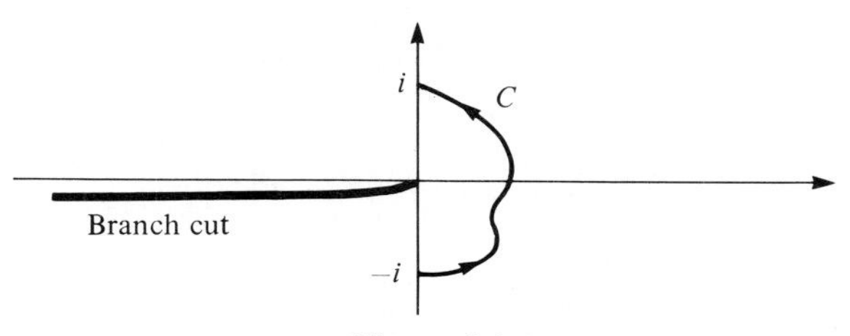

Figure 4.4–3

where k is an integer, while at $+i$ the argument becomes $\pi/2 + 2k\pi$, where k must have the *same value* in both cases. Hence,

$$\int_{-i}^{+i} \frac{1}{z}\,dz = i\left(\frac{\pi}{2} + 2k\pi\right) - i\left(-\frac{\pi}{2} + 2k\pi\right) = \pi i. \quad ◀$$

Comment: The solution of the problems contained in Example 2 depends on our finding or knowing a function $F(z)$ that is analytic at every point on the contour of integration and that satisfies $dF/dz = f(z)$ everywhere on the contour. Here $f(z)$ is the function to be integrated. *If $f(z)$ fails to be analytic at one or more points on the contour of integration, then it is impossible to find an $F(z)$ satisfying the required equation.* This fact is not now obvious but is proved in Section 4.5, where it is shown that if $F(z)$ is analytic at a point then all its derivatives must be analytic functions at this point. It is therefore futile to seek an analytic function $F(z)$ whose derivative $f(z)$ is not analytic. If in Example 2 we replace $1/z$ by $1/\bar{z}$ (which is nowhere analytic) the technique of Example 2 would not be applicable. To complete the solution we would need to know a mathematical formula for the contour C. The integration could then, in principle, be carried out with the aid of Eq. (4.2–5).

The reader of this text should recall the fundamental theorem of (real) calculus, i.e., if $f(x)$ is a continuous function of x, then

$$\frac{d}{dx}\int_a^x f(w)\,dw = f(x).$$

Here w is a dummy variable and a is a constant. If $F(x) = \int_a^x f(w)\,dw$, then $dF/dx = f(x)$. The theorem relates integration and differentiation, asserting that integration is the inverse operation of differentiation.

There is a corresponding statement for integrations involving *analytic* functions of a complex variable. Let w be a (dummy) complex variable and $f(w)$ be analytic in a simply connected domain D in the w-plane. Let a and z be two points in D. We regard z as a variable. From Theorem 4, $F(z) = \int_a^z f(w)\,dw$ is a function that is independent of the contour C used to connect a and z provided C lies in D. We will show that $dF/dz = f(z)$.

Refer to Fig. 4.4–4. Now $F(z) = \int_a^z f(w)\,dw$, where the integration is along C. Furthermore, $F(z + \Delta z) = \int_a^{z+\Delta z} f(w)\,dw$, where the path goes from a to z along C; then from z to $z + \Delta z$ along the straight line path, that is, along the vector Δz. Consider now

$$g(\Delta z) = \left|\frac{F(z + \Delta z) - F(z)}{\Delta z} - f(z)\right|. \tag{4.4–5}$$

Recall that

$$\frac{dF}{dz} = \lim_{\Delta z \to 0} \frac{F(z + \Delta z) - F(z)}{\Delta z}.$$

If we can show that $\lim_{\Delta z \to 0} g(\Delta z) = 0$ in Eq. (4.4–5), then it must be true that $dF/dz = f(z)$.

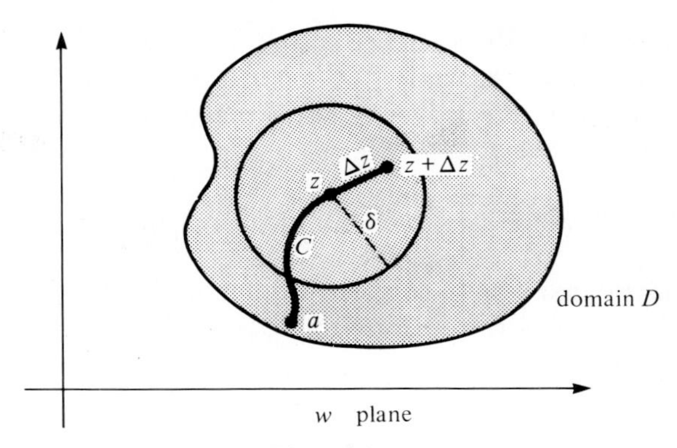

Figure 4.4–4

Notice that

$$F(z + \Delta z) - F(z) = \int_z^{z+\Delta z} f(w)\, dw, \tag{4.4–6}$$

where the integration is along Δz in Fig. 4.4–4. With Eq. (4.4–6) used in Eq. (4.4–5) we obtain

$$g(\Delta z) = \left|\frac{1}{\Delta z}\right| \left| \int_z^{z+\Delta z} f(w)\, dw - f(z)\Delta z \right|. \tag{4.4–7}$$

We have factored out $1/|\Delta z|$. Since $\Delta z = \int_z^{z+\Delta z} dw$ we can rewrite Eq. (4.4–7) as

$$g(\Delta z) = \left|\frac{1}{\Delta z}\right| \left| \int_z^{z+\Delta z} f(w)\, dw - f(z) \int_z^{z+\Delta z} dw \right|,$$

or

$$g(\Delta z) = \left|\frac{1}{\Delta z}\right| \left| \int_z^{z+\Delta z} [f(w) - f(z)]\, dw \right|. \tag{4.4–8}$$

Because $f(z)$ is analytic it must be continuous. Thus (see Section 2.2) if we have any positive number ε, there must exist a circle of radius δ centered at z such that, for values of w inside this circle,

$$|f(w) - f(z)| < \varepsilon. \tag{4.4–9}$$

We can choose Δz small enough so that the point $z + \Delta z$ lies inside this circle. Along the straight line path of integration in Fig. (4.4–4), Eq. (4.4–9) is satisfied.

We now apply the ML inequality (Section 4.2) to the integral of Eq. (4.4–8). From Eq. (4.4–9) we see that we can let $M = \varepsilon$. The length of the path $L = |\Delta z|$. Thus

$$g(\Delta z) \le \left|\frac{1}{\Delta z}\right| \varepsilon |\Delta z| = \varepsilon.$$

Since ε can be made arbitrarily small (we shrink $|\Delta z|$ to keep $z + \Delta z$ inside the circle of Fig. 4.4–4), it must follow that $\lim_{\Delta z \to 0} g(\Delta z) = 0$, and as Eq. (4.4–5) now indicates, we have $dF/dz = f(z)$. We have proved not only that dF/dz exists but have also found its value. In summary:

THEOREM 6 Fundamental theorem of the calculus of analytic functions

If $f(w)$ is analytic in a simply connected domain D of the w-plane, then the integral $\int_a^z f(w)\,dw$ performed along any contour in D defines an analytic function of z satisfying

$$\frac{d}{dz}\int_a^z f(w)\,dw = f(z). \quad \square \tag{4.4–10}$$

One says, as in real calculus, that if $dF/dz = f(z)$, then $F(z)$ is an *antiderivative* of $f(z)$. Thus, from Eq. (4.4–10), $\int_a^z f(w)\,dw$ is an antiderivative of $f(z)$. Of course if $dF/dz = f(z)$, then $F_1(z) = F(z) + C$ (a constant) will also have derivative $f(z)$. Hence $f(z)$ has an infinite number of antiderivatives. They differ by constant values. The indefinite integral $\int f(z)\,dz$ is used to mean all the possible antiderivatives of $f(z)$. It contains, as in real calculus, an arbitrary additive constant. For example, since $(d/dz)(\sin z + C) = \cos z$, all the antiderivatives of $\cos z$ are contained in the statement $\int \cos z\,dz = \sin z + C$; i.e., the antiderivatives are of the form $\sin z + C$.

The value of the constant for a specific antiderivative $\int_a^z f(w)\,dw$ is established by the lower limit of integration, as shown in Example 3 to follow.

Note that the identity involving antiderivatives that the reader learned in real calculus,

$$\int u\,dv = uv - \int v\,du \tag{4.4–11}$$

(integration by parts), applies equally well in complex variable theory. The identity is derived from the formula for the derivative of the product uv, and that expression holds for both real and complex functions and variables.

EXAMPLE 3

a) Find the antiderivatives of ze^z.
b) Use the result of (a) to find $\int_i^z we^w\,dw$.
c) Verify Theorem 6 for the integral in part (b).
d) Use the result of (a) to find $\int_i^1 ze^z\,dz$.

Solution

Part (a): To determine $\int ze^z\,dz$ we use integration by parts. Thus, applying Eq. (4.4–11) with $u = z$, $dv = e^z\,dz$, $v = e^z$, we have $\int ze^z\,dz = ze^z - \int e^z\,dz = ze^z - e^z + C = F(z)$.

Part (b): Using the result of (a) we have $\int_i^z we^w\,dw = ze^z - e^z + C$. To evaluate C we observe that the left side of this equation is zero when $z = i$. The right side will agree with the left at $z = i$ if we put $C = -ie^i + e^i$. Thus

$$\int_i^z we^w\,dw = ze^z - e^z - ie^i + e^i.$$

Part (c): Theorem 6 asserts that

$$\frac{d}{dz}\left(\int_i^z we^w\,dw\right) = ze^z.$$

Using the value for the preceding integral, $ze^z - e^z - ie^i + e^i$, and differentiating with respect to z, we see that this is true. Since ze^z is entire (we^w is entire in the w-plane), the theorem also tells us that $\int_i^z we^w\,dw$ is analytic throughout the z-plane, and indeed $ze^z - e^z - ie^i + e^i$ is an entire function.

Part (d): With $F(z) = ze^z - e^z + C$ from part (a) we may now use Theorem 5 directly. Since $dF/dz = ze^z$ throughout the z-plane, we have

$$\int_i^1 ze^z\,dz = ze^z - e^z + C\Big|_i^1 = -ie^i + e^i.$$ ◀

EXERCISES

1. In Fig. 4.4–5 contour C_1 (solid line) and contour C_2 (broken line) each connect points z_1 and z_2. The contours also intersect at one other point, designated z_3. Let $f(z)$ be analytic in a simply connected domain containing C_1 and C_2. Show that

$$\underset{\text{along } C_1}{\int_{z_1}^{z_2} f(z)\,dz} = \underset{\text{along } C_2}{\int_{z_1}^{z_2} f(z)\,dz}.$$

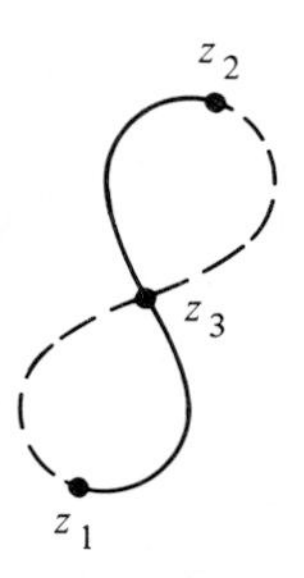

Figure 4.4–5

Evaluate the following integrals along the curve $y = \sqrt{x}$.

2. $\displaystyle\int_{1+i}^{9+3i} e^{2z}\,dz$ **3.** $\displaystyle\int_{1+i}^{9+3i} (z^3 + z)\,dz$

4. $\displaystyle\int_{1+i}^{9+3i} e^{z}\sin(e^{z})\,dz$ **5.** $\displaystyle\int_{1+i}^{9+3i} z\cos(z)\,dz$

6. $\displaystyle\int_{1+i}^{9+3i} e^{z}\cos(z)\,dz$ **7.** $\displaystyle\int_{1+i}^{9+3i} z^{-2}\,dz$

8. a) What, if anything, is incorrect about the following two integrations? The integrals are both along the line $y = x$.

$$\int_{0+i0}^{1+i} z\,dz = \frac{z^2}{2}\bigg|_{0+i0}^{1+i} = \frac{(1+i)^2}{2} = i$$

$$\int_{0+i0}^{1+i} \bar{z}\,dz = \frac{\bar{z}^2}{2}\bigg|_{0+i0}^{1+i} = \frac{(1-i)^2}{2} = -i$$

b) What is the correct numerical value of each of the above integrals?

9. Find the value of $\int_e^i \operatorname{Log} z\,dz$ taken along the line connecting $z = e$ with $z = i$. Why is it necessary to specify the contour?

10. Find $\int_{1+i}^{-1-i} \operatorname{Log} z/z\,dz$, where the integral is along a contour not intersecting the branch cut for $\operatorname{Log} z$.

11. Find $\int_1^i z^{1/2}\,dz$. The principal branch of $z^{1/2}$ is used. The contour does not pass through any point satisfying $y = 0$, $x \le 0$.

12. Find $\int_1^i z^{1/2}\,dz$. The branch of $z^{1/2}$ used equals -1 when $z = 1$. The branch cut lies along $y = 0$, $x \le 0$, and the contour does not pass through the branch cut.

13. Consider contours C_1 and C_2 shown in Fig. 4.4–6. We can use the result derived in Example 2 of this section to show that along C_1, $\int_{-i}^{+i} 1/z\,dz = \pi i$. Can we use the principle of path independence to show that along C_2, $\int_{-i}^{+i} 1/z\,dz = \pi i$? If not, what is the correct value of this integral?

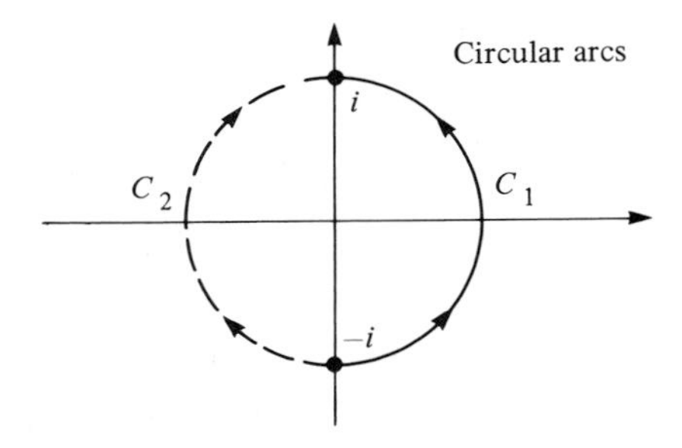

Figure 4.4–6

14. a) Find $\int_0^{2i} dz/(z - i)$ taken along the arc satisfying $|z - i| = 1$, $\text{Re}\, z \geq 0$.

b) Repeat part (a) with the same limits, but use the arc $|z - i| = 1$, $\text{Re}\, z \leq 0$.

15. Let z_1 and z_2 be a pair of arbitrary points in the complex plane. Contours C_1 and C_2 each connect points z_1 and z_2. The contours do not otherwise intersect, and neither passes through $z = 0$. Explain why

$$\underset{\text{along } C_1}{\int_{z_1}^{z_2} \frac{1}{z^2}\, dz} = \underset{\text{along } C_2}{\int_{z_1}^{z_2} \frac{1}{z^2}\, dz}.$$

Consider two cases:

a) $z = 0$ does not belong to the domain whose boundaries are C_1 and C_2 (see Fig. 4.4–7).

b) $z = 0$ does belong to the domain whose boundaries are C_1 and C_2 (see Fig. 4.4–8).

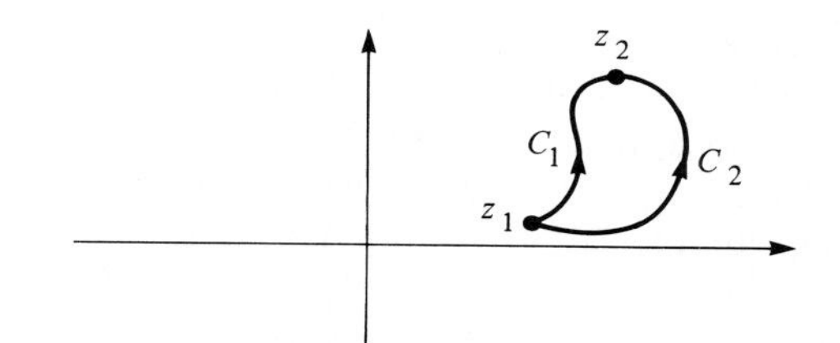

Figure 4.4–7

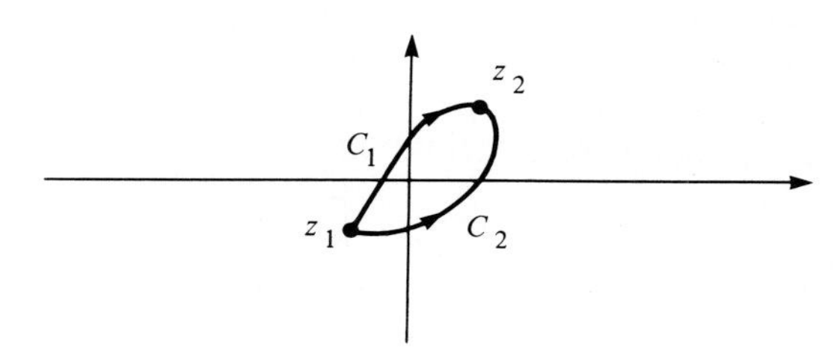

Figure 4.4–8

16. In elementary calculus the reader learned the *Mean Value Theorem*: If $f(x)$ is continuous for $a \leq x \leq b$, then there exists a number x_1, where $a < x_1 < b$, such that

$$\int_a^b f(x)\, dx = f(x_1)(b - a).$$

Show that this theorem does not have a counterpart for complex line integrals by doing the following:

a) Show that $\int_1^i (1/z^2)\, dz = 1 + i$, where the integral is along $x + y = 1$.

b) Show that there is no point z_1 along the contour of integration satisfying $(1/z_1^2)(i - 1) = 1 + i$.

17. a) Find the antiderivatives of $1/(z - i)^2$ in the domain $\operatorname{Re} z > 0$.

b) Find the specific antiderivative that is zero when $z = 1 + i$.

18. a) Find the antiderivatives of $1/(z^2 + 1)$ in the domain $|\operatorname{Im} z| < 1$.

b) Find the specific antiderivative that equals $\pi/4$ when $z = 1$.

19. a) Show that any branch of z^α (α is any number, real or complex) has antiderivative $zz^\alpha/(\alpha + 1)$ in the domain of analyticity of z^α. See Section 3.6.

b) Using principal values for all functions in the integrand, find $\int_{-i}^{i} (z^i - i^z)\, dz$. Employ a contour of integration not passing through $z = 0$ or the negative real axis.

4.5 THE CAUCHY INTEGRAL FORMULA AND ITS EXTENSION

Consider a simple closed contour C (see, for example, Fig. 4.5–1). Starting with the Cauchy integral theorem, we will derive a remarkable fact. Suppose $f(z)$ is a function that is analytic on C and at all points belonging to the interior of C. Then, if we know the values of $f(z)$ only on C, we have sufficient information to find $f(z)$ at any point z_0 in the interior of C. The means for finding $f(z_0)$ is the Cauchy integral formula, which we will now obtain. The impatient reader may wish to skip directly to the simple and nonrigorous formal derivation outlined in Exercise 1 at the end of this section.

To derive the result rigorously we proceed as follows: Let C_0 be a circle, centered at z_0, of radius r. The value of r is sufficiently small so that C_0 lies entirely within C. The function $f(z)/(z - z_0)$ is analytic at all points for which $f(z)$ is analytic except $z = z_0$. Thus $f(z)/(z - z_0)$ is analytic on C_0, C, and at all points lying outside C_0 but inside C. Using the principle of deformation of contours, we can assert that

$$\oint_C \frac{f(z)}{z - z_0}\, dz = \oint_{C_0} \frac{f(z)}{z - z_0}\, dz. \tag{4.5–1}$$

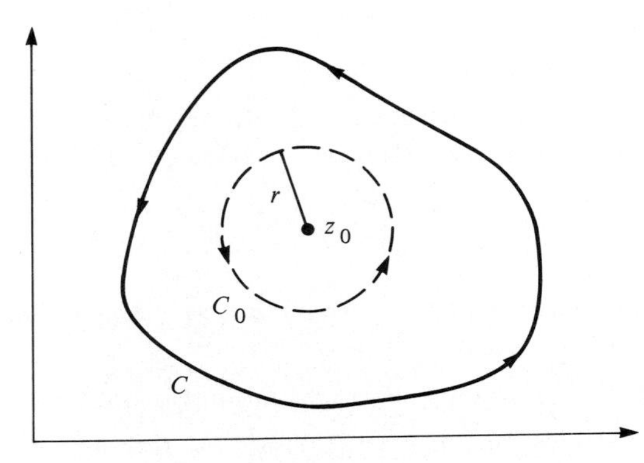

Figure 4.5–1

Let us show that the expression on the right equals $2\pi i f(z_0)$. Recall from Eq. (4.3–10) that $\oint_{C_0} dz/(z - z_0) = 2\pi i$. Thus

$$2\pi i f(z_0) = \oint_{C_0} \frac{dz}{(z - z_0)} f(z_0) = \oint_{C_0} \frac{f(z_0)}{z - z_0}\, dz. \tag{4.5–2}$$

Since the quantity $f(z_0)$ is a constant, it was taken under the integral sign on the right in Eq. (4.5–2). If we subtract both the left side and the right side of Eq. (4.5–2) from the right-hand integral in Eq. (4.5–1), we obtain

$$\oint_{C_0} \frac{f(z)}{z - z_0}\, dz - 2\pi i f(z_0) = \oint_{C_0} \frac{f(z) - f(z_0)}{z - z_0}\, dz. \tag{4.5–3}$$

Our goal is to show that the integral on the right is zero. The value of this integral, although still unknown, must, by the principle of deformation of contours, be independent of r, the radius of C_0.

The ML inequality of Section 4.2 can be applied to obtain a bound on the magnitude of the right-hand side of Eq. (4.5–3). The length of path, L, is here merely the circumference of the circle C_0, that is, $2\pi r$. The quantity M must have the property that

$$\frac{|f(z) - f(z_0)|}{|z - z_0|} \le M, \qquad \text{for } z \text{ on } C_0.$$

On the contour of integration we have $|z - z_0| = r$.

Since $f(z)$ is continuous at z_0, we can apply the definition of continuity (see Section 2.2) and assert that, given a positive number ε, there exists a positive number δ such that $|f(z) - f(z_0)| < \varepsilon$ for $|z - z_0| < \delta$. If the radius r of C_0 is chosen to be less than δ, it follows that on C_0 we have $|f(z) - f(z_0)| < \varepsilon$. Hence, we have on C_0 that

$$\left|\frac{f(z) - f(z_0)}{z - z_0}\right| = \left|\frac{f(z) - f(z_0)}{r}\right| < \frac{\varepsilon}{r},$$

and we can take M in Eq. (4.5–4) as ε/r.

Now, knowing M and L, we apply the ML inequality to the right side of Eq. (4.5–3) and obtain

$$\left|\oint_{C_0} \frac{f(z) - f(z_0)}{r}\, dz\right| \le \frac{\varepsilon}{r} 2\pi r = 2\pi\varepsilon. \tag{4.5–5}$$

Since ε on the right in Eq. (4.5–5) can be made arbitrarily small, the absolute value of the integral on the left can likewise be made arbitrarily small. Reducing ε merely implies that we must shrink the radius r of C_0.

We observed earlier that the value of the integral within the absolute magnitude signs in Eq. (4.5–5) must be independent of r. Since the absolute value of this integral can be made as small as we please, we conclude that the actual value of the integral is zero.

Because we have shown that the right side of Eq. (4.5–3) is zero, we can rearrange the left side to yield

$$2\pi i f(z_0) = \oint_{C_0} \frac{f(z)}{(z - z_0)} dz. \tag{4.5–6}$$

Now Eq. (4.5–6) shows that the right side of Eq. (4.5–1) is $2\pi i f(z_0)$. Dividing both sides of Eq. (4.5–1) by $2\pi i$, we obtain the Cauchy integral formula.

THEOREM 7 Cauchy integral formula

Let $f(z)$ be analytic on and in the interior of a simple closed contour C. Let z_0 be a point in the interior of C. Then

$$f(z_0) = \frac{1}{2\pi i} \oint_C \frac{f(z)\, dz}{(z - z_0)}. \quad \square \tag{4.5–7}$$

EXAMPLE 1

a) Find $\oint_C (\cos z)/(z - 1)\, dz$, where C is the triangular contour shown in Fig. 4.5–2.
b) Find $\oint_C (\cos z)/(z + 1)\, dz$, where C is the same as in part (a).

Solution

Part (a): Since $\cos z$ is an entire function and the point $z_0 = 1$ lies within C, we can apply Eq. (4.5–7). Thus

$$\frac{1}{2\pi i} \oint_C \frac{\cos z}{z - 1} dz = \cos 1, \quad \text{or} \quad \oint_C \frac{\cos z}{(z - 1)} dz = 2\pi i \cos 1 \doteq 3.39i.$$

Part (b): The integrand can be written as $(\cos z)/(z - (-1))$. Employing Eq. (4.5–7) we find $z_0 = -1$ lies outside the contour of integration. The Cauchy integral formula *does not* apply here. However, because $(\cos z)/(z + 1)$ is analytic both on C and at all points in the interior of C, the Cauchy–Goursat theorem does apply. Hence, the value of the given integral is zero. ◀

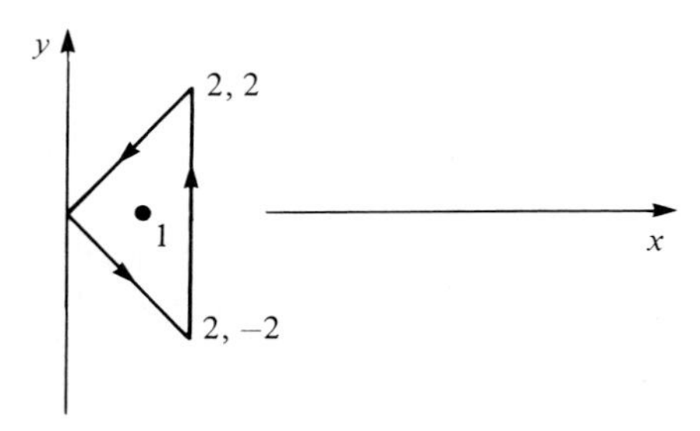

Figure 4.5–2

EXAMPLE 2

Find $(1/2\pi i) \oint_C (\cos z)/(z^2 + 1)\, dz$, where C is the circle $|z - 2i| = 2$.

Solution

It is not immediately apparent whether the Cauchy integral formula or the Cauchy–Goursat theorem is applicable here. Factoring the denominator, we have

$$\frac{1}{2\pi i} \oint \frac{\cos z}{(z-i)(z+i)}\, dz.$$

Notice that the factor $(z - i)$ goes to zero within the contour of integration (at $z = i$), and $z + i$ remains nonzero both on and inside the contour (see Fig. 4.5–3). Writing the given integral as

$$\frac{1}{2\pi i} \oint \frac{\left[\dfrac{\cos z}{z+i}\right]}{(z-i)}\, dz,$$

we see that because $\cos z/(z + i)$ is analytic both on and inside $|z - 2i| = 2$, the Cauchy integral formula is applicable. In Eq. (4.5–7) we take $f(z) = \cos z/(z + i)$ and $z_0 = i$. Hence, the value of the given integral is

$$\left(\frac{\cos z}{z+i}\right)_{z=i} = \frac{\cos i}{2i} = \frac{-i}{2} \cosh 1. \blacktriangleleft$$

An integration such as

$$\oint_{|z|=2} \frac{\cos z}{(z^2+1)}\, dz = \oint_{|z|=2} \frac{\cos z}{(z-i)(z+i)}\, dz,$$

where $z^2 + 1$ goes to zero *at two* points inside the contour of integration cannot be directly evaluated by means of the Cauchy integral formula. However, the formula can be adapted to deal with problems of this type as is shown in Exercise 16 of this section.

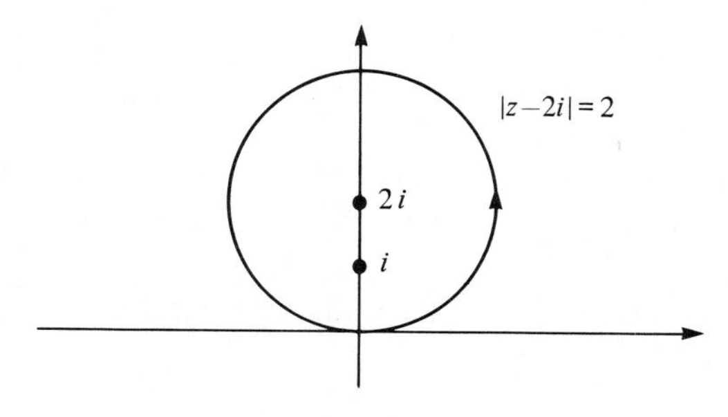

Figure 4.5–3

The Cauchy integral formula yields the value of an analytic function at a point when we know the values assumed by that function on a surrounding simple closed contour. This formula can be extended. With the "extended formula" the derivatives of any order of the function at this same point are obtainable provided we again know the function everywhere along a surrounding curve.

The extended formula is obtainable by the following series of manipulations, which do not constitute a proof. With $f(z)$ analytic on and interior to a simple closed contour C and with z_0 a point inside C, we have, from Eq. (4.5–7),

$$f(z_0) = \frac{1}{2\pi i}\oint_C \frac{f(z)}{z - z_0}\,dz. \tag{4.5–8}$$

We now regard $f(z_0)$ as a function of the variable z_0, and we differentiate both sides of Eq. (4.5–8) with respect to z_0. We will assume that it is permissible to take the d/dz_0 operator under the integral sign. Thus

$$\begin{aligned}\frac{d}{dz_0}f(z_0) &= \frac{d}{dz_0}\frac{1}{2\pi i}\oint_C \frac{f(z)}{z - z_0}\,dz = \frac{1}{2\pi i}\oint_C f(z)\frac{d}{dz_0}\left(\frac{1}{z - z_0}\right)dz \\ &= \frac{1}{2\pi i}\oint_C \frac{f(z)}{(z - z_0)^2}\,dz.\end{aligned} \tag{4.5–9}$$

In effect, we have assumed that Leibnitz's rule† applies not only to real integrals but also to contour integrals. When the second and nth derivatives are found in this way, we have

$$f^{(2)}(z_0) = \frac{1}{2\pi i}\oint_C f(z)\frac{d^2}{dz_0^2}\left(\frac{1}{z - z_0}\right)dz = \frac{2}{2\pi i}\oint_C \frac{f(z)}{(z - z_0)^3}\,dz, \tag{4.5–10}$$

$$f^{(n)}(z_0) = \frac{1}{2\pi i}\oint_C f(z)\frac{d^n}{dz_0^n}\left(\frac{1}{z - z_0}\right)dz = \frac{n!}{2\pi i}\oint_C \frac{f(z)}{(z - z_0)^{n+1}}\,dz. \tag{4.5–11}$$

The formulas obtained in Eqs. (4.5–9) through (4.5–11) can, in fact, be rigorously justified. To properly obtain $f'(z_0)$ we use the definition of the first derivative and the Cauchy integral formula Eq. (4.5–8):

$$\begin{aligned}f'(z_0) &= \lim_{\Delta z_0 \to 0} \frac{f(z_0 + \Delta z_0) - f(z_0)}{\Delta z_0} \\ &= \lim_{\Delta z_0 \to 0} \frac{1}{\Delta z_0}\left[\frac{1}{2\pi i}\oint_C \frac{f(z)\,dz}{z - (z_0 + \Delta z_0)} - \frac{1}{2\pi i}\oint_C \frac{f(z)}{(z - z_0)}\,dz\right] \\ &= \lim_{\Delta z_0 \to 0} \frac{1}{2\pi i \Delta z_0}\oint_C \left[\frac{1}{z - (z_0 + \Delta z_0)} - \frac{1}{z - z_0}\right] f(z)\,dz \\ &= \lim_{\Delta z_0 \to 0} \frac{1}{2\pi i}\oint_C \frac{f(z)\,dz}{[z - (z_0 + \Delta z_0)](z - z_0)}.\end{aligned} \tag{4.5–12}$$

---

† See W. Kaplan, *Advanced Calculus*, 4th ed. (Reading, MA: Addison-Wesley, 1991) p. 266.

If we could interchange the order of the $\lim_{\Delta z_0 \to 0}$ operation and the integration in the last term in Eq. (4.5–12), we would obtain the expression for $f'(z_0)$ presented in Eq. (4.5–9). However, we have no obvious means of justifying this step. Instead, what can be done is to show that the absolute value of the difference between

$$\frac{1}{2\pi i}\oint_C \frac{f(z)\,dz}{[z-(z_0+\Delta z_0)](z-z_0)} \quad \text{and} \quad \frac{1}{2\pi i}\oint_C \frac{f(z)\,dz}{(z-z_0)^2}$$

goes to zero as $\Delta z_0 \to 0$. This would establish the validity of Eq. (4.5–9) for $f'(z_0)$. The procedure is quite straightforward and involves the ML inequality in a manner similar to that used in deriving the Cauchy integral formula. The reader can find the details outlined in Exercise 15 at the end of this section.

Once the validity of $f'(z_0) = (1/2\pi i) \oint_C f(z)/(z-z_0)^2\,dz$ is established, one can, by a similar, rigorous procedure justify the formula for $f^{(2)}(z_0)$ given in Eq. (4.5–10). Since the derivative of $f'(z_0)$ with respect to z_0 not only exists but exists in any domain in which $f(z_0)$ is an analytic function, we can assert that $f'(z_0)$ is itself an analytic function of z_0. The preceding procedure can be carried out any number of times so as to yield any derivative of $f(z_0)$. A formula for the nth derivative is obtained and is given in Eq. (4.5–11). Summarizing these results, we have the following theorem.

THEOREM 8 Extension of Cauchy integral formula

If a function $f(z)$ is analytic within a domain, then it possesses derivatives of all orders in that domain. These derivatives are themselves analytic functions in the domain. If $f(z)$ is analytic on and in the interior of a simple closed contour C and if z_0 is inside C, then

$$f^{(n)}(z_0) = \frac{n!}{2\pi i}\oint_C \frac{f(z)}{(z-z_0)^{n+1}}\,dz. \quad \square \tag{4.5–13}$$

Note that if we interpret $f^{(0)}(z_0)$ as $f(z_0)$, and $0! = 1$, then Eq. (4.5–13) contains the Cauchy integral formula for $f(z_0)$.

Let $f(z)$ be defined throughout a neighborhood of z_0. If $f(z)$ fails to be analytic at z_0, then it is impossible to find a function $F(z)$ such that $dF/dz = f(z)$ will be satisfied throughout this neighborhood. If $F(z)$ existed, then it would be analytic, and according to Theorem 8, its second derivative df/dz would exist throughout this neighborhood. Thus $f(z)$ would be analytic at z_0, which is a contradiction.

As an illustration of this, the function $f(z) = x^2 + iy^2$, which is nowhere analytic (see Example 2 in Section 2.4), cannot be expressed as the derivative of a function $F(z)$ in any domain.

If an analytic function $f(z)$ is expressed in the form $u(x, y) + iv(x, y)$, then the various derivatives of $f(z)$ can be written in terms of the partial derivatives of u and v (see Section 2.3 and Exercise 19 at the end of that section). For example,

$$f'(z) = \frac{\partial u}{\partial x} + i\frac{\partial v}{\partial x} = \frac{\partial v}{\partial y} - i\frac{\partial u}{\partial y}, \tag{4.5–14}$$

$$f''(z) = \frac{\partial^2 u}{\partial x^2} + i\frac{\partial^2 v}{\partial x^2} = -\frac{\partial^2 u}{\partial y^2} - i\frac{\partial^2 v}{\partial y^2}. \tag{4.5–15}$$

The extension of the Cauchy integral formula tells us that if $f(z)$ is analytic it possesses derivatives of all orders. Since these derivatives are defined by Eqs. (4.5–14) and (4.5–15) as well as similar equations involving higher-order partial derivatives, we see that the partial derivatives of u and v of all orders must exist. Since a harmonic function can be regarded as the real (or imaginary) part of an analytic function we can assert Theorem 9.

THEOREM 9

A function that is harmonic in a domain will possess partial derivatives of *all* orders in that domain. □

EXAMPLE 3

Determine the value of $\oint_C [(z^3 + 2z + 1)/(z - 1)^3]\, dz$, where C is the contour $|z| = 2$.

Solution

Considering the form of the denominator in the integrand, we will use Eq. (4.5–13) with $n = 2$. We then have

$$f^{(2)}(z_0) = \frac{2!}{2\pi i} \oint_C \frac{f(z)}{(z - z_0)^3}\, dz.$$

With $z_0 = 1$ and a simple multiplication the preceding equation becomes

$$\frac{2\pi i}{2} f^{(2)}(1) = \oint_C \frac{f(z)}{(z - 1)^3}\, dz.$$

This formula, with $f(z) = z^3 + 2z + 1$, yields the value of the given integral. Thus

$$\oint_C \frac{z^3 + 2z + 1}{(z - 1)^3}\, dz = \pi i \frac{d^2}{dz^2}(z^3 + 2z + 1)\bigg|_{z=1} = \pi i(6z)\bigg|_{z=1} = 6\pi i.$$

Notice that if the contour C were $|z| = 1/2$, we would apply the Cauchy integral theorem. This is because $(z^3 + 2z + 1)/(z - 1)^3$ is analytic on and inside this circle. ◀

EXAMPLE 4

Find $\oint_C [\cos z/((z - 1)^3(z - 5)^2)]\, dz$, where C is the circle $|z - 4| = 2$.

Solution

Let us examine the two factors in the denominator. The term $(z - 1)^3$ is nonzero both inside and on the contour of integration. However, $(z - 5)^2$ does become zero at the point $z = 5$ inside C. We therefore rewrite the integral as

$$\oint_C \frac{\left(\dfrac{\cos z}{(z - 1)^3}\right)}{(z - 5)^2}\, dz$$

and apply Eq. (4.5–13) with $n = 1$, $z_0 = 5$, $f(z) = \cos z/(z-1)^3$. Thus

$$\frac{1}{2\pi i}\oint_C \frac{\left(\dfrac{\cos z}{(z-1)^3}\right)}{(z-5)^2}\,dz = \frac{d}{dz}\frac{\cos z}{(z-1)^3}\bigg|_{z=5} = \frac{-64\sin 5 - 48\cos 5}{4^6}.$$

The value of the *given* integral is $2\pi i$ times the preceding result. ◀

EXERCISES

1. To arrive at a formal (nonrigorous) derivation of the Cauchy integral formula let $f(z)$ be analytic on and inside a simple closed contour C, let z_0 lie inside C, and let C_0 be a circle centered at z_0 and lying completely inside C. From the principle of deformation of contours we then have

$$\oint_C \frac{f(z)}{(z-z_0)}\,dz = \oint_{C_0} \frac{f(z)}{(z-z_0)}\,dz.$$

a) Rewrite the integral on the right by means of the change of variables $z = z_0 + re^{i\theta}$, where r is the radius of C_0 and θ increases from 0 to 2π (see Fig. 4.3–8). Note that $dz/d\theta = ire^{i\theta}$.

b) For the integral obtained in part (a) let $r \to 0$ in the integrand. Now perform the integration and use your result to show that

$$\oint_C \frac{f(z)}{(z-z_0)}\,dz = 2\pi i f(z_0).$$

c) What makes this derivation nonrigorous?

Evaluate the following integrals. Use the Cauchy integral formula (or its extension) or the Cauchy integral theorem where appropriate.

2. $\displaystyle\oint \frac{dz}{e^z(z-2)}$ around $\dfrac{x^2}{9} + \dfrac{y^2}{16} = 1$

3. $\displaystyle\frac{1}{2\pi i}\oint \frac{(\cos z + \sin z)}{(z^2+25)(z+1)}\,dz$ around $\dfrac{x^2}{9} + \dfrac{y^2}{16} = 1$

4. $\displaystyle\oint \frac{\cosh z}{z^2+z+1}\,dz$ around $(x-1)^2 + (y-1)^2 = 1$

5. $\displaystyle\frac{1}{2\pi i}\oint \frac{\sinh z}{z^2+z+1}\,dz$ around $|z-2i| = 2$

6. $\displaystyle\frac{1}{2\pi i}\oint \frac{\cos(\sinh z)}{z^2+z+1}\,dz$ around $|z-1-i| = \dfrac{3}{2}$

7. $\displaystyle\oint \frac{[\mathrm{Log}(z/e)]^2}{z^2-6z+5}\,dz$ around $|z-1| = \dfrac{1}{2}$

8. $\oint \frac{\sin(e^z + \cos z)}{(z-1)^2(z+3)}\,dz$ around $\frac{x^2}{2} + y^2 = 1$

9. $\oint \frac{e^{3z}}{(z-2i)(z-1)^2}\,dz$ around $\frac{x^2}{2} + y^2 = 1$

10. $\oint \frac{\cos(2z)}{z^{20}}\,dz$ around $|z| = 1$

11. $\oint \frac{\cos(2z)}{z^{21}}\,dz$ around $|z| = 1$

12. A student is attempting to perform the integration $\int_{0+i0}^{1+i} \bar{z}\,dz$ along the line $y = \sqrt{\sin(\frac{\pi}{2}x)}$. He studies Theorem 5 in Section 4.4 and reasons that if he can find a function $F(z)$ satisfying $dF/dz = \bar{z}$ in a domain containing the path of integration, then he can evaluate the integral as $F(1+i) - F(0+i0)$ without having to use the path of integration. Explain why this will not work.

13. a) Use the extension of the Cauchy integral formula to show that $\oint e^{az}/(z^{n+1})\,dz = a^n 2\pi i/n!$, where the integration is performed around $|z| = 1$.

b) Rewrite the integral of part (a) using the substitution $z = e^{i\theta}$ $(0 \le \theta \le 2\pi)$ when z lies on the unit circle. Integrating on θ show that, when a is real, $\int_0^{2\pi} e^{a\cos\theta}\cos(a\sin\theta - n\theta)\,d\theta = 2\pi a^n/n!$, and $\int_0^{2\pi} e^{a\cos\theta}\sin(a\sin\theta - n\theta)\,d\theta = 0$.

14. a) If a is a real number and $|a| < 1$ show that

$$\int_0^{2\pi} \frac{1 - a\cos\theta}{1 - 2a\cos\theta + a^2}\,d\theta = 2\pi.$$

Hint: Consider $\oint dz/(z-a)$ around $|z| = 1$. What is the value of this integral? Now rewrite this integral using $z = e^{i\theta}$, $0 \le \theta \le 2\pi$, for z on the unit circle.

b) Suppose a is a real number, but $a > 1$. Now what is the value of

$$\int_0^{2\pi} \frac{1 - a\cos\theta}{1 - 2a\cos\theta + a^2}\,d\theta?$$

15. The rigorous proof of the extended Cauchy integral formula for the first derivative, which was begun on page 184, requires for its completion our showing that

$$\lim_{\Delta z_0 \to 0} \frac{1}{2\pi}\left|\oint_C \frac{f(z)\,dz}{(z - (z_0 + \Delta z_0))(z - z_0)} - \oint_C \frac{f(z)\,dz}{(z - z_0)^2}\right| = 0.$$

Complete the proof.

Hint: Let b equal the shortest distance from z_0 to any point on the contour C, let m be the maximum value of $|f(z)|$ on C, let L be the length of C, and assume $|\Delta z_0| \le b/2$. Show that you can rewrite the preceding limit with a single integral:

$$\lim_{\Delta z_0 \to 0} \frac{1}{2\pi}\left|\oint_C \frac{f(z)}{(z - z_0)^2}\left(\frac{\Delta z_0}{z - (z_0 + \Delta z_0)}\right) dz\right|.$$

Apply the ML inequality to this integral using m, b, L, etc., and then pass to the limit indicated.

Contour C (solid line)

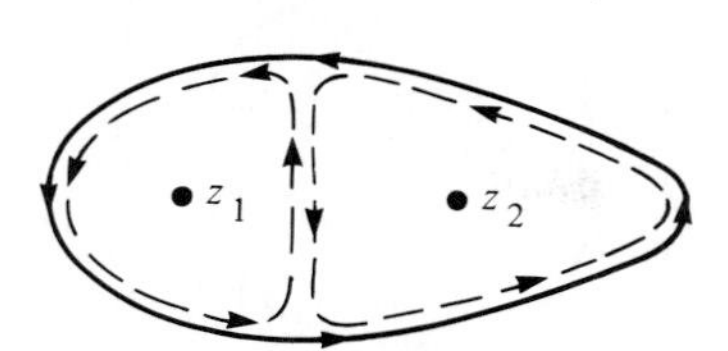

Figure 4.5–4

16. Let $f(z)$ be analytic on and inside a simple closed contour C. Let z_1 and z_2 lie inside C (see Fig. 4.5–4).

a) Show that

$$\frac{1}{2\pi i}\oint_C \frac{f(z)}{(z-z_1)(z-z_2)}\,dz = \frac{f(z_1)}{z_1-z_2} + \frac{f(z_2)}{z_2-z_1}.$$

Hint: Integrate around the two contours shown by the broken line in Fig. 4.5–4 and combine the results.

b) Let $f(z)$ have the same properties as in part (a), and let $z_1, z_2, \ldots, z_n$ lie inside C. Assume that $z_1, z_2, \ldots, z_n$ are numerically distinct, i.e., no two values are the same. Extend the method used in part (a) to show that

$$\begin{aligned}\frac{1}{2\pi i}\oint \frac{f(z)\,dz}{(z-z_1)(z-z_2)\cdots(z-z_n)} &= \frac{f(z_1)}{(z_1-z_2)(z_1-z_3)\cdots(z_1-z_n)} \\ &\quad + \frac{f(z_2)}{(z_2-z_1)(z_2-z_3)\cdots(z_2-z_n)} \\ &\quad + \cdots + \frac{f(z_n)}{(z_n-z_1)(z_n-z_2)\cdots(z_n-z_{n-1})}.\end{aligned}$$

The following problems require either the results derived in the previous problem for their solution, or an extension of the methods employed there.

17. $\dfrac{1}{2\pi i}\displaystyle\oint \frac{\cos(z-1)}{(z+1)(z-2)}\,dz$ around $|z| = 3$

18. $\displaystyle\oint \frac{dz}{e^z(z^2-1)}$ around the square with corners at $z = \pm 2$, and $z = \pm 2i$

19. $\displaystyle\oint \frac{\operatorname{Log} z}{z^2 - z + 1/2}\,dz$ around $|z-1| = 8/9$

20. $\displaystyle\oint \frac{dz}{e^z(z^2-1)^2}$ around the contour of Exercise 18

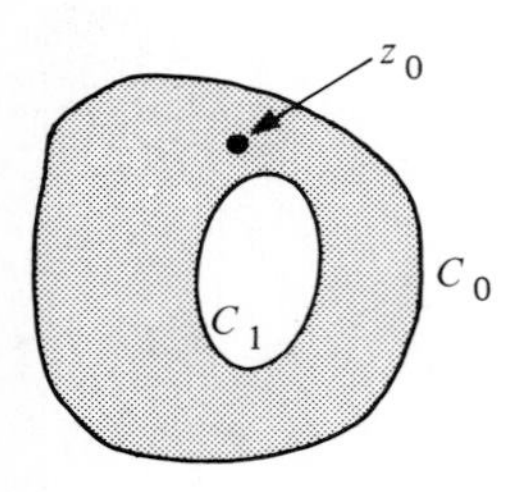

Figure 4.5–5

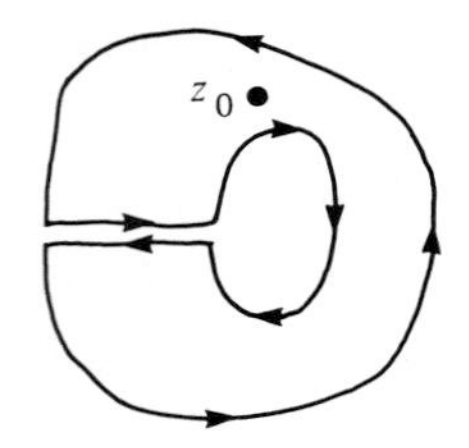

Figure 4.5–6

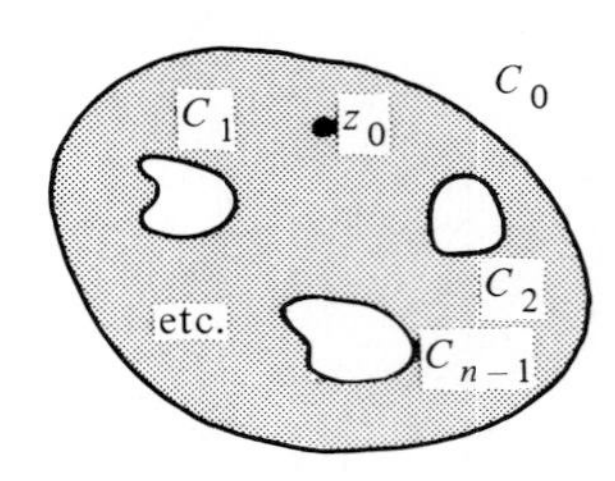

Figure 4.5–7

21. a) Let D be a doubly connected domain bounded by the simple closed contours C_0 and C_1 as shown in Fig. 4.5–5. Let $f(z)$ be analytic in D and on its boundaries, and let z_0 lie in D. Note that $f(z)$ is not necessarily analytic inside C_1. Show that

$$\frac{1}{2\pi i}\oint_{C_0}\frac{f(z)}{z-z_0}\,dz = f(z_0) + \frac{1}{2\pi i}\oint_{C_1}\frac{f(z)}{(z-z_0)}\,dz.$$

This is the Cauchy integral formula for doubly connected domains.

Hint: Do the integration $(1/2\pi i)\oint f(z)/(z-z_0)\,dz$ around the simple closed contour shown in Fig. 4.5–6. What portions of the integral cancel?

b) Use the preceding result to show that

$$\frac{1}{2\pi i}\oint_{|z|=2}\frac{dz}{(z-1)\sin z} = \frac{1}{\sin 1} + \frac{1}{2\pi i}\oint_{|z|=1/2}\frac{dz}{(z-1)\sin z}.$$

c) Let D be an n-tuply connected domain bounded by the closed contours $C_0, C_1, \ldots, C_{n-1}$ as shown in Fig. 4.5–7. Let $f(z)$ be analytic in D and on its boundaries, and let z_0 lie in D. Show that

$$\frac{1}{2\pi i}\oint_{C_0}\frac{f(z)}{(z-z_0)}\,dz = f(z_0) + \frac{1}{2\pi i}\oint_{C_1}\frac{f(z)}{(z-z_0)}\,dz + \cdots + \frac{1}{2\pi i}\oint_{C_{n-1}}\frac{f(z)}{(z-z_0)}\,dz.$$

This is the Cauchy integral formula for n-tuply connected domains.

4.6 SOME APPLICATIONS OF THE CAUCHY INTEGRAL FORMULA

In this and the following section we will explore a few consequences of the Cauchy integral formula and its extension. Certain of these results can be used to solve physical problems, while others are purely mathematical. We will begin with an easily derived result that is one of the interesting consequences of the Cauchy integral formula.

THEOREM 10 Gauss's mean value theorem

Let $f(z)$ be analytic in a simply connected domain. Consider any circle lying in this domain. The value assumed by $f(z)$ at the center of the circle equals the average of the values assumed by $f(z)$ on its circumference. □

To establish this fact let the circle be centered at z_0 and have radius r. A glance at Fig. 4.6–1 shows that any point z on the circle can be expressed in the form $z = z_0 + re^{i\theta}$, where $0 \leq \theta \leq 2\pi$. Note that $dz = re^{i\theta} i\, d\theta$. With these substitutions for z and dz made in Eq. (4.5–7), we obtain

$$f(z_0) = \frac{1}{2\pi i}\int_0^{2\pi} \frac{f(z_0 + re^{i\theta})}{re^{i\theta}} ire^{i\theta}\, d\theta.$$

With some obvious cancellations this becomes

$$f(z_0) = \frac{1}{2\pi}\int_0^{2\pi} f(z_0 + re^{i\theta})\, d\theta. \tag{4.6–1}$$

The expression on the right in Eq. (4.6–1) is the arithmetic mean (average value) of $f(z)$ on the circumference of the circle. Equation (4.6–1) is the mathematical statement of Gauss's mean value theorem.

If the function $f(z)$ is expressed in terms of its real and imaginary parts, $f(z) = u(x, y) + iv(x, y)$, we can recast Eq. (4.6–1) as follows:

$$u(z_0) + iv(z_0) = \frac{1}{2\pi}\int_0^{2\pi} u(z_0 + re^{i\theta})\, d\theta + \frac{i}{2\pi}\int_0^{2\pi} v(z_0 + re^{i\theta})\, d\theta. \tag{4.6–2}$$

Taking $z_0 = x_0 + iy_0$ and equating corresponding parts of each side of Eq. (4.6–2), we obtain

$$u(x_0, y_0) = \frac{1}{2\pi}\int_0^{2\pi} u(z_0 + re^{i\theta})\, d\theta, \tag{4.6–3a}$$

$$v(x_0, y_0) = \frac{1}{2\pi}\int_0^{2\pi} v(z_0 + re^{i\theta})\, d\theta. \tag{4.6–3b}$$

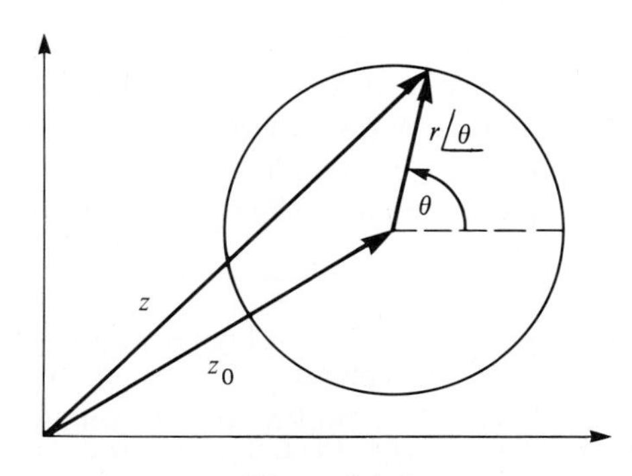

Figure 4.6–1

We see from Eq. (4.6–3a) that the real part u of the analytic function evaluated at the center of the circle is equal to u averaged over the circumference of the circle. A corresponding statement contained in Eq. (4.6–3b) applies to the imaginary part v.

If the value of u is known at certain discrete points along the circumference of a circle, these values can be used to determine, approximately, the value of the integral on the right in Eq. (4.6–3a). In this way an approximation can be obtained for the value of u at the center of the circle (see Exercise 7 in this section). Typically, four uniformly spaced points on the circumference might be used. This technique forms the basis of a numerical procedure called the *finite difference method*, which is used to evaluate a harmonic function at the interior points of a domain when the values of the function are known on the boundaries. The method is used to solve physical problems in such specialities as electrostatics, heat transfer, and fluid mechanics.[†]

Gauss's mean value theorem can be used to establish an important property of analytic functions:

THEOREM 11 Maximum modulus theorem

Let a nonconstant function $f(z)$ be continuous throughout a closed bounded region R. Let $f(z)$ be analytic at every interior point of R. Then the maximum value of $|f(z)|$ in R must occur on the boundary of R. □

Loosely stated, the theorem asserts that the maximum value of the modulus of $f(z)$ occurs on the boundary of a region.

To prove the theorem we must employ without proof the following property of analytic functions: If an analytic function fails to assume a constant value over all the interior points of a region, then it is not constant in any neighborhood of any interior point of that region.[‡] From Exercise 19, Section 2.4 we see, in addition, that $|f(z)|$ will not be constant in any such neighborhood.

Returning to the maximum modulus theorem, let us assume that the maximum value of $|f(z)|$ in R occurs at z_0, an interior point of R. At z_0 we have $|f(z_0)| = m$ (the maximum value). We assume R to be the set of points on and inside the contour C of Fig. 4.6–2.

Consider a circle C_0 of radius r centered at z_0 and lying entirely within C. Since $|f(z)|$ is not constant in any neighborhood of z_0, we can choose r so that C_0 passes through at least one point where $|f(z)| < m$. If we describe C_0 by the equation $z = z_0 + re^{i\theta}$, $0 \le \theta \le 2\pi$, we have at this point $|f(z_0 + re^{i\theta})| < m$.

Because $f(z)$ is a continuous function, there must be a finite segment of arc on C_0 along which $|f(z_0 + re^{i\theta})| \le m - b$. Here b is a positive constant such that $b < m$. For simplicity let the arc in question extend from $\theta = 0$ to $\theta = \beta$. Along the remainder

---

[†] For an application to electrostatics see W. H. Hayt, *Engineering Electromagnetics*, 4th ed. (New York: McGraw-Hill, 1981), Chapter 6.

[‡] See, for example, R. V. Churchill, J. Brown, and R. Verhey, *Complex Variables and Applications* (New York: McGraw-Hill, 1974), pp. 135, 284.

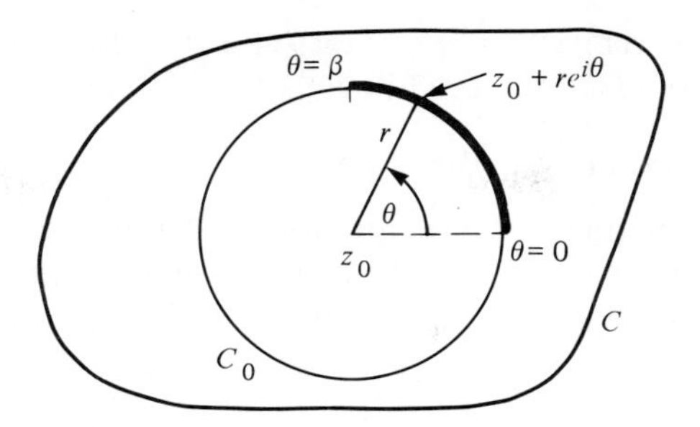

Figure 4.6–2

of the arc, $\beta \le \theta \le 2\pi$, we have $|f(z_0 + re^{i\theta})| \le m$ since $|f(z_0)| = m$ is the maximum value of $|f(z)|$.

Now we refer to Eq. (4.6–1) and write the integral around C_0 in two parts:

$$f(z_0) = \frac{1}{2\pi}\int_0^\beta f(z_0 + re^{i\theta})\,d\theta + \frac{1}{2\pi}\int_\beta^{2\pi} f(z_0 + re^{i\theta})\,d\theta.$$

Let us take the absolute magnitude of both sides of this equation and also apply a triangle inequality. We then have

$$|f(z_0)| \le \frac{1}{2\pi}\left|\int_0^\beta f(z_0 + re^{i\theta})\,d\theta\right| + \frac{1}{2\pi}\left|\int_\beta^{2\pi} f(z_0 + re^{i\theta})\,d\theta\right|.$$

We can apply the ML inequality to each of these integrals. For the first integral on the right we know that $|f(z_0 + re^{i\theta})| \le m - b$, and for the second we can assert that $|f(z_0 + re^{i\theta})| \le m$. The quantity L in each case is just the interval of integration: β and $2\pi - \beta$, respectively. Thus

$$|f(z_0)| \le \frac{1}{2\pi}(m - b)\beta + \frac{m}{2\pi}(2\pi - \beta).$$

Adding the terms on the right side of this equation we obtain

$$|f(z_0)| \le m - \frac{b\beta}{2\pi}.$$

The quantity on the left, $|f(z_0)|$, is m, the maximum value of $|f(z)|$. But m cannot be less than $m - b\beta/2\pi$. We have obtained a contradiction. Our assumption that z_0 is an interior point of R must be false. Since $|f(z)|$ must be a maximum *somewhere* in R (see Theorem 2, Chapter 2) this maximum must be at a boundary point. For the region R of Fig. 4.6–2 such a point would be on the boundary C.

There is a similar theorem, which is proved in Exercise 8 of this section, pertaining to the minimum value of $|f(z)|$ achieved in R:

THEOREM 12 Minimum modulus theorem

Let a nonconstant function $f(z)$ be continuous and nowhere zero throughout a closed bounded region R. Let $f(z)$ be analytic at every interior point of R. Then the minimum value of $|f(z)|$ in R must occur on the boundary of R. □

Note the additional requirement $f(z) \neq 0$.[†]

We will see in Exercises 13, 14, 15, and 16 of this section that the maximum and minimum modulus theorems can tell us some useful properties about the behavior of harmonic functions in bounded regions. These properties have direct physical application to problems involving heat conduction (see, for example, Exercise 16) and electrostatics.

EXAMPLE 1

Consider $f(z) = e^z$ in the region $|z| \leq 1$. Find the points in this region where $|f(z)|$ achieves its maximum and minimum values.

Solution

Because e^z is an entire function and e^z is never 0 in the given region, both the maximum and minimum modulus theorems should be confirmed by our result. We have $|f(z)| = |e^z| = |e^{x+iy}| = |e^x||e^{iy}| = |e^x| = e^x$. Because e^x is nonnegative, we were able to drop the absolute magnitude signs. Now $|f(z)|$ is maximum at the point in the region where e^x achieves its largest value, that is, at $x = 1$, $y = 0$, and $|f(z)|$ is minimum where e^x is smallest, that is, at $x = -1$, $y = 0$. Both points are on the boundary of R (see Fig. 4.6–3). ◀

The maximum and minimum modulus theorems came to us from the Cauchy integral formula, with Gauss's mean value theorem as an intermediate step. The extension of the Cauchy integral formula can be used to establish this novel result:

THEOREM 13 Liouville's theorem[‡]

An entire function whose absolute value is bounded (that is, does not exceed some constant) throughout the z-plane is a constant. □

[†] There are other versions of the maximum and minimum modulus theorems. They are called the "local" versions and are stated as follows: a) Let $f(z)$ be analytic and not constant in a neighborhood N of z_0. Then there are points in N lying arbitrarily close to z_0 where $|f(z)| > |f(z_0)|$, that is, $|f(z)|$ cannot have a local maximum at z_0. b) If, in addition, $f(z_0) \neq 0$, there are points in N lying arbitrarily close to z_0, where $|f(z)| < |f(z_0)|$, that is, $|f(z)|$ cannot have a local minimum at z_0.

[‡] Named for Joseph Liouville (1809–1882), a French mathematician.

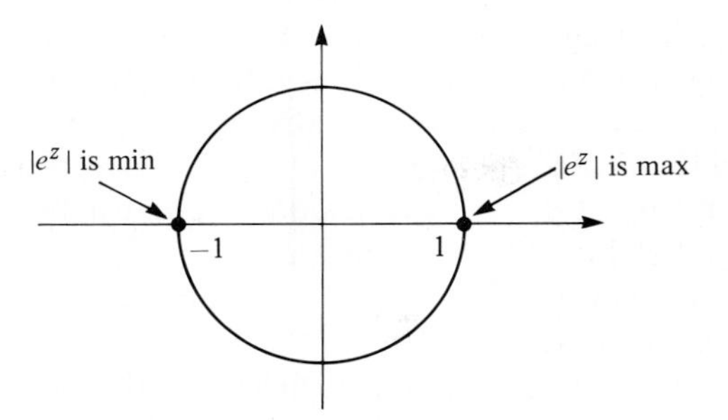

Figure 4.6–3

To prove this theorem consider a circle C, of radius r, centered at z_0. Since $f(z)$ is everywhere analytic, we can use Eq. (4.5–13) with $n = 1$ to integrate $f(z)/(z - z_0)^2$ around C. Thus

$$f'(z_0) = \frac{1}{2\pi i} \oint_C \frac{f(z)}{(z - z_0)^2}\, dz.$$

Taking magnitudes we have

$$|f'(z_0)| = \frac{1}{2\pi} \left| \oint_C \frac{f(z)}{(z - z_0)^2}\, dz \right|.$$

We now apply the ML inequality to the integral and take L as the circumference of C, that is, $2\pi r$. Thus

$$|f'(z_0)| = \frac{1}{2\pi} \left| \oint_C \frac{f(z)}{(z - z_0)^2}\, dz \right| \le \frac{1}{2\pi} M 2\pi r, \tag{4.6–4}$$

where M is a constant satisfying

$$\left| \frac{f(z)}{(z - z_0)^2} \right| \le M$$

along C. Because $|z - z_0| = r$ on C, the preceding can be rewritten

$$\frac{|f(z)|}{r^2} \le M. \tag{4.6–5}$$

We have assumed that $|f(z)|$ is bounded throughout the z-plane. Thus there is a constant m such that $|f(z)| \le m$ for all z. Dividing both sides of this inequality by r^2 results in

$$\frac{|f(z)|}{r^2} \le \frac{m}{r^2}. \tag{4.6–6}$$

A comparison of Eqs. (4.6–5) and (4.6–6) shows that we can take M as m/r^2. Rewriting Eq. (4.6–4) with this choice of M, we have $|f'(z_0)| \le m/r$.

The preceding inequality can be applied to any circle centered at z_0. Because we can consider circles of arbitrary large radius r, the right-hand side of this inequality can be made arbitrarily small. Thus the derivative of $f(z)$ at z_0, which is some specific number, has a magnitude of zero. Hence $f'(z_0) = 0$. Since the preceding argument can be applied at any point z_0, the expression $f'(z)$ must be zero throughout the z-plane. This is only possible if $f(z)$ is constant.

Liouville's Theorem can be used to prove the fundamental theorem of algebra. Although the reader has used algebra for many years he or she is perhaps unacquainted with its fundamental theorem which states that the polynomial equation $a_n z^n + a_{n-1}z^{n-1} + \cdots + a_0 = 0$ has at least one solution in the complex plane. We assume that $n \geq 1$ and that $a_n \neq 0$.

Take $p(z) = a_n z^n + a_{n-1}z^{n-1} + \cdots + a_0$. Let us consider two regions in the complex plane. R_1 is the disc $|z| \leq r$, while R_2 is the remainder of the plane: $|z| > r$. Assume now that $p(z) = 0$ has no roots in the complex plane (i.e., the polynomial equation has no solutions).

This means that $1/p(z)$ is a continuous function in R_1. According to Theorem 2, part (d), in Chapter 2, there exists a constant M such that $|1/p(z)| \leq M$ when z lies in the bounded region R_1.

Now we study $p(z)$ and $1/p(z)$ in R_2. Recall the triangle inequality $|f + g| \geq |f| - |g|$ (when $|f| \geq |g|$) from Eq. (1.3–20). Taking $f = a_n z^n$ and $g = a_{n-1}z^{n-1} + \cdots + a_0$ (note $p(z) = f + g$), we see that it is certainly possible to take r large enough so that in R_2, where $|z| > r$, we have $|f| \geq |g|$. Hence from our triangle inequality we have in R_2,

$$|p(z)| \geq |a_n z^n| - |a_{n-1}z^{n-1} + \cdots + a_0|. \tag{4.6–7}$$

From another triangle inequality (see Eq. 1.3–8), we have that

$$|a_{n-1}z^{n-1} + a_{n-2}z^{n-2} + \cdots + a_0| \leq |a_{n-1}||z|^{n-1} + \cdots + |a_0|. \tag{4.6–8}$$

Combining the inequalities in Eqs. (4.6–7) and (4.6–8) we obtain

$$|p(z)| \geq |a_n||z|^n - (|a_{n-1}||z|^{n-1} + \cdots + |a_0|), \tag{4.6–9}$$

where we again assume that $|z| > r$ is large enough so that the right side remains positive.

Factoring $|z|^{n-1}$ on the right in Eq. (4.6–9) we get

$$|p(z)| \geq |a_n||z|^n - |z|^{n-1}\left[|a_{n-1}| + \frac{|a_{n-2}|}{|z|} + \cdots + \frac{|a_0|}{|z|^{n-1}}\right]. \tag{4.6–10}$$

Let A be the largest of the numbers $|a_{n-1}|, |a_{n-2}|, \ldots, |a_0|$. Then, if $|z| > 1$, we obtain

$$|a_{n-1}| + \frac{|a_{n-2}|}{|z|} + \cdots + \frac{|a_0|}{|z|^{n-1}} \leq A + \frac{A}{|z|} + \cdots + \frac{A}{|z|^{n-1}} \leq nA. \tag{4.6–11}$$

Combining the inequalities in Eqs. (4.6–11) and (4.6–10) we find

$$|p(z)| \geq |a_n||z|^n - |z|^{n-1}nA = |z|^n\left[|a_n| - \frac{nA}{|z|}\right], \tag{4.6–12}$$

where we again assume that $|z|$ is large enough to render the right side positive throughout R_2.

Inverting the inequality of Eq. (4.6–12) we see that

$$\frac{1}{|p(z)|} \le \frac{1}{|z|^n[|a_n| - nA/|z|]}. \tag{4.6–13}$$

The right side of Eq. (4.6–13) will achieve its maximum value in R_2, at those points where $|z|$ is smallest. Since $|z| > r$, we have, in R_2,

$$\frac{1}{|p(z)|} \le \frac{1}{r^n[|a_n| - nA/r]}. \tag{4.6–14}$$

Recall that in R_1 we have $1/|p(z)| \le M$. Letting M' be the larger of M and the right side of Eq. (4.6–14) we have, therefore, throughout the z-plane,

$$1/|p(z)| \le M'.$$

With the aid of Liouville's Theorem and the preceding inequality we have that the bounded analytic function $1/p(z)$ is a constant, or equivalently, $p(z)$ is a constant. Since $p(z)$ is clearly not a constant, it must *not* be true that $p(z) = 0$ fails to have a root in the complex plane. This completes the proof.

The reader first encountered complex numbers in trying to solve quadratic equations like $az^2 + bz + c = 0$. Linear problems like $az + b = 0$ did not require complex numbers for their solution if a and b were real. We see now that cubic ($az^3 + bz^2 + cz + d = 0$), quartic, etc., equations do not require the use of anything beyond the complex number system for their solution. Actually, once we have shown that $p(z) = 0$ has one root in the complex plane, it is not hard (see Exercise 18) to show that it has n roots.

The first proof of the fundamental theorem of algebra was published by Carl Friedrich Gauss around 1798 as part of his doctoral dissertation. Although he employed complex numbers he did not use the method presented here. There are numerous other proofs of the theorem. One is given in Exercise 9 of Section 7.3.

EXERCISES

Use Gauss's mean value theorem in various versions (see Eqs. 4.6–1 and 4.6–3) to evaluate the following integrals.

1. $(1/2\pi)\int_0^{2\pi} e^{(e^{i\theta})}\, d\theta$

2. $\int_0^{2\pi} \cos(\cos\theta + i\sin\theta)\, d\theta$

3. $\int_{-\pi}^{\pi} \cos(\cos\theta)\cosh(\sin\theta)\, d\theta$

4. $\displaystyle\int_{-\pi}^{\pi} \frac{a + \cos(n\theta)}{a^2 + 1 + 2a\cos(n\theta)}\, d\theta$, $a > 1$, n integer

Hint: $f(z) = \dfrac{1}{z^n + a}$

5. $\int_0^{2\pi} \text{Log}[a^2 + 1 + 2a\cos(n\theta)]\, d\theta$, $a > 1$, n integer

Hint: $a^2 + 1 + 2a\cos(n\theta) = |a + e^{in\theta}|^2$

6. Show that $\int_0^{2\pi} \text{Log}(a + b\cos\theta)\, d\theta = 2\pi\, \text{Log}[(a + \sqrt{a^2 - b^2})/2]$, where $a > b \geq 0$.

7. a) Let $u(x, y)$ be a harmonic function. Let u_0 be the value of u at the center of the circle, of radius r, shown in Fig. 4.6–4. The values of u at four equally spaced points on the circumference are u_1, u_2, u_3, u_4. Refer to Eq. (4.6–3a) and use an approximation to the integral to show that

$$u_0 \approx \frac{u_1 + u_2 + u_3 + u_4}{4}.$$

b) Use a pocket calculator to evaluate the harmonic function $e^x \cos y$ at the four points (1.1, 1), (0.9, 1), (1, 1.1), and (1, 0.9). Compare the average of these results with $e^x \cos y$ evaluated at (1, 1).

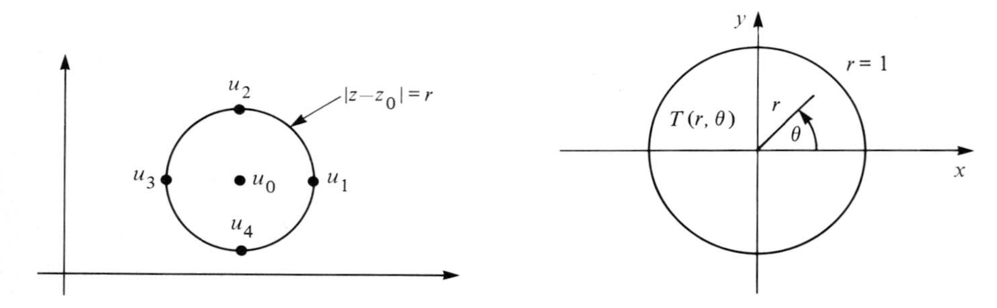

Figure 4.6–4

Figure 4.6–5

8. Let $f(z)$ be a nonconstant function that is continuous and nonzero throughout a closed bounded region R. Let $f(z)$ be analytic at every interior point of R. Show that the minimum value of $|f(z)|$ in R must occur on the boundary of R.

Hint: Consider $g(z) = 1/f(z)$ and recall the maximum modulus theorem.

For the following closed regions R and functions $f(z)$ find the values of z in R where $|f(z)|$ achieves its maximum and minimum values. If your answers do not lie on the boundary of R, give an explanation. Give the values of $|f(z)|$ at its maximum and minimum in R.

9. $f(z) = z$, R is $|z - 1 - i| \leq 1$

10. $f(z) = z^2$, R is $|z - 1 - i| \leq 2$

11. $f(z) = e^z$, R same as in Exercise 9.

12. $f(z) = 1/(z + 1)$, R same as in Exercise 9.

13. Let $u(x, y)$ be real, nonconstant, and continuous in a closed bounded region R. Let $u(x, y)$ be harmonic in the interior of R. Prove that the maximum value of $u(x, y)$ in this region occurs on the boundary. This is known as the *maximum principle*.

Hint: Consider $F(z) = u(x, y) + iv(x, y)$, where v is the harmonic conjugate of u. Let $f(z) = e^{F(z)}$. Explain why $|f(z)|$ has its maximum value on the boundary. How does it follow that $u(x, y)$ has its maximum value on the boundary?

14. For $u(x, y)$ described in Exercise 13 show that the minimum value of this function occurs on the boundary. This is known as the *minimum principle.*

Hint: Follow the suggestions given in Exercise 13 but show that $|f(z)|$ has its minimum value on the boundary.

15. Consider the closed bounded region R given by $0 \le x \le 1, 0 \le y \le 1$. Now $u = (x^2 - y^2)$ is harmonic in R. Find the maximum and minimum values of u in R and state where they are achieved.

16. A long cylinder of unit radius, shown in Fig. 4.6–5, is filled with a heat-conducting material. The temperature inside the cylinder is described by the harmonic function $T(r, \theta)$ (see Section 2.6). The temperature on the surface of the cylinder is known and is given by $\sin\theta\cos^2\theta$. Since $T(r, \theta)$ is continuous for $0 \le r \le 1$, $0 \le \theta \le 2\pi$, we require that $T(1, \theta) = \sin\theta\cos^2\theta$. Use the results derived in Exercises 13 and 14 to establish upper and lower bounds on the temperature inside the cylinder.

17. In this problem we derive one of the four Wallis formulas. They allow one to evaluate $\int_0^{\pi/2} [f(\theta)]^m \, d\theta$ where $m \ge 0$ is an integer and $f(\theta) = \sin\theta$ or $\cos\theta$. The cases of odd and even m must be considered separately. We will consider m even.

a) Show using the binomial theorem that

$$z^{-1}\left(z + \frac{1}{z}\right)^{2n} = \sum_{k=0}^{2n} \frac{(2n)! z^{2n-2k-1}}{(2n-k)!k!}, \qquad n = 0, 1, 2, \ldots .$$

b) Using the above result, a term-by-term integration, and the extended Cauchy integral formula or Eq. (4.3–10), show that

$$\oint z^{-1}\left(z + \frac{1}{z}\right)^{2n} dz = 2\pi i \frac{(2n)!}{(n!)^2},$$

where the integration is around $|z| = 1$.

c) With $z = e^{i\theta}$, $0 \le \theta \le 2\pi$, on the unit circle, show from (b) that

$$\int_0^{2\pi} (2\cos\theta)^{2n} \, d\theta = 2\pi \frac{(2n)!}{(n!)^2}.$$

d) Noting the symmetry of $\cos\theta$, and that $2n$ is even ($n = 0, 1, 2, \ldots$), explain why

$$\int_0^{\pi/2} (\cos\theta)^{2n} \, d\theta = \frac{\pi}{2} \frac{(2n)!}{(n!)^2 2^{2n}}.$$

This is one of Wallis's formulas. Wallis (1616–1703) was one of the great English mathematicians of the seventeenth century. He was also chaplain to Charles II. He did not employ complex variables in the derivation of the formulas named after him. The symbol ∞ (for infinity) was invented by Wallis.

e) Find

$$\int_0^{\pi/2} (\sin \theta)^{2n}\, d\theta,$$

where $n = 0, 1, 2, \ldots$.

18. The fundamental theorem of algebra shows that $p(z) = a_n z^n + a_{n-1} z^{n-1} + \cdots + a_0 = 0$ has at least one root z_0 in the complex plane. We show in this exercise that by a simple extension this equation has n roots $z_0, z_1, \ldots, z_{n-1}$.

a) Show that $z^n - z_0^n$ can be written in the form $(z^n - z_0^n) = (z - z_0)R(z)$, where $R(z) = z^{n-1} + z_0 z^{n-2} + z_0^2 z^{n-3} + \cdots + z_0^{n-2} z + z_0^{n-1}$, is a polynomial of degree $n - 1$ in z.

b) If z_0 is the root of $p(z) = 0$ given by the fundamental theorem, explain why $p(z) = a_n(z^n - z_0^n) + a_{n-1}(z^{n-1} - z_0^{n-1}) + \cdots + a_1(z - z_0)$.

Hint: Consider $p(z) - p(z_0)$, and combine terms.

c) Using the results in (b) and (a) show that $p(z) = a_n(z - z_0)R_{n-1} + a_{n-1}(z - z_0)R_{n-2} + \cdots + a_1(z - z_0)$, where R_j is a polynomial of degree j in z.

d) Use the result derived in (c) to show that $p(z) = (z - z_0)A(z)$, where $A(z)$ is a polynomial of degree $n - 1$ in z.

Comment: The polynomial equation $A(z) = 0$ has, from the fundamental theorem, a root z_1 in the complex plane. Thus $A(z) = (z - z_1)B(z)$, where $B(z)$ is a polynomial of degree $n - 2$ in z. Hence $p(z) = (z - z_0)(z - z_1)B(z)$. We can then extract a multiplicative factor from $B(z)$ and continue this procedure until we have $p(z) = (z - z_0)(z - z_1) \cdots (z - z_{n-1})K$, where K is a constant (polynomial of degree zero). Some of these roots may be identical, and such roots are termed multiple or repeated roots.

4.7 INTRODUCTION TO DIRICHLET PROBLEMS—THE POISSON INTEGRAL FORMULA FOR THE CIRCLE AND HALF PLANE

In previous sections we have seen the close relationship that exists between harmonic functions and analytic functions. In this section we continue exploring this connection and, in so doing, will solve some physical problems whose solutions are harmonic functions.

An important type of mathematical problem, with physical application, is the *Dirichlet problem.* Here, an unknown function must be found that is harmonic within a domain and that also assumes preassigned values on the boundary of the domain.†

---

† The function being sought should be continuous in the region consisting of the domain and its boundary except that discontinuities are permitted at those boundary points where the given boundary condition is itself discontinuous.

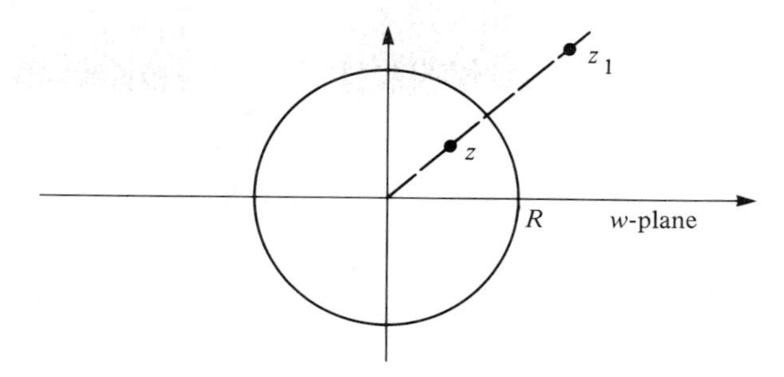

Figure 4.7–1

For an example of such a problem refer to Exercise 16 of the previous section. In this exercise the temperature inside the cylinder $T(r, \theta)$ is a harmonic function. We know what the temperature is on the boundary. If we try to find $T(r, \theta)$ subject to the requirement that on the boundary it agrees with the given function $\sin \theta \cos^2 \theta$, we are solving a Dirichlet problem.

Dirichlet problems, such as the one just discussed, in which the boundaries are of simple geometrical shape, are frequently solved by *separation of variables*, a technique discussed in most textbooks on partial differential equations. Another method, which can sometimes be used for such simple domains as well as for those of more complicated shapes, is *conformal mapping*. This subject is considered at some length in Chapter 8. An approach is discussed in this section that is applicable when the boundary of the domain is a circle or an infinite straight line.

Nowadays most Dirichlet problems involving relatively complicated boundaries are solved through the use of numerical techniques, which are implemented on a digital computer. Even desk top microcomputers are used in this capacity. The answers obtained are approximations.† The analytical methods given in this book for relatively simple boundaries can provide some physical insight as well as approximate checks to the solutions of those problems that must be solved on a computer.

The Dirichlet Problem for a Circle

The Cauchy integral formula is helpful in solving the Dirichlet problem when we have a circular boundary. Consider a circle of radius R whose center lies at the origin of the complex w-plane (see Fig. 4.7–1). Let $f(w)$ be a function that is analytic on and throughout the interior of this circle.

The variable z locates some arbitrary point inside the circle. Applying the Cauchy integral formula on this circular contour and using w as the variable of integration,

† See for example the computer software *Electromagnetism—Physics Simulations II* by Blas Cabrera and the Faculty Author Development Program, Stanford U. This is designed for the Macintosh Computer and is available from Intellimation Library for the Macintosh, 130 Cremona Drive, Santa Barbara, CA 93116.

we have

$$f(z) = \frac{1}{2\pi i} \oint \frac{f(w)}{w - z} \, dw. \tag{4.7–1}$$

Suppose we write $f(z) = U(x, y) + iV(x, y)$. We would like to use the preceding integral to obtain explicit expressions for U and V.

We begin by considering the point z_1 defined by $z_1 = R^2/\bar{z}$. Note that

$$|z_1| = \frac{R^2}{|\bar{z}|} = \frac{R}{|z|} R.$$

Since $|z| < R$, the preceding shows that $|z_1| > R$, that is, the point z_1 lies outside the circle in Fig. 4.7–1. It is easy to show that $\arg z_1 = \arg z$. The function $f(w)/(w - z_1)$ is analytic in the w-plane on and inside the given circle. Hence, from the Cauchy integral theorem,

$$0 = \frac{1}{2\pi i} \oint \frac{f(w)}{w - z_1} \, dw = \frac{1}{2\pi i} \oint \frac{f(w)}{w - \dfrac{R^2}{\bar{z}}} \, dw. \tag{4.7–2}$$

Subtracting Eq. (4.7–2) from Eq. (4.7–1), we obtain

$$f(z) = \frac{1}{2\pi i} \oint f(w) \left[\frac{1}{w - z} - \frac{1}{w - \dfrac{R^2}{\bar{z}}} \right] dw$$

$$= \frac{1}{2\pi i} \oint f(w) \left[\frac{z - \dfrac{R^2}{\bar{z}}}{(w - z)\left(w - \dfrac{R^2}{\bar{z}}\right)} \right] dw. \tag{4.7–3}$$

Because we are integrating around a circular contour, we switch to polar coordinates. Let $w = Re^{i\phi}$ and $z = re^{i\theta}$. Thus $\bar{z} = re^{-i\theta}$. Along the path of integration $dw = Re^{i\phi} i \, d\phi$, and ϕ ranges from 0 to 2π. Rewriting the right side of Eq. (4.7–3), we have

$$f(r, \theta) = \frac{1}{2\pi i} \int_0^{2\pi} f(R, \phi) \left[\frac{re^{i\theta} - \dfrac{R^2}{re^{-i\theta}}}{(Re^{i\phi} - re^{i\theta})\left(Re^{i\phi} - \dfrac{R^2}{re^{-i\theta}}\right)} \right] Re^{i\phi} i \, d\phi$$

$$= \frac{1}{2\pi} \int_0^{2\pi} f(R, \phi) \left[\frac{\left(re^{i\theta} - \dfrac{R^2}{r} e^{i\theta}\right) Re^{i\phi}}{(Re^{i\phi} - re^{i\theta})\left(Re^{i\phi} - \dfrac{R^2}{r} e^{i\theta}\right)} \right] d\phi.$$

If we multiply the two terms in the denominator of the preceding integral together and then multiply numerator and denominator by $(-r/R)e^{-i(\theta+\phi)})$, we can show, with the aid of Euler's identity, that

$$f(r, \theta) = \frac{1}{2\pi}\int_0^{2\pi} \frac{f(R, \phi)(R^2 - r^2)\, d\phi}{R^2 + r^2 - 2\,Rr\cos(\phi - \theta)}. \tag{4.7–4}$$

The analytic function $f(z)$ will now be represented in terms of its real and imaginary parts U and V. Thus $f(R, \phi) = U(R, \phi) + iV(R, \phi)$, $f(r, \theta) = U(r, \theta) + iV(r, \theta)$ and Eq. (4.7–4) becomes

$$U(r, \theta) + iV(r, \theta) = \frac{1}{2\pi}\int_0^{2\pi} \frac{[U(R, \phi) + iV(R, \phi)][R^2 - r^2]\, d\phi}{R^2 + r^2 - 2\,Rr\cos(\phi - \theta)}. \tag{4.7–5}$$

By equating the real parts on either side of this equation, we obtain the following formula:

POISSON INTEGRAL FORMULA (FOR INTERIOR OF A CIRCLE)

$$U(r, \theta) = \frac{1}{2\pi}\int_0^{2\pi} \frac{U(R, \phi)(R^2 - r^2)\, d\phi}{R^2 + r^2 - 2\,Rr\cos(\phi - \theta)}. \tag{4.7–6}$$

A corresponding expression relates $V(r, \theta)$ and $V(R, \phi)$ and is obtained by our equating imaginary parts in Eq. (4.7–5).

Equation (4.7–6), the Poisson integral formula, is important. The formula yields the value of the harmonic function $U(r, \theta)$ everywhere inside a circle of radius R, provided we know the values $U(R, \phi)$ assumed by U on the circumference of the circle.

Since we required that $f(z)$ be analytic inside, and on, the circle of radius R, the reader must assume that the function $U(R, \phi)$ in Eq. (4.7–6) is continuous. In fact, this condition can be relaxed† to allow $U(R, \phi)$ to have a finite number of finite "jump" discontinuities. The Poisson integral formula will remain valid.

In Exercise 4 we develop a formula comparable to Eq. (4.7–6) that works outside the circle; i.e., if the value of a harmonic function is known on the circumference of a circle, the formula will tell us the value of this function everywhere outside the circle. The formula presumes that the harmonic function sought is bounded (its magnitude is $\leq$ a constant) in the domain external to the circle.

All the work in this section is based on the writings of a Frenchman, Siméon-Denis Poisson, who lived from 1781 to 1840. The reader has perhaps encountered his name in connection with probability theory (the Poisson distribution) or electrostatics (the Poisson equation). He is credited with helping to bring mathematical analysis to bear on the subjects of electricity, magnetism, and elasticity.

---

† See R. V. Churchill and J. W. Brown, *Complex Variables and Applications*, 5th ed. (New York: McGraw-Hill, 1990), Section 95.

EXAMPLE 1

An electrically conducting tube of unit radius is separated into two halves by means of infinitesimal slits. The top half of the tube ($R = 1, 0 < \phi < \pi$) is maintained at an electrical potential of 1 volt while the bottom half ($R = 1, \pi < \phi < 2\pi$) is at -1 volt. Find the potential at an arbitrary point r, θ inside the tube (see Fig. 4.7–2). Assume there is a dielectric material inside the tube.

Solution

Since the electrostatic potential is a harmonic function (see Section 2.6), the Poisson integral formula is applicable. From Eq. (4.7–6), with $R = 1$, we have

$$U(r, \theta) = \frac{1}{2\pi}\int_0^{\pi} \frac{(1 - r^2)\, d\phi}{1 + r^2 - 2r\cos(\phi - \theta)} - \frac{1}{2\pi}\int_{\pi}^{2\pi} \frac{(1 - r^2)\, d\phi}{1 + r^2 - 2r\cos(\phi - \theta)}. \tag{4.7–7}$$

In each integral we make the change of variables $x = \phi - \theta$; from a standard table of integrals we find the following formula, which is valid for $a^2 > b^2 \geq 0$:

$$\int \frac{dx}{a + b\cos x} = \frac{2}{\sqrt{a^2 - b^2}} \tan^{-1}\left[\frac{\sqrt{a^2 - b^2}\tan(x/2)}{a + b}\right].$$

Using this formula in Eq. (4.7–7) with $a = 1 + r^2$, $b = -2r$, we obtain

$$\begin{aligned} U(r, \theta) = \frac{1}{\pi}\Bigg[& 2\tan^{-1}\left(\frac{1 + r}{1 - r}\tan\left(\frac{\pi}{2} - \frac{\theta}{2}\right)\right) - \tan^{-1}\left(\frac{1 + r}{1 - r}\tan\left(\pi - \frac{\theta}{2}\right)\right) \\ & - \tan^{-1}\left(\frac{1 + r}{1 - r}\tan\left(-\frac{\theta}{2}\right)\right)\Bigg]. \end{aligned} \tag{4.7–8}$$

Since the arctangent is a multivalued function, some care must be taken in applying this formula. Recalling that the values assumed by U on the boundaries are ± 1, we

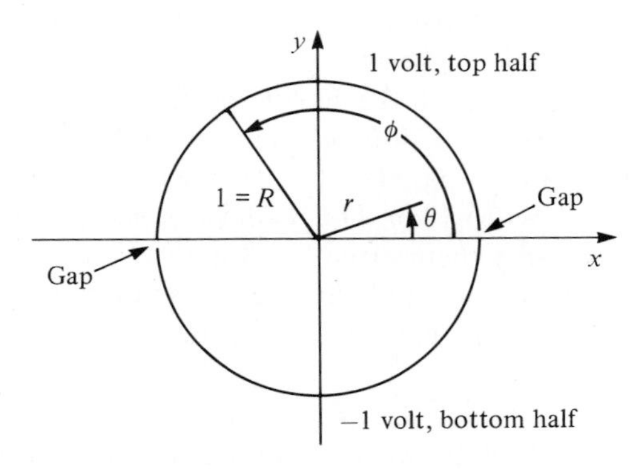

Figure 4.7–2

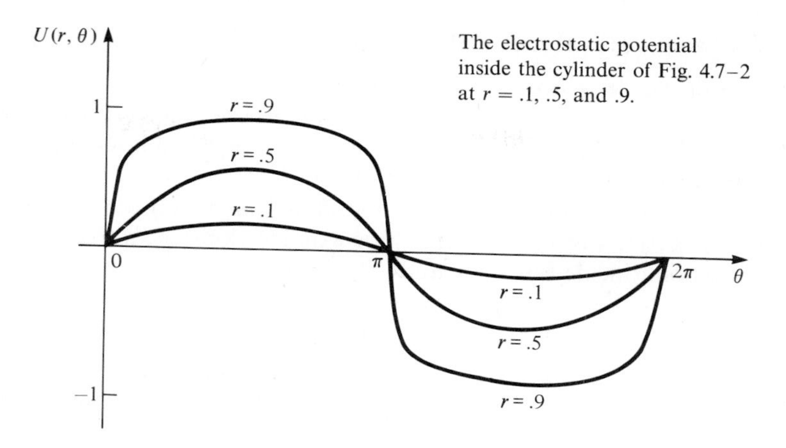

Figure 4.7–3

can use physical reasoning[†] to conclude that $-1 \leq U(r, \theta) \leq 1$ when $r \leq 1$. Moreover, the values of the arctangents must be chosen so that $U(r, \theta)$ is continuous for all $r < 1$, and $U(1, \theta)$ is discontinuous only at the slits $\theta = 0$ and $\theta = \pi$. ◀

For purposes of computation, it is convenient to have Eq. (4.7–8) in a different form. Recalling that $\tan(n\pi + \alpha) = \tan \alpha$, where α is any angle and n any integer, and that the arctangent and tangent are both odd functions,[‡] we can recast Eq. (4.7–8) as follows:

$$U(r, \theta) = \frac{2}{\pi}\left[\tan^{-1}\left[\frac{1+r}{1-r}\tan\left(\frac{\pi}{2} - \frac{\theta}{2}\right)\right] + \tan^{-1}\left[\frac{1+r}{1-r}\tan\frac{\theta}{2}\right] \pm \frac{\pi}{2}\right], \tag{4.7–9}$$

where the minus sign is to be used in front of $\pi/2$ when $0 < \theta < \pi$ and the plus sign when $\pi < \theta < 2\pi$. All values of the arctangents are evaluated so as to satisfy $-\pi/2 \leq \tan^{-1}(\ ) \leq \pi/2$. This is the convention used in most pocket calculators and in the FORTRAN and BASIC computer languages. With Eq. (4.7–9) and a simple computer program we have evaluated $U(r, \theta)$ for various values of r and plotted them in Fig. 4.7–3.

[†] This same conclusion can also be reached through the maximum and minimum principles (Exercises 13 and 14, Section 4.6), although strictly speaking these principles only apply when the voltage in the region is continuous. Our boundary voltage has two points of discontinuity.

[‡] An odd function $f(x)$ is one that satisfies $f(x) = -f(-x)$.

The Dirichlet Problem for a Half Plane (the Infinite Line Boundary)

As in the case of the circle, we will state our new Dirichlet problem in the w-plane. Our problem is to find a function $\phi(u, v)$ that is harmonic in the upper half plane (the domain $v > 0$). In addition, $\phi(u, v)$ must satisfy a prescribed boundary condition $\phi(u, 0)$ on the line $v = 0$.

Let $f(w) = \phi(u, v) + i\psi(u, v)$ be a function that is analytic for $v \geq 0$. Consider the closed semicircle C (see Fig. 4.7–4) whose base extends along the u-axis from $-R$ to $+R$. Let z be a point inside this semicircle. Then, from the Cauchy integral formula,

$$f(z) = \frac{1}{2\pi i}\oint_C \frac{f(w)}{w - z}\,dw, \tag{4.7–10}$$

where the integral is taken along the base and arc of the given contour. Now, since z lies inside the semicircle, observe that $\bar{z}$ must lie in the space $v < 0$ and is therefore outside the semicircle. Hence, the function $f(w)/(w - \bar{z})$ is analytic on and interior to the contour C. Thus, from the Cauchy–Goursat theorem,

$$0 = \frac{1}{2\pi i}\oint_C \frac{f(w)}{(w - \bar{z})}\,dw. \tag{4.7–11}$$

Let us subtract Eq. (4.7–11) from Eq. (4.7–10):

$$f(z) = \frac{1}{2\pi i}\oint_C f(w)\left(\frac{1}{w - z} - \frac{1}{w - \bar{z}}\right)dw = \frac{1}{2\pi i}\oint_C \frac{(z - \bar{z})f(w)}{(w - z)(w - \bar{z})}\,dw. \tag{4.7–12}$$

We break the integral along C into two parts: along the base ($v = 0$, $-R \leq u \leq R$), which we symbolize with $\rightarrow$; and along the arc of radius R, which we symbolize with $\curvearrowleft$. Thus

$$f(z) = \frac{1}{2\pi i}\int_{\rightarrow} \frac{(z - \bar{z})f(w)}{(w - z)(w - \bar{z})}\,dw + \frac{1}{2\pi i}\int_{\curvearrowleft} \frac{(z - \bar{z})f(w)}{(w - z)(w - \bar{z})}\,dw. \tag{4.7–13}$$

Consider the Cartesian representations $z = x + iy$, $\bar{z} = x - iy$, and $w = u + iv$. We see that $z - \bar{z} = 2iy$ and that $w = u$ along the base. Hence, the first integrand on the right in the preceding equation can be rewritten as

$$\frac{(z - \bar{z})f(w)}{(w - z)(w - \bar{z})} = \frac{2iyf(u)}{[u - (x + iy)][u - (x - iy)]} = \frac{2iyf(u)}{(u - x)^2 + y^2},$$

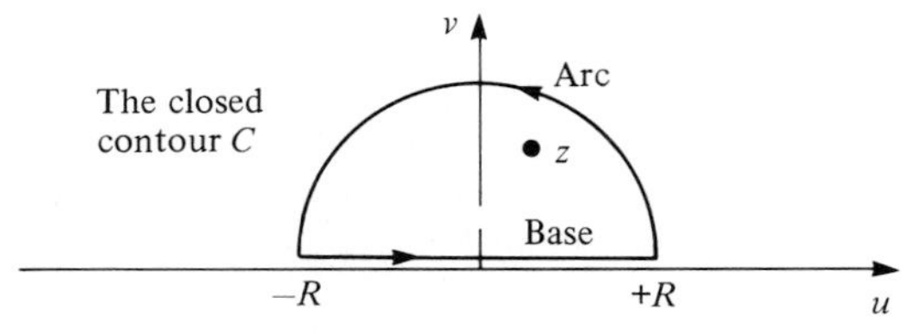

Figure 4.7–4

so that Eq. (4.7–13) becomes

$$f(z) = \frac{y}{\pi} \int_{-R}^{+R} \frac{f(u)\, du}{(u-x)^2 + y^2} + \frac{y}{\pi} \int \frac{f(w)\, dw}{(w-z)(w-\bar{z})}. \tag{4.7–14}$$

In Exercise 8 of this section we discover that, as the radius R of the arc tends to infinity, the value of the integral along the arc in Eq. (4.7–14) goes to zero. The proof requires our assuming the existence of a constant m such that $|f(w)| \le m$ for all Im $w \ge 0$, that is, $|f(w)|$ is bounded in the upper half plane. Passing to the limit $R \to \infty$, we find that Eq. (4.7–14) simplifies to

$$f(z) = \frac{y}{\pi} \int_{-\infty}^{+\infty} \frac{f(u)\, du}{(u-x)^2 + y^2}. \tag{4.7–15}$$

If $f(w)$ is known on the whole real axis of the w-plane, (that is, along $w = u$), this formula will yield the value of the function at any arbitrary point $w = z$, provided Im $z > 0$.

Let us now rewrite $f(z)$ and $f(w)$ explicitly in terms of real and imaginary parts. With $f(z) = \phi(x, y) + i\psi(x, y)$ and $f(w) = \phi(u, v) + i\psi(u, v)$ we obtain, from Eq. (4.7–15),

$$\phi(x, y) + i\psi(x, y) = \frac{y}{\pi} \int_{-\infty}^{+\infty} \frac{\phi(u, 0) + i\psi(u, 0)}{(u-x)^2 + y^2}\, du.$$

Equating the real parts in this equation, we arrive at the following formula:

POISSON INTEGRAL FORMULA (FOR THE UPPER HALF PLANE)

$$\phi(x, y) = \frac{y}{\pi} \int_{-\infty}^{+\infty} \frac{\phi(u, 0)\, du}{(u-x)^2 + y^2} \tag{4.7–16}$$

A corresponding equation, relating $\psi(x, y)$ and $\psi(u, 0)$ is obtained if we equate imaginary parts.

Equation (4.7–16), called the Poisson Integral Formula for the Upper Half Plane, will yield the value of a harmonic function $\phi(x, y)$ anywhere in the upper half plane provided ϕ is already completely known over the entire real axis. It can be shown that this is the only solution to the Dirichlet problem that is *bounded* in the upper half plane. Without this restriction other solutions can be found.

In our derivation we assumed that $\phi(u, v)$ is the real part of a function $f(u, v)$, which is analytic for Im $v \ge 0$. This would require that $\phi(u, 0)$ in Eq. (4.7–16) be continuous for $-\infty < u < \infty$. Actually, this requirement can be relaxed to permit $\phi(u, 0)$ to have a finite number of finite jumps. We then can still use Eq. (4.7–16).

EXAMPLE 2

As indicated in Fig. 4.7–5, the upper half space Im $w > 0$ is filled with a heat-conducting material. The boundary $v = 0$, $u > 0$ is maintained at a temperature of 0 while the boundary $v = 0$, $u < 0$ is kept at temperature T_0. Find the steady state distribution of temperature $\phi(x, y)$ throughout the conducting material.

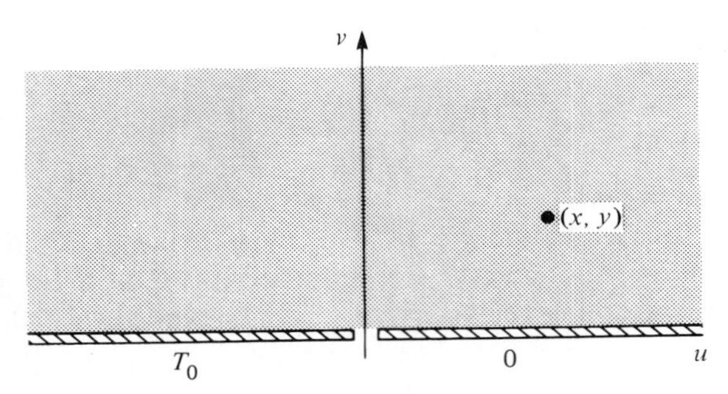

Figure 4.7–5

Solution

Since, as shown in Section 2.6, the temperature is a harmonic function, the Poisson integral formula is directly applicable. We have $\phi(u, 0) = T_0$, $u < 0$; and $\phi(u, 0) = 0$, $u > 0$. Thus

$$\phi(x, y) = \frac{y}{\pi} \int_{-\infty}^{0} \frac{T_0\, du}{(u - x)^2 + y^2} + \frac{y}{\pi} \int_{0}^{\infty} \frac{0\, du}{(u - x)^2 + y^2}.$$

The second integral is zero. In the first we make the change of variables $p = x - u$. Thus

$$\phi(x, y) = \frac{T_0 y}{\pi} \int_{x}^{\infty} \frac{dp}{p^2 + y^2} = \frac{T_0}{\pi} \tan^{-1} \frac{p}{y} \bigg|_{x}^{\infty} = \frac{T_0}{\pi} \left[\frac{\pi}{2} - \tan^{-1} \frac{x}{y} \right]. \qquad (4.7\text{–}17)$$

From the trigonometric identity $\tan^{-1} s = \pi/2 - \tan^{-1}(1/s)$ we see that the expression in the brackets on the right side of Eq. (4.7–17) is $\tan^{-1}(y/x) = \theta$, where θ is the polar angle associated with the point (x, y). From physical considerations we require that $0 \le \phi(x, y) \le T_0$, that is, the maximum and minimum temperatures are on the boundary. This is satisfied if we take θ as the principal polar angle in the space $y \ge 0$. Thus

$$\phi(x, y) = \frac{T_0}{\pi} \theta, \qquad 0 \le \theta \le \pi.$$

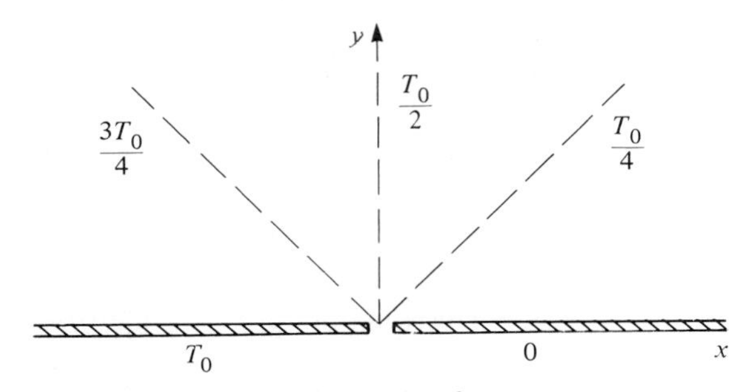

Figure 4.7–6

Some surfaces on which the temperature displays constant values are exhibited as rays in Fig. 4.7–6. ◀

EXERCISES

1. a) In Fig. 4.7–2 (Example 1) explain physically why the potential should be zero along the ray going from the origin to $x = 1$, $y = 0$. Show that our answer, Eq. (4.7–9), confirms this result by considering the limit $\theta \to 0+$ (θ shrinks to zero through positive values). Be sure to use the correct formula (with the minus sign).

b) Verify that Eq. (4.7–9) does satisfy the following boundary conditions:

$$\lim_{r\to 1} U(r, \theta) = \begin{cases} 1, & 0 < \theta < \pi, \\ -1, & \pi < \theta < 2\pi. \end{cases}$$

c) With the aid of a programmable calculator or a computer, obtain the numerical data and plot curves like those in Fig. 4.7–3 for $r = 0.25$ and $r = 0.75$.

2. a) The temperature of the surface of a cylinder of radius 5 is maintained as shown in Fig. 4.7–7. Show that the steady state temperature inside the cylinder, $U(r, \theta)$, is given by

$$U(r, \theta) = \frac{100}{\pi}\left[\tan^{-1}\left(\frac{5+r}{5-r}\tan\left(\frac{\pi}{2} - \frac{\theta}{2}\right)\right) + \tan^{-1}\left(\frac{5+r}{5-r}\tan\frac{\theta}{2}\right) + C\right],$$

where $C = 0$ for $0 < \theta < \pi$ and $C = \pi$ for $\pi < \theta < 2\pi$. In each case the arctangent satisfies $-\pi/2 \le \tan^{-1}(\ldots) \le \pi/2$

b) Verify that the following boundary conditions are satisfied by the preceding formula:

$$\lim_{r\to 5} U(r, \theta) = \begin{cases} 100, & 0 < \theta < \pi, \\ 0, & \pi < \theta < 2\pi. \end{cases}$$

c) Using a programmable calculator or a computer to obtain the numerical data, plot $U(r, \theta)$, $0 \le \theta \le 2\pi$, for $r = 1$, 2, and 4.

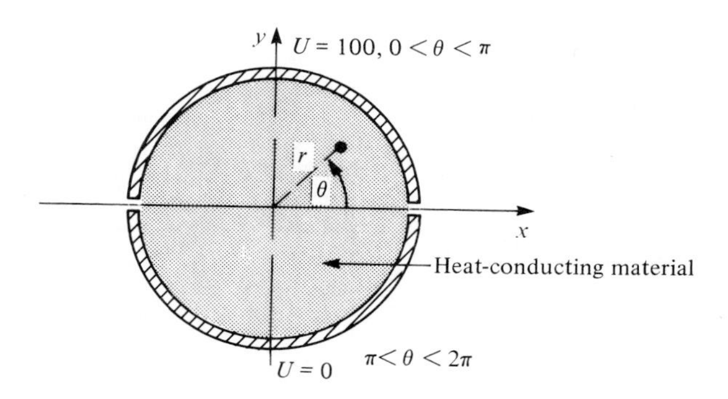

Figure 4.7–7

3. Use the Poisson integral formula for the circle to show that if the electrostatic potential on the surface of any cylinder is constant and equal to V_0, then the potential everywhere inside is equal to V_0.

4. The purpose of this exercise is to obtain a formula for a function that is harmonic in the unbounded domain external to a circle. The function is required to achieve prescribed values on the circumference of the circle and to be bounded in the domain outside the circle. This is known as the *external Dirichlet problem* for a circle.

a) Consider a function $f(w)$ that is analytic at all points in the w-plane that satisfy $|w| \geq R$. Place two circles, as shown in Fig. 4.7–8, in the w-plane centered at $w = 0$. Their radii are R and R', while $R < R'$. Let $z = re^{i\theta}$ be a point lying within the annular domain formed by the circles. Thus $R < r < R'$. Show that

$$f(z) = \frac{1}{2\pi i} \oint_{|w|=R'} \frac{f(w)}{w - z}\, dw + \frac{1}{2\pi i} \oint_{|w|=R} \frac{f(w)}{w - z}\, dw.$$

Note the direction of integration around the two circles.

Hint: See Exercise 21a in Section 4.5.

b) Let $z_1 = R^2/\bar{z}$. Note that this point lies inside the inner circle. Show that

$$0 = \frac{1}{2\pi i} \oint_{|w|=R'} \frac{f(w)}{w - z_1}\, dw + \frac{1}{2\pi i} \oint_{|w|=R} \frac{f(w)}{w - z_1}\, dw.$$

c) Subtract the formula of part (b) from that derived in part (a) and show that

$$f(z) = \frac{1}{2\pi i} \oint_{|w|=R'} \frac{f(w)(z - R^2/\bar{z})}{(w - z)(w - R^2/\bar{z})}\, dw + \frac{1}{2\pi i} \oint_{|w|=R} \frac{f(w)(z - R^2/\bar{z})}{(w - z)(w - R^2/\bar{z})}\, dw.$$

d) Assume that $|f(w)| \leq m$ (a constant) when $|w| > R$. Let $R' \to \infty$. Show that in the limit the integral around $|w| = R'$ goes to zero.

Hint: Use the ML inequality.

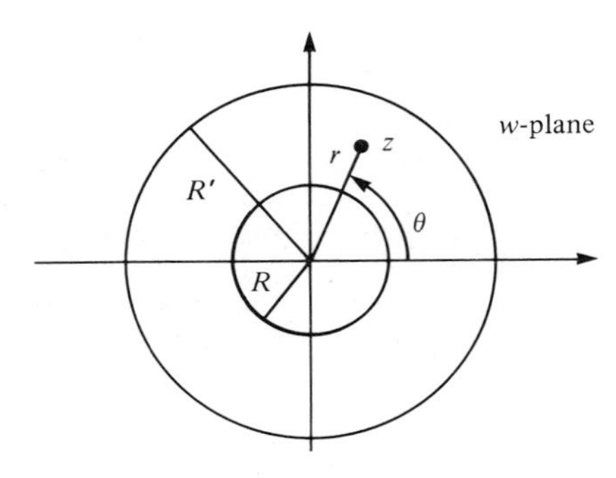

Figure 4.7–8

e) Rewrite the remaining integral in part (c) by using polar coordinates $z = re^{i\theta}$, $w = Re^{i\phi}$. Put $f(z) = U(r, \theta) + iV(r, \theta)$. Show that

$$U(r, \theta) = \frac{1}{2\pi}\int_0^{2\pi} \frac{U(r, \phi)(r^2 - R^2)\, d\phi}{R^2 + r^2 - 2\, Rr\cos(\phi - \theta)}, \qquad r > R. \qquad (4.7\text{–}18)$$

Hint: Study the derivation of the Poisson integral formula for the interior of a circle.

5. In this exercise we have an external Dirichlet problem with a circular boundary.

a) Consider the configuration shown in Exercise 2. Assume that the temperature distribution along the cylindrical surface is the same as was given in that exercise but that now the region external to the cylinder is filled with a heat-conducting material. Use Eq. (4.7–18), derived in Exercise 4, to show that $U(r, \theta)$, the temperature distribution, is given for $r > 5$ by

$$U(r, \theta) = \frac{100}{\pi}\left[\tan^{-1}\left(\frac{5 + r}{r - 5}\tan\left(\frac{\pi}{2} - \frac{\theta}{2}\right)\right) + \tan^{-1}\left(\frac{5 + r}{r - 5}\tan\frac{\theta}{2}\right) + C\right],$$

where C and the arctangent are defined as in Exercise 2.

b) Verify that the expression derived in part (a) fulfills the boundary conditions

$$\lim_{r \to 5} U(r, \theta) = 100, \qquad 0 < \theta < \pi, \qquad \text{and} \qquad \lim_{r \to 5} U(r, \theta) = 0, \qquad \pi < \theta < 2\pi.$$

c) Plot $U(r, \pi/2)$ for $5 \le r \le 50$. What temperature is created at $r = \infty$ by the configuration?

6. a) An electrically conducting metal sheet is perpendicular to the y-axis and passes through $y = 0$, as shown in Fig. 4.7–9. The right half of the sheet, $x > 0$, is maintained at an electrical potential of V_0 volts while the left half, $x < 0$, is maintained at a voltage $-V_0$. Show that in the half space, $y \ge 0$, the electrostatic potential is given by

$$\phi(x, y) = V_0 - \frac{2V_0}{\pi}\tan^{-1}\frac{y}{x} = V_0 - \frac{2V_0}{\pi}\,\mathrm{Im}(\mathrm{Log}\, z),$$

where $0 \le \tan^{-1}(y/x) \le \pi$.

b) Sketch the equipotential lines (or surfaces) on which $\phi(x, y) = V_0/2$, $\phi(x, y) = 0$, $\phi(x, y) = -V_0/2$.

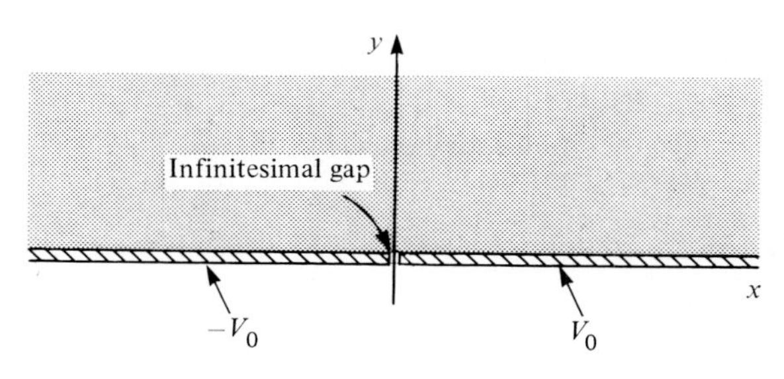

Figure 4.7–9

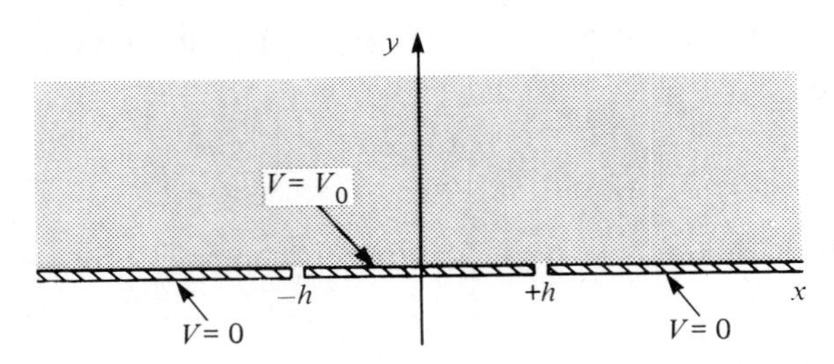

Figure 4.7–10

c) Find the components of the electric field E_x and E_y at $x = 1$, $y = 1$, and draw a vector representing the field at this point (see Section 2.6).

7. a) The surface $y = 0$ is maintained at an electrostatic potential $V(x)$ described by

$$-\infty < x < -h, \quad V(x) = 0; \qquad -h < x < h, \quad V(x) = V_0; \qquad h < x < \infty, \quad V(x) = 0.$$

This potential distribution is shown in Fig. 4.7–10. Show that the electrostatic potential in the space $y > 0$ is given by

$$\phi(x, y) = \frac{V_0}{\pi}\left[-\tan^{-1}\frac{x-h}{y} + \tan^{-1}\frac{x+h}{y}\right].$$

b) Consider the limit $y = 0+$ in the result for part (a). Show that the boundary conditions are satisfied for the cases $x < -h$, $-h < x < h$, and $x > h$ when we evaluate the arctangent according to $-\pi/2 \le \tan^{-1}(\ldots) \le \pi/2$.

c) Show that along the line $x = 0$, we have, when $y \gg h$,

$$\phi(0, y) \approx \frac{V_0}{\pi}\frac{2h}{y}.$$

Hint: For small arguments $\tan^{-1} w \approx w$.

d) Let $h = 1$. Plot $\phi(x, 0.5)$ for $-5 \le x \le 5$. Let $V_0 = 1$.

8. Complete the proof of the Poisson integral formula for the upper half plane by showing that the integral over the arc (of radius R) in Eq. (4.7–14) goes to zero as $R \to \infty$.

Hint: Explain why $|w - z| \ge |w| - |z|$ and $|w - \bar{z}| \ge |w| - |\bar{z}|$ in the integrand. We must assume that $|f(w)| \le m$ for $\operatorname{Im} w \ge 0$. Now explain why $|f(w)/[(w - z)(w - \bar{z})]| \le m/(R - |z|)^2$ on the path of integration. Calling the right side of this inequality M, show that the magnitude of the integral on the arc is $\le M\pi R$ if we ignore the factor y/π in Eq. (4.7–14). Allow $R \to \infty$.

9. Derive a Poisson integral formula analogous to Eq. (4.7–16) that applies in the *lower* half plane.

Hint: Begin with the contour of integration shown in Fig. 4.7–11.

Answer:

$$\phi(x, y) = -\frac{y}{\pi}\int_{-\infty}^{+\infty} \frac{\phi(u, 0)\, du}{(u - x)^2 + y^2}, \qquad y < 0.$$

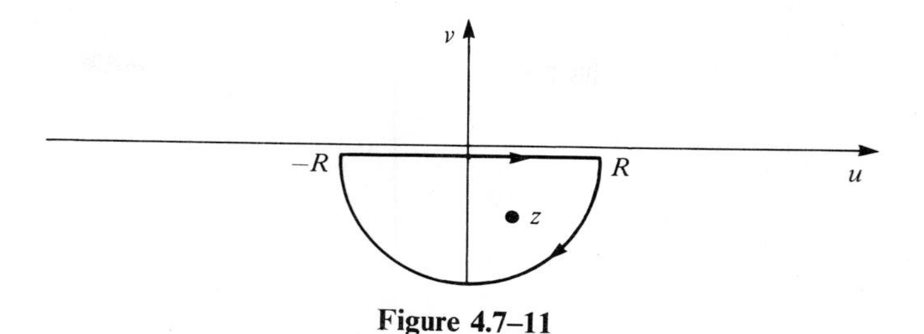

Figure 4.7–11

APPENDIX TO CHAPTER 4

GREEN'S THEOREM IN THE PLANE†

Let us prove our theorem for a simple closed contour C that has this property: If a straight line is drawn parallel to either the x- or y-axis, it will intersect C at two points at most. Such a curve is shown in Fig. A4–1. The points A and B are the pair of points on C having the smallest and largest x-coordinates. These coordinates are a and b, respectively. Now consider

$$\iint_R \frac{-\partial P}{\partial y}\,dx\,dy,$$

where the integral is taken over the region R consisting of the contour C and its interior. The function $P(x, y)$ is assumed to be continuous and to have continuous first partial derivatives in R.

The contour C creates two distinct paths connecting A and B. They are given by the equations $y = g(x)$ and $y = f(x)$ (see Fig. A4–1). Thus

$$\iint_R \frac{-\partial P}{\partial y}\,dx\,dy = -\int_{x=a}^{x=b}\left[\int_{y=f(x)}^{y=g(x)} \frac{\partial P}{\partial y}\,dy\right]dx = -\int_a^b P(x, y)\Big|_{y=f(x)}^{y=g(x)}\,dx$$

$$= \int_a^b [P(x, f(x)) - P(x, g(x))]\,dx$$

$$= \int_a^b P(x, f(x))\,dx + \int_b^a P(x, g(x))\,dx.$$

† See Section 4.3.

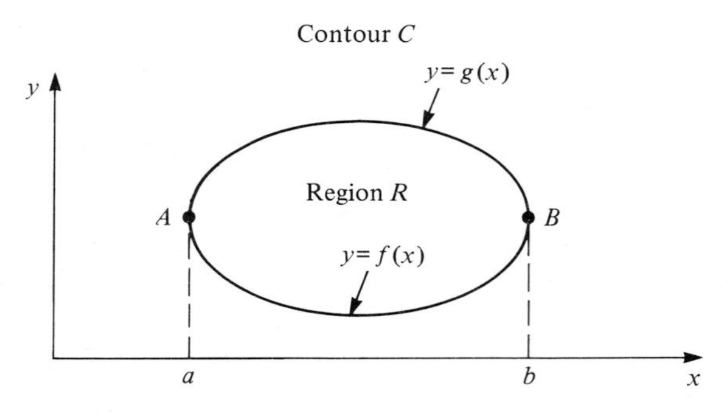

Figure A4–1

This final pair of integrals, one from a to b, the other from b to a, together form the line integral $\int P(x, y)\, dx$ taken around the contour C in the positive (counterclockwise) direction. Thus

$$\iint_R -\frac{\partial P}{\partial y}\, dx\, dy = \oint_C P(x, y)\, dx. \tag{A4–1}$$

In a similar way (refer to Fig. A4–2) we have for a function $Q(x, y)$, which has the same properties of continuity as $P(x, y)$:

$$\iint_R \frac{\partial Q}{\partial x}\, dx\, dy = \int_e^d [Q(n(y), y) - Q(m(y), y)]\, dy$$

$$= \int_e^d Q(n(y), y)\, dy + \int_d^e Q(m(y), y)\, dy = \oint_C Q(x, y)\, dy. \tag{A4–2}$$

Adding Eqs. (A4–1) and (A4–2) we obtain our desired result:

$$\oint_C P(x, y)\, dx + Q(x, y)\, dy = \iint_R \left(\frac{\partial Q}{\partial x} - \frac{\partial P}{\partial y}\right) dx\, dy.$$

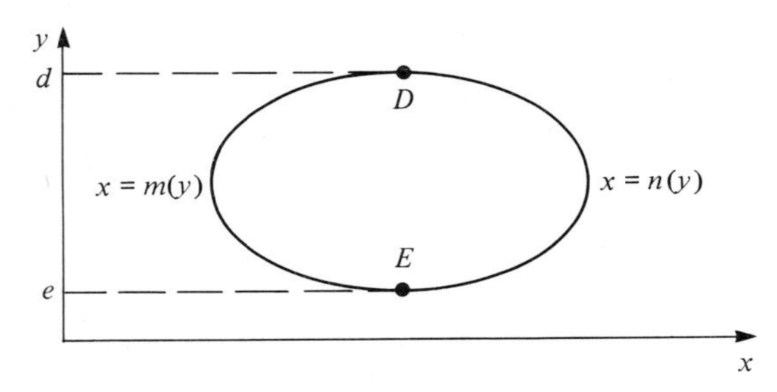

Figure A4–2

The condition that straight lines drawn parallel to the x- or y-axes intersect C at two points, at most, is easily relaxed. A slightly more complicated proof is required.

The following theorem, related to Green's theorem, is of use in complex variable theory. It enables one to prove a converse of the Cauchy–Goursat theorem (see Exercise 11, Section 4.3).

THEOREM 14 Let $P(x, y)$, $Q(x, y)$, $\partial P/\partial y$ and $\partial Q/\partial x$ be continuous in a simply connected domain D. Suppose $\oint P\,dx + Q\,dy = 0$ around every simple closed contour in D. Then, $\partial Q/\partial x = \partial P/\partial y$ in D. □

To prove this theorem suppose that $\partial Q/\partial x - \partial P/\partial y > 0$ at the point x_0, y_0 in D. Then, since both these derivatives are continuous, we can find in D a circle C centered at x_0, y_0 such that $\partial Q/\partial x - \partial P/\partial y > 0$ inside and on C. Applying Green's theorem to this circle, we have

$$\oint_C P\,dx + Q\,dy = \iint_{\text{interior of } C} \left(\frac{\partial Q}{\partial x} - \frac{\partial P}{\partial y}\right) dx\,dy. \qquad \text{(A4–3)}$$

Since the integrand on the right is positive, the integral on the right must result in a positive number. The integral on the left is, by hypothesis, zero. Hence, we have obtained a contradiction.

Assuming $\partial Q/\partial x - \partial P/\partial y < 0$, we can go through a similar argument and also obtain a contradiction. Thus since $\partial Q/\partial x$ is neither less than nor greater than $\partial P/\partial y$, we have, at (x_0, y_0), $\partial Q/\partial x = \partial P/\partial y$.

CHAPTER 5

Infinite Series Involving a Complex Variable

5.1 INTRODUCTION AND REVIEW OF REAL SERIES

The reader doubtless already knows something about infinite series involving functions of a real variable. For example, the equations

$$e^x = \sum_{n=0}^{\infty} \frac{x^n}{n!} = 1 + x + \frac{x^2}{2!} + \frac{x^3}{3!} + \cdots \tag{5.1–1}$$

and

$$\frac{1}{1-x} = \sum_{n=0}^{\infty} \frac{(x+1)^n}{2^{n+1}} = \frac{1}{2} + \frac{(x+1)}{4} + \frac{(x+1)^2}{8} + \cdots \tag{5.1–2}$$

should look at least slightly familiar. The infinite sums appearing here are called power series because each term is of the form $(x - x_0)^n$, where x_0 is a real constant and $n \geq 0$ is an integer. When the real variable x assumes certain allowable values, the infinite sums correctly represent the functions on the left.

In this chapter we will learn to obtain power series representations of functions of a complex variable z. These series will contain terms of the form $(z - z_0)^n$ instead of $(x - x_0)^n$. Here z_0 is a complex constant. We are interested in such series for several reasons. First, there is a close connection between the analyticity of a function and its ability to be represented by a power series. Secondly, just as for real series, complex series are useful for numerical approximation. Without the benefit of either tables or a pocket calculator we could use the first three terms in the sum in Eq. (5.1–1) to establish that $e^{0.2} \doteq 1 + 0.2 + 0.02 = 1.22$, a result accurate to better than 0.15%. Even better accuracy is obtained with more terms. Similarly, without resorting to Eq. (3.1–1) we can use a power series in the variable z to obtain a good approximation to $e^{0.2+0.1i}$. Power series are used to obtain numerical approximations for the value of an integral that cannot be evaluated in terms of standard functions. For example, given the problem of determining $\int_0^{0.2} (e^x - 1)/x \, dx$, the student might first seek, in a table of integrals, the antiderivative of $(e^x - 1)/x$. None will be found. However, an approximate evaluation of the integral can be had by replacing e^x in the integrand by perhaps the first three terms in Eq. (5.1–1). Having done so, we find that we must now determine $\int_0^{0.2} (1 + x/2) \, dx$, which is an easy matter. Given a series expansion for e^z, we could proceed in a similar manner to find $\int_0^{0.2+0.1i} (e^z - 1)/z \, dz$ taken along some path. Power series are also used extensively in the solution of differential equations.

Toward the end of this chapter we will consider Laurent series, which contain $(z - z_0)$ raised to positive and negative powers. These series lead directly to the subject of residues (see Chapter 6). Residues are enormously useful in the rapid evaluation of contour integrals taken along closed contours.

A power series expansion of the real function $f(x)$ takes the form

$$f(x) = \sum_{n=0}^{\infty} c_n(x - x_0)^n = c_0 + c_1(x - x_0) + c_2(x - x_0)^2 + \cdots. \tag{5.1–3}$$

Here, c_0, c_1, etc. are called the coefficients of the expansion. For finding the coefficients the reader probably recalls the straightforward procedure

$$c_n = \frac{f^{(n)}(x_0)}{n!}. \tag{5.1–4}$$

When coefficients obtained in this way are used in Eq. (5.1–3), we say that Eq. (5.1–3) is the Taylor series expansion of $f(x)$ about x_0. The coefficients in Eqs. (5.1–1) and (5.1–2) can be derived through the use of Eq. (5.1–4). However, attempting the expansion

$$x^{1/2} = \sum_{n=0}^{\infty} c_n x^n, \tag{5.1–5}$$

we land in trouble since $x^{1/2}$ does not possess a first- or higher-order derivative at $x = 0$. Thus not all functions of x have a Taylor series expansion.

Let us consider another difficulty posed by Taylor series. If in Eq. (5.1–2) we substitute $x = -1/2$ in the series as well as in the function on the left, we have

$$\frac{2}{3} = \frac{1}{2} + \frac{1}{8} + \frac{1}{32} + \frac{1}{128} + \cdots. \tag{5.1–6}$$

Adding the first four terms on the right and getting 0.664, we become convinced that the infinite sum approaches $2/3 = 0.6666\ldots$. However, with $x = 2$ on both sides in

Eq. (5.1–2) we obtain

$$-1 = \frac{1}{2} + \frac{3}{4} + \frac{9}{8} + \cdots. \tag{5.1–7}$$

Clearly, the infinite sum will not yield the numerical value -1. What we have seen–that there are values of x for which the series is not valid—is typical of Taylor series for functions of real variables.

The problems just discussed also occur when we study power series expansions of functions of a complex variable. We will thus be concerned with the question of which functions $f(z)$ can be represented by a power series, how the series is obtained, and for what values of z the series actually does represent or "converge to" $f(z)$. We must also carefully consider the meaning of the term "converge" when applied to series generally. In studying these matters we will also achieve a better understanding of the behavior of real power series.

EXERCISES†

Use Eq. (5.1–4) to obtain the indicated Taylor series. Give the first four nonzero terms. Put each series in the form $\sum_{n=0}^{\infty} u_n(x)$, where terms that are identically zero are not included. State the general term $u_n(x)$. Review the "ratio test" in an elementary calculus book, and use this test to establish those values of x for which each series is convergent. Notice that in certain of these problems the ratio test cannot establish convergence or divergence at the endpoints of the interval of convergence.

1. $1/(1 - x)$ about $x = 0$

2. $2/(1 - x)$ about $x = 2$

3. e^x about $x = 1$

4. $\sin x$ about $x = 0$

5. $\sin x$ about $x = \pi/4$

6. $\operatorname{Log} x$ about $x = e$

7. Recall the "nth term test" from elementary calculus. It asserts that a necessary condition for a series to converge is that the general term $u_n(x)$ (nth term) of the series must go to zero as $n \to \infty$. If the limit fails to exist, or if it exists but is nonzero, the series diverges. The test cannot be used to establish convergence. Use this test to prove that the series obtained in Exercise 1 diverges at $x = \pm 1$ and that the series obtained in Exercise 2 diverges at $x = 1$ and $x = 3$. Can the test be used on the series obtained in Exercise 6 at $x = 0$ and $x = 2e$? Explain.

5.2 CONVERGENCE OF COMPLEX SERIES

Before we enter a discussion of power series involving a complex variable we must try to define precisely the term "convergent" as applied to series in general. The definition is also applicable to real series. In addition, we should know something

---

† These exercises are primarily for review.

about the properties of convergent series and how to test for convergence.

Let

$$u_1(z) + u_2(z) + \cdots = \sum_{j=1}^{\infty} u_j(z) \tag{5.2–1}$$

be a series with an infinite number of terms in which the members $u_1(z), u_2(z), \ldots$ are functions of the complex variable z. A function $u_j(z)$ is assumed to exist for each positive[†] value of the integer index j. Examples of such series are $e^z + e^{2z} + e^{3z} + \cdots$ and $(z-1)/1! + (z-1)^2/2! + (z-1)^3/3! + \cdots$. Thus for the first series, $u_j(z) = e^{jz}$, while for the second, $u_j(z) = (z-1)^j/j!$.

We define

$$\begin{aligned} S_1(z) &= u_1(z), \\ S_2(z) &= u_1(z) + u_2(z), \\ S_3(z) &= u_1(z) + u_2(z) + u_3(z), \end{aligned} \tag{5.2–2}$$

and so forth.

A sum

$$S_n(z) = \sum_{j=1}^{n} u_j(z) \tag{5.2–3}$$

involving the first n terms in Eq. (5.2–1) is called the *nth partial sum* of the infinite series. Partial sums can be arranged to create a sequence of functions of the form

$$S_1(z), S_2(z), S_3(z), \ldots, S_n(z), \ldots. \tag{5.2–4}$$

For example, if the series shown in Eq. (5.2–1) were $\sum_{j=1}^{\infty} e^{jz}$, then the sequence of partial sums would be $e^z, e^z + e^{2z}, e^z + e^{2z} + e^{3z}, \ldots$.

Let us assume that the sequence of partial sums in Eq. (5.2–4) has a limit $S(z)$ as $n \to \infty$. Mathematically we can define this as stated below.

DEFINITION Limit of a sequence

For the sequence of functions $S_1(z), S_2(z), \ldots, S_n(z), \ldots$ we say that there is a limit $S(z)$ as $n \to \infty$ if, given a number $\varepsilon > 0$, we determine that there exists an integer N such that

$$|S(z) - S_n(z)| < \varepsilon, \qquad \text{for all } n > N. \tag{5.2–5}$$

We then write $\lim_{n\to\infty} S_n(z) = S(z)$. □

A sequence may or may not have a limit. Typically, a limit will exist only when z is confined to some set of points in the complex plane.

[†] Sometimes we will use series that begin with $u_0(z)$ or $u_N(z)$, where $N \neq 1$. The discussion presented here applies substantially unchanged to series of this sort. If necessary, such series can be reindexed to begin with $j = 1$.

The concept of a limit of a sequence is not confined to sequences formed from the partial sums of a series. (In Appendix A of this chapter we discuss fractals—an interesting topic involving sequences of complex numbers.) However when $S_1(z)$, $S_2(z), \dots$ are partial sums we can make the following definition.

DEFINITION Ordinary convergence

If the sequence of partial sums in Eq. (5.2–2) formed from the infinite series in Eq. (5.2–1) has limit $S(z)$ as $n \to \infty$, we say that the infinite series in Eq. (5.2–1) *converges* and has sum $S(z)$. The set of all values of z for which the series converges is called its *region of convergence.*[†] For z lying in this region we write

$$S(z) = \sum_{j=1}^{\infty} u_j(z). \tag{5.2–6}$$

If the sequence of partial sums does not have a limit, we say that the series in Eq. (5.2–1) *diverges.* □

In informal language the series in Eq. (5.2–1) "converges to $S(z)$" if in Eq. (5.2–2) the succession of approximations to the whole series shown in Eq. (5.2–1) yields $S(z)$. Equation (5.2–5) says that we can make the difference between the sum of the series and the partial sum of the series, $S_n(z)$, be smaller than any preassigned number ε by taking enough terms in the partial sum. We must make n, the number of terms used to form the partial sum, exceed a number N. Typically, N depends on ε and z.

Occasionally we will deal with series of the form $\sum_{j=1}^{\infty} c_j$, where c_1, $c_2, \dots$ are complex constants. The definition of convergence of such series is identical to that just given for a series of functions, except that the sum and partial sums are independent of z. The convergence or divergence of such series likewise is independent of z.

EXAMPLE 1

Show that

$$\sum_{j=1}^{\infty} z^{j-1} = 1 + z + z^2 + \cdots = S(z) = \frac{1}{1-z}, \qquad |z| < 1. \tag{5.2–7}$$

Solution

This result should look plausible since, if we replace z by x in Eq. (5.2–7), we obtain a familiar real geometric series and its sum.

The nth partial sum of our series is $S_n(z) = 1 + z + z^2 + \cdots + z^{n-1}$. Notice that $S_n(z) - zS_n(z) = (1 + z + \cdots + z^{n-1}) - (z + z^2 + \cdots + z^n) = 1 - z^n$, so that $S_n(z)(1 - z) = 1 - z^n$ or, for $z \neq 1$,

$$S_n(z) = \frac{1 - z^n}{1 - z} = 1 + z + z^2 + \cdots + z^{n-1}. \tag{5.2–8}$$

[†] The region of convergence is often, but not always, a region in the sense in which this word was defined in Chapter 1.

Since the sum in Eq. (5.2–7) is $S(z) = 1/(1 - z)$, we have

$$|S(z) - S_n(z)| = \left|\frac{1 - (1 - z^n)}{1 - z}\right| = \frac{|z|^n}{|1 - z|}. \tag{5.2–9}$$

Referring to Eq. (5.2–5), we require for convergence that

$$\frac{|z|^n}{|1 - z|} < \varepsilon, \qquad \text{for } n > N, \tag{5.2–10}$$

or that

$$\left|\frac{1}{z}\right|^n > \frac{1}{\varepsilon|1 - z|}.$$

Taking logarithms of the preceding, we obtain

$$n \operatorname{Log}\left|\frac{1}{z}\right| > \operatorname{Log}\frac{1}{\varepsilon|1 - z|}. \tag{5.2–11}$$

Inside the disc $|z| < 1$ we have $|1/z| > 1$ and $\operatorname{Log}|1/z| > 0$. The inequality in (5.2–11) can be rearranged as

$$n > \frac{\operatorname{Log}\dfrac{1}{\varepsilon|1 - z|}}{\operatorname{Log}\left|\dfrac{1}{z}\right|} = \frac{\operatorname{Log}\varepsilon|1 - z|}{\operatorname{Log}|z|}. \tag{5.2–12}$$

If we choose N as a positive integer that equals or exceeds the right side of Eq. (5.2–12) and take $n > N$, then Eq. (5.2–10) is satisfied. Hence, $|S(z) - S_n(z)|$ in Eq. (5.2–9) will be $< \varepsilon$. Thus Eq. (5.2–5) is satisfied and, according to our definition of convergence, Eq. (5.2–7) has been proved. ◀

With Theorem 2 (the nth term test for complex series), which follows in a few pages, we will readily prove that the series in Eq. (5.2–7) diverges for $|z| \geq 1$. Thus the relationship described in Eq. (5.2–7) holds only if $|z| < 1$.

EXAMPLE 2

Use the known sum of a geometric series, shown in Eq. (5.2–7), to sum $\sum_{n=0}^{\infty} e^{inz} = 1 + e^{iz} + e^{i2z} + \cdots$. State the region of convergence.

Solution

We replace z by e^{iz} in Eq. (5.2–7) and obtain

$$1 + e^{iz} + e^{i2z} + \cdots = \frac{1}{1 - e^{iz}}, \qquad |e^{iz}| < 1.$$

Now

$$|e^{iz}| = |e^{i(x+iy)}| = |e^{ix}e^{-y}| = |e^{ix}||e^{-y}| = e^{-y}.$$

To justify the final step on the right, recall that $|e^{ix}| = 1$ and that e^{-y} is not negative. The requirement for convergence of our given series $|e^{iz}| < 1$ now becomes $e^{-y} < 1$. A sketch of e^{-y} against y reveals that the inequality is satisfied only if $y > 0$, that is, $\text{Im } z > 0$. ◀

Comment: In this example we used a familiar series and its known sum (Eq. 5.2–7) to obtain a new series and its sum. We achieved this by making a change of variable in the original series. The technique is a useful one, and we will employ it on other occasions.

Often the functions $u_j(z)$ in an infinite series appear in the form

$$u_j(z) = R_j(x, y) + iI_j(x, y), \tag{5.2–13}$$

where R_j and I_j are the real and imaginary parts of u_j. Thus

$$\sum_{j=1}^{\infty} u_j(z) = \sum_{j=1}^{\infty} R_j(x, y) + iI_j(x, y). \tag{5.2–14}$$

THEOREM 1 The convergence of both the real series $\sum_{j=1}^{\infty} R_j(x, y)$ and $\sum_{j=1}^{\infty} I_j(x, y)$ is a necessary and sufficient condition for the convergence of $\sum_{j=1}^{\infty} u_j(z)$, where $u_j(z) = R_j(x, y) + iI_j(x, y)$. If $\sum_{j=1}^{\infty} R_j(x, y)$ and $\sum_{j=1}^{\infty} I_j(x, y)$ converge to the functions $R(x, y)$ and $I(x, y)$, respectively, then $\sum_{j=1}^{\infty} u_j(z)$ converges to $S(z) = R(x, y) + iI(x, y)$. Conversely if $\sum_{j=1}^{\infty} u_j(z)$ converges to $S(z) = R(x, y) + iI(x, y)$, then $\sum_{j=1}^{\infty} R_j(x)$ converges to $R(x, y)$ and $\sum_{j=1}^{\infty} I_j(x)$ converges to $I(x, y)$. □

The rather simple proofs will not be presented here.

EXAMPLE 3

Consider the series

$$1 + e^{-y} \cos x + e^{-2y} \cos 2x + \cdots,$$

which is obtained by taking the real part of each term in the series of Example 2. Find the sum of this new series.

Solution

The series of Example 2 converges to $1/(1 - e^{iz})$ in the domain $\text{Im } z > 0$. Thus the series of the present example converges to $\text{Re}[1/(1 - e^{iz})]$ in this domain. We have

$$\text{Re}\left[\frac{1}{1 - e^{iz}}\right] = \text{Re}\left[\frac{e^{-iz/2}}{e^{-iz/2} - e^{iz/2}}\right] = \text{Re}\left[\frac{\cos(z/2) - i\sin(z/2)}{-2i\sin(z/2)}\right]$$

$$= \text{Re}\left[\frac{i}{2}\cot\left(\frac{z}{2}\right) + \frac{1}{2}\right].$$

Now

$$\cot\left(\frac{z}{2}\right) = \frac{\sin x - i\sinh y}{\cosh y - \cos x}$$

(Exercise 29, Section 3.2). Thus the sum of our series is

$$\frac{\sinh y}{2(\cosh y - \cos x)} + \frac{1}{2}.$$ ◀

We should recall that two convergent real series can be added term by term. The resulting series converges to a function obtained by adding the sums of the two original series. Convergent complex series can also be added in this way. Subtraction of series is also performed in an analogous manner.

The nth term test, derived for real series in elementary calculus, also applies to complex series, and is described by the following theorem.

THEOREM 2 nth term test

The series $\sum_{n=1}^{\infty} u_n(z)$ diverges if

$$\lim_{n\to\infty} u_n(z) \neq 0 \tag{5.2–15a}$$

or, equivalently, if

$$\lim_{n\to\infty} |u_n(z)| \neq 0. \quad \square \tag{5.2–15b}$$

Loosely speaking, if the terms of a series do not ultimately start shrinking to zero, then the series cannot converge. Notice that the phrase "only if" does not appear in this theorem; there are divergent series whose nth term goes to zero as $n \to \infty$. The test can, however, be used to establish the divergence of some series, as the following example illustrates.

EXAMPLE 4

Use Theorem 2 to show that the series of Example 1, $\sum_{j=1}^{\infty} z^{j-1}$, diverges for $|z| \geq 1$.

Solution

We take $u_n(z) = z^{n-1}$ and $|u_n(z)| = |z^{n-1}| = |z|^{n-1}$. If $|z| = 1$, then $\lim_{n\to\infty} |u_n(z)| = \lim_{n\to\infty} 1^{n-1} = 1$. Since this limit is nonzero, the series diverges if $|z| = 1$. For $|z| > 1$, $\lim_{n\to\infty} |z|^{n-1} = \infty$, which is clearly nonzero. The series again diverges.

Notice that with $|z| < 1$ we have $\lim_{n\to\infty} |z|^{n-1} = 0$. However, this is of no use in proving that the series converges for $|z| < 1$. ◀

The notions of absolute and conditional convergence that apply to real series also apply to complex series, as we see from the following definition.

DEFINITION Absolute and conditional convergence

The series $\sum_{j=1}^{\infty} u_j(z)$ is called *absolutely convergent* if the series $\sum_{j=1}^{\infty} |u_j(z)|$ is convergent. □

Thus a series is absolute convergent if the series formed by taking the magnitude of each of its terms is convergent. It is possible that the series consisting of the magnitude of each term diverges while the original series converges. We then have the following:

DEFINITION Conditional convergence

The series $\sum_{j=1}^{\infty} u_j(z)$ is called *conditionally convergent* if it converges but $\sum_{j=1}^{\infty} |u_j(z)|$ diverges. □

An absolutely convergent complex series has some useful properties that we will now list without proof. In each case the proof is similar to that for a corresponding property of absolute convergent real series. The reader is referred to standard texts on the calculus of real variables.

THEOREM 3 An absolutely convergent series converges in the ordinary sense. □

THEOREM 4 The sum of an absolutely convergent series is independent of the order in which the terms are added. □

THEOREM 5 Two absolutely convergent series can be multiplied together in the same way as one multiplies two polynomials. The resulting series is absolutely convergent. Its sum, which is independent of how the terms are arranged, is the product of the sums of the two original series. □

Theorem 5 has the following implications: Suppose we have two series that are both absolutely convergent when z is confined to some region. They are $\sum_{j=1}^{\infty} u_j(z) = S(z)$ and $\sum_{j=1}^{\infty} v_j(z) = T(z)$. Recalling how we might multiply two polynomials, we consider the product $(u_1 + u_2 + \cdots)\cdot(v_1 + v_2 + \cdots)$. According to the theorem, we have, in our region,

$$(u_1 v_1) + (u_1 v_2 + u_2 v_1) + (u_1 v_3 + u_2 v_2 + u_3 v_1) + \cdots = S(z)T(z). \quad (5.2\text{–}16)$$

The series on the left is absolutely convergent. The parentheses can be dropped without affecting the result. This particular way of multiplying series is called the Cauchy product. If c_1 is $u_1 v_1$ (the terms in the first set of parentheses), c_2 is $u_1 v_2 + u_2 v_1$, etc., then we have, for the set of terms in the nth set of parentheses,

$$c_n(z) = \sum_{j=1}^{n} u_j v_{n-j+1},$$

and we can rewrite Eq. (5.2–16) as

$$\sum_{n=1}^{\infty} c_n(z) = S(z)T(z).$$

The sum of the resulting series is the product of the sums of the two original series.

The ratio test, which is used to establish the absolute convergence of a real infinite series, also applies to complex series.

THEOREM 6 Ratio test

For the series $\sum_{j=1}^{\infty} u_j(z)$ consider

$$\Gamma = \lim_{j\to\infty} \left| \frac{u_{j+1}(z)}{u_j(z)} \right|; \tag{5.2–17}$$

then

a) the series converges if $\Gamma < 1$, and the convergence is absolute;

b) the series diverges if $\Gamma > 1$;

c) Eq. (5.2–17) provides no information about the convergence of the series if the indicated limit fails to exist or if $\Gamma = 1$. □

It is an easy matter to use Eq. (5.2–17) to show that the series of Example 1, $\sum_{j=1}^{\infty} z^{j-1}$, is absolutely convergent for $|z| < 1$. Instead we shall consider a slightly harder example as follows.

EXAMPLE 5

Use the ratio test and the nth term test to investigate the convergence of

$$\sum_{j=1}^{\infty} (-1)^j j 2^{j+1} z^{2j} = -4z^2 + 16z^4 - 48z^6 + \cdots.$$

Solution

$$u_j = (-1)^j j 2^{j+1} z^{2j} \quad \text{and} \quad u_{j+1} = (-1)^{j+1}(j+1)2^{j+2} z^{2(j+1)}.$$

Thus

$$\left| \frac{u_{j+1}}{u_j} \right| = \left| \frac{(-1)^{j+1}(j+1)2^{j+2}z^{2j+2}}{(-1)^j j 2^{j+1} z^{2j}} \right| = \left| \frac{j+1}{j} 2z^2 \right|.$$

In the preceding equation we put $j \to \infty$ on the right side and notice that $(j+1)/j$ equals 1 in this limit. From Eq. (5.2–17) we have, for our series,

$$\Gamma = 2|z^2|.$$

Now, we use part (a) of Theorem 6 and set $\Gamma < 1$. This requires that

$$2|z^2| < 1 \quad \text{or} \quad |z| < \frac{1}{\sqrt{2}}.$$

Thus our series converges absolutely if z lies inside a circle of radius $1/\sqrt{2}$ centered at the origin. Using part (b) of the same theorem we readily show that the series diverges for $|z| > 1/\sqrt{2}$, that is, when z lies outside the circle just mentioned.

On $|z| = 1/\sqrt{2}$ we have $\Gamma = 1$, which provides no information about convergence. However, observe that on $|z| = 1/\sqrt{2}$ we have

$$|u_j(z)| = j2^{j+1}\left(\frac{1}{\sqrt{2}}\right)^{2j} = j\frac{2^{j+1}}{2^j} = 2j.$$

Clearly, as $j \to \infty$, we do not have $|u_j| \to 0$. Thus according to Theorem 2, the series diverges on $|z| = 1/\sqrt{2}$. ◀

EXERCISES

Use the nth term test to establish the divergence of the following series in the indicated regions.

1. $\sum_{n=1}^{\infty} nz^n$ for $|z| \geq 1$

2. $\sum_{n=1}^{\infty} \left(\frac{n}{n+1}\right)(2z)^n$ for $|z| \geq 1/2$

3. $\sum_{n=0}^{\infty} e^{inz}$ for $\operatorname{Im} z \leq 0$

4. $\sum_{j=1}^{\infty} \frac{1}{(z-1)^j}$ for $|z-1| \leq 1$

Use the ratio test to prove that the following series are absolutely convergent in the indicated domains.

5. $\sum_{n=0}^{\infty} n! e^{in^2 z}$ for $\operatorname{Im} z > 0$

6. $\sum_{n=1}^{\infty} \frac{1}{z^n n!}$ for $|z| > 0$

7. $\sum_{j=1}^{\infty} \frac{2^{-2j}(z+i)^{j+1}}{j}$ for $|z+i| < 4$

8. $\sum_{j=0}^{\infty} \frac{2^j}{(z-1-i)^{2j}}$ for $|z-1-i| > \sqrt{2}$

9. $\sum_{n=1}^{\infty} n!\left(\frac{z}{n}\right)^n$ for $|z| < e$

Hint: Recall, from elementary calculus, $\lim_{n\to\infty}(1 + 1/n)^n = e$.

10. Show that the sum of the series $e^{-y}\sin x + e^{-2y}\sin 2x + e^{-3y}\sin 3x + \cdots$, for $y > 0$, is

$$\frac{e^{-y}\sin x}{1 - 2e^{-y}\cos x + e^{-2y}}.$$

In Exercises 11 and 12 find the sum of the series by making a suitable change of variables in Eq. (5.2–7). In each case state where in the complex plane the series converges to the sum.

11. $1 + (z-1)^2 + (z-1)^4 + (z-1)^6 + \cdots$

12. $1 + 1/z + 1/z^2 + 1/z^3 + \cdots$

13. a) Prove that $\sum_{n=1}^{\infty} nz^{n-1} = 1/(1-z)^2$ for $|z| < 1$ by using series multiplication, i.e., Theorem 5, and the result $\sum_{j=1}^{\infty} z^{j-1} = 1/(1-z)$ for $|z| < 1$.

b) Using the result derived in part (a), and an additional series multiplication, show that

$$\sum_{n=1}^{\infty} \frac{n(n+1)}{2} z^{n-1} = \frac{1}{(1-z)^3}.$$

The identity $\sum_{j=1}^{n} j = n(n+1)/2$ can be helpful here.

14. Consider the series $\sum_{k=1}^{\infty} kz^{k-1} = 1 + 2z + 3z^2 + \cdots$. Show that its nth partial sum, $1 + 2z + 3z^2 + \cdots + nz^{n-1}$ is given by

$$\frac{z^n[n(z-1)-1]+1}{(1-z)^2} \qquad \text{for } z \neq 1.$$

Hint: Refer to Eq. (5.2–8), which gives the nth partial sum of the series studied in Example 1. Notice that if we differentiate the series in that equation we will obtain the series for the $(n-1)$st partial sum in the present problem.

15. Using the kind of argument presented in Example 1, prove that $\sum_{k=1}^{\infty} kz^{k-1} = 1/(1-z)^2$ for $|z| < 1$. Use the nth partial sum derived in Exercise 14.

5.3 UNIFORM CONVERGENCE OF SERIES

In Example 1 of the previous section we showed that the series $\sum_{j=1}^{\infty} z^{j-1}$ converges to $1/(1-z)$ for $|z| < 1$. To accomplish this we found a number N such that $|S(z) - S_n(z)| < \varepsilon$ for $n > N$. Here S_n was the nth partial sum, and $S(z)$ the sum of the given series. We should recall (see Eq. 5.2–12) that our N depended on both ε and z.

Suppose, however, that in the course of establishing the convergence of a series we find, when z lies in some region R, an expression for N that is independent of z. Such a series has special properties and is called *uniformly convergent* in R. More precisely:

DEFINITION Uniform convergence

The series $\sum_{j=1}^{\infty} u_j(z)$ whose nth partial sum is $S_n(z)$ is said to converge uniformly to $S(z)$ in a region R if, for any $\varepsilon > 0$, there exists a number N *independent* of z so that for all z in R

$$|S(z) - S_n(z)| < \varepsilon, \qquad \text{for all } n > N. \quad \square \qquad (5.3\text{–}1)$$

There are various ways to show that a series is uniformly convergent in a region. In Exercise 8 of this section, for example, the reader will encounter one method. The series $\sum_{j=1}^{\infty} z^{j-1}$ is shown to be uniformly convergent in the disc $|z| \leq r$ (where $r < 1$) since we are able to find the required value of N in Eq. (5.3–1) that depends on only r and ε. This approach is time consuming. It is often easier to establish uniform convergence with the Weierstrass M test, which is described as follows:

THEOREM 7 Weierstrass M test

Let $\sum_{j=1}^{\infty} M_j$ be a convergent series whose terms $M_1, M_2, \ldots$ are all positive constants. The series $\sum_{j=1}^{\infty} u_j(z)$ converges uniformly in a region R if

$$|u_j(z)| \leq M_j, \qquad \text{for all } z \text{ in } R. \quad \square \qquad (5.3\text{–}2)$$

The test asserts that a series of complex functions $u_1(z) + u_2(z) + \cdots$ is uniformly convergent in a region R if there exists a convergent series of positive constants $M_1 + M_2 + \cdots$ each of whose terms $M_1, M_2, \ldots$ equals or exceeds the magnitude of the corresponding term $|u_1|, |u_2|, \ldots$ throughout R.

If Eq. (5.3–2) is satisfied, then $\sum_{j=1}^{\infty} u_j(z)$ is also absolutely convergent in R. This follows from the "comparison test" that the reader encountered for real series; it also applies to complex series.

Theorem 7, the M test, has a counterpart in the theory of real series. The proof of the real case, which is not difficult, is virtually identical to that for the complex case. The reader is referred to standard texts on real calculus for the derivation.†

Karl Weierstrass (1815–1897), the inventor of the test, was a German mathematician known for his contributions to infinite series and the modern theory of functions.

EXAMPLE 1

Use the M test to show that $\sum_{j=1}^{\infty} z^{j-1}$ is uniformly convergent in the disc $|z| \le 3/4$.

Solution

From Eq. (5.2–7) with $z = 3/4$ or from a previous knowledge of real geometric series, we know that if $M_j = (3/4)^{j-1}$, then

$$\sum_{j=1}^{\infty} M_j = 1 + \frac{3}{4} + \left(\frac{3}{4}\right)^2 + \cdots = \frac{1}{1 - \frac{3}{4}}. \tag{5.3–3}$$

Now, with $u_j = z^{j-1}$ we have the given series:

$$\sum_{j=1}^{\infty} u_j = \sum_{j=1}^{\infty} z^{j-1} = 1 + z + z^2 + \cdots. \tag{5.3–4}$$

If $|z| \le 3/4$, then the magnitude of each term of the series in Eq. (5.3–4) is less than or equal to the corresponding term in Eq. (5.3–3), for example, $|z^2| \le (3/4)^2$, $|z^3| \le (3/4)^3$, etc., so that $|u_j| \le M_j$ and the M test is satisfied in the given region.

Comment: We can, by an identical argument, show that $\sum_{j=1}^{\infty} z^{j-1}$ is uniformly convergent in any circular region $|z| \le r$ provided $r < 1$. The proof involves replacing 3/4 by r in the preceding example. ◀

We have dwelt on uniform convergence because series with this feature have some useful properties, which we will now list. The scope of this text does not allow for a derivation of all these properties. Most are derived as part of the exercises at the end of this section, and the reader is referred to more advanced texts for justification of the others.‡

---

† See, for example, W. Kaplan, *Advanced Calculus*, 4th ed. (Reading, MA: Addison-Wesley, 1991), Section 6.13.

‡ See for example: T. Apostol, *Mathematical Analysis*, 2nd ed. (Reading, MA: Addison-Wesley, 1974), Chapter 9.

THEOREM 8 Let $\sum_{j=1}^{\infty} u_j(z)$ converge uniformly in a region R to $S(z)$. Let $f(z)$ be bounded in R, that is, $|f(z)| \leq k$ (k is constant) throughout R. Then, in R,

$$\sum_{j=1}^{\infty} f(z)u_j(z) = f(z)u_1(z) + f(z)u_2(z) + \cdots = f(z)S(z).$$

The series converges uniformly to $f(z)S(z)$. □

For example, since $\sum_{j=1}^{\infty} z^{j-1}$ converges uniformly for $|z| \leq r$, where $r < 1$, $\sum_{j=1}^{\infty} e^z z^{j-1}$ converges uniformly in the same region. (Recall that since e^z is continuous in the disc $|z| \leq r$ it must be bounded in this region.) From Eq. (5.2–7) and the preceding theorem we have, in this region, $\sum_{j=1}^{\infty} e^z z^{j-1} = e^z/(1 - z)$.

THEOREM 9 Let $\sum_{j=1}^{\infty} u_j(z)$ be a series converging uniformly to $S(z)$ in R. If all the functions $u_1(z), u_2(z), \ldots$ are continuous in R, then so is the sum $S(z)$. □

For example, all the terms in the series $1 + z + z^2 + \cdots$ are continuous in the z-plane. We showed previously that this series is uniformly convergent if $|z| \leq r$ $(r < 1)$. Thus the sum must be continuous in $|z| \leq r$. A glance at the sum $1/(1 - z)$ (see Example 1 in the preceding section) reveals this to be true.

THEOREM 10 Term-by-term integration

Let $\sum_{j=1}^{\infty} u_j(z)$ be a series that is uniformly convergent to $S(z)$ in R and let all the terms $u_1(z), u_2(z), \ldots$ be continuous in R. If C is a contour in R, then

$$\int_C S(z)\, dz = \sum_{j=1}^{\infty} \int_C u_j(z)\, dz = \int_C u_1(z)\, dz + \int_C u_2(z)\, dz + \cdots, \tag{5.3–5}$$

that is, when a uniformly convergent series of continuous functions is integrated term by term the resulting series has a sum that is the integral of the sum of the original series. □

To illustrate this theorem we again consider

$$\frac{1}{1 - z} = 1 + z + z^2 + \cdots, \qquad |z| \leq r, \quad \text{and} \quad r < 1.$$

We integrate this series term by term along a contour C that lies entirely inside the disc $|z| \leq r$. The contour is assumed to connect the points $z = 0$ and $z = z'$. Thus, from Eq. (5.3–5),

$$\int_0^{z'} \frac{1}{1 - z}\, dz = \int_0^{z'} dz + \int_0^{z'} z\, dz + \int_0^{z'} z^2\, dz + \cdots. \tag{5.3–6}$$

The left side involves an integrand $1/(1 - z)$, which is the derivative of an analytic branch of a multivalued function. Such integrations were considered in Section 4.4.

We have

$$\int_0^{z'} \frac{1}{1-z}\,dz = -\text{Log}(1-z)\Big|_0^{z'} = \text{Log}\left(\frac{1}{1-z'}\right), \tag{5.3–7}$$

where we have elected to use the principal branch of $\log(1-z)$ since it is analytic in the disc under consideration. Notice this branch satisfies $-\text{Log}\, w = \text{Log}(1/w)$.

The result in Eq. (5.3–7) can be used on the left in Eq. (5.3–6); the integrations on the right in Eq. (5.3–6) are readily performed. We have, finally,

$$\text{Log}\frac{1}{(1-z')} = z' + \frac{(z')^2}{2} + \frac{(z')^3}{3} + \cdots = \sum_{j=1}^{\infty} \frac{(z')^j}{j}, \qquad |z'| \le r, \quad r < 1. \tag{5.3–8}$$

As a practical matter, the restriction on z' in this equation can be written simply $|z'| < 1$.

As shown in the exercises, Theorem 10 can be used to establish the following theorem.

THEOREM 11 Analyticity of the sum of a series

If $\sum_{j=1}^{\infty} u_j(z)$ converges uniformly to $S(z)$ for all z in R and if $u_1(z), u_2(z), \ldots$ are all analytic in R, then $S(z)$ is analytic in R. □

The preceding theorem guarantees the existence of the derivative of the sum of a uniformly convergent series of analytic functions. We have a way to arrive at this derivative:

THEOREM 12 Term-by-term differentiation

Let $\sum_{j=1}^{\infty} u_j(z)$ converge uniformly to $S(z)$ in a region R. If $u_1(z), u_2(z), \ldots$ are all analytic in R, then at any interior point of this region

$$\frac{dS}{dz} = \sum_{j=1}^{\infty} \frac{du_j(z)}{dz}. \quad \square \tag{5.3–9}$$

The theorem states that when a uniformly convergent series of analytic functions is differentiated term by term, we obtain the derivative of the sum of the original series.

We illustrate the preceding with our geometric series. Since $1/(1-z) = \sum_{j=1}^{\infty} z^{j-1} = 1 + z + z^2 + \cdots$, where convergence is uniform for $|z| \le r$ (with $r < 1$), we have upon differentiation

$$\frac{d}{dz}\frac{1}{1-z} = \frac{1}{(1-z)^2} = \frac{d}{dz}(1 + z + z^2 + \cdots) = 1 + 2z + 3z^2 + \cdots,$$

or

$$\frac{1}{(1-z)^2} = \sum_{j=1}^{\infty} jz^{j-1}, \qquad |z| < r, \quad r < 1. \tag{5.3–10}$$

This result was obtained relatively painlessly with the use of Theorem 12. Without this theorem the same result could be had from the more difficult manipulation required in Exercises 13, 14, and 15 of Section 5.2.

EXERCISES

Use the Weierstrass M test to establish the uniform convergence of the following series in the indicated regions.

1. $\displaystyle\sum_{n=0}^{\infty} \frac{e^{in^2x}}{n!}$ for $|z| \leq r$, $r < \infty$

Hint: Recall the infinite series for e.

2. $\displaystyle\sum_{n=0}^{\infty} \frac{z^{4n}}{n+1}$ for $|z| \leq 0.99$

3. $\displaystyle\sum_{n=1}^{\infty} \frac{(1+ni)z^n}{n^3}$ for $|z| \leq 1$

Hint: The ratio test tells us that $\sum_{n=1}^{\infty} c/n^2$ is convergent, where c is any constant.

4. $\displaystyle\sum_{n=1}^{\infty} \frac{e^{nz}}{n^2}$ for $\text{Re}\, z \leq 0$ (see the hint for Exercise 3)

5. $\displaystyle\sum_{n=1}^{\infty} (-n + e^{-n})z^n$ for $|z| \leq \rho$, $\rho < 1$

Hint: See Eq. (5.3–10).

6. Use the fifth partial sum (that is, the first five terms) in Eq. (5.3–8) to find an approximate value of Log 2. Sum the terms with a pocket calculator. Compare your result, to three significant digits, with the exact value.

7. Observe that $\pi = 6 \text{ Im Log}((\sqrt{3} + i)/2)$. Use the first three terms of the series in Eq. (5.3–8) to make an approximate evaluation of π. How good is your result?

8. In this exercise we show that the series $\sum_{j=1}^{\infty} z^{j-1}$ converges uniformly to $1/(1-z)$ in the disc $|z| \leq r$, where $r < 1$. The proof requires that we obtain a value for N satisfying Eq. (5.3–1).

a) Explain why $\text{Log}(1/|z|) \geq \text{Log}(1/r)$ in the disc.

b) Prove $1/|1-z| \leq 1/(1-r)$ and $\text{Log}[1/(\varepsilon|1-z|)] \leq \text{Log}[1/(\varepsilon(1-r))]$ for $|z| \leq r$. Take $\varepsilon > 0$.

c) Assume that $0 < \varepsilon < 1/2$. Show that in the disc $|z| \leq r$ we have

$$\frac{\text{Log}\,\dfrac{1}{\varepsilon|1-z|}}{\text{Log}\,\dfrac{1}{|z|}} \leq \frac{\text{Log}\,\dfrac{1}{\varepsilon(1-r)}}{\text{Log}\,\dfrac{1}{r}}.$$

d) Observe that the left side of the preceding inequality is equal to the right side of Eq. (5.2–12). Explain why we can take N as any positive integer greater than or equal to

$$\frac{\text{Log}\,\dfrac{1}{\varepsilon(1-r)}}{\text{Log}\,\dfrac{1}{r}},$$

and Eq. (5.3–1) will be satisfied. Observe that N is independent of z.

9. We will prove Theorem 10, that is, establish that in a region R,

$$\sum_{j=1}^{\infty} \int_C u_j(z)\, dz = \int_C S(z)\, dz,$$

where $\sum_{j=1}^{\infty} u_j(z)$ is assumed to converge uniformly to $S(z)$. From the definition of convergence we must prove that, given $\varepsilon_1 > 0$, there exists a number N such that

$$\left| \int_C S(z)\, dz - \sum_{j=1}^{n} \int_C u_j(z)\, dz \right| < \varepsilon_1 \qquad \text{for } n > N. \tag{5.3–11}$$

a) Notice that for a finite sum

$$\sum_{j=1}^{n} \int_C u_j(z)\, dz = \int_C \sum_{j=1}^{n} u_j(z)\, dz$$

since u_1, u_2, etc. are assumed to be continuous (see Eq. 4.2–7c). Explain why the following is true:

$$\left| \int_C S(z)\, dz - \sum_{j=1}^{n} \int_C u_j(z)\, dz \right| = \left| \int_C S(z)\, dz - \int_C \sum_{j=1}^{n} u_j(z)\, dz \right|$$
$$= \left| \int_C \left[S(z) - \sum_{j=1}^{n} u_j(z) \right] dz \right|.$$

b) Given $\varepsilon > 0$, determine that there exists an N such that

$$\left| \int_C \left[S(z) - \sum_{j=1}^{n} u_j(z) \right] dz \right| < \varepsilon L, \qquad \text{for } n > N, \tag{5.3–12}$$

where L is the length of C. (Recall the definition of a uniformly convergent series and use the ML inequality.) Now observe that if we take $\varepsilon = \varepsilon_1/L$ in Eq. (5.3–12) we have proved Eq. (5.3–11).

10. Morera's Theorem states that if, in a simply connected domain D, $\oint f(z)\, dz = 0$ for every possible simple closed contour of integration in D, then $f(z)$ is analytic in D. A proof is given in Exercise 11, Section 4.3. Use this theorem as well as Theorem 10 to prove Theorem 11.

11. Prove Theorem 12.

Hint: Consider the series $\sum_{j=1}^{\infty} u_j(z') = S(z')$, where convergence is uniform in a region R of the z'-plane. Let z be any point in R except a boundary point. Consider a simple closed contour C lying in R and enclosing z. Then, dividing the preceding series by $(z' - z)^2$ and invoking Theorem 8, we have

$$\frac{S(z')}{2\pi i(z' - z)^2} = \frac{u_1(z')}{2\pi i(z' - z)^2} + \frac{u_2(z')}{2\pi i(z' - z)^2} + \frac{u_3(z')}{2\pi i(z' - z)^2} + \cdots,$$

where z' is now assumed to lie on C. Note that $1/(z' - z)^2$ is bounded on C. Integrate the left side of the preceding equation around C, and make a term-by-term integration of the right side around the same contour. Use Theorems 8 and 10 for justification. Evaluate each integral by using an extension of the Cauchy integral formula and thus obtain Eq. (5.3–9).

5.4 POWER SERIES AND TAYLOR SERIES

As we noted earlier, a power series is a sum of the form $\sum_{n=0}^{\infty} c_n(z - z_0)^n$. Part of our task in this section is to see when the theorems of the previous section, especially those on uniform convergence, are applicable to power series. The series notation $\sum_{j=1}^{\infty} u_j(z)$ used in the previous section can be used to generate a power series with the substitution

$$u_j(z) = c_{j-1}(z - z_0)^{j-1}. \tag{5.4–1}$$

We begin by discussing two theorems that apply specifically to power series.

THEOREM 13 If $\sum_{n=0}^{\infty} c_n(z - z_0)^n$ converges when $z = z_1$, then this series converges for all z satisfying $|z - z_0| < |z_1 - z_0|$. The convergence is absolute for these values of z. □

To understand this theorem, imagine a circle in the z-plane centered at z_0, as shown in Fig. 5.4–1. Suppose the given series is known to converge for $z = z_1$. Then the series will converge for any z lying within the solid circle in the figure.

The proof of Theorem 13, which involves a comparison test, is not difficult; in fact, it will not be presented here because it is sufficiently similar to the proof to be presented for Theorem 14.

THEOREM 14 Uniform convergence and analyticity of power series

If $\sum_{n=0}^{\infty} c_n(z - z_0)^n$ converges when $z = z_1$, where $z_1 \neq z_0$, then the series converges uniformly for all z in the disc $|z - z_0| \leq r$, where $r < |z_1 - z_0|$. The sum of the series is an analytic function for $|z - z_0| \leq r$. □

We assume in this theorem that $z_1 \neq z_0$; therefore, the distance $|z_1 - z_0|$ is nonzero. The theorem asserts that if the power series converges at z_1 in Fig. 5.4–1, then the series converges to an analytic function on and inside the broken circle shown in this figure.

The proof of Theorem 14 involves the Weierstrass M test. To begin, we consider the convergent series

$$\sum_{n=0}^{\infty} c_n(z_1 - z_0)^n = c_0 + c_1(z_1 - z_0) + c_2(z_1 - z_0)^2 + \cdots. \tag{5.4–2}$$

For a convergent series of constants, such as the preceding one, we can find a number m that equals or exceeds the magnitude of any of the terms.† Thus

$$|c_n(z_1 - z_0)^n| \leq m, \qquad n = 0, 1, 2, \ldots. \tag{5.4–3}$$

---

† A convergent series satisfies the nth term test (Theorem 2). If this test is satisfied it is an easy matter to show that a bound must exist on the magnitudes of the terms of the series.

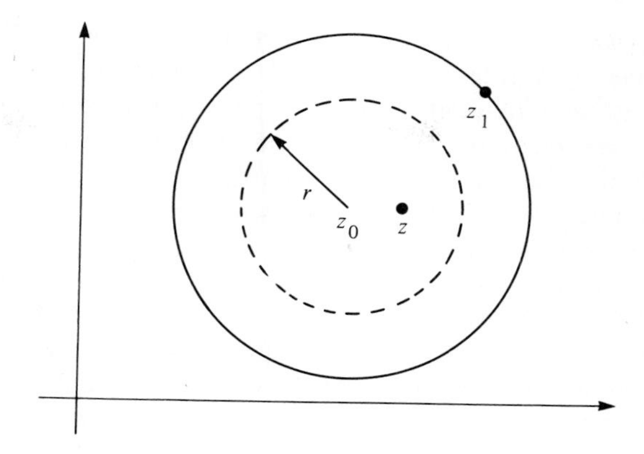

Figure 5.4–1

Now consider the original series

$$\sum_{n=0}^{\infty} c_n(z - z_0)^n = c_0 + c_1(z - z_0) + c_2(z - z_0)^2 + \cdots, \tag{5.4–4}$$

where we take $|z - z_0| \leq r$ and $r < |z_1 - z_0|$. Notice that the terms in Eq. (5.4–4) can be written

$$c_n(z - z_0)^n = c_n(z_1 - z_0)^n\left(\frac{z - z_0}{z_1 - z_0}\right)^n.$$

Taking magnitudes yields

$$|c_n(z - z_0)^n| = |c_n(z_1 - z_0)^n|\left|\frac{z - z_0}{z_1 - z_0}\right|^n. \tag{5.4–5}$$

Let $p = r/|z_1 - z_0|$, where, by hypothesis, $p < 1$. Since $|z - z_0| \leq r$, we have

$$\left|\frac{z - z_0}{z_1 - z_0}\right| < p. \tag{5.4–6}$$

Simultaneously applying this inequality, as well as Eq. (5.4–3), to the right side of Eq. (5.4–5), we obtain

$$|c_n(z - z_0)^n| < mp^n. \tag{5.4–7}$$

Let $M_n = mp^n$. From Eq. (5.3–7) we have

$$|c_n(z - z_0)^n| < M_n. \tag{5.4–8}$$

The summation

$$\sum_{n=0}^{\infty} M_n = \sum_{n=0}^{\infty} mp^n = m\sum_{n=0}^{\infty} p^n, \qquad p < 1 \tag{5.4–9}$$

involves a convergent geometric series of real constants (see Eq. 5.2–7).

The inequality shown in Eq. (5.4–8), the convergence of Eq. (5.4–9), and Theorem 7 together guarantee the uniform convergence $\sum_{n=0}^{\infty} c_n(z - z_0)^n$ for $|z - z_0| \le r$. Because the individual terms $c_n(z - z_0)^n$ in this series are each analytic functions, it follows (see Theorem 11) that the sum of this series is an analytic function in $|z - z_0| \le r$. The proof of Theorem 14 is complete.

Now consider all the possible values of z for which $\sum_{n=0}^{\infty} c_n(z - z_0)^n$ is convergent. Suppose we find that value of z lying farthest from z_0 for which this series converges. Calling this value z_2 and taking $|z_2 - z_0| = \rho$, we see from Theorem 13 that $|z - z_0| < \rho$ describes the largest disc centered at z_0 within which our power series is convergent. By Theorem 14 this series converges uniformly to an analytic function on and inside any circle centered at z_0 whose radius is less than ρ. A circle such as the one just described is known as the *circle of convergence* of a power series.

DEFINITION Circle of convergence

The largest circle centered at z_0 inside which the series $\sum_{n=0}^{\infty} c_n(z - z_0)^n$ converges everywhere is called the *circle of convergence* of this series. The radius ρ of the circle is called the *radius of convergence* of the series. The center of the circle z_0 is called the *center of expansion* of the series. □

It is possible for the radius ρ to be as large as infinity in some cases. There are also series such that $\sum_{n=0}^{\infty} c_n(z - z_0)^n$ converges *only* when $z = z_0$. In such a case ρ is zero. Theorem 14 now does not apply, and we cannot assert that the sum of the series is an analytic function. An example of such a series is $\sum_{n=0}^{\infty} (n + 1)!z^n$. An application of the ratio test shows that this series converges only at $z = 0$.

Given an analytic function, is it always possible to find a power series whose sum, in some domain, is that function? In other words, is an analytic function always representable by a power series? The answer is yes, as we will see by proving the following theorem.

THEOREM 15 Taylor series

Let $f(z)$ be analytic at z_0. Let C be the largest circle centered at z_0, inside which $f(z)$ is everywhere analytic, and let $a > 0$ be the radius of C. Then there exists a power series $\sum_{n=0}^{\infty} c_n(z - z_0)^n$, which converges to $f(z)$ in C; that is,

$$f(z) = \sum_{n=0}^{\infty} c_n(z - z_0)^n, \qquad |z - z_0| < a, \tag{5.4–10}$$

where

$$c_n = \frac{f^{(n)}(z_0)}{n!}. \tag{5.4–11}$$

This power series is called the Taylor series expansion of $f(z)$ about z_0. In the special case $z_0 = 0$ we call the Taylor series a Maclaurin series. □

Notice that the preceding theorem makes no guarantees concerning the convergence of the series to $f(z)$ when z lies on C. Here, each series, and each value of z on C, must be examined on an individual basis.

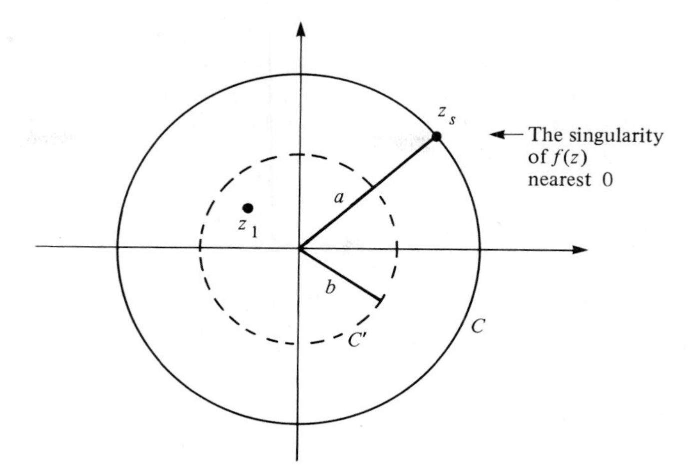

Figure 5.4–2

For simplicity we will prove the theorem by making an expansion about $z_0 = 0$, and we will then indicate how to extend our work to the case $z_0 \neq 0$. We thus begin with a function $f(z)$ that is analytic at $z = 0$. Let z_s be that singularity of $f(z)$ lying closest to $z = 0$. We construct a circle, as shown in Fig. 5.4–2, that is centered at the origin and that passes through z_s.† The radius of the circle $a = |z_s|$. Let z_1 lie within this contour. We enclose z_1 by a second circle C' centered at the origin but having a radius less than that of C. Since the radius of C' is b, we have $|z_1| < b < a$. By the Cauchy integral formula,

$$f(z_1) = \frac{1}{2\pi i} \oint_{C'} \frac{f(z)}{(z - z_1)} dz = \frac{1}{2\pi i} \oint_{C'} \frac{f(z)\, dz}{z\left(1 - \dfrac{z_1}{z}\right)}. \tag{5.4–12}$$

Now consider

$$\frac{1}{1 - \dfrac{z_1}{z}} = 1 + \frac{z_1}{z} + \left(\frac{z_1}{z}\right)^2 + \left(\frac{z_1}{z}\right)^3 + \cdots. \tag{5.4–13}$$

This is just the series of Example 1 of Section 5.3 with z replaced by z_1/z. According to that example, the series of Eq. (5.4–13) is uniformly convergent when

$$\left|\frac{z_1}{z}\right| \leq r, \qquad \text{where } r < 1. \tag{5.4–14}$$

---

† If there is no one singularity closest to the origin, but there are two or more singularities that are closest, then the circle C will pass through them. If $f(z)$ has no singularities, then the radius of C is infinite.

If z is confined to the contour C' (see Fig. 5.4–2), we observe that $|z_1/z| < 1$, and we can readily find a value of r satisfying Eq. (5.4–14).

The function $f(z)/z$ is bounded on C'. By invoking Theorem 8 we can formally multiply both sides of Eq. (5.4–13) by $f(z)/z$ and obtain

$$\frac{f(z)}{z\left(1-\frac{z_1}{z}\right)} = f(z)\frac{1}{z} + f(z)\frac{z_1}{z^2} + f(z)\frac{z_1^2}{z^3} + \cdots, \tag{5.4–15}$$

which is uniformly convergent in some region containing C'.

Notice that the right side of Eq. (5.4–15) is a series expansion of the integrand in Eq. (5.4–12). Each term of this series is continuous on C'. Thus from Theorem 10 a term-by-term integration of the series in Eq. (5.4–15) is possible around C'. Therefore from Eqs. (5.4–12) and (5.4–15)

$$f(z_1) = \frac{1}{2\pi i}\oint_{C'} \frac{f(z)}{z}\,dz + \frac{z_1}{2\pi i}\oint_{C'} \frac{f(z)}{z^2}\,dz + \frac{z_1^2}{2\pi i}\oint_{C'} \frac{f(z)}{z^3}\,dz + \cdots. \tag{5.4–16}$$

Since z_1 is a fixed point, it was brought out from under each integral sign.

There are an infinite number of integrals on the right in Eq. (5.4–16), each of which can be evaluated with the extended Cauchy integral formula. With $z_0 = 0$ in Eq. (4.5–13) we have

$$\frac{1}{2\pi i}\oint_{C'} \frac{f(z)}{z^{n+1}}\,dz = \frac{f^{(n)}(0)}{n!}, \qquad n = 0, 1, 2, \ldots. \tag{5.4–17}$$

Rewriting Eq. (5.4–16) with this formula we obtain

$$f(z_1) = \sum_{n=0}^{\infty} c_n z_1^n = c_0 + c_1 z_1 + c_2 z_1^2 + \cdots, \tag{5.4–18}$$

for $|z_1| < b < a$, where

$$c_n = \frac{f^{(n)}(0)}{n!}. \tag{5.4–19}$$

Replacing what is now the dummy variable z_1 in Eq. (5.4–18) by z, we find that we have derived Eqs. (5.4–10 and (5.4–11) for the special case $z_0 = 0$. The constraint on $|z_1|$ now becomes $|z| < b < a$, and because b can be made arbitrarily close to a, we will write this as $|z| < a$. Thus with $z_0 = 0$, z is constrained to lie inside the largest circle, centered at the origin, within which $f(z)$ is analytic.

The more general result $z_0 \neq 0$ described in Theorem 15, is obtained by a derivation much like the one just given. The contours C and C' in Fig. 5.4–2 become circles centered at z_0. Again, z_1 lies inside C'. Equation (5.4–12) still holds, but the integrand is now written

$$\frac{f(z)}{z-z_1} = \frac{f(z)}{(z-z_0)\left[1-\frac{(z_1-z_0)}{(z-z_0)}\right]},$$

and a series expansion is made in powers of $(z_1 - z_0)/(z - z_0)$. The reader can supply the additional details in Exercise 2 of this section.

Theorem 15 is enormously useful. It tells us that any function $f(z)$, analytic at z_0, is the sum of a power series, called a Taylor series, containing powers of $(z - z_0)$. In Eq. (5.4–11) the theorem tells us how to obtain the coefficients for the Taylor series. Since all the derivatives of an analytic function exist (see Section 4.5), all the coefficients are defined. The procedure for getting the coefficients is identical to that used in Taylor expansions of functions of a real variable (see Eq. 5.1–4). Finally, the theorem guarantees that as long as z lies within a certain circle, centered at z_0, the Taylor series converges to $f(z)$. The radius of this circle is precisely the distance from z_0 to the nearest singularity of $f(z)$ in the complex plane. Some examples of Taylor series follow.

A corollary to Theorem 15 is known as Taylor's theorem. It states that if $f(z)$ satisfies the conditions described in Theorem 15, then $f(z)$ can be represented within the domain $|z - z_0| < b$ (where $b < a$) by the sum of a power series with a finite number of terms plus a remainder, i.e.,

$$f(z) = \sum_{n=0}^{N-1} c_n(z - z_0)^n + R_N(z).$$

Here c_n is again $f^n(z_0)/n!$, while $R_N(z)$ is expressed as a contour integration around the circle $|z - z_0| = b$. We can often establish an upper bound on $|R_N(z)|$ and thereby establish a bound on the error made if we use only the finite series $\Sigma_{n=0}^{N-1} c_n(z - z_0)^n$ to approximate $f(z)$. The proof of Taylor's theorem as well as an expression for $R_N(z)$ are obtained in Exercise 31 for the special case $z_0 = 0$.

Taylor's series is named for an Englishman, Brook Taylor (1685–1731), who derived this expansion for real functions of a real variable and published his findings in 1715. He gave scant attention to the question of convergence and was also unaware that a Scotsman, James Gregory (1638–1675), had obtained this result 40 years earlier. The Maclaurin series is named for another Scotsman, Colin Maclaurin (1698–1746), who did not invent this expansion (it was part of Taylor's earlier work) but who demonstrated its usefulness in a publication of 1742. Some examples of Taylor and Maclaurin series follow.

EXAMPLE 1

Expand e^z in (a) a Maclaurin series and (b) a Taylor series about $z = i$.

Solution

Part (a):

$$e^z = c_0 + c_1 z + c_2 z^2 + \cdots.$$

From Eq. (5.4–11), with $z_0 = 0$,

$$c_n = \frac{\left.\dfrac{d^n}{dz^n} e^z\right|_{z=0}}{n!} = \left.\frac{e^z}{n!}\right|_{z=0} = \frac{1}{n!}.$$

Thus

$$e^z = \sum_{n=0}^{\infty} \frac{1}{n!} z^n. \tag{5.4–20}$$

Because e^z is analytic for all finite z, Theorem 15 guarantees that the series in Eq. (5.4–20) converges to e^z everywhere in the complex plane. Putting $z = x$ in Eq. (5.4–20) yields a familiar expansion for the real function e^x.

Part (b):

$$e^z = c_0 + c_1(z - i) + c_2(z - i)^2 + \cdots.$$

From Eq. (5.4–11), with $z_0 = i$,

$$c_n = \frac{\left.\frac{d^n}{dz^n} e^z\right|_{z=i}}{n!} = \left.\frac{e^z}{n!}\right|_{z=i} = \frac{e^i}{n!};$$

thus

$$e^z = \sum_{n=0}^{\infty} \frac{e^i}{n!}(z - i)^n.$$

The series representation is again valid throughout the complex plane. ◀

Other useful Maclaurin series expansions besides Eq. (5.4–20) are

$$\begin{aligned} \sin z &= z - \frac{z^3}{3!} + \frac{z^5}{5!} - \cdots, \\ \cos z &= 1 - \frac{z^2}{2!} + \frac{z^4}{4!} - \cdots, \\ \sinh z &= z + \frac{z^3}{3!} + \frac{z^5}{5!} + \cdots, \\ \cosh z &= 1 + \frac{z^2}{2!} + \frac{z^4}{4!} + \cdots. \end{aligned} \tag{5.4–21}$$

Because the four preceding functions on the left are analytic for $|z| < \infty$, the series representations are valid throughout the z-plane.†

The question of convergence is of consequence in the following example.

EXAMPLE 2

Expand

$$f(z) = \frac{1}{(1 - z)}$$

† Series expansions for the inverse functions $\sin^{-1}(z)$ and $\cos^{-1}(z)$ are developed in Exercises 25 and 26 of the following section.

in the Taylor series $\sum_{n=0}^{\infty} c_n(z+1)^n$. For what values of z must the series converge to $f(z)$?

Solution

The expansion is about $z_0 = -1$. From Eq. (5.4–11), with $f(z) = 1/(1-z)$, we find $c_0 = 1/2$, $c_1 = 1/4$, and, in general, $c_n = 1/2^{n+1}$. Thus

$$f(z) = \frac{1}{1-z} = \sum_{n=0}^{\infty} \frac{1}{2^{n+1}}(z+1)^n. \tag{5.4–22}$$

To study the validity of this series representation we must see where the singularities of $1/(1-z)$ lie in the complex plane and determine which one lies closest to $z_0 = -1$.

Since $f(z)$ is analytic except at $z = 1$, Theorem 15 *guarantees* that the series in Eq. (5.4–22) will converge to $f(z)$ for all z lying inside a circle centered at -1 having radius 2. We will soon see that it is impossible for the series to converge to $f(z)$ outside this circle. ◀

Given any analytic function $f(z)$ we know that it can be represented in a Taylor series about the point z_0. Might there be some other power series using powers of $(z - z_0)$ that converges to $f(z)$ in a neighborhood of z_0? The answer is no.

Let

$$f(z) = b_0 + b_1(z - z_0) + b_2(z - z_0)^2 + \cdots, \qquad |z - z_0| \leq r. \tag{5.4–23}$$

We can show that this must be the Taylor expansion of $f(z)$ about z_0. Invoking Theorem 14 and Theorem 12, we differentiate Eq. (5.4–23) and find that

$$f'(z) = b_1 + 2b_2(z - z_0) + \cdots, \qquad |z - z_0| < r. \tag{5.4–24}$$

This series can be differentiated again:

$$f''(z) = 2b_2 + 3\cdot 2b_3(z - z_0) + \cdots \tag{5.4–25}$$

and again.

Setting $z = z_0$ in Eqs. (5.4–23), (5.4–24), and (5.4–25) we get

$$b_0 = f(z_0), \qquad b_1 = f'(z_0), \qquad b_2 = \frac{f''(z_0)}{2},$$

and, in general, one readily shows that

$$b_n = \frac{f^{(n)}(z_0)}{n!}, \qquad n = 0, 1, 2, \ldots.$$

But these coefficients are precisely those used in the Taylor expansion (see Eq. 5.4–11). Thus we conclude:

THEOREM 16 The Taylor series expansion about z_0 of a function $f(z)$ is the *only* power series using powers of $(z - z_0)$ that will converge to $f(z)$ everywhere in a circular domain centered at z_0. □

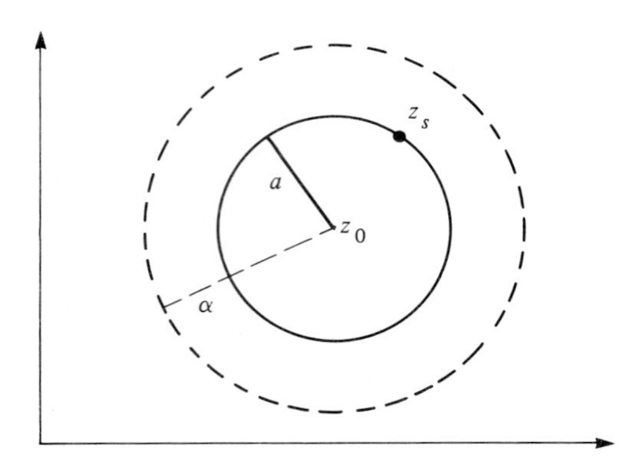

Figure 5.4–3

According to Theorem 15, when a function $f(z)$ is expanded in a Taylor series about z_0, the resulting series *must* converge to $f(z)$ whenever z resides within a certain circle centered at z_0.

Theorem 15 explains how to find the radius of the circle. We can prove that this is the largest circle, centered at z_0, in which the Taylor series converges to $f(z)$.

Let z_s be that singularity of $f(z)$ lying closest to z_0. The circle shown in solid line in Fig. 5.4–3 is the one described in Theorem 15. Assume that the Taylor expansion of $f(z)$ converges to $f(z)$ in the disc $|z - z_0| < \alpha$, where $\alpha > a = |z_s - z_0|$. We thus have a power series that converges in the disc $|z - z_0| < \alpha$. Now, according to Theorem 14, such a power series converges to a function that is analytic throughout a disc that is larger than, and contains, the disc of radius a shown in Fig. 5.4–3. This larger disc contains the point z_s, where $f(z)$ is known to be nonanalytic. Thus our assumption that the Taylor expansion converges to $f(z)$ in a circle larger than $|z - z_s| = a$ must be false. To summarize:

THEOREM 17 Let $f(z)$ be expanded in a Taylor series about z_0. The largest circle within which this series converges to $f(z)$ at each point is $|z - z_0| = a$, where a is the distance from z_0 to the nearest singular point of $f(z)$. □

Notice that this theorem *does not* assert that the Taylor series fails to converge outside $|z - z_0| = a$. It asserts that this is the largest circle throughout which the series converges to $f(z)$.

The circle in which the Taylor series $\sum_{n=0}^{\infty} [f^{(n)}(z_0)/n!](z - z_0)^n$ is everywhere convergent to $f(z)$ and the circle throughout which this series converges are not necessarily the same. The second circle could be larger. This fact is considered in Exercise 29 of this section. It can be shown, however, that when the singularity z_s lying nearest z_0 is one where $|f(z)|$ becomes infinite, then the two circles are identical. This is the case in most of the examples that we will consider.

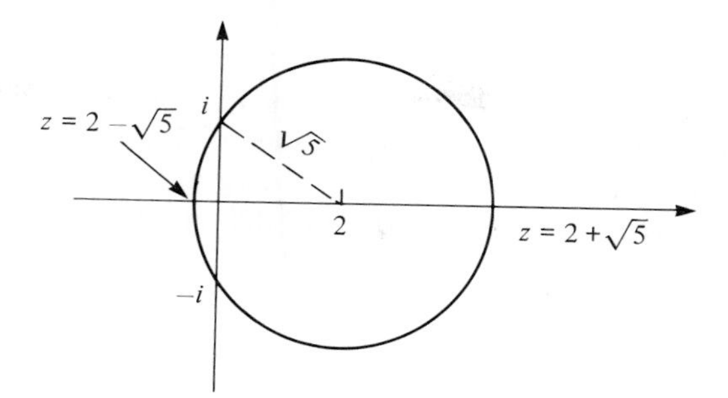

Figure 5.4–4

EXAMPLE 3

Without actually obtaining the Taylor series give the largest circle throughout which the indicated expansion is valid:

$$f(z) = \frac{1}{z^2 + 1} = \sum_{n=0}^{\infty} c_n(z - 2)^n. \tag{5.4–26}$$

Solution

Convergence takes place within a circle centered at $z = 2$, as shown in Fig. 5.4–4. The singularities of $f(z)$ lie at $\pm i$. The nearest singularity to $z = 2$ is, in this case, either $+i$ or $-i$. They are equally close. The distance from $z = 2$ to these points is $\sqrt{5}$. Thus the Taylor series converges to $f(z)$ throughout the circular domain $|z - 2| < \sqrt{5}$. ◀

EXAMPLE 4

Consider the real Taylor series expansion

$$\frac{1}{x^2 + 1} = \sum_{n=0}^{\infty} c_n(x - 2)^n. \tag{5.4–27}$$

Determine the largest interval along the x-axis inside which the series converges to $1/(x^2 + 1)$.

Solution

By requiring z to be a real variable ($z = x$) in Eq. (5.4–26), we will obtain the series of the present problem. If z is a real variable and if, in addition, the series on the right in Eq. (5.4–26) is to converge to the function on the left, not only must z lie on the real axis in Fig. 5.4–4, it must be inside the indicated circle as well. Thus we require $2 - \sqrt{5} < x < 2 + \sqrt{5}$ for convergence. Whether the series in Eq. (5.4–27) will converge to $1/(x^2 + 1)$ at either of the endpoints of this interval, that is, $x = 2 \pm \sqrt{5}$, cannot be determined from Theorem 17. ◀

REMARKS ON ANALYTICITY

We have seen from Theorem 15 that there is an intimate connection between analyticity and Taylor series. Summarizing what we know about analyticity from Theorems 15 and 16, the extended Cauchy integral formula, and our work in Section 2.4, we have the following equally valid definitions of analyticity. Functions satisfying any one of the following satisfy the others. A function $f(z)$ is analytic in a domain D if

a) $f'(z)$ exists throughout D;

b) $f(z)$ has derivatives of all orders throughout D;

c) $f(z)$ has a Taylor series expansion valid in a neighborhood of each point in D;

d) $f(z)$ is the sum of a convergent power series in a neighborhood of each point in D.

EXERCISES

1. Derive the Maclaurin series expansions in Eq. (5.4–21).

2. Theorem 15 on Taylor series was derived for expansions about the origin $z_0 = 0$. Follow the suggestions given in that derivation and give a proof valid for any z_0.

State the first three nonzero terms in the following Taylor series expansions. The function to be expanded and the point about which you are expanding are indicated. Give the nth term (general term) of the series, and state the circle within which the expansion is valid.

3. $1/z$, $z = 1 + i$

4. $\text{Log } z$, $z = ie$

5. $1/(z + i)^2$, $z = i$

6. e^z, $z = \pi i$

7. $1/(z - 1)^2$, $z = 1 + i$

8. z^i (principal branch), $z = 1$

9. $\sinh z - \sin z$, $z = 0$

10. $z^2 + z + 1$, $z = i$

11. a) Explain why $z^{1/2}$ (principal branch) cannot be expanded in a Maclaurin series. Also, explain why the series expansion sought in Eq. (5.1–5) does not exist.

b) Explain whether this same branch of $z^{1/2}$ can be expanded in a Taylor series about 1. If so, find the first three terms and state the circle within which the expansion is valid.

12. Consider the two infinite series, $1 + z + z^2 + z^3 + \cdots$ and $1 + (z/\text{Log } 2) + (z^2/\text{Log } 3) + (z^3/\text{Log } 4) + \cdots$. Both of these series will converge to $1/(1 - z)$ at $z = 0$. Yet the second series is not the Taylor expansion of $1/(1 - z)$. Does this not contradict Theorem 16, which asserts that the Taylor series expansion of a function is the power series expansion of the function? Explain.

Without actually obtaining the series determine the center and radius of the circle within which the indicated Taylor series converges to the given function.

13. $\dfrac{1}{z^4 + 1} = \sum_{n=0}^{\infty} c_n(z - 1)^n$

14. $\dfrac{1}{\sin z} = \sum_{n=0}^{\infty} c_n(z - 1 - i)^n$

15. $\dfrac{1}{z^4 + z^2 + 1} = \sum_{n=0}^{\infty} c_n(z - i)^n$

16. $\dfrac{1}{1 - \tan z}$ expanded about $1 + 2i$

17. $\dfrac{1}{z^{3/2} + 8i}$ expanded about $z = 5$ (use principal branch $z^{3/2}$)

18. $\tanh z$ expanded about $z = i$

Without obtaining the series determine the interval along the x-axis for which the indicated real Taylor series converges to the given real function. Convergence at the endpoints of the interval need not be considered.

19. $\dfrac{1}{1 - x}$, expanded about $x = -1$. How do your findings confirm Eqs. (5.1–2), (5.1–6), and (5.1–7)?

20. $\dfrac{1}{x^2 + x + 1}$ expanded about $x = 1$

21. $1/\text{Log}\, x$ expanded about $x = 3/4$

22. $1/\text{Log}\, x$ expanded about $x = 1/4$

23. $\sqrt{x + 1}$ expanded about $x = 0$

24. $\sqrt{x}$ expanded about $x = 2$

25. $\dfrac{1}{1 - \sqrt{x}}$ expanded about $x = 3/2$

26. Use Eq. (5.4–20) and a triangle inequality to prove that in the disc $|z| \le 1$ we have $|e^z - 1| \le (e - 1)|z|$.

27. Using Taylor series, express $z^5 - 1$ in the form $\sum_{n=0}^{5} c_n(z - 1)^n$, valid for all z. State the coefficients.

28. a) Let $z^N - z_0^N = \sum_{n=1}^{N} c_n(z - z_0)^n$, valid for all z. N is a positive integer. Show that $c_n = N! z_0^{N-n}/[n!(N - n)!]$.

    b) Replace z in the above with $z + z_0$, and show that

$$(z + z_0)^N = \sum_{n=0}^{N} \frac{N! z_0^{N-n}}{n!(N - n)!} z^n.$$

    This is the familiar binomial expansion.

29. Let a function $f(z)$ be expanded in a Taylor series about z_0. The circle, centered at z_0, within which this series converges is in certain cases larger than, but concentric with, the circle in which the series converges to $f(z)$. We will investigate this possibility in one particular case.

    a) Let $f(z) = \text{Log}(z)$, where the principal branch of the logarithm is used. Thus $f(z)$ is defined by means of the branch cut $y = 0$, $x \le 0$.

    Show that

$$f(z) = \sum_{n=0}^{\infty} c_n(z + 1 - i)^n,$$

where

$$c_0 = \text{Log}\,\sqrt{2} + \frac{i3\pi}{4} \quad \text{and} \quad c_n = \frac{(-1)^{n+1}e^{-i(3\pi/4)n}}{n(\sqrt{2})^n}, \qquad n \neq 0.$$

b) What is the radius of the largest circle centered at $-1 + i$ within which the series of part (a) converges to $f(z)$?

Hint: Draw the branch cut mentioned in part (a).

c) Use the ratio test to establish that the series of part (a) converges inside $|z - (-1 + i)| = \sqrt{2}$ and diverges outside this circle. Compare this circle with the one in part (b).

30. a) Let $f(z)$ be analytic in a domain containing $z = 0$. Assume that $f(z)$ is an *even function* of z in this domain, i.e., $f(z) = f(-z)$. Show that in the Maclaurin expansion $f(z) = \sum_{n=0}^{\infty} c_n z^n$ the coefficients of odd order, $c_1, c_3, c_5, \ldots$, must be zero.

b) Show that for an *odd function*, $f(z) = -f(-z)$, analytic in the same kind of domain, the coefficients of even order, $c_0, c_2, c_4, \ldots$, are zero.

c) What coefficients vanish in the Maclaurin expansions of $z \sin z$, $z^2 \tan z$, and $\cosh z/(1 + z^2)$?

31. This problem investigates Taylor series with a *remainder*, which is also known as Taylor's theorem. In numerical calculations we often use the first n terms of a Taylor series instead of the entire (infinite) expansion. The difference between the sum $f(z)$ of the infinite series and the sum of the finite series actually used constitutes an error and is called the *remainder*. The size of the remainder is of interest. An upper bound for its magnitude can be determined with the help of Taylor's theorem. We will derive this theorem for the special case of an expansion about the origin.

a) Refer to the proof of Theorem 15 and to Fig. 5.4–2. Use Eq. (5.4–12) to show that

$$f(z_1) = \frac{1}{2\pi i}\oint_{C'} f(z)\left[\frac{1}{z} + \frac{z_1}{z^2} + \cdots + \frac{z_1^{n-1}}{z^n} + \left(\frac{z_1}{z}\right)^n \frac{1}{(z - z_1)}\right] dz.$$

Hint: Refer to Eq. (5.2–8), which implies that

$$\frac{1}{1 - z} = 1 + z + z^2 + \cdots + z^{n-1} + \frac{z^n}{1 - z},$$

and replace z by z_1/z.

b) Use the expression for $f(z_1)$ given in part (a) to show, after integration, that

$$f(z_1) = f(0) + f'(0)z_1 + \frac{f''(0)}{2}z_1^2 + \cdots + \frac{f^{(n-1)}(0)}{(n-1)!}z_1^{n-1} + R_n, \tag{5.4–28}$$

where

$$R_n = \frac{z_1^n}{2\pi i}\oint_{C'} \frac{f(z)}{z^n(z - z_1)}\,dz. \tag{5.4–29}$$

We see that R_n, the remainder, represents the difference between $f(z_1)$ and the first n terms of its Maclaurin expansion.

c) We can place an upper bound on the remainder in Eq. (5.4–29). Assume $|f(z)| \le m$ everywhere on $|z| = b$ (the contour C'). Use the ML inequality to show that

$$|R_n| \le \left|\frac{z_1}{b}\right|^n \frac{mb}{b - |z_1|}. \tag{5.4–30}$$

Hint: Note that for z lying on contour C'

$$\frac{1}{|z - z_1|} \le \frac{1}{b - |z_1|}.$$

Why?

In passing, we notice that since $|z_1/b| < 1$, the remainder R_n in Eq. (5.4–30) tends to zero as $n \to \infty$. Using this limit we find that the right side of Eq. (5.4–28) becomes the Maclaurin series of $f(z_1)$. This constitutes a derivation of the Maclaurin expansion shown in Eq. (5.4–18) not requiring the use of uniform convergence. A similar derivation applies for the Taylor expansion.

d) Suppose we wish to determine the approximate value of $\cosh i$ by the finite series $i^0 + i^2/2! + \cdots + i^{10}/10!$ (see Eq. (5.4–21)). Taking the contour C' in Eq. (5.4–29) as $|z| = 2$, show by using Eq. (5.4–30) that the error made cannot exceed $(\cosh 2)/2^{10} \doteq 3.67 \times 10^{-3}$.

Hint: Observe that $|\cosh z| \le \cosh x$ and use this fact to find m.

5.5 TECHNIQUES FOR OBTAINING TAYLOR SERIES EXPANSIONS

Equation (5.4–11) allows us, in principle, to obtain the coefficients for the Taylor series expansion of any analytic function. Alternatively, there are sometimes shortcuts that can be applied that will relieve us of the tedium of taking high-order derivatives to obtain the coefficients. We will explore some of these techniques in this section.

Term-by-Term Differentiation and Integration

In our derivation of Theorem 16 we observed that the Taylor series of $f(z)$ can be differentiated term by term. If the original series converged to $f(z)$ inside a circle having center z_0 and radius r, the series obtained through differentiation converges to $f'(z)$ inside this circle. The procedure can be repeated indefinitely to yield a series for any derivative of $f(z)$.

Finally, Theorem 10 permits us to integrate the Taylor series for $f(z)$ term by term along a path lying inside the circle of radius r. The resulting series converges to the integral of $f(z)$ taken along this path.

EXAMPLE 1

Use term-by-term differentiation and the expansion of $1/z$ about $z = 1$ to obtain the expansion of $1/z^2$ about the same point.

Solution

From Theorem 15, with $f(z) = 1/z$ and the use of Eqs. (5.4–10) and (5.4–11), we find that

$$\frac{1}{z} = 1 - (z - 1) + (z - 1)^2 + \cdots, \qquad |z - 1| < 1. \tag{5.5–1}$$

Differentiating both sides with respect to z and multiplying by (-1), we obtain

$$\frac{1}{z^2} = 1 - 2(z - 1) + 3(z - 1)^2 + \cdots = \sum_{n=0}^{\infty} (-1)^n(n + 1)(z - 1)^n, \tag{5.5–2}$$

valid for $|z - 1| < 1$. ◀

Term-by-term integration is illustrated in the following example.

EXAMPLE 2

Obtain the Maclaurin expansion of

$$\text{Si}(z) = \int_0^z f(z')\,dz', \tag{5.5–3}$$

where

$$f(z') = \frac{\sin z'}{z'}, \qquad z' \neq 0, \tag{5.5–4a}$$

$$f(0) = 1, \qquad z' = 0. \tag{5.5–4b}$$

The function Si(z) is called the *sine integral* and cannot be evaluated in terms of elementary functions. It appears often in problems involving electromagnetic radiation.

Solution

From Eq. (5.4–21) we have

$$\sin z' = z' - \frac{(z')^3}{3!} + \frac{(z')^5}{5!} + \cdots,$$

so that

$$\frac{\sin z'}{z'} = 1 - \frac{(z')^2}{3!} + \frac{(z')^4}{5!} + \cdots.$$

Notice that this series converges to $f(z')$ in Eq. (5.5–4) for $z' \neq 0$ and $z' = 0$. We now integrate as follows:

$$\int_0^z \frac{\sin z'}{z'}\,dz' = \int_0^z dz' + \int_0^z \frac{-(z')^2}{3!}\,dz' + \int_0^z \frac{(z')^4}{5!}\,dz' + \cdots$$
$$= z - \frac{z^3}{3 \cdot 3!} + \frac{z^5}{5 \cdot 5!} + \cdots.$$

Thus

$$\mathrm{Si}(z) = \sum_{n=0}^{\infty} c_n z^{(2n+1)}, \tag{5.5–5}$$

where

$$c_n = \frac{(-1)^n}{(2n+1)!(2n+1)}.$$

The expansion is valid throughout the z-plane. ◀

Series Expansions of Branches of Multivalued Functions

An analytic branch of a multivalued function can be expanded in a Taylor series about any point within the domain of analyticity of the branch provided one takes care to use this branch consistently in obtaining the coefficients of the series. The following problem illustrates this.

EXAMPLE 3

Find the Maclaurin expansion of $f(z) = (z+1)^{1/2}$, where the principal branch of the function is used. Where is the expansion valid?

Solution

Recall that the branch in question is identical to $e^{1/2\,\mathrm{Log}(z+1)}$ (see Section 3.8) and that its derivative is given by

$$e^{1/2\,\mathrm{Log}(z+1)} \frac{1}{2(z+1)} = \frac{(z+1)^{1/2}}{2(z+1)}.$$

We may of course differentiate indefinitely and thus have

$$f^{(1)}(z) = \frac{1}{2}(z+1)^{1/2-1}, \qquad f^{(2)}(z) = \frac{1}{2}\left(\frac{1}{2} - 1\right)(z+1)^{1/2-2},$$
$$f^{(3)}(z) = \frac{1}{2}\left(\frac{1}{2} - 1\right)\left(\frac{1}{2} - 2\right)(z+1)^{1/2-3}, \qquad \text{etc.}$$

In general,

$$f^{(n)}(z) = \frac{1}{2}\left(\frac{1}{2} - 1\right)\left(\frac{1}{2} - 2\right)\cdots\left(\frac{1}{2} - (n-1)\right)(z+1)^{1/2-n}. \tag{5.5–6}$$

Note that $(z+1)^{1/2-n}$ must be interpreted as

$$\frac{(z+1)^{1/2}}{(z+1)^n} = \frac{e^{1/2\,\mathrm{Log}(z+1)}}{(z+1)^n}.$$

When $z = 0$ this function equals $e^{1/2\,\mathrm{Log}\,1}/1^n = 1$. With this result and Eqs. (5.5–6) and (5.4–11) we have finally

$$(1+z)^{1/2} = \sum_{n=0}^{\infty} c_n z^n, \tag{5.5–7a}$$

where

$$c_0 = 1,$$
$$c_n = \frac{1}{n!}\left[\left(\frac{1}{2}\right)\left(\frac{1}{2}-1\right)\left(\frac{1}{2}-2\right)\cdots\left(\frac{1}{2}-(n-1)\right)\right], \qquad n \geq 1. \tag{5.5–7b}$$

The singularity of $(z+1)^{1/2}$ nearest the origin is the branch point at $z = -1$. Thus Eq. (5.5–7) is valid in the domain $|z| < 1$. ◀

Multiplication and Division of Series

Let $g(z)$ and $h(z)$ be analytic functions. Then a Taylor series expansion of $f(z) = g(z)h(z)$ can be obtained, in principle, by multiplying together the Taylor expansions for $g(z)$ and $h(z)$. This is because Taylor series are absolutely convergent and, as discussed in Section 5.2, the product of two absolutely convergent series is an absolutely convergent series whose sum is the product of the sums of the two original series. To obtain the Taylor series for $f(z)$, the method of series multiplication that should be used is the Cauchy product (see Eq. 5.2–16). Of course, both of the original series must employ the same center of expansion z_0. The procedure is readily extended to finding the Taylor expansion of the product of more than two functions.

Multiplication of series is often a tedious procedure, especially if we want a general formula for the nth coefficient in the resulting series. However, if we need only the first few terms in the result, it is easy to use.

EXAMPLE 4

a) Using series multiplication, obtain the Maclaurin expansion of $f(z) = e^z/(1-z)$.

b) Use your result to obtain the value of the 10th derivative of $f(z)$ at $z = 0$.

Solution

Part (a): We are fortunate that in this case a general formula for the nth coefficient can be found. With $e^z = \sum_{n=0}^{\infty} z^n/n!$ (valid for all z) and $1/(1-z) = \sum_{n=0}^{\infty} z^n$ (for $|z| < 1$), we have

$$f(z) = \left(1 + z + \frac{z^2}{2!} + \frac{z^3}{3!} + \cdots\right)(1 + z + z^2 + \cdots)$$
$$= 1 + (1+1)z + \left(1 + 1 + \frac{1}{2!}\right)z^2$$
$$+ \left(1 + 1 + \frac{1}{2!} + \frac{1}{3!}\right)z^3 + \cdots.$$

Or, equivalently,

$$\frac{e^z}{1-z} = \sum_{n=0}^{\infty} c_n z^n, \tag{5.5–8a}$$

where

$$c_n = \sum_{j=0}^{n} \frac{1}{j!}. \tag{5.5–8b}$$

The expansion for $1/(1-z)$ is valid for $|z| < 1$, while that for e^z holds for all z. The more restrictive condition, $|z| < 1$, applies to Eq. (5.5–8a) since it is in this domain that the two series used to obtain this result are simultaneously valid.

Part (b): It is a little tedious to obtain the 10th derivative of $f(z)$ by differentiating this function 10 times. Note, however, that in the Maclaurin expansion $f(z) = \sum_{n=0}^{\infty} c_n z^n$ we have $c_n = f^{(n)}(0)/n!$ (see Eq. 5.4–11). Thus using the result of part (a) and taking $n = 10$, we find

$$f^{(10)}(0) = 10! \sum_{j=0}^{10} \frac{1}{j!} = 10!\left(1 + \frac{1}{1!} + \frac{1}{2!} + \cdots + \frac{1}{10!}\right),$$

which with the aid of a pocket calculator turns out to be 9,864,101. ◀

Suppose $f(z)$ and $g(z)$ are both analytic at z_0. Then, if $g(z_0) \neq 0$, the quotient

$$h(z) = \frac{f(z)}{g(z)} \tag{5.5–9}$$

is analytic at z_0 and can be expanded in a Taylor series about this point. Let us use the series $h(z) = \sum_{n=0}^{\infty} c_n(z-z_0)^n$, $f(z) = \sum_{n=0}^{\infty} a_n(z-z_0)^n$, $g(z) = \sum_{n=0}^{\infty} b_n(z-z_0)^n$, where a_n and b_n are presumed known (we can obtain them, in principle, by differentiation) and the coefficients c_n are unknown. Now from Eq. (5.5–9) we have $h(z)g(z) = f(z)$. Using the Taylor series for each term in the product, and multiplying the two series appearing on the left side of the equation in accordance with the Cauchy product (Section 5.2), we have

$$\sum_{n=0}^{\infty} c_n(z-z_0)^n \sum_{n=0}^{\infty} b_n(z-z_0)^n = \sum_{n=0}^{\infty} a_n(z-z_0)^n$$

and

$$c_0b_0 + (c_0b_1 + c_1b_0)(z-z_0)^1 + (c_0b_2 + c_1b_1 + c_2b_0)(z-z_0)^2 + \cdots$$
$$= a_0 + a_1(z-z_0) + a_2(z-z_0)^2 + \cdots.$$

Equating coefficients of corresponding powers of $(z-z_0)$, we have

$$c_0b_0 = a_0, \tag{5.5–10a}$$

$$c_0b_1 + c_1b_0 = a_1, \tag{5.5–10b}$$

$$c_0b_2 + c_1b_1 + c_2b_0 = a_2. \tag{5.5–10c}$$

From the first equation, we have

$$c_0 = a_0/b_0, \tag{5.5–11a}$$

and with this result used in the second we find

$$c_1 = \frac{a_1}{b_0} - \frac{a_0 b_1}{b_0^2}. \tag{5.5–11b}$$

The preceding allows us to solve Eq. (5.5–10c) for c_2, with the result

$$c_2 = \frac{a_2}{b_0} - \frac{a_1 b_1 + a_0 b_2}{b_0^2} + \frac{a_0 b_1^2}{b_0^3}. \tag{5.5–11c}$$

The process can be repeated to yield any coefficient c_n, where n is as large as we wish. This is an example of a recursive procedure; i.e., we use those values $c_1, c_2, \ldots, c_{n-1}$ that have already been determined to find the next unknown c_n.

In Exercise 20, the reader will verify that the same coefficients are obtained through a formal division of the Taylor series for $f(z)$ by the series for $g(z)$. This method is used in the following example.

EXAMPLE 5

Obtain the Maclaurin expansion of $(e^z - 1)/\cos z$ from the Maclaurin series for $e^z - 1$ and $\cos z$.

Solution

From Eqs. (5.4–20) and (5.4–21), we have

$$e^z - 1 = z + \frac{z^2}{2!} + \frac{z^3}{3!} + \cdots, \tag{5.5–12}$$

$$\cos z = 1 - \frac{z^2}{2!} + \frac{z^4}{4!} - \cdots. \tag{5.5–13}$$

We divide these series as follows:

$$
\begin{array}{r|l}
 & z + \frac{z^2}{2!} + z^3\left(\frac{1}{3!} + \frac{1}{2!}\right) + \cdots \\
\cline{2-2}
1 - \frac{z^2}{2!} + \frac{z^4}{4!} - \cdots & z + \frac{z^2}{2!} + \frac{z^3}{3!} + \frac{z^4}{4!} + \cdots \\
 & z \qquad - \frac{z^3}{2!} \qquad\qquad + \cdots \\
\cline{2-2}
 & \frac{z^2}{2!} + z^3\left(\frac{1}{3!} + \frac{1}{2!}\right) + \frac{z^4}{4!} + \cdots \\
 & \frac{z^2}{2!} \qquad\qquad\qquad - \frac{z^4}{(2!)^2} + \cdots \\
\cline{2-2}
 & z^3\left(\frac{1}{3!} + \frac{1}{2!}\right) + z^4\left(\frac{1}{4!} + \frac{1}{(2!)^2}\right) \\
 & + \cdots .
\end{array}
$$

Recalling that $\cos z = 0$ for $z = \pm\pi/2$, we have

$$\frac{e^z - 1}{\cos z} = \sum_{n=0}^{\infty} c_n z^n,$$

valid for $|z| < \pi/2$. Our division shows $c_0 = 0$, $c_1 = 1$, $c_2 = 1/2!$, $c_3 = (1/3! + 1/2!) = 2/3$.

As a check, we use Eqs. (5.5–11). From Eq. (5.5–12), we have $a_0 = 0$, $a_1 = 1$, $a_2 = 1/2!$, and from Eq. (5.5–13), $b_0 = 1$, $b_1 = 0$, $b_2 = -1/2!$. From Eq. (5.5–11a), we can confirm that $c_0 = 0$, from Eq. (5.5–11b) that $c_1 = 1$, and from Eq. (5.5–11c) that $c_2 = 1/2$. ◀

The Method of Partial Fractions

Consider a rational algebraic function

$$f(z) = \frac{P(z)}{Q(z)},$$

where P and Q are polynomials in z. If $Q(z_0) \neq 0$, then $f(z)$ has a Taylor expansion about z_0. The coefficients in this series can, in principle, be obtained through differentiation of $f(z)$, but the process is often tedious.

When the degree of Q (its highest power of z) exceeds the degree of P, the use of partial fractions provides a systematic procedure for obtaining the coefficients in the Taylor series. When the degree of P equals or exceeds that of Q, the method to be presented also helps, provided we first perform a simple division (see Exercise 19 in this section).

The reader should review the techniques learned in elementary calculus for decomposing real rational functions into partial fractions. The method works equally well for complex functions—the algebraic manipulations are the same. The form of partial fraction expansions is governed by the following rules.

RULE I Nonrepeated factors

Let $P(z)/Q(z)$ be a rational function where the polynomial $P(z)$ is of lower degree than the polynomial $Q(z)$. If $Q(z)$ can be factored into the form

$$Q(z) = C(z - a_1)(z - a_2)\cdots(z - a_n), \tag{5.5–14}$$

where $a_1, a_2, \ldots$ are all different constants and C is a constant, then

$$\frac{P(z)}{Q(z)} = \frac{A_1}{(z - a_1)} + \frac{A_2}{(z - a_2)} + \cdots + \frac{A_n}{(z - a_n)}, \tag{5.5–15}$$

where $A_1, A_2, \ldots$ are constants. Equation (5.5–15), called the partial fraction expansion of $P(z)/Q(z)$, is valid for all $z \neq a_j$ $(j = 1, 2, \ldots, n)$. □

This rule does not apply to the function $z/[(z + 1)(z^2 - 1)]$ since $Q(z) = (z + 1)^2(z - 1)$. The first factor here appears raised to the second power and not to the first as required by Eq. (5.5–14). Instead, we use the following rule.

RULE II Repeated factors

Let $Q(z)$ be factored as in Eq. (5.5–14), except that $(z - a_1)$ appears raised to the m_1 power, $(z - a_2)$ to the m_2 power, etc. Then, $P(z)/Q(z)$ can be decomposed as in Eq. (5.5–15), except that for each factor of $Q(z)$ of the form $(z - a_j)^{m_j}$, where $m_j \geq 2$, we replace $A_j/(z - a_j)$ in Eq. (5.5–15) by

$$\frac{A_{j1}}{(z - a_j)} + \frac{A_{j2}}{(z - a_j)^2} + \cdots + \frac{A_{jm_j}}{(z - a_j)^{m_j}}. \quad \square$$

Thus Rule I tells us that

$$\frac{z}{(z - 1)(z + 1)} = \frac{A_1}{(z - 1)} + \frac{A_2}{(z + 1)},$$

whereas Rule II establishes

$$\frac{z}{(z - 1)^2(z + 1)} = \frac{A_{11}}{(z - 1)} + \frac{A_{12}}{(z - 1)^2} + \frac{A_2}{(z + 1)}.$$

The utility of partial fractions in the generation of series and various procedures for obtaining the coefficients are illustrated in Examples 6 and 7.

First, however, for future reference we state four Maclaurin expansions:

$$\frac{1}{1 - w} = 1 + w + w^2 + \cdots, \qquad |w| < 1; \tag{5.5–16a}$$

$$\frac{1}{1 + w} = 1 - w + w^2 - w^3 + \cdots, \qquad |w| < 1; \tag{5.5–16b}$$

$$\frac{1}{(1 - w)^2} = 1 + 2w + 3w^2 + \cdots, \qquad |w| < 1; \tag{5.5–16c}$$

$$\frac{1}{(1 + w)^2} = 1 - 2w + 3w^2 - \cdots, \qquad |w| < 1. \tag{5.5–16d}$$

Equation (5.5–16a) is actually Eq. (5.2–7) with z replaced by w. Equation (5.5–16c) is similarly derived from Eq. (5.3–10). Equations (5.5–16b) and (5.5–16d) are obtained if we substitute $-w$ for w in Eqs. (5.5–16a) and (5.5–16c), respectively.

A general expansion for $1/(1 \pm w)^N$, which one sometimes needs, can be obtained from Exercise 3 in this section.

EXAMPLE 6

Expand

$$f(z) = \frac{z}{z^2 - z - 2} = \frac{z}{(z + 1)(z - 2)}$$

in a Taylor series about the point $z = 1$.

Solution

We have, from Rule I,

$$\frac{z}{(z+1)(z-2)} = \frac{a}{(z+1)} + \frac{b}{(z-2)}. \tag{5.5–17}$$

(It is simpler to use a and b here instead of the subscript notation A_1 and A_2.)
Clearing the fractions in Eq. (5.5–17) yields

$$z = a(z-2) + b(z+1).$$

We can find a and b by letting z in the above equation equal -1 and then 2. This type of procedure is useful and can be generalized to yield the required partial fractions whenever $Q(z)$ has any number of nonrepeated factors (see Exercise 10 in this section). For another approach we rearrange the previous equation as

$$z = (a+b)z + (-2a+b).$$

The coefficients of like powers of z on each side of the equation must be in agreement. We then have

$$1 = a + b \qquad (z^1 \text{ coefficients});$$
$$0 = (-2a + b) \qquad (z^0 \text{ coefficients});$$

whose solution is $a = 1/3$, $b = 2/3$.
Thus, from Eq. (5.5–17),

$$\frac{z}{(z+1)(z-2)} = \frac{1/3}{(z+1)} + \frac{2/3}{(z-2)}. \tag{5.5–18}$$

To expand $z/[(z+1)(z-2)]$ in powers of $(z-1)$ we expand each fraction on the right in Eq. (5.5–18) in these powers. Thus

$$\frac{1/3}{(z+1)} = \frac{1/3}{(z-1)+2} = \frac{1/6}{1+\frac{(z-1)}{2}} = 1/6\left[1 - \frac{(z-1)}{2} + \frac{(z-1)^2}{4} - \cdots\right],$$
$$\text{for } |z-1| < 2. \tag{5.5–19}$$

The preceding series is obtained with the substitution $w = (z-1)/2$ in Eq. (5.5–16b). The requirement $|z-1| < 2$ is identical to the constraint $|w| < 1$.
Similarly,

$$\frac{2/3}{(z-2)} = \frac{2/3}{(z-1)-1} = \frac{-2/3}{1-(z-1)} = -\frac{2}{3}[1 + (z-1) + (z-1)^2 + \cdots],$$
$$\text{for } |z-1| < 1, \tag{5.5–20}$$

where we have used Eq. (5.5–16a) and taken $w = z - 1$. The series in Eqs. (5.5–19) and (5.5–20) are now substituted in the right side of Eq. (5.5–18). Thus

$$\frac{z}{(z+1)(z-2)} = \frac{1}{6}\underbrace{\left[1 - \frac{(z-1)}{2} + \frac{(z-1)^2}{4} - \cdots\right]}_{|z-1|<2}$$
$$- \frac{2}{3}\underbrace{[1 + (z-1) + (z-1)^2 + \cdots]}_{|z-1|<1}.$$

In the domain $|z - 1| < 1$ *both* series converge and their terms can be combined. Thus

$$\frac{z}{(z+1)(z-2)} = \left(\frac{1}{6} - \frac{2}{3}\right) + \left(-\frac{1}{12} - \frac{2}{3}\right)(z-1) + \left(\frac{1}{24} - \frac{2}{3}\right)(z-1)^2 + \cdots,$$

or

$$\frac{z}{(z+1)(z-2)} = \sum_{n=0}^{\infty} c_n(z-1)^n, \qquad |z-1| < 1, \tag{5.5–21}$$

where

$$c_n = \frac{1}{6}\left(-\frac{1}{2}\right)^n - \frac{2}{3}.$$

We could have obtained the constraint $|z - 1| < 1$ by studying the location of the singularities of $z/[(z + 1)(z - 2)]$ to see which one ($z = 2$) lies closest to $(1, 0)$. ◀

EXAMPLE 7

Expand

$$f(z) = \frac{z}{(z+1)^2(z-2)}$$

in a Maclaurin series.

Solution

Since $(z + 1)$ is raised to the second power, we must follow Rule II and seek a partial fraction expansion of the form

$$\frac{z}{(z+1)^2(z-2)} = \frac{A}{(z+1)} + \frac{B}{(z+1)^2} + \frac{C}{(z-2)}. \tag{5.5–22}$$

Clearing fractions we obtain

$$z = A(z+1)(z-2) + B(z-2) + C(z+1)^2, \tag{5.5–23}$$

or

$$z = (A + C)z^2 + (-A + B + 2C)z + (-2A - 2B + C). \tag{5.5–24}$$

By putting $z = 2$ and then $z = -1$ in Eq. (5.5–23), we discover that $C = 2/9$ and $B = 1/3$. Note that z^2 does not appear on the left in Eq. (5.5–24), which means z^2 must not appear on the right; hence, $A = -C = -2/9$. Thus from Eq. (5.5–22)

$$\frac{z}{(z+1)^2(z-2)} = \frac{-2/9}{z+1} + \frac{1/3}{(z+1)^2} + \frac{2/9}{z-2}. \tag{5.5–25}$$

We now expand each fraction in powers of z. Taking $w = z$, we have, from Eq. (5.5–16b),

$$\frac{-2/9}{1+z} = -\frac{2}{9}[1 - z + z^2 - \cdots], \qquad |z| < 1,$$

and, from Eq. (5.5–16d),

$$\frac{1/3}{(1+z)^2} = \frac{1}{3}[1 - 2z + 3z^2 - 4z^3 + \cdots], \qquad |z| < 1.$$

With $w = z/2$ in Eq. (5.5–16a) we obtain

$$\frac{2/9}{z-2} = \frac{-1/9}{1-z/2} = -\frac{1}{9}\left[1 + \frac{z}{2} + \frac{z^2}{4} + \cdots\right], \qquad |z| < 2.$$

The substitution of the three preceding series on the right in Eq. (5.5–25) yields

$$\frac{z}{(z+1)^2(z-2)} = -\frac{2}{9}\underbrace{[1 - z + z^2 - \cdots]}_{|z|<1} + \frac{1}{3}\underbrace{[1 - 2z + 3z^2 - \cdots]}_{|z|<1}$$
$$-\frac{1}{9}\underbrace{\left[1 + \frac{z}{2} + \frac{z^2}{4} + \cdots\right]}_{|z|<2}.$$

Inside $|z| = 1$ we can add the three series together and obtain

$$\frac{z}{(z+1)^2(z-2)} = \sum_{n=0}^{\infty} c_n z^n, \qquad |z| < 1, \tag{5.5–26}$$

where

$$c_n = (-1)^{n+1}\frac{2}{9} + \frac{(-1)^n}{3}(n+1) - \frac{1}{9}\left(\frac{1}{2}\right)^n. \qquad \blacktriangleleft$$

EXERCISES

1. Differentiate the series of Eq. (5.5–2) to show that

$$\frac{1}{z^3} = 1 - \frac{3 \cdot 2}{2}(z-1) + \frac{4 \cdot 3}{2}(z-1)^2 - \frac{5 \cdot 4}{2}(z-1)^3 + \cdots, \qquad |z-1| < 1.$$

2. By differentiating the series of Eq. (5.2–7) several times, find c_n in the expansion $1/(1-z)^4 = \sum_{n=0}^{\infty} c_n z^n$, $|z| < 1$.

3. Use the series in Eq. (5.2–7), and successive differentiation, to show that, for $N \geq 0$,

$$\frac{1}{(1-z)^N} = \sum_{n=0}^{\infty} c_n z^n, \qquad c_n = \frac{(N-1+n)!}{n!(N-1)!}, \qquad |z| < 1.$$

4. a) Show that

$$\frac{1}{1+z^2} = 1 - z^2 + z^4 - z^6 + \cdots, \qquad |z| < 1.$$

b) Integrate the above series, and its sum, along a contour from the origin to an arbitrary point z to show that

$$\tan^{-1} z = \sum_{n=0}^{\infty} (-1)^n \frac{z^{2n+1}}{(2n+1)}, \qquad |z| < 1. \tag{5.5–27}$$

5. a) Consider $\mathrm{Si}(z)$ discussed in Example 2. Show that $f(z) = \int_0^z \mathrm{Si}(z')\,dz'$ can be expressed as $f(z) = \sum_{n=1}^{\infty} c_n z^{2n}$ (for all z). State c_n.

b) Evaluate approximately $f(2i)$ by using the first four terms of this series.

6. The Fresnel integrals $C(P)$ and $S(P)$ are used in optics and in the design of microwave antennas. They are defined by

$$C(P) = \int_0^P \cos\left(\frac{\pi t^2}{2}\right) dt$$

and

$$S(P) = \int_0^P \sin\left(\frac{\pi t^2}{2}\right) dt,$$

where $P \geq 0$ is a real number and t a real variable. In Exercise 26 of Section 6.6 the reader can prove that when $P = \infty$, $C = S = 0.5$. For finite P both C and S must be evaluated numerically. Notice that

$$F(P) = C(P) + iS(P) = \int_0^P e^{i\pi t^2/2}\,dt.$$

Why is this so?

a) Expand the preceding integrand in a Maclaurin series and integrate to show that

$$C(P) + iS(P) = \sum_{n=0}^{\infty} \frac{\left(\dfrac{i\pi}{2}\right)^n P^{2n+1}}{n!(2n+1)}.$$

b) Allow P to assume the values 0, 0.2, 0.4, 0.6, and 0.8. Use the first four terms of your series and a pocket calculator to compute the corresponding values of $F(P)$. Plot these values of $F(P)$ as points in a complex plane whose real axis is $C(P)$ and whose imaginary axis is $S(P)$. Connect the points by a smooth curve. This locus is, approximately, a portion of the "Cornu spiral."† The entire spiral can be used to determine $F(P) = C(P) + iS(P)$ for any real number P.

7. a) By considering the first and second derivatives of the geometric series in Eq. (5.2–7) show that $\sum_{n=1}^{\infty} n^2 z^n = (z + z^2)/(1 - z)^3$ for $|z| < 1$.

b) Use your result to evaluate $\sum_{n=1}^{\infty} n^2/2^n$.

Use series multiplication to find a formula for c_n in these Maclaurin expansions. In what circle is each series valid?

8. $\dfrac{\cosh z}{1 - z} = \sum_{n=0}^{\infty} c_n z^n$

9. $\dfrac{\text{Log}(1 - z)}{1 + z} = \sum_{n=0}^{\infty} c_n z^n$

10. Assume $P(z)/Q(z)$ satisfies the requirements of Rule I for partial fractions. Thus $Q(z)$ has no repeated factors and

$$\frac{P(z)}{Q(z)} = \frac{P}{C(z - a_1)(z - a_2)\cdots(z - a_n)} = \frac{A_1}{z - a_1} + \frac{A_2}{z - a_2} + \cdots + \frac{A_n}{z - a_n}.$$

a) Multiply both sides of the preceding equation by $(z - a_1)(z - a_2)\cdots(z - a_n)$ and cancel common factors to show that

$$\frac{P(z)}{C} = A_1(z - a_2)(z - a_3)\cdots(z - a_n) + A_2(z - a_1)(z - a_3)\cdots(z - a_n) + \cdots + A_n(z - a_1)(z - a_2)\cdots(z - a_{n-1}).$$

b) Show how to obtain any coefficient A_j $(j = 1, 2, \ldots, n)$ by setting $z = a_j$ in the previous equation.

c) Show that the result obtained in part (b) is identical to

$$A_j = \lim_{z \to a_j} \left[(z - a_j)\frac{P(z)}{Q(z)} \right] = \frac{P(a_j)}{Q'(a_j)}.$$

Hint: Consider L'Hôpital's Rule.

d) Expand $z/[(z^2 + 1)(z - 2)]$ in partial fractions by using the results of part (b) or (c).

---

† See, e.g., J. D. Kraus, *Antennas*, 2nd ed. (New York: McGraw-Hill, 1988), p. 185.

Obtain the following Taylor expansions. Give a general formula for the nth coefficient, and state the circle within which your expansion is valid.

11. $\dfrac{z}{(z-1)(z+2)}$ expanded about $z = 0$

12. $\dfrac{z}{(z+1)(z+2)}$ expanded about $z = 1$

13. $\dfrac{1}{z^2}$ expanded about $z = 1 + i$

14. $\dfrac{1}{z^3}$ expanded about $z = i$

15. $\dfrac{z+1}{(z-1)^2(z+2)}$ expanded about $z = 2$

16. $\dfrac{1}{(z-1)^2(z+1)^2}$ expanded about $z = 2$

17. $\dfrac{e^z}{(z-2)(z+1)}$ expanded about $z = 0$

18. Use the answer to Exercise 16 to find the value of the 10th derivative of $1/[(z-1)^2(z+1)^2]$ at $z = 2$.

19. Expand $\dfrac{z^3 + 2z^2 + z - 1}{z^2 - 4}$ about $z = 1$.

Hint: The denominator is not of higher degree than the numerator; thus we cannot immediately make a partial fraction decomposition. Show by a long division that the given function can be written as

$$(z + 2) + \frac{5z + 7}{z^2 - 4} = (z - 1) + 3 + \frac{5z + 7}{z^2 - 4}.$$

Now apply the method of partial fractions.

20. Let $h(z) = f(z)/g(z)$, where $f(z) = \sum_{n=0}^{\infty} a_n(z - z_0)^n$ and $g(z) = \sum_{n=0}^{\infty} b_n(z - z_0)^n$, and $g(z_0) = b_0 \neq 0$. We seek a Taylor expansion of $h(z)$ of the form $h(z) = \sum_{n=0}^{\infty} c_n(z - z_0)^n$. Find c_0, c_1, c_2 by long division of the series for $f(z)$ by the series for $g(z)$. Show that you obtain results identical to Eq. (5.5–11).

21. Find the coefficients c_0, c_1, c_2, c_3 in the Maclaurin expansion $\text{Log}(1 + z)/\cos z = \sum_{n=0}^{\infty} c_n z^n$, $|z| < 1$, by the series division of Exercise 20, or by the technique used in deriving Eq. (5.5–11).

22. Obtain all the coefficients in the following Maclaurin expansion by doing a long division:

$$\frac{1 + z}{1 + z + z^2 + z^3 + \cdots} = \sum_{n=0}^{\infty} c_n z^n, \ |z| < 1.$$

Explain why there are only two nonzero coefficients in your result.

23. a) The Bernoulli numbers $B_0, B_1, B_2, \ldots$ are defined by†

$$B_n = n!c_n,$$

where

$$f(z) = \begin{Bmatrix} \dfrac{z}{e^z - 1}, & z \neq 0 \\ 1, & z = 0 \end{Bmatrix} = \sum_{n=0}^{\infty} c_n z^n.$$

Note that $f(z)$ is analytic at $z = 0$ since, for all z,

$$\frac{z}{e^z - 1} = \frac{z}{z + \dfrac{z^2}{2!} + \dfrac{z^3}{3!} + \cdots} = \frac{1}{1 + \dfrac{z}{2!} + \dfrac{z^2}{3!} + \cdots}.$$

Perform long division on the right-hand quotient to show that $B_0 = 1$, $B_1 = -1/2$, $B_2 = 1/6$.

b) Show that the coefficients of odd order beyond 1, i.e., $B_3, B_5, B_7, \ldots$, are all zero.

Hint: $f(z) + z/2 = (z/2)\cosh(z/2)/\sinh(z/2)$ is an even function of z. See Exercise 30 of the previous section.

c) Where is the series expansion of part (a) valid?

24. a) Consider the Maclaurin expansion

$$\frac{1}{\cosh z} = \sum_{n=0}^{\infty} \frac{E_n}{n!} z^n,$$

where E_n are known as the Euler numbers. What is the radius of convergence of this series? Show that $E_0 = 1$, $E_2 = -1$, $E_4 = 5$, $E_6 = -61$, by a long division for the Maclaurin series of cosh z. Explain why $E_1, E_3, E_5, \ldots$ are zero. The Euler numbers are tabulated in handbooks.‡

b) Show that

$$\frac{1}{\cos z} = E_0 - \frac{E_2}{2!} z^2 + \frac{E_4}{4!} z^4 - \frac{E_6}{6!} z^6 + \cdots.$$

---

† The Bernoulli numbers also appear in other expansions. Tables of the numbers can be found in various handbooks, e.g., M. Abramowitz and I. Stegun, *Handbook of Mathematical Functions* (New York: Dover Publications, 1965), p. 810.

‡ Abramowitz and Stegun, *op. cit.* Other handbooks may use a slightly different definition of these numbers.

c) Multiply the Maclaurin series for $\sin z$ by the series obtained in part (b) to show that

$$\tan z = \frac{E_0}{0!}z - \left(\frac{E_2}{1!2!} + \frac{E_0}{3!1!}\right)z^3 + \left(\frac{E_4}{1!4!} + \frac{E_2}{3!2!} + \frac{E_0}{5!0!}\right)z^5 - \left(\frac{E_6}{1!6!} + \frac{E_4}{3!4!} + \frac{E_2}{5!2!} + \frac{E_0}{7!0!}\right)z^7 + \cdots, \qquad |z| < \pi/2$$

25. a) Let α be any complex number except zero or a positive integer. Using the branch of $(1+z)^\alpha$ defined by $e^{\alpha \operatorname{Log}(1+z)}$ (principal branch), show that, for $|z| < 1$,

$$(1+z)^\alpha = 1 + \alpha z + \frac{\alpha(\alpha-1)z^2}{2!} + \frac{\alpha(\alpha-1)(\alpha-2)z^3}{3!} + \cdots = 1 + \sum_{n=1}^{\infty} c_n z^n,$$

where $c_n = (1/n!)[\alpha(\alpha-1)(\alpha-2)\cdots(\alpha-(n-1))]$. Follow the method of Example 3.

b) Show that if α is a positive integer, then $(1+z)^\alpha = 1 + \sum_{n=1}^{\alpha} c_n z^n$, where c_n is given in part (a). Where is this expansion valid?

26. a) Use the result derived in Exercise 25(a) and a change of variable to show that

$$\frac{1}{(1-z)^{1/2}} = 1 + \frac{z}{2} + \left(\frac{1}{2}\right)\left(\frac{3}{2}\right)\frac{z^2}{2!} + \left(\frac{1}{2}\right)\left(\frac{3}{2}\right)\left(\frac{5}{2}\right)\frac{z^3}{3!} + \cdots, \qquad |z| < 1.$$

Use the first four terms of this series to evaluate approximately $\sqrt{2}$. Compare this with the value obtained from your pocket calculator.

b) Show that

$$\frac{1}{(1-z^2)^{1/2}} = 1 + \frac{1}{2}z^2 + \frac{\left(\frac{1}{2}\right)\left(\frac{3}{2}\right)z^4}{2!} + \frac{\left(\frac{1}{2}\right)\left(\frac{3}{2}\right)\left(\frac{5}{2}\right)z^6}{3!} + \cdots, \qquad |z| < 1.$$

c) Use the preceding result and a term-by-term integration to show that

$$\sin^{-1}(z) = z + \frac{z^3}{2\cdot 3\cdot 1!} + \frac{1\cdot 3z^5}{2^2\cdot 5\cdot 2!} + \frac{1\cdot 3\cdot 5z^7}{2^3\cdot 7\cdot 3!} + \cdots, \qquad |z| < 1,$$

where this branch of $\sin^{-1}(z)$ is analytic inside the unit circle, and $\sin^{-1}(0) = 0$. Note that $\cos^{-1}(z) = \pi/2 -$ (the series on the above right), provided $|z| < 1$.

d) Use the series for $\sin^{-1}(z)$ to obtain a numerical series for $\pi/6$. Use the first four terms of your result to evaluate approximately $\pi/6$.

5.6 LAURENT SERIES

A Taylor series expansion of a function $f(z)$ contains only terms of the form $(z - z_0)$ raised to nonnegative integer powers. The Laurent series,[†] which is related to the Taylor series, is defined as follows:

DEFINITION Laurent series

The Laurent series expansion of a function $f(z)$ is an expansion of the form

$$f(z) = \sum_{n=-\infty}^{\infty} c_n(z - z_0)^n = \cdots + c_{-2}(z - z_0)^{-2} + c_{-1}(z - z_0)^{-1}$$
$$+ c_0 + c_1(z - z_0) + \cdots, \qquad (5.6\text{–}1)$$

where the series converges to $f(z)$ in a region or domain. □

Thus a Laurent expansion, unlike a Taylor expansion, can contain one or more terms with $(z - z_0)$ raised to a negative power. It can also contain positive powers of $(z - z_0)$.

Examples of Laurent series are often obtained from some simple manipulations on Taylor series. Thus

$$e^u = 1 + u + \frac{u^2}{2!} + \cdots, \qquad \text{all finite } u.$$

Putting $u = (z - 1)^{-1}$ in the preceding equation, we have

$$e^{1/(z-1)} = 1 + (z - 1)^{-1} + \frac{(z - 1)^{-2}}{2!} + \frac{(z - 1)^{-3}}{3!} + \cdots, \qquad z \neq 1.$$

We can reverse the order of the terms on the right to comply with the form of Eq. (5.6–1) and obtain

$$e^{1/(z-1)} = \cdots + \frac{(z - 1)^{-3}}{3!} + \frac{(z - 1)^{-2}}{2!} + \frac{(z - 1)^{-1}}{1!} + 1, \qquad z \neq 1. \qquad (5.6\text{–}2)$$

This is a Laurent series with no positive powers of $(z - 1)$. Multiplying both sides of Eq. (5.6–2) by $(z - 1)^2$, we have

$$(z - 1)^2 e^{1/(z-1)} = \cdots + \frac{(z - 1)^{-1}}{3!} + \frac{1}{2!} + (z - 1) + (z - 1)^2, \qquad z \neq 1. \qquad (5.6\text{–}3)$$

This is a Laurent series with both negative and positive powers of $(z - 1)$.

A knowledge of Laurent series is necessary for an understanding of the calculus of residues. Residues, treated in Chapter 6, are an invaluable tool in the evaluation of many types of integrals.

In Appendix B of this chapter another use of Laurent series is given. The **z** transformation, used in applied mathematics and various branches of engineering, is

---

[†] Named for its discoverer, the French mathematician Paul Mathieu Hermann Laurent (1841–1908).

based directly upon the Laurent series representation of an analytic function. After finishing this section the reader may wish to turn to this appendix, where various properties of the transformation are developed and applied.

What kinds of functions can be represented by Laurent series, and in what region of the complex plane will the representation be valid? The answer is contained in the following theorem, which we will soon prove.

THEOREM 18 Laurent's theorem

Let $f(z)$ be analytic in D, an annular domain $r_1 < |z - z_0| < r_2$. If z lies in D, $f(z)$ can be represented by the Laurent expansion

$$f(z) = \sum_{n=-\infty}^{\infty} c_n(z - z_0)^n = \cdots + c_{-2}(z - z_0)^{-2} + c_{-1}(z - z_0)^{-1}$$
$$+ c_0 + c_1(z - z_0) + c_2(z - z_0)^2 + \cdots. \tag{5.6–4}$$

The coefficients are given by

$$c_n = \frac{1}{2\pi i} \oint_C \frac{f(z)}{(z - z_0)^{n+1}}\, dz, \tag{5.6–5}$$

where C is any simple closed contour lying in D and enclosing the inner boundary $|z - z_0| = r_1$. The series is uniformly convergent in any annular region centered at z_0 and lying in D. □

The theorem asserts that if $f(z)$ is analytic in a "washer-shaped" domain, like the one shown in Fig. 5.6–1, then it can be represented by a Laurent series throughout this domain. The coefficients can be found by a line integral (see Eq. 5.6–5) taken around a loop C, such as the one shown in the figure.

For simplicity we consider a proof in which z_0 is zero; that is, we seek an expansion in an annulus centered at the origin. The annulus, having inner and outer radii r_1 and r_2, is shown in Fig. 5.6–2(a).

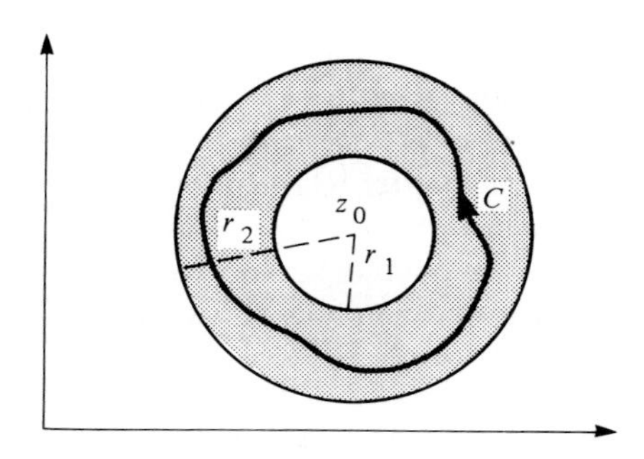

Figure 5.6–1

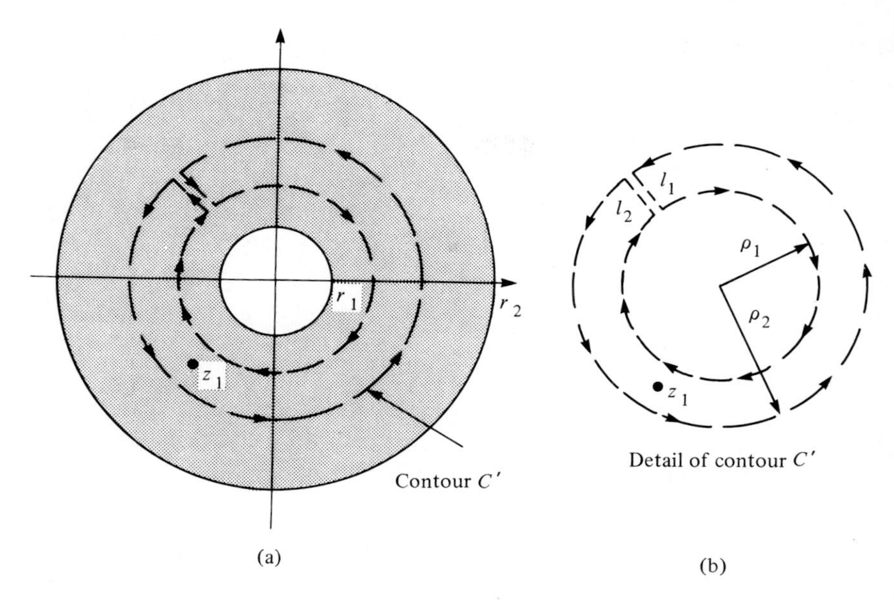

Figure 5.6–2

Now, using the contour C', which lies in the annulus, we apply the Cauchy integral formula. Observe that C' encloses the point z_1 and that $f(z)$ is analytic on and inside this contour. Thus

$$f(z_1) = \frac{1}{2\pi i} \oint_{C'} \frac{f(z)}{(z - z_1)}\, dz. \tag{5.6–6}$$

The portions of the preceding integral taken along the contiguous lines l_1 and l_2 (see Fig. 5.6–2b) cancel because of the opposite directions of integration. Thus Eq. (5.6–6) becomes

$$f(z_1) = I_A + I_B, \tag{5.6–7}$$

where

$$I_A = \frac{1}{2\pi i} \oint_{|z|=\rho_2} \frac{f(z)}{(z - z_1)}\, dz \tag{5.6–8}$$

and

$$I_B = \frac{1}{2\pi i} \oint_{|z|=\rho_1} \frac{f(z)}{(z - z_1)}\, dz. \tag{5.6–9}$$

Note the directions for each of these two integrations. The integral I_A is dealt with in the same manner as the integral in Eq. (5.4–12) in our derivation of the Taylor series. We have

$$I_A = \frac{1}{2\pi i}\oint_{|z|=\rho_2} \frac{f(z)}{z\left(1 - \dfrac{z_1}{z}\right)}\, dz = \frac{1}{2\pi i}\oint_{|z|=\rho_2} \frac{f(z)}{z}\left(1 + \frac{z_1}{z} + \left(\frac{z_1}{z}\right)^2 + \cdots\right) dz$$

$$= \sum_{n=0}^{\infty} c_n z_1^n, \tag{5.6–10}$$

where

$$c_n = \frac{1}{2\pi i}\oint_{|z|=\rho_2} \frac{f(z)}{z^{n+1}}\, dz, \qquad n = 0, 1, 2, \ldots. \tag{5.6–11}$$

In Eq. (5.6–10) we require that $|z_1/z| < 1$ or $|z_1| < \rho_2$ (since $|z| = \rho_2$).

In the integral I_B (see Eq. 5.6–9) we reverse the direction of integration and compensate with a minus sign in the integrand. Thus

$$I_B = \frac{1}{2\pi i}\oint_{|z|=\rho_1} \frac{f(z)}{(z_1 - z)}\, dz = \frac{1}{2\pi i}\oint_{|z|=\rho_1} \frac{f(z)}{z\left(1 - \dfrac{z}{z_1}\right)}\, dz. \tag{5.6–12}$$

Now

$$\frac{1}{1 - \dfrac{z}{z_1}} = 1 + \frac{z}{z_1} + \left(\frac{z}{z_1}\right)^2 + \cdots, \qquad \text{if } \left|\frac{z}{z_1}\right| < 1 \text{ or } |z| < |z_1|.$$

This series converges uniformly in a region containing the circle $|z| = \rho_1$ (since $|z| = \rho_1 < |z_1|$). Using this series in Eq. (5.6–12) and integrating we obtain

$$I_B = \frac{1}{2\pi i}\oint_{|z|=\rho_1} \frac{f(z)}{z_1}\left(1 + \frac{z}{z_1} + \left(\frac{z}{z_1}\right)^2 + \cdots\right) dz = \frac{z_1^{-1}}{2\pi i}\oint_{|z|=\rho_1} f(z)\, dz$$

$$+ \frac{z_1^{-2}}{2\pi i}\oint_{|z|=\rho_1} z f(z)\, dz + \frac{z_1^{-3}}{2\pi i}\oint_{|z|=\rho_1} z^2 f(z)\, dz + \cdots. \tag{5.6–13}$$

We have moved the constant z_1 outside the integral signs. We may rewrite Eq. (5.6–13) more succinctly as

$$I_B = \sum_{n=-\infty}^{-1} c_n z_1^n, \qquad |z_1| > \rho_1, \tag{5.6–14}$$

where

$$c_n = \frac{1}{2\pi i}\oint_{|z|=\rho_1} z^{-n-1} f(z)\, dz = \frac{1}{2\pi i}\oint_{|z|=\rho_1} \frac{f(z)}{z^{n+1}}\, dz, \qquad n = \cdots, -2, -1. \tag{5.6–15}$$

Let us compare Eqs. (5.6–11) and (5.6–15). The former equation gives the coefficients for a series representation of the integral I_A (see Eq. 5.6–10). The index n in

Eq. (5.6–11) is zero or positive. Equation (5.6–15) gives the coefficients for a series expansion of the integral I_B (see Eq. 5.6–14). The index n in Eq. (5.6–15) is negative. Both Eq. (5.6–11) and Eq. (5.6–15) are identical in form except that the paths of integration are circles of different radii. Observe, however, that $f(z)/z^{n+1}$ is analytic throughout the annular domain $r_1 < |z| < r_2$. Thus, by the principle of deformation of contours (see Section 4.3), we can perform the integrations in both Eq. (5.6–11) and Eq. (5.6–15) around any simple closed contour C lying in this domain and encircling the inner boundary $|z| = r_1$. The same contour can be used in both formulas.

Substituting series expansions, as shown in Eqs. (5.6–14) and (5.6–10), for the two integrals on the right in Eq. (5.6–7), we have

$$f(z_1) = \underbrace{\sum_{n=0}^{\infty} c_n z_1^n}_{|z_1| < \rho_2} + \underbrace{\sum_{n=-\infty}^{-1} c_n z_1^n}_{|z_1| > \rho_1}, \tag{5.6–16}$$

where

$$c_n = \frac{1}{2\pi i} \oint_C \frac{f(z)}{z^{n+1}}\, dz, \qquad n = 0, \pm 1, \pm 2, \ldots. \tag{5.6–17}$$

We can rewrite Eq. (5.6–16) as a single summation,

$$f(z_1) = \sum_{n=-\infty}^{+\infty} c_n z_1^n, \tag{5.6–18}$$

that is valid when z_1 satisfies $\rho_1 < |z_1| < \rho_2$. Since ρ_1 can be brought arbitrarily close to r_1 and ρ_2 arbitrarily close to r_2 (see Fig. 5.6–2) this restriction can be relaxed to $r_1 < |z_1| < r_2$. Replacing z_1 by z in Eq. (5.6–18), we find that we have derived Eq. (5.6–4) for the special case $z_0 = 0$. A derivation valid for an arbitrary z_0 is developed in Exercise 19 of this section.

The M test (see Theorem 7) can be used to study the uniform convergence of each series on the right in Eq. (5.6–16). It is easily established that the overall series shown in Eq. (5.6–18) is uniformly convergent in any closed annular region contained in, and concentric with, the domain $r_1 < |z| < r_2$ (see Fig. 5.6–2). Just as is the case for a Taylor series, the uniform convergence of a Laurent series permits term-by-term integration and differentiation. New series are obtained that converge to the integral and derivative, respectively, of the sum of the original series.

The imaginative reader may compare Eq. (5.6–17) with the extended Cauchy integral formula (see Eq. 4.5–13) and conclude that the coefficients for our Laurent series with $z_0 = 0$ are given by

$$c_n = \frac{f^{(n)}(0)}{n!}, \qquad \text{for } n \geq 0. \tag{5.6–19}$$

In fact, this very step was taken in the derivation of the Taylor series (see Eq. 5.4–19). This manuever is not permitted here. The Cauchy integral formula and its extension apply only when $f(z)$ in Eq. (5.6–17) is analytic not only on C but throughout its interior. We have made no such assumption concerning $f(z)$. Our derivation admits the possibility of $f(z)$ having singularities inside the circle $|z| = r_1$ in Fig. 5.6–2. If $f(z)$ were analytic throughout the disc $|z| \le r_1$, as well as in the annulus $r_1 < |z| < r_2$ shown in Fig. 5.6–2, then $c_0, c_1, c_2, \ldots$ would indeed be given by Eq. (5.4–19). Moreover, according to Eq. (5.6–17) and the Cauchy integral theorem, the coefficients c_{-1}, $c_{-2}, \ldots$ would be zero. A special kind of Laurent series, namely, a Taylor expansion of $f(z)$, is obtained. The preceding discussion is easily altered to deal with Eq. (5.6–5); in other words, that integral, in general, is *not* $f^n(z_0)/n!$.

Although the coefficients in a Laurent expansion can, in principle, be derived from Eq. (5.6–5), this formula is rarely used. In practice, the coefficients are obtained from manipulations involving known series, as in the derivation of Eq. (5.6–2), and with partial fraction decompositions. The techniques are illustrated in the following examples. A useful corollary to Theorem 18, which we will not prove, is that in a given annulus the Laurent series for a function is unique. Hence, if we find a Laurent expansion of $f(z)$ valid in an annular domain, we have found the only Laurent expansion of $f(z)$ in this domain. We will learn from Examples 2 and 3 in this section, that a given $f(z)$ can have different Laurent expansions valid in different annuli sharing the same center z_0.

A discussion of Laurent series and of analytic functions sometimes involves the notion of an isolated singular point, which is defined as follows:

DEFINITION Isolated singular point

The point z_p is an *isolated singular point* of $f(z)$ if $f(z)$ is not analytic at z_p but is analytic in a deleted neighborhood of z_p. □

For example, $1/[(z-1)(z-2)^3]$ has isolated singular points at $z = 1$ and $z = 2$ since we can find a disc, centered at each of these points, in which this function is everywhere analytic except for the center.

Let a function $f(z)$ be expanded in a Laurent series involving powers of $(z - z_0)$, where z_0 happens to be an isolated singular point of $f(z)$. We can find a series that converges to $f(z)$ in an annular domain centered at z_0. The inner radius r_1 of the domain can be made arbitrarily small but cannot be made zero since the point z_0 must, according to Laurent's theorem, be excluded from the domain. A series representation in such a domain (a disc with the center removed) is valid in a deleted neighborhood of z_0. An instance of this occurs in Example 3 below. In our work on residues in the next chapter we will be especially concerned with such series.

Finally, if $f(z)$ is analytic at all points in the z-plane lying outside some circle, then it is possible to find a Laurent expansion for $f(z)$ valid in an annulus whose outer radius r_2 is infinitely large. Equation (5.6–2) shows this possibility. Here we have a Laurent series expansion that is valid in a deleted neighborhood of $z = 1$. The outer radius of the domain in which the expansion holds is infinite.

EXAMPLE 1

Expand

$$f(z) = \frac{1}{(z-3)}$$

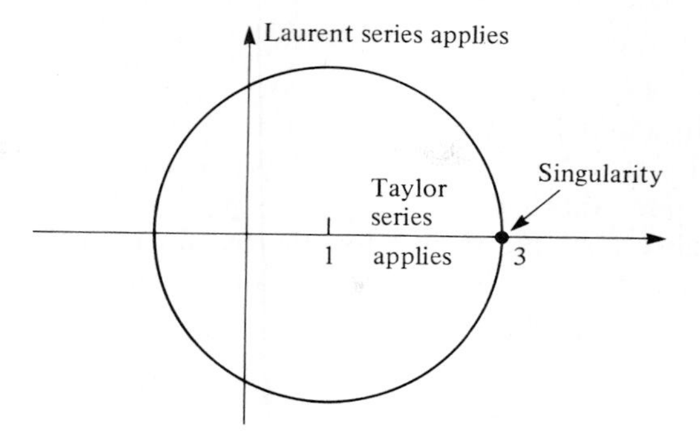

Figure 5.6–3

in a Laurent series in powers of $(z - 1)$. State the domain in which the series converges to $f(z)$.

Solution

Noting that $f(z)$ has its only singularity at $z = 3$, we see that a Taylor series representation of $f(z)$ is valid in the domain $|z - 1| < 2$ (see Fig. 5.6–3). According to Theorem 18, with $z_0 = 1$, we can represent $f(z)$ in a Laurent series in the domain $|z - 1| > 2$. We proceed by recalling Eq. (5.5–16a).

$$\frac{1}{1 - w} = 1 + w + w^2 + \cdots, \qquad |w| < 1. \tag{5.6–20}$$

Now

$$\frac{1}{z - 3} = \frac{1}{(z - 1) - 2} = \frac{1/(z - 1)}{1 - 2/(z - 1)}. \tag{5.6–21}$$

Comparing Eqs. (5.6–20) and (5.6–21) and taking $w = 2/(z - 1)$, we obtain our Laurent series. Thus

$$\frac{1}{z - 3} = \frac{1}{(z - 1)}\left[1 + \frac{2}{(z - 1)} + \frac{4}{(z - 1)^2} + \cdots\right]$$
$$= (z - 1)^{-1} + 2(z - 1)^{-2} + 4(z - 1)^{-3} + \cdots.$$

The condition $|w| < 1$ in Eq. (5.6–20) becomes $|2/(z - 1)| < 1$, or $|z - 1| > 2$. We anticipated that our Laurent series would be valid in this domain. ◀

EXAMPLE 2

Expand

$$f(z) = \frac{1}{(z + 1)(z + 2)}$$

in a Laurent series in powers of $(z - 1)$ valid in an annular domain containing the point $z = 7/2$. State the domain in which the series converges to $f(z)$.

Solution

From Theorem 18 we know that a Laurent series in powers of $(z - 1)$ is capable of representing $f(z)$ in annular domains centered at $z_0 = 1$. Refer to Fig. 5.6–4. Since $f(z)$ has singularities at -2 and -1, we see that one such domain is D_1 defined by $2 < |z - 1| < 3$, while another is D_2 given by $|z - 1| > 3$. A Taylor series representation is also available in the domain D_3 described by $|z - 1| < 2$. Since $z = 7/2$ lies in D_1, it is the Laurent expansion valid in this domain that we seek.

We break $f(z)$ into partial fractions. Thus

$$\frac{1}{(z+1)(z+2)} = \frac{1}{(z+1)} - \frac{1}{(z+2)}. \tag{5.6–22}$$

Because we wish to generate powers of $(z - 1)$, we rewrite the first fraction as

$$\frac{1}{(z+1)} = \frac{1}{(z-1)+2} = \frac{1/2}{1+(z-1)/2} \tag{5.6–23}$$

or, alternatively, as

$$\frac{1}{z+1} = \frac{1}{(z-1)+2} = \frac{1/(z-1)}{1+2/(z-1)}. \tag{5.6–24}$$

Recall now Eq. (5.5–16b):

$$\frac{1}{1+w} = 1 - w + w^2 - w^3 + \cdots, \qquad |w| < 1.$$

With $w = (z - 1)/2$ we expand Eq. (5.6–23) to obtain

$$\frac{1}{z+1} = \frac{1}{2}\left[1 - \frac{(z-1)}{2} + \frac{(z-1)^2}{4} + \cdots\right], \qquad \text{if } \left|\frac{z-1}{2}\right| < 1, \text{ or } |z-1| < 2. \tag{5.6–25}$$

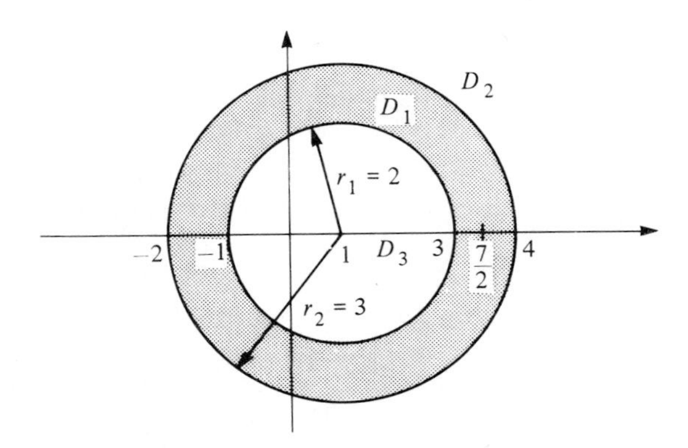

Figure 5.6–4

Taking $w = 2/(z-1)$, we can expand Eq. (5.6–24) as follows:

$$\frac{1}{z+1} = \frac{1}{(z-1)}\left[1 - \frac{2}{(z-1)} + \frac{4}{(z-1)^2} - \cdots\right] = (z-1)^{-1} - 2(z-1)^{-2}$$
$$+ 4(z-1)^{-3} - \cdots, \qquad \text{if } \left|\frac{2}{z-1}\right| < 1, \text{ or } |z-1| > 2. \tag{5.6–26}$$

We have expressed $1/(z+1)$ as a Taylor series and a Laurent series, both in powers of $(z-1)$. A similar procedure can be applied to the remaining partial fraction in Eq. (5.6–22). Thus, with $w = (z-1)/3$,

$$-\frac{1}{(z+2)} = \frac{-1}{(z-1)+3} = \frac{-1/3}{1 + \left(\dfrac{z-1}{3}\right)}$$
$$= -\frac{1}{3}\left[1 - \frac{(z-1)}{3} + \frac{(z-1)^2}{9} - \cdots\right], \qquad |z-1| < 3, \tag{5.6–27}$$

and, with $w = 3/(z-1)$,

$$-\frac{1}{z+2} = \frac{-1}{(z-1)+3} = \frac{-1/(z-1)}{1 + 3/(z-1)}$$
$$= -\frac{1}{(z-1)}\left[1 - \frac{3}{(z-1)} + \frac{9}{(z-1)^2} - \cdots\right]$$
$$= -(z-1)^{-1} + 3(z-1)^{-2} - 9(z-1)^{-3} + \cdots, \qquad |z-1| > 3. \tag{5.6–28}$$

In the domain D_1 the series in Eqs. (5.6–25) and (5.6–28) are of no use. However, the series in Eqs. (5.6–26) and (5.6–27) converge to their respective functions in this domain. Using these equations, we replace each fraction on the right in Eq. (5.6–22) by a series and obtain

$$\frac{1}{(z+1)(z+2)} = \underbrace{(z-1)^{-1} - 2(z-1)^{-2} + 4(z-1)^{-3} - \cdots}_{|z-1|>2}$$
$$\underbrace{-\frac{1}{3} + \frac{1}{9}(z-1) - \frac{1}{27}(z-1)^2 + \cdots}_{|z-1|<3}, \tag{5.6–29}$$

which, when written more succinctly, reads

$$\frac{1}{(z+1)(z+2)} = \sum_{n=-\infty}^{+\infty} c_n (z-1)^n, \tag{5.6–30}$$

where

$$c_n = \left(-\frac{1}{3}\right)^{n+1}, \qquad n \geq 0, \tag{5.6–31a}$$

and

$$c_n = (-1)^{n+1}2^{-n-1}, \qquad n \le -1. \tag{5.6–31b}$$

We see from Eq. (5.6–29) that the Laurent series in Eq. (5.6–30) is composed of two series that simultaneously converge to their respective partial fractions only in the annulus $2 < |z - 1| < 3$.

Comment: A Laurent series expansion of $f(z)$ in the domain $|z - 1| > 3$, that is, D_2, is possible. We represent the partial fractions in Eq. (5.6–22) by the series shown in Eqs. (5.6–26) and (5.6–28). Both are valid in D_2. Adding these series we have

$$\frac{1}{(z+1)(z+2)} = \sum_{n=-\infty}^{-2} c_n(z-1)^n, \qquad |z - 1| > 3,$$

where

$$c_n = (-1)^n[3^{-n-1} - 2^{-n-1}], \qquad n = \ldots, -3, -2.$$ ◀

EXAMPLE 3

Expand

$$f(z) = \frac{1}{(z)(z-1)}$$

in a Laurent series that is valid in a deleted neighborhood of $z = 1$. State the domain throughout which the series is valid.

Solution

Observe that $f(z)$ has singularities at $z = 1$ and $z = 0$. The annulus $0 < |z - 1| < 1$ is the largest deleted neighborhood of $z = 1$ that excludes both singularities of $f(z)$. Hence, we take $z_0 = 1$ in Theorem 18 and seek a Laurent expansion in powers of $(z - 1)$. Decomposing $f(z)$ into partial fractions, we obtain

$$\frac{1}{(z)(z-1)} = -\frac{1}{z} + \frac{1}{z-1}. \tag{5.6–32}$$

This equality breaks down at $z = 0$ and $z = 1$. The second fraction, $(z - 1)^{-1}$, is already expanded in powers of $(z - 1)$. It is a one-term Laurent series. No other expansion of this fraction in powers of $(z - 1)$ is possible. For the fraction $-1/z$ we have the choice of two series containing powers of $(z - 1)$. Thus

$$-\frac{1}{z} = \frac{-1}{1 + (z-1)} = -(1 - (z-1) + (z-1)^2 - \cdots), \qquad |z - 1| < 1, \tag{5.6–33}$$

and

$$-\frac{1}{z} = \frac{-1/(z-1)}{1 + 1/(z-1)} = -(z-1)^{-1}\left(1 - \frac{1}{(z-1)} + \frac{1}{(z-1)^2} - \cdots\right)$$
$$= -(z-1)^{-1} + (z-1)^{-2} - (z-1)^{-3} + \cdots, \qquad |z - 1| > 1. \tag{5.6–34}$$

Using Eq. (5.6–33) on the right in Eq. (5.6–32) to represent $-1/z$, we get

$$\frac{1}{(z)(z-1)} = \underbrace{-1 + (z-1) - (z-1)^2 + \cdots}_{|z-1|<1} + \underbrace{(z-1)^{-1}}_{z \neq 1},$$

or, more neatly,

$$\frac{1}{(z)(z-1)} = \sum_{n=-1}^{\infty} (-1)^{n+1}(z-1)^n, \qquad 0 < |z-1| < 1.$$

Comment: Had we used Eq. (5.6–34) instead of Eq. (5.6–33) to represent $-1/z$ on the right in Eq. (5.6–32), we would have obtained the Laurent expansion

$$\frac{1}{z(z-1)} = (z-1)^{-2} - (z-1)^{-3} + (z-1)^{-4} - \cdots.$$

This expansion is valid in the same annulus as the series in Eq. (5.6–34), that is, $|z-1| > 1$, which is not the required deleted neighborhood of $z = 1$. ◀

The preceding examples have shown how to obtain Laurent expansions when the function to be expanded is a ratio of polynomials. In Eqs. (5.6–2) and (5.6–3) we showed how the transcendental function $(z-1)^2 e^{1/(z-1)}$ could be expanded in a Laurent series about $z = 1$ if we make a change of variable in the Maclaurin expansion of e^z. Laurent series for transcendental functions are sometimes obtained either by division of Taylor series or by a recursive procedure equivalent to series division, as in the following example.

EXAMPLE 4

Expand $1/\sin z$ in a Laurent series valid in a deleted neighborhood of the origin. Where in the complex plane will your series converge to this function?

Solution

Recall that $\sin z = 0$ is satisfied when $z = 0, \pm\pi, \pm 2\pi, \ldots$. Thus $z = 0$, $z = -\pi$, and $z = \pi$ are isolated singular points of $1/\sin z$. A Laurent expansion of this function, employing powers of z, is thus possible in the punctured disc $0 < |z| < \pi$.

We seek a series expansion of the form $1/\sin z = \sum_{n=-\infty}^{\infty} c_n z^n$. It is helpful to now show that many of the coefficients in the series are zero. Note that

$$\frac{z}{\sin z} = z \sum_{n=-\infty}^{\infty} c_n z^n = \cdots + c_{-3}z^{-2} + c_{-2}z^{-1} + c_{-1} + c_0 z + c_1 z^2 + \cdots. \tag{5.6–35a}$$

Now from L'Hôpital's Rule we have

$$\lim_{z\to 0} \frac{z}{\sin z} = \lim_{z\to 0} \frac{1}{\cos z} = 1.$$

If the series on the right in Eq. (5.6–35a) is to produce this same limit, we require that $c_{-2}, c_{-3}, c_{-4}, \ldots$ all be zero. Otherwise the terms $c_{-2}z^{-1}$, $c_{-3}z^{-2}$, $c_{-4}z^{-3}$, etc., on the right would become infinite as $z \to 0$.

Having eliminated all c_n for $n \leq -2$, we have

$$\frac{1}{\sin z} = c_{-1}z^{-1} + c_0 + c_1 z + c_2 z^2 + \cdots, \qquad 0 < |z| < \pi.$$

We now follow the recursive procedure described in Section 5.5 to obtain the above coefficients. Multiplying both sides of the preceding equation by $\sin z$ and using the expansion $\sin z = z - z^3/3! + z^5/5! + \cdots$, we have

$$1 = \left(z - \frac{z^3}{3!} + \frac{z^5}{5!} + \cdots\right)$$
$$\cdot(c_{-1}z^{-1} + c_0 + c_1 z + c_2 z^2 + c_3 z^3 + \cdots).$$

Now multiplying the series on the above right and equating the coefficients of the various powers of z to the corresponding coefficients on the left, we find

$$1 = c_{-1} \qquad [z^0 \text{ term}],$$
$$0 = c_0 \qquad [z^1 \text{ term}],$$
$$0 = c_1 - \frac{c_{-1}}{3!} \qquad [z^2 \text{ term}],$$
$$0 = c_2 - \frac{c_0}{3!} \qquad [z^3 \text{ term}],$$
$$0 = c_3 - \frac{c_1}{3!} + \frac{c_{-1}}{5!} \qquad [z^4 \text{ term}],$$

etc.

From the preceding it becomes apparent that the coefficients of all even powers, c_0, $c_2, c_4, \ldots$, are zero. For the odd coefficients, $c_{-1} = 1$, $c_1 = 1/6$, $c_3 = -1/5! + (1/3!)/3! = 7/360$. A general formula for any odd coefficient is given in Exercise 23. Thus we have

$$\frac{1}{\sin z} = \frac{1}{z} + \frac{z}{6} + \frac{7z^3}{360} + \cdots, \qquad 0 < |z| < \pi. \tag{5.6–35b}$$

Comment: The Laurent series in Eq. (5.6–35b) converges to $1/\sin z$ in domain D_1 shown in Fig. 5.6–5. From the location of the singularities of $1/\sin z$ we see that it should be possible to obtain another Laurent series, in powers of z, valid in D_2 of the same figure, i.e.,

$$\frac{1}{\sin z} = \sum_{n=-\infty}^{\infty} d_n z^n, \qquad \pi < |z| < 2\pi.$$

Similarly, there is a third Laurent series valid in the domain D_3 described by $2\pi < |z| < 3\pi$. In fact, an infinite number of such Laurent series are possible, each

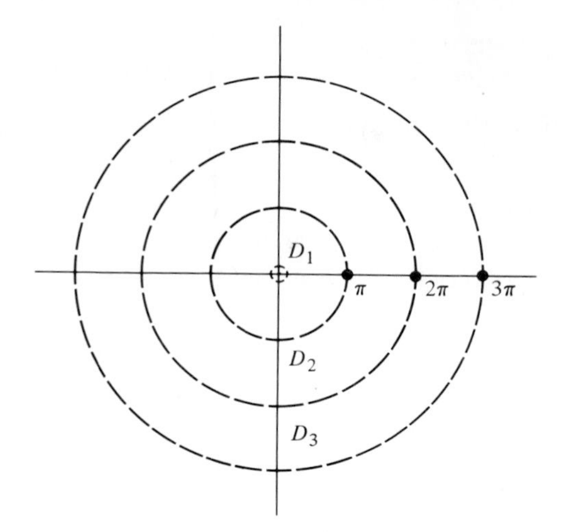

Figure 5.6–5

valid in a different annular region centered at the origin. The coefficients in these new series cannot be obtained by a straightforward division of the Maclaurin series for $\sin z$, as was just done in this last example. We will postpone the problem of finding the Laurent series in each ring until we consider the subject of residues, in the next chapter. In principle, we could find each coefficient by an application of Eq. (5.6–5), but for now we see no easy way to evaluate these integrals. ◀

EXERCISES

Obtain the following Laurent expansions. State explicitly the nth term in the series, and give the largest annular domain in which your series is valid.

1. $\dfrac{1}{z^4}\cos z$ expanded in powers of z

2. $z^4 \cos\dfrac{1}{z}$ expanded in powers of z

3. $ze^{1/(z-1)}$ expanded in powers of $z - 1$

4. $\sin[(z-1)/z]$ expanded in powers of z

5. $\dfrac{\sin(z-1)}{z}$ expanded in powers of $z - 1$

Obtain the indicated Laurent expansions of $1/(z + 2)$. State the nth term of the series.

6. An expansion valid for $|z| > 2$.

7. An expansion valid for $|z + 1| > 1$.

Expand the following functions in a Laurent series valid in a domain whose outer radius is infinite. State the center and inner radius of the domain. Give the nth term of the series.

8. $1/(z-1)$ expanded in powers of $z+3$

9. $1/(z+2)$ expanded in powers of $z-i$

10. $z/(z-i)$ expanded in powers of $z-1$

11. a) Consider $f(z) = 1/[z(z-1)(z+3)]$. This function is expanded in three different Laurent series involving powers of z. State the three domains in which Laurent series are available.

b) Find each series and give an explicit formula for the nth term.

For the following functions, find the Laurent series valid in an annular domain that contains the point $z = 2 + 2i$. The center of the annulus is at $z = 1$. State the domain in which each series is valid, and give an explicit formula for the nth term of your series.

12. $f(z) = 1/[z(z-2)]$

13. $f(z) = 1/[z(z-4)]$

14. $f(z) = 1/[(z-1)(z-3)]$

15. $f(z) = \dfrac{z-i}{z-1}$

16. $f(z) = \dfrac{1}{(z-1)^3} + \dfrac{1}{z}$

17. $f(z) = \dfrac{1}{(z-1)^3} + z^3$

18. The exponential integral $E_1(a)$ is defined by the improper integral

$$E_1(a) = \int_a^\infty \frac{e^{-x}}{x}\,dx, \qquad a > 0.$$

Thus

$$E_1(a) - E_1(b) = \int_a^b \frac{e^{-x}}{x}\,dx.$$

Use a Laurent expansion for e^{-z}/z and a term-by-term integration to show that

$$E_1(a) - E_1(b) = \text{Log}\frac{b}{a} - (b-a) + \frac{b^2 - a^2}{(2!)(2)} - \frac{(b^3 - a^3)}{(3!)(3)} + \cdots.$$

19. A derivation of Laurent's theorem was given for the case $z_0 = 0$. Derive this theorem for the more general case where z_0 is not necessarily 0.

Hint: Redraw Fig. 5.6–2 with all circles centered at $z_0 \neq 0$. Notice that

$$f(z_1) = \frac{1}{2\pi i}\oint_{|z-z_0|=\rho_2} \frac{f(z)\,dz}{(z-z_0)-(z_1-z_0)} + \frac{1}{2\pi i}\oint_{|z-z_0|=\rho_1} \frac{f(z)\,dz}{(z_1-z_0)-(z-z_0)}.$$

Now expand each integral in either positive or negative powers of $(z_1 - z_0)$.

20. Explain why $\text{Log}\, z$ (principal branch) cannot be expanded in a Laurent series involving powers of z.

21. Explain why $(\sin z)/z^{1/2}$ cannot be expanded in a Laurent series valid in a deleted neighborhood of $z = 0$. Use the principal branch of $z^{1/2}$.

22. Obtain the Laurent expansion of $z^{1/2}/(z-1)$ valid in a punctured disc centered at $z = 1$. Give an explicit formula for the nth term, and state the maximum outer radius of the disc. Use the principal branch of $z^{1/2}$.

Hint: Use the result contained in Exercise 25(a) of the previous section.

23. a) Extend the work of Example 4 to show that in the expansion

$$\frac{1}{\sin z} = \sum_{n=-1}^{\infty} c_n z^n, \qquad 0 < |z| < \pi,$$

we can get c_n from the recursion formula

$$c_n = \left[\frac{c_{n-2}}{3!} - \frac{c_{n-4}}{5!} + \frac{c_{n-6}}{7!} + \cdots \pm \frac{c_{-1}}{(n+2)!}\right]$$

when n is odd. Recall that $c_n = 0$ if n is even, and that $c_{-1} = 1$.

b) Find c_5 for the series.

24. Consider the expansion $1/\text{Log}\, z = \sum_{n=-m}^{\infty} c_n(z-1)^n$, $0 < |z-1| < r$.

a) Using the method of Example 4 show that $m = 1$.

b) Find r (maximum value).

c) Find a recursion formula for the nth coefficient c_n like that given in Exercise 23. What is c_{-1}? What is c_4?

25. a) Obtain the expansion

$$\begin{aligned}\cot z &= \frac{B_0}{z} - \frac{B_2 2^2 z}{2!} + \frac{B_4 2^4 z^3}{4!} - \frac{B_6 2^6 z^5}{6!} + \cdots \\ &= \frac{1}{z} - \frac{z}{3} - \frac{z^3}{45} - \frac{2z^5}{945} - \cdots, \qquad 0 < |z| < \pi,\end{aligned}$$

where B_n are the Bernoulli numbers defined in Exercise 23 of the previous section.

Hint: Replace z in that problem with $2iz$.

b) Check the first three terms of the preceding result by multiplying the Maclaurin series for $\cos z$ by the Laurent series in Eq. (5.6–35b).

26. One way of defining the Bessel functions of the first kind is by means of an integral:

$$J_n(w) = \frac{1}{2\pi}\int_{-\pi}^{+\pi} \cos(n\theta - w\sin\theta)\, d\theta,$$

where n is an integer. The number n is called the order of the Bessel function. There is a connection between this integral and the coefficients of z in a Laurent expansion of $e^{w(z-1/z)/2}$.

Let

$$e^{(w/2)(z-z^{-1})} = \sum_{n=-\infty}^{+\infty} c_n z^n, \qquad |z| > 0. \tag{5.6–36}$$

Show using Laurent's theorem that

$$c_n = J_n(w). \tag{5.6–37}$$

Hint: Refer to Eq. (5.6–5). Take as a contour $|z| = 1$. Make a change of variables to polar coordinates ($z = e^{i\theta}$). Then use Euler's identity and symmetry to argue that a portion of your result is zero.

The expression $e^{(w/2)(z-z^{-1})}$ is called a *generating function* for these Bessel functions.

27. Refer to Eqs. (5.6–36) and (5.6–37). Show that

$$J_n(w) = \sum_{k=0}^{\infty} \frac{(-1)^k (w/2)^{n+2k}}{k!(n+k)!}, \qquad n = 0, 1, 2, \ldots.$$

Hint: The left side of Eq. (5.6–36) is $e^{wz/2}e^{-w/2z}$. Multiply the Maclaurin series for the first term by a Laurent series for the second term.

5.7 SOME PROPERTIES OF ANALYTIC FUNCTIONS RELATED TO TAYLOR SERIES

In this section we will consider a few of the many properties of analytic functions related to their power series expansions. These properties were selected because of their usefulness to us in later chapters.

Isolation of Zeros

Suppose the function $f(z)$ satisfies $f(z_0) = 0$. We say the z_0 *is a zero of* $f(z)$. Thus the zeros of $z^2 + 1$ are at $\pm i$, while the zeros of $\sin z$ are at $n\pi$, n is any integer.

An *isolated zero* of $f(z)$ is one that has this property:

DEFINITION Isolated zero

If $f(z_0) = 0$, z_0 is said to be an isolated zero of $f(z)$ if there exists a deleted neighborhood of z_0 throughout which $f(z) \neq 0$. □

Thus $f(z) = (z - 1)(z - 3)$ has an isolated zero at $z = 1$ since any punctured disc centered at $z = 1$ with radius less than 2 will be a domain in which $f(z) \neq 0$. There is also an isolated zero at $z = 3$.

If a function is analytic at z_0, and if $f(z_0) = 0$, we can make this generalization: either z_0 is an isolated zero of $f(z)$ or $f(z) = 0$ *throughout* a neighborhood of z_0. For our proof we notice that since $f(z)$ is analytic at z_0 we have a Taylor expansion:

$$f(z) = \sum_{n=0}^{\infty} c_n(z - z_0)^n = c_0 + c_1(z - z_0) + c_2(z - z_0)^2 + \cdots.$$

Since $f(z_0) = 0$ we require $c_0 = 0$. Now it is *possible* that $c_1 = 0$, $c_2 = 0$, etc. If *all* the coefficients c_n $(n = 0, 1, \ldots, \infty)$ are zero, then the Taylor series representation of $f(z)$ about z_0 is zero, and according to Theorem 15, Section 5.4, and Eq. (5.4–10), $f(z)$ must be zero throughout a disc of nonzero radius centered at z_0.

Suppose, however, that not all the coefficients are zero. Let c_m be the first nonzero coefficient in the series. Thus $f^{(m)}(z_0) \neq 0$, and we have

$$f(z) = c_m(z - z_0)^m + c_{m+1}(z - z_0)^{m+1} + \cdots.$$

We now factor out $(z - z_0)^m$ on the right and get

$$f(z) = (z - z_0)^m \phi(z), \tag{5.7–1}$$

where

$$\phi(z) = c_m + c_{m+1}(z - z_0) + \cdots = \sum_{n=m}^{\infty} c_n(z - z_0)^{n-m}. \tag{5.7–2}$$

Note that $\phi(z_0) = c_m \neq 0$. This leads us to the following.

DEFINITION Order of a zero

Let m be an integer ≥ 1. A function $f(z)$ that is both analytic and zero at z_0 and has throughout a neighborhood of z_0 the form $(z - z_0)^m \phi(z)$, where $\phi(z_0) \neq 0$, has a *zero of order m at z_0*. □

We have shown that when $f(z_0) = 0$ the order of the zero is the order of the first nonvanishing coefficient in the Taylor expansion of $f(z)$ about z_0. Of course, if $f(z_0) \neq 0$ it makes no sense to define the order of the zero.

The function $\phi(z)$ of Eq. (5.7–1) has a Taylor series expansion about z_0 (see Eq. 5.7–2) and is therefore analytic and hence continuous at z_0. Thus, given an $\varepsilon > 0$, there must exist a neighborhood of z_0, say $|z - z_0| < \delta$, throughout which

$$|\phi(z) - \phi(z_0)| < \varepsilon. \tag{5.7–3}$$

(The reader may wish to review the concept of continuity in Section 2.2.)

Suppose now in Eq. (5.7–3) we choose $\varepsilon = |\phi(z_0)/2|$. Thus there exists a neighborhood of z_0 in which

$$|\phi(z) - \phi(z_0)| < |\phi(z_0)/2|. \tag{5.7–4}$$

It is clear that $\phi(z) \neq 0$ throughout this neighborhood, because if $\phi(z) = 0$, then Eq. (5.7–4) would require that $|-\phi(z_0)| < |\phi(z_0)/2|$ at some point—an impossibility.

In this same neighborhood we have $f(z) = (z - z_0)^m \phi(z)$. Since $\phi(z) \neq 0$, it follows that z_0 is the *only* zero of $f(z)$ in the neighborhood. The preceding argument required that not every coefficient in the Taylor expansion of $f(z)$ about z_0 be zero. Taking into account this possibility, and recalling its consequences, we have the following:

THEOREM 19 Isolation of zeros

Let $f(z)$ be analytic at z_0, and let $f(z_0) = 0$. Then either there is a neighborhood of z_0 in which $f(z) = 0$ is satisfied only at z_0, or there is a neighborhood of z_0 in which $f(z) = 0$ everywhere. □

Recall that the latter possibility in this theorem holds only when $f(z)$ and all its derivatives vanish at z_0. *Thus a zero of $f(z)$ at z_0 is isolated unless there is a disc centered at z_0 in which $f(z) = 0$ everywhere.* The radius of this disc is determined by the distance from z_0 to the nearest singular point of $f(z)$.

EXAMPLE 1

Give the location and order of the zeros of $f(z) = \sin(1/z)$. Show that each zero is isolated.

Solution

We know that the solutions of $\sin z = 0$ are $z = n\pi$, where n is any integer (see Section 3.2, Exercise 18). Thus $\sin(1/z) = 0$ for $z = 1/(n\pi)$, $n = \pm 1, \pm 2, \ldots$. As n increases in magnitude, the zeros of $\sin(1/z)$ become increasingly close to $z = 0$, as shown in Fig. 5.7–1. At $z = 0$, $\sin(1/z)$ is undefined and thus not analytic.

Consider a zero at $1/(n\pi)$ where $n \geq 2$. The two closest neighboring zeros are at $1/((n-1)\pi)$ and $1/((n+1)\pi)$, of which the latter is closest to $1/(n\pi)$ (this is easy to prove). Thus a neighborhood of $1/(n\pi)$ having radius δ, where $\delta < (1/n\pi - 1/(n+1)\pi)$ is one in which $\sin(1/z) = 0$ is satisfied only at the center. A similar discussion applies to zeros at $1/(n\pi)$, $n \leq -2$, and also to zeros at $\pm 1/\pi$.

Notice that $f'(z) = -z^{-2}\cos(1/z)$ and that $f'(1/(n\pi)) = -(n\pi)^2\cos(n\pi) = -(n\pi)^2(-1)^n$, which is nonzero. Since the first derivative of $\sin(1/z)$ is nonvanishing at the zeros of $\sin(1/z)$, it follows that the zeros are all of first order.

It is interesting that every neighborhood of the origin contains zeros of $\sin(1/z)$; thus $\sin(1/z) = 0$ somewhere in every disc centered at $z = 0$. Indeed the origin is a limit point (this term is defined in Section 1.5) of the set of zeros of $\sin(1/z)$. This does not contradict our assertion that the zeros of $\sin(1/z)$ are isolated. The point $z = 0$ is not a zero of $\sin(1/z)$ but rather a singular point of this function. ◀

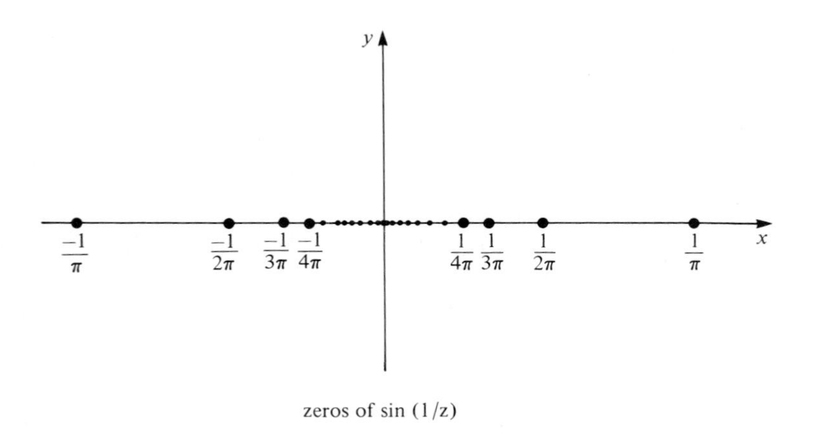

Figure 5.7–1

A relationship such as $\sin^2 x = 1/2 - 1/2\cos(2x)$, which we already know is valid when x is real, can, with the aid of Theorem 19, be shown to be true throughout the complex plane, i.e., $\sin^2 z = 1/2 - 1/2\cos(2z)$.

Consider $F(z) = \sin^2 z + 1/2\cos(2z) - 1/2$. When $z = x$ we know that $F(z) = 0$. Thus $F(z) = 0$ all along the real axis. The zeros of $F(z)$ on the real axis are thus not isolated. Hence, from Theorem 19, each point on the real axis has a neighborhood throughout which $F(z) = 0$. Expanding $F(z)$ in a Taylor series about such a point, we find that every coefficient vanishes. Because $F(z)$ is analytic throughout the complex plane, this Taylor series, which converges to $F(z)$, is valid everywhere in the complex plane. Thus $F(z) = \sin^2 z + 1/2\cos(2z) - 1/2 = 0$ for all z, and therefore $\sin^2 z = 1/2 - 1/2\cos(2z)$ for all z.

We have shown that the form of the equation $\sin^2 x = 1/2 - 1/2\cos(2x)$ is preserved (i.e., is still valid) when x is replaced by z. It is apparent that other relationships can be extended in this way from the real axis into the complex plane. Indeed, if two entire functions $f(z)$ and $g(z)$ are equal everywhere along a segment of the real axis $a < x < b$, they will be equal throughout the complex plane.

Analytic Continuation

Suppose we are given a function $f(z)$ that is analytic at a point z_0. The function is not described by a simple formula like $\sin z$ or e^z but instead is given by a convergent power series involving powers of $(z - z_0)$. We know that this series $\sum_{n=0}^{\infty} c_n(z - z_0)^n = f(z)$ will converge inside a disc centered at z_0 whose radius is determined by that singularity of $f(z)$ lying closest to z_0.

In Fig. 5.7–2(a) we show this disc, which we call region R. For any z lying in R we evaluate $f(z)$ by summing all the terms in the infinite series. The disc-shaped region R_1 depicted in Fig. 5.7–2(b) lies partly inside and partly outside R. Is it possible to find a function $g(z)$ that is analytic in R_1 and has this additional property: for

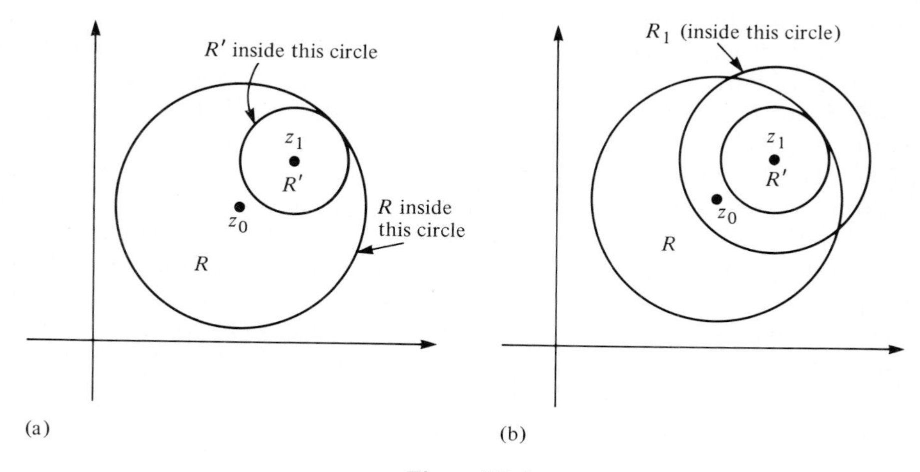

Figure 5.7–2

each z belonging to both† R and R_1, the function $g(z)$ assumes the same values as $f(z)$? Often the answer is yes. Under these circumstances, we say that $g(z)$ is an *analytic continuation* of $f(z)$ into the region R_1.

As an example of an analytic continuation, we are given a function $f(z)$ defined by a series: $f(z) = 1 + z + z^2 + \cdots$. Application of the ratio and nth term tests shows that the series converges for $|z| < 1$ and diverges for $|z| \geq 1$. Thus $f(z)$ is analytic inside a unit circle centered at the origin. Is there an analytic continuation of $f(z)$ beyond this circle? We should recognize that our power series is the Maclaurin expansion of $1/(1 - z)$. Since $1/(1 - z)$ agrees with our given $f(z)$ for $|z| < 1$, and is analytic for $z \neq 1$, it is apparent that $1/(1 - z)$ is the analytic continuation of $f(z)$ into the entire complex plane with the point $z = 1$ deleted. Thus for the present example $g(z) = 1/(1 - z)$.

We were fortunate in this example that we could find an analytic continuation of $f(z)$ by recognizing that the given series has sum $1/(1 - z)$. If we cannot establish a closed form expression ("a formula") for the sum of the series, analytic continuation may still be possible. Refer now to Fig. 5.7–2(a). Suppose we are given $f(z)$ defined by means of a series in powers of $(z - z_0)$. The series is valid in R. To find $g(z)$ we now proceed to expand $f(z)$ in a Taylor series about z_1, which lies in R; i.e., we obtain the series $\sum_{n=0}^{\infty} d_n(z - z_1)^n$, where $d_n = f^{(n)}(z_1)/n!$. Each coefficient d_n is obtained by repeatedly differentiating and summing $\sum_{n=0}^{\infty} c_n(z - z_0)^n$. Thus

$$d_0 = \sum_{n=0}^{\infty} c_n(z_1 - z_0)^n, \qquad d_1 = \sum_{n=1}^{\infty} c_n n(z_1 - z_0)^{n-1}, \qquad \text{etc.}$$

According to Theorem 15, the series $\sum_{n=0}^{\infty} d_n(z - z_1)^n$ must converge to $f(z)$ inside the circular region R' centered at z_1 and shown in Fig. 5.7–2(a). R' lies inside R. However, it is *possible* that this series will converge in the larger region R_1 that extends beyond R as shown in Fig. 5.7–2(b). If this is the case we will define $g(z)$ as $\sum_{n=0}^{\infty} d_n(z - z_1)^n$. Recalling that a power series converges to an analytic function, we see that $g(z)$ must be analytic in R_1 and, as such, is an analytic continuation of $f(z)$ into a region that has points lying outside R. We note that although we are guaranteed by Theorem 15 that $g(z)$ agrees with $f(z)$ for those values of z in R_1 that also belong to both R and R', an additional proof is required to show that $g(z)$ and $f(z)$ are identical for values of z simultaneously belonging to both R_1 and R, but lying *outside* R'. This relatively simple proof will not be given here.

The procedure just used may sometimes be repeated to provide an analytic continuation of $g(z)$ itself. This would provide a further analytic continuation of $f(z)$. Refer to Fig. 5.7–3. We select a point z_2 lying in R_1 and expand $g(z)$ in a Taylor series here. If this series converges in R_2, which lies in part outside R_1, we have another analytic continuation. Sometimes we can by means of series and a chain of circles like that shown in Fig. 5.7–3 obtain an analytic continuation of $f(z)$ into most of the complex plane. In practice this is not done, and the analytic continuation of

---

† These values of z lie in the *intersection* of R and R_1 (see Exercises 18–21, Section 1.5).

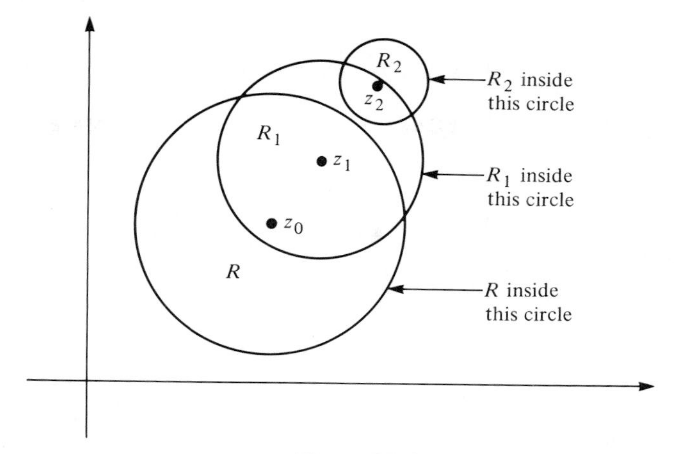

Figure 5.7–3

a function is performed by means of a relatively simple formula, as in our original example where we recognized that our given series converged to $1/(1 - z)$.

The questions of when this "circle chain" method of analytic continuation can be performed and whether the functions obtained are unique (do we obtain the same function by approaching a region via two different chains?) cannot be dealt with here. The reader is referred to more advanced texts for their consideration.†

So far we have been concerned with finding the analytic continuation of a function defined by an infinite series. Recalling that an integral is the limit of a sum, it is not surprising to us that analytic continuations of functions created by integrals are possible.

In Chapter 7 we will be studying an important integral that creates functions, i.e., $\int_0^\infty f(t)e^{-zt}\,dt$, defined as $\lim_{L\to\infty}\int_0^L f(t)e^{-zt}\,dt$. Here t is a real variable and z is a complex variable. If the integral exists it defines a function $F(z)$, which we call the Laplace transform of $f(t)$. Typically the integral exists if $\operatorname{Re} z > x_0$, where x_0 is some real constant whose value depends on $f(t)$. Thus $F(z)$ is defined by the integral only when z lies in a half plane lying to the right of a vertical line in the complex plane.

Let us study $F(z)$ when $f(t) = 1$, for $t \geq 0$. We have

$$F(z) = \int_0^\infty e^{-zt}\,dt = \lim_{L\to\infty}\int_0^L e^{-zt}\,dt = \lim_{L\to\infty}\frac{1}{z}[1 - e^{-zL}]$$

$$= \lim_{L\to\infty}\frac{1}{z}[1 - e^{-(x+iy)L}] = \lim_{L\to\infty}\frac{1}{z}[1 - e^{-xL}\operatorname{cis}(-yL)].$$

If $x > 0$, then $\lim_{L\to\infty} e^{-xL}\operatorname{cis}(-yL) = 0$; while if $x \leq 0$, no limit exists as $L \to \infty$. Thus the Laplace transform of $f(t) = 1$ is $F(z) = 1/z$ for $\operatorname{Re} z > 0$, while $F(z)$ is

† See, e.g., J. Marsden and M. Hoffman, *Basic Complex Analysis*, 2nd ed. (New York: W. H. Freeman, 1987), Section 6.1.

undefined for $\operatorname{Re} z \leq 0$. However, it should be apparent that $1/z$ is an analytic continuation of $F(z)$ (the function defined by our integral) into the entire complex plane with the origin deleted.

EXAMPLE 2

A student is given the function $f(z)$ defined by the Maclaurin series $1 - z + z^2 - z^3 + \cdots$. Using the ratio test he concludes correctly that this series defines an analytic function inside $|z| = 1$. He now seeks to expand $f(z)$ in a Taylor series about $z = 1/2$ and thus wants the coefficients c_n in the following:

$$\sum_{n=0}^{\infty} c_n\left(z - \frac{1}{2}\right)^n = f(z) = 1 - z + z^2 - z^3 + \cdots. \tag{5.7–5}$$

He knows nothing about the sum of the series on the above right, but he does have a pocket calculator. Using $c_n = f^{(n)}(1/2)/n!$ to obtain his Taylor coefficients, he finds that $c_0 = 1 - 1/2 + 1/4 - 1/8 + \cdots$. Summing a large number of terms he concludes that $c_0 = 0.6666$, to which he assigns the value 2/3. Differentiating Eq. (5.7–5) and putting $z = 1/2$ he obtains $c_1 = -1 + 1 - 3/4 + 4/8 - \cdots$. Again using a calculator he concludes that $c_1 = -0.444$, which he calls $-4/9$. Continuing in this way, he decides correctly that $c_n = (-1)^n(2/3)^{n+1}$. (Of course he could have obtained this result more easily had he realized that for $|z| < 1$ his series converges to $1/(1 + z)$.) Does his Taylor expansion $\sum_{n=0}^{\infty}(-1)^n(2/3)^{n+1}(z - 1/2)^n$ represent an analytic continuation of $f(z)$, and, if so, into what region?

Solution

Consider the function $g(z) = \sum_{n=0}^{\infty}(-1)^n(2/3)^{n+1}(z - 1/2)^n$. Using the ratio test we find that the series is absolutely convergent inside the circle $|z - 1/2| = 3/2$. Thus $g(z)$ is an analytic function inside the circle C' (see Fig. 5.7–4) and is an analytic

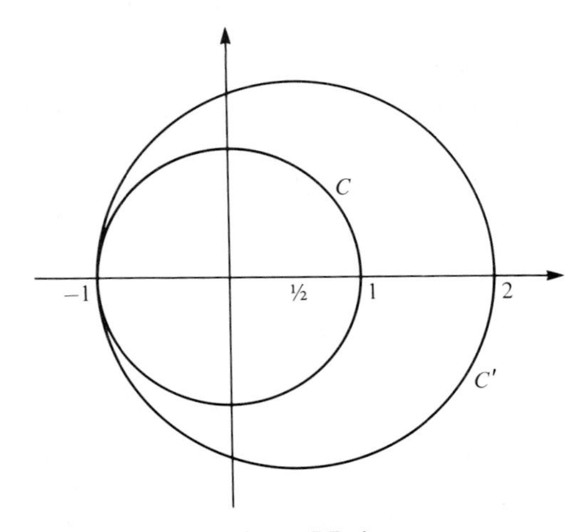

Figure 5.7–4

continuation of $f(z)$ into that region. Recall that $f(z)$ was defined only inside the circle C of the same figure. The analytic continuation of $f(z)$ into $|z - 1/2| < 3/2$ provides us with an analytic function defined in the crescent shape lying between C and C'. Here $f(z)$ is undefined.

Comment: Taylor series expansions about other points inside $|z| = 1$ can result in functions defined in other crescents lying outside $|z| = 1$. It will be found in each case that the values assumed by each series are identical to those of the function $1/(1 + z)$. ◀

EXERCISES

1. Consider $f(z) = z^3 - x^3 + 3xy^2 + i(y^3 - 3x^2y)$, where $z = x + iy$.

a) Show that the zeros of this function on the axis $y = 0$, $-\infty < x < \infty$, are not isolated.

b) Does the result of (a) contradict the statement that the zeros of an analytic function are isolated? Explain.

2. a) Consider $f(z) = e^z - e^{iy}$. Show that this function is zero everywhere on the y (imaginary) axis of the complex z-plane.

b) Can you conclude from part (a) and Theorem 19 that every point on the y-axis has a neighborhood throughout which $f(z) = 0$? Explain.

3. a) Are the zeros of $\sin(\pi/(z^2 + 1))$, in the domain $|z| < 1$, isolated?

b) Find all the zeros of this function in the domain.

c) Identify any limit (accumulation) points of the set, and state whether they belong to the given domain.

Find the order of the zeros of the following functions at the points indicated.

4. $\cos z$ at $z = n\pi + \pi/2$, n any integer

5. $\operatorname{Log} z$ at $z = 1$

6. $(z^4 - 1)^2/z$ at $z = i$

7. $z^3 \sin z$ at $z = 0$ and also $z = \pi$

8. Show that if $f(z)$ has a zero of order n at z_0, then $[f(z)]^m$ (m is an integer) has a zero of order nm at z_0.

Use the preceding result to find the order of the zeros in the following.

9. $(\operatorname{Log} z - 1)^2$ at $z = e$

10. $(\sin z)^4$ at $z = \pi$

11. $(z^3 \sin z)^2$ at $z = 0$

12. a) Let $f(z) = 1 + z + z^2 + \cdots$, $|z| < 1$. Expand this function in a Taylor series about $z = -3/4$; i.e., state c_n in its expansion $\Sigma_{n=0}^{\infty} c_n(z + 3/4)^n$. You may use your knowledge of the sum of the given series.

b) Does the Taylor series found in (a) produce an analytic continuation of $f(z)$ into a region extending beyond $|z| = 1$? Explain.

13. a) Find in closed form the function defined by $\int_0^\infty e^{2t}e^{-zt}\,dt$ for $\text{Re}\, z > 2$.

b) What is the analytic continuation of this function? What is the largest region in which this continuation is valid?

14. a) Show that the function $f(z) = \int_0^z (2 + 3\cdot 2w + 4\cdot 3w^2 + 5\cdot 4w^3 + \cdots)\,dw$ is analytic in the disc $|z| < 1$, and is undefined for $|z| > 1$.

b) What is the analytic continuation of $f(z)$ beyond this disc? Give a closed form expression by integrating the sum of the series under the integral sign.

APPENDIX A

SEQUENCES, FRACTALS, AND THE MANDELBROT SET

Until now, every topic covered in this text was known to mathematicians by the start of the twentieth century. With the exception of some material in Chapters 6 and 7, the subjects covered in the remaining chapters were understood by the end of the nineteenth century. It is therefore refreshing to treat here a topic that was not only developed mostly in this century but is still a lively subject of current research. It is moreover a field for which new applications are being found in the solution of physical problems.

The word *fractal* is related to the word fraction, meaning broken or not whole. It was coined fairly recently by Prof. Benoit Mandelbrot† to define sets of numbers that describe objects in space whose dimensionality is, in a certain sense, fractional. The set of numbers satisfying $|z| = 1$ will, when plotted in the complex plane, produce the outline of a two-dimensional object—a disc. However, when a fractal set of numbers is drawn in the same plane the object delineated can have a fractional dimension lying between one and two. Space does not permit our discussing the precise mathematical meaning of this statement. For a brief and interesting explanation, see the book by Ivars Peterson.‡

Mandelbrot has developed a fractal geometry of nature to describe objects that cannot be described by the ordinary straight lines and smooth arcs familiar to us from Euclidean geometry. To see the need for a different geometry, imagine a trip along the seacoast of Maine by three travelers: a motorist driving along the coastal highway; a pedestrian walking along the shore who follows, at the high-water mark, all the peninsulas and inlets; and finally an ant who follows a path like that of the pedestrian but who, because of his much smaller body, is more aware of tiny fluctuations in the coast line and is able to follow them.

---

† B. B. Mandelbrot, *Fractals: Form, Chance, and Dimension* (San Francisco: W. H. Freeman, 1977).

‡ Ivars Peterson, *The Mathematical Tourist* (San Francisco: W. H. Freeman, 1988), pp. 116–123.

Obviously the first distance covered is much less than the second, which, in turn, is much less than the third. An animal smaller than the ant, one whose body is the size of a grain of sand, could follow variations in the coast that we might notice only with a magnifying glass. This distance traversed by this fourth traveler would be yet greater than the others. If we are at liberty to choose voyagers whose sensitivity to imperfections in the coastline is arbitrarily fine, we can perhaps conclude that the coast is infinitely long. (Of course this statement ignores the atomic structure of matter, and we will stop short of that level of magnification.)

Mandelbrot's fractal geometry can be used to make mathematical models of physical structures such as coastlines. This kind of boundary, which never looks smooth under any degree of magnification, and which nowhere has a definable tangent, is also found on surfaces—e.g., the surface of a chunk of metal at a fracture, or the outlines of clouds in the sky. In addition, fractals appear in the analysis of the chaos that can appear in systems with vibrations or fluid flow.†

Before we study fractals we must extend our work in Section 5.2 on limits of sequences. Refer to Eqs. (5.2–4) and (5.2–5) and the definition of a limit of a sequence.

Let $b_0, b_1, b_2, \ldots, b_n, \ldots$ be a sequence of complex numbers defined for each integer $n \geq 0$. An example might be $1, (1+i), (1+i)^2, \ldots,$ where $b_n = (1+i)^n$. We are dealing here with sequences like those defined in Eq. (5.2–4) except that now the elements are constants, and they have been reindexed to begin with $n = 0$. The sequence $b_0, b_1, b_2, \ldots$ has limit b as $n \to \infty$ if, given any $\varepsilon > 0$, there exists an N such that

$$|b - b_n| < \varepsilon \qquad \text{for } n > N. \tag{A5–1}$$

This is just Eq. (5.2–5) with b and b_n replacing S and S_n, respectively. In other words, if the sequence has limit b, then at some point in the sequence the terms must start getting closer and closer to b.

If there is no value of b satisfying Eq. (A5–1) the given sequence diverges. For example, if $b_n = i^n$ we have the divergent sequence $1, i, -1, -i, 1, \ldots,$ which (the reader should verify) cannot satisfy Eq. (A5–1) for any b if we are given $0 < \varepsilon \leq 1$. The sequence $b_n = (1+i)^n$, that is, $1, (1+i), (1+i)^2, (1+i)^3, \ldots$ also diverges. Equation (A5–1) cannot be satisfied for any b because $|b_n| = (\sqrt{2})^n$ and the expression $|b - b_n|$ is unbounded as n increases beyond any preassigned N. This second sequence not only diverges, it exhibits a behavior called divergence to infinity, defined as follows.

DEFINITION Divergence to infinity

The sequence $b_0, b_1, \ldots, b_n, \ldots$ *diverges to infinity* if, given any $\gamma > 0$, there exists an integer N such that

$$|b_n| > \gamma \qquad \text{for all } n > N. \tag{A5–2}$$

† See *Engineering: Cornell Quarterly*, Vol. 20, No. 3, Winter 1986, Ithaca, N.Y., for a variety of articles devoted to the subject of fractals and their use in modeling.

Thus if a sequence diverges to infinity, the magnitude of its terms, beyond the Nth, will become larger than any preassigned number γ. Typically N depends on γ. The sequence $b_n = i^n$, although divergent, does not diverge to infinity. The magnitude of each term never exceeds 1.

The *Mandelbrot set*, which we sometimes simply call M, is a set of numbers whose definition requires the concept of divergence to infinity. Suppose we are given

$$f(z) = z^2 + c, \tag{A5–3}$$

where $z = x + iy$ and c is a complex number. We define the sequence $z_0, z_1, z_2, \ldots$ as follows. We take $z_0 = 0$, $z_1 = f(z_0) = z_0^2 + c$, $z_2 = f(z_1) = z_1^2 + c$, etc. Thus each element of the sequence is found by our evaluating $f(z)$ at the previous value, i.e.,

$$z_n = z_{n-1}^2 + c \qquad \text{(where } z_0 = 0\text{)}. \tag{A5–4}$$

The set M is generated by our considering different values of c. Those values of c for which the resulting sequence $z_0, z_1, z_2, \ldots$ *does not diverge to infinity* are in M. The elements of the Mandelbrot set (i.e., these values of c) are typically plotted as points in the complex c-plane. Figure A5–1 shows, in black, the set M.

We can quickly see that certain numbers do or do not lie in M. Taking $c = 0$, we have, from Eq. (A5–4) and the starting value $z_0 = 0$, that $z_1 = 0$, $z_2 = 0$, etc. Since $z_n = 0$ for all $n \geq 0$ the sequence converges to zero. Thus $c = 0$ is in the Mandelbrot set, as Fig. A5–1 shows.

Taking $c = -2$, we have $z_0 = 0$, $z_1 = -2$, $z_2 = 2$, $z_3 = 2$, and $z_n = 2$ for $n \geq 2$. The sequence converges to 2, and so $c = -2$ is also in M.

In a similar way we take $c = i$ and find the sequence $0, i, -1 + i, -i, -1 + i, -i, \ldots$. Except for the first two terms, the elements of the sequence alternate between $-1 + i$ and $-i$. This is a divergent sequence, but it is not a sequence diverging to

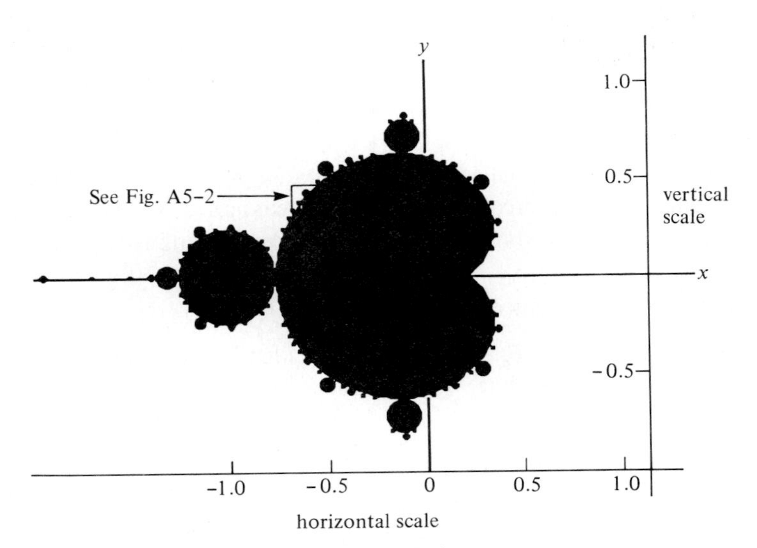

Figure A5–1

infinity since no term in the sequence has a magnitude greater than $\sqrt{2}$. Thus $c = i$ is in M and lies in the black region of Fig. A5–1—a fact not obvious from the figure.

If $c = 1$ we generate the sequence 0, 1, 2, 5, 26, 677,. . . . The terms rapidly grow larger and larger, without a bound—it is clear that the sequence diverges to infinity and that $c = 1$ is not in the set.

For most values of c, however, it is not immediately evident whether the resulting sequence $z_0, z_1, z_2, \ldots$ is divergent to infinity. For example, we know that $c = i$ is in M. What about the nearby point $c = 0.99i$? With the aid of a simple computer program we can compute the terms of the resulting sequence. Table A1 shows approximate decimal values for the first 11 terms. It is not clear if this sequence will converge or diverge, and, if it does diverge, whether divergence is to infinity. We will see shortly that Table A1 has enough information to show that the sequence does diverge to infinity. Thus $c = 0.99i$ is not in the Mandelbrot set. There are in fact other points lying equally close to $c = i$ that lie outside M. The reader may begin to appreciate that the shape of M is very complicated, much more so than Fig. A5–1 suggests.

One might guess that the point $c = i$ is isolated from the rest of M, i.e., it has a deleted neighborhood whose every point is outside M. This guess is wrong—a fact known only since 1982 when two mathematicians, J. Hubbard and A. Douady, proved that M is a connected closed set.

The boundary of the Mandelbrot set is its most fascinating aspect. The region shown in the box in Fig. A5–1, which lies close to the boundary, is described by $-0.68 < x < -0.58$, $0.388 < y < 0.46$. This set of points is shown magnified in Fig. A5–2.

The points of this region belonging to M are drawn in black; other points outside M are in white or grey. The meaning of this shading will be explained shortly. The boundary contains countless numbers of inlets and peninsulas with yet smaller "fingers" of land extending from each peninsula. A further magnification of a portion

Table A1

z_n and $|z_n|$ when $c = 0.99i$

z_n	$|z_n|$
$z_0 = 0.000 + 0.000i$	$|z_0| = 0.000$
$z_1 = 0.000 + 0.990i$	$|z_1| = 0.990$
$z_2 = -0.980 + 0.990i$	$|z_2| = 1.393$
$z_3 = -0.020 - 0.951i$	$|z_3| = 0.951$
$z_4 = -0.903 + 1.027i$	$|z_4| = 1.368$
$z_5 = -0.239 - 0.865i$	$|z_5| = 0.898$
$z_6 = -0.692 + 1.404i$	$|z_6| = 1.565$
$z_7 = -1.492 - 0.952i$	$|z_7| = 1.770$
$z_8 = 1.319 + 3.831i$	$|z_8| = 4.052$
$z_9 = -12.940 + 11.093i$	$|z_9| = 17.044$
$z_{10} = 44.390 - 286.100i$	$|z_{10}| = 289.523$

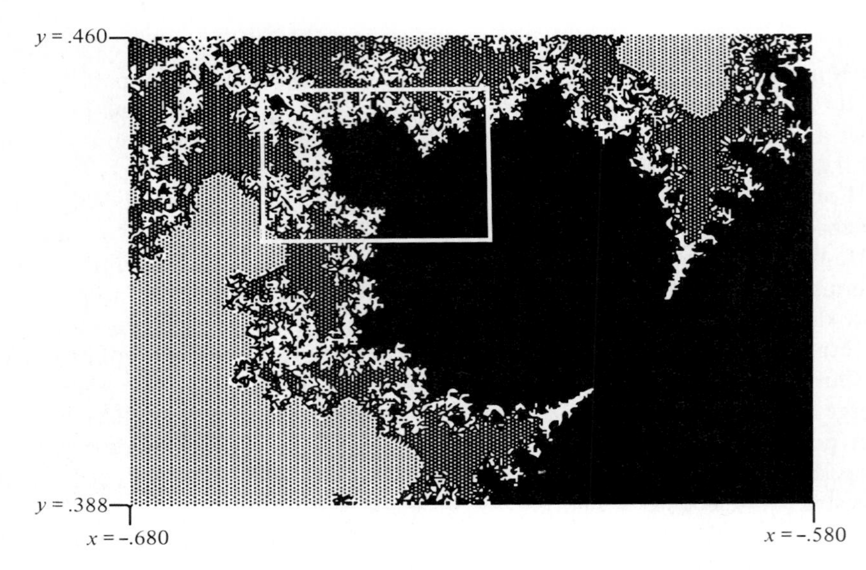

Figure A5–2

$(-0.66 < x < -0.628,\ 0.429 < y < 0.452)$ of the preceding rectangle is depicted in Fig. A5–3 and shows strands of land coming from the fingers. More magnification would reveal even additional fine structure, and at no degree of magnification would the boundary of any portion of the Mandelbrot set appear to be composed of the smooth arcs and straight lines of Euclidean geometry. No wonder that Hubbard, an expert on this subject, calls the Mandelbrot set "the most complicated object in mathematics." It is remarkable that this shape came from such a simple formula and procedure. Incidentally Hubbard's finding that M is connected is not contradicted by any black "islands" appearing in Fig. A5–3. Any apparent lack of connectedness in M appears because of the limitations of the computer program and video display used to obtain the image. This also explains why $z = i$, which we noticed is in the Mandelbrot set, does not appear in the black part of Fig. A5–1. This point is connected to the main area of the set by a thread so thin that it could not be displayed by the computer screen.

The set of boundary points of M is an example of a fractal set. The resemblance of the behavior of this boundary to the coastline of a real country should be apparent. There are other fractal sets, besides M, that can be used in this way, as well as to model other physical phenomena.

Figures A5–1, A5–2, and A5–3 were obtained from a computer program called "Mandelzoom," which runs on the Apple Macintosh computer.† The program employs

---

† This is an excellent piece of inexpensive software, and it is available, together with other fractal generating programs, on disc #113 from Educorp, 7434 Trade St., San Diego, CA 92121.

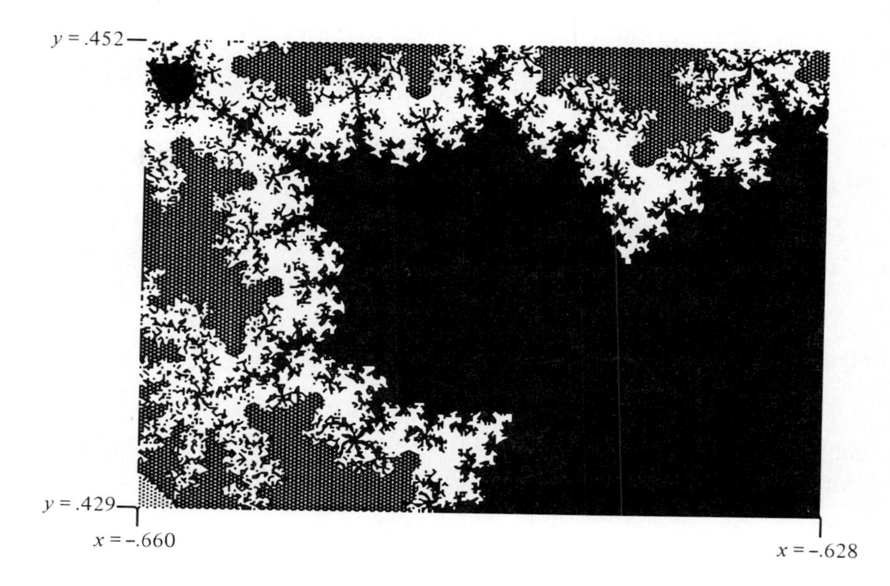

Figure A5–3

a procedure outlined in an interesting article on fractals in the magazine *Scientific American.*[†]

Let us see how a computer program might determine whether a point belongs to M. We need not check the whole complex c-plane, point by point. First we show that

The Mandelbrot set lies in and on a circle of radius 2 in the complex c-plane.

To prove the preceding, first assume that $|c| > 2$. We can show that the resulting sequence $z_0, z_1, z_2, \ldots$ diverges to infinity. From Eq. (A5–3) we have $z_1 = f(z_0) = f(0) = c$, and so $|z_1| = |c| > 2$. Now $z_2 = f(z_1)$ and so

$$|z_2| = |f(z_1)| = |z_1^2 + c| = |z_1||z_1 + c/z_1|. \tag{A5–5}$$

Now recall the inequality derived in Exercise 30, Section 1.3:

$$|u + v| \geq |u| - |v| \geq 0 \qquad \text{(provided } |u| \geq |v|\text{)}. \tag{A5–6}$$

Applying this to Eq. (A5–5) we obtain

$$|z_2| \geq |z_1|(|z_1| - |c|/|z_1|).$$

Since $|z_1| = |c|$ we have

$$|z_2| \geq |c|(|c| - 1). \tag{A5–7}$$

---

† A. K. Dewdney, Computer Recreations, *Scientific American*, Vol. 253, No. 2, August 1985, pp. 16–24.

Continuing in this way we obtain

$$|z_3| = |f(z_2)| = |z_2^2 + c| = |z_2||z_2 + c/z_2|$$

and

$$|z_3| \geq |z_2|(|z_2| - |c|/|z_2|). \tag{A5–8}$$

Since $(|c| - 1) > 1$, a moment's study of Eq. (A5–7) reveals that $|z_2| > |c|$, and so $|c|/|z_2| < 1$. Thus $|z_2| - |c|/|z_2| > |c| - 1$, and we have, from Eq. (A5–8),

$$|z_3| > |z_2|(|c| - 1).$$

Combining Eq. (A5–7) with the preceding we have

$$|z_3| > |c|(|c| - 1)(|c| - 1) = |c|(|c| - 1)^2.$$

Continuing in this fashion we can show that $|z_4| > |c|(|c| - 1)^3$ and, in general, for $n \geq 3$ we obtain

$$|z_n| > |c|(|c| - 1)^{n-1}. \tag{A5–9}$$

Since the right side of this inequality tends to infinity as $n \to \infty$, the magnitude of the terms in the sequence $z_0, z_1, z_2, \ldots$ grow arbitrarily large, and the sequence diverges to infinity. Thus no points in the Mandelbrot set involve $|c| > 2$.

Now selecting $|c| \leq 2$ we begin to generate the sequence $z_0, z_1, z_2, \ldots$. If we encounter an element z_n such that $|z_n| > 2$ we can prove that we can stop our procedure. We can conclude that the sequence diverges to infinity and that our chosen c is not in M. To prove this, assume $|z_n| > 2$. We then have

$$|z_{n+1}| = |z_n^2 + c| = |z_n||z_n + c/z_n| \geq |z_n|[|z_n| - |c|/|z_n|], \tag{A5–10}$$

where we have employed Eq. (A5–6). Now since $|c/z_n| < 1$ it is apparent that $(|z_n| - |c|/|z_n|) > (|z_n| - 1)$, which we use in Eq. (A5–10) to yield

$$|z_{n+1}| > |z_n|[|z_n| - 1]. \tag{A5–11}$$

Since $|z_n| > 2$, the preceding shows that $|z_{n+1}| > |z_n| > 2$.

Now, proceeding much as before,

$$\begin{aligned} |z_{n+2}| &= |z_{n+1}^2 + c| = |z_{n+1}||z_{n+1} + c/z_{n+1}| \\ &\geq |z_{n+1}|[|z_{n+1}| - |c|/|z_{n+1}|]. \end{aligned} \tag{A5–12}$$

Since $|c|/|z_{n+1}| < 1$ and $|z_{n+1}| > |z_n|$ we have, from Eq. (A5–12),

$$|z_{n+2}| > |z_{n+1}|[|z_n| - 1]. \tag{A5–13}$$

Employing Eq. (A5–11) in the above yields

$$|z_{n+2}| > |z_n|[|z_n| - 1][|z_n| - 1] = |z_n|[|z_n| - 1]^2.$$

Continuing in this way we find that $|z_{n+3}| > |z_n|[|z_n| - 1]^3$ and that in general $|z_{n+k}| > |z_n|[|z_n| - 1]^k$. Thus as $k \to \infty$ the elements of the sequence have magnitudes that become arbitrarily large, and the sequence diverges to infinity.

Equation (A5–13) shows that $|z_{n+2}| > |z_{n+1}|$, and we earlier proved that $|z_{n+1}| > |z_n|$. This can easily be generalized. When $|z_n| > 2$ we have $|z_n| < |z_{n+1}| < |z_{n+2}| < |z_{n+3}| < \cdots$. Thus as soon as we encounter $|z_n| > 2$ the magnitudes of the subsequent terms get progressively larger. An example of this occurs in Table A1. Having evaluated $|z_8| = 4.05$ we can conclude that the sequence diverges. Note that $|z_8| < |z_9| < |z_{10}|$.

We have seen that when we seek to determine if a given c lies in M we can stop our calculation upon encountering $|z_n| > 2$. We conclude that c is not in M.

Suppose, however, we hit upon a value of c such that the sequence $z_0, z_1, \ldots$ does not result in $|z_n| > 2$ after many iterations of $f(z)$. According to Hubbard, who is quoted by Dewdney, if we reach z_{1000} and $|z_{1000}| \leq 2$, there is only a very, very small chance that the sequence will diverge to infinity. We can thus, with great safety, assign this value of c to the Mandelbrot set. We can even make this decision sooner, say at z_{100}, and have very few errors.

A computer program displaying the Mandelbrot set does not investigate the sequence $z_0, z_1, \ldots$ at every single point in any region of the complex plane. An infinite number of points would have to be considered. A tightly spaced grid is placed in the region and the behavior of the sequence is evaluated at each grid intersection. When a point is judged to lie in M, a black dot is placed on the screen at the corresponding grid point. In this way the black portions of Figs. A5–1, A5–2, and A5–3 were obtained.

The values of c outside the Mandelbrot set are of interest. The computer screen can be used to display these values of c with a color assigned to each point that indicates how rapidly the sequence $z_0, z_1, \ldots$ diverges to infinity, i.e., how soon the condition $|z_n| > 2$ is reached. In the absence of a screen displaying colors, a shade of gray (including white) might be employed. This technique is used in Figs. A5–2 and A5–3. Those points colored white result in sequences that diverge to infinity relatively slowly, while those in the middle gray correspond to sequences moving to infinity more rapidly. Obviously, finer gradations can be displayed if we use more shades of gray. A color video terminal can also be used to great advantage.

The Mandelbrot set is by no means the only set of numbers in the complex plane leading to fractals. The *Julia sets*, to cite one example, can as well.[†] An example of these sets can be generated by a simple procedure. We return to $f(z) = z^2 + c$ and assign a value to c. Next we give a value to z—we will call it z_0—and compute $z_1 = f(z_0)$, $z_2 = f(z_1)$, etc., as before. Note that z_0 is not necessarily zero. If the resulting sequence $z_0, z_1, z_2, \ldots$ does not diverge to infinity, the value z_0 is said to lie in a filled Julia set. Now we try a new value for z_0 and repeat the procedure, getting $z_1, z_2, \ldots$ to see if this new z_0 lies in the filled Julia set. For any given c there is a filled Julia set: it consists of all the values z_0 such that the sequence $z_0, z_1, z_2, \ldots$ does not diverge to infinity.

[†] Other examples of fractals can be found in M. Barnsley, *Fractals Everywhere* (New York: Academic Press, 1988). Julia sets were first described by two Frenchmen, Gaston Julia and Pierre Fatou, who worked in this subject in the period 1900–1930.

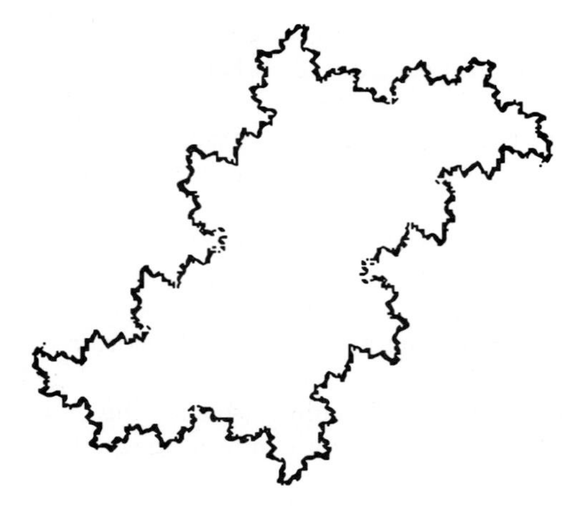

Figure A5–4

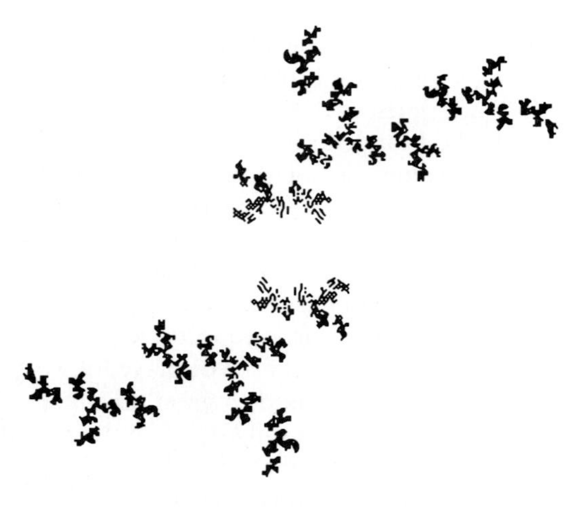

Figure A5–5

Other Julia sets can be generated through iterations of more complicated polynomials in z than $f(z) = z^2 + c$. For example, Linda Keen shows in a recent publication an interesting filled Julia set obtained by iteration of $z^2 - \lambda z$, where $\lambda = 0.7373688 + i0.6754903$.[†]

The boundary of a filled Julia set is usually a fractal set—i.e., it exhibits a complicated structure under any degree of magnification and describes an object

---

[†] Keen's article appears on page 68 of an attractive survey text: R. Devaney and L. Keen, eds., *Chaos and Fractals, The Mathematics Behind the Computer Graphics* (Providence, RI: American Mathematical Society, 1989).

whose dimensionality is not a whole number. The set of boundary points of the filled Julia set is often simply called a Julia set.

It was shown only in 1980 by Mandelbrot that a filled Julia set obtained from $z^2 + c$ is connected if and only if the value of c used to obtain it is a member of the Mandelbrot set. The Mandelbrot set is thus the set of all values of c that lead to connected filled Julia sets when Julia sets are obtained from $f(z) = z^2 + c$.

Inexpensive computer software is also available for generating Julia sets. Examples of Julia sets obtained from the previously mentioned Educorp software are shown in Figs. A5–4 and A5–5. Again $z^2 + c$ is employed. The values of c are $0.60i$ and $0.90i$, respectively. The first is in M, the second outside. For the first value there is a corresponding connected filled Julia set. It occupies the region on and inside the irregular closed curve of Fig. A5–4. The curve looks as though it could be the border of an actual country.

EXERCISES

1. Consider the sequence $b_0, b_1, b_2, \ldots$.

a) If $b_n = n^2 i^n$, prove that this sequence diverges to infinity.

b) If $b_n = n^2 \cos(n\pi/2)$, prove that this sequence diverges and prove that it does not diverge to infinity.

2. Prove that if a value of c is in the Mandelbrot set, then $\bar{c}$ must also be in the set.

3. Find a value of c in the Mandelbrot set such that $-c$ is not in the set. It is easiest to look for a real value.

4. You are testing to see whether a value of c belongs to the Mandelbrot set M. You generate the sequence $z_0, z_1, z_2, \ldots$ and find two values z_n and z_{n+p} such that $z_n = z_{n+p}$ $(p \neq 0)$. Explain why c must be in M. Give an example of $c \neq 0$ that leads to this situation.

5. a) Write a simple computer program that will find $z_n = z_{n-1}^2 + c$ for any $n \geq 1$. Assume $z_0 = 0$ and c is any complex constant. Use your program to verify Table A1.

b) Use your program to show that $c = 1.01i$ is not in the Mandelbrot set M. For what smallest n is $|z_n| > 2$?

c) If $|c| \leq 2$ and $|z_{1000}| \leq 2$ you may assume with little risk that c is in M. Use your program to show that $-1.25 + 0.01i$ is in M but that $-1.25 + 0.1i$ is not in M.

In Exercises 6–9 assume that the Julia sets under discussion are obtained by iteration of $f(z) = z^2 + c$.

6. If $z = z_0$ is in a filled Julia set, prove that $z = -z_0$ is in the same set.

7. What is the filled Julia set corresponding to $c = 0$? Note the boundary is not a fractal set.

8. Prove that $z = 0$ lies in all connected filled Julia sets. Use Mandelbrot's result concerning connected filled Julia sets.

9. Why must a filled Julia set corresponding to value c contain the two points given by the values of $[1 + (1 - 4c)^{1/2}]/2$?

APPENDIX B
THE z TRANSFORMATION

The **z** transformation is a mathematical procedure in which a sequence of numbers is used to create an analytic function. A Laurent series whose coefficients are the elements of the sequence is the vehicle used to create the function.

The transformation is much used in the analysis of sampled data systems. Here the values assumed by an electrical signal that varies with time are recorded at uniform discrete time intervals. These values (or samples) form a sequence of numbers; this sequence is fed into a computer that alters it in various useful ways. This procedure is called digital signal processing and is often used in communications links and radar systems.

The output from the computer is found to be the solution of a difference equation that, as the name suggests, involves the differences between the values assumed by the samples. The **z** transformation is a useful tool in the solution of such equations.

Difference equations also appear in problems in economics, population growth, and biology. Not surprisingly, the **z** transformation is also applied in these subjects. In this section we will learn the definition of the transformation and see how it can be used to solve some simple difference equations of the kind that might arise in these disciplines.

Consider a function $f(t)$, defined for $t \geq 0$, such as the one sketched in Fig. B5–1. The values (samples) of the function taken at intervals of T, beginning at $t = 0$, produce the sequence $f(0)$, $f(T)$, $f(2T)$, ... from which we make the definition:

DEFINITION **z** transform

The **z** transform of the function $f(t)$, that is, $\mathbb{Z}[f(t)]$, is given by

$$\mathbb{Z}[f(t)] = \sum_{n=0}^{\infty} f(nT)z^{-n} = f(0) + f(T)z^{-1} + f(2T)z^{-2} + \cdots, \quad \text{(B5–1)}$$

where $T > 0$. The function so defined is called $F(z)$. We say that $\mathbb{Z}[f(t)] = F(z)$. □

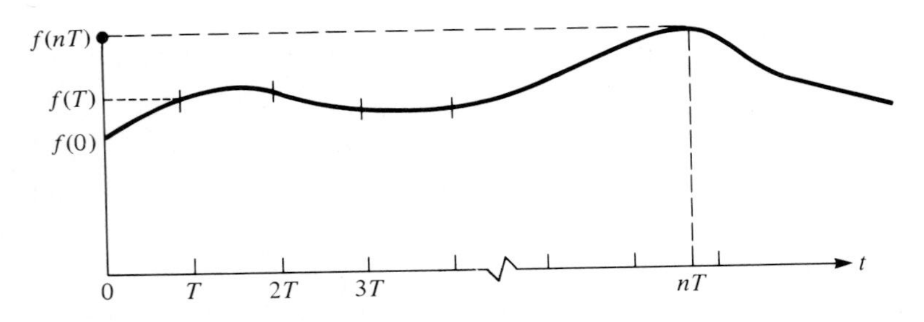

Figure B5–1

Usually lowercase letters like f and g will be reserved for functions of t, while the corresponding uppercase letter (here F and G) will be the associated **z** transform.

If the series of Eq. (B5–1) converges in a domain, it is a Laurent series with *no positive exponent* in any term. Notice $F(z)$ depends not only on $f(t)$, the function being transformed, but also on the parameter T, which measures how often the samples of $f(t)$ are taken. Observe that to perform the transformation, $f(t)$ need be defined only for $t = nT$, $n = 0, 1, 2, \ldots$. The transformation is the conversion of a sequence of numbers $c_n = f(nT)$ $(n = 0, 1, 2, \ldots)$ to a function of z by means of $\sum_{n=0}^{\infty} c_n z^{-n}$. Two different functions of t can have, for some value of T, identical **z** transforms. For example, the functions $\sin(\pi t)$ and $\sin^2(\pi t)$ will have **z** transforms that are identically zero if we choose $T = 1$, since $\sin(n\pi)$ and $\sin^2(n\pi)$ are zero when n is an integer. This lack of uniqueness will not be an impediment to us since we will be using the **z** transform to solve problems where $f(t)$ is needed only for $t = nT$.

Let us set $w = 1/z$ in Eq. (B5–1), which we assume contains a convergent Laurent series. The function on the right now becomes $\sum_{n=0}^{\infty} c_n w^n$, where $c_n = f(nT)$. This function of w is a power series, which, as we know, converges inside some circle of radius ρ centered at $w = 0$. Assuming $\rho > 0$, we have from Theorem 14 that $\sum_{n=0}^{\infty} c_n w^n$ is an analytic function of w for $|w| \le r$, where $r < \rho$. Thus $\sum_{n=0}^{\infty} c_n (1/z)^n = F(z)$ is an analytic function of z for $|1/z| \le r$ or $|z| \ge r$. Therefore $F(z)$, the **z** transform of $f(t)$, is a function, defined by a Laurent series, that is analytic in the z-plane in an annular domain whose outer radius is infinite. From Eq. (B5–1) we have that $\lim_{z\to\infty} F(z) = f(0)$, since all the terms involving $z^{-1}, z^{-2}, z^{-3}, \ldots$ vanish in this limit.

There are two recurring problems when one deals with **z** transforms. One is to obtain the transform of a given function of t. Of course we could simply state each term of the resulting infinite series; in practice we would prefer a closed form expression (a formula) for the sum of the series. This expression is the analytic function whose Laurent expansion, valid for all $T > 0$, is $\sum_{n=0}^{\infty} f(nT) z^{-n}$. Sometimes we will be lucky enough to recognize the closed form expression by studying the series; otherwise we must be content with using the series as the transformation.

The other problem is to obtain $f(t)$ when we are given $F(z)$. This is not strictly speaking possible since, as noted, two different functions of t can have the same **z** transform. What we obtain, given $F(z)$, is the set of numbers $f(nT)$, $n = 0, 1, 2, \ldots$. We then say

$$\mathbb{Z}^{-1}[F(z)] = f(nT).$$

If $F(z)$ is stated as a Laurent series, as on the right side of Eq. (B5–1), $f(nT)$ is obtained merely by identifying the coefficients of the terms $z^0, z^{-1}, \ldots$. If $F(z)$ is given as a closed form expression, we must first obtain its Laurent expansion valid in an annulus centered at the origin. The outer radius of the annulus must be infinite. Another approach is to perform a contour integration. In Exercise 17 it is shown that if $F(z)$ is analytic for $|z| > R$, then

$$f(nT) = \frac{1}{2\pi i} \oint_C F(z) z^{n-1}\, dz, \qquad n = 0, 1, 2, \ldots, \tag{B5–2}$$

where C is any circle centered at the origin with radius greater than R. For some simple functions $F(z)$ we can evaluate the previous integral by means of the Cauchy integral formula or its extension. For some complicated functions the integral can often be evaluated by the method of residues, which is discussed in the next chapter.

To take the **z** transform of a function it is not necessary to know the value for any $t < 0$. It is convenient to define all the functions with which we will be dealing here as *being zero for negative t*. One such function, which is very handy, is $u(t)$, the *unit step function*, given by

$$u(t) = 1, \qquad t \geq 0, \tag{B5–3a}$$

$$u(t) = 0, \qquad t < 0. \tag{B5–3b}$$

Notice that $u(t - \tau) = 1$ when $t \geq \tau$, and $u(t - \tau) = 0$ for $t < \tau$. Given a function $g(t)$ we have $g(t)u(t - \tau) = g(t)$ when $t \geq \tau$, and $g(t)u(t - \tau) = 0$ for $t < \tau$.

EXAMPLE 1

Find $\mathbb{Z}[u(t)]$, the transform of the unit step function.

Solution

From Eq. (B5–3) it is obvious that $u(nT) = 1$ for $n = 0, 1, 2, \ldots$. Thus from Eq. (B5–1), $\mathbb{Z}[u(t)] = \mathbb{Z}[1] = \sum_{n=0}^{\infty} z^{-n} = 1 + 1/z + 1/z^2 + \cdots$. Recalling that $1/(1 - w) = 1 + w + w^2 + \cdots$ for $|w| < 1$, we see that with $w = 1/z$ we have

$$1 + \frac{1}{z} + \frac{1}{z^2} + \cdots = \frac{1}{1 - 1/z} = \frac{z}{z - 1},$$

which is valid for $|1/z| < 1$, or $|z| > 1$. Thus

$$\mathbb{Z}[u(t)] = \mathbb{Z}[1] = \frac{z}{z - 1}, \qquad |z| > 1.$$

We were fortunate to obtain a closed form expression. ◀

EXAMPLE 2

Find the **z** transform of $f(t) = tu(t)$. (Notice that using $u(t)$ saves us the trouble of saying $f(t) = 0$, $t < 0$.)

Solution

Here $f(nT) = nT$, $n = 0, 1, 2, \ldots$. Thus

$$\mathbb{Z}[tu(t)] = \sum_{n=0}^{\infty} (nT)z^{-n} = T\left[\frac{1}{z} + \frac{2}{z^2} + \frac{3}{z^3} + \cdots\right]. \tag{B5–4}$$

The right side of the preceding can be evaluated with help from Eq. (5.5–16c). Multiplying that equation by w yields

$$\frac{w}{(1 - w)^2} = w + 2w^2 + 3w^3 + \cdots, \qquad |w| < 1.$$

If we replace w with $1/z$ in the preceding and multiply both sides by T we have

$$\frac{T(1/z)}{(1-1/z)^2} = T\left[\frac{1}{z} + \frac{2}{z^2} + \frac{3}{z^3} + \cdots\right], \qquad |z| > 1.$$

Comparing this equation with Eq. (B5–4) we obtain

$$\mathbb{Z}[tu(t)] = \frac{T(1/z)}{(1-1/z)^2} = \frac{Tz}{(z-1)^2}.$$ ◀

Other transforms are developed in the exercises. An extensive table of **z** transforms can be found in various texts.†

Linearity of the Transformation

The operation that creates $F(z)$ from $f(t)$ is linear, i.e., $\mathbb{Z}[cf(t)] = c\mathbb{Z}[f(t)]$, where c is any constant. Furthermore if $f(t)$ and $g(t)$ are both defined for $t = nT$, $n \geq 0$, then

$$\mathbb{Z}[f(t) + g(t)] = \mathbb{Z}[f(t)] + \mathbb{Z}[g(t)] = F(z) + G(z), \tag{B5–5}$$

where $F(z)$ and $G(z)$ are the transforms of $f(t)$ and $g(t)$. The equation is valid when z is confined to a domain in which both $F(z)$ and $G(z)$ exist. The proof is easy and follows from the definition of the transformation.

In a similar way the inverse transformation is linear, i.e.,

$$\mathbb{Z}^{-1}[F(z) + G(z)] = \mathbb{Z}^{-1}F(z) + \mathbb{Z}^{-1}G(z) = f(nT) + g(nT).$$

One can verify this from Eq. (B5–2), which provides a means for performing the inverse transformation.

To illustrate Eq. (B5–5) we can combine the results of Examples 1 and 2 as follows:

$$\mathbb{Z}[(1+t)u(t)] = \frac{z}{z-1} + \frac{Tz}{(z-1)^2} = \frac{z^2 - z + Tz}{(z-1)^2}.$$

Some examples of inverse transformations follows.

EXAMPLE 3

If $F(z) = (z+1)/z^2$, find $\mathbb{Z}^{-1}[F(z)] = f(nT)$.

Solution

Rewriting $F(z)$ as a two-term Laurent series we have $F(z) = 1/z + 1/z^2$. A glance at Eq. (B5–1) shows that $f(0) = 0$, $f(T) = f(2T) = 1$, and $f(nT) = 0$, $n \geq 3$. ◀

EXAMPLE 4

If $F(z) = (z+1)/(z-1)$, find $\mathbb{Z}^{-1}[F(z)]$.

---

† See, e.g., E. L. Jury, *Theory and Applications of the z Transform Method* (New York: John Wiley, 1964), Appendix.

Solution

$F(z)$ can be expanded in a Laurent series valid for $|z| > 1$. We have

$$F(z) = \frac{z+1}{z-1} = \frac{(z-1)+2}{z-1} = 1 + \frac{2}{z-1}.$$

Now $2/(z-1) = (2/z)/(1-1/z) = 2[1 + 1/z + 1/z^2 + \cdots]/z$ for $|z| > 1$. Thus, $F(z) = 1 + 2/z + 2/z^2 + \cdots$, $|z| > 1$. Studying the coefficients and using Eq. (B5–1), we conclude that $f(0) = 1$, $f(nT) = 2$, $n \geq 1$. ◄

Comment: A given $F(z)$ does not necessarily have an inverse **z** transform. If $F(z)$ cannot be expanded in a Laurent series of the form $\sum_{n=0}^{\infty} c_n z^{-n}$, no inverse transformation is possible. The functions $(z^3 - 1)/z^2$ and $\sin z$ have no inverse transforms. Neither function has a limit as $z \to \infty$. Functions having the desired Laurent expansion must have a limit as $z \to \infty$. This limit is c_0.

Two features of the **z** transformation, which we call translation properties, are useful for solving difference equations. These allow us to determine $\mathbb{Z}[f(t \pm kT)]$ (k is an integer) when $\mathbb{Z}[f(t)]$ is known. When $k \geq 0$, a graph of $f(t - kT)$ is identical to a graph of $f(t)$ but is translated kT units to the right. Similarly a graph of $f(t + kT)$ involves a shift of kT units to the left.

Let $\mathbb{Z}[f(t)] = F(z) = \sum_{n=0}^{\infty} f(nT)z^{-n}$. Now, $\mathbb{Z}[f(t - kT)] = \sum_{n=0}^{\infty} f(nT - kT)z^{-n}$. Recalling that $f(t) = 0$, $t < 0$, we see that $f(nT - kT) = 0$ when $n < k$. We can thus rewrite the previous sum and get

$$\mathbb{Z}[f(t - kT)] = \sum_{n=k}^{\infty} f(nT - kT)z^{-n} = \sum_{n=k}^{\infty} f((n-k)T)z^{-n}.$$

We now reindex this summation using $m = n - k$. Thus

$$\mathbb{Z}[f(t - kT)] = \sum_{m=0}^{\infty} f(mT)z^{-(m+k)} = z^{-k} \sum_{m=0}^{\infty} f(mT)z^{-m}.$$

This last sum is $F(z)$, and we have the *first translation formula*:

$$\mathbb{Z}[f(t - kT)] = z^{-k}F(z), \tag{B5–6}$$

which is valid when $\mathbb{Z}[f(t)] = F(z)$, $k \geq 0$, and $f(t) = 0$, $t < 0$.

Let us study the second property. Consider $\mathbb{Z}[f(t + kT)]$ when $k = 1$. We have

$$\begin{aligned}\mathbb{Z}[f(t + T)] &= \sum_{n=0}^{\infty} f(nT + T)z^{-n} = \sum_{n=0}^{\infty} f((n+1)T)z^{-n} \\ &= f(T) + f(2T)z^{-1} + f(3T)z^{-2} + \cdots.\end{aligned}$$

Adding and subtracting $f(0)z$ in this last series we have

$$\mathbb{Z}[f(t + T)] = [f(0)z + f(T)z^0 + f(2T)z^{-1} + \cdots] - f(0)z.$$

The expression in the brackets is $zF(z)$. Thus

$$\mathbb{Z}[f(t + T)] = zF(z) - zf(0). \tag{B5–7}$$

As a further example,

$$\mathbb{Z}[f(t+2T)] = \sum_{n=0}^{\infty} f(nT+2T)z^{-n} = f(2T) + f(3T)z^{-1} + f(4T)z^{-2} + \cdots$$
$$= [f(0)z^2 + f(T)z + f(2T) + f(3T)z^{-1} + f(4T)z^{-2} + \cdots]$$
$$- z^2 f(0) - zf(T).$$

The bracketed expression is $z^2F(z)$. Thus

$$\mathbb{Z}[f(t) + 2T)] = z^2F(z) - z^2f(0) - zf(T). \tag{B5–8}$$

The preceding results, Eqs. (B5–7) and (B5–8), are examples of our *second translation formula.* We can show by the same technique that in general

$$\mathbb{Z}[f(t+kT)] = z^kF(z) - z^kf(0) - z^{k-1}f(T) - z^{k-2}f(2T)$$
$$- \cdots - zf((k-1)T),$$

where $k \geq 0$.

EXAMPLE 5

In Exercise 1 you will show that if $f(t) = e^{at}u(t)$, then $F(z) = z/(z - e^{aT})$ for $|z| > |e^{aT}|$. Use this result to find $\mathbb{Z}[g(t)]$, where $g(t) = e^{a(t-T)}u(t-T)$. Also find $\mathbb{Z}[h(t)]$, where $h(t) = e^{a(t+T)}u(t+T)$. A sketch of $f(t)$, $g(t)$, and $h(t)$ are given in Fig. B.5–2, where we assume $a > 0$.

Solution

Since $g(t) = f(t - T)$ we use Eq. (B5–6) with $k = 1$ to get $G(z)$. Thus

$$G(z) = z^{-1}\frac{z}{z - e^{aT}} = \frac{1}{z - e^{aT}} \qquad \text{for } |z| > e^{aT}.$$

Since $h(t) = f(t + T)$ we use Eq. (B5–7) to get $H(z)$. Noting that $f(0) = 1$ we have

$$\mathbb{Z}[h(t)] = \frac{z^2}{z - e^{aT}} - z = \frac{ze^{aT}}{z - e^{aT}}.$$ ◀

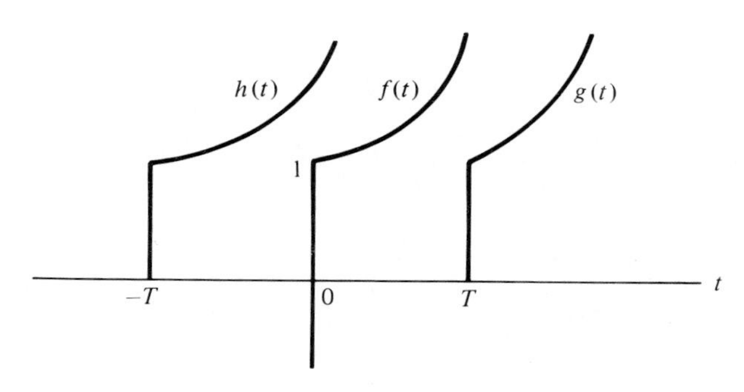

Figure B5–2

z Transforms of Products of Functions

Let $\mathbb{Z}[f(t)] = \sum_{n=0}^{\infty} c_n z^{-n} = F(z)$ and $\mathbb{Z}[g(t)] = \sum_{n=0}^{\infty} d_n z^{-n} = G(z)$, where $c_n = f(nT)$ and $d_n = g(nT)$. Suppose we seek $\mathbb{Z}[f(t)g(t)]$. By definition $\mathbb{Z}[f(t)g(t)] = \sum_{n=0}^{\infty} f(nT)g(nT)z^{-n}$. Thus

$$\mathbb{Z}[f(t)g(t)] = \sum_{n=0}^{\infty} c_n d_n z^{-n}. \tag{B5–9}$$

Thus if we have the **z** transforms of $f(t)$ and $g(t)$ as Laurent series, the **z** transform of $f(t)g(t)$ is easily obtained as a Laurent series.

If the transforms of $f(t)$ and $g(t)$ are presented to us in closed form, we can still find $\mathbb{Z}[f(t)g(t)]$ without first obtaining Laurent expansions of $F(z)$ and $G(z)$. However, we must be prepared to evaluate a contour integral, which we now derive.

Let $F(z)$ and $G(z)$ both be analytic in the domain $|z| > R$. From Laurent expansions we have $F(w) = \sum_{m=0}^{\infty} c_m w^{-m}$, $|w| > R$, and $G(z/w) = \sum_{n=0}^{\infty} d_n (z/w)^{-n} = \sum_{n=0}^{\infty} d_n w^n z^{-n}$. This last expansion is valid for $|z/w| > R$, or $|z| > R|w|$. Multiplying our series we have

$$F(w)G(z/w) = \sum_{m=0}^{\infty} \sum_{n=0}^{\infty} c_m d_n w^{n-m} z^{-n}, \tag{B5–10}$$

where we choose $|w| > R$ and $|z| > R|w|$.

Now consider a circle of radius ρ centered at the origin in the complex w-plane. (Refer to Fig. B5–3.) We take $\rho > R$, and we place our variable w on this circle so that $|w| = \rho$. In Eq. (B5–10) we will require that $|z| > R\rho$. The Laurent expansion in Eq. (B5–10) is according to Theorem 18 uniformly convergent in a domain in the w-plane containing the circle $|w| = \rho$. According to Theorem 8 the following Laurent

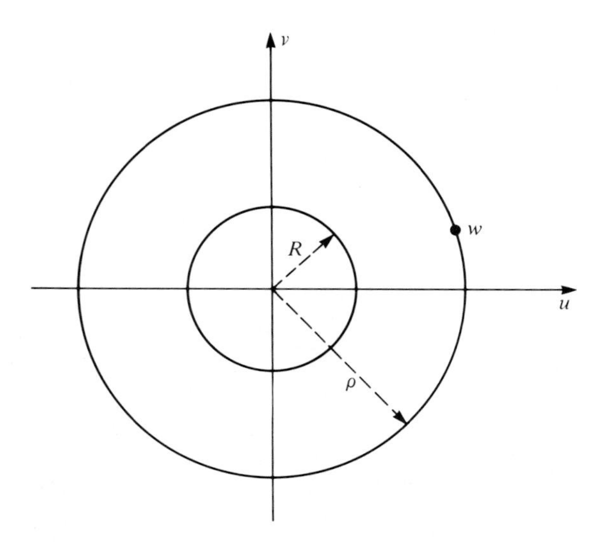

Figure B5–3

expansion is also uniformly convergent in this domain:

$$\frac{1}{2\pi i w}F(w)G(z/w) = \frac{1}{2\pi i}\sum_{m=0}^{\infty}\sum_{n=0}^{\infty} c_m d_n \frac{z^{-n}w^{n-m}}{w}.$$

We can thus integrate this series term by term around $|w| = \rho$, so that

$$\frac{1}{2\pi i}\oint_{|w|=\rho}\frac{F(w)G(z/w)}{w}\,dw = \frac{1}{2\pi i}\sum_{m=0}^{\infty}\sum_{n=0}^{\infty}\oint_{|w|=\rho} c_m d_n z^{-n}\frac{w^{n-m}}{w}\,dw. \tag{B5–11}$$

Recall that if k is an integer $\oint_{|w|=\rho} w^k\,dw$ is zero or $2\pi i$ according to whether $k \neq -1$ or $k = -1$. We notice that the integrals on the right in Eq. (B5–11) are zero except when $n = m$. Then $\oint_{|w|=\rho} z^{-n}w^{n-m}/w\,dw = 2\pi i z^{-n}$. Thus Eq. (B5–11) becomes

$$\frac{1}{2\pi i}\oint_{|w|=\rho}\frac{F(w)G(z/w)}{w}\,dw = \sum_{n=0}^{\infty} c_n d_n z^{-n}. \tag{B5–12}$$

Comparing the above with Eq. (B5–9) we have our desired result:

$$\mathbb{Z}[f(t)g(t)] = \frac{1}{2\pi i}\oint_{|w|=\rho}\frac{F(w)G(z/w)}{w}\,dw. \tag{B5–13}$$

In this integral we require that $|z| > R\rho$, where $\rho > R$. Recall that R is such that $F(w)$ and $G(w)$ are analytic for $|w| > R$.

EXAMPLE 6

Find $\mathbb{Z}[te^{at}u(t)]$ from Eq. (B5–13).

Solution

Let $f(t) = tu(t)$ and $g(t) = e^{at}u(t)$. From Example 2, $\mathbb{Z}[f(t)] = Tz/(z-1)^2 = F(z)$. In Exercise 1 it is shown that $\mathbb{Z}[g(t)] = z/(z - e^{aT}) = G(z)$. Notice that $F(z)$ is analytic except at $z = 1$, while $G(z)$ is analytic except at $z = e^{aT}$. Substituting in Eq. (B5–13) we find

$$\begin{aligned}\mathbb{Z}[te^{at}u(t)] &= \frac{1}{2\pi i}\oint_{|w|=\rho}\frac{Tw}{w(w-1)^2}\frac{z/w}{(z/w - e^{aT})}\,dw\\ &= \frac{zT}{2\pi i}\oint_{|w|=\rho}\frac{1}{(w-1)^2}\frac{1}{z - we^{aT}}\,dw.\end{aligned} \tag{B5–14}$$

We require $\rho > R$, where R is the radius of a circle in the w-plane outside which $F(w)$ and $G(w)$ are analytic. Thus, $R > 1$ and also $R > |e^{aT}|$. Hence ρ is larger than the greater of 1 and $|e^{aT}|$.

The integrand of the last integral fails to be analytic at $w = 1$, a point enclosed by the contour of integration. The integrand is also not analytic where $we^{aT} = z$. However, this point lies outside $|w| = \rho$, as we shall see. Recall that (B5–13) is valid for $R\rho < |z|$. We have, for w lying on or inside the contour $|w| = \rho$,

$$|we^{aT}| = |w||e^{aT}| \le \rho|e^{aT}| < \rho R < |z|.$$

The preceding, $|we^{aT}| < |z|$, tells us that $z - we^{aT} = 0$ cannot be satisfied on and inside our contour of integration.

We now evaluate Eq. (B5–14) using the extended Cauchy integral formula. Thus

$$\mathbb{Z}[te^{at}u(t)] = zT\frac{d}{dw}\left[\frac{1}{z - we^{aT}}\right]_{w=1} = \frac{zTe^{aT}}{(z - e^{aT})^2}. \blacktriangleleft \qquad \text{(B5–15)}$$

Inverse z Transform of a Product of Two Functions

If we are given functions $F(z)$ and $G(z)$, whose inverse transforms $f(nT)$ and $g(nT)$ are known, can we find directly $\mathbb{Z}^{-1}[F(z)G(z)]$? The answer is yes. To prove this we need a definition—the convolution of $f(t)$ with $g(t)$, written $f(t) * g(t)$.

DEFINITION Convolution

$$f(t) * g(t) = \sum_{k=0}^{\infty} f(kT)g((n-k)T), \qquad n = 0, 1, 2, \ldots. \qquad \text{(B5–16)}$$

It is easy to show that $g(t) * f(t) = \sum_{k=0}^{\infty} g(kT)f((n-k)T) = f(t) * g(t)$ (convolution is commutative) when $f(t)$ and $g(t)$ are zero for $t < 0$. Notice that the preceding sum for $g(t) * f(t)$ as well as the sum in Eq. (B5–16) for $f(t) * g(t)$ need be carried out only from $k = 0$ to $k = n$.

The function $f(t) * g(t)$ defines a function of the variable nT, where $n = 0, 1, 2, \ldots$. Let $h(t) = f(t) * g(t) = \sum_{k=0}^{\infty} f(kT)g((n-k)T)$, which defines $h(t)$ for $t = nT$. Now

$$\mathbb{Z}[h(t)] = \sum_{n=0}^{\infty}\left[\sum_{k=0}^{\infty} f(kT)g((n-k)T)\right]z^{-n}.$$

The inner sum needs to be carried out only as far as n. Taking $f(kT) = a_k$ and $g(jT) = b_j$ we have

$$\mathbb{Z}[h(t)] = \sum_{n=0}^{\infty}\sum_{k=0}^{n} a_k b_{n-k} z^{-n}. \qquad \text{(B5–17)}$$

Now $\mathbb{Z}[f(t)] = \sum_{k=0}^{\infty} a_k z^{-k} = F(z)$ and $\mathbb{Z}[g(t)] = \sum_{j=0}^{\infty} b_j z^{-j} = G(z)$. Thus

$$\begin{aligned} F(z)G(z) &= \sum_{k=0}^{\infty} a_k z^{-k} \sum_{j=0}^{\infty} b_j z^{-j} \\ &= (a_0 + a_1/z + a_2/z^2 + \cdots)(b_0 + b_1/z + b_2/z^2 + \cdots) \\ &= a_0b_0 + (a_0b_1 + a_1b_0)z^{-1} + (a_0b_2 + a_1b_1 + a_2b_0)z^{-2} + \cdots. \end{aligned}$$

(Recall from Theorem 5 in Section 5.2 that two absolutely convergent series can be multiplied in this fashion.) Studying the coefficients in the preceding series we see that $F(z)G(z) = \sum_{n=0}^{\infty} c_n z^{-n}$, where $c_n = \sum_{k=0}^{n} a_k b_{n-k}$. Comparing this series with Eq. (B5–17) we have, finally,

$$\mathbb{Z}[h(t)] = \mathbb{Z}[f(t) * g(t)] = F(z)G(z). \qquad \text{(B5–18)}$$

Thus

*The **z** transform of the convolution of two functions is the product of the **z** transforms of each function,*

and, conversely,

*The inverse **z** transform of the product of two functions is the convolution of the inverse transform of each function.*

EXAMPLE 7

Using the concept of convolution, find

$$\mathbb{Z}^{-1}\left[\frac{z^2}{(z - e^{aT})(z - 1)}\right] = h(nT).$$

Solution

Rewriting the expression in the brackets and using the inverse of Eq. (B5–18) we have

$$\mathbb{Z}^{-1}\left[\frac{z}{z - e^{aT}}\frac{z}{(z - 1)}\right] = f(t) * g(t),$$

where

$$f(nT) = \mathbb{Z}^{-1}\left[\frac{z}{z - e^{aT}}\right] \quad \text{and} \quad g(nT) = \mathbb{Z}^{-1}\left[\frac{z}{z - 1}\right].$$

We could obtain $f(nT)$ and $g(nT)$ from a standard table of transforms. However, from Example 1 we have that $\mathbb{Z}^{-1}[z/(z - 1)] = u(t)$, and from Exercise 1, $\mathbb{Z}^{-1}[z/(z - e^{aT})] = e^{at}u(t)$, where $t = nT$ in both cases. Taking $f(nT) = e^{anT}u(nT)$ and $g(nT) = u(nT)$ and performing their convolution, we get

$$h(nT) = \sum_{k=0}^{\infty} e^{akT}u((n - k)T).$$

Now $u((n - k)T) = 0$ for $k > n$ and $u((n - k)T) = 1$ for $n \geq k$. We can thus rewrite the preceding as

$$h(nT) = \sum_{k=0}^{n} e^{akT} = \sum_{k=0}^{n} (e^{aT})^k.$$

Recalling the sum for a finite geometric series, $\sum_{k=0}^{n} p^k = (1 - p^{n+1})/(1 - p)$, and taking $p = e^{aT}$ we have

$$h(nT) = \frac{1 - e^{a(n+1)T}}{1 - e^{aT}} = \mathbb{Z}^{-1}\left[\frac{z^2}{(z - e^{aT})(z - 1)}\right].$$ ◀

Difference Equations and the z Transform

Here is an example of a problem involving a *difference equation.* Let $f(nT)$ be a function defined for $n = 0, 1, 2, \ldots$ and assume $T > 0$. Obtain a closed form expression for the solution of the equation

$$f((n+1)T) - 2f(nT) = 0,$$

given that $f(0) = 1$.

To solve this equation we might put $n = 0$, $f(0) = 1$ and obtain $f(T) = 2$. Then putting $n = 1$, $f(T) = 2$, we get $f(2T) = 4$. Continuing in this way we find $f(nT) = 2^n$, $n = 0, 1, 2, \ldots$.

A more elegant method, which is useful in a wide range of problems, employs the **z** transform. We perform a **z** transformation on both sides of the given equation taking $t = nT$, $\mathbb{Z}[0] = 0$, $2\mathbb{Z}[f(nT)] = 2F(z)$. From our translation formula, Eq. (B5–7), with $f(0) = 1$, we have $\mathbb{Z}[f((n+1)T)] = zF(z) - z$. Thus the transformed equation, $\mathbb{Z}[f[(n+1)T]] - 2\mathbb{Z}[f(nT)] = \mathbb{Z}[0]$, becomes $zF(z) - z - 2F(z) = 0$. Solving for $F(z)$ we have $F(z) = z/(z-2)$. To obtain $f(nT)$ we have

$$F(z) = \frac{1}{1 - 2/z} = 1 + \frac{2}{z} + \frac{4}{z^2} + \frac{8}{z^3} + \cdots = \sum_{n=0}^{\infty} f(nT)z^{-n}.$$

Thus $f(nT) = 2^n$.

The equation we solved is a linear difference equation.[†] The general form of the linear difference equation is

$$a_0 f(t + NT) + a_1 f(t + (N-1)T) + a_2 f(t + (N-2)T) + \cdots + a_N f(t) = g(t). \tag{B5–19}$$

Here t is constrained to equal nT, $n = 0, 1, 2, \ldots$, and $g(t)$ must be defined for these values of t. N is an integer ≥ 1. The coefficients $a_0, a_1, \ldots, a_N$ can be known functions of t. In the problems considered here they will all be assumed to be constant. Such equations turn up in a variety of disciplines, and **z** transforms can be used to solve them.

EXAMPLE 8

The *Fibonacci sequence* of numbers was first described in the early thirteenth century by the Italian mathematician Leonardo Fibonacci.[‡] The sequence is 0, 1, 1, 2, 3, 5, 8, 13, 21, 34, ... Each element of the sequence is the sum of the two preceding elements. Fibonacci described these numbers in the solution of a problem in the growth of a rabbit population. The numbers arise also in plant growth, puzzles, and in theories of aesthetics. For $n \geq 0$, the nth element of the sequence, $f(n)$, satisfies

---

[†] For an explanation of why the word "difference" is applied to these equations and for an extensive discussion of them, see P. V. O'Neil, *Advanced Engineering Mathematics*, 3rd ed. (Belmont, CA: Wadsworth Publishing Co., 1991), Chapter 10.

[‡] See, e.g., N. N. Vorob'ev, *Fibonacci Numbers*, translated by H. Mors (New York: Blaisdell, 1961).

the difference equation $f(n+2) = f(n+1) + f(n)$, or

$$f(n+2) - f(n+1) - f(n) = 0. \tag{B5–20}$$

The preceding is of the form shown in Eq. (B5–19) if we take $T = 1$, $N = 2$, $a_0 = 1$, $a_1 = -1$, $a_2 = -1$. Note that $f(0) = 0$, $f(1) = 1$, $f(2) = 1$, etc. Our problem is to find a closed form solution of Eq. (B5–20) by using **z** transforms.

Solution

Taking the **z** transformation of Eq. (B5–20) we have

$$\mathbb{Z}[f(n+2)] - \mathbb{Z}[f(n+1)] - \mathbb{Z}[f(n)] = 0. \tag{B5–21}$$

With $T = 1$, $f(0) = 0$, $f(1) = 1$, we obtain from Eqs. (B5–7) and (B5–8) that $\mathbb{Z}[f(n+1)] = zF(z)$ and $\mathbb{Z}[f(n+2)] = z^2F(z) - z$. Substituting these into our transformed equation we have $z^2F(z) - z - zF(z) - F(z) = 0$, from which we obtain $F(z) = z/(z^2 - z - 1)$.

To obtain $f(n)$ we expand the preceding in a Laurent series containing z to only nonpositive powers. Partial fractions are handy here. Thus

$$F(z) = \frac{z}{z^2 - z - 1} = \frac{1}{\sqrt{5}}\left[\frac{(1+\sqrt{5})/2}{z - (1+\sqrt{5})/2} - \frac{(1-\sqrt{5})/2}{z - (1-\sqrt{5})/2}\right].$$

Each fraction can be expanded in negative powers of z, and we obtain

$$F(z) = \sum_{n=0}^{\infty} c_n z^{-n}, \qquad |z| > (1+\sqrt{5})/2,$$

where

$$c_n = \frac{1}{\sqrt{5}\,2^n}[(1+\sqrt{5})^n - (1-\sqrt{5})^n].$$

Since $c_n = f(n)$, the problem is solved. We can determine any term in the sequence, e.g., with a reasonably good pocket calculator. We find that the 20th Fibonacci number ($n = 20$) is 6765. ◀

EXERCISES

1. Show that $\mathbb{Z}[e^{at}] = \dfrac{z}{z - e^{aT}}$, $|z| > |e^{aT}|$, a is complex.

2. Show that $\mathbb{Z}[b^t] = \dfrac{z}{z - b^T}$, $|z| > |b^T|$, b is complex, b^t is a principal value.

Establish the following formulas by using the result of Exercise 1, suitable values for a, and the linearity property of the **z** transform. For example, take $a = \pm i\alpha$ in Exercise 3.

3. $\mathbb{Z}[\sin(\alpha t)] = \dfrac{z\sin(\alpha T)}{z^2 - 2z\cos(\alpha T) + 1}$, $|z| > 1$, α is real

4. $\mathbb{Z}[\cos(\alpha t)] = \dfrac{z(z - \cos(\alpha T))}{z^2 - 2z\cos(\alpha T) + 1}$, $|z| > 1$, α is real

5. $\mathbb{Z}[\sinh(\alpha t)] = \dfrac{z\sinh(\alpha T)}{z^2 - 2z\cosh(\alpha T) + 1}$, $|z| > e^{|\alpha| T}$, α is real

6. $\mathbb{Z}[\cosh(\alpha t)] = \dfrac{z(z - \cosh(\alpha T))}{z^2 - 2z\cosh(\alpha T) + 1}$, $|z| > e^{|\alpha| T}$, α is real

7. If $\mathbb{Z}[f(t)] = F(z)$ show that $\mathbb{Z}[tf(t)] = -zT\,dF/dz$.

8. a) Using $\mathbb{Z}[u(t)]$ from Example 1, and the result of Exercise 7, show that

$$\mathbb{Z}[tu(t)] = \frac{zT}{(z-1)^2}, \qquad |z| > 1.$$

b) In a similar way, use the preceding result to show that

$$\mathbb{Z}[t^2u(t)] = \frac{zT^2(z+1)}{(z-1)^3}, \qquad |z| > 1.$$

Use the result $\mathbb{Z}[u(t)] = z/(z-1)$ as well as the translational and linearity properties of the z transform to establish the following:

9. $\mathbb{Z}[u(t - T)] = 1/(z - 1)$

10. $\mathbb{Z}[u(t) - u(t - T)] = 1$

11. $\mathbb{Z}[u(t - T) - u(t - 2T)] = 1/z$.

12. Show that $\operatorname{Log}(z/(z-1))$ is analytic in a cut plane defined by the branch cut $y = 0$, $0 \le x \le 1$. Expand this function in a Laurent series valid for $|z| > 1$, and use your result to show that

$$\mathbb{Z}\left[\frac{T}{t}u(t - T)\right] = \operatorname{Log}\left(\frac{z}{z-1}\right).$$

We define $u(t - T)/t = 0$ when $t = 0$.

Find $\mathbb{Z}^{-1}[F(z)] = f(nT)$ for these functions:

13. $F(z) = \dfrac{1}{(z-1)^2}$ 14. $F(z) = \dfrac{1}{z^4(1 - z)}$ 15. $F(z) = e^{1/z}$

16. a) If $\mathbb{Z}[f(t)] = F(z)$ show that

$$\mathbb{Z}[e^{\beta t}f(t)] = F(ze^{-\beta T}).$$

b) Use the preceding result and the result of Exercise 3 to show that

$$\mathbb{Z}[e^{\beta t}\sin(\alpha t)] = \frac{ze^{\beta T}\sin(\alpha T)}{z^2 - 2ze^{\beta T}\cos(\alpha T) + e^{2\beta T}},$$

α, β real, $|z| > e^{\beta T}$.

17. If $\mathbb{Z}[f(t)] = F(z)$, where $F(z)$ is analytic for $|z| > R$, show that

$$f(nT) = \frac{1}{2\pi i} \oint_C F(z) z^{n-1}\, dz,$$

where C is the circle $|z| = R_0$, $R_0 > R$. C can also be any closed contour into which $|z| = R_0$ can be deformed, by the principle of deformation of contours.

18. Show that

a) $\mathbb{Z}^{-1}[z/(z-1)^2] = h(nT) = n$ by using the convolution of the inverse transforms of $z/(z-1)$ and $1/(z-1)$.

b) Obtain the preceding result by using the contour integration derived in Exercise 17 and the extended Cauchy integral formula.

19. The gamma function and the **z** transform:

a) Consider the gamma function $\Gamma(z) = \lim_{L\to\infty} \int_0^L u^{z-1} e^{-u}\, du$, commonly written $\int_0^\infty u^{z-1} e^{-u}\, du$. Here u is a real variable, z a complex variable, and $u^{z-1} = e^{(z-1)\,\mathrm{Log}\, u}$. It can be shown† that $\Gamma(z)$ is analytic for $\mathrm{Re}\, z > 0$. Do an integration by parts to show that

$$\Gamma(z+1) = z\Gamma(z).$$

b) Show that $\Gamma(1) = 1$, $\Gamma(2) = 1$, $\Gamma(3) = 2$. Taking $n \geq 0$ as an integer show by induction that $\Gamma(n+1) = n!$.

c) Show that $\mathbb{Z}[1/\Gamma(t/T + 1)] = e^{1/z}$, $|z| > 0$.

20. a) Use the result derived in Exercise 19(c), the transform derived in Exercise 1, and Eq. (B5–13) to show that

$$\mathbb{Z}\left[\frac{e^{at}}{\Gamma(t/T+1)}\right] = e^{e^{aT}/z}.$$

b) Derive this same formula by using the results of Exercises 16(a) and 19(c).

21. Use the **z** transform to find $f(nT)$ satisfying the difference equation $f(t+T) - 2f(t) = 0$, where $t = nT$ and $f(0) = 2$.

Solve these difference equations for $f(n)$, $n \geq 0$.

22. $f(n+2) = f(n+1) - f(n)$, where $f(0) = f(1) = 1$

23. $f(n+2) - f(n+1) - 2f(n) = 0$, where $f(0) = 0$, $f(1) = 1$

24. Consider the sequence 1, 1, 2, 4, 7, 11, 16,.... Let $f(n)$ be the nth term in the sequence, which begins with $n = 0$. Show that $f(n+1) - f(n) = n$, and use **z** transforms to prove that $f(n) = (n^2 - n + 2)/2$.

25. Following the method of the previous problem, show that the nth term in the sequence 0, 0, 1, 5, 14, 30, 55,... is

$$\frac{1}{6}\left(\frac{(n+1)!}{(n-2)!} + \frac{n!}{(n-3)!}\right) \qquad \text{for } n \geq 3.$$

---

† G. Polya and G. Latta, *Complex Variables* (New York: John Wiley, 1974), pp. 235–237.

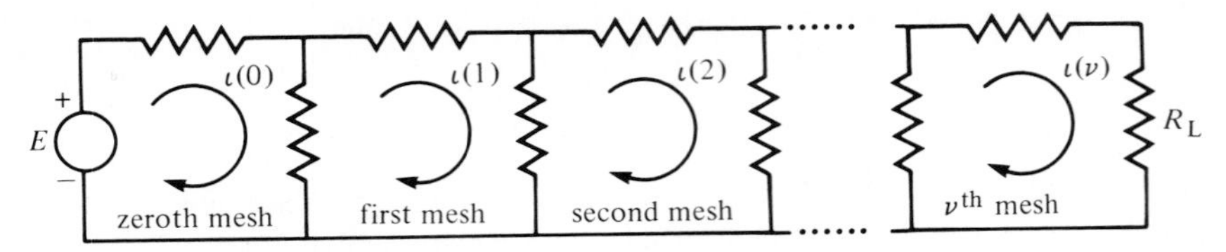

Figure B5–4

26. Shown in Fig. B5–4 is a ladder network containing $\nu + 1$ meshes. All resistors are 1 ohm except the one on the far right, which is R_L ohms. The generator on the left has a potential of E volts. Let $\imath(n) = \imath_n$ be the current in the nth mesh. From Kirchhoff's voltage law we have, in the zeroth mesh, $E = 2\imath_0 - \imath_1$, while in the last (νth) mesh $0 = \imath_\nu(2 + R_L) - \imath_{\nu-1}$. Writing Kirchhoff's voltage law for the $(n + 1)$ mesh we have $0 = -\imath_n + 3\imath_{n+1} - \imath_{n+2}$, where we avoid the first and last meshes by the restriction $n = 0, 1, 2, \ldots, \nu - 2$.

a) Take the **z** transform of the preceding equation (with $t = nT$, $T = 1$) to show

$$I(z) = \frac{z[z\imath_0 - 3\imath_0 + \imath_1]}{z^2 - 3z + 1}, \qquad \text{where } I(z) = \mathbb{Z}[\imath(n)] = \mathbb{Z}[\imath_n].$$

b) Use the equation for the zeroth mesh to eliminate $\imath_1$ from $I(z)$ in part (a). Show that

$$I(z) = \frac{z[z - (1 + (E/\imath_0))]\imath_0}{z^2 - 3z + 1}.$$

c) Show that the preceding equation can be written

$$I(z) = \frac{z(z - 3/2) + z(1/2 - E/\imath_0)}{z^2 - 2(3/2)z + 1}\imath_0.$$

Now use the result of Exercises 5 and 6 to show that

$$\imath_n = \imath_0\left[\cosh(na) + \frac{(1/2 - E/\imath_0)\sinh(na)}{(\sqrt{5}/2)}\right],$$

where $a = \cosh^{-1}(3/2)$ and $a > 1$. Note that $\sinh a = \sqrt{5}/2$.

d) If $\imath_0$ has the proper value in the preceding equation, then $\imath_n$ can correctly describe the current in all the meshes. Use the above equation† to obtain expressions for $\imath_n$ and $\imath_{n-1}$. Substitute these expressions in our equation obtained by writing Kirchhoff's voltage law for the νth mesh. Show that

$$\frac{E}{\imath_0} = \frac{1}{2} + \frac{(\sqrt{5}/2)[(2 + R_L)\cosh(\nu a) - \cosh((\nu - 1)a)]}{(2 + R_L)\sinh(\nu a) - \sinh((\nu - 1)a)}.$$

This is the resistance "seen" by the generator. One can use this expression to eliminate $E/\imath_0$ in the result of part (c). Thus a formula is obtained for the current in any mesh.

---

† Although the difference equation used to derive $\imath_n$ in (c) is not valid in the zeroth and νth meshes, the resulting expression for $\imath_n$ is valid for $n = \nu$. There is an analogous situation in the solution of linear differential equations where the differential equation is not satisfied on a boundary but the resulting solution is valid on the boundary.

CHAPTER 6

Residues and Their Use in Integration

6.1 DEFINITION OF THE RESIDUE

In the previous chapter we devoted nearly equal attention to Laurent series and Taylor series. Although some uses were demonstrated for Taylor series (for example, in the numerical evaluation of integrals and transcendental functions), almost none were demonstrated for Laurent series. Now we will remedy this imbalance and show that a Laurent series (in particular, one of its terms) can be used in the evaluation of many kinds of integrals. Indeed the majority of this chapter will be concerned with the evaluation of integrals whose values would be difficult if not impossible to obtain without the techniques of complex calculus. At the end of the chapter, in the appendix, we see how the sums of certain infinite series can be found with the same methods.

Let z_0 be an isolated singular point of the analytic function $f(z)$. Consider any simple closed contour enclosing z_0 and no other singularity of $f(z)$. The integral $(1/2\pi i)\oint f(z)\,dz$ taken around C is typically nonzero. However, by the principle of deformation of contours (see Section 4.3) its value is independent of the precise shape of C, that is, all closed curves that contain z_0 and no other singular point of $f(z)$ will lead to the same value for the integral. This leads us to create the following definition.

DEFINITION Residue

Let $f(z)$ be analytic on a simple closed contour C and at all points interior to C except for the point z_0. Then the residue of $f(z)$ at z_0, written $\text{Res}[f(z), z_0]$, is defined by

$$\text{Res}[f(z), z_0] = \frac{1}{2\pi i}\oint_C f(z)\,dz. \quad \square \tag{6.1–1}$$

The connection between $\text{Res}[f(z), z_0]$ and a Laurent series for $f(z)$ will soon be apparent. Because z_0 is an isolated singular point of $f(z)$ a Laurent expansion

$$f(z) = \cdots + c_{-2}(z - z_0)^{-2} + c_{-1}(z - z_0)^{-1} + c_0 + c_1(z - z_0) + c_2(z - z_0)^2 + \cdots, \qquad 0 < |z - z_0| < R, \tag{6.1–2}$$

of $f(z)$ about z_0 is possible. This series converges to $f(z)$ at all points (except z_0) within a circle of radius R centered at z_0, i.e., in a deleted neighborhood of z_0.

Now, to evaluate $\text{Res}[f(z), z_0]$ we take as C in Eq. (6.1–1) a circle of radius r centered at z_0. We choose $r < R$, which means that $f(z)$ in Eq. (6.1–1) can be represented by means of the Laurent series in Eq. (6.1–2), and a term-by-term integration is possible. Thus

$$\begin{aligned}\text{Res}[f(z), z_0] &= \frac{1}{2\pi i}\oint_{|z-z_0|=r} \sum_{n=-\infty}^{\infty} c_n(z - z_0)^n \, dz \\ &= \frac{1}{2\pi i}\sum_{n=-\infty}^{\infty} c_n \oint_{|z-z_0|=r} (z - z_0)^n \, dz.\end{aligned} \tag{6.1–3}$$

It is now helpful to take note of Eq. (4.3–10).

$$\oint_{|z-z_0|=r} (z - z_0)^n \, dz = \begin{cases} 0, & n \neq -1, \\ 2\pi i, & n = -1. \end{cases} \tag{6.1–4}$$

Thus all the integrals on the right side of Eq. (6.1–3) have the value zero except the one for which $n = -1$. We then have

$$\text{Res}[f(z), z_0] = c_{-1}. \tag{6.1–5}$$

The result contained in this equation is extremely important and is summarized in the following theorem.

THEOREM 1 The residue of the function $f(z)$ at the isolated singular point z_0 is the coefficient of $(z - z_0)^{-1}$ in the Laurent series representing $f(z)$ in an annulus $0 < |z - z_0| < R$. □

The term "residue," meaning "that which is left over," seems particularly appropriate when applied to Eqs. (6.1–1) and (6.1–5). When a valid Laurent series for $f(z)$ is used in Eq. (6.1–1) and the integration is performed term by term, all that remains is a particular coefficient in the series. An alternative derivation of Eq. (6.1–5) is given in Exercise 9 of this section.

EXAMPLE 1

Let

$$f(z) = \frac{1}{(z)(z - 1)}.$$

Find, using Theorem 1, $(1/2\pi i) \oint_C f(z)\, dz$, where C is the contour shown in Fig. 6.1–1.

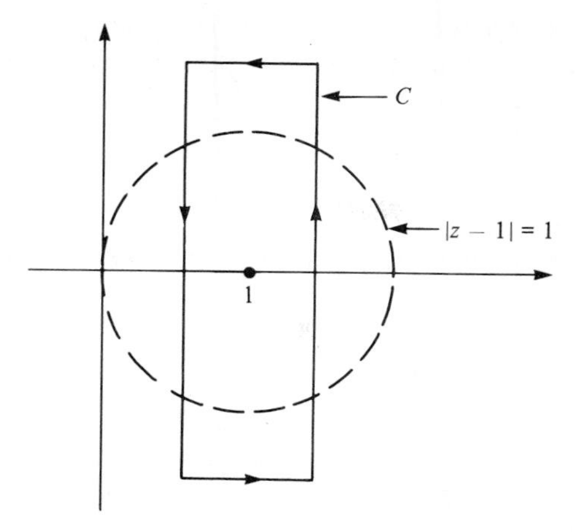

Figure 6.1–1

Solution

Note that $f(z)$ is analytic on C and at all points inside C except $z = 1$, which is an isolated singularity. Thus Theorem 1 is applicable. In Example 3 of Section 5.6 we considered two Laurent series expansions of $f(z)$ that converge in annular regions centered at $z = 1$:

$$\frac{1}{z(z-1)} = (z-1)^{-1} - 1 + (z-1) - (z-1)^2 + \cdots, \qquad 0 < |z-1| < 1; \tag{6.1–6}$$

$$\frac{1}{z(z-1)} = (z-1)^{-2} - (z-1)^{-3} + (z-1)^{-4} - \cdots, \qquad |z-1| > 1. \tag{6.1–7}$$

The series of Eq. (6.1–6), which applies around the point $z = 1$, is of use to us here. The coefficient of $(z-1)^{-1}$ is 1, which means that

$$\operatorname{Res}\left[\frac{1}{z(z-1)}, 1\right] = 1.$$

Thus

$$\frac{1}{2\pi i}\oint_C f(z)\,dz = 1.$$

Observe that the contour C lies in part outside the domain in which the Laurent series of Eq. (6.1–6) correctly represents $f(z)$ (see Fig. 6.1–1).

However, contour C can legitimately be deformed into a circle, centered at $z = 1$ and lying within the domain $0 < |z-1| < 1$. It is around this circle that the term-by-term integration, leading to our final result, can be applied.

Comment: If we change the previous example so that the contour C now encloses only the singular point $z = 0$ (for example, C is $|z| = 1/2$), then our solution would require that we extract the residue at $z = 0$ from the expansion

$$\frac{1}{(z)(z-1)} = -z^{-1} - 1 - z - z^2 - \cdots, \qquad 0 < |z| < 1,$$

which the reader should confirm. The required residue, at $z = 0$, is -1. Thus for this new contour

$$\frac{1}{2\pi i}\oint_C \frac{dz}{z(z-1)} = -1.$$

◀

The preceding example could have been solved without recourse to residues. In fact, the Cauchy integral formula (see Section 4.5) could have yielded the answers more quickly. However, we will soon be dealing with more difficult integrations that can be performed only with residue calculus, as in the following example.

EXAMPLE 2

Find $(1/2\pi i)\oint_C z\sin(1/z)\,dz$ integrated around $|z| = 2$.

Solution

The point $z = 0$ is an isolated singularity of $\sin(1/z)$ and lies inside the given contour of integration. We require a Laurent expansion of $z\sin(1/z)$ about this point. From Eq. (5.4–21), with $1/z$ substituted for z, we can obtain

$$z\sin\left(\frac{1}{z}\right) = 1 - \frac{\left(\frac{1}{z}\right)^2}{3!} + \frac{\left(\frac{1}{z}\right)^4}{5!} - \cdots, \qquad |z| > 0.$$

Since the coefficient $(1/z)$ in the preceding series is zero, we have

$$\operatorname{Res}\left[\left(z\sin\frac{1}{z}\right), 0\right] = 0.$$

We see that a function can have a singularity at a point and possess a residue of zero there. The value of the given integral is thus zero. ◀

Thus far we have used residues to evaluate only those contour integrals whose path of integration encloses one isolated singularity of the integrand. Theorem 2 enables us to use residue calculus to evaluate integrals when more than one isolated singularity is enclosed.

THEOREM 2 Residue theorem

Let C be a simple closed contour and let $f(z)$ be analytic on C and at all points inside C except for isolated singularities at $z_1, z_2, \ldots, z_n$. Then

$$\frac{1}{2\pi i}\oint_C f(z)\,dz = \operatorname{Res}[f(z), z_1] + \operatorname{Res}[f(z), z_2] + \cdots + \operatorname{Res}[f(z), z_n],$$

which is more neatly written

$$\oint_C f(z)\,dz = 2\pi i \sum_{k=1}^{n} \text{Res}[f(z), z_k]. \quad \square \tag{6.1–8}$$

Thus the integral of $f(z)$ around C is $2\pi i$ times the sum of the residues of $f(z)$ inside C.

To prove the residue theorem, we first surround each of the singularities in C by circles $C_1, C_2, \ldots, C_n$ that intersect neither each other nor C (see Fig. 6.1–2a). A set of paths, illustrated with broken lines, is then drawn connecting $C, C_1, \ldots, C_n$ as shown. Two simple closed contours, C_U and C_L can then be formed as shown in Fig. 6.1–2(b). The function $f(z)$ is analytic on and inside C_U and C_L. Hence, from the Cauchy integral theorem,

$$\frac{1}{2\pi i}\oint_{C_U} f(z)\,dz = 0, \qquad \frac{1}{2\pi i}\oint_{C_L} f(z)\,dz = 0.$$

We now add these two expressions:

$$\frac{1}{2\pi i}\left[\oint_{C_U} f(z)\,dz + \oint_{C_L} f(z)\,dz\right] = 0. \tag{6.1–9}$$

Note that those portions of the integral along C_U that take place along the paths illustrated with broken lines are exactly canceled by those portions of the integral along C_L that take place along the same path. Cancellation is due to the opposite directions of integration. What remains on the left side of Eq. (6.1–9) is the integral of $f(z)$ taken around C in the positive (counterclockwise) sense, plus the integrals around $C_1, C_2, \ldots, C_n$ in the negative direction. Hence

$$\frac{1}{2\pi i}\left[\oint_{C} f(z)\,dz + \oint_{C_1} f(z)\,dz + \oint_{C_2} f(z)\,dz + \cdots + \oint_{C_n} f(z)\,dz\right] = 0.$$

We can rearrange this as

$$\frac{1}{2\pi i}\oint_{C} f(z)\,dz = \frac{1}{2\pi i}\oint_{C_1} f(z)\,dz + \frac{1}{2\pi i}\oint_{C_2} f(z)\,dz + \cdots + \frac{1}{2\pi i}\oint_{C_n} f(z)\,dz, \tag{6.1–10}$$

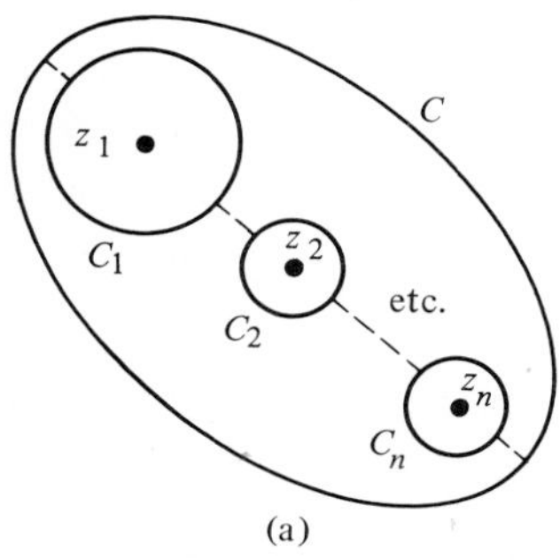

Figure 6.1–2(a)

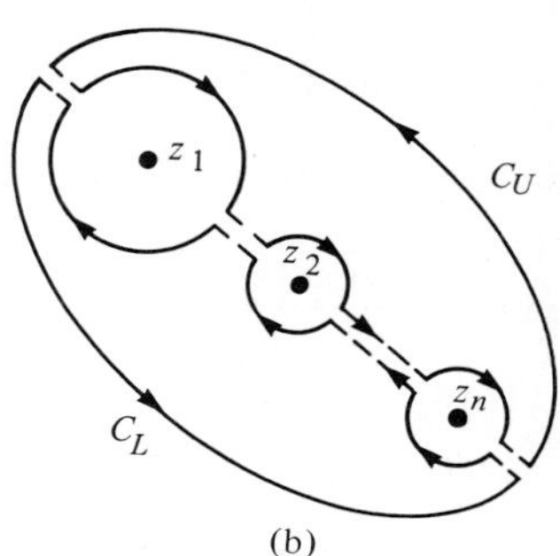

Figure 6.1–2(b)

where all the integrations are now performed in the positive sense. Each of the integrals on the right in Eq. (6.1–10) is taken around an isolated singularity and is numerically equal to the residue of $f(z)$ evaluated at that singularity. Hence

$$\frac{1}{2\pi i}\oint_C f(z)\,dz = \text{Res}[f(z), z_1] + \text{Res}[f(z), z_2] + \cdots + \text{Res}[f(z), z_n].$$

Multiplying both sides of the equation by $2\pi i$, we obtain Eq. (6.1–8). Note that in the summation of Eq. (6.1–8) we include residues at only those singularities inside C.

EXAMPLE 3

Find $\oint_C 1/[(z)(z-1)]\,dz$, where C is the circle $|z-1| = 6$, by means of the residue theorem.

Solution

The contour C encloses the isolated singularities at $z = 1$ and $z = 0$. The residues of $1/[z(z-1)]$ at 1 and 0 were derived in Example 1 and are 1 and -1, respectively. Applying Eq. (6.1–8), we see that since the sum of these residues is zero, the value of the given integral is zero. ◀

EXERCISES

Evaluate the following integrals using residues, i.e., follow the methods of Examples 1 or 3.

1. $\displaystyle\oint_{|z|=2} \frac{dz}{(z-1)(z+3)}$ **2.** $\displaystyle\oint_{|z|=4} \frac{dz}{(z-1)(z+3)}$

3. $\displaystyle\oint_{|z|=3} \frac{dz}{z^2-1}$

Evaluate the following integrals by the method of residues. Where necessary use Laurent expansions about singular points to obtain the residue.

4. $\displaystyle\oint \sum_{n=0}^{\infty} \frac{1}{(z-i)^{n-4}} \frac{\sqrt{n}}{n!}\,dz$ around $|z+i| = 3$

5. $\displaystyle\oint z^3 \cos(1/z)\,dz$ around $|z+1+i| = 4$

6. $\displaystyle\oint \frac{\sin z - \sinh z}{z^8}\,dz$ around $|z| = 1$

7. $\displaystyle\oint \frac{\sin z}{z \sinh z}\,dz$ around $|z| = 1$

8. $\displaystyle\oint \frac{dz}{\text{Log}\, z - 1}$ around $|z - e| = 1$

9. Show how the result in Eq. (6.1–5), which relates the residue to a particular coefficient in a Laurent series, can be derived through the use of Eq. (5.6–5) and the definition of the residue shown in Eq. (6.1–1).

10. Use residue calculus to show that if $n \geq 1$ is an integer, then

$$\oint_C \left(z + \frac{1}{z}\right)^n dz = \begin{cases} \dfrac{2\pi i n!}{\left(\dfrac{n-1}{2}\right)!\left(\dfrac{n+1}{2}\right)!}, & n \text{ odd}, \\ 0, & n \text{ even}, \end{cases}$$

where C is any simple closed contour encircling the origin.

Hint: Use the binomial theorem.

11. We wish to evaluate

$$\oint_{\substack{|z|=R \\ R>1}} (z^2 - 1)^{1/2}\, dz,$$

where we employ a branch of the integrand defined by a straight branch cut connecting $z = 1$ and $z = -1$, and $(z^2 - 1)^{1/2} > 0$ on the line $y = 0$, $x > 1$. Note that the singularities enclosed by the path of integration are not isolated.

a) Show that $(z^2 - 1)^{1/2} = z(1 - 1/z^2)^{1/2} = z - 1/(2z) + \cdots$, $|z| > 1$.

Hint: Consider $(1 + w)^{1/2} = \sum_{n=0}^{\infty} c_n w^n$, $|w| < 1$, let $w = -1/z^2$.

b) Evaluate the given integral by a term-by-term integration of the Laurent series found in part (a).

12. Use the technique of Exercise 11 to evaluate $\oint 1/(z^2 - 1)^{1/2}\, dz$, where the same contour of integration and branch of $(z^2 - 1)^{1/2}$ are used as in Exercise 11.

6.2 ISOLATED SINGULARITIES

From the previous section it should be apparent that when $f(z)$ has an isolated singular point at z_0, it is useful to know the coefficient of $(z - z_0)^{-1}$ in the Laurent expansion about that point. In this section we do some preliminary work that will allow us, in Section 6.3, to often find this coefficient without obtaining the Laurent series.

Kinds of Isolated Singularities

Let $\sum_{n=-\infty}^{\infty} c_n(z - z_0)^n$ be the Laurent expansion of $f(z)$ about the isolated singular point z_0. We have

$$f(z) = \cdots + c_{-2}(z - z_0)^{-2} + c_{-1}(z - z_0)^{-1} + c_0 + c_1(z - z_0)^1 + \cdots.$$

DEFINITION Principal part

That portion of the Laurent series containing only the negative powers of $(z - z_0)$ is known as the *principal part.* □

We now distinguish among three different kinds of principal parts.

a) A principal part with a *finite number* of nonzero terms. Since the number of terms in the principal part is neither zero nor infinite it takes the form

$$c_{-N}(z - z_0)^{-N} + c_{-(N-1)}(z - z_0)^{-(N-1)} + \cdots + c_{-1}(z - z_0)^{-1},$$

where $c_{-N} \neq 0$. The most negative power of $(z - z_0)$ in the principal part is $-N$, where $N > 0$. Thus we arrive at the following definition.

DEFINITION Pole of order N

A function whose Laurent expansion about a singular point z_0 has a principal part, in which the most negative power of $(z - z_0)$ is $-N$, is said to have a *pole of order* N at z_0. □

The function $f(z) = 1/[(z)(z - 1)^2]$ has singularities at $z = 0$ and $z = 1$. The Laurent expansions about these points are

$$f(z) = z^{-1} + 2 + 3z + 4z^2 + \cdots, \qquad 0 < |z| < 1$$

and

$$f(z) = (z - 1)^{-2} - (z - 1)^{-1} + 1 - (z - 1) + (z - 1)^2 + \cdots, \qquad 0 < |z - 1| < 1.$$

The first series reveals that $f(z)$ has a pole of order 1 at $z = 0$, while the second shows a pole of order 2 at $z = 1$.

A function possessing a pole of order 1, at some point, is said to have a *simple pole* at that point. Some discussion of the term "pole" is given at the end of this section.

b) There are an *infinite number* of nonzero terms in the principal part. Unlike the case just described we are now unable to find in the principal part a nonzero term $C_{-N}(z - z_0)^{-N}$ containing the most negative power of $(z - z_0)$.

DEFINITION Isolated essential singularity

A function, whose Laurent expansion about the isolated singular point z_0 contains an infinite number of nonzero terms in the principal part, is said to have an *isolated essential singularity* at z_0. □

We will usually delete the word "isolated" and simply say essential singularity.†

Transcendental functions defined in terms of exponentials (sine, cosh, etc.) can exhibit this behavior when their arguments become infinite. For example, from

† There is a kind of nonisolated singular point called a *nonisolated essential singularity.* An example is given in Exercise 29. Because we will rarely encounter this kind of singularity it is convenient to use "essential singularity" to mean an isolated essential singularity.

Eq. (5.4–21), with z replaced by $1/z$, we have

$$\sin\frac{1}{z} = z^{-1} - \frac{z^{-3}}{3!} + \frac{z^{-5}}{5!} + \cdots, \qquad z \neq 0, \tag{6.2–1}$$

which shows that $\sin(1/z)$ has an essential singularity at $z = 0$.

Another example is $e^{1/(z-1)}/(z-1)^2$. Using $e^u = 1 + u + u^2/2! + \cdots$, and putting $u = (z-1)^{-1}$, we find that

$$\frac{e^{1/(z-1)}}{(z-1)^2} = (z-1)^{-2} + (z-1)^{-3} + \frac{(z-1)^{-4}}{2!} + \cdots, \qquad z \neq 1, \tag{6.2–2}$$

which shows that the given function has an essential singularity at $z = 1$.

c) There are functions possessing an isolated singular point z_0 such that a sought after Laurent expansion about z_0 will be found to have *no terms* in the principal part, that is, all the coefficients involving negative powers of $(z - z_0)$ are zero. In fact, a Taylor series is obtained. In these cases it is found that the singularity exists because the function is undefined at z_0 or defined so as to create a discontinuity. By properly defining $f(z)$ at z_0 the singularity is removed.

DEFINITION Removable singular point

When a singularity of a function $f(z)$ at z_0 can be removed by suitably defining $f(z)$ at z_0, we say that $f(z)$ has a *removable singular point* at z_0. □

One example of the preceding is $f(z) = \sin z/z$, which is undefined at $z = 0$. Since $\sin z = z - z^3/3! + z^5/5! - \cdots$, $|z| < \infty$, we have

$$\frac{\sin z}{z} = 1 - \frac{z^2}{3!} + \frac{z^4}{5!} - \cdots.$$

Because $\sin 0/0$ is undefined, the function on the left in this equation possesses a singular point at $z = 0$. The Taylor series on the right represents an analytic function everywhere inside its circle of convergence (the entire z-plane). The value of the series on the right, at $z = 0$, is 1. By defining $f(0) = 1$, we obtain a function

$$f(z) = \begin{cases} \dfrac{\sin z}{z}, & z \neq 0, \\ 1, & z = 0, \end{cases}$$

which is analytic for all z. The singularity of $f(z)$ at $z = 0$ has been removed by an appropriate definition of $f(0)$. This value could also have been obtained by evaluating $\lim_{z\to 0} \sin z/z$ with L'Hôpital's rule. (See Section 2.4.) Since we require continuity at $z = 0$, this limit must equal $f(0)$.

Establishing the Nature of the Singularity

When a function $f(z)$ possesses an essential singularity at z_0, the only means we have for obtaining its residue there is to use the Laurent expansion about this point and pick out the appropriate coefficient. Thus from Eq. (6.2–1) we see that $\text{Res}[\sin 1/z, 0] = 1$

(the coefficient of z^{-1}) while Eq. (6.2–2) shows that

$$\operatorname{Res}\left[\frac{e^{1/(z-1)}}{(z-1)^2}, 1\right] = 0,$$

which is the coefficient of $(z-1)^{-1}$.

If however a function has a pole singularity at z_0, we need not obtain the entire Laurent expansion about z_0 in order to find the one coefficient in the series that we actually need. Provided we know that the singularity is a pole, there are a variety of techniques open to us. Furthermore, finding the residue is made easier by our first knowing the order of the pole. We will now find some rules for doing this.

Let $f(z)$ have a pole of order N at $z = z_0$. Then

$$f(z) = c_{-N}(z-z_0)^{-N} + c_{-(N-1)}(z-z_0)^{-(N-1)} + \cdots + c_0 + c_1(z-z_0) + \cdots, \tag{6.2–3}$$

where $c_{-N} \neq 0$. The above expansion is valid in a deleted neighborhood of z_0. Multiplying both sides by $(z-z_0)^N$, we have

$$(z-z_0)^N f(z) = c_{-N} + c_{-(N-1)}(z-z_0) + \cdots + c_0(z-z_0)^N + c_1(z-z_0)^{N+1} + \cdots. \tag{6.2–4}$$

From the preceding we have

$$\lim_{z\to z_0} [(z-z_0)^N f(z)] = c_{-N}. \tag{6.2–5}$$

Since $|(z-z_0)^N| \to 0$ as $z \to z_0$, Eq. (6.2–5) shows us that $\lim_{z\to z_0}|f(z)| \to \infty$, that is, *if $f(z)$ has a pole at z_0 then $|f(z)|$ is unbounded as $z \to z_0$.*

The function $f(z) = \sinh z/z$ does not have a pole at $z = 0$. To see this we apply L'Hôpital's rule to evaluate $\lim_{z\to 0} f(z)$. We obtain $\lim_{z\to 0}\cosh z/1 = 1$. In fact, $f(z)$ has a removable singularity at $z = 0$.

A function $f(z)$ having an essential singularity at z_0 does not have a limit of ∞ as $z \to z_0$. For example, in the case of $e^{1/z}$, if we approach the origin along the line $y = 0$, $x > 0$, we find that $f(z) = e^{1/x}$ becomes unbounded as $x \to 0$. However, if we approach the origin along the line $x = 0$, we have $f(z) = e^{1/iy} = \cos(y^{-1}) - i\sin(y^{-1})$, which is a complex number of modulus 1 for all y.

Analytic branches of some multivalued functions, such as $\log z$ and $1/(z-1)^{1/2}$ have moduli that become infinite at their singular points (in these examples at $z = 0$ and $z = 1$, respectively). However, these singular points are not poles but branch points. A function that "blows up" at a point does not necessarily have a pole there. However, a function that has a limit of ∞ at an *isolated* singular point does have a pole at that point. The following pair of rules, based on Eqs. (6.2–3) and (6.2–5), are useful in establishing the existence of a pole and its order. To establish the second rule we must multiply Eq. (6.2–3) by $(z-z_0)^n$.

RULE I Let z_0 be an isolated singular point of $f(z)$. If $\lim_{z\to z_0}(z-z_0)^N f(z)$ exists and if this limit is neither zero nor infinity, then $f(z)$ has a pole of order N at z_0. □

RULE II If N is the order of the pole of $f(z)$ at z_0, then

$$\lim_{z\to z_0}(z-z_0)^n f(z) = \begin{cases} 0, & n > N, \\ \infty, & n < N. \end{cases} \quad \square \tag{6.2–6}$$

EXAMPLE 1

Discuss the singularities of

$$f(z) = \frac{z\cos z}{(z-1)(z^2+1)^2(z^2+3z+2)}.$$

Solution

This function possesses only isolated singularities, and they occur only where the denominator becomes zero. Factoring the denominator we have

$$f(z) = \frac{z\cos z}{(z-1)(z+i)^2(z-i)^2(z+2)(z+1)}.$$

There is a pole of order 1 (simple pole) at $z = 1$ since

$$\lim_{z\to 1}[(z-1)f(z)] = \lim_{z\to 1}\frac{(z-1)z\cos z}{(z-1)(z+i)^2(z-i)^2(z+2)(z+1)} = \frac{\cos 1}{24},$$

which is finite and nonzero.

There is a second-order pole at $z = -i$ since

$$\lim_{z\to -i}[(z+i)^2 f(z)] = \lim_{z\to -i}\frac{(z+i)^2 z\cos z}{(z-1)(z+i)^2(z-i)^2(z+2)(z+1)}$$
$$= \frac{-i\cos(-i)}{8(2-i)}.$$

Similarly, there is a pole of order 2 at $z = i$ and poles of order 1 at -2 and -1. ◀

EXAMPLE 2

Discuss the singularities of

$$f(z) = \frac{e^z}{\sin z}.$$

Solution

Wherever $\sin z = 0$, that is, for $z = k\pi$, $k = 0, \pm 1, \pm 2, \ldots, f(z)$ has isolated singularities.

Assuming these are simple poles, we evaluate $\lim_{z\to k\pi}[(z-k\pi)e^z/\sin z]$. This indeterminate form is evaluated from L'Hôpital's rule and equals

$$\lim_{z\to k\pi}\frac{(z-k\pi)e^z + e^z}{\cos z} = \frac{e^{k\pi}}{\cos k\pi}.$$

Because this result is finite and nonzero, the pole at $z = k\pi$ is of first order.

Had this result been infinite, we would have recognized that the order of the pole exceeded 1, and we might have investigated $\lim_{z\to k\pi}(z - k\pi)^2 f(z)$, etc. On the other hand, had our result been zero, we would have concluded that $f(z)$ had a removable singularity at $z = k\pi$. ◀

Problems like the one discussed in Example 2, in which we must investigate the order of the pole for a function of the form $f(z) = g(z)/h(z)$, are so common that we will give them some special attention.

If $g(z)$ and $h(z)$ are analytic at z_0, with $g(z_0) \neq 0$ and $h(z_0) = 0$, then $f(z)$ will have an isolated singularity at z_0.† With these assumptions we expand $h(z)$ in a Taylor series about z_0.

Let

$$h(z) = a_N(z - z_0)^N + a_{N+1}(z - z_0)^{N+1} + \cdots.$$

The leading term, that is, the one containing the lowest power of $(z - z_0)$, is $a_N(z - z_0)^N$, where $a_N \neq 0$ and $N \geq 1$. Recall from Section 5.7 that $h(z)$ has a *zero of order* N at z_0.

Rewriting our expression for $f(z)$ by using the series for $h(z)$, we have

$$f(z) = \frac{g(z)}{a_N(z - z_0)^N + a_{N+1}(z - z_0)^{N+1} + \cdots}.$$

To show that this expression has a pole of order N at z_0, consider

$$\lim_{z\to z_0}[(z - z_0)^N f(z)] = \lim_{z\to z_0}\frac{(z - z_0)^N g(z)}{a_N(z - z_0)^N + a_{N+1}(z - z_0)^{N+1} + \cdots}$$

$$= \lim_{z\to z_0}\frac{g(z)}{a_N + a_{N+1}(z - z_0) + \cdots} = \frac{g(z_0)}{a_N}.$$

Since this limit is finite and nonzero, we may conclude the following rule.

RULE I Quotients

If $f(z) = g(z)/h(z)$, where $g(z)$ and $h(z)$ are analytic at z_0, and if $h(z_0) = 0$ and $g(z_0) \neq 0$, then the order of the pole of $f(z)$ at z_0 is identical to the order of the zero of $h(z)$ at this point. □

The preceding procedure can be modified to deal with the case $g(z_0) = 0$, $h(z_0) = 0$. Under these conditions if $\lim_{z\to z_0} f(z) = \lim_{z\to z_0}[g(z)/h(z)]$ is infinite, then $f(z)$ has a pole at z_0, whereas if the limit is finite, $f(z)$ has a removable singularity at z_0. L'Hôpital's rule is often useful in finding the limit.

If there is a pole, and we want its order, we might expand both $g(z)$ and $h(z)$ in Taylor series about z_0. This would establish the order of the zeros of $g(z)$ and $h(z)$ at z_0. Then, as is shown in Exercise 10 of this section, the following rule applies.

---

† Recall from Section 5.7 that the zeros of an analytic function are isolated, that is, every zero has some neighborhood containing no other zero. Thus a zero appearing in a denominator creates an isolated singularity.

RULE II Quotients

The order of the pole of $f(z) = g(z)/h(z)$ at z_0 is the order of the zero of $h(z)$ at this point less the order of the zero of $g(z)$ at the same point. □

The number found from this rule must be positive, otherwise there would be no pole.

EXAMPLE 3

Find the order of the pole of $(z^2 + 1)/(e^z + 1)$ at $z = i\pi$.

Solution

With $g(z) = (z^2 + 1)$ and $h(z) = (e^z + 1)$ we verify that $g(i\pi) = -\pi^2 + 1 \neq 0$, and $h(i\pi) = e^{i\pi} + 1 = 0$.

To find the order of the zero of $(e^z + 1)$ at $z = i\pi$, we make the Taylor expansion

$$h(z) = e^z + 1 = c_0 + c_1(z - i\pi) + c_2(z - i\pi)^2 + \cdots.$$

Note that $c_0 = 0$ because $h(i\pi) = 0$. Since

$$c_1 = \frac{d}{dz}(e^z + 1)\bigg|_{z = i\pi} = -1,$$

which is nonzero, we see that $h(z)$ has a zero of order 1 at $z = i\pi$. Thus by our Rule I, $f(z)$ has a pole of order 1 at $z = i\pi$. ◀

EXAMPLE 4

Find the order of the pole of

$$f(z) = \frac{\sinh z}{\sin^5 z}, \qquad \text{at } z = 0.$$

Solution

With $g(z) = \sinh z$ and $h(z) = \sin^5 z$ we find that $g(0) = 0$, and $h(0) = 0$.

Because $\sinh z = z + z^3/3! + z^5/5! + \cdots$, we see that $g(z)$ has a zero of order 1 at $z = 0$. Since $(\sin z)^5 = (z - z^3/3! + z^5/5! + \cdots)^5$, we see that the lowest power of z in the Maclaurin series for $\sin^5 z$ is 5. Thus $(\sin z)^5$ has a zero of order 5 at $z = 0$. The order of the pole of $f(z)$ at $z = 0$ is, by Rule II, the order of the zero of $(\sin z)^5$ less the order of the zero of $\sinh z$, that is, $5 - 1 = 4$. ◀

EXAMPLE 5

Find the poles and establish their order for the function

$$f(z) = \frac{1}{(\text{Log } z + i\pi)(z^{1/2} - 1)}.$$

Use the principal branch of $z^{1/2}$.

Solution

Note that the principal branch of the logarithm is also being used. Referring to Sections 3.5 and 3.8 we recall that both the principal branch of $z^{1/2}$ and $\operatorname{Log} z$ are analytic in the cut plane defined by the branch cut $y = 0$, $-\infty \le x \le 0$ (see Fig. 3.5–2). This cut plane is the domain of analyticity of $f(z)$.

Where are the singularities of $f(z)$ in this domain? If $\operatorname{Log} z + i\pi = 0$, we have $\operatorname{Log} z = -i\pi$, or $z = e^{-i\pi} = -1$. This condition cannot occur in the domain. Alternatively we can say that $z = -1$ is not an isolated singular point of $f(z)$ since it lies on the branch cut containing all the nonisolated singular points of $f(z)$.

Consider $z^{1/2} - 1 = 0$, or $z^{1/2} = 1$. Squaring, we get $z = 1$. Since the principal value of $1^{1/2}$ is 1, we see that $f(z)$ has an isolated singular point at $z = 1$. Now

$$z^{1/2} - 1 = \sum_{n=0}^{\infty} c_n(z-1)^n.$$

We readily find that $c_0 = 0$ and $c_1 = [(1/2)/z^{1/2}]_{z=1}$, or $c_1 = 1/2$. This shows that $z^{1/2} - 1$ has a zero of order 1 at $z = 1$.

Since

$$f(z) = \frac{\left(\dfrac{1}{\operatorname{Log} z + i\pi}\right)}{z^{1/2} - 1}$$

and the numerator of this expression is analytic at $z = 1$ while the denominator $z^{1/2} - 1$ has a zero of order 1 at the same point, the given $f(z)$ must, by Rule I, have a simple pole at $z = 1$. ◀

Comment on the term "pole": To the lay reader the word "pole" might well suggest a narrow cylinder protruding into the air from the ground. There is a connection between this colloquial definition and the mathematical one. We have seen that if $f(z)$ has a pole at z_0 then $\lim_{z \to z_0} f(z) = \infty$. Equivalently $|f(z)|$ becomes unbounded as z approaches z_0. Thus a three-dimensional plot showing $|f(z)|$ as a function of x and y would create a surface rising to a peak of infinite height as z approaches z_0. The relationship of this behavior to the conventional meaning of the word pole should be obvious.

Suppose $f(z)$ has a pole of order N at z_0. Dividing both sides of Eq. (6.2–4) by $(z - z_0)^N$ we have this representation of $f(z)$ in a deleted neighborhood of z_0:

$$f(z) = \psi(z)/(z - z_0)^N,$$

where

$$\psi(z) = c_{-N} + c_{-(N-1)}(z - z_0) + c_{-(N-2)}(z - z_0)^2 + \cdots$$

and $c_{-N} \neq 0$.

Thus as z approaches z_0 we have

$$f(z) \approx \frac{\psi(z_0)}{(z - z_0)^N} = \frac{c_{-N}}{(z - z_0)^N},$$

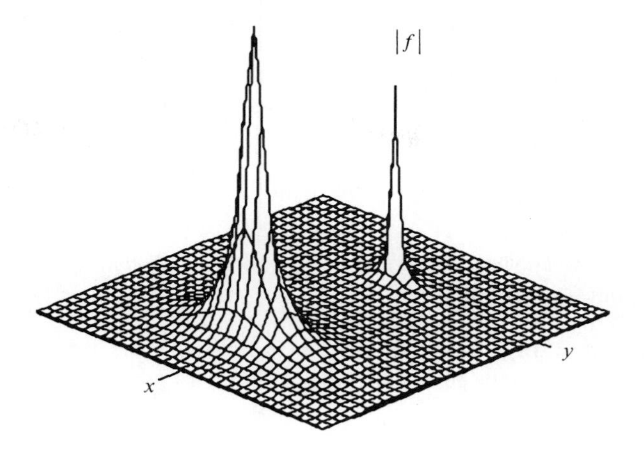

Figure 6.2–1

and so

$$|f(z)| \approx \frac{|c_{-N}|}{|(z - z_0)|^N}.$$

We see that the higher the order N of the pole the more steeply will the surface depicting $|f(z)|$ rise as $z \to z_0$ (compare $1/|z|$ with $1/|z|^4$ for small $|z|$).

The preceding principle can be studied with the aid of a desk top computer. In Fig. 6.2–1 we have plotted, with the aid of a Macintosh computer, $|f(z)| = |1/z(z - 2)^2|$. Now $f(z)$ has poles of order 1 at $z = 0$ and order 2 at $z = 2$.

We see from the figure that the pole at $z = 2$ (which is in the left foreground) causes a more rapid rise in the surface than does the one at $z = 0$ (which is in the right background). The surface should in theory rise to infinity at both poles. In both cases the actual plot has been leveled off at a finite but large value to facilitate display on the computer screen. The appropriateness of the word "pole" should be apparent.

EXERCISES

Show, by means of a Laurent expansion, that the following functions possess essential singularities at the points stated. Give an explicit formula for the nth term in the expansion. State the residue at the singularity.

1. $\cosh(1/z)$ at $z = 0$

2. $ze^{1/(z-1)}$ at $z = 1$

3. $e^{1/(z-1)}\cos[1/(z - 1)]$ at $z = 1$

4. Does $e^{\text{Log } 1/(z-1)}$ have an essential singularity at $z = 1$? Explain.

Use series expansions or L'Hôpital's rule to show that the following functions possess removable singularities at the indicated singular points; that is, show that $\lim_{z-z_0} f(z)$ exists and is finite at each such point. How, in each case, should $f(z_0)$ be defined in order to remove the singularity?

5. $\sin z/\sinh z$ at $z = 0$

6. $(z^2 + \pi^2)\sin z/\sinh z$ at $z = i\pi$

7. $(z^2 + z + 1)/(2z + 1 - i\sqrt{3})$ at the singular point

8. $\sin(z - 1)/\text{Log}\, z$ at $z = 1$

9. a) Let both $g(z)$ and $h(z)$ be analytic and have zeros of order m at z_0. Let $f(z) = g(z)/h(z)$. Show that $\lim_{z\to z_0} f(z) = g^{(m)}(z_0)/h^{(m)}(z_0)$.

Hint: Begin with a Taylor series expansion of both $g(z)$ and $h(z)$. Note that by putting $m = n + 1$ we get the generalization of L'Hôpital's rule mentioned in Section 2.4.

b) Explain why $f(z)$ has a removable singularity at z_0. How should $f(z)$ be defined to remove its singularity?

10. a) Let $f(z) = g(z)/h(z)$, where $g(z) = b_M(z - z_0)^M + b_{M+1}(z - z_0)^{M+1} + \cdots$ has a zero of order M at $z = z_0$ and $h(z) = a_N(z - z_0)^N + a_{N+1}(z - z_0)^{N+1} + \cdots$, which has a zero of order N at $z = z_0$.

b) Show that if $N > M$, $f(z)$ has a pole of order $N - M$ at $z = z_0$.

c) If $N \le M$, how should $f(z_0)$ be defined so that the singularity of $f(z)$ at z_0 is eliminated? Consider the cases $N = M$, $N < M$.

State the locations of all the poles for each of the following functions, and give the order of each pole. Use the principal branch of the given functions.

11. $\dfrac{1}{z^4 - 1}$

12. $\dfrac{z^2 + 1}{(z^4 - 1)^3}$

13. $\dfrac{2z + 1 - i\sqrt{3}}{(z^2 + z + 1)^2}$

14. $\dfrac{z^2 + 3z + 2}{(\sin z)(z + 1)}$

15. $\dfrac{\sin z}{z^{10}(z + 1)}$

16. $\dfrac{1}{(\cosh z) - ae^z}$, where a is real

17. $\dfrac{1}{10^z - e^z}$

18. $\dfrac{\sinh z}{z \sin z}$

19. $\dfrac{1}{(e^z + 1)^4}$

20. $\dfrac{1}{\sin(\pi z/e)[\text{Log}(z) - 1]}$

21. $\dfrac{\sin 1/z}{(z + 1/z)^3}$

22. $\dfrac{\sin(z - i\pi/4)}{e^{2z} - i}$

23. $\dfrac{1}{z^{1/2}\sinh^4 z}$

24. $\dfrac{1}{1 + z^{1/2}}$

25. $\dfrac{1}{(1 - z^{1/2})^4}$

Let $f(z)$ have a pole of order m at z_0, and let $g(z)$ have a pole of order n at z_0.

26. Prove that (fg) has a pole of order $m + n$ at z_0.

27. If $m \neq n$ prove that the order of the pole at z_0 of $(f + g)$ is the greater of m and n.

28. If $m = n$ prove that the order of the pole at z_0 of $(f + g)$ need not be m.

29. A *nonisolated essential singular point* of a function is a singular point whose every neighborhood contains an infinite number of isolated singular points of the function. An example is $f(z) = 1/\sin(1/z)$, whose nonisolated essential singular point is at $z = 0$.

a) Show that in the domain $|z| < \varepsilon$ $(\varepsilon > 0)$, there are poles at $z = \pm 1/n\pi$, $\pm 1/(n+1)\pi$, $\pm 1/(n+2)\pi, \ldots$, where n is an integer such that $n > 1/(\pi\varepsilon)$.

b) Is there a Laurent expansion of $f(z)$ in a deleted neighborhood of $z = 0$?

c) Find a different function with a nonisolated essential singular point at $z = 0$. Prove that it has an infinite number of isolated singular points inside $|z| = \varepsilon$.

6.3 FINDING THE RESIDUE

When a function $f(z)$ is known to possess a pole at z_0, there is a straightforward method for finding its residue at this point that does not entail our obtaining a Laurent expansion about z_0. When the order of the pole is known, the technique involved is even easier.

We have found in the preceding section (see Eq. (6.2–4)) that when $f(z)$ has a pole of order N at z_0, that $\psi(z) = (z - z_0)^N f(z)$ has the following Taylor series expansion about z_0:

$$\psi(z) = (z - z_0)^N f(z) = c_{-N} + c_{-(N-1)}(z - z_0) + \cdots + c_0(z - z_0)^N + \cdots. \tag{6.3–1}$$

Suppose $f(z)$ has a simple pole at $z = z_0$. Then $N = 1$. Hence

$$\psi(z) = c_{-1} + c_0(z - z_0) + c_1(z - z_0)^2 + \cdots. \tag{6.3–2}$$

The residue c_{-1} is easily seen to be $\lim_{z\to z_0} \psi(z) = \lim_{z\to z_0}[(z - z_0)f(z)]$. Hence, we arrive at Rule I.

RULE I Residues

If $f(z)$ has a pole of order 1 at $z = z_0$, then

$$\operatorname{Res}[f(z), z_0] = \lim_{z\to z_0} [(z - z_0)f(z)]. \quad \square \tag{6.3–3}$$

Suppose $f(z)$ has a pole of order 2 at $z = z_0$. We then have, from Eq. (6.3–1),

$$\psi(z) = c_{-2} + c_{-1}(z - z_0) + c_0(z - z_0)^2 + \cdots.$$

We notice that

$$\frac{d\psi}{dz} = c_{-1} + 2c_0(z - z_0) + \cdots,$$

so that

$$c_{-1} = \lim_{z \to z_0} \frac{d}{dz}(\psi(z)) = \lim_{z \to z_0} \frac{d}{dz}[(z - z_0)^2 f(z)],$$

from which we obtain Rule II.

RULE II Residues

If $f(z)$ has a pole of order 2 at $z = z_0$,

$$\text{Res}[f(z), z_0] = \lim_{z \to z_0} \frac{d}{dz}[(z - z_0)^2 f(z)]. \quad \square \tag{6.3–4}$$

The method can be generalized, the result of which is Rule III.

RULE III Residues

If $f(z)$ has a pole of order N at $z = z_0$, then

$$\text{Res}[f(z), z_0] = \lim_{z \to z_0} \frac{1}{(N-1)!} \frac{d^{N-1}}{dz^{N-1}}[(z - z_0)^N f(z)]. \quad \square \tag{6.3–5}$$

Rule II is contained in Rule III (put $N = 2$), as is Rule I if we take $0! = 1$, and $d^{N-1}/dz^{N-1}|_{N=1} = 1$.

If the order of the pole of $f(z)$ at z_0 is known, the application of Eq. (6.3–5) yields the residue directly. If the order is unknown, we might seek to determine its order by means of the methods suggested in Section 6.2. Another possibility is the following method, which is proved in Exercise 36 of this section.

> Guess the order of the pole and use Eq. (6.3–5), taking N as the conjectured value. If the guessed N is less than the actual order, an infinite result, and not the residue, is obtained. However, if this N equals or exceeds the order of the pole, the residue is correctly obtained.

The problem of finding the residue at a pole of first order for a quotient of the form $f(z) = g(z)/h(z)$ occurs so often that we will derive a special formula for this case.

Let us assume that $f(z)$ has a simple pole at z_0 and that $g(z_0) \neq 0$. Thus $h(z)$ has a zero of first order at z_0. Applying Rule I just given,

$$\text{Res}[f(z), z_0] = \lim_{z \to z_0} (z - z_0)\frac{g(z)}{h(z)},$$

which results in the indeterminate form 0/0. Using L'Hôpital's rule, we obtain

$$\lim_{z \to z_0} \frac{(z - z_0)g(z)}{h(z)} = \lim_{z \to z_0} \frac{(z - z_0)g'(z) + g(z)}{h'(z)} = \frac{g(z_0)}{h'(z_0)}.$$

Since $h(z)$ has a zero of order 1 at z_0, $h'(z_0) \neq 0$. We summarize the preceding steps as follows:

RULE IV Residues

The residue of $f(z) = g(z)/h(z)$ at a simple pole, where $g(z_0) \neq 0$, $h(z_0) = 0$, is given by

$$\text{Res}[f(z), z_0] = \frac{g(z_0)}{h'(z_0)}. \quad \square \tag{6.3–6}$$

If $f(z) = g(z)/h(z)$ has a pole of order two at z_0 and $g(z_0) \neq 0$, there is a formula like Eq. (6.3–6) that yields the residue at z_0. It is derived in Exercise 15.

EXAMPLE 1

Find the residue of

$$f(z) = \frac{e^z}{(z^2+1)z^2}$$

at all poles.

Solution

We rewrite $f(z)$ with a factored denominator:

$$f(z) = \frac{e^z}{(z+i)(z-i)z^2},$$

which shows that there are simple poles at $z = \pm i$ and a pole of order 2 at $z = 0$.
From Rule I we obtain the residue at i:

$$\text{Res}[f(z), i] = \lim_{z \to i} \frac{(z-i)e^z}{(z+i)(z-i)z^2} = \frac{e^i}{(2i)(-1)}.$$

The residue at $-i$ could be similarly calculated. Instead, for variety, let's use Rule IV. Taking $g(z) = e^z/z^2$ (which is nonzero at $-i$) and $h(z) = z^2 + 1$, so that $h'(z) = 2z$, we have

$$\text{Res}[f(z), -i] = \left[\frac{e^z/z^2}{2z}\right]_{z=-i} = \frac{e^{-i}}{2i}.$$

Notice that we could also have taken $g(z) = e^z$, $h(z) = z^2(z^2+1)$ and the same result would ultimately be obtained.
The residue at $z = 0$ is computed from Rule II as follows:

$$\text{Res}[f(z), 0] = \lim_{z \to 0} \frac{d}{dz} \frac{z^2 e^z}{(z^2+1)(z^2)} = \lim_{z \to 0} \frac{e^z(z^2+1) - 2ze^z}{(z^2+1)^2} = 1 \quad ◀$$

EXAMPLE 2

Find the residue of

$$f(z) = \frac{\tan z}{z^2 + z + 1}$$

at all singularities of tan z.

Solution

Rewriting $f(z)$ as $\sin z/[(\cos z)(z^2 + z + 1)]$, we see that there are poles of $f(z)$ for z satisfying $\cos z = 0$, that is, $z = \pi/2 + k\pi$, $k = 0, \pm 1, \pm 2, \ldots$. We can show that these are simple poles by expanding $\cos z$ in a Taylor series about the point $z_0 = \pi/2 + k\pi$. We obtain

$$\cos z = a_1(z - z_0) + a_2(z - z_0)^2 + \cdots, \qquad \text{where } a_1 = -\cos k\pi \neq 0.$$

Since $\cos z$ has a zero of order 1 at z_0, $f(z)$ must have a pole of order 1 there.

Let us apply Rule IV, taking $g(z) = \sin z/(z^2 + z + 1)$ and $h(z) = \cos z$, so that $h'(z) = -\sin z$. Thus

$$\begin{aligned} \text{Res}[f(z), \pi/2 + k\pi] &= -\frac{1}{z^2 + z + 1}\bigg|_{\pi/2 + k\pi} \\ &= \frac{-1}{(\pi/2 + k\pi)^2 + (\pi/2 + k\pi) + 1}, \qquad k = 0, \pm 1, \pm 2, \ldots. \end{aligned}$$

Instead of first determining the order of the poles at $z = k\pi + \pi/2$, we might have just assumed that they were of first order and then applied Rule I or Rule IV. The finite result thus obtained would justify our guess.

Comment: There are also poles of $f(z)$, for z satisfying $z^2 + z + 1 = 0$. The roots of this quadratic are $z_1 = -1/2 + i\sqrt{3}/2$ and $z_2 = -1/2 - i\sqrt{3}/2$. Because the roots are distinct, the quadratic expression is a product of *nonrepeated* factors $(z - z_1)(z - z_2)$, and the poles of $f(z)$ at z_1 and z_2 are thus of first order. The residues at these poles can be found from Rule IV. We take $g(z) = \tan z$, and $h(z) = z^2 + z + 1$. The residue at z_1 is $[\tan(-1/2 + i\sqrt{3}/2)]/(i\sqrt{3})$, and the residue at z_2 is $[\tan(-1/2 - i\sqrt{3}/2)]/(-i\sqrt{3})$, which the reader should verify. ◀

EXAMPLE 3

Find the residue of

$$f(z) = \frac{z^{1/2}}{z^3 - 4z^2 + 4z}$$

at all poles. Use the principal branch of $z^{1/2}$.

Solution

We factor the denominator and obtain

$$f(z) = \frac{z^{1/2}}{z(z - 2)^2}.$$

It appears that there is a simple pole at $z = 0$. This is wrong. A pole is an isolated singularity, and $f(z)$ does not have an isolated singularity at $z = 0$. The factor $z^{1/2}$ has a branch point at this value of z that in turn causes $f(z)$ to have a branch point there.

However, $f(z)$ does have a pole of order 2 at $z = 2$. Applying Rule II, we find

$$\text{Res}[f(z), 2] = \lim_{z \to 2} \frac{d}{dz}\left[\frac{(z-2)^2 z^{1/2}}{z(z-2)^2}\right] = \frac{-1}{4(2)^{1/2}},$$

where, because we are using the principal branch of the square root, $2^{1/2}$ is chosen positive. ◀

EXAMPLE 4

Find the residue of

$$f(z) = \frac{e^{1/z}}{1-z}$$

at all singularities.

Solution

Obviously, there is a simple pole at $z = 1$. The residue there, from Rule I, is found to be $-e$. Since

$$e^{1/z} = 1 + z^{-1} + \frac{z^{-2}}{2!} + \cdots$$

has an essential singularity at $z = 0$, this will also be true of $f(z) = e^{1/z}/(1 - z)$.

The residue of $f(z)$ at $z = 0$ is calculable only if we find the Laurent expansion about this point and extract the appropriate coefficient. Since

$$\frac{1}{1-z} = 1 + z + z^2 + \cdots, \qquad |z| < 1,$$

we have

$$\frac{e^{1/z}}{1-z} = (1 + z + z^2 + z^3 + \cdots)\left(1 + z^{-1} + \frac{z^{-2}}{2!} + \cdots\right)$$
$$= \cdots + c_{-2}z^{-2} + c_{-1}z^{-1} + c_0 + \cdots.$$

Our interest is in c_{-1}. If we multiply the two series together and confine our attention to products resulting in z^{-1}, we have

$$c_{-1}z^{-1} = \left[1 + \frac{1}{2!} + \frac{1}{3!} + \cdots\right]z^{-1}.$$

Recalling the definition $e = 1 + 1 + 1/2! + 1/3! + \cdots$, we see that $c_{-1} = e - 1 = \text{Res}[f(z), 0]$. ◀

EXAMPLE 5

Find the residue of

$$f(z) = \frac{e^z - 1}{\sin^3 z} \qquad \text{at } z = 0.$$

Solution

Both numerator and denominator of the given function vanish at $z = 0$. To establish the order of the pole, we will expand both these expressions in Maclaurin series by the usual means.

$$e^z - 1 = z + \frac{z^2}{2!} + \frac{z^3}{3!} + \cdots, \qquad \sin^3 z = z^3 - \frac{z^5}{2} + \cdots.$$

Thus

$$\frac{e^z - 1}{\sin^3 z} = \frac{z + \dfrac{z^2}{2!} + \dfrac{z^3}{3!} + \cdots}{z^3 - \dfrac{z^5}{2} + \cdots}.$$

Since the numerator has a zero of order 1 and the denominator has a zero of order 3, the quotient has a pole of order 2.

To find the residue of $f(z)$ at $z = 0$, we could apply Rule II. However, it will be found, after performing the differentiations, that the expression obtained is indeterminate at $z = 0$. The required limit as $z \to 0$ is found only after successive applications of L'Hôpital's rule—a tedious procedure.

Instead, the quotient of the two series appearing above is expanded in a Laurent series by means of long division. We need proceed only far enough to obtain the term containing z^{-1}. Thus

$$\begin{array}{r} z^{-2} + \dfrac{z^{-1}}{2} + \cdots \\ z^3 - \dfrac{z^5}{2} + \cdots \overline{\Big) z + \dfrac{z^2}{2!} + \dfrac{z^3}{3!} + \cdots,} \end{array}$$

from which we see that the residue is 1/2. ◀

EXERCISES

1. Let $f(z) = g(z) + h(z)$. Prove that the residue of $f(z)$ at z_0 is the sum of the residues of $g(z)$ and $h(z)$ at z_0. Assume z_0 is an isolated singular point of both $g(z)$ and $h(z)$.
2. Can a function have a residue of zero at a simple pole? Can a function have a residue of zero at a higher-order pole? Can a function have a residue of zero at an essential singularity? Explain.

For each of the following functions state the location and order of each pole and find the corresponding residue. Use the principal branch of any multivalued function given below.

3. $\dfrac{e^{2z}}{z^2 - z + 1}$

4. $\dfrac{1}{z^3 + i}$

5. $\dfrac{1}{z^{1/2}(z - i)^3}$

6. $\dfrac{\text{Log } z}{z^4(z - 1)^2}$

7. $\dfrac{1}{(\text{Log } z)(z^2 + 1)^2}$

8. $\dfrac{\sin z - z}{z \sinh z}$

9. $\dfrac{z^8 + 1}{z^4}$

10. $\dfrac{1}{(\text{Log}(z/e) - 1)^2}$

11. $\dfrac{1}{\sin z^2}$

12. $\dfrac{1}{10^z - e^z}$

13. $\dfrac{\cos 1/z}{\sin z}$

14. $\dfrac{1}{e^{2z} + e^z + 1}$

15. a) Consider the analytic function $f(z) = g(z)/h(z)$, having a pole at z_0. Let $g(z_0) \neq 0$, $h(z_0) = h'(z_0) = 0$, $h''(z_0) \neq 0$. Thus $f(z)$ has a pole of second order at $z = z_0$. Show that

$$\text{Res}[f(z), z_0] = \frac{2g'(z_0)}{h''(z_0)} - \frac{2}{3}\frac{g(z_0)h'''(z_0)}{[h''(z_0)]^2}. \tag{6.3–7}$$

Hint: Write down the Taylor series expansion, about z_0, for $g(z)$ and $h(z)$, taking note of which coefficients are zero. Divide the two series using long division and so obtain the Laurent expansion of $f(z)$ about z_0.

b) Use the formula of part (a) to obtain

$$\text{Res}\left[\frac{\cos z}{(\text{Log } z - 1)^2}, e\right].$$

Find the residue of the following functions at the indicated point.

16. $\dfrac{z - 1}{z} e^{1/z}$ at 0

17. $\dfrac{e^{1/z^2}}{1 - z}$ at 0

18. $\dfrac{1}{(z + i)^5}$ at $-i$

19. $\dfrac{\sin z}{(z + i)^5}$ at $-i$

20. $\dfrac{z^{12}}{(z - 1)^{10}}$ at 1

21. $\dfrac{1}{\sinh(2 \text{ Log } z)}$ at i

22. $\dfrac{1}{\cos\left(\dfrac{\pi}{2} e^z + \sin z\right)}$ at 0

23. $\dfrac{1}{\sin[z(e^z - 1)]}$ at 0

24. $\dfrac{\cos(z - 1)}{z^{10}} + \dfrac{2}{z - 1}$ at $z = 1$

25. $\dfrac{\cos(z - 1)}{z^{10}} + \dfrac{2}{z - 1}$ at $z = 0$

26. a) Let $n \geq 1$ be an integer. Show that the n poles of

$$\frac{1}{z^n + z^{n-1} + z^{n-2} + \cdots + 1}$$

are at $\text{cis}(2k\pi/(n + 1))$, $k = 1, 2, \ldots, n$.

Hint for Part (a): Let $P(z) = z^n + z^{n-1} + \cdots + 1$. Show that $P(z)(z - 1) = z^{n+1} - 1$. Thus $P(z) = (z^{n+1} - 1)/(z - 1)$ for $z \neq 1$. Now explain why the n roots of $P(z) = 0$ are the possible values of $1^{1/(n+1)}$ excluding the value at 1.

b) Show that the poles are simple.

c) Show that the residue at $\text{cis}(2k\pi/(n + 1))$ is

$$\frac{\text{cis}\left(\dfrac{2k\pi}{n+1}\right) - 1}{(n + 1)\,\text{cis}\left(\dfrac{2k\pi n}{n+1}\right)}.$$

Use residues to evaluate the following integrals. Use the principal branch of multivalued functions.

27. $\displaystyle\oint \frac{dz}{\sin z}$ around $|z - 6| = 4$

28. $\displaystyle\oint \frac{\sinh 1/z}{z - 1}\,dz$ around $|z| = 2$

29. $\displaystyle\oint \frac{\sin z}{\sinh^2 z}\,dz$ around $|z| = 3$

30. $\displaystyle\oint \frac{dz}{[\text{Log}(\text{Log}\, z) - 1]}$ around $|z - 16| = 5$

31. $\displaystyle\oint \frac{e^{1/z}}{z^2 - 1}\,dz$ around $|z - 1| = 3/2$

32. $\displaystyle\oint \frac{dz}{\sinh z - 2e^z}$ around $|z + 1| = 2$

33. $\displaystyle\oint \frac{dz}{\sin(z^{1/2})}$ around $|z - 9| = 5$

34. $\displaystyle\oint \frac{\text{Log}\, z}{\sin(z^{1/2} - 1)}\,dz$ around $|z - 11| = 10$

35. In Section 5.6 we used long division to obtain the Laurent expansion of $1/\sin z$ in the domain $0 < |z| < \pi$. We observed that a Laurent expansion in the annular domain $\pi < |z| < 2\pi$ is also possible as well as expansions in other rings centered at the origin. Here we use residues to obtain

$$\frac{1}{\sin z} = \sum_{n=-\infty}^{+\infty} c_n z^n \qquad \text{for } \pi < |z| < 2\pi.$$

a) Use Eq. (5.6–5) to show that

$$c_n = \frac{1}{2\pi i} \oint \frac{1}{z^{n+1} \sin z}\, dz,$$

where the integral can be around $|z| = R$, $\pi < R < 2\pi$.

b) Show that

$$c_n = d_n + e_n + f_n,$$

where

$$d_n = \operatorname{Res}\left[\frac{1}{z^{n+1} \sin z}, 0\right],$$

$$e_n = \operatorname{Res}\left[\frac{1}{z^{n+1} \sin z}, \pi\right],$$

$$f_n = \operatorname{Res}\left[\frac{1}{z^{n+1} \sin z}, -\pi\right].$$

c) Show that for $n \le -2$, $d_n = 0$.

d) Show that $e_n + f_n = 0$ when n is even, and $e_n + f_n = -2/\pi^{n+1}$ for n odd.

e) Show that $c_n = 0$ for even n and that $c_{-1} = -1$, $c_1 = -2/\pi^2 + 1/6$, $c_3 = -2/\pi^4 + 7/360$, and $c_n = -2/\pi^{n+1}$ for $n \le -3$.

36. This problem proves the assertion made earlier that if we guess the order of a pole we can use Eq. (6.3–5) to compute the corresponding residue provided we have either guessed correctly or guessed too high. If we guess too low, Eq. (6.3–5) yields infinity in the limit. Let $f(z)$ have a pole of order m at $z = z_0$, so that, about the point z_0, we have the Laurent expansion

$$f(z) = c_{-m}(z - z_0)^{-m} + c_{-(m-1)}(z - z_0)^{-(m-1)} + \cdots.$$

a) Consider $\psi(z) = (z - z_0)^N f(z)$. Suppose $N \ge m$. What is the Taylor expansion for $\psi(z)$ about z_0? Show that

$$\lim_{z \to z_0} \frac{1}{(N-1)!} \frac{d^{N-1}}{dz^{N-1}}[(z - z_0)^N f(z)] = c_{-1} = \operatorname{Res}[f(z), z_0].$$

b) Suppose $1 \le N < m$. Show that $\psi(z)$ has a Laurent expansion about z_0. Show that

$$\lim_{z \to z_0} \frac{1}{(N-1)!} \frac{d^{N-1}}{dz^{N-1}}[(z - z_0)^N f(z)] = \infty.$$

6.4 EVALUATION OF REAL INTEGRALS WITH RESIDUE CALCULUS, I

Real definite integrals of the type $\int_0^{2\pi} R(\sin\theta, \cos\theta)\, d\theta$, where R is a *rational function* of $\sin\theta$ and/or $\cos\theta$ are frequently very difficult to evaluate with the methods of elementary calculus.[†] However, the calculus of residues can give the result in a straightforward manner.

[†] Recall that these functions R are quotients of polynomials in $\sin\theta$ and $\cos\theta$.

An example of such an integral is $\int_0^{2\pi} 1/(2 + \sin\theta)\, d\theta$. Integrals like this occur in Dirichlet problems for the circle solved by the Poisson integral formula (see Section 4.7).

To evaluate all integrals of the form $\int_0^{2\pi} R(\sin\theta, \cos\theta)\, d\theta$, the approach is the same. The given expression is converted into a line integration in the complex z-plane by the following change of variables:

$$z = e^{i\theta}, \qquad dz = e^{i\theta} i\, d\theta$$

so that

$$\begin{aligned} d\theta &= \frac{dz}{iz}, \\ \sin\theta &= \frac{e^{i\theta} - e^{-i\theta}}{2i} = \frac{z - z^{-1}}{2i}, \\ \cos\theta &= \frac{e^{i\theta} + e^{-i\theta}}{2} = \frac{z + z^{-1}}{2}. \end{aligned} \tag{6.4–1}$$

As θ ranges from 0 to 2π, or over any interval of 2π, the point representing $z = \cos\theta + i\sin\theta$ proceeds in the counterclockwise direction around the *unit circle* in the complex z-plane. The contour integral on this circle is evaluated with residue theory.

The method fails if the integrand for the contour integration has pole singularities *on* the unit circle. However, this can occur only if $\int_0^{2\pi} R(\sin\theta, \cos\theta)\, d\theta$ is an improper integral, that is, the rational function $R(\sin\theta, \cos\theta)$ exhibits a vanishing denominator on the interval $0 \le \theta \le 2\pi$.

EXAMPLE 1

Find

$$I = \int_0^{2\pi} \frac{d\theta}{k + \sin\theta}, \qquad k > 1,$$

by using residues.

Solution

With the change of variables suggested by Eq. (6.4–1) we have

$$I = \oint_{|z|=1} \frac{\dfrac{dz}{iz}}{k + \dfrac{z - z^{-1}}{2i}} = \oint_{|z|=1} \frac{2\, dz}{z^2 + 2ikz - 1}.$$

We now examine the integrand on the right for poles. From the quadratic formula we find that $z^2 + 2ikz - 1 = 0$ has roots at $z_1 = i(-k + \sqrt{k^2 - 1})$ and $z_2 = -i(k + \sqrt{k^2 - 1})$. Now recall that the product of the two roots of the general quadratic expression $az^2 + bz + c$ is c/a. In the present case, $z_1 z_2 = -1$, so that $|z_1| = 1/|z_2|$. If one root lies outside the circle $|z| = 1$, the other must lie inside. For $k > 1$ it is obvious that $z_2 = -i(k + \sqrt{k^2 + 1})$ is outside the unit circle. Thus z_1 is inside and it is here that we require the residue. Using Rule IV of the previous section, we have

$$I = \oint \frac{2\,dz}{z^2 + 2ikz - 1} = \frac{4\pi i}{2z + 2ik}\bigg|_{z = i[-k + \sqrt{k^2 - 1}]} = \frac{2\pi}{\sqrt{k^2 - 1}}.$$

Thus

$$\int_0^{2\pi} \frac{d\theta}{k + \sin\theta} = \frac{2\pi}{\sqrt{k^2 - 1}}$$

for $k > 1$. Putting $k = a/b$, where $a > b > 0$, we have

$$\int_0^{2\pi} \frac{d\theta}{a/b + \sin\theta} = \frac{2\pi}{\sqrt{a^2/b^2 - 1}},$$

or

$$\int_0^{2\pi} \frac{d\theta}{a + b\sin\theta} = \frac{2\pi}{\sqrt{a^2 - b^2}},$$

which is a well-known identity. ◀

Functions of the form $\cos n\theta$ and $\sin n\theta$, where n is an integer, are expressible in terms of sums and differences of integral powers of $\cos\theta$ and $\sin\theta$ and are therefore rational functions of $\cos\theta$ and $\sin\theta$. Integrals containing rational expressions in $\cos n\theta$ and $\sin n\theta$ are readily evaluated by the method just discussed. We still take $z = e^{i\theta}$ and use the substitution

$$\cos n\theta = \frac{e^{in\theta} + e^{-in\theta}}{2} = \frac{z^n + z^{-n}}{2}, \qquad \sin n\theta = \frac{e^{in\theta} - e^{-in\theta}}{2i} = \frac{z^n - z^{-n}}{2i}.$$

EXAMPLE 2

Find

$$I = \int_0^{2\pi} \frac{\cos 2\theta}{5 - 4\sin\theta}\,d\theta.$$

Solution

With the substitutions

$$\cos 2\theta = \frac{z^2 + z^{-2}}{2}, \qquad \sin\theta = \frac{z - z^{-1}}{2i}, \qquad d\theta = \frac{dz}{iz},$$

we have

$$I = \oint_{|z|=1} \frac{\dfrac{z^2 + z^{-2}}{2}}{5 - \dfrac{2}{i}(z - z^{-1})}\left(\frac{dz}{iz}\right) = \oint_{|z|=1} \frac{(z^4 + 1)\,dz}{2iz^2[2iz^2 + 5z - 2i]}.$$

There is a second-order pole at $z = 0$. Solving $2iz^2 + 5z - 2i = 0$, we find simple poles at $i/2$ and $2i$. The pole at $2i$ is outside the circle $|z| = 1$ and can be ignored. Thus

$$I = 2\pi i \sum_{\text{res}} \frac{z^4 + 1}{2iz^2[2iz^2 + 5z - 2i]}, \qquad \text{at } z = 0 \text{ and } i/2.$$

From Eq. (6.3–4) we find the residue at $z = 0$:

$$\frac{1}{2i}\frac{d}{dz}\left[\frac{z^4 + 1}{2iz^2 + 5z - 2i}\right]\Bigg|_{z=0} = \frac{-5}{8}i;$$

and from Eq. (6.3–6), the residue at $i/2$:

$$\frac{1}{(2i)\left(\dfrac{i}{2}\right)^2} \frac{\left(\dfrac{i}{2}\right)^4 + 1}{\dfrac{d}{dz}[2iz^2 + 5z - 2i]}\Bigg|_{z=i/2} = \frac{17i}{24}.$$

Thus

$$I = 2\pi i\left(\frac{-5i}{8} + \frac{17i}{24}\right) = \frac{-\pi}{6}.$$

◀

EXERCISES

Using residue calculus, establish the following identities:

1. $\displaystyle\int_0^{2\pi} \frac{d\theta}{k - \sin\theta} = \frac{2\pi}{\sqrt{k^2 - 1}}$ for $k > 1$. Does your result hold for $k < -1$? Explain.

2. $\displaystyle\int_{-\pi}^{\pi} \frac{d\theta}{a + b\cos\theta} = \frac{2\pi}{\sqrt{a^2 - b^2}}$ for $a > b \geq 0$

3. $\displaystyle\int_{-\pi/2}^{3\pi/2} \frac{\cos\theta}{a + b\cos\theta}\,d\theta = \frac{2\pi}{b}\left[1 - \frac{a}{\sqrt{a^2 - b^2}}\right]$ for $a > b > 0$

4. $\displaystyle\int_0^{2\pi} \sin^4\theta\,d\theta = 3\pi/4$

5. $\displaystyle\int_0^{2\pi} \cos^m\theta\,d\theta = \frac{2\pi}{2^m}\frac{m!}{\left[\left(\frac{m}{2}\right)!\right]^2}$ for $m \geq 0$ even.

Show that the preceding integral is zero when m is odd.

6. $\displaystyle\int_0^{2\pi} \frac{d\theta}{(a + b\sin\theta)^2} = \frac{2\pi a}{(\sqrt{a^2 - b^2})^3}$ for $a > b \geq 0$

7. $\displaystyle\int_0^{2\pi} \frac{d\theta}{a + \sin^2\theta} = \frac{2\pi}{\sqrt{a(a+1)}}$ for $a > 0$

8. $\displaystyle\int_{-\pi}^{+\pi} \frac{\cos\theta}{1 - 2a\cos\theta + a^2}\, d\theta = \frac{2\pi}{a(a^2 - 1)}$ for a real, $|a| > 1$

9. $\displaystyle\int_0^{2\pi} \frac{\cos\theta}{1 - 2a\cos\theta + a^2}\, d\theta = \frac{2\pi a}{1 - a^2}$ for a real, $|a| < 1$

10. $\displaystyle\int_0^{2\pi} \frac{\cos n\theta\, d\theta}{\cosh a + \cos\theta} = \frac{2\pi(-1)^n e^{-na}}{\sinh a}$ $n \geq 0$ is an integer, $a > 0$

11. $\displaystyle\int_0^{2\pi} \frac{d\theta}{a^2\sin^2\theta + b^2\cos^2\theta} = \frac{2\pi}{ab}$ for a, b real, $ab > 0$

Evaluate the following integrals. Where necessary, use the periodic or symmetric properties of the integrand to convert the following expressions to integrals over an interval of 2π. In Exercise 12, for example, $\int_0^\pi = 1/2\int_0^{2\pi}$. Why? Evaluate the resulting expression.

12. $\displaystyle\int_0^{\pi} \frac{\cos\theta}{5 + 4\cos\theta}\, d\theta$

13. $\displaystyle\int_{-\pi/2}^{+\pi/2} \frac{\sin\theta}{5 - 4\sin\theta}\, d\theta$

14. $\displaystyle\int_0^{\pi} \sin^5 x \sin 5x\, dx$

15. $\displaystyle\int_{-\pi}^{+\pi} \frac{\sin 2\theta}{5 - 4\sin\theta}\, d\theta$

16. $\displaystyle\int_0^{\pi} \frac{\cos 2\theta}{2 - \cos\theta}\, d\theta$

17. $\displaystyle\int_0^{\pi/2} \frac{\sin^2\theta}{5 + 4\cos^2\theta}\, d\theta$

6.5 EVALUATION OF INTEGRALS, II

In previous work in mathematics and physics the reader has probably encountered "improper" integrals in which one or both limits are infinite, that is, expressions of the form

$$\int_k^{\infty} f(x)\, dx, \qquad \int_{-\infty}^{k} f(x)\, dx, \qquad \int_{-\infty}^{+\infty} f(x)\, dx,$$

where $f(x)$ is a real function of x, and k is a real constant.

Integrals of the first two types are defined in terms of proper integrals (Riemann sums) as follows:

$$\int_k^{\infty} f(x)\, dx = \lim_{R\to\infty} \int_k^{R} f(x)\, dx, \tag{6.5–1}$$

$$\int_{-\infty}^{k} f(x)\, dx = \lim_{R\to\infty} \int_{-R}^{k} f(x)\, dx, \tag{6.5–2}$$

provided the indicated limits exist.

Such improper integrals do not always exist. Thus, for example,

a) $$\int_1^\infty \frac{1}{1+x^2}\,dx = \lim_{R\to\infty}\int_1^R \frac{dx}{1+x^2} = \lim_{R\to\infty}(\arctan R - \arctan 1) = \frac{\pi}{2} - \frac{\pi}{4}$$

exists; however,

b) $$\int_1^\infty \frac{1}{x}\,dx = \lim_{R\to\infty}(\log R - \log 1)$$

fails to exist, as does

c) $$\int_0^\infty \cos x\,dx = \lim_{R\to\infty}\sin R.$$

In case (b) as x increases, the curve $y = 1/x$ does not fall to zero fast enough for the area under the curve to approach a finite limit. In case (c) a sketch of $y = \cos x$ shows that, along the positive x-axis, the total area under this curve has no meaning.

We define an improper integral with two infinite limits by the following equation.

CAUCHY PRINCIPAL VALUE

$$\int_{-\infty}^{+\infty} f(x)\,dx = \lim_{R\to\infty}\int_{-R}^{+R} f(x)\,dx \qquad (6.5\text{–}3)$$

Integrals between $-\infty$ and $+\infty$ are often defined in another, more restrictive way. The definition given in Eq. (6.5–3) is known as the *Cauchy principal value* of the improper integral. A different definition of an integral between these limits, the standard or ordinary value, is considered in Exercise 9 of this section. It is shown there that if the ordinary value exists, it agrees with the Cauchy principal value, and that there are instances where the Cauchy principal value exists and the ordinary value does not. Unless otherwise stated, we will be using Cauchy principal values of integrals having infinite limits.

Now, if $f(x)$ is an odd function, that is, $f(x) = -f(-x)$, we have $\int_{-R}^{+R} f(x)\,dx = 0$ since the area under the curve $y = f(x)$ to the left of $x = 0$ cancels the area to the right of $x = 0$. Thus, from Eq. (6.5–3),

$$\int_{-\infty}^{+\infty} f(x)\,dx = 0, \qquad \text{if } f(x) \text{ is odd}, \qquad (6.5\text{–}4)$$

for the Cauchy principal value of this integral. To illustrate:

$$\int_{-\infty}^{+\infty} \frac{x^3}{x^4+1}\,dx = 0, \qquad \int_{-\infty}^{+\infty} \frac{x}{x^2+1}\,dx = 0, \qquad \int_{-\infty}^{+\infty} x\,dx = 0.$$

When $f(x)$ is an even function of x, we have $f(x) = f(-x)$. Because of the symmetry of $y = f(x)$ about $x = 0$,

$$\int_{-R}^{+R} f(x)\,dx = 2\int_0^R f(x)\,dx.$$

From Eqs. (6.5–1) and (6.5–3) we thus obtain

$$2\int_{0}^{\infty} f(x)\,dx = \int_{-\infty}^{\infty} f(x)\,dx, \qquad \text{if } f(x) \text{ is even.} \tag{6.5–5}$$

To illustrate:

$$2\int_{0}^{\infty} \frac{1}{x^2+1}\,dx = \int_{-\infty}^{+\infty} \frac{dx}{x^2+1}.$$

Let us see, with an example, how residue calculus enables us to find the Cauchy principal value of a real integral taken between $-\infty$ and $+\infty$.

ound the closed contour C (see Fig. 6.5–1) $-R \le x \le R$ and the semicircle $|z| = R$, means that C encloses all the poles of iated u.h.p.).

at all poles in u.h.p.

two parts: an integral along the real axis icircular arc C_1 in the upper half plane.

$$i\sum_{\text{res}} \frac{z^2}{z^4+1}, \qquad \text{at all poles in u.h.p.} \tag{6.5–6}$$

xtreme left becomes the Cauchy principal For $R \to \infty$ we can show that the second iblish this we use the ML inequality (see

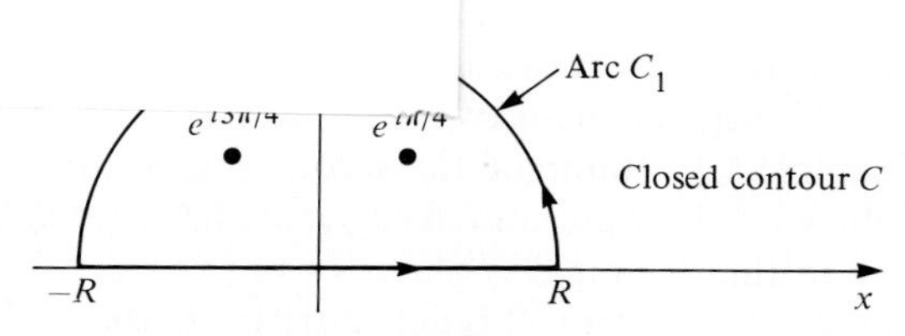

Figure 6.5–1

Section 4.2) and arrive at

$$\left|\int_{C_1} \frac{z^2}{z^4+1}\,dz\right| \le ML = M\pi R, \tag{6.5–7}$$

where $L = \pi R$ is the length of the semicircle C_1.

We require $|z^2/(z^4+1)| \le M$ on C_1. Since $|z| = R$ on this contour, we can instead require that $R^2/|z^4+1| \le M$. By a triangle inequality (see Eq. 1.3–20) $|z^4+1| \ge |z^4| - 1 = R^4 - 1$. Hence, $R^2/|z^4+1| \le R^2/(R^4-1)$. Thus we can put $M = R^2/(R^4-1)$ and use it in Eq. (6.5–7) with the result that

$$\left|\int_{C_2} \frac{z^2}{z^4+1}\,dz\right| \le \frac{\pi R^3}{R^4-1}.$$

As $R \to \infty$, the right side of this equation goes to zero, which means that the integral on the left must also become zero.

Armed with this fact, we put $R \to \infty$ in Eq. (6.5–6). The first integral on the left is the desired Cauchy principal value, the second disappears, and the right side remains unchanged. Thus

$$\int_{-\infty}^{+\infty} \frac{x^2}{x^4+1}\,dx = 2\pi i \sum \operatorname{res} \frac{z^2}{z^4+1}, \qquad \text{at all poles in u.h.p.}$$

The equation $z^4 = -1$ has solutions $e^{i\pi/4}$, $e^{i3\pi/4}$, $e^{-i\pi/4}$, $e^{-i3\pi/4}$, of which only the first two lie in the upper half plane. The residues at the simple poles $e^{i\pi/4}$ and $e^{i3\pi/4}$ are easily found from Eq. (6.3–6) to be $(1/4)e^{-i\pi/4}$ and $(1/4)e^{-i3\pi/4}$, respectively. Thus

$$\int_{-\infty}^{+\infty} \frac{x^2}{x^4+1}\,dx = \frac{2\pi i}{4}\left[e^{-i\pi/4} + e^{-i3\pi/4}\right] = \frac{\pi}{\sqrt{2}}.$$

Because $x^2/(x^4+1)$ is an even function, we have, as a bonus,

$$\int_0^{\infty} \frac{x^2}{x^4+1}\,dx = \frac{1}{2}\frac{\pi}{\sqrt{2}}.$$ ◀

We can solve the problem just considered by using a contour of integration containing a semicircular arc in the lower half plane (abbreviated l.h.p.). Referring to Fig. 6.5–2, we have

$$\int_{-R}^{+R} \frac{x^2}{x^4+1}\,dx + \int_{C_2} \frac{z^2}{z^4+1}\,dz = -2\pi i \sum \operatorname{res} \frac{z^2}{z^4+1}, \qquad \text{at all poles in l.h.p.}$$

Note the minus sign on the right. It arises because the closed contour in Fig. 6.5–2 is being negotiated in the negative (clockwise) sense. We again let $R \to \infty$ and apply the arguments of Example 1 to eliminate the second integral on the left. The reader should sum the residues on the right and verify that the same value is obtained for the integral evaluated in that example.

The technique involved in Example 1 is not restricted to the problem just presented but has wide application in the evaluation of other integrals taken between infinite

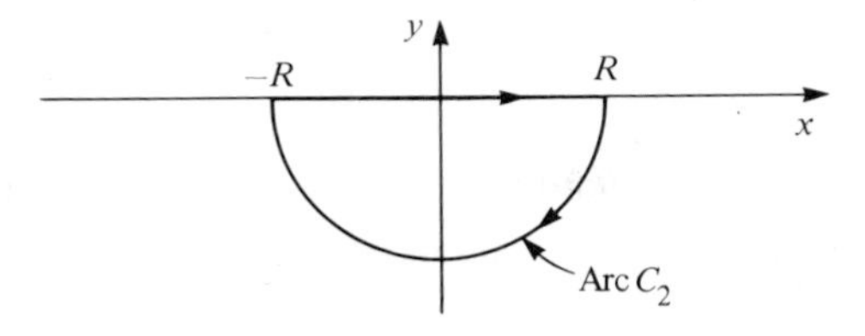

Figure 6.5–2

limits. In all cases we must be able to argue that the integral taken over the arc becomes zero as $R \to \infty$. Theorem 3 is of use in asserting that this is so.

THEOREM 3 Let $f(z)$ have the following property in the half plane $\text{Im } z \geq 0$. There exist constants $k > 1$, R_0, and μ such that

$$|f(z)| \leq \frac{\mu}{|z|^k}, \qquad \text{for all } |z| \geq R_0 \text{ in this half plane.}$$

Then, if C_1 is the semicircular arc $Re^{i\theta}$, $0 \leq \theta \leq \pi$, and $R > R_0$, we have

$$\lim_{R\to\infty} \int_{C_1} f(z)\, dz = 0. \quad \square \tag{6.5–8}$$

The preceding merely says that if $|f(z)|$ falls off more rapidly than the reciprocal of the radius of C_1, then the integral of $f(z)$ around C_1 will vanish as the radius of C_1 becomes infinite. A corresponding theorem can be stated for a contour in the lower half plane.

The proof of Eq. (6.5–8) is simple. Assuming $R > R_0$, we apply the ML inequality as follows:

$$\left|\int_{C_1} f(z)\, dz\right| \leq ML = M\pi R,$$

where $L = \pi R$ is the length of C_1. We require $|f(z)| \leq M$ on C_1. By hypothesis $|f(z)| \leq \mu/|z|^k = \mu/R^k$ on C_1. Thus taking $M = \mu/R^k$ in the ML inequality, we have

$$\left|\int_{C_1} f(z)\, dz\right| \leq \frac{\pi R \mu}{R^k}, \qquad \text{where } k > 1.$$

As $R \to \infty$, the right side of the preceding inequality goes to zero; thus, since the integral over C_1 must also go to zero, the theorem is proved.

Consider the rational function

$$\frac{P(z)}{Q(z)} = \frac{a_n z^n + a_{n-1}z^{n-1} + \cdots + a_0}{b_m z^m + b_{m-1}z^{m-1} + \cdots + b_0},$$

where m, the degree of the denominator Q, is assumed to exceed n, the degree of the numerator P. As $|z|$ grows without limit, the leading terms in the numerator and denominator become dominant. We therefore see intuitively that

$$\frac{P(z)}{Q(z)} \approx \frac{a_n}{b_m}\frac{z^n}{z^m} = \frac{a_n}{b_m}\frac{1}{z^d}, \qquad \text{where } d = m - n.$$

Thus it should seem plausible that for sufficiently large $|z|$ there must exist constants μ and R_0 such that

$$\left|\frac{P(z)}{Q(z)}\right| \le \frac{\mu}{|z|^d} \qquad \text{for } |z| \ge R_0. \tag{6.5–9}$$

The proof can be found in Exercise 25. Thus when $d \ge 2$, the function $f(z) = P(z)/Q(z)$ will satisfy Eq. (6.5–8), and we can assert that

$$\lim_{R\to\infty} \int_{C_1} \frac{P(z)}{Q(z)}\,dz = 0, \tag{6.5–10}$$

where P, Q are polynomials and degree Q − degree $P \ge 2$.

Now, integrating $P(z)/Q(z)$ around contour C of Fig. 6.5–1, and taking R sufficiently large, we have from residue theory

$$\int_{-R}^{+R} \frac{P(x)}{Q(x)}\,dx + \int_{C_1} \frac{P(z)}{Q(z)}\,dz = 2\pi i \sum \operatorname{res} \frac{P(z)}{Q(z)}, \qquad \text{at all poles in u.h.p.}$$

On the left z has been set equal to x on the straight portion of the path. Passing to the limit $R \to \infty$, we use Eq. (6.5–10) to eliminate the integral over the arc C_1 and then obtain the following theorem:

THEOREM 4 Let $P(x)$ and $Q(x)$ be polynomials in x, and let the degree of $Q(x)$ exceed that of $P(x)$ by two or more. Let $Q(x)$ be nonzero for all real values of x. Then

$$\int_{-\infty}^{+\infty} \frac{P(x)}{Q(x)}\,dx = 2\pi i \sum \operatorname{res} \frac{P(z)}{Q(z)}, \qquad \text{at all poles in u.h.p.} \quad \square \tag{6.5–11}$$

The requirement $Q(x) \ne 0$ assures us that the integrand in Eq. (6.5–11) is finite for all x. The question of how to evaluate integrals in which $Q(x) = 0$, for some x, is dealt with later in this chapter.

EXAMPLE 2

Find $\int_{-\infty}^{+\infty} x^2/(x^4 + x^2 + 1)\,dx$.

Solution

Equation (6.5–11) can be used directly since the degree of the denominator, which is 4, differs from that of the numerator by 2. A difference of at least two is required. Thus

$$\int_{-\infty}^{+\infty} \frac{x^2\,dx}{x^4 + x^2 + 1} = 2\pi i \sum \operatorname{res} \frac{z^2}{z^4 + z^2 + 1}, \qquad \text{at all poles in u.h.p.}$$

Using the quadratic formula, we can solve $z^4 + z^2 + 1 = 0$ for z^2 and obtain

$$z^2 = \frac{-1 \pm i\sqrt{3}}{2} = e^{i2\pi/3},\ e^{-i2\pi/3}.$$

Taking square roots yields $z = e^{i\pi/3}, e^{-i2\pi/3}, e^{-i\pi/3}, e^{i2\pi/3}$. Thus $z^2/(z^4 + z^2 + 1)$ has simple poles in the u.h.p. at $e^{i\pi/3}$ and $e^{i2\pi/3}$. Evaluating the residues at these two poles in the usual way (see, for example, Eq. 6.3–6), we find that the value of the given integral is

$$2\pi i \sum_{\text{res}} \frac{z^2}{z^4 + z^2 + 1} = \frac{\pi}{\sqrt{3}}, \qquad \text{in u.h.p.}$$ ◀

EXERCISES

Which of the following integrals exist?

1. $\int_0^\infty e^{-2x}\, dx$ **2.** $\int_0^\infty e^{2x}\, dx$

3. $\int_0^\infty xe^{-2x}\, dx$ **4.** $\int_0^\infty \frac{x}{x^2+1}\, dx$

For which of the following integrals does the Cauchy principal value exist?

5. $\int_{-\infty}^{+\infty} e^{-x}\, dx$ **6.** $\int_{-\infty}^{+\infty} e^{-x^2}\, dx$†

7. $\int_{-\infty}^{+\infty} \frac{x^2+x}{1+x^2}\, dx$ **8.** $\int_{-\infty}^{+\infty} \frac{x-1}{1+x^2}\, dx$

9. The standard or ordinary definition of $\int_{-\infty}^{+\infty} f(x)\, dx$ is given by

$$\int_{-\infty}^{+\infty} f(x)\, dx = \lim_{b\to\infty} \int_0^b f(x)\, dx + \lim_{a\to\infty} \int_{-a}^0 f(x)\, dx,$$

where the two limits must exist independently of one another. Work the following without using complex variables.

a) Show that $\int_{-\infty}^{+\infty} \sin x\, dx$ fails to exist according to the standard definition.

b) Show that the Cauchy principal value of the preceding integral does exist and is zero.

c) Show that $\int_{-\infty}^{+\infty} dx/(1 + x^2) = \pi$ for both the standard definition and the Cauchy principal value.

d) Show that if the ordinary value of $\int_{-\infty}^{+\infty} f(x)\, dx$ exists, then the Cauchy principal value must also exist and that the two results agree.

---

† *Hint:* Look up the comparison test for improper integrals. See, for example, W. Kaplan, *Advanced Calculus*, 4th ed. (Reading, MA: Addison-Wesley, 1991), Section 6.22.

Evaluate the following integrals by means of residue calculus. Use the Cauchy principal value where appropriate.

10. $\displaystyle\int_{-\infty}^{+\infty} \frac{dx}{x^2 - x + 1}$

11. $\displaystyle\int_{-\infty}^{+\infty} \frac{dx}{(x^2 - x + 1)(x^2 + x + 1)}$

12. $\displaystyle\int_{0}^{\infty} \frac{x^4 + 1}{x^6 + 1}\, dx$

13. $\displaystyle\int_{-\infty}^{+\infty} \frac{x^2 + x + 1}{x^4 + x^2 + 1}\, dx$

14. Consider

$$\int_{-\infty}^{+\infty} \frac{x^3 + x^2}{(x^2 + 1)(x^2 + 4)}\, dx.$$

Does Theorem 4 apply directly to this integral? Evaluate this integral by evaluating the sum of two Cauchy principal values.

15. When a and b are positive, prove that

$$\int_{0}^{\infty} \frac{dx}{(x^2 + a^2)(x^2 + b^2)} = \frac{\pi}{2} \frac{1}{(b + a)ab}$$

for both $a \neq b$ and $a = b$.

Show that for a, b, c real, and $b^2 < 4ac$, the following hold.

16. $\displaystyle\int_{-\infty}^{+\infty} \frac{dx}{ax^2 + bx + c} = \frac{2\pi}{\sqrt{4ac - b^2}}$

17. $\displaystyle\int_{-\infty}^{+\infty} \frac{dx}{(ax^2 + bx + c)^2} = \frac{4\pi a}{[\sqrt{4ac - b^2}]^3}$

Obtain the above result using residues, but check the answer by differentiating both sides of the equation in Exercise 16 with respect to c. You may differentiate under the integral sign.

18. Find $\displaystyle\int_{0}^{\infty} \frac{dx}{x^{100} + 1}$. Answer: $\displaystyle\frac{\left(\frac{\pi}{100}\right)}{\sin\left(\frac{\pi}{100}\right)}$

19. Find $\displaystyle\int_{-\infty}^{+\infty} \frac{dx}{x^4 + x^3 + x^2 + x + 1}$.

Hint: See Exercise 26, Section 6.3 to locate the required poles.

20. Restate Eq. (6.5–11) so as to employ only residues in the lower half plane. State the conditions on P and Q.

21. a) Solve Example 1 by using only residues in the lower half plane.

b) Solve Example 2 by using residues in the lower half plane.

22. Evaluate $\int_{-\infty}^{+\infty} du/\cosh u$ by making the change of variable $x = e^u$ and then applying residues. Answer: π.

23. Let $f(z) = P(z)/Q(z)$, where P and Q are polynomials in z with the property that degree Q − degree $P \geq 2$.

a) Show that

$$\sum_{\text{res}} \frac{P(z)}{Q(z)} = 0, \qquad \text{all poles.}$$

Hint: Consider $\lim_{R\to\infty} \oint_{|z|=R} f(z)\,dz$. Use Eq. (6.5–10) and its counterpart in the lower half plane.

b) Verify the result of part (a) by summing the residues of $f(z) = z/(z^3 + 1)$.

24. a) Explain why $\int_0^\infty x/(x^4 + 1)\,dx$ cannot be evaluated through the use of a closed semicircular contour in the upper or lower half plane (see Fig. 6.5–1 or Fig. 6.5–2).

b) Consider the quarter circle contour shown in Fig. 6.5–3. C_1 is the arc of radius $R > 1$. Show that

$$\int_0^R \frac{x}{x^4+1}\,dx - \int_R^0 \frac{y}{y^4+1}\,dy + \int_{C_1} \frac{z\,dz}{z^4+1} = 2\pi i \sum_{\text{res}} \frac{z}{z^4+1}, \qquad \text{at all poles in first quadrant.}$$

c) Let $R \to \infty$ and show that $\int_0^\infty x/(x^4 + 1)\,dx = \pi/4$.

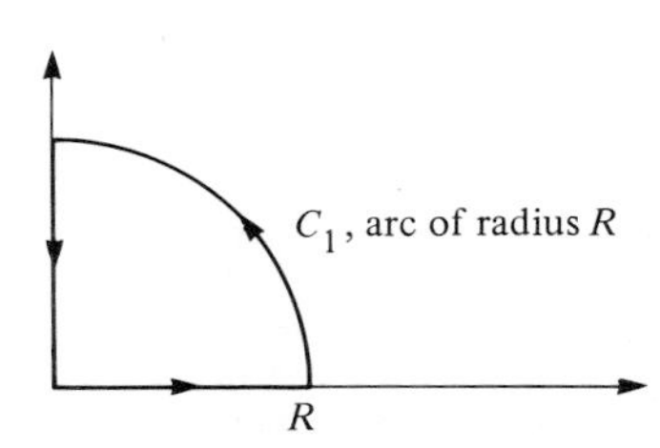

Figure 6.5–3

25. Let $P(z) = a_n z^n + a_{n-1} z^{n-1} + \cdots + a_0$ and $Q(z) = b_m z^m + b_{m-1} z^{m-1} + \cdots + b_0$ be polynomials in z, with $m > n$. In this exercise we show that if $f(z) = P(z)/Q(z)$ then there exist constants μ and R_0 such that $|f(z)| \leq \mu/|z|^d$ for all $|z| \geq R_0$, where $d = m - n$. Thus if $d \geq 2$, $f(z) = P(z)/Q(z)$ fulfills the requirements of Theorem 3.

a) Let A be the largest of the numbers $|a_n|, |a_{n-1}|, \ldots, |a_0|$. Using an elementary triangle inequality (Section 1.3) show that $|P(z)| \leq (n + 1)A|z|^n$ for $|z| \geq 1$.

b) Note that $Q(z) = b_m z^m g(z)$, where

$$g(z) = 1 + \frac{b_{m-1}}{b_m z} + \frac{b_{m-2}}{b_m z^2} + \cdots + \frac{b_0}{b_m z^m}.$$

Recall the triangle inequality $|s + t| \geq |s| - |t| \geq 0$, which holds if $|s| \geq |t|$. Now let B be the largest of the numbers $|b_{m-1}/b_m|, |b_{m-2}/b_m|, \ldots, |b_0/b_m|, 1$. Assume $m \geq 1$. Show that

$$|g(z)| \geq 1 - \left|\frac{b_{m-1}}{b_m z} + \frac{b_{m-2}}{b_m z^2} + \cdots + \frac{b_0}{b_m z^m}\right| \geq 0$$

for $|z| \geq 2mB$. Now use a triangle inequality to establish an upper bound on

$$\left|\frac{b_{m-1}}{b_m z} + \frac{b_{m-2}}{b_m z^2} + \cdots + \frac{b_0}{b_m z^m}\right|$$

in the region $|z| \geq 2mB$, and show that $|g(z)| \geq 1 - m/2m = 1/2$ for $|z| \geq 2mB$.

c) Use the preceding result to show that $|Q(z)| \geq |b_m||z|^m/2$ for $|z| \geq 2mB$.

d) Show that

$$\left|\frac{P(z)}{Q(z)}\right| \leq \frac{2(n+1)A}{|b_m||z|^d}$$

for $|z| \geq R_0$ where $R_0 = 2mB$. This completes the proof.

26. Show that

$$\int_0^\infty \frac{x^m}{x^n + 1}\,dx = \frac{\pi}{n \sin[\pi(m+1)/n]},$$

where n and m are nonnegative integers and $n - m \geq 2$.

Hint: Use the method employed in Exercise 24 above, but change to the contour of integration in Fig. 6.5–4.

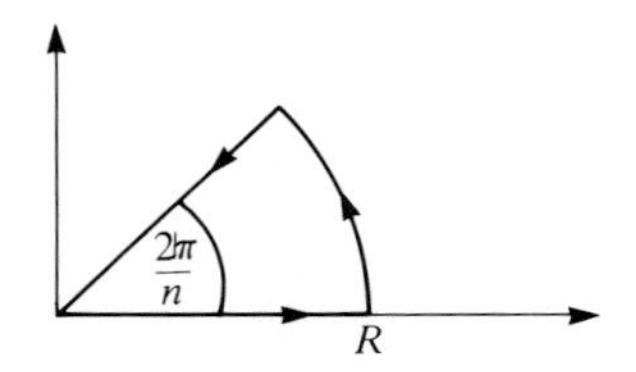

Figure 6.5–4

27. a) Show that

$$\int_0^\infty \frac{u^{1/l}}{u^k + 1}\,du = \frac{\pi}{k \sin[\pi(l+1)/(lk)]},$$

where k and l are integers, $l > 0$, which satisfy $l(k-1) \geq 2$. Take $u^{1/l}$ as a nonnegative real function in the interval of integration.

Hint: First work Exercise 26 above. Then, in the present problem, make the change of variable $x = u^{1/l}$ and use the result of Exercise 26.

b) What is $\int_0^\infty u^{1/4}/(u^5 + 1)\,du$?

6.6 EVALUATION OF INTEGRALS, III

Integrals of the type $\int_{-\infty}^{+\infty} f(x) \cos px\,dx$ and $\int_{-\infty}^{+\infty} f(x) \sin px\,dx$, where $f(x)$ is a rational function of x, and p is a real constant are often evaluated by methods similar to that just presented. These integrals appear in the theory of Fourier transforms, which is discussed in Section 6.9. Generally, we will determine the Cauchy principal value of such integrals and ignore the question of whether the ordinary values exist (see Exercise 9, Section 6.5).

To give some insight into the method discussed in this section, we try to evaluate $\int_{-\infty}^{+\infty} \cos(3x)/((x-1)^2+1)\,dx$ using the technique of the preceding section. We integrate $\cos 3z/((z-1)^2+1)$ around the closed semicircular contour of Fig. 6.5–1 and evaluate the result with residues. Thus

$$\int_{-R}^{+R} \frac{\cos 3x}{(x-1)^2+1}\,dx + \int_{C_1} \frac{\cos 3z\,dz}{(z-1)^2+1} = 2\pi i \sum_{\text{res}} \frac{\cos 3z}{(z-1)^2+1}, \quad \text{in u.h.p.}$$

As before, C_1 is an arc of radius R in the upper half plane. Although the preceding equation is valid for sufficiently large R it is of no use to us. We would like to show that as $R \to \infty$, the integral over C_1 goes to zero. However,

$$\cos 3z = \frac{e^{3iz} + e^{-3iz}}{2} = \frac{e^{i3x-3y} + e^{-i3x+3y}}{2}.$$

As $R \to \infty$, the y-coordinates of points on C_1 become infinite and the term $e^{-i3x+3y}$, whose magnitude is e^{3y}, becomes unbounded. The integral over C_1 thus does not vanish with increasing R.

The correct approach in solving the given problem is to begin by finding $\int_{-\infty}^{+\infty} e^{3ix}/((x-1)^2+1)\,dx$. Its value *can* be determined if we use the technique of the previous section, that is, we integrate $\int e^{3iz}/((z-1)^2+1)\,dz$ around the closed contour of Fig. 6.5–1 and obtain

$$\int_{-R}^{+R} \frac{e^{3ix}}{(x-1)^2+1}\,dx + \int_{C_1} \frac{e^{3iz}}{(z-1)^2+1}\,dz = 2\pi i \sum_{\text{res}} \frac{e^{3iz}}{(z-1)^2+1}, \quad \text{in u.h.p.} \tag{6.6–1}$$

Assuming we can argue that the integral over arc C_1 vanishes as $R \to \infty$ (the troublesome e^{-3iz} no longer appears), we have, in this limit,

$$\int_{-\infty}^{+\infty} \frac{e^{3ix}}{(x-1)^2+1}\,dx = 2\pi i \sum_{\text{res}} \frac{e^{3iz}}{(z-1)^2+1}, \quad \text{in u.h.p.}$$

Putting $e^{3ix} = \cos 3x + i \sin 3x$ and rewriting the integral on the left as two separate expressions, we have

$$\int_{-\infty}^{+\infty} \frac{\cos 3x}{(x-1)^2+1}\,dx + i\int_{-\infty}^{+\infty} \frac{\sin 3x}{(x-1)^2+1}\,dx = 2\pi i \sum_{\text{res}} \frac{e^{3iz}}{(z-1)^2+1}, \quad \text{in u.h.p.}$$

When we equate corresponding parts (reals and imaginaries) in this equation, the values of two real integrals are obtained:

$$\int_{-\infty}^{+\infty} \frac{\cos 3x}{(x-1)^2+1}\,dx = \text{Re}\left[2\pi i \sum_{\text{res}} \frac{e^{3iz}}{(z-1)^2+1}\right], \quad \text{at all poles in u.h.p.} \tag{6.6–2}$$

$$\int_{-\infty}^{+\infty} \frac{\sin 3x}{(x-1)^2+1}\,dx = \text{Im}\left[2\pi i \sum_{\text{res}} \frac{e^{3iz}}{(z-1)^2+1}\right], \quad \text{at all poles in u.h.p.} \tag{6.6–3}$$

Equation (6.6–2) contains the result being sought while the integral in Eq. (6.6–3) has been evaluated as a bonus.

Solving the equation $(z-1)^2 = -1$ and finding that $z = 1 \pm i$, we see that, on the right sides of Eqs. (6.6–2) and (6.6–3), we must evaluate a residue at the simple pole $z = 1 + i$. From Eq. (6.3–6) we obtain

$$2\pi i \operatorname{Res}\left(\frac{e^{3iz}}{(z-1)^2+1}, 1+i\right) = 2\pi i \lim_{z\to(1+i)} \frac{e^{3iz}}{2(z-1)}$$
$$= \pi e^{-3+3i} = \pi e^{-3}[\cos 3 + i \sin 3].$$

Using the result in Eqs. (6.6–2) and (6.6–3), we have finally

$$\int_{-\infty}^{+\infty} \frac{\cos 3x}{(x-1)^2+1}\, dx = \pi e^{-3} \cos 3 \quad \text{and} \quad \int_{-\infty}^{+\infty} \frac{\sin 3x}{(x-1)^2+1}\, dx = \pi e^{-3} \sin 3.$$

Recall now that we still have the task of showing that the second integral on the left in Eq. (6.6–1) becomes zero as $R \to \infty$. Rather than supply the details, we instead prove the following theorem and lemma. These not only perform our required task but many similar ones that we will encounter.

THEOREM 5 Let $f(z)$ have the following property in the half plane $\operatorname{Im} z \geq 0$. There exist constants, $k > 0$, R_0, and μ such that

$$|f(z)| \leq \frac{\mu}{|z|^k}, \qquad \text{for all } |z| \geq R_0 \text{ in this half plane.}$$

Then, if C_1 is the semicircular arc $Re^{i\theta}$, $0 \leq \theta \leq \pi$, and $R > R_0$, we have

$$\lim_{R\to\infty} \int_{C_1} f(z) e^{ivz}\, dz = 0, \qquad \text{when } v > 0. \quad \square \tag{6.6–4}$$

When $v < 0$, there is a corresponding theorem that applies in the lower half plane.

Equation (6.6–4) should be compared with Eq. (6.5–8). Notice that when the factor e^{ivz} is not present, as happens in Eq. (6.5–8), we require $k > 1$, whereas the validity of Eq. (6.6–4) requires the less-restrictive condition $k > 0$.

To prove Eq. (6.6–4) we rewrite the integral on the left, which we call I, in terms of polar coordinates; taking $z = Re^{i\theta}$, $dz = Re^{i\theta} i\, d\theta$, we have

$$I = \int_{C_1} f(z) e^{ivz}\, dz = \int_0^{\pi} f(Re^{i\theta}) e^{ivRe^{i\theta}} iRe^{i\theta}\, d\theta. \tag{6.6–5}$$

Recall now the inequality

$$\left| \int_a^b g(\theta)\, d\theta \right| \leq \int_a^b |g(\theta)|\, d\theta$$

derived in Exercise 17 of Section 4.2. Applying this to Eq. (6.6–5) and recalling that $|e^{i\theta}| = 1$, we have

$$|I| \leq R \int_0^{\pi} |f(Re^{i\theta})| |e^{ivRe^{i\theta}}|\, d\theta. \tag{6.6–6}$$

We see that

$$|e^{ivRe^{i\theta}}| = |e^{ivR(\cos\theta + i\sin\theta)}| = |e^{-vR\sin\theta}||e^{ivR\cos\theta}|.$$

Now

$$|e^{ivR\cos\theta}| = 1,$$

and since $e^{-vR\sin\theta} > 0$, we find that

$$|e^{ivRe^{i\theta}}| = e^{-vR\sin\theta}.$$

Rewriting Eq. (6.6–6) with the aid of the previous equation we have

$$|I| \le R\int_0^{\pi} |f(Re^{i\theta})|e^{-vR\sin\theta}\,d\theta.$$

With the assumption $|f(z)| = |f(Re^{i\theta})| \le \mu/R^k$ it should be clear that

$$|I| \le R\int_0^{\pi} \frac{\mu}{R^k}e^{-vR\sin\theta}\,d\theta = \frac{\mu}{R^{k-1}}\int_0^{\pi} e^{-vR\sin\theta}\,d\theta. \tag{6.6–7}$$

Since $\sin\theta$ is symmetric about $\theta = \pi/2$ (see Fig. 6.6–1), we can perform the integration on the right in Eq. (6.6–7) from 0 to $\pi/2$ and then double the result. Hence,

$$|I| \le \frac{2\mu}{R^{k-1}}\int_0^{\pi/2} e^{-vR\sin\theta}\,d\theta. \tag{6.6–8}$$

Figure 6.6–1 also shows that over the interval $0 \le \theta \le \pi/2$ we have $\sin\theta \ge 2\theta/\pi$. Thus when $v \ge 0$ we find $e^{-vR\sin\theta} \le e^{-vR\theta 2/\pi}$ for $0 \le \theta \le \pi/2$.

Making use of this inequality in Eq. (6.6–8), we have

$$|I| \le \frac{2\mu}{R^{k-1}}\int_0^{\pi/2} e^{-vR\theta 2/\pi}\,d\theta = \frac{\pi\mu}{vR^k}[1 - e^{-vR}].$$

With $R \to \infty$ the right-hand side of this equation becomes zero, which implies $I \to 0$ in the same limit. Thus

$$\lim_{R\to\infty}\int_{C_1} f(z)e^{ivz}\,dz = 0.$$

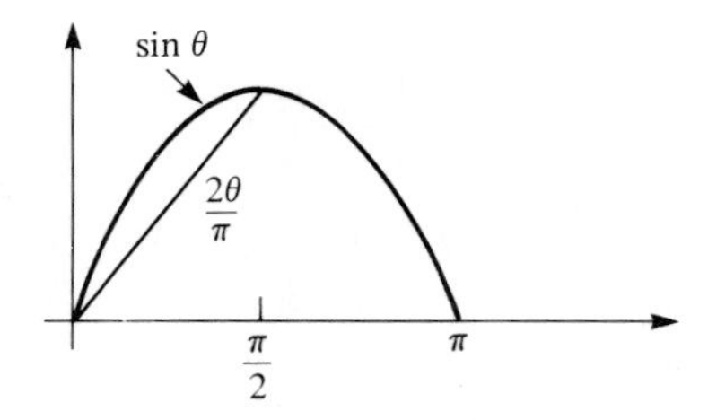

Figure 6.6–1

Any rational function $f(z) = P(z)/Q(z)$, where the degree of the polynomial $Q(z)$ exceeds that of the polynomial $P(z)$ by one or more, will fulfill the requirements of the theorem just presented (see Eq. 6.5–9) and leads us to the following lemma.

JORDAN'S LEMMA

$$\lim_{R\to\infty}\int_{C_1}\frac{P(z)}{Q(z)}e^{ivz}\,dz = 0, \qquad \text{if } v > 0, \text{ degree } Q - \text{degree } P \geq 1 \quad (6.6\text{–}9)$$

Jordan's lemma can be used to assert that the integral over C_1 in Eq. (6.6–1) becomes zero as $R \to \infty$. This was a required step in the derivation of Eqs. (6.6–2) and (6.6–3). We can use this lemma to develop a general formula for evaluating many other integrals involving polynomials and trigonometric functions.

Let us evaluate $\int e^{ivz}P(z)/Q(z)\,dz$ around the closed contour of Fig. 6.5–1 by using residues. All zeros of $Q(z)$ in the u.h.p. are assumed enclosed by the contour, and we also assume $Q(x) \neq 0$ for all real values of x. Therefore,

$$\int_{-R}^{+R}\frac{P(x)}{Q(x)}e^{ivx}\,dx + \int_{C_1}\frac{P(z)}{Q(z)}e^{ivz}\,dz = 2\pi i\sum_{\text{res}}\frac{P(z)}{Q(z)}e^{ivz}, \qquad \text{at all poles in u.h.p.} \quad (6.6\text{–}10)$$

Now, provided the degrees of Q and P are as described in Eq. (6.6–9), we put $R \to \infty$ in Eq. (6.6–10) and discard the integral over C_1 in this equation by invoking Jordan's lemma. We obtain the following:

$$\int_{-\infty}^{-\infty}\frac{P(x)}{Q(x)}e^{ivx}\,dx = 2\pi i\sum_{\text{res}}\frac{P(z)}{Q(z)}e^{ivz}, \qquad \text{in u.h.p.} \quad (6.6\text{–}11)$$

The derivation of Eq. (6.6–11) requires that $v > 0$, $Q(x) \neq 0$ for $-\infty < x < \infty$, and the degree of Q exceed the degree of P by at least one.

We now apply Euler's identity on the left in Eq. (6.6–11) and obtain

$$\int_{-\infty}^{+\infty}(\cos vx + i\sin vx)\frac{P(x)}{Q(x)}\,dx = 2\pi i\sum_{\text{res}}e^{ivz}\frac{P(z)}{Q(z)}, \qquad \text{in u.h.p.}$$

Now, assume that $P(x)$ and $Q(x)$ are *real* functions of x, that is, the coefficients of x in these polynomials are real numbers. We can then equate corresponding parts (reals and imaginaries) on each side of the preceding equation with the result that

$$\int_{-\infty}^{+\infty}\cos vx\,\frac{P(x)}{Q(x)}\,dx = \text{Re}\left[2\pi i\sum_{\text{res}}e^{ivz}\frac{P(z)}{Q(z)}\right], \qquad \text{in u.h.p.} \quad (6.6\text{–}12a)$$

$$\int_{-\infty}^{+\infty}\sin vx\,\frac{P(x)}{Q(x)}\,dx = \text{Im}\left[2\pi i\sum_{\text{res}}e^{ivz}\frac{P(z)}{Q(z)}\right], \qquad \text{in u.h.p.,} \quad (6.6\text{–}12b)$$

where degree Q − degree $P \geq 1$, $Q(x) \neq 0$, $-\infty < x < \infty$, $v > 0$.

These equations are useful in the rapid evaluation of integrals like those appearing in the Exercises below and in Eqs. (6.6–2) and (6.6–3). When $v = 0$, the integral on the left in Eq. (6.6–12b) is zero while that on the left in Eq. (6.6–12a) is evaluated from Eq. (6.5–11) if the degree of Q exceeds that of P by two or more.

When v is negative, we do not use Eqs. (6.6–11), (6.6–12a) or (6.6–12b). It is instructive and useful for later work to have formulas valid for $v < 0$. The reader should refer to Exercise 13 below where these are stated and derived.

EXERCISES

Evaluate the following integrals by means of residue calculus. Use the Cauchy principal value where appropriate.

1. $\displaystyle\int_{-\infty}^{+\infty} \frac{\cos 3x}{x^2 + 4}\, dx$

2. $\displaystyle\int_{0}^{\infty} \frac{x \sin 3x}{(x^2 + 4)^2}\, dx$

3. $\displaystyle\int_{-\infty}^{+\infty} \frac{\sin x}{x^2 + x + 1}\, dx$

4. $\displaystyle\int_{0}^{\infty} \frac{x \sin x}{x^4 + 1}\, dx$

5. $\displaystyle\int_{-\infty}^{+\infty} \frac{\cos x}{(x^2 + x + 1)^2}\, dx$

6. $\displaystyle\int_{-\infty}^{+\infty} \frac{\sin x}{x^4 + x^2 + 1}\, dx$

7. $\displaystyle\int_{0}^{\infty} \frac{\cos x}{x^4 + x^2 + 1}\, dx$

8. $\displaystyle\int_{-\infty}^{+\infty} \frac{\sin x}{[(x-1)^2 + 1]^2}\, dx$

Prove the following where $a > 0$.

9. $\displaystyle\int_{0}^{\infty} \frac{\sin^2 x}{x^2 + a^2}\, dx = \frac{\pi}{4a}(1 - e^{-2a})$

Hint: $\sin^2 x = \frac{1}{2}(1 - \cos 2x)$.

10. $\displaystyle\int_{0}^{\infty} \frac{\sin mx \sin nx}{a^2 + x^2}\, dx = \frac{\pi}{2a} e^{-ma} \sinh na \text{ for } m \ge n \ge 0$

Hint: Express $\sin mx \sin nx$ as a sum involving $\cos(m + n)x$ and $\cos(m - n)x$.

11. $\displaystyle\int_{0}^{\infty} \frac{x \sin mx \cos nx}{x^2 + a^2}\, dx = \frac{-\pi}{2} e^{-na} \sinh ma \text{ for } n > m \ge 0$

Hint: See hint for Exercise 10 but use $\sin(n \pm m)x$.

12. Explain why even though $\int_0^\infty (\cos x)/(x^2 + 1)\, dx$ can be evaluated with the aid of Eq. (6.6–12a), $\int_0^\infty (\sin x)/(x^2 + 1)\, dx$ cannot be evaluated with the help of Eq. (6.6–12b). This latter integral can be evaluated approximately with a numerical table or a computer program.

13. a) Refer to the contour shown in Fig. 6.6–2. Let C_2 be the arc of radius R in the lower half plane (l.h.p.). Show that

$$\lim_{R\to\infty} \int_{C_2} \frac{P(z)}{Q(z)} e^{ivz}\, dz = 0$$

if $v < 0$, where Q and P are polynomials such that degree Q – degree $P \ge 1$. This is Jordan's lemma in the l.h.p.

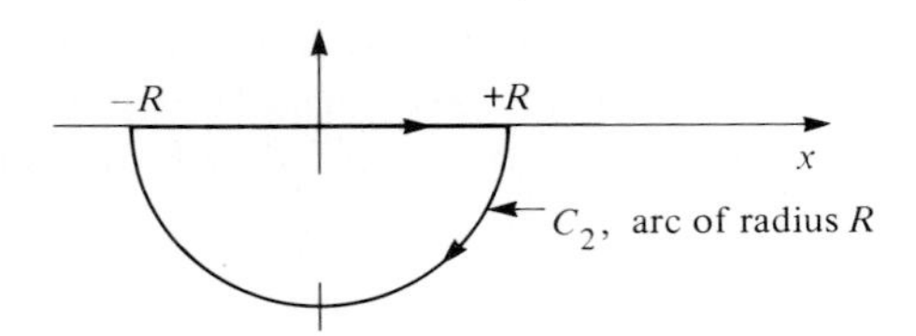

Figure 6.6–2

Hint: Begin by finding a formula analogous to Eq. (6.6–4) that applies when $v < 0$ and the contour is a semicircular arc in the l.h.p.

b) Perform an integration of $e^{ivz}P(z)/Q(z)$ around the closed contour in Fig. 6.6–2, allow $R \to \infty$, and use the result of part (a) to show that

$$\int_{-\infty}^{\infty} \frac{P(x)}{Q(x)} e^{ivx}\, dx = -2\pi i \sum_{\text{res}} \frac{P(z)}{Q(z)} e^{ivz}, \qquad \text{in l.h.p.} \tag{6.6–13}$$

if $v < 0$ and $Q(x) \neq 0$ for $-\infty < x < \infty$. Why is there a minus sign in Eq. (6.6–13) that does not appear in Eq. (6.6–11)?

c) Assume that $P(x)$ and $Q(x)$ are real functions. Use Eq. (6.6–13) to show that

$$\int_{-\infty}^{\infty} \cos vx \frac{P(x)}{Q(x)}\, dx = -\text{Re}\left[2\pi i \sum_{\text{res}} \frac{P(z)}{Q(z)} e^{ivz}\right], \qquad \text{in l.h.p. for } v < 0, \tag{6.6–14a}$$

$$\int_{-\infty}^{\infty} \sin vx \frac{P(x)}{Q(x)}\, dx = -\text{Im}\left[2\pi i \sum_{\text{res}} \frac{P(z)}{Q(z)} e^{ivz}\right], \qquad \text{in l.h.p. for } v < 0. \tag{6.6–14b}$$

Evaluate the following integrals by direct use of Eq. (6.6–13).

14. $\displaystyle\int_{-\infty}^{+\infty} \frac{e^{-ix}}{(x+1)^2+1}\, dx$ **15.** $\displaystyle\int_{-\infty}^{+\infty} \frac{(x^3+1)e^{-ix}}{x^4+1}\, dx$

16. a) Explain why $\int_{-\infty}^{+\infty} (\sin 2x)/(x-i)\, dx$ cannot be evaluated by means of Eq. (6.6–12b).

b) Evaluate this integral through the use of Eqs. (6.6–11) and (6.6–13).

Hint: Express $\sin 2x$ in terms of e^{i2x} and e^{-i2x}. Write the given integral as the sum of two integrals and evaluate each by using residues.

Using a method suggested by the previous exercise, evaluate these integrals:

17. $\displaystyle\int_{-\infty}^{+\infty} \frac{\cos x}{(x-1)^2+i}\, dx$

18. $\displaystyle\int_{-\infty}^{+\infty} \frac{e^{i\omega x}\cos x}{x^2+1}\, dx$ Consider the cases $\omega \leq -1$, $-1 \leq \omega \leq 1$, $\omega \geq 1$.

19. $\displaystyle\int_{-\infty}^{+\infty} \frac{e^{i\omega x}\sin x}{x^2+1}\, dx$ Consider ω as in Exercise 18.

20. To establish the well-known result

$$\int_0^\infty \frac{\sin x}{x}\,dx = \frac{\pi}{2}$$

we proceed as follows:

a) Show that

$$f(z) = \frac{e^{iz} - 1}{z}$$

has a removable singularity at $z = 0$. How should $f(0)$ be defined to remove the singularity?

b) Using the contour of Fig. 6.5–1, prove that

$$\int_{-R}^{+R} \frac{e^{ix} - 1}{x}\,dx + \int_{C_1} \frac{e^{iz} - 1}{z}\,dz = 0$$

and also

$$\int_{-R}^{+R} \frac{\cos x - 1}{x}\,dx + i\int_{-R}^{+R} \frac{\sin x}{x}\,dx = \int_{C_1} \frac{1}{z}\,dz - \int_{C_1} \frac{e^{iz}}{z}\,dz.$$

c) Evaluate the first integral on the above right by using the polar representation of C_1: $z = Re^{i\theta}$, $0 \le \theta \le \pi$. Pass to the limit $R \to \infty$ and explain why the second integral on the right goes to zero. Thus prove that

$$\int_{-\infty}^{+\infty} \frac{\cos x - 1}{x}\,dx = 0$$

and

$$\int_{-\infty}^{+\infty} \frac{\sin x}{x}\,dx = \pi,$$

and finally that

$$\int_0^\infty \frac{\sin x}{x}\,dx = \frac{\pi}{2}.$$

21. The expression

$$g(a) = \int_0^\infty \frac{\cos x}{x + a}\,dx, \qquad a > 0,$$

called the *auxiliary cosine integral*, must be evaluated numerically for every value of a. Using a computer and a suitable program, we might determine

$$\int_0^R \frac{\cos x}{x + a}\,dx,$$

where R is chosen "large" in the sense that a further increase in R yields a negligible change in the numerical result. Thus for a given value of a, we arrive at an approximation to $g(a)$. The difficulty encountered is that the magnitude of the integrand falls to zero slowly with increasing x. Thus a large R and hence a long interval of integration must be employed. Also, because of the oscillations of $\cos x$, there is a tendency for the contributions to the integral over the intervals $0 \le x \le \pi$, $\pi \le x \le 2\pi$, etc., to nearly cancel, and a high degree of accuracy in the numerical evaluation of the integral becomes difficult to obtain. With the aid of the Cauchy Integral Theorem we may find an equivalent integral whose numerical evaluation does not have the problems described.

a) Explain why

$$\oint \frac{e^{iz}}{z+a}\,dz = 0,$$

where the contour of integration is shown in Fig. 6.5–3.

b) Show that as $R \to \infty$ the portion of the preceding integral taken over the 90° arc C_1 vanishes. (The proof is similar to that for Theorem 5.)

c) Using the results of parts (a) and (b) show that

$$\int_0^{\infty} \frac{\cos x + i \sin x}{x+a}\,dx = \int_0^{\infty} \frac{e^{-y} i}{iy+a}\,dy.$$

d) Use the preceding equation to show that

$$g(a) = \int_0^{\infty} \frac{\cos x}{x+a}\,dx = \int_0^{\infty} \frac{ye^{-y}}{y^2+a^2}\,dy$$

and also

$$f(a) = \int_0^{\infty} \frac{\sin x}{x+a}\,dx = a\int_0^{\infty} \frac{e^{-y}}{y^2+a^2}\,dy,$$

where $f(a)$ is known as the *auxiliary sine integral.* Notice that each of the integrands in the variable y is nonoscillatory and decays exponentially with increasing y. Their numerical integration is readily accomplished.

e) With the aid of a computer or a programmable pocket calculator evaluate numerically $g(1)$ by using the result of part (d). Use as your upper limit of integration $y = 5$, $y = 10$, and $y = 20$. Give results to three significant figures and see whether your answers are sensitive to your choice of upper limit.
Answers: 0.342, 0.343, 0.343.

22. Consider the problem of evaluating

$$I = \int_{-\infty}^{+\infty} \frac{\cos x}{\cosh x}\,dx = \mathrm{Re}\int_{-\infty}^{+\infty} \frac{e^{ix}}{\cosh x}\,dx.$$

If we try to evaluate the preceding integral by employing the contour of Fig. 6.5–1 and the methods leading to Eq. (6.6–12a) we get into difficulty because the function $e^{iz}/\cosh z$ has an infinite number of poles in the upper half plane, i.e., at $z = i(n\pi + \pi/2)$, $n = 0, 1, 2, \ldots,$

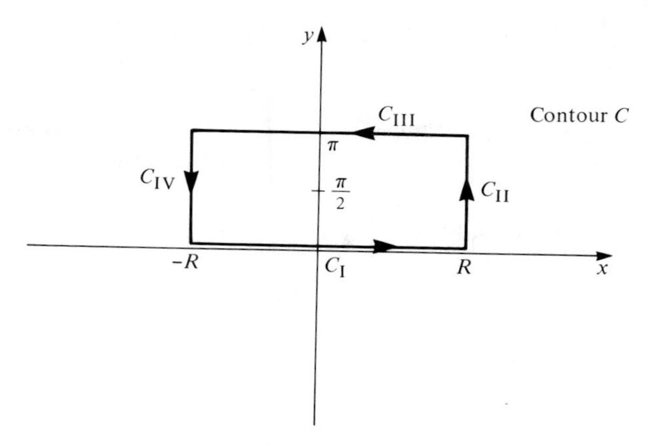

Figure 6.6–3

which are the zeros of cosh z. Thus Theorem 5 is inapplicable. However, we can determine I with the aid of the contour C shown in Fig. 6.6–3.

a) Using residues show that

$$\oint_C \frac{e^{iz}}{\cosh z}\,dz = 2\pi e^{-\pi/2}.$$

b) Using the appropriate values of z in the integrals along the top and bottom portions of the rectangle show that

$$\int_{-R}^{+R} \frac{e^{ix}}{\cosh x}\,dx + \int_{+R}^{-R} \frac{e^{i(x+i\pi)}}{\cosh(x+i\pi)}\,dx + \int_{C_{II}} \frac{e^{iz}}{\cosh z}\,dz + \int_{C_{IV}} \frac{e^{iz}}{\cosh z}\,dz = 2\pi e^{-\pi/2}.$$

c) Combine the first two integrals on the left in the preceding equation to show that

$$\int_{-R}^{+R} \frac{e^{ix}}{\cosh x}\,dx(1+e^{-\pi}) + \int_{C_{II}} \frac{e^{iz}}{\cosh z}\,dz + \int_{C_{IV}} \frac{e^{iz}}{\cosh z}\,dz = 2\pi e^{-\pi/2}.$$

d) Let $R \to \infty$ in the preceding. Using the ML inequality argue that the integrals along C_{II} and C_{IV} are zero in this limit.

Hint: Recall that $|\cosh z| = \sqrt{\sinh^2 x + \cos^2 y}$ (Exercise 17, Section 3.3).

Thus on C_{II} and C_{IV} we have

$$\left|\frac{e^{iz}}{\cosh z}\right| \le \frac{1}{\sinh R}.$$

Prove finally that

$$\int_{-\infty}^{+\infty} \frac{\cos x}{\cosh x}\,dx = \frac{2\pi e^{-\pi/2}}{1+e^{-\pi}} = \frac{\pi}{\cosh \pi/2},$$

and explain why

$$\int_0^\infty \frac{\cos x}{\cosh x}\,dx = \frac{\pi}{2\cosh \pi/2}.$$

Using a technique similar to that employed in Exercise 22 prove the following:

23. $\displaystyle\int_{-\infty}^{+\infty} \frac{\cosh x}{\cosh ax}\,dx = \frac{\pi}{a\cos(\pi/2a)}$ for $a > 1$

Hint: Use a rectangle like that in Fig. 6.6–3 but having height π/a. Take $\cosh z/\cosh az$ as the integrand, and prove that the integrations along segments C_{II} and C_{IV} each go to zero as $R \to \infty$.

24. $\displaystyle\int_{-\infty}^{+\infty} \frac{e^x}{1+e^{ax}}\,dx = \frac{\pi}{a\sin(\pi/a)}$ for $a > 1$

Hint: Use a rectangle like that in Fig. 6.6–3 but having height $2\pi/a$.

25. The result

$$\int_{-\infty}^{+\infty} e^{-a^2x^2}\,dx = \frac{\sqrt{\pi}}{a}, \qquad a > 0,$$

is derived in many standard texts on real calculus.† Use this identity to show that

$$\int_{-\infty}^{+\infty} e^{-m^2x^2}\cos bx\,dx = \frac{\sqrt{\pi}}{m}e^{-b^2/4m^2},$$

where b is a real number and $m > 0$.

Hint: Assume $b > 0$. Integrate $e^{-m^2z^2}$ around a contour similar to that in Fig. 6.6–3. Take $b/2m^2$ as the height of the rectangle. Argue that the integrals along C_{II} and C_{IV} vanish as $R \to \infty$. Why does your result apply to the case $b \leq 0$?

---

† To derive the result for $a = 1$ note that

$$I = \int_{-\infty}^{+\infty} e^{-x^2}\,dx = \int_{-\infty}^{+\infty} e^{-y^2}\,dy$$

and

$$I^2 = \int_{-\infty}^{+\infty} e^{-x^2}\,dx \int_{-\infty}^{+\infty} e^{-y^2}\,dy$$

$$= \int_{-\infty}^{+\infty}\int_{-\infty}^{+\infty} e^{-(x^2+y^2)}\,dx\,dy,$$

which you can interpret as an area integral and evaluate with a switch to polar variables, so that $x = r\cos\phi$ and $y = r\sin\phi$. Thus

$$I^2 = \int_0^{2\pi}\int_0^{\infty} e^{-r^2} r\,dr\,d\phi = \pi, \qquad \text{or } I = \sqrt{\pi}.$$

For $a \neq 1$ you can make a change of variable in the preceding result.

26. In Exercise 6 of Section 5.5 the reader studied the Fresnel integrals

$$C(P) = \int_0^P \cos\left(\frac{\pi}{2}t^2\right) dt$$

and

$$S(P) = \int_0^P \sin\left(\frac{\pi}{2}t^2\right) dt,$$

where $P \geq 0$. We can evaluate the integrals in closed form in the limit $P \to \infty$.

a) Consider $\int e^{iz^2}\, dz$ taken around the contour of Fig. 6.5–4. The angle $2\pi/n$ should be chosen as $\pi/4$, and $z = R \operatorname{cis} \theta$ on the arc. Prove that

$$\int_0^R \cos x^2\, dx + i \int_0^R \sin x^2\, dx + \int_0^{\pi/4} e^{iR^2 \operatorname{cis} 2\theta} iRe^{i\theta}\, d\theta = (1 + i) \int_0^{R/\sqrt{2}} e^{-2x^2}\, dx.$$

b) Show that the preceding integral on θ goes to zero as $R \to \infty$.

Hint: Use Eq. (4.2–14b) to show that

$$\left| \int_0^{\pi/4} e^{iR^2 \operatorname{cis} 2\theta} iRe^{i\theta}\, d\theta \right| \leq R \int_0^{\pi/4} e^{-R^2 \sin 2\theta}\, d\theta.$$

Rewrite the integral on the right with the change of variable $\phi = 2\theta$. Use the inequality $\sin \phi \geq 2\phi/\pi$ for $0 \leq \phi \leq \pi/2$ to argue that the resulting integral goes to zero as $R \to \infty$ (see the derivation of Theorem 5).

c) With $R \to \infty$ show that the equation derived in part (a) now yields

$$\int_0^\infty \cos x^2\, dx = \int_0^\infty \sin x^2\, dx = \int_0^\infty e^{-2x^2}\, dx.$$

Now use the result given in Exercise 25(a) and a change of variables to show that

$$\int_0^\infty \cos\left(\frac{\pi}{2}x^2\right) dx = \int_0^\infty \sin\left(\frac{\pi}{2}x^2\right) dx = \frac{1}{2}.$$

Make a change of variables in the preceding result and use symmetry arguments to show that, for b real and nonzero,

$$\int_{-\infty}^{+\infty} \cos bu^2\, du = \frac{1}{\sqrt{|b|}}\sqrt{\frac{\pi}{2}}, \tag{6.6–15a}$$

$$\int_{-\infty}^{+\infty} \sin bu^2\, du = \frac{(\pm 1)}{\sqrt{|b|}}\sqrt{\frac{\pi}{2}}, \tag{6.6–15b}$$

where $(\pm)$ is chosen to conform to the sign of b.

6.7 INTEGRALS INVOLVING INDENTED CONTOURS

In previous sections we gave particular attention to the evaluation of integrals of the form $\int_{-\infty}^{+\infty} P(x)/Q(x)\,dx$ and $\int_{-\infty}^{+\infty} P(x)/Q(x)(\cos vx$ or $\sin vx)\,dx$. We required $Q(x) \neq 0$ for all x. In this section we dispense with this requirement, which means that the integrand can become infinite somewhere in the interval of integration. We will find that when the limiting process involved in the integration is appropriately defined, the integral can sometimes be evaluated.

We begin with a brief and ultimately useful digression. Let the function $f(z)$ possess a simple pole at the point z_0, and let C be a circle of radius r centered at z_0. Suppose c_{-1} is the residue of $f(z)$ at z_0. Assuming that $f(z)$ has no other singularities on and inside C, we know immediately that

$$\oint_C f(z)\,dz = 2\pi i c_{-1}. \tag{6.7–1}$$

The reader may wonder if an integration taken only halfway around C would yield $2\pi i c_{-1}/2$ and if an integration performed 1/4 of the way around C would yield $2\pi i c_{-1}/4$. A specific example (see Exercise 1 of this section) shows this is a naive expectation. What is true, however, is that an integral taken around a fraction of C can be evaluated in the *limit* as the radius of C shrinks to zero by our using only the corresponding fraction of the residue of $f(z)$ at z_0. To be more specific, consider Theorem 6.

THEOREM 6 Let $f(z)$ have a simple pole at z_0. An arc C_0 of radius r is constructed using z_0 as its center. The arc subtends an angle α at z_0 (see Fig. 6.7–1). Then

$$\lim_{r\to 0}\int_{C_0} f(z)\,dz = 2\pi i\left[\frac{\alpha}{2\pi}\operatorname{Res}[f(z), z_0]\right], \tag{6.7–2}$$

where the integration is done in the counterclockwise direction. (For a clockwise integration, a factor of -1 is placed on the right in Eq. 6.7–2). □

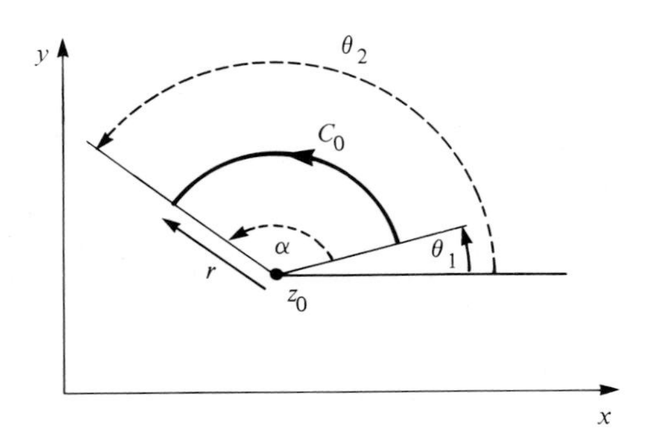

Figure 6.7–1

Note that when the integration in Theorem 6 is performed around an entire circle, α equals 2π and Eq. (6.7–2) yields a familiar result.[†]

To prove this theorem, we first expand $f(z)$ in a Laurent series about z_0. Because of the simple pole at z_0, the series assumes the form

$$f(z) = \frac{c_{-1}}{(z - z_0)} + \sum_{n=0}^{\infty} c_n(z - z_0)^n = \frac{c_{-1}}{(z - z_0)} + g(z),$$

where $g(z) = \sum_{n=0}^{\infty} c_n(z - z_0)^n$, which is the sum of a Taylor series, is analytic at z_0, and where $c_{-1} = \text{Res}[f(z), z_0]$. We now integrate the series expansion of $f(z)$ along C_0 in Fig. 6.7–1. Thus

$$\int_{C_0} f(z)\,dz = \int_{C_0} \frac{c_{-1}}{(z - z_0)}\,dz + \int_{C_0} g(z)\,dz. \tag{6.7–3}$$

Because $g(z)$ is continuous at z_0, we assert that $|g(z)|$ is bounded in a neighborhood of z_0; that is, there is a real constant M such that $|g(z)| \le M$ in this neighborhood. The radius r of C_0 is taken sufficiently small so that C_0 lies entirely in the neighborhood in question. Applying the ML inequality to the second integral on the right in Eq. (6.7–3), we have

$$\left| \int_{C_0} g(z)\,dz \right| \le Mr\alpha, \tag{6.7–4}$$

where $r\alpha$ is the length of contour C_0. From Eq. (6.7–4) we see that

$$\lim_{r \to 0} \int_{C_0} g(z)\,dz = 0. \tag{6.7–5}$$

The first integral on the right in Eq. (6.7–3) can be rewritten with a switch to polar variables. With $z = z_0 + re^{i\theta}$, $dz = ire^{i\theta}\,d\theta$, and with the limits on θ indicated in Fig. 6.7–1, we have

$$\int_{C_0} \frac{c_{-1}}{(z - z_0)}\,dz = \int_{\theta_1}^{\theta_1 + \alpha} \frac{c_{-1} ire^{i\theta}\,d\theta}{re^{i\theta}} = c_{-1}\alpha i = 2\pi i \frac{\alpha}{2\pi} c_{-1}. \tag{6.7–6}$$

Thus, passing to the limit $r \to 0$ in Eq. (6.7–3) and using Eqs. (6.7–5) and (6.7–6), we prove the theorem at hand. Applications of this theorem will now be discussed.

For integrals of the form $\int_a^b f(x)\,dx$, we sometimes find that $f(x)$ becomes infinite at some point, let us say p, that lies between a and b. Let us assume that $f(x)$ is continuous at all other points in the interval $a \le x \le b$. We thus have an improper integral. Previously we have encountered improper integrals involving infinite limits. The term "improper" is used in both cases because neither integral is expressible as

---

[†] In Eq. (6.7–2) we should strictly write the limit using the special notation $\lim_{r \to 0+}$ to signify that the limit is evaluated as r shrinks to zero through *positive* values. This kind of limit will often be used without special notation throughout the remainder of this book.

the usual limit of a sum. Some improper integrals of the type we are considering here are

$$\int_{-1}^{+1} \frac{1}{x}\,dx, \quad \int_{1}^{3} \frac{1}{(x-2)}\,dx, \quad \text{and} \quad \int_{-\pi/2}^{\pi/2} \frac{1}{\sin x}\,dx.$$

In colloquial language each of the integrands "blows up" somewhere between the limits of integration.

To evaluate such expressions we require a suitable definition. The term "Cauchy principal value," which we have used before (with another meaning) is applied to the following definition of this kind of improper integral.

DEFINITION Cauchy principal value of an integral

$$\int_a^b f(x)\,dx = \lim_{\varepsilon\to 0}\left[\int_a^{p-\varepsilon} f(x)\,dx + \int_{p+\varepsilon}^{b} f(x)\,dx\right], \tag{6.7–7}$$

where $f(x)$ is continuous for $a \le x < p$ and $p < x \le b$ and ε shrinks to zero through *positive* values. □

In both integrals appearing in the brackets the troublesome point $x = p$ is excluded from the interval of integration. In the first integral on the right the point p is approached from the left on the x-axis while in the second integral this same point is approached from the right. The preceding definition is meaningless if the limit in Eq. (6.7–7) does not exist.

EXAMPLE 1

Find the Cauchy principal value of $\int_{-1}^{2} 1/x\,dx$.

Solution

Applying Eq. (6.7–7), we take $a = -1$, $b = 2$, and $p = 0$, (since $1/x$ has a discontinuity at $x = 0$). Thus for the Cauchy principal value we have

$$\int_{-1}^{2} \frac{1}{x}\,dx = \lim_{\varepsilon\to 0}\left[\int_{-1}^{-\varepsilon} \frac{1}{x}\,dx + \int_{\varepsilon}^{2} \frac{1}{x}\,dx\right]. \tag{6.7–8}$$

Using the indefinite integral $\int 1/x\,dx = \operatorname{Log}|x|$, we obtain

$$\int_{-1}^{2} \frac{1}{x}\,dx = \lim_{\varepsilon\to 0}\left[\operatorname{Log}\frac{|-\varepsilon|}{|-1|} + \operatorname{Log}\frac{2}{\varepsilon}\right] = \lim_{\varepsilon\to 0}\,[\operatorname{Log}\varepsilon + \operatorname{Log} 2 - \operatorname{Log}\varepsilon] = \operatorname{Log} 2.$$

Notice that in the limit $\varepsilon \to 0$, neither of the integrals in Eq. (6.7–8) exists separately. However, because of the cancellation of negative and positive areas about $x = 0$ (see Fig. 6.7–2), the sum of these integrals does possess a finite limit as $\varepsilon \to 0$. In the exercises we will see that there are integrals whose Cauchy principal value does not exist since the limit in Eq. (6.7–7) does not exist. ◀

Presently, we will be dealing with integrals that are improper not only because of discontinuities in the integrand but also because of infinite limits of integration.

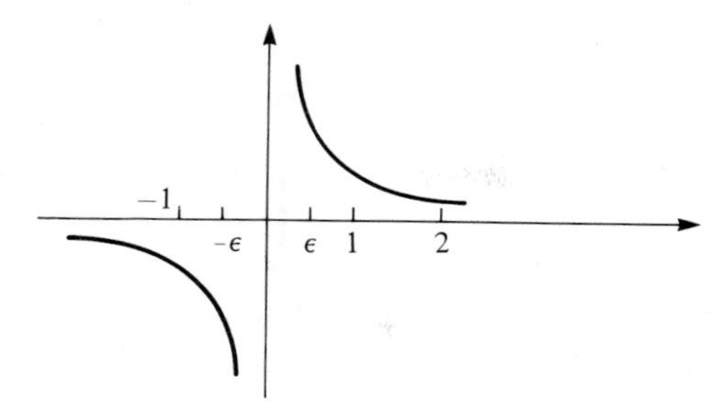

Figure 6.7–2

Thus both definitions of the Cauchy principal value are required; for example,

$$\int_{-\infty}^{+\infty} \frac{\cos x}{(x-1)}\,dx = \lim_{\substack{\varepsilon\to 0\\ R\to\infty}} \left[\int_{-R}^{1-\varepsilon} \frac{\cos x}{(x-1)}\,dx + \int_{1+\varepsilon}^{R} \frac{\cos x}{(x-1)}\,dx\right].$$

The point $x = 1$ is approached in a symmetric fashion, as are the limits at infinity. The concept of the Cauchy principal value can readily be extended to cover cases where the integrand has two or more points of discontinuity, as, for example, in this integral:

$$\int_{-\infty}^{\infty} \frac{\cos x}{(x+2)(x-3)}\,dx = \lim_{\substack{\varepsilon\to 0\\ \delta\to 0\\ R\to\infty}} \left[\int_{-R}^{-2-\delta} \frac{\cos x}{(x+2)(x-3)}\,dx \right.$$

$$\left. + \int_{-2+\delta}^{3-\varepsilon} \frac{\cos x}{(x+2)(x-3)}\,dx + \int_{3+\varepsilon}^{R} \frac{\cos x}{(x+2)(x-3)}\,dx\right].$$

The previous ideas, as well as the theorem just presented, can be combined in order to evaluate integrals of the form $\int_{-\infty}^{+\infty} f(x)\,dx$, where the discontinuities of $f(x)$ at real values of x coincide with *simple poles* of the analytic function $f(z)$. The method to be presented is known as *indentation of contours*, a term whose meaning will become clear with an example.

EXAMPLE 2

Find the Cauchy principal value of $\int_{-\infty}^{+\infty} (\cos 3x)/(x-1)\,dx$.

Solution

If we were to proceed according to the methods of Section 6.6, we would consider $\int e^{i3z}/(z-1)\,dz$ integrated along a closed contour like that of Fig. 6.5–1. We use here a contour like that one but with a modification. Because $e^{i3z}/(z-1)$ has a pole at $z = 1$, this point must be avoided by means of a semicircular indentation of radius ε in the contour. The closed contour actually used is shown in Fig. 6.7–3. Notice that $e^{i3z}/(z-1)$ is analytic at all points lying on, and interior to, this contour. Integrating this function around the path shown and putting $z = x$ where appropriate,

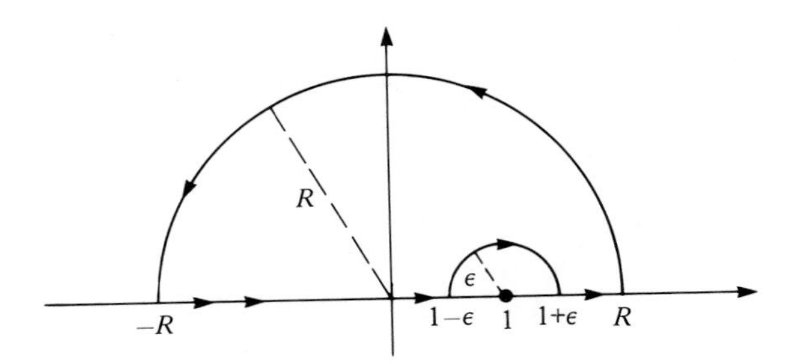

Figure 6.7–3

we have

$$\int_{-R}^{1-\varepsilon} \frac{e^{i3x}}{(x-1)}\,dx + \int_{\substack{\curvearrowright \\ |z-1|=\varepsilon}} \frac{e^{i3z}}{z-1}\,dz + \int_{1+\varepsilon}^{R} \frac{e^{i3x}}{(x-1)}\,dx + \int_{\substack{\curvearrowleft \\ |z|=R}} \frac{e^{i3x}}{(z-1)}\,dz = 0. \tag{6.7–9}$$

Allowing $R \to \infty$ and invoking Jordan's lemma (see Eq. 6.6–9) we can easily argue that the integral around the semicircle of radius R goes to zero.

Taking the limit $\varepsilon \to 0$, we can evaluate the integral over the semicircular indentation at $z = 1$. Using Eq. (6.7–2) with $\varepsilon = r$, $f(z) = e^{i3z}/(z-1)$, $z_0 = 1$, and $\alpha = \pi$ (for a semicircle), we find that the second integral on the left becomes in the limit $-i\pi e^{3i}$. The minus sign appears because of the clockwise direction of integration. With $R \to \infty$ and $\varepsilon \to 0$ we rewrite Eq. (6.7–9) as

$$\lim_{\substack{R\to\infty \\ \varepsilon\to 0}} \left[\int_{-R}^{1-\varepsilon} \frac{e^{i3x}}{x-1}\,dx + \int_{1+\varepsilon}^{R} \frac{e^{i3x}}{x-1}\,dx \right] - i\pi e^{i3} = 0.$$

The sum of the two integrals in the bracket becomes, with the limits indicated, the Cauchy principal value of

$$\int_{-\infty}^{+\infty} \frac{e^{i3x}}{(x-1)}\,dx = \int_{-\infty}^{+\infty} \frac{\cos 3x + i \sin 3x}{(x-1)}\,dx.$$

Thus

$$\int_{-\infty}^{+\infty} \frac{\cos 3x}{(x-1)}\,dx + i \int_{-\infty}^{+\infty} \frac{\sin 3x}{(x-1)}\,dx = i\pi e^{i3} = i\pi[\cos 3 + i \sin 3].$$

Equating real and imaginary parts on either side, we have

$$\int_{-\infty}^{\infty} \frac{\cos 3x}{(x-1)}\,dx = -\pi \sin 3 \quad \text{and} \quad \int_{-\infty}^{+\infty} \frac{\sin 3x}{(x-1)}\,dx = \pi \cos 3. \qquad ◀$$

Comment: In this example the real integral evaluated had one discontinuity in its integrand, at $x = 1$. The integral was evaluated by means of a contour integration with the contour indented about the value of z corresponding to this point. In some of the following exercises we consider integrals in which the integrand possesses more than one point of discontinuity. Here it becomes necessary to employ contours of integration indented around *each* such point. For example, to find

$$\int_{-\infty}^{+\infty} \frac{\cos x}{x^2 - 9}\, dx$$

we must have indentations at $z = \pm 3$.

EXERCISES

1. a) Find $\oint_{|z|=1} (z + 1)/z\, dz$.

b) Find $\int_C (z + 1)/z\, dz$, where C is the semicircle $|z| = 1$, $0 \leq \arg z \leq \pi$. Integrate in the counterclockwise sense.

c) In part (b) you integrated halfway around the circle used in part (a). Is the answer to part (b) half that of part (a)? Explain.

2. a) Evaluate $\int_{1-\varepsilon}^{1+\varepsilon} (z + 1)/(z - 1)\, dz$ around the semicircular arc of radius ε shown in Fig. 6.7–4.

b) In the answer to (a) let $\varepsilon \to 0$. Obtain $-2\pi i(1/2)\operatorname{Res}[(z + 1)/(z - 1), 1]$.

3. Obtain the Cauchy principal value required in Example 2 by using, instead of Fig. 6.7–3, the contour shown in Fig. 6.7–5. Notice that a pole singularity is now enclosed.

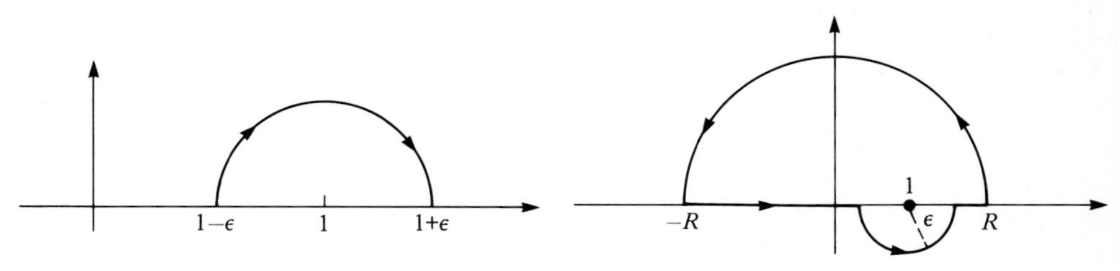

Figure 6.7–4 **Figure 6.7–5**

Find the Cauchy principal value of each of the following integrals:

4. $\displaystyle\int_{-\infty}^{+\infty} \frac{\sin 2x}{x + 4}\, dx$

5. $\displaystyle\int_{-\infty}^{+\infty} \frac{\cos 2x}{x^2 - 16}\, dx$

6. $\displaystyle\int_{-\infty}^{+\infty} \frac{\sin x}{(x - \pi/2)(x - \pi)}\, dx$

7. $\displaystyle\int_{-\infty}^{+\infty} \frac{\cos x}{(x - \pi/2)(x^2 + 1)}\, dx$

8. $\displaystyle\int_{-\infty}^{+\infty} \frac{\cos\left(\frac{\pi}{2}x\right)}{(x - 1)^2}\, dx$

Hint: Evaluate $\displaystyle\int_{-\infty}^{+\infty} \frac{e^{i\pi x/2} - i}{(x - 1)^2}\, dx.$

Prove the following, where the Cauchy principal value is used:

9.
$$\int_{-\infty}^{+\infty} \frac{\cos mx}{ax^2 + bx + c}\, dx = \frac{-2\pi \cos\dfrac{mb}{2a} \sin\dfrac{m\sqrt{b^2 - 4ac}}{2a}}{\sqrt{b^2 - 4ac}},$$

where $m \geq 0$, a, b, c are real, $b^2 > 4ac$, and $a \neq 0$.

10. $\displaystyle\int_{-\infty}^{+\infty} \frac{\cos mx}{x^4 - b^4}\, dx = \frac{-\pi}{2b^3} \sin mb - \frac{\pi e^{-mb}}{2b^3}$, where $m \geq 0$, $b > 0$

11. $\displaystyle\int_{-\infty}^{+\infty} \frac{\sin bx}{\sinh ax}\, dx = \frac{\pi}{a} \tanh\frac{\pi b}{2a}$, where $a > 0$

Hint: See technique discussed in Exercise 22, Section 6.6.

12. In Exercise 20, Section 6.6 you showed that

$$\int_{-\infty}^{+\infty} \frac{\sin x}{x}\, dx = \pi.$$

Using the concept of indented contours obtain the same result by an easier method, i.e., consider

$$\int_{-\infty}^{+\infty} \frac{e^{ix}}{x}\, dx.$$

13. Show that

$$\int_{-\infty}^{+\infty} \frac{\sin^2 x}{x^2}\, dx = \pi.$$

Hint: $\sin^2 x = \dfrac{1}{2} - \dfrac{1}{2} \cos 2x = \text{Re}\left[\dfrac{1 - e^{2ix}}{2}\right]$.

14. Show that

$$\int_{0}^{\infty} \left(\frac{\cos t - e^{-t}}{t}\right) dt = 0.$$

Hint: Integrate e^{iz}/z around the contour shown in Fig. 6.7–6 and allow $\varepsilon \to 0$, $R \to \infty$.

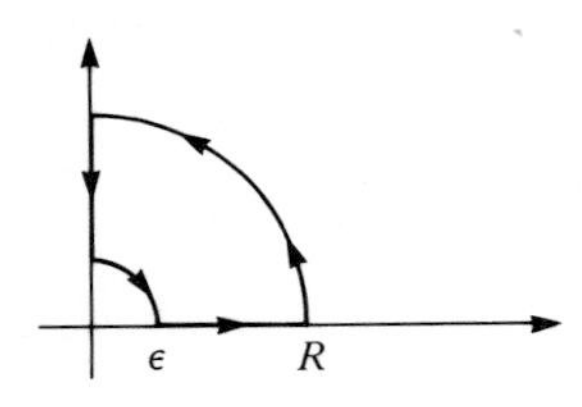

Figure 6.7–6

15. Show that

$$\int_{-\infty}^{+\infty} \frac{\cos ax - \cos bx}{x^2}\, dx = \pi(b - a), \qquad \text{where } b > 0,\ a > 0.$$

Hint: The integrand equals $\operatorname{Re}\left[\dfrac{e^{iax} - e^{ibx}}{x^2}\right]$.

6.8 CONTOUR INTEGRATIONS INVOLVING BRANCH POINTS AND BRANCH CUTS

Up to now we have been using contour integrations to evaluate integrals whose integrands are either rational functions or products of rational functions and trigonometric functions. In this section we will show how contour integration can often be used to evaluate integrals whose integrands are single-valued branches of multivalued functions; for example, $\int_0^\infty 1/[\sqrt{x}(x + 1)]\, dx$, where we have chosen the positive square root of x along the interval of integration.

Unlike the previous integrals considered, the present ones cannot be evaluated through a prescribed list of rules. What makes these problems more difficult is that their solution involves contour integrations taken along branch cuts and around branch points. Some of the techniques involved are illustrated in the following examples.

EXAMPLE 1

Find $\int_0^\infty (\operatorname{Log} x)/(x^2 + 4)\, dx$. Notice that $\operatorname{Log} x$ has a discontinuity at $x = 0$. This integral is thus defined as

$$\lim_{\substack{\varepsilon \to 0 \\ R \to \infty}} \int_\varepsilon^R \frac{\operatorname{Log} x\, dx}{x^2 + 4}.$$

Solution

Following earlier reasoning, we try to evaluate $\int_C \log z/(z^2 + 4)\, dz$ around a closed contour C, a portion of which in some limit will coincide with the positive x-axis. We will use the principal branch of $\log z$ since it agrees with $\operatorname{Log} x$ on the positive x-axis, and the integral along this portion of C becomes identical to that of the given problem. Other branches providing such agreement can also be used. The contour C is shown in Fig. 6.8–1.

The integral along the negative real axis is taken along the "upper side" of the branch cut.† Since we want the integrand to be analytic on and inside C, the branch point $z = 0$ is avoided by means of a semicircle of radius ε.

---

† Strictly speaking, we should use a contour of the shape shown in Fig. 6.8–2 and allow $\alpha \to 0+$. In this way we can define what is meant by the "upper side" of the branch cut.

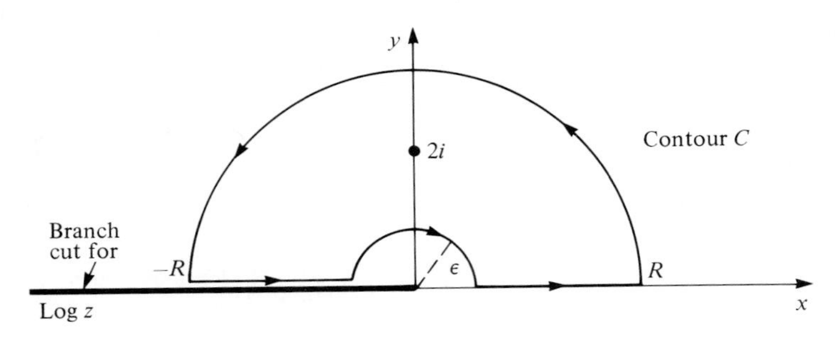

Figure 6.8–1

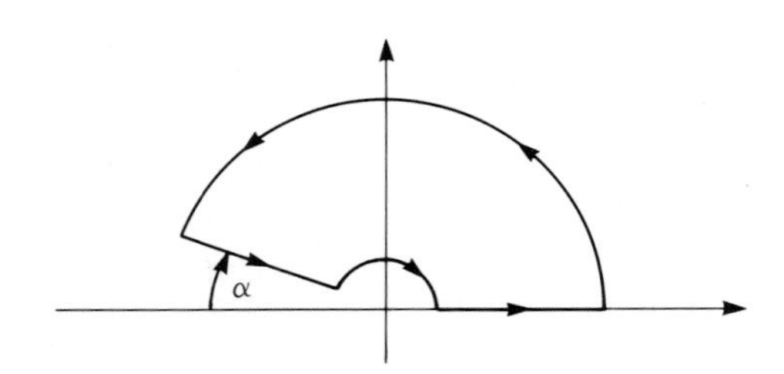

Figure 6.8–2

We now express $\oint_C \operatorname{Log} z/(z^2+4)\,dz$ in terms of integrals taken around the various parts of C. Note that the integrand has a simple pole at $z = 2i$. Thus

$$\int_{-R}^{-\varepsilon} \frac{\operatorname{Log} z}{z^2+4}\,dz + \int\limits_{\substack{|z|=\varepsilon \\ \curvearrowright}} \frac{\operatorname{Log} z}{z^2+4}\,dz + \int_{\varepsilon}^{R} \frac{\operatorname{Log} x}{x^2+4}\,dx + \int\limits_{\substack{|z|=R \\ \curvearrowleft}} \frac{\operatorname{Log} z}{z^2+4}\,dz$$

$$= 2\pi i \operatorname{Res}\left[\frac{\operatorname{Log} z}{z^2+4}, 2i\right]. \tag{6.8–1}$$

Consider the first integral on the left. For the principal branch of the logarithm $\operatorname{Log} z = \operatorname{Log}|z| + i \arg z$, $-\pi < \arg z \le \pi$. Since we are integrating along the upper side of the branch cut in this integral, we see that $\arg z$ is π while $|z| = |x|$. Thus

$$\int_{-R}^{-\varepsilon} \frac{\operatorname{Log} z\,dz}{z^2+4} = \int_{-R}^{-\varepsilon} \frac{\operatorname{Log}|x|}{x^2+4}\,dx + \int_{-R}^{-\varepsilon} \frac{i\pi}{x^2+4}\,dx. \tag{6.8–2}$$

The first integration on the right can be taken between the limits ε and R (the integrand is an even function) and $\operatorname{Log}|x|$ can be replaced by $\operatorname{Log} x$. Therefore

$$\int_{-R}^{-\varepsilon} \frac{\operatorname{Log} z\,dz}{z^2+4} = \int_{\varepsilon}^{R} \frac{\operatorname{Log} x\,dx}{x^2+4} + \int_{-R}^{-\varepsilon} \frac{i\pi}{x^2+4}\,dx. \tag{6.8–3}$$

Using Eq. (6.8–3) or the far left in Eq. (6.8–1) and combining two identical integrals we have

$$i\pi \int_{-R}^{-\varepsilon} \frac{dx}{x^2+4} + 2\int_{\varepsilon}^{R} \frac{\operatorname{Log} x}{x^2+4}\,dx + \int_{\substack{|z|=\varepsilon \\ \curvearrowright}} \frac{\operatorname{Log} z}{z^2+4}\,dz + \int_{\substack{|z|=R \\ \curvearrowleft}} \frac{\operatorname{Log} z}{z^2+4}\,dz$$

$$= 2\pi i \operatorname{Res}\left[\frac{\operatorname{Log} z}{z^2+4}, 2i\right]. \tag{6.8–4}$$

We need to show that the integrals over the semicircles of radii ε and R go to zero in the limits $\varepsilon \to 0$ and $R \to \infty$, respectively. We will present the first result; the derivation of the second is similar. The third integral on the left in Eq. (6.8–4) is

$$I = \int_{|z|=\varepsilon} \frac{\operatorname{Log} z}{z^2+4}\,dz, \tag{6.8–5}$$

where $z = \varepsilon e^{i\theta}$, $0 \le \theta \le \pi$, on the path of integration. We apply the ML inequality to I, where $L = \pi\varepsilon$ (the path length) and M is a constant satisfying

$$\left|\frac{\operatorname{Log} z}{z^2+4}\right| = \left|\frac{\operatorname{Log} \varepsilon e^{i\theta}}{\varepsilon^2 e^{2i\theta}+4}\right| \le M, \qquad 0 \le \theta \le \pi. \tag{6.8–6}$$

Thus

$$|I| \le M\pi\varepsilon. \tag{6.8–7}$$

If $0 \le \varepsilon \le 1$, then $|\varepsilon^2 e^{2i\theta} + 4| \ge 3$, and

$$\left|\frac{1}{\varepsilon^2 e^{i2\theta}+4}\right| \le \frac{1}{3}. \tag{6.8–8}$$

Also, notice that

$$|\operatorname{Log}(\varepsilon e^{i\theta})| = |\operatorname{Log}(\varepsilon) + i\theta| \le [|\operatorname{Log} \varepsilon| + \pi], \tag{6.8–9}$$

where we have recalled that $0 \le \theta \le \pi$. Combining Eqs. (6.8–9) and (6.8–8), we see that

$$\left|\frac{\operatorname{Log}(\varepsilon e^{i\theta})}{\varepsilon^2 e^{i2\theta}+4}\right| \le \frac{[|\operatorname{Log} \varepsilon| + \pi]}{3}. \tag{6.8–10}$$

A glance at Eq. (6.8–6) shows that the right side of Eq. (6.8–10) can be identified as M. The right side of Eq. (6.8–7) becomes $(\pi\varepsilon/3)[|\operatorname{Log} \varepsilon| + \pi]$. As $\varepsilon \to 0$, this expression goes to zero.† The integral in Eq. (6.8–5) and the integral over the semicircle of radius ε in Eq. (6.8–4) thus become zero as $\varepsilon \to 0$.

The residue on the right in Eq. (6.8–4) is easily found to be $(\operatorname{Log} 2 + i\pi/2)/(4i)$. Passing to the limits $\varepsilon \to 0$ and $R \to \infty$ in Eq. (6.8–4) and using the computed residue,

† To evaluate $\lim_{\varepsilon \to 0+} \varepsilon \operatorname{Log} \varepsilon$, let $-x = \operatorname{Log} \varepsilon$, $\varepsilon = e^{-x} = 1/e^x$. Consider $\lim_{x\to\infty} -x/e^x = 0$.

we have

$$i\pi \int_{-\infty}^{0} \frac{dx}{x^2+4} + 2\int_0^{\infty} \frac{\operatorname{Log} x}{x^2+4}\,dx = \frac{\pi}{2}\left[\operatorname{Log} 2 + \frac{i\pi}{2}\right].$$

Identifying real and imaginary parts in the above we have

$$\int_0^{\infty} \frac{\operatorname{Log} x}{x^2+4}\,dx = \frac{\pi}{4}\operatorname{Log} 2 \quad \text{and} \quad \int_{-\infty}^{0} \frac{dx}{x^2+4} = \frac{\pi}{4}.$$

The right-hand result is of course more easily found with the method shown in Section 6.5. ◀

EXAMPLE 2

Evaluate $\int_0^\infty dx/[x^{1/\alpha}(x+1)]$, where $\alpha > 1$ and $x^{1/\alpha} = \sqrt[\alpha]{x}$ for $0 \le x \le \infty$. The integrand is thus real and nonnegative within the limits of integration. Notice that as in the previous problem the integrand has a discontinuity at $x = 0$.

Solution

We will consider $\oint_C dz/[z^{1/\alpha}(z+1)]$, where the contour of integration C lies partly along the x-axis. When we pass to appropriate limits, the integration along this part of the contour reduces to the given integral. In this calculation we must use a specific branch of $z^{1/\alpha}$. With the polar representation $z = re^{i\theta}$ we choose $z^{1/\alpha} = \sqrt[\alpha]{r}\, e^{i\theta/\alpha}$, where $0 \le \theta < 2\pi$. This branch is analytic in a domain defined by a branch cut along $y = 0$, $x \ge 0$. The contour of integration C, which lies in this domain, is shown in Fig. 6.8–3. The circular path of radius ε is necessary in order to exclude the branch point of $z^{1/\alpha}$ from the path of integration.

We express $\oint_C dz/[z^{1/\alpha}(z+1)]$ as integrals along the four paths shown in the figure. Along path I, $z = r$, $dz = dr$, $z^{1/\alpha} = \sqrt[\alpha]{r}\, e^{i\theta/\alpha}$, where $\theta = 0$. Since $r = x$ on I, we have $z = x$, $dz = dx$, $z^{1/\alpha} = \sqrt[\alpha]{x}$. Along path III, $z = re^{i2\pi} = r$, $z^{1/\alpha} = \sqrt[\alpha]{r}\, e^{i\theta/\alpha}$, where

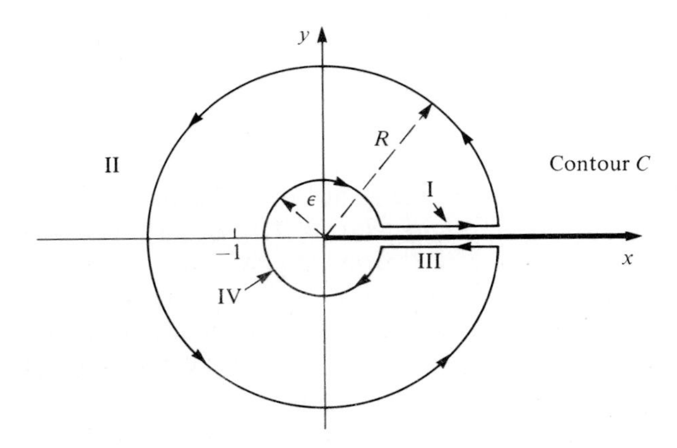

Figure 6.8–3

$\theta = 2\pi$. Since $r = x$ on III, this becomes $z = x$, $dz = dx$, $z^{1/\alpha} = \sqrt[\alpha]{x}\, e^{i2\pi/\alpha}$. Along path II, $z = Re^{i\theta}$, $z^{1/\alpha} = \sqrt[\alpha]{R}\, e^{i\theta/\alpha}$, $dz = iRe^{i\theta}\, d\theta$. And along path IV, $z = \varepsilon e^{i\theta}$, $z^{1/\alpha} = \sqrt[\alpha]{\varepsilon}\, e^{i\theta/\alpha}$, and $dz = i\varepsilon e^{i\theta}\, d\theta$.

The contour C encloses the simple pole of $1/[z^{1/\alpha}(z+1)]$ at $z = -1$. We have

$$\operatorname{Res}\left[\frac{1}{z^{1/\alpha}(z+1)}, -1\right] = \lim_{z \to -1} \frac{z+1}{z^{1/\alpha}(z+1)} = \left(\frac{1}{z^{1/\alpha}}\right)_{z=-1}$$
$$= \left[\frac{1}{\sqrt[\alpha]{r}\, e^{i\theta/\alpha}}\right]_{r=1,\, \theta=\pi} = \frac{1}{e^{i\pi/\alpha}}. \tag{6.8–11}$$

We have been careful to use the particular value of $(-1)^{1/\alpha}$ belonging to the chosen branch of $z^{1/\alpha}$. The integral around C, expressed in terms of integrals along the four paths mentioned and evaluated with Eq. (6.8–11), yields the equation

$$\int_{\varepsilon \atop \mathrm{I}}^{R} \frac{dx}{\sqrt[\alpha]{x}(x+1)} + \int_{0 \atop \mathrm{II}}^{2\pi} \frac{Re^{i\theta} i\, d\theta}{\sqrt[\alpha]{R}\, e^{i\theta/\alpha}(Re^{i\theta}+1)} + \int_{R \atop \mathrm{III}}^{\varepsilon} \frac{dx}{\sqrt[\alpha]{x}\, e^{i2\pi/\alpha}(x+1)}$$
$$+ \int_{2\pi \atop \mathrm{IV}}^{0} \frac{\varepsilon e^{i\theta} i\, d\theta}{\sqrt[\alpha]{\varepsilon}\, e^{i\theta/\alpha}[\varepsilon e^{i\theta}+1]} = \frac{2\pi i}{e^{i\pi/\alpha}}. \tag{6.8–12}$$

Let I be the integral along path II. Using the ML inequality we can show that I tends to zero as $R \to \infty$. Since $z = Re^{i\theta}$ we have $|I| \le ML$, where

$$\left|\frac{1}{z^{1/\alpha}(z+1)}\right| = \frac{1}{\sqrt[\alpha]{R}\,|Re^{i\theta}+1|} \le M$$

and $L = 2\pi R$. Notice that, for $R > 1$,

$$\frac{1}{\sqrt[\alpha]{R}\,|Re^{i\theta}+1|} \le \frac{1}{\sqrt[\alpha]{R}(R-1)}.$$

Thus we can take

$$M = \frac{1}{\sqrt[\alpha]{R}(R-1)},$$

and we observe that

$$\lim_{R\to\infty} (ML) = \lim_{R\to\infty} \frac{2\pi}{\sqrt[\alpha]{R}(1-1/R)} = 0.$$

Thus the integral over path II in Eq. (6.8–12) goes to zero as $R \to \infty$. A similar discussion demonstrates that the integral over path IV in the same equation becomes zero as $\varepsilon \to 0$. Taking the limits $R \to \infty$, $\varepsilon \to 0$ in Eq. (6.8–12), we now have

$$\int_0^\infty \frac{dx}{\sqrt[\alpha]{x}(x+1)} + e^{-i2\pi/\alpha} \int_\infty^0 \frac{dx}{\sqrt[\alpha]{x}(x+1)} = 2\pi i e^{-i\pi/\alpha}. \tag{6.8–13}$$

Reversing the limits on the second integral on the left, compensating with a minus sign, and multiplying both sides of Eq. (6.8–13) by $e^{i\pi/\alpha}$, we get

$$\int_0^\infty \frac{dx}{\sqrt[\alpha]{x}\,(x+1)}(e^{i\pi/\alpha} - e^{-i\pi/\alpha}) = 2\pi i. \tag{6.8–14}$$

The exponentials inside the parentheses sum to $2i\sin(\pi/\alpha)$. Dividing by this factor we have

$$\int_0^\infty \frac{dx}{x^{1/\alpha}(x+1)} = \frac{\pi}{\sin(\pi/\alpha)}, \qquad \alpha > 1.$$

Comment: We might try to solve this problem by means of a closed semicircular contour like that used in Example 1. However, because the pole of $1/[z^{1/\alpha}(z+1)]$ at $z = -1$ lies along this contour, an indentation must be made around this point. Such an approach is investigated in Exercise 7 below. ◀

EXERCISES

By employing the contour in Fig. 6.8–1 prove the following for $a > 0$.

1. $\displaystyle\int_0^\infty \frac{\operatorname{Log} x}{x^2 + a^2}\,dx = \frac{\pi}{2a}\operatorname{Log} a$

2. $\displaystyle\int_0^\infty \frac{\operatorname{Log} x}{(x^2 + a^2)^2}\,dx = \frac{\pi}{4a^3}\operatorname{Log}(a/e)$

3. $\displaystyle\int_0^\infty \frac{x^2 \operatorname{Log} x}{x^4 + a^4}\,dx = \frac{\pi\sqrt{2}}{4a}\operatorname{Log} a + \frac{\pi^2\sqrt{2}}{16a}$

4. Use a contour like Fig. 6.8–1 with additional semicircular indentations at $z = \pm a$ to establish the following Cauchy principal value:

$$\int_0^\infty \frac{\operatorname{Log} x}{x^2 - a^2}\,dx = \frac{\pi^2}{4a}, \qquad a > 0.$$

Hint: Your branch cut should extend from $z = 0$ into the lower half plane.

By employing the contour in Fig. 6.8–3 prove the following for $a > 0$ and $x^\beta \geq 0$.

5. $\displaystyle\int_0^\infty \frac{x^\beta}{(x + a)^2}\,dx = \frac{\pi}{a}\,\frac{\beta a^\beta}{\sin\beta\pi}, \quad -1 < \beta < 1,\ \beta \neq 0$

6. $\displaystyle\int_0^\infty \frac{x^\beta}{x^2 + a^2}\,dx = \frac{\pi a^\beta}{2a\cos\beta\pi/2}, \quad -1 < \beta < 1$

7. a) Evaluate the integral of Example 2 by using the indented semicircular contour C shown in Fig. 6.8–4 and passing to appropriate limits.

Hint: Consider $\int_C dz/[z^{1/\alpha}(z + 1)]$. Take the required limits for R and ε. Evaluate the integral by employing a residue. Equate real and imaginary parts on both sides of the resulting equation.

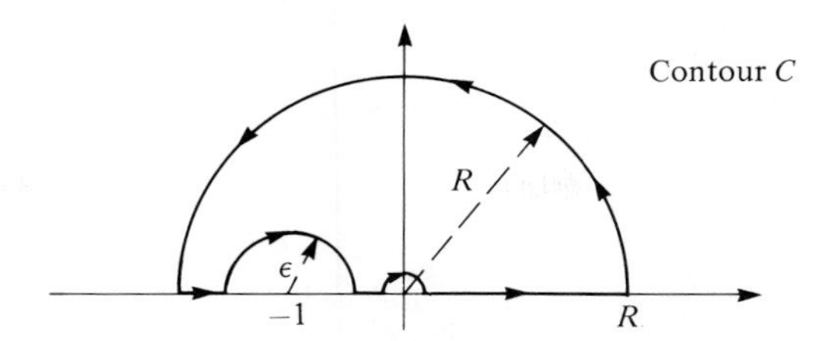

Figure 6.8–4

Solve the resulting pair of equations simultaneously for an unknown integral. Check your result using Example 2.

b) Use a method similar to that of part (a) to show that

$$\int_0^\infty \frac{du}{u^{1/\alpha}(u-1)} = \pi \cot\frac{\pi}{a} \qquad \text{(Cauchy principal value),}$$

where $\alpha > 1$ and $u^{1/\alpha} \geq 0$.

8. Show that

$$\int_0^\infty \frac{x^{1/\alpha}\,dx}{x^2 - a^2} = \frac{\pi}{2a}\,\frac{a^{1/\alpha}}{\sin\left(\dfrac{\pi}{\alpha}\right)}\left[1 - \cos\frac{\pi}{\alpha}\right] \qquad \text{(Cauchy principal value),}$$

where $a > 0$, $x^{1/\alpha} = \sqrt[\alpha]{x}$, $-1 < 1/\alpha < 1$.

9. Use the contour of Fig. 6.8–1 to prove, as parts of the same problem, that, for $a > 0$,

$$\int_0^\infty \frac{\sqrt{x}\,\operatorname{Log} x}{x^2 + a^2}\,dx = \frac{\pi}{\sqrt{2a}}\left(\operatorname{Log} a + \frac{\pi}{2}\right)$$

and

$$\int_0^\infty \frac{\sqrt{x}}{x^2 + a^2}\,dx = \frac{\pi}{\sqrt{2a}}.$$

10. Show that for $a > 0$, $\beta > 1$, $x^{1/\beta} = \sqrt[\beta]{x} \geq 0$, we have

$$\int_0^\infty \frac{\operatorname{Log} x}{x^{1/\beta}(x + a)}\,dx = \frac{\pi}{\sqrt[\beta]{a}\,\sin \pi/\beta}\left(\pi \cot\frac{\pi}{\beta} + \operatorname{Log} a\right).$$

Hint: Use the contour of Fig. 6.8–3. Employ a branch of $\log z$ that assumes values identical to $\operatorname{Log} x$ on top of the branch cut.

11. Show that, for $a > 0$ and $v > 0$,

$$\int_0^\infty \operatorname{Log}\frac{a^2 + x^2}{x^2}\cos vx\,dx = \frac{\pi}{v}(1 - e^{-av}).$$

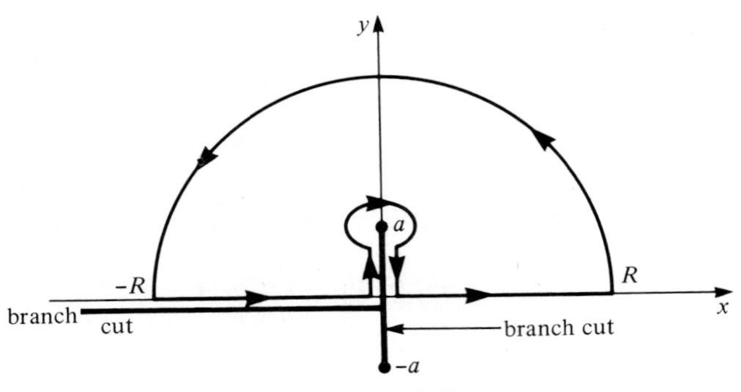

Figure 6.8–5

Hint: Evaluate $\int \operatorname{Log}[(a^2 + z^2)/z^2]e^{ivz}\,dz$ around the contour shown in Fig. 6.8–5. The integrals along each side of the branch cut ($x = 0$, $-a \le y \le a$) are easy. To argue that as $R \to \infty$ the integral around the semicircle $|z| = R$ tends to zero, expand $\operatorname{Log}(1 + a^2/z^2)$ in a Laurent series valid for $|z| > a$. Integrate term by term, and apply Jordan's Lemma to each integration.

12. Show that, for $a > 0$ and $v > 0$,

$$\int_0^\infty \tan^{-1}\left(\frac{a}{x}\right) \sin vx\,dx = \frac{\pi}{2v}(1 - e^{-va}).$$

Take $0 \le \tan^{-1}(a/x) \le \pi/2$.

Hint: Consider

$$\int \operatorname{Log}\frac{z + ia}{z} e^{ivz}\,dz$$

around a contour like that in Fig. 6.8–1. The branch cut for $\operatorname{Log}[(z + ia)/z]$ is a line from $z = 0$ to $z = -ia$. Use a technique from Exercise 11 to eliminate the integral on the large semicircle as $R \to \infty$. The result of Exercise 11 is also needed to complete the proof.

13. The modified Bessel function of the second kind, of order zero, $K_0(w)$, is defined for $w > 0$ by

$$K_0(w) = \int_0^\infty \frac{\cos wx}{\sqrt{x^2 + 1}}\,dx = \frac{1}{2}\int_{-\infty}^{+\infty} \frac{e^{iwx}}{\sqrt{x^2 + 1}}\,dx.$$

This function occurs in problems involving radiation. The integral must be evaluated numerically. If $w \gg 1$ the numerical evaluation becomes difficult because of the rapid oscillation of the integrand.

a) Using a contour integration in the upper half of the complex z-plane show that an equivalent form is

$$K_0(w) = \int_1^\infty \frac{e^{-wy}}{\sqrt{y^2 - 1}}\,dy.$$

Hint: Use branch cuts along $x = 0$, $|y| \geq 1$.

The preceding integral is more amenable to numerical integration.

b) Explain why if $w \gg 1$ we have

$$K_0(w) \approx \int_1^{\infty} \frac{e^{-wy}}{\sqrt{2}\sqrt{y-1}}\,dy.$$

c) Make a change of variable in the preceding expression and show that

$$K_0(w) \approx \frac{e^{-w}}{\sqrt{2}} \int_0^{\infty} \frac{e^{-wt}}{\sqrt{t}}\,dt.$$

d) Let $x^2 = t$ in the preceding integral. The resulting integral is of known value (see Exercise 25, Section 6.6). Thus prove that

$$K_0(w) \approx \sqrt{\frac{\pi}{2w}}\,e^{-w} \qquad \text{for } w \gg 1.$$

6.9 INTEGRATION AROUND INFINITY AS A TOOL FOR EVALUATING DEFINITE INTEGRALS

We remarked in Section 1.5 that we will sometimes think of infinity as a complex number and that there is a point corresponding to this number that we can display on the Riemann number sphere (but not on the Cartesian plane). Some definite integrals can be evaluated fairly easily if we now reconsider the point at infinity and introduce the notion of integration along contours that enclose infinity.

As z approaches infinity, the quantity $|z|$ grows without bound. The moduli of such analytic functions as $f(z) = z$ and $f(z) = z^2 + 1$ become unbounded in a way reminiscent of the behavior of analytic functions near poles. Indeed, when we think of infinity as a point in the extended complex plane, we can speak of the behavior of $f(z)$ in the neighborhood of the point and define the nature of the singularity of $f(z)$ at $z = \infty$.

Let $z = 1/w$. As $w \to 0$, $|z| \to \infty$. The behavior displayed by $f(z)$ near $z = \infty$ corresponds to the behavior of $F(w) = f(1/w)$ near $w = 0$. Thus we have the following:

DEFINITION Behavior at infinity

The function $f(z)$ is said to have a singularity at $z = \infty$ if the function $F(w)$ has a singularity at $w = 0$, where $F(w) = f(1/w)$. If $F(w)$ is analytic at $w = 0$, then $f(z)$ is said to be analytic at $z = \infty$.

The kind of singularity (pole, branch point, removable, etc.) possessed by $f(z)$ at $z = \infty$ is defined as being identical to the kind of singularity possessed by $F(w)$ at $w = 0$. □

In order to study the behavior of a function in the extended complex plane, it is convenient to employ the rules introduced in Section 1.5 for using ∞ in various calculations. The reader should review these rules since they will be employed in this section.

From our definition $f(z) = z^2$ has a pole of order two at ∞, since $F(w) = 1/w^2$ has a pole of order two at the origin. Also $f(z) = e^z$ has an essential singularity at ∞, since $e^{1/w}$ has the same singularity at $w = 0$; similarly $z^{1/2}$ has a branch point singularity at ∞ because $1/w^{1/2}$ has a branch point at $w = 0$.

Now consider $f(z) = 1/z$. We have $F(w) = 1/(1/w)$. Putting $w = 0$ we find by using the rules of Section 1.5 that $F(0) = 1/\infty = 0$. Note that for $w \neq 0$ we have $F(w) = w$. Thus $f(z) = 1/z$ is analytic at $z = \infty$, since $F(w) = w$ is analytic at $w = 0$.

Finally consider $f(z) = z \sin 1/z$. We have $F(w) = (1/w) \sin w$, which we recognize as having a removable singularity at $w = 0$ (we take $F(0) = 1$). Thus $f(z)$ has a removable singularity at $z = \infty$.

Presently we will define the meaning of the "residue at infinity," but it is perhaps not too soon to warn the reader not to be quick to use those formulas learned for obtaining the residues at finite points in the complex plane. The residue of $f(z)$ at $z = \infty$ *cannot* in general be computed by finding the residue of $F(w) = f(1/w)$ at $w = 0$.

Recall that the residue at a finite point in the complex plane was defined by means of a contour integral (see Eq. (6.1–1)). We integrated along the contour in the positive sense. The contour was the boundary of a bounded domain containing the singular point in question and no other singularities of the function. An analogous definition provides the residue at infinity.

Suppose $f(z)$ is analytic except possibly at $z = \infty$ and at points lying within a bounded domain. Then recalling the principle of deformation of contours (see Section 4.3), we see that all integrals of $f(z)$ taken in the same sense around simple closed contours whose interiors contain all the singularities of $f(z)$ lying in the finite z-plane have the same value. Thus we make the following definition:

DEFINITION Residue of $f(z)$ at ∞

$$\text{Res}[f(z), \infty] = \frac{1}{2\pi i} \oint_C f(z)\,dz, \tag{6.9–1}$$

where C is a simple closed contour such that all singularities of $f(z)$ in the finite plane are in the domain inside C. In addition $f(z)$ is assumed analytic outside C except that $f(z)$ is permitted to have a singularity at infinity. □

Note the sense of integration around C—the opposite of our usual direction. The situation is shown in Fig. 6.9–1. The direction of integration is such that the unbounded domain lying outside C and containing $z = \infty$ is on our left as we negotiate the contour. We should imagine that we are enclosing the domain containing the point $z = \infty$ in the positive sense.

This can be appreciated if we project a simple closed curve C onto the Riemann number sphere (Section 1.5) as shown in Fig. 6.9–2. A loop is formed on the sphere and two domains now exist on the sphere; their boundaries are the loop. One domain, the one above the loop, contains the north pole, which, we recall, corresponds to $z = \infty$ in the x,y-plane. In traveling around the projection of C on the sphere we

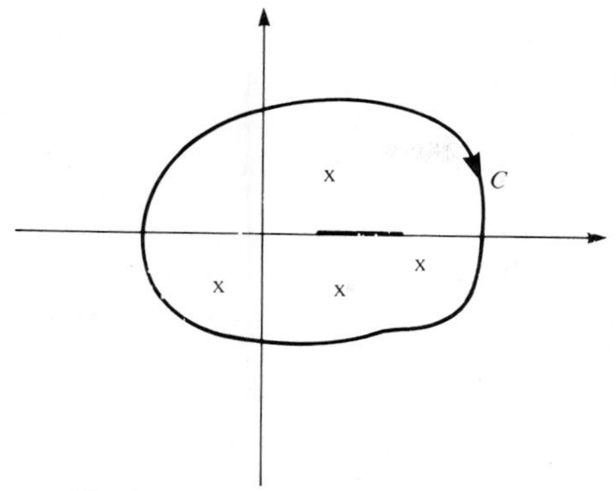

The singular points of $f(z)$ in the finite plane are inside the loop. A branch cut is shown as well as isolated singular points.

Figure 6.9–1

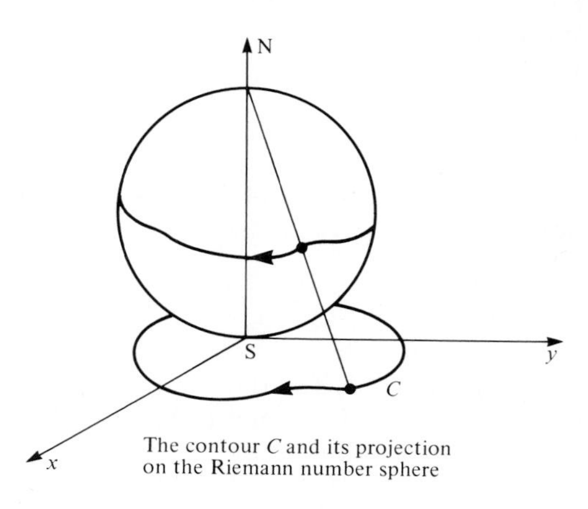

The contour C and its projection on the Riemann number sphere

Figure 6.9–2

encircle both domains, one in the positive sense and the other in the negative sense. Reversing the direction of travel reverses the sense in which each domain is enveloped.

EXAMPLE 1

Using Eq. (6.9–1), find the residues of the following functions at $z = \infty$. State the nature of the singularity at ∞.

a) $f(z) = k$ (a constant)

b) $f(z) = z$

c) $f(z) = 1/z$

d) $f(z) = 1/(z^2 + 1)$

Solution

Part (a): Any constant has derivative zero and is analytic throughout the extended complex plane. Substituting k for $f(z)$ in Eq. (6.9–1) we obtain as our residue $(1/2\pi i)\oint k\,dz$. Any simple closed contour can be used, and the residue is zero by the Cauchy-Goursat Theorem.

Part (b): Since $f(z) = z$, we have $F(w) = 1/w$ in our definition of the singularity at ∞. Because $F(w)$ has a simple pole at $w = 0$ we have that $f(z)$ has a simple pole at $z = \infty$. According to Eq. (6.9–1) we have $\text{Res}[z, \infty] = (1/2\pi i) \oint z\,dz$, where any closed contour can be used since $f(z)$ has no singularities in the finite complex plane. The value of this integral is zero by the Cauchy-Goursat Theorem. Thus we see that *a function with a pole at infinity can have a residue of zero at infinity.* Note that a function having a simple pole at a point other than infinity cannot have residue of zero at this pole (recall the form of the Laurent expansion about this pole).

Part (c): Earlier in this section we showed that $f(z) = 1/z$ is analytic at ∞. From Eq. (6.9–1) we have $\text{Res}[1/z, \infty] = (1/2\pi i) \oint 1/z\,dz$, where our contour can be $|z| = r$, $r > 0$. Now $1/z$ has a simple pole at $z = 0$ and our result is -1 (the clockwise direction of integration gives the minus sign). Thus *a function can be analytic at infinity and have a nonzero residue at infinity.*

Part (d): If $f(z) = 1/(z^2 + 1)$, then $F(w) = 1/(1/w^2 + 1)$. If $w \neq 0$ this function is identical to $w^2/(w^2 + 1)$, while the rules of Section 1.5 show that $F(0) = 0$. Thus $F(w)$ is analytic at $w = 0$ and $f(z)$ is analytic at ∞. From Eq. (6.9–1) we have

$$\text{Res}\left[\frac{1}{z^2+1}, \infty\right] = \frac{1}{2\pi i}\oint \frac{1}{z^2+1}\,dz,$$

where our contour can be $|z| = r$, $r > 1$. The right side equals the negative of the sum of the residues of $1/(z^2 + 1)$ at $\pm i$ and is found to be zero. ◄

There is a direct connection between the residue of a function at ∞ and a coefficient in a Laurent expansion. Assume that all the singularities of $f(z)$ in the finite complex plane lie in the disc $|z| \le r$ and that $f(z)$ is analytic throughout the remainder of the finite plane. Thus there exists a Laurent expansion of $f(z)$ of the form

$$f(z) = \cdots + c_{-2}z^{-2} + c_{-1}z^{-1} + c_0 + c_1 z + c_2 z^2 + \cdots, \tag{6.9–2}$$

valid for $|z| > r$. When $z = \infty$ we cannot evaluate the terms of this series having positive exponents. Thus we say that the preceding series is valid in a *deleted neighborhood of infinity.*

Using Eq. (6.9–2) in Eq. (6.9–1) and integrating term by term along C we obtain the following.

RULE I Residue at ∞

$$\text{Res}[f(z), \infty] = -c_{-1}, \tag{6.9–3}$$

where c_{-1} is the coefficient of z^{-1} in a Laurent expansion of $f(z)$ valid in a domain $|z| > r$, where r is a positive real constant.

The details are not supplied here because they are virtually identical to the derivation of Eq. (6.1–5), which yields the residue at a point in the finite plane. Note that the minus sign on the right in Eq. (6.9–3) is due to the clockwise sense of integration employed.

In Exercise 27 we easily establish the following useful rule, which follows from Eq. (6.9–3):

RULE II Residue at ∞

If $f(z)$ is analytic at ∞ or has a removable singularity at ∞, and if $\lim_{z\to\infty} f(z) = 0$, then $\text{Res}[f(z), \infty] = \lim_{z\to\infty} -zf(z)$. □

Thus, for example, to get the residue of $1/(1 + z)$ at ∞ we have

$$\lim_{z\to\infty} \frac{-z}{1+z} = \lim_{z\to\infty} \frac{-1}{1+1/z} = -1.$$

If we make the change of variables $w = 1/z$ in Eq. (6.9–2) we obtain the Laurent expansion

$$F(w) = f(1/w) = \cdots + c_{-2}w^2 + c_{-1}w + c_0 + c_1 w^{-1} + c_2 w^{-2} + \cdots, \quad (6.9\text{–}4)$$

which is valid for $|1/w| > r$. We must also avoid $w = 0$, where the terms with negative exponents become undefined. Thus Eq. (6.9–4) is a valid expansion of $F(w)$ for $0 < |w| < 1/r$. Now dividing Eq. (6.9–4) by w^2 we have

$$\frac{1}{w^2}F(w) = \cdots + c_{-2} + c_{-1}w^{-1} + c_0 w^{-2} + c_1 w^{-3} + c_2 w^{-4} + \cdots.$$

The preceding is a Laurent expansion of $F(w)/w^2$ valid in a deleted neighborhood of $w = 0$. Now from the definition of the residue we have from the above $\text{Res}[(1/w^2)F(w), 0] = c_{-1}$. Since $c_{-1} = -\text{Res}[f(z), \infty]$ (see Eq. (6.9–3)), we have the following.

RULE III Residue at ∞

$$\text{Res}[f(z), \infty] = -\text{Res}[w^{-2}F(w), 0], \quad (6.9\text{–}5)$$

where $F(w) = f(1/w)$. □

Equation (6.9–5) allows us to use our rules for getting residues in the finite plane to get residues at infinity. Notice that Rule II, which requires that $\lim_{z\to\infty} f(z) = 0$ for its application, is more restrictive than Rules I and III, which do not have this requirement.

EXAMPLE 2

Find $\text{Res}[(z - 1)/(z + 1), \infty]$ by Rule III and Rule I.

Solution

Since $f(z) = (z - 1)/(z + 1)$, we have

$$F(w) = f(1/w) = \frac{1/w - 1}{1/w + 1} = \frac{1 - w}{1 + w}.$$

Now from Eq. (6.9–5), Rule III, we have

$$\operatorname{Res}[f(z), \infty] = \operatorname{Res}\left[-\frac{1}{w^2}\left(\frac{1-w}{1+w}\right), 0\right].$$

This function has a second order pole at $w = 0$. From Eq. (6.3–4) we find its residue to be

$$\lim_{w\to 0} -\frac{d}{dw}\left(\frac{1-w}{1+w}\right) = \lim_{w\to 0}\frac{2}{(1+w)^2} = 2,$$

which is the answer.

Now we apply Rule I. We have

$$\frac{z-1}{z+1} = \frac{1-1/z}{1+1/z} = \left(1-\frac{1}{z}\right)\left(1-\frac{1}{z}+\frac{1}{z^2}-\frac{1}{z^3}+\cdots\right)$$
$$= \left(1-\frac{2}{z}+\frac{2}{z^2}-\cdots\right).$$

The Laurent expansion on the above right is valid for $|z| > 1$. Since the coefficient of z^{-1} in the expansion is -2 we have from Rule I that the required residue is 2. ◀

The following useful theorem can easily be proved:

THEOREM 7 Let $f(z)$ be analytic throughout the complex plane except at a finite number of isolated singular points. Then the sum of the residues of $f(z)$ at all singular points in the finite plane plus the residue at infinity is zero. □

For our proof, we consider the integral $\oint f(z)\,dz$ around a circle $|z| = R$, where R is so large that all singularities of $f(z)$ in the finite z-plane are enclosed in the positive sense. The integral equals $2\pi i$ times the sum of the residues of $f(z)$ at its singular points in the finite plane. Since $f(z)$ is analytic outside $|z| = R$ except perhaps at $z = \infty$, we see from Eq. (6.9–1) that our integral $\oint f(z)\,dz = -2\pi i \operatorname{Res}[f(z), \infty]$. Thus

$$-2\pi i \operatorname{Res}[f(z), \infty] = 2\pi i \sum \operatorname{Res}[f(z), \text{all isolated singularities, finite plane}]$$

Adding $2\pi i \operatorname{Res}[f(z), \infty]$ to both sides of the preceding equation yields the theorem.

The preceding theorem applies to the rational function

$$f(z) = \frac{a_n z^n + a_{n-1}z^{n-1} + \cdots + a_0}{b_m z^m + b_{m-1}z^{m-1} + \cdots + b_0}, \tag{6.9–6}$$

where n and m are integers. The singularities in the finite plane occur at the m roots of the denominator. There can also be an isolated singularity at infinity.

The following conclusion applies to some rational functions and is derived in Exercise 30.

RULE IV Residue at ∞

Let a rational function $f(z)$ consist of the quotient of two polynomials in z. If m is the degree of the denominator and n the degree of the numerator, then

$$\text{Res}[f(z), \infty] = 0 \qquad \text{if } m - n \geq 2. \quad \square$$

EXAMPLE 3

Find

$$\oint \frac{z^5 + z + 1}{z^6 + 1}\, dz$$

around $|z| = 3$ by using a residue at ∞.

Solution

The denominator $z^6 + 1$ is zero at six points inside $|z| = 3$, and the value of the integral is $2\pi i$ times the sum of the residues at these six locations. The calculation is a little tedious. The integrand has perhaps a nonzero residue at ∞. By the previous theorem, the sum of the residues inside $|z| = 3$ is the negative of the residue at ∞. Calling the given integral I, we have

$$I = -2\pi i \,\text{Res}\left[\frac{z^5 + z + 1}{z^6 + 1}, \infty\right].$$

Applying Eq. (6.9–5) we obtain

$$I = 2\pi i \,\text{Res}\left[\frac{\dfrac{1}{w^2}\left(\dfrac{1}{w^5} + \dfrac{1}{w} + 1\right)}{\dfrac{1}{w^6} + 1}, 0\right] = 2\pi i \,\text{Res}\left[\frac{1 + w^4 + w^5}{w(1 + w^6)}, 0\right] = 2\pi i.$$

(Note that the final expression has a simple pole at $w = 0$.) ◀

Comment: If we change the preceding problem and consider $\oint (z^4 + z + 1)/(z^6 + 1)\, dz$ around the same contour, we can conclude immediately that the integral is zero. Following the same procedure as in the example, we find the result is $-2\pi i \,\text{Res}[(z^4 + z + 1)/(z^6 + 1), \infty]$. By Rule IV this residue is zero.

In Section 6.7 we learned that if we integrate an analytic function along an arc that "partially" encloses a simple pole of that function, then, as the radius of the arc shrinks to zero, the integral equals in this limit some fraction of the residue of the function (at the pole) times $2\pi i$. The fraction of the residue to be used is the fraction of a full circle described by the arc. A similar statement applies with some modification to integration around the point at infinity, i.e., the following.

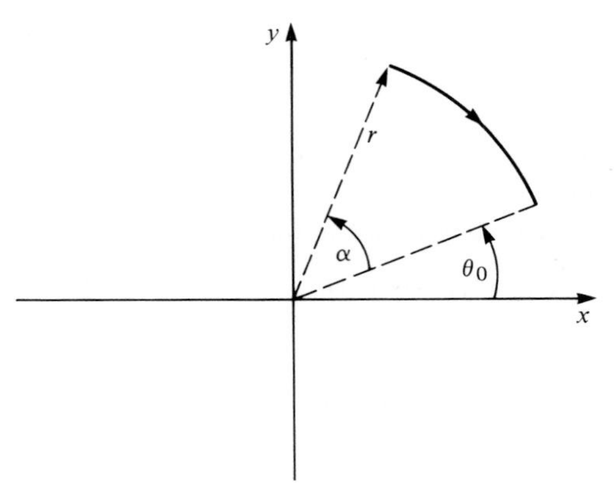

Figure 6.9–3

THEOREM 8 Let $f(z)$ be analytic at ∞ with $f(\infty) = 0$. Let C be the arc $|z| = r$, $\theta_0 \le \arg z \le \theta_0 + \alpha$ (see Fig. 6.9–3). Then

$$\lim_{r\to\infty} \int_C f(z)\, dz = \frac{\alpha}{2\pi} 2\pi i \operatorname{Res}[f(z), \infty] = \alpha i \operatorname{Res}[f(z), \infty], \qquad (6.9\text{–}7)$$

where the integration is done *clockwise* on C. □

The theorem is somewhat restrictive—it requires analyticity at ∞. We do not integrate around a pole at ∞. If a function has a removable singularity at ∞ the theorem can still be applied if the singularity can be removed by defining $f(\infty) = 0$. When the theorem is applicable it tells us in particular the value of a contour integral on a semicircle half enclosing infinity in the positive sense. As the radius of the arc tends to zero, the integral equals $2\pi i$ times half the residue at ∞.

The proof of the theorem is quite similar to the proof of Theorem 6 of Section 6.7, where we integrate along an arc that goes part way around a pole. We obtain only a fraction of the residue at that pole when the radius of the arc shrinks to zero. To make the present proof resemble that one, we first make the change of variable $z = 1/w$ on the left in Eq. (6.9–7) so that the radius of the resulting contour of integration will shrink to zero as $|z| = r \to \infty$. The remaining details of the proof are given in Exercise 33.

EXAMPLE 4

Using the concept of the residue at infinity, evaluate the Cauchy principal value† of

$$\int_{-\infty}^{+\infty} \frac{x}{x^2 + x + 1}\, dx.$$

† Because for $|x| \gg 1$ the integrand varies as $1/x$ this integral does not exist in the ordinary sense but *must* be evaluated as a Cauchy principal value (see Section 6.5).

Solution

The degree of the denominator less the degree of the numerator is 1, which is less than 2. Thus we cannot apply Theorem 4 of Section 6.5. To evaluate the given integral, consider $\oint z/(z^2+z+1)\,dz$ around the semicircular contour C of Fig. 6.5–1. The equation $z^2+z+1=0$ has roots at $-1/2 \pm i\sqrt{3}/2$, and we take R large enough so that C encloses them. Thus following the method leading up to Eq. (6.5–6), we obtain

$$\int_{-R}^{+R} \frac{x}{x^2+x+1}\,dx + \int_{C_1} \frac{z}{z^2+z+1}\,dz = 2\pi i \operatorname{Res}\left[\frac{z}{z^2+z+1}, -\frac{1}{2}+\frac{i\sqrt{3}}{2}\right]. \tag{6.9–8}$$

In Section 6.5 we let $R \to \infty$ and argued that the integral over C_1 becomes zero in this limit, but that analysis does not apply here. Instead we again let $R \to \infty$ but use Theorem 8 to show that, along C_1,

$$\lim_{R\to\infty} \int \frac{z}{z^2+z+1}\,dz = -\pi i \operatorname{Res}\left[\frac{z}{z^2+z+1}, \infty\right].$$

The minus sign arises from the counterclockwise sense of integration.

Combining this result with Eq. (6.9–8) (with $R \to \infty$) we have

$$\int_{-\infty}^{+\infty} \frac{x}{x^2+x+1}\,dx = 2\pi i \operatorname{Res}\left[\frac{z}{z^2+z+1}, -\frac{1}{2}+\frac{i\sqrt{3}}{2}\right] + \pi i \operatorname{Res}\left[\frac{z}{z^2+z+1}, \infty\right]$$

The first residue on the right is found from Eq. (6.3–6) to be $(-1/2 + i\sqrt{3}/2)/(i\sqrt{3})$, while the second is found from Rule II of the present section to be

$$\lim_{z\to\infty} \frac{-z^2}{z^2+z+1} = \lim_{z\to\infty} \frac{-1}{1+1/z+1/z^2} = -1.$$

Thus

$$\int_{-\infty}^{+\infty} \frac{x}{x^2+x+1}\,dx = \frac{2\pi}{\sqrt{3}}\left(-\frac{1}{2}+\frac{i\sqrt{3}}{2}\right) - \pi i = \frac{-\pi}{\sqrt{3}}.$$ ◀

Sometimes we will encounter functions $f(z)$ having nonisolated singularities that lie in a bounded region of the complex plane. The function is analytic outside this region except perhaps at a finite number of isolated singular points, one of which can be infinity. An example of such a function is the branch of $z^{1/2}(z-1)^{1/2}$ discussed in Example 3, part (c), in Section 3.8. The branch cut employed is the line segment $y=0$, $0 \le x \le 1$ (see Fig. 3.8–6c). This function has no other singularities except for a pole at infinity (which the reader should verify). Thus all the nonisolated singularities lie along the branch cut of finite length.

It is convenient to be able to perform integrations around a simple closed contour whose interior contains the region with the nonisolated singular points. The value of the integral is determined by the following theorem, which we shall prove.

THEOREM 9 Let $f(z)$ be analytic on a simple closed contour C, and let $f(z)$ be analytic in the unbounded domain D *outside* C except possibly at a finite number of isolated singular points. Then

$$\oint_C f(z)\,dz = -2\pi i \sum \text{all residues of } f(z) \text{ in } D. \tag{6.9–9}$$

The residue at ∞, if any, must be included in Eq. (6.9–9). □

Recall that $f(z)$ may have a nonzero residue at ∞ even if $f(z)$ is analytic at ∞.

To prove this theorem, refer to Fig. 6.9–4(a). The points $s_1, s_2, \ldots, s_n$ refer to the n singularities of $f(z)$ lying outside C in the finite plane. The circle C_0, centered at $z = 0$, has a radius so large that $s_1, s_2, \ldots, s_n$ are enclosed. We draw two lines, shown with dashes, from C to C_0. The lines are constructed so as not to intersect any of the singular points of $f(z)$, and they form two simple closed contours C_1 and C_2 as shown in Fig. 6.9–4(b). Because of cancellation of the integrations along the contiguous straight line segments shown in the figure, we have

$$\oint_{C_1} f(z)\,dz + \oint_{C_2} f(z)\,dz = \oint_C f(z)\,dz + \oint_{C_0} f(z)\,dz.$$

Note the directions of the arrows. The two integrals on the left can be evaluated by the residue theorem (Theorem 2, Section 6.1). Their sum is $-2\pi i \sum$ residues of $f(z)$ at $s_1, s_2, \ldots, s_n$. The minus sign is from the clockwise sense of integration. The integral on the far right is by definition $2\pi i \operatorname{Res}[f(z), \infty]$. Thus with a little rearranging we have

$$\oint_C f(z)\,dz = -2\pi i \sum \text{residues of } f(z) \text{ at infinity and all isolated singular points.}$$

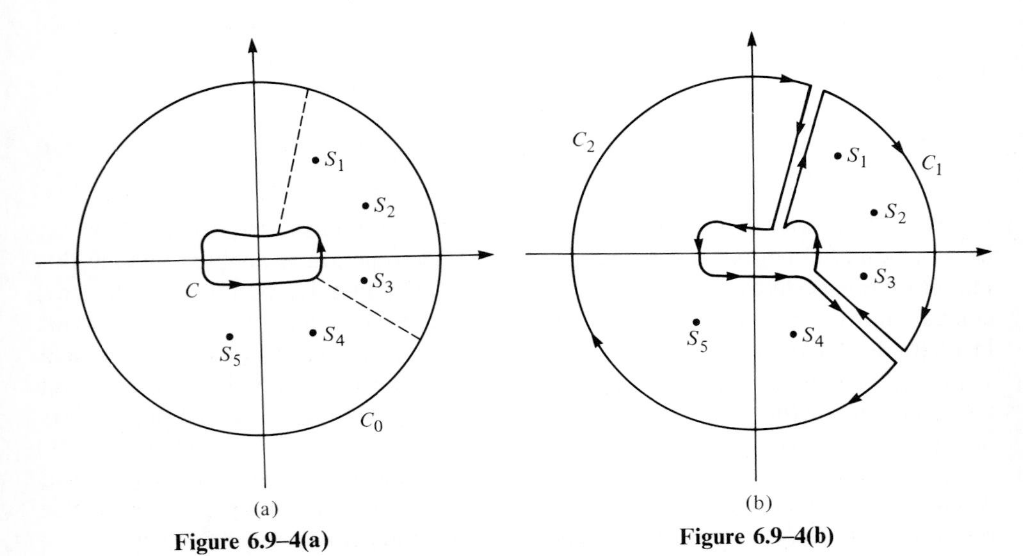

Figure 6.9–4(a) Figure 6.9–4(b)

Reversing the sense of integration we have, equivalently,

$$\oint_C f(z)\,dz = 2\pi i \sum \text{residues of } f(z) \text{ at infinity and all isolated singular points.} \tag{6.9–10}$$

EXAMPLE 5

Find

$$\int_0^1 \frac{\sqrt{x}\sqrt{1-x}}{x^2+1}\,dx$$

by using the preceding theorem.

Solution

We consider the function

$$f(z) = \frac{z^{1/2}(1-z)^{1/2}}{z^2+1},$$

which has branch points at $z = 0$ and $z = 1$. A branch of $z^{1/2}(1-z)^{1/2}$ that does not involve a branch cut passing through infinity can be created by using a straight line cut of unit length connecting $z = 0$ and $z = 1$ (see Example 3, part (c), Section 3.8 for a similar example). The branch cut is illustrated in Fig. 6.9–5(a). Note that $f(z)$ has poles at $\pm i$ and that any singularity at ∞ will be isolated—a requirement of the preceding theorem.

To define a branch of $z^{1/2}(1-z)^{1/2}/(1+z^2)$ that is analytic except at $\pm i$ (and possibly at ∞) in the cut plane of Fig. 6.9–5(a) we consider the vector representation of $z^{1/2}$ and $(1-z)^{1/2}$ derivable from Fig. 6.9–5(b). We take $\arg z = \alpha$, where $\alpha = 0$ when z is on the line $y = 0$, $x > 1$. Also we take $\arg(1-z) = \beta$, where $\beta = -\pi$ when z lies on the same line. Thus we can write

$$z^{1/2}(1-z)^{1/2} = \sqrt{|z|}\,e^{i\alpha/2}\sqrt{|1-z|}\,e^{i\beta/2}.$$

Note that if we evaluate $f(z)$ just above the branch cut, e.g., at p in Fig. 6.9–5(c), we have $\alpha = 0$, $\beta = 0$, and $f(z) = \sqrt{x}\sqrt{1-x}/(1+x^2)$. Similarly if we evaluate $f(z)$ just below the cut, e.g., at q in the same figure, we can take $\alpha = 2\pi$, $\beta = 0$. Then

$$f(z) = \frac{\sqrt{x}\,e^{i2\pi/2}\sqrt{1-x}}{1+x^2} = \frac{-\sqrt{x}\sqrt{1-x}}{1+x^2}.$$

Thus if p and q lie on opposite sides of the branch cut but at identical values of x, the values assumed by $f(z)$ at these two points are negatives of each other.

We now consider $\oint z^{1/2}(1-z)^{1/2}/(z^2+1)\,dz$ around C, the dumbbell shaped curve of Fig. 6.9–5(a). Because of the branch points at $z = 0$ and $z = 1$ we make circular detours of radius ε around these two points. We can let $\varepsilon \to 0+$ and easily argue (see Section 6.8) that in this limit the integrals along these two paths vanish. Notice also that the integrands on the straight line segments along the top and bottom sides of

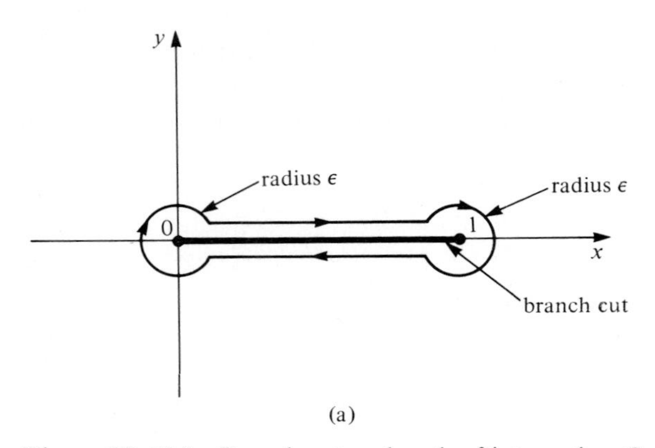

Figure 6.9–5(a) Branch cut and path of integration C

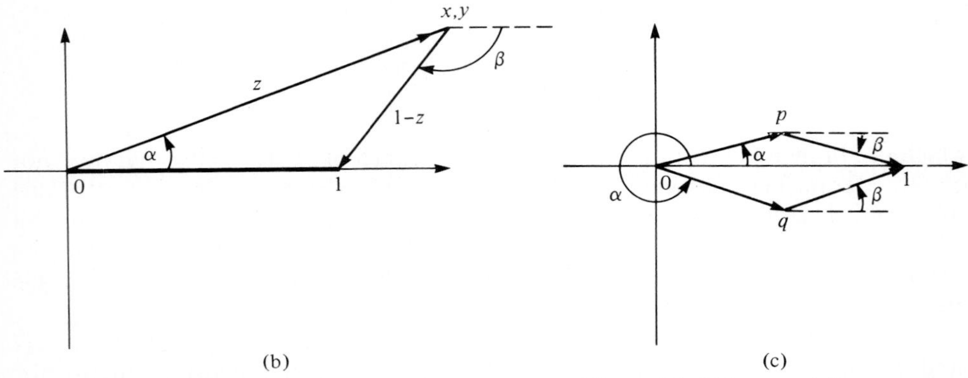

Figure 6.9–5(b) Definitions of α and β

Figure 6.9–5(c) α and β at p and q

the cut are negatives of each other at identical values of x. Since the directions of integration are opposite in each case, the contributions to the contour integration around C are identical along these two straight paths. With this fact, and the appropriate limit for ε, we have

$$\oint_C \frac{z^{1/2}(1-z)^{1/2}}{1+z^2}\,dz = 2\int_0^1 \frac{\sqrt{x}\sqrt{1-x}}{1+x^2}\,dx. \tag{6.9–11}$$

We have employed twice the integral along the top of the cut.

From the preceding theorem and Eq. (6.9–10) we get

$$\oint_C \frac{z^{1/2}(1-z)^{1/2}}{1+z^2}\,dz = 2\pi i \sum \text{Res}[f(z), \text{ at } \pm i \text{ and } \infty].$$

The residue at infinity can be computed from Eq. (6.9–5). We must find the residue at $w = 0$ of

$$-\frac{1}{w^2} f(1/w) = -\frac{1}{w^2}\,\frac{1}{w^{1/2}}\,\frac{(1-1/w)^{1/2}}{1/w^2+1} = (-1/w)\,\frac{(w-1)^{1/2}}{w^2+1}.$$

This function has a simple pole at $w = 0$. The residue is $-(w-1)^{1/2}$ evaluated at $w = 0$. We must decide whether to use i or $-i$. To extract the appropriate root, recall that $w = 1/z$, so that

$$-(w-1)^{1/2} = -\left(\frac{1}{z} - 1\right)^{1/2} = -\frac{z^{1/2}(1-z)^{1/2}}{z}.$$

If w approaches zero, z approaches infinity. Moving to infinity along any line in the cut plane of Fig. 6.9–5(a) and using the previously described branch of $z^{1/2}(1-z)^{1/2}$, we find the argument of $-z^{1/2}(1-z)^{1/2}/z$ tends to $\pi/2 + 2k\pi$ (k is any integer). The reader can easily verify this by proceeding to infinity along the line $y = 0$, $x > 1$ in Fig. 6.9–5(b) and using $\alpha = 0$, $\beta = -\pi$. Thus the residue at ∞ of $f(z)$ is i.

The residue of $f(z)$ at $z = \pm i$ is $z^{1/2}(1-z)^{1/2}/2z$ evaluated at $\pm i$. Some study of Fig. 6.9–5(b) and our branch of $z^{1/2}(1-z)^{1/2}$ shows that since $\alpha = \pi/2$ when $z = i$ we have here $z^{1/2} = \operatorname{cis} \pi/4$. Because $\beta = -\pi/4$ at the same point we find $(1-z)^{1/2} = \sqrt{\sqrt{2}} \operatorname{cis}(-\pi/8)$. Thus

$$\operatorname{Res}[f(z), i] = \frac{\sqrt{\sqrt{2}}}{2i} \operatorname{cis}\frac{\pi}{4} \operatorname{cis}\left(-\frac{\pi}{8}\right) = \frac{\sqrt{\sqrt{2}}}{2i} \operatorname{cis}\frac{\pi}{8}.$$

In a similar fashion, at $z = -i$ we have $\alpha = 3\pi/2$, $\beta = \pi/4$, and we find

$$\operatorname{Res}[f(z), -i] = \frac{\sqrt{\sqrt{2}}}{2i} \operatorname{cis}\left(-\frac{\pi}{8}\right).$$

The sum of the residues at ∞ and $\pm i$ is thus

$$i + \frac{\sqrt{\sqrt{2}}}{2i}\left[\operatorname{cis}\frac{\pi}{8} + \operatorname{cis} -\frac{\pi}{8}\right] = i\left(1 - \sqrt{\sqrt{2}}\cos\frac{\pi}{8}\right).$$

Multiplying the preceding by $2\pi i$ we evaluate the integral on the left in Eq. (6.9–11). A division by two yields the integral on the right. Hence

$$\int_0^1 \frac{\sqrt{x}\sqrt{1-x^2}}{1+x^2}\,dx = \pi\left[\sqrt{\sqrt{2}}\cos\frac{\pi}{8} - 1\right].$$

Comment: To some extent the choice of branch employed in this problem is arbitrary. The branch cut used must be that in Fig. 6.9–5(a). The branch employed must make $z^{1/2}(1-z)^{1/2}$ real on the top and bottom sides of the cut. Beyond that there are no restrictions. ◀

EXERCISES

Determine whether the following functions have a singularity at $z = \infty$ or are analytic at $z = \infty$. If the function has a singularity, state the nature of the singularity.

1. $z^3 + z$ **2.** $z^{-3} + z$ **3.** $\operatorname{Log} z$

4. $\operatorname{Log}[z/(z+1)]$ **5.** $(\sin z)/z$ **6.** $e^{1/z}$

7. $z^{1/3}$, principal branch

8. $z^{1/3}(z-1)^{2/3}$, branch is positive real for $y = 0$, $x > 1$, branch cut in Fig. 6.9–5(a)

9. $z^2 \cos 1/z$ **10.** $z \tan 1/z$ **11.** $z + e^z$

12. $z \operatorname{Log}(1 + 1/z)$

Find the residues of the following functions at infinity, and state which functions are analytic at infinity.

13. e^z **14.** $e^{1/z}$ **15.** $(z + 1/z)e^{1/z}$

16. $(z + 1/z)^4$ **17.** $\dfrac{z}{z^2 + z + 1}$ **18.** $\dfrac{z^2}{z^2 + z + 1}$

19. $\dfrac{z^3}{z^2 + z + 1}$ **20.** $\sin z$ **21.** $\displaystyle\sum_{n=-5}^{5} \frac{(z-1)^n}{|n|!}$

22. $(z^2 - 1)^{1/2}$, function is positive real for $y = 0, x > 1$, branch cut goes from $(-1, 0)$ to $(1, 0)$

23. $(z^2 - 1)^{-1/2}$, branch defined as in Exercise 22

Find $\oint z^n/(z^{10} + 1)\, dz$ around $|z| = 3$ by using the concept of the residue at ∞. Consider:

24. $n = 8$ **25.** $n = 9$ **26.** $n = 10$

27. Proof of Rule II.

a) Use Eq. (6.9–2) to show that if $f(z)$ is analytic at $z = \infty$ it must have a Laurent expansion of the form $f(z) = c_0 + c_{-1}z^{-1} + c_{-2}z^{-2} + \cdots$ for $|z| > r$.

Hint: Replace z with $1/w$ in Eq. (6.9–2) and argue that if $F(w) = f(1/w)$ is analytic at $w = 0$ then certain coefficients must be zero.

b) Using the series from part (a) and Rule I, show that if $\lim_{z\to\infty} f(z) = 0$, then $\operatorname{Res}[f(z), \infty] = \lim_{z\to\infty} -zf(z)$.

28. Evaluate all the residues of $z/(z^2 + z + 1)$ including the one at infinity and add them to show that their sum is zero. What theorem does this illustrate?

29. Compute the residue at ∞ of $z^2/(z^4 + 1)$ by using Rule III and show that the answer agrees with the residue obtained by Rule IV.

30. Assume that $f(z)$ is a rational function as described in Eq. (6.9–6), and assume that $m - n \geq 2$. Show that $\operatorname{Res}[f(z), \infty] = 0$. This proves Rule IV.

Hint: Consider $F(w) = f(1/w)$. Show that $(-1/w^2)F(w)$ is analytic at $w = 0$. Now use Rule III.

31. Using the residue at infinity, determine the Cauchy principal value of

$$\int_{-\infty}^{+\infty} \frac{x^3}{(x^2 + 1)[(x-1)^2 + 1]}\, dx.$$

32. Using the residue at infinity, prove that the following integral has the stated value. To find the required residues in the finite plane refer to Exercise 26, Section 6.3.

$$\int_{-\infty}^{+\infty} \frac{x^3}{x^4 + x^3 + x^2 + x + 1}\, dx = \frac{-2\pi}{5}\left(\sin\frac{2\pi}{5} + \sin\frac{4\pi}{5}\right)$$

33. Derive Eq. (6.9–7) (Theorem 8).

Hint: Since $f(z)$ is analytic at $z = \infty$, and $f(\infty) = 0$, show using Eq. (6.9–2) that $F(w) = f(1/w)$ has a Taylor expansion of the form $\sum_{n=1}^{\infty} c_{-n} w^n$. Show that, under the change of variables $w = 1/z$,

$$\int_C f(z)\,dz = \int_{C'} F(w)(-1/w^2)\,dw,$$

where C is defined in the theorem and C' is an arc of radius $1/r$ on which we integrate counterclockwise from $\arg w = -(\theta_0 + \alpha)$ to $\arg w = -\theta_0$. Show that $(-1/w^2)F(w) = -c_{-1}/w + g(w)$, where $g(w)$ is analytic at $w = 0$. Now show that

$$\lim_{r\to\infty} \int_{C'} F(w)(-1/w^2)\,dw = -c_{-1}\alpha i.$$

You must show that the integral of $g(w)$ along C' vanishes in the same limit (see Section 6.7). This completes the proof since $-c_{-1} = \operatorname{Res}[f(z), \infty]$.

Following the method used in Example 5, prove the following:

34. $\displaystyle\int_0^1 x^3 \sqrt{x}\sqrt{1-x}\,dx = \frac{\pi}{5!}\frac{1\cdot 3\cdot 5\cdot 7}{2^5}$

35. $\displaystyle\int_0^1 \frac{\sqrt{x}\sqrt{1-x}}{1+x}\,dx = \pi\left(\frac{3}{2} - \sqrt{2}\right)$

36. $\displaystyle\int_{-1}^{+1} \frac{\sqrt{1-x^2}}{1+x^2}\,dx = \pi(\sqrt{2} - 1)$

6.10 RESIDUE CALCULUS APPLIED TO FOURIER TRANSFORMS

The theory of Fourier transforms is a branch of mathematics with wide physical application.† We do not have the space here to delve into this theory. However, we will see how residue calculus is useful in the evaluation of integrals that arise when one is using Fourier transforms.

A few definitions are first required.

DEFINITION Absolute integrability

A function $f(t)$ of a real variable is *absolutely integrable* if

$$\int_{-\infty}^{\infty} |f(t)|\,dt \text{ exists. } \square \qquad (6.10\text{–}1)$$

---

† See, for example, J. R. Hanna and J. H. Rowland, *Fourier Series, Transforms, and Boundary Value Problems* (New York: John Wiley, 1990).

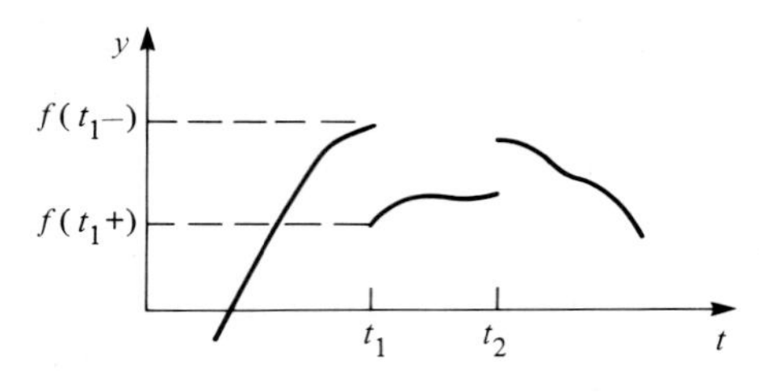

Figure 6.10–1

Next, we require the notion of piecewise continuity:

DEFINITION Piecewise continuity

The function $f(t)$ is *piecewise continuous* over an interval on the t-axis if this interval can be divided into a finite number of subintervals in which $f(t)$ is continuous. For each subinterval $f(t)$ has a finite limit as the ends are approached from the interior. □

An example of a real function $y = f(t)$ that is piecewise continuous is shown in Fig. 6.10–1. Note that the only discontinuities experienced by $f(t)$ are "jumps" of finite size. At a typical jump, say t_1, $f(t)$ has finite right- and left-hand limits defined by $\lim_{\delta\to 0+} f(t_1 + \delta) = f(t_1+)$ and $\lim_{\delta\to 0+} f(t_1 - \delta) = f(t_1-)$, respectively. The symbol $\lim_{\delta\to 0+}$ means that δ shrinks to zero only through positive values. Thus in the first case t_1 is approached from the right while in the second case it is approached from the left. A piecewise continuous complex function of t, for example, $\phi(t) + i\psi(t)$, can have jump discontinuities that occur in both $\phi(t)$ and $\psi(t)$.

Suppose we have an absolutely integrable function $f(t)$ that is piecewise continuous over every finite interval along the t-axis. Then, we can define a new function, $F(\omega)$, called the Fourier transform of $f(t)$, which is given by the following definition.

DEFINITION Fourier transform

$$F(\omega) = \frac{1}{2\pi}\int_{-\infty}^{+\infty} f(t)e^{-i\omega t}\,dt, \qquad -\infty < \omega < \infty \quad \square \qquad (6.10\text{–}2)$$

Note that ω is real. A comparison test[†] guarantees the existence of $F(\omega)$. In this section all improper integrals are to be regarded as Cauchy principal values. Usually, we will use a lowercase letter (like f) to denote a function of t and the corresponding uppercase letter (here F) to denote its Fourier transform. It is well to note that we have stated *sufficient* conditions for the existence of $F(\omega)$. There are functions that fail to satisfy Eq. (6.10–1) but that do have Fourier transforms (see Exercise 16 of this section).

[†] See W. Kaplan, *Operational Methods for Linear Systems* (Reading, MA: Addison-Wesley, 1962), Chapter 5.

It can be shown that, except for one limitation, the following formula permits us to recover or find $f(t)$ when its Fourier transform is known:

$$f(t) = \int_{-\infty}^{+\infty} F(\omega)e^{i\omega t}\, d\omega, \tag{6.10–3}$$

where the integral is a Cauchy principal value. The limitation on Eq. (6.10–3) is that this formula correctly yields $f(t)$ *except* for values of t where $f(t)$ is discontinuous. Here the formula gives the *average* of the right- and left-hand limits of $f(t)$, that is, $(1/2)f(t+) + (1/2)f(t-)$. The function $f(t)$ and its corresponding function $F(\omega)$ are known as Fourier transform pairs. Equation (6.10–3) is called the Fourier integral representation of $f(t)$. We will often regard the variable t as meaning time.

EXAMPLE 1

For the function

$$f(t) = \begin{cases} e^{-t}, & t \geq 0, \\ 0, & t < 0, \end{cases} \tag{6.10–4}$$

find the Fourier transform and verify the Fourier integral representation shown in Eq. (6.10–3).

Solution

From Eq. (6.10–2) we obtain

$$F(\omega) = \frac{1}{2\pi}\int_0^{\infty} e^{-t}e^{-i\omega t}\, dt = \frac{1}{2\pi}\int_0^{\infty} e^{-(1+i\omega)t}\, dt$$
$$= \frac{1}{2\pi}\frac{e^{-(1+i\omega)t}}{-1-i\omega}\Bigg|_0^{\infty} = \frac{1}{2\pi}\frac{1}{1+i\omega}.$$

Substituting this $F(\omega)$ in Eq. (6.10–3), we have

$$f(t) = \frac{1}{2\pi}\int_{-\infty}^{+\infty} \frac{e^{i\omega t}}{1+i\omega}\, d\omega = \frac{1}{2\pi}\int_{-\infty}^{+\infty} \frac{e^{ixt}}{1+ix}\, dx. \tag{6.10–5}$$

We have replaced ω by x in order to evaluate our integral with a contour integration in the more familiar z-plane. In Example 2, and thereafter, we dispense with this step.

With $t > 0$, Eq. (6.10–5) is readily evaluated from Eq. (6.6–11) and equals $i\,\text{Res}[e^{izt}/(1+iz), i] = e^{-t}$. With $t < 0$, Eq. (6.10–5) is evaluated from Eq. (6.6–13) and found to be zero since $e^{izt}/(1+iz)$ has no poles in the lower half of the z-plane.

The case $t = 0$ in Eq. (6.10–5) is considered separately. Evaluating, we find that

$$\frac{1}{2\pi}\int_{-\infty}^{+\infty} \frac{dx}{1+ix} = \frac{1}{2\pi}\int_{-\infty}^{+\infty} \frac{1-ix}{x^2+1}\, dx = \frac{1}{2\pi}\int_{-\infty}^{+\infty} \frac{dx}{x^2+1} - \frac{i}{2\pi}\int_{-\infty}^{+\infty} \frac{x\, dx}{x^2+1}. \tag{6.10–6}$$

The last integral on the right in Eq. (6.10–6) does not exist in the ordinary sense. However, because the integrand is an odd function, its Cauchy principal value is

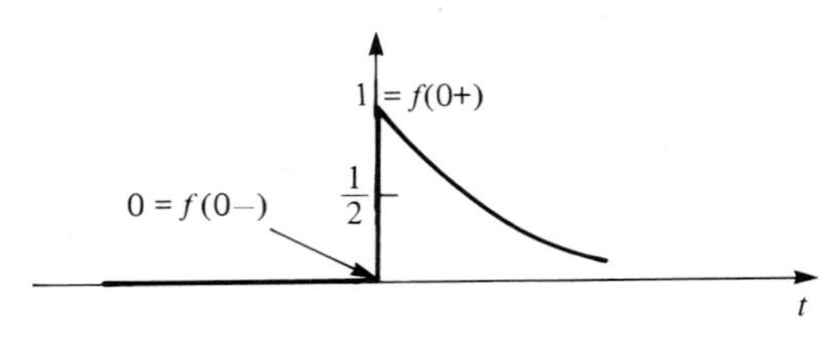

Figure 6.10–2

zero. The remaining integral on the right in Eq. (6.10–6) is readily evaluated with residues as follows:

$$\frac{1}{2\pi}\int_{-\infty}^{+\infty}\frac{dx}{x^2+1}=\frac{2\pi i}{2\pi}\operatorname{Res}\left[\frac{1}{z^2+1}, i\right]=\frac{1}{2}.$$

To summarize:

$$\int_{-\infty}^{+\infty}\frac{1}{2\pi}\frac{e^{-i\omega t}}{1+i\omega}\,d\omega=\begin{cases}e^{-t}, & t>0,\\ \frac{1}{2}, & t=0,\\ 0, & t<0.\end{cases}\tag{6.10–7}$$

Note that the function of t defined by Eq. (6.10–7) agrees with the given $f(t)$ in Eq. (6.10–4) for all t except $t = 0$. Here, the Fourier integral yields 1/2, whereas $f(0) = e^{-t}|_0 = 1$. The discrepancy occurs because $f(t)$ in Eq. (6.10–4) is discontinuous at $t = 0$. The right- and left-hand limits of $f(t)$ at $t = 0$ are 1 and 0 (see Fig. 6.10–2). The average of these quantities is 1/2, which is the value produced by the Fourier integral shown in Eq. (6.10–3). ◀

The Fourier integral representation of $f(t)$ given in Eq. (6.10–3) probably reminds us of the complex phasors discussed in the appendix to Chapter 3. Equation (A3–1),

$$f(t) = \operatorname{Re}[Fe^{st}] = \operatorname{Re}[Fe^{(\sigma+i\omega)t}],$$

where F is the complex phasor corresponding to $f(t)$, is in a sense analogous to Eq. (6.10–3). A similarity exists between the Fourier transform $F(\omega)$ and the complex phasor F. Phasor representations are limited to functions of the form $e^{\sigma t}\cos(\omega t + \theta)$, that is, functions exhibiting a single complex frequency $\sigma + i\omega$. The Fourier integral, which represents a function of time by means of an integration over all frequencies, is not limited to the representation of functions possessing a single frequency.

Fourier transforms are used in the solution of differential equations in much the same way as are phasors. The transform of the sum of two or more functions is the sum of their transforms. A property of Fourier transforms analogous to property (5) of phasors in the appendix to Chapter 3 would be useful. Thus given the relationship between $f(t)$ and $F(\omega)$ described by Eq. (6.10–2), we want a quick method for finding the Fourier transform of df/dt, that is, for finding $(1/2\pi)\int_{-\infty}^{+\infty}(df/dt)\,e^{-i\omega t}\,dt$. Suppose df/dt is piecewise continuous and $f(t)$ is continuous over every finite interval on the t-axis, and suppose that $f(t)$ and its derivatives are absolutely integrable. Assume again that $\lim_{t\to\pm\infty} f(t) = 0$. Integrating by parts we have

$$\frac{1}{2\pi}\int_{-\infty}^{+\infty} \frac{df}{dt} e^{-i\omega t}\, dt = \left.\frac{e^{-i\omega t}f(t)}{2\pi}\right|_{-\infty}^{+\infty} + \frac{1}{2\pi}\int_{-\infty}^{+\infty} f(t) i\omega\, e^{-i\omega t}\, dt.$$

The first term on the right becomes zero at the limits $\pm\infty$. Hence, we obtain

$$\frac{1}{2\pi}\int_{-\infty}^{+\infty} \frac{df}{dt} e^{-i\omega t}\, dt = \frac{i\omega}{2\pi}\int_{-\infty}^{+\infty} f(t)\, e^{-i\omega t}\, dt = i\omega F(\omega), \tag{6.10–8}$$

which shows that if $f(t)$ has Fourier transform $F(\omega)$, then df/dt has Fourier transform $i\omega F(\omega)$. This result is also obtainable from formal differentiation of Eq. (6.10–3); the operator d/dt is placed under the integral sign.

With certain restrictions this procedure can be repeated again and again. Thus, if $d^n f/dt^n$ is piecewise continuous over every interval on the t-axis and if the lower order derivatives $d^{n-1}f/dt^{n-1}$, $d^{n-2}f/dt^{n-2}$, etc. (including $f(t)$), are continuous for $-\infty < t < \infty$, then the Fourier transform of $d^n f/dt^n$ is $(i\omega)^n F(\omega)$, i.e.,

$$(i\omega)^n F(\omega) = \frac{1}{2\pi}\int_{-\infty}^{+\infty} \frac{d^n f}{dt^n} e^{-i\omega t}\, dt,$$

provided $d^n f/dt^n$ and the lower order derivatives (including $f(t)$) are absolutely integrable and all these functions vanish as $t \to \pm\infty$. Eq. (6.10–8) is analogous to Property 5 for phasors mentioned in the appendix to Chapter 3.

Let $g(t) = \int_c^t f(x)\, dx$, where, for some choice of the constant c, $g(t)$ is absolutely integrable. Note that $dg/dt = f(t)$, which implies that $i\omega G(\omega) = F(\omega)$ or $G(\omega) = F(\omega)/i\omega$. This is the counterpart to Property 6 for phasors in Chapter 3.

The utility of the Fourier transform in the solution of physical problems is demonstrated in the following example.

EXAMPLE 2

Consider the series electric circuit in Fig. 6.10–3 containing a resistor r and inductance L. The voltage $v(t)$ supplied by the generator is a sine function that is turned on for only two cycles. What is the current $\imath(t)$?

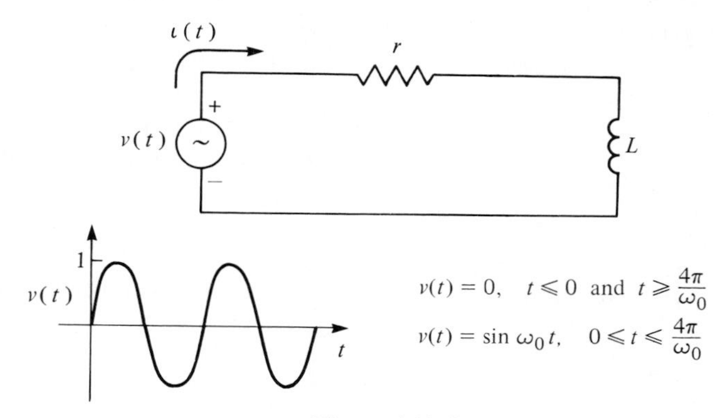

Figure 6.10–3

Solution

Applying the Kirchhoff voltage law around the circuit we have

$$v(t) = L\frac{d\imath}{dt} + \imath r. \tag{6.10–9}$$

Unlike Example 1 in the appendix to Chapter 3, the voltage in this problem fails to be harmonic for *all* time and therefore does not possess a phasor. Nonetheless, $v(t)$ does have a Fourier transform. Applying Eq. (6.10–2) to $v(t)$, using the exponential form of $\sin \omega_0 t$, and integrating we have

$$V(\omega) = \frac{1}{2\pi}\int_0^{4\pi/\omega_0} \frac{[e^{i\omega_0 t} - e^{-i\omega_0 t}]}{2i} e^{-i\omega t}\,dt = \frac{1}{2\pi}[1 - e^{-i4\pi\omega/\omega_0}]\frac{\omega_0}{\omega_0^2 - \omega^2}. \tag{6.10–10}$$

Transforming both sides of Eq. (6.10–9) according to the rules just described we get

$$V(\omega) = i\omega L I(\omega) + I(\omega) r. \tag{6.10–11}$$

Using Eq. (6.10–10) in the preceding equation, we solve for $I(\omega)$ and obtain

$$I(\omega) = \frac{1}{2\pi}[1 - e^{-i4\pi\omega/\omega_0}]\frac{\omega_0}{(i\omega L + r)(\omega_0^2 - \omega^2)}. \tag{6.10–12}$$

The desired time function $\imath(t)$ is now produced from Eqs. (6.10–3) and (6.10–12). Thus

$$\imath(t) = \imath_1(t) - \imath_2(t), \tag{6.10–13}$$

where

$$\imath_1(t) = \frac{\omega_0}{2\pi}\int_{-\infty}^{+\infty} \frac{e^{i\omega t}}{(\omega_0^2 - \omega^2)(i\omega L + r)}\,d\omega, \tag{6.10–14}$$

and

$$\imath_2(t) = \frac{\omega_0}{2\pi}\int_{-\infty}^{+\infty} \frac{e^{i\omega(t - 4\pi/\omega_0)}}{(\omega_0^2 - \omega^2)(i\omega L + r)}\,d\omega. \tag{6.10–15}$$

We first evaluate $\imath_1(t)$ for $t > 0$ by means of a contour integration in the complex $\boldsymbol{\omega}$-plane.† Because of singularities at $\omega = \pm\omega_0$, we determine the Cauchy principal value of the integral. The contour used is shown below in Fig. 6.10–4. Notice that indentations of radius ε are used around $-\omega_0$ and ω_0. We have

$$\int_{-R}^{-\omega_0-\varepsilon} \cdots + \int_{\substack{\frown \\ |\boldsymbol{\omega}+\omega_0|=\varepsilon}} \cdots + \int_{-\omega_0+\varepsilon}^{\omega_0-\varepsilon} \cdots + \int_{\substack{\frown \\ |\boldsymbol{\omega}-\omega_0|=\varepsilon}} \cdots + \int_{\omega_0+\varepsilon}^{R} \cdots$$

$$+ \int_{\substack{\frown \\ |\boldsymbol{\omega}|=R}} \frac{\omega_0}{2\pi}\frac{e^{i\boldsymbol{\omega} t}}{(\omega_0^2 - \boldsymbol{\omega}^2)(i\boldsymbol{\omega} L + r)}\,d\boldsymbol{\omega} = 2\pi i \operatorname{Res}\left[\frac{\omega_0}{2\pi}\frac{e^{i\boldsymbol{\omega} t}}{(\omega_0^2 - \boldsymbol{\omega}^2)(i\boldsymbol{\omega} L + r)}, \frac{ir}{L}\right]. \tag{6.10–16}$$

---

† The symbol $\boldsymbol{\omega}$ (boldface) will refer to a complex variable whose real part is ω. Thus $\operatorname{Re}(\boldsymbol{\omega}) = \omega$.

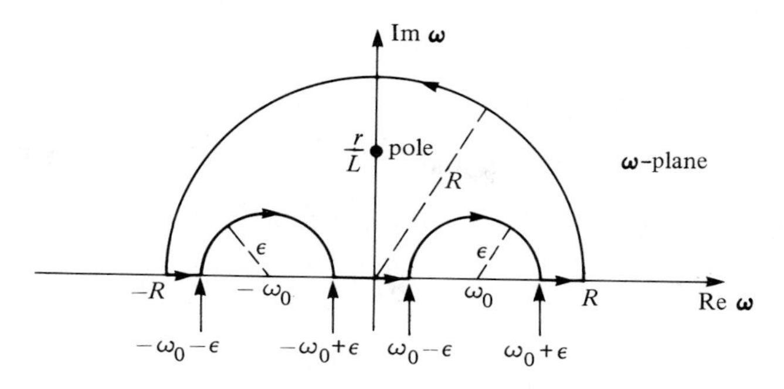

Figure 6.10–4

Only one singularity of the integrand is enclosed by the contour of integration. This is the pole where $i\omega L + r = 0$ or $\omega = ir/L$. We let $R \to \infty$ in Eq. (6.10–16) and invoke Jordan's lemma (see Eq. 6.6–9) to set the integral over the large semicircle to zero. We allow $\varepsilon \to 0$ and evaluate the integrals over the semicircular indentations of radius ε, in this limit, by using Eq. (6.7–2) with $\alpha = \pi$. The result is

$$\imath_1(t) = \left[\frac{r \sin(\omega_0 t) - \omega_0 L \cos(\omega_0 t)}{2(r^2 + \omega_0^2 L^2)} + \frac{\omega_0 L e^{-(r/L)t}}{r^2 + \omega_0^2 L^2}\right], \qquad t \geq 0. \qquad (6.10\text{–}17)$$

Although our derivation of Eq. (6.10–17) presupposed $t > 0$, we have indicated $t \geq 0$ in Eq. (6.10–17). The case $t = 0$ in Eq. (6.10–14) can be treated with the contour of Fig. 6.10–4. We use Eq. (6.5–10) to argue that the integral over the large semicircle vanishes. It is found that Eq. (6.10–17), with $t = 0$, gives the correct result.

For $t < 0$ we must evaluate $\imath_1(t)$ by means of an integration over a semicircular contour lying in the lower half of the ω-plane. The contour used is obtained by reflecting the one in Fig. 6.10–4 about the real axis. The integrand employed is the same as in Eq. (6.10–16). The new contour does not encircle the pole at ir/L. We can use Eq. (6.6–13) to argue that the integral over the semicircle of radius R vanishes as $R \to \infty$. We ultimately obtain (see Exercise 10 of this section)

$$\imath_1(t) = \frac{\omega_0 L \cos(\omega_0 t) - r \sin(\omega_0 t)}{2(r^2 + \omega_0^2 L^2)}, \qquad t < 0. \qquad (6.10\text{–}18)$$

To evaluate $\imath_2(t)$ we first consider the case $t \geq 4\pi/\omega_0$. The contour of integration used is identical to Fig. 6.10–4, and the integrand is the same as in Eq. (6.10–16), except $(t - 4\pi/\omega_0)$ is substituted for t. We also make this substitution in Eq. (6.10–17) to obtain $\imath_2(t)$. Note that $\sin[(\omega_0)(t - 4\pi/\omega_0)] = \sin \omega_0 t$ and $\cos[(\omega_0)(t - 4\pi/\omega_0)] = \cos \omega_0 t$. Thus

$$\imath_2(t) = \left[\frac{r \sin(\omega_0 t) - \omega_0 L \cos(\omega_0 t)}{2(r^2 + \omega_0^2 L^2)} + \frac{\omega_0 L e^{(-r/L)(t - 4\pi/\omega_0)}}{r^2 + \omega_0^2 L^2}\right], \qquad t \geq \frac{4\pi}{\omega_0}.$$

To obtain $\imath_2(t)$ for $t < 4\pi/\omega_0$, we follow a procedure analogous to the derivation of Eq. (6.10–18). The contour of integration used is the same, and t in the integrand is replaced by $t - 4\pi/\omega_0$. With this change in Eq. (6.10–18) we get

$$\imath_2(t) = \frac{\omega_0 L\cos(\omega_0 t) - r\sin(\omega_0 t)}{2(r^2 + \omega_0^2 L^2)}, \qquad t < \frac{4\pi}{\omega_0}. \tag{6.10–20}$$

Combining the last four equations according to Eq. (6.10–13), we have

$$\imath(t) = \frac{\omega_0 L e^{-rt/L}}{r^2 + \omega_0^2 L^2}[1 - e^{4\pi r/(\omega_0 L)}], \qquad t \geq \frac{4\pi}{\omega_0}; \tag{6.10–21}$$

$$\imath(t) = \left[\frac{r\sin(\omega_0 t) - \omega_0 L\cos(\omega_0 t)}{(r^2 + \omega_0^2 L^2)} + \frac{\omega_0 L e^{-rt/L}}{(r^2 + \omega_0^2 L^2)}\right], \qquad 0 \leq t \leq \frac{4\pi}{\omega_0}; \tag{6.10–22}$$

$$\imath(t) = 0, \qquad t \leq 0. \tag{6.10–23}$$ ◀

Although there are physical problems, like the preceding one, that are unsolvable with phasors and solvable with Fourier transforms, these transforms are inconvenient or useless in many situations. The Fourier transform technique is incapable of accounting for the response of a system due to any initial conditions that exist at the instant of excitation. Suppose, in the problem just given, a known current is already flowing in the circuit at $t = 0$. The Fourier transformation solution given takes no account of how the response of the system is affected by this current. The student of differential equations will perhaps realize that to obtain this portion of the response one must make a separate solution of the homogeneous differential equation describing the network.

The method of Laplace transforms, which is probably familiar to the reader, takes direct account of the initial conditions imposed on a system. In the next chapter we discuss Laplace transforms in their relationship to complex variable theory.

EXERCISES

Use Eq. (6.10–3) and contour integration to establish the functions $f(t)$ corresponding to each of the following $F(\omega)$. Consider *all* real values of t. Assume $a > 0$, $b > 0$, and use Cauchy principal values for Eq. (6.10–3) where appropriate.

1. $\dfrac{1}{\omega^2 + a^2}$ **2.** $\dfrac{-i}{\omega - ia}$

3. $\dfrac{e^{-ib\omega}}{\omega^2 + a^2}$ **4.** $\dfrac{2}{(\omega - ia)^2}$

5. $\dfrac{1}{\omega^2 - a^2}$ **6.** $\dfrac{\sin a\omega}{\omega}$

7. $\dfrac{\cos a\omega}{\omega^2 + b^2}$

Use Eq. (6.10–3) to show that for the following functions $F(\omega)$ the corresponding $f(t)$ is as indicated.

8. $F(\omega) = e^{-a^2\omega^2}$, $f(t) = \dfrac{\sqrt{\pi}}{a} e^{-t^2/4a^2}$, $a > 0$

Hint: Consider a contour in the complex ω-plane like that in Fig. 6.6–3. The height of the rectangle is $t/(2a^2)$. Integrate $e^{-a^2\omega^2}$ around the rectangle and let $R \to \infty$. On the bottom of the contour the integrand is $e^{-a^2\omega^2}$ and on the top it is $e^{-a^2(\omega + it/2a^2)^2}$. Recall (see Exercise 25, Section 6.6) that

$$\int_{-\infty}^{+\infty} e^{-a^2x^2}\, dx = \frac{\sqrt{\pi}}{a}, \qquad a > 0.$$

9. $F(\omega) = \dfrac{e^{a\omega}}{\sinh b\omega}$,

$$f(t) = \frac{\pi}{b}\left[\frac{\sin\left(\frac{\pi a}{b}\right) + i \sinh\left(\frac{\pi}{b}t\right)}{\cosh\left(\frac{\pi}{b}t\right) + \cos\left(\frac{\pi a}{b}\right)}\right], \ b > |a| \geq 0, \ a \text{ real}$$

Hint: In the complex ω-plane use a contour like that in Fig. 6.6–3. The height of the rectangle is π/b. Semicircular indentations are required at $\omega = 0$ and $\omega = i\pi/b$. Integrate $e^{i\omega t}e^{a\omega}/\sinh \omega b$ around the contour. See Exercises 22–24 of Section 6.6.

10. Supply the necessary details for the derivation of Eqs. (6.10–17) and (6.10–18).

11. If $f(t)$ has Fourier transform $F(\omega)$, what is the Fourier transform of $f(t - \tau)$? Answer: $e^{-i\omega\tau}F(\omega)$.

12. Let

$$f(t) = \begin{cases} 1 & 0 \leq t \leq T, \\ 0, & t < 0, \\ 0, & t > T. \end{cases}$$

Find the Fourier transform $F(\omega)$. Verify the Fourier integral theorem by obtaining $f(t)$ from $F(\omega)$. Consider all possible real values of t. Use Eq. (6.10–3).

13. Use the results of Exercises 11 and 12 but no new integrations to obtain the Fourier transform of the function $f(t)$ given by

$$f(t) = \begin{cases} 1, & 5 \leq t \leq 6, \\ 0, & t < 5, \\ 0, & t > 6. \end{cases}$$

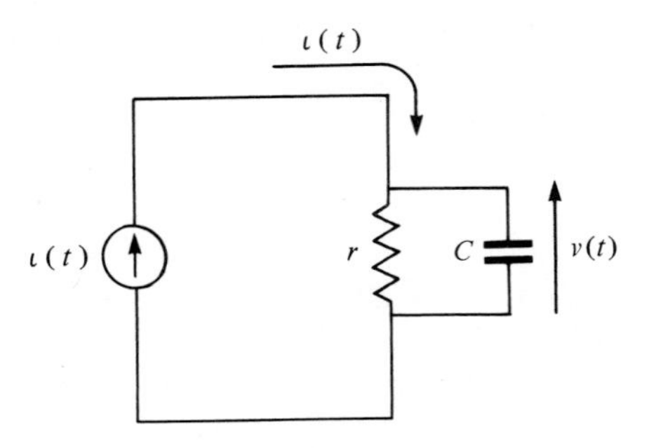

Figure 6.10–5

14. In Fig. 6.10–5 let $v(t)$ be the voltage across this parallel r, C circuit, and let $\imath(t)$ be the current supplied by the generator. Then, from Kirchhoff's current law,

$$\frac{v(t)}{r} + C\frac{dv}{dt} = \imath(t).$$

Assume that for $t < 0$, $v(t) = 0$. Let $\imath(t)$ be the function of time defined in Exercise 12. Use the method of Fourier transforms to find $V(\omega)$ and $v(t)$. What is the "system function" $V(\omega)/I(\omega)$?

15. a) Use residues to obtain the Fourier transform of

$$f(t) = \frac{\sin\left(\frac{2\pi t}{T}\right)}{t}, \qquad t \neq 0, \qquad f(0) = \frac{2\pi}{T}.$$

b) Sketch $F(\omega)$ and $f(t)$. What are the effects on $f(t)$ and $F(\omega)$ of increasing the period of oscillation T?

16. In Exercise 15 the Fourier transform of $f(t) = (\sin(2\pi t/T))/t$ was found. Show that $\int_{-\infty}^{+\infty} |f(t)|\, dt$ fails to exist.

Hint: When A is any nonzero constant, $\sum_{n=1}^{\infty} A/n$ is known to diverge. Make a comparison test† in which you show that, for an appropriate choice of A, the terms in this series are smaller than corresponding terms in the sequence of integrals

$$\int_0^T \frac{\left|\sin\frac{2\pi}{T}\right|}{t}\, dt, \quad \int_T^{2T} \frac{\left|\sin\frac{2\pi t}{T}\right|}{t}\, dt, \ldots .$$

Thus we see that Eq. (6.10–1) provides a sufficient but not a necessary condition for the existence of a Fourier transform.

---

† See W. Kaplan, *Advanced Calculus*, 4th ed. (Reading, MA: Addison-Wesley, 1991), Sections 6.6, 6.22.

17. In Chapter 2 we discussed the equation governing the two-dimensional steady-state flow of heat. Under transient (nonsteady-state) conditions Eq. (2.6–5) must be altered. The temperature $\phi(x, y)$ now satisfies

$$\frac{\partial^2 \phi}{\partial x^2} + \frac{\partial^2 \phi}{\partial y^2} = \frac{1}{a^2}\frac{\partial \phi}{\partial t}, \tag{6.10–24}$$

where $t =$ time and a is a constant characteristic of the heat-conducting material.

Consider an infinite rod extending along the x-axis as shown in Fig. 6.10–6. The rod is assumed to be insulated along all its faces, so that the temperature varies only with x and t. Assume that at $t = 0$ the temperature $\phi_0(x)$ along the rod is known. In this exercise we obtain $\phi(x, t)$ for $t > 0$. We use the Fourier transformation

$$\tilde{\phi}(\omega, t) = \frac{1}{2\pi}\int_{-\infty}^{+\infty} \phi(x, t)e^{-i\omega x}\, dx.$$

a) Take the Fourier transformation of both sides of Eq. (6.10–24) (ϕ is independent of y) to show that

$$-\omega^2 \tilde{\phi}(\omega, t) = \frac{1}{a^2}\frac{d\tilde{\phi}}{dt}(\omega, t).$$

You may assume that the order of a differentiation and an integration can be reversed.

b) Solve the preceding equation and show that

$$\tilde{\phi}(\omega, t) = A(\omega)e^{-\omega^2 a^2 t}, \qquad t \geq 0,$$

where

$$A(\omega) = \frac{1}{2\pi}\int_{-\infty}^{+\infty} \phi_0(x)e^{-i\omega x}\, dx = \frac{1}{2\pi}\int_{-\infty}^{+\infty} \phi_0(\zeta)e^{-i\omega\zeta}\, d\zeta,$$

and a change of variable is made from x to ζ for convenience.

c) Show that

$$\phi(x, t) = \int_{-\infty}^{+\infty}\left[\frac{1}{2\pi}\int_{-\infty}^{+\infty} \phi_0(\zeta)e^{-i\omega\zeta}\, d\zeta e^{-\omega^2 a^2 t}\right]e^{i\omega x}\, d\omega.$$

d) Show that

$$\phi(x, t) = \frac{1}{2a\sqrt{\pi t}}\int_{-\infty}^{+\infty} \phi_0(\zeta)e^{-(x-\zeta)^2/(4a^2 t)}\, d\zeta. \tag{6.10–25}$$

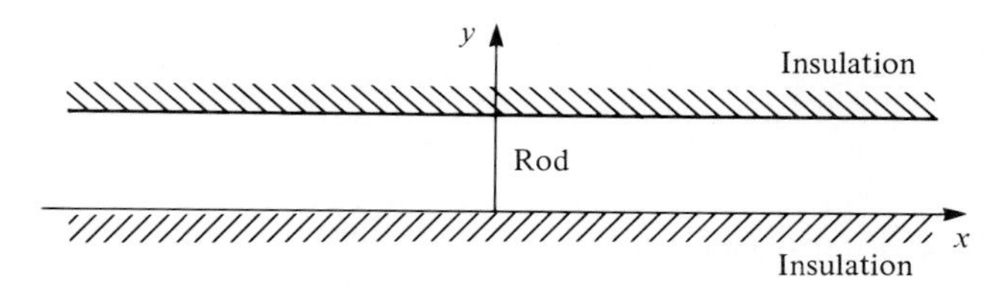

Figure 6.10–6

Hint: One can interchange the order of integration of the double integral in part (c). To do the integration on ω, one uses the formula[†]

$$\int_{-\infty}^{+\infty} e^{-m^2u^2}\cos(bu)\,du = \frac{\sqrt{\pi}}{m}e^{-b^2/(4m^2)}, \qquad m > 0.$$

e) Assume $a^2 = 1/2$; $\phi_0(\zeta) = 1$, $-1 \le \zeta \le 1$; and $\phi_0(\zeta) = 0$, $|\zeta| > 1$. Plot $\phi(x, t)$ as a function of x for $t = 0.1$ and also $t = 1$. To evaluate Eq. (6.10–25) either use a table of the error function[‡] or write a computer program.

18. Suppose a taut vibrating string lies along the x-axis except for small displacements $y(x, t)$ parallel to the y-axis (see Fig. 6.10–7). Let the deviation from the axis at any time be described by $y(x, t)$, where t is time. Assuming that the amplitude of vibrations is small enough so that each part of the string moves only in the y-direction and that gravitational forces are negligible, one can show that

$$\frac{\partial^2 y}{\partial x^2} = \frac{1}{c^2}\frac{\partial^2 y}{\partial t^2}, \tag{6.10–26}$$

where c is the velocity of propagation of waves along the string. We will use Fourier transforms to solve Eq. (6.10–26) for an infinitely long string that, at $t = 0$, is subject to the displacement $y_0(x) = y(x, 0)$ and velocity $v_0(x) = \partial y(x, 0)/\partial t$. The behavior of the string for $t > 0$ is sought. We use the transformation

$$Y(\omega, t) = \frac{1}{2\pi}\int_{-\infty}^{+\infty} y(x, t)e^{-i\omega x}\,dx.$$

a) Transform both sides of Eq. (6.10–26) and show that

$$\frac{d^2 Y}{dt^2}(\omega, t) + \omega^2 c^2 Y(\omega, t) = 0.$$

Hint: Assume that the operation $\partial/\partial t$ and the Fourier transformation can be performed in any order.

b) Show that $Y(\omega, t)$ must be of the form

$$Y(\omega, t) = A(\omega)\cos(\omega ct) + B(\omega)\sin(\omega ct). \tag{6.10–27}$$

c) By putting $t = 0$ in the preceding show that

$$A(\omega) = \frac{1}{2\pi}\int_{-\infty}^{+\infty} y_0(x)e^{-i\omega x}\,dx.$$

[†] Derived in Exercise 25, Section 6.6.

[‡] See M. Abramowitz and I. Stegun, *Handbook of Mathematical Functions* (New York: Dover 1970), p. 297.

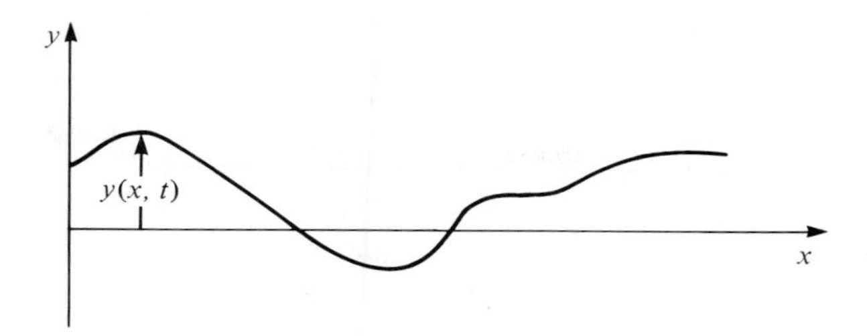

Figure 6.10–7

d) Differentiate Eq. (6.10–27) with respect to time; put $t = 0$, and show that

$$B(\omega) = \frac{1}{\omega c 2\pi} \int_{-\infty}^{+\infty} v_0(x) e^{-i\omega x}\, dx.$$

e) Show that

$$y(x, t) = \int_{-\infty}^{+\infty} [A(\omega)\cos(\omega ct) + B(\omega)\sin(\omega ct)] e^{i\omega x}\, d\omega, \qquad (6.10\text{–}28)$$

where $A(\omega)$ and $B(\omega)$ are given in parts (c) and (d).

f) Assume that $v_0(x) = 0$ and that $y_0(x) = \Delta e^{-|x|}$, where $\Delta > 0$ is a constant. Using residue calculus to evaluate the integral in Eq. (6.10–28), find $y(x, t)$ for $t > 0$. Consider the three cases: $x > ct$, $-ct < x < ct$, $x < -ct$. Check your answer by considering $\lim_{t\to 0} y(x, t)$.

APPENDIX TO CHAPTER 6

THE USE OF RESIDUES TO SUM CERTAIN SERIES

Numerical Series

In the present chapter we have seen numerous instances where the calculus of residues can be used to evaluate definite integrals. In this appendix we will first see that residues can also sometimes be used to determine the sums of infinite series of constants, e.g., series like $\sum_{n=-\infty}^{+\infty} 1/(n^2+1)$ or $\sum_{n=-\infty}^{+\infty} (-1)^n n^2/(n^4+1)$, etc. In particular, we will learn a procedure to sum any series of the form

$$\sum_{n=-\infty}^{+\infty} \frac{P(n)}{Q(n)} \quad \text{and} \quad \sum_{n=-\infty}^{+\infty} (-1)^n \frac{P(n)}{Q(n)},$$

where $P(n)$ and $Q(n)$ are polynomials in n, the degree of Q exceeds that of P by two or more, and $Q(n) \neq 0$ for all integer n.

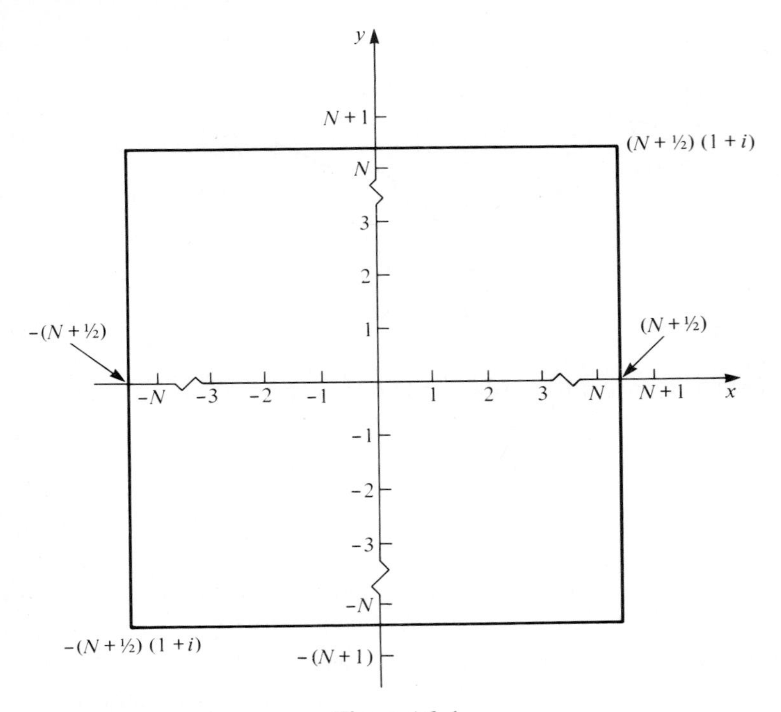

Figure A6–1

We begin by considering $\oint \pi \cot \pi z\, f(z)\, dz$ taken around the contour C_N shown in Fig. A6–1. C_N is one of a family of square contours, centered at the origin, with corners at $\pm(N + 1/2)(1 \pm i)$. Here $N \geq 0$ is an integer. To apply the ML inequality to this integral, we first seek an upper bound for $|\cot \pi z|$ when z lies on C_N.

At an arbitrary point on the right side of the contour we have $z = (N + 1/2) + iy$, where $|y| \leq N + 1/2$. Here

$$\cot \pi z = \frac{\cos[\pi(N + 1/2) + i\pi y]}{\sin[\pi(N + 1/2) + i\pi y]}. \tag{A6–1}$$

With the aid of Eqs. (3.2–9) and (3.2–10) and the identities $\cos[\pm\pi(N + 1/2)] = 0$, $|\sin[\pm\pi(N + 1/2)]| = 1$ we have, from Eq. (A6–1),

$$|\cot \pi z| = \left|\frac{\sinh \pi y}{\cosh \pi y}\right| = |\tanh \pi y|. \tag{A6–2}$$

Since $|\tanh \pi y|$ increases monotonically with $|y|$, the largest values achieved by this expression on the right side of C_N are at $y = \pm(N + 1/2)$, i.e., the two corners. Thus on the right side $|\cot \pi z| \leq |\tanh \pi(N + 1/2)|$. Since the hyperbolic tangent of any finite real number has magnitude less than 1 (the reader can verify this with a pocket calculator) we can say that, on the right side of C_N,

$$|\cot \pi z| \leq 1. \tag{A6–3}$$

On the left side of C_N, $z = -(N + 1/2) + iy$. The preceding argument can again be applied and Eq. (A6–3) is found to be valid.

On the top side of C_N we have $z = x + iy$, where $y = N + 1/2$ and $|x| \le N + 1/2$. Again using Eqs. (3.2–9) and (3.2–10) we have

$$|\cot \pi z| = \left|\frac{\cos \pi x \cosh \pi y - i \sin \pi x \sinh \pi y}{\sin \pi x \cosh \pi y + i \cos \pi x \sinh \pi y}\right|. \tag{A6–4}$$

Since $|\sinh \pi y| < |\cosh \pi y|$, the numerator in Eq. (A6–4) satisfies $|\cos \pi x \cosh \pi y - i \sin \pi x \sinh \pi y| < |\cos \pi x \cosh \pi y - i \sin \pi x \cosh \pi y| = \cosh \pi y |\cos \pi x - i \sin \pi x| = \cosh \pi y$. Similarly for the denominator, $|\sin \pi x \cosh \pi y + i \cos \pi x \sinh \pi y| > |\sin \pi x \sinh \pi y + i \cos \pi x \sinh \pi y| = |\sinh \pi y||\sin \pi x + i \cos \pi x| = |\sinh \pi y|$. Thus returning to Eq. (A6–4) with these inequalities we have

$$|\cot \pi z| \le \frac{\cosh \pi y}{|\sinh \pi y|} = |\coth \pi y|. \tag{A6–5}$$

The right side of the preceding equation decreases monotonically with increasing $|y|$. If we consider all possible contours C_N in our family, the one in which $|y|$ is smallest on the top side is the case $N = 0$. Here $y = 1/2$. Thus on the top side of all contours we have $|\cot \pi z| \le \coth \pi/2 \approx 1.09$. The preceding argument can also be applied to the bottom side of all contours C_N, and the preceding inequality again derived. Therefore, we can say that

$$|\cot \pi z| \le \coth \pi/2 \tag{A6–6}$$

is satisfied on every contour C_N.

Now return to $\oint_{C_N} \pi \cot \pi z\, f(z)\, dz$, where $f(z)$ is assumed analytic throughout the complex plane except at a finite number of poles, none of which is an integer. Assume also that there exist reals $k > 1$, m, and R such that

$$|f(z)| \le m/|z|^k \qquad \text{for } |z| \ge R. \tag{A6–7}$$

Let us take N sufficiently large so that C_N encloses all poles of $f(z)$. Now the singularities of $\pi \cot \pi z\, f(z)$ are the poles of $f(z)$ and also the poles of $\cot \pi z$. The poles of $\cot \pi z$ are the zeros of $\sin \pi z$ and lie at $z = 0, \pm 1, \pm 2, \ldots$. Thus the poles of $\cot \pi z$ enclosed by C_N are those lying at $z = 0, \pm 1, \pm 2, \ldots, \pm N$. The remaining poles are outside C_N. To summarize: the singularities of $\pi \cot \pi z\, f(z)$ enclosed by C_N are all the poles of $f(z)$ and the points $z = 0, \pm 1, \pm 2, \ldots, \pm N$. Thus from the residue theorem,

$$\begin{aligned} &\oint_{C_N} \pi \cot \pi z\, f(z)\, dz \\ &\quad = 2\pi i \sum_{\text{residues}} \pi \cot \pi z\, f(z) \text{ at all poles of } f(z) \\ &\qquad + 2\pi i \sum_{\text{residues}} \pi \cot \pi z\, f(z) \text{ at } z = 0, \pm 1, \pm 2, \ldots, \pm N. \end{aligned} \tag{A6–8}$$

From Rule IV, Section 6.3 we have

$$\operatorname{Res}[\pi \cot \pi z\, f(z), n] = \operatorname{Res}\left[\pi \frac{\cos \pi z}{\sin \pi z} f(z), n\right] = f(n), \tag{A6–9}$$

where n is any integer.

Now we allow $N \to \infty$; i.e., we consider a sequence of increasingly large contours, and we will argue that the integral on C_N vanishes. The length L of the contour is $4(2N + 1)$. The sides of the contour are taken sufficiently far from $z = 0$ so that Eq. (A6–7) is valid on the path. Thus combining Eq. (A6–7) with Eq. (A6–6), we have on C_N that

$$|\pi \cot \pi z\, f(z)| \le \pi \coth(\pi/2)(m/|z|^k).$$

On C_N, $|z|^k \ge (N + 1/2)^k$, so that $|1/z|^k \le 1/(N + 1/2)^k$. Hence on C_N,

$$|\pi \cot \pi z\, f(z)| \le \pi \coth(\pi/2)m/(N + 1/2)^k. \tag{A6–10}$$

Now applying the ML inequality to the integral on the left in Eq. (A6–8), and taking $L = 4(2N + 1)$ and M as the right side of Eq. (A6–10), we have

$$\left|\int_{C_N} \pi \cot \pi z\, f(z)\, dz\right| \le \frac{\pi \coth(\pi/2)\, m}{(N + 1/2)^k} 4(2N + 1).$$

Passing to the limit $N \to \infty$ in the preceding we find (since $k > 1$) that the right side goes to zero. Thus $\lim_{N\to\infty} \oint_{C_N} \pi \cot \pi z\, f(z)\, dz = 0$. As $N \to \infty$ the contour C_N grows in size so as to enclose all the singularities of $\pi \cot \pi z\, f(z)$ at $z = 0, \pm 1, \pm 2, \ldots$ in the complex plane. With $N \to \infty$ in Eq. (A6–8) we thus obtain

$$0 = 2\pi i \sum_{\text{residues}} \pi \cot \pi z\, f(z) \text{ at all poles of } f(z)$$

$$+ 2\pi i \sum_{\text{residues}} \pi \cot \pi z\, f(z) \text{ at } 0, \pm 1, \pm 2, \ldots, \infty.$$

Using Eq. (A6–9) to evaluate the residues at the integer values of z on the right, we have finally

$$\sum_{n=-\infty}^{+\infty} f(n) = -\pi \sum_{\text{residues}} \cot \pi z\, f(z) \text{ at all poles of } f(z). \tag{A6–11}$$

To summarize, the preceding is valid if $f(z)$ is analytic except at a finite number of poles none of which is an integer and if $|f(z)| \le m/|z|^k$ $(k > 1)$ for $|z| > R$. This can be satisfied by $f(z) = P(z)/Q(z)$ where P and Q are polynomials in z, with the degree of Q exceeding the degree of P by two or more (see the discussion leading to Theorem 4, Section 6.5, also Exercise 25, Section 6.5). Thus we have the following theorem:

THEOREM 10 Let $P(z)$ and $Q(z)$ be polynomials in z such that the degree of Q exceeds that of P by two or more. Assume that $Q(z) = 0$ has no solution for integer z. Then

$$\sum_{n=-\infty}^{+\infty} \frac{P(n)}{Q(n)} = -\pi \sum_{\text{residues}} \cot \pi z \frac{P(z)}{Q(z)} \text{ at all poles of } \frac{P(z)}{Q(z)}. \quad \square \qquad \text{(A6–12)}$$

EXAMPLE 1

Find $S = \sum_{n=0}^{\infty} 1/(n^2+1) = 1 + 1/2 + 1/5 + 1/10 + \cdots$.

Solution

The summation given is not precisely of the form shown on the left in Theorem 10. However, $S_0 = \sum_{n=-\infty}^{+\infty} 1/(n^2+1) = \cdots + 1/5 + 1/2 + 1 + 1/2 + 1/5 + \cdots$ does have the form of the theorem. Notice that $(S_0 + 1)/2 = S$.

To find S_0 we use Eq. (A6–12) taking $P(n) = 1$, $Q(n) = n^2 + 1$, $P(z) = 1$, $Q(z) = z^2 + 1$. As required the roots of $z^2 + 1 = 0$, which are $\pm i$, are not integers. Applying Eq. (A6–12) we have

$$S_0 = -\pi \sum_{\text{residues}} \frac{\cot \pi z}{z^2+1} \text{ at } \pm i.$$

The poles at $\pm i$ are simple. From Eq. (6.3–6) we have

$$S_0 = -\pi\left[\frac{\cot \pi i}{2i} + \frac{\cot(-\pi i)}{-2i}\right] = \frac{-\pi}{i}\cot \pi i = \frac{-\pi}{i}\frac{\cos \pi i}{\sin \pi i} = \frac{-\pi}{i}\frac{\cosh \pi}{i \sinh \pi}$$

$$= \pi \coth \pi = \sum_{n=-\infty}^{+\infty} \frac{1}{n^2+1}.$$

Our desired result is $S = (\pi \coth \pi + 1)/2 \approx 2.077$. As a check, we approximate S by the first 100 terms in our given series, $\sum_{n=0}^{99} 1/(n^2+1)$ and obtain, with the help of a programmable calculator, 2.067. ◀

The technique used in deriving Eq. (A6–11) and Theorem 10 can be repeated with some modification to obtain a related result. If $f(z)$ satisfies the same requirements as for Eq. (A6–11) we can show that

$$\sum_{n=-\infty}^{+\infty} (-1)^n f(n) = -\pi \sum_{\text{residues}} \frac{1}{\sin \pi z} f(z) \text{ at all poles of } f(z), \qquad \text{(A6–13)}$$

from which we obtain the following:

THEOREM 11 If $P(z)$ and $Q(z)$ satisfy the same conditions as in Theorem 10 we have

$$\sum_{n=-\infty}^{\infty} (-1)^n \frac{P(n)}{Q(n)} = -\pi \sum_{\text{residues}} \frac{1}{\sin \pi z}\frac{P(z)}{Q(z)} \text{ at all poles of } \frac{P(z)}{Q(z)}. \quad \square$$

(A6–14)

The details of the derivation of Eqs. (A6–13) and (A6–14) as well as some applications are given in the exercises.

The Watson Transformation and Summation of Fourier Series

We have already seen in this appendix how complex variable theory can sometimes be used to obtain the sum of a numerical series. The reader probably has some familiarity with Fourier series and has had occasion to expand functions in Fourier series.†

Recall that if a function $f(\phi)$ is piecewise continuous on the interval $0 \leq \phi \leq 2\pi$ it can be represented in the form

$$f(\phi) = \sum_{n=0}^{\infty} a_n \cos n\phi + b_n \sin n\phi, \qquad 0 \leq \phi \leq 2\pi,$$

where a_n and b_n are constants. The value of b_0 is immaterial and we will take it as zero. The series representation of $f(\phi)$ is valid except where $f(\phi)$ is discontinuous. Using the exponential forms of $\cos n\phi$ and $\sin n\phi$ in the preceding equation and reindexing the summation, we have

$$f(\phi) = \sum_{n=-\infty}^{+\infty} g(n)e^{in\phi}, \qquad 0 \leq \phi \leq 2\pi, \tag{A6–15}$$

where $g(n) = (a_n - ib_n)/2$ for $n \geq 0$, $g(n) = (a_n + ib_n)/2$ for $n \leq 0$.

Invoking the orthogonal properties of $\cos n\phi$ and $\sin n\phi$ (and by extension $e^{in\phi}$) we can readily show that

$$g(n) = \frac{1}{2\pi}\int_0^{2\pi} f(\phi)e^{-in\phi}\,d\phi, \qquad n = 0, \pm 1, \pm 2, \ldots. \tag{A6–16}$$

Thus obtaining the complex Fourier coefficients $g(n)$ is relatively straightforward.

Given the series on the right in Eq. (A6–15) we might wish to obtain numerical values of $f(\phi)$. This can be done by placing specific values of ϕ in the finite sum $\sum_{-N}^{+N} g(n)e^{in\phi}$ and adding the numbers with a computer or calculator. Some series converge slowly and the number N that must be used in the summation to achieve a reasonable accuracy can be very large. If many values of ϕ are of interest, this procedure can be time-consuming, even on a computer.

Ideally we would like to derive a closed form expression for $f(\phi)$ when we are given the Fourier coefficients $g(n)$. What is available, however, is the *Watson transformation*, which yields $f(\phi)$ in terms of a contour integral, provided we can find an analytic function $g(z)$ that agrees with the coefficients $g(n)$ when z is an integer. This is usually the case. We can sometimes evaluate the integral using residue calculus. Even if this is not possible, the integral may be more amenable to numerical evaluation than the original series.

The method is named for the British mathematician G. N. Watson (1886–1965) who invented the technique around 1919 to sum a series in his analysis of the

---

† For a review or an introduction to Fourier series see, e.g., J. R. Hanna and J. H. Rowland, *Fourier Series, Transforms, and Boundary Value Problems* (New York: John Wiley, 1990).

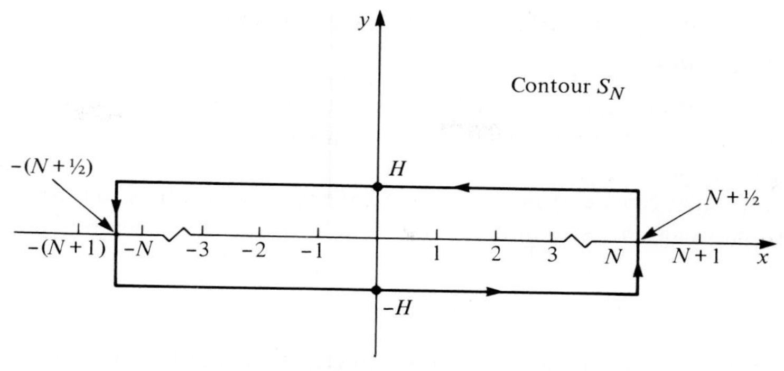

Figure A6–2

propagation of electromagnetic waves around the earth. It is still widely used in problems involving the scattering of waves.

Consider a function $g(z)$ that is analytic throughout the complex plane except possibly at poles or along branch cuts. Assume that none of the singular points of $g(z)$ is on the real axis. We will also assume that $g(n)$ (n is any integer) is the set of Fourier coefficients in a given complex Fourier series. Consider

$$\frac{1}{2i}\oint_{S_N} \frac{e^{iz\phi}e^{-i\pi z}g(z)}{\sin \pi z}\,dz,$$

where S_N is the rectangular contour shown in Fig. A6–2. We assume that $g(z)$ is analytic on and inside S_N. The sides of this contour are at $x = \pm(N + 1/2)$ (N is an integer) and at $y = \pm H$. We will take $0 < \phi < 2\pi$. The only singularities of the integrand inside S_N are the zeros of $\sin \pi z$. These occur at the integers $z = 0, \pm 1, \pm 2, \ldots, \pm N$. The residue of

$$\frac{1}{2i}\frac{e^{iz\phi}e^{-i\pi z}g(z)}{\sin \pi z}$$

at $z = n$ is

$$\frac{1}{2\pi i}\frac{e^{in\phi}e^{-in\pi}g(n)}{\cos n\pi} = \frac{1}{2\pi i}e^{in\phi}g(n).$$

Summing these residues at the poles inside S_N and applying the residue theorem, Eq. (6.1–8), we have

$$\frac{1}{2i}\oint_{S_N} \frac{e^{iz\phi}e^{-i\pi z}g(z)}{\sin \pi z}\,dz = \sum_{-N}^{+N} g(n)e^{in\phi}. \tag{A6–17}$$

Passing to the limit $N \to \infty$, we obtain the

WATSON TRANSFORMATION

$$\lim_{N\to\infty}\frac{1}{2i}\oint_{S_N} \frac{e^{iz\phi}e^{-i\pi z}g(z)}{\sin \pi z}\,dz = \sum_{-\infty}^{+\infty} g(n)e^{in\phi}, \tag{A6–18}$$

where $g(z)$ must be analytic on the real axis and the integral is assumed to exist in the limit. □

The expression on the right in Eq. (A6–18) is our given complex Fourier series. There are other transformations† similar in form to this one. They are useful when we use the real form of the Fourier series.

If the singularities of $g(z)$ in Eq. (A6–18) are limited to poles that are not on the real axis we can evaluate the integral on the left, in the desired limit, provided $g(z)$ satisfies, for $|z| \geq R$,

$$|g(z)| \leq m/|z|^k, \tag{A6–19}$$

where $k > 0$, $m > 0$. Let us assume for simplicity that $g(z)$ has a simple pole at z_k. Using the principle of deformation of contours, we conclude that the integral around the contour S_N can be converted to the integral around S'_N shown in Fig. A6–3. Here S'_N is a square contour like that in Fig. A6–1 except that a detour has been made around the pole at z_K. (We assume that N is large enough so that this detour is necessary.) The pole is encircled in the negative (clockwise) sense. The integrals along the paths P_1 and P_2 cancel, while the integral around the circle of radius ε has the value

$$-2\pi i \operatorname{Res} \frac{1}{2i}\left[\frac{e^{iz\phi}e^{-i\pi z}g(z)}{\sin \pi z}\right] \text{ at } z_K.$$

Thus

$$\begin{aligned}
\frac{1}{2i}\oint_{S_N} &\frac{e^{iz\phi}e^{-i\pi z}g(z)}{\sin \pi z}\,dz \\
&= \frac{1}{2i}\oint_{S'_N} \frac{e^{iz\phi}e^{-i\pi z}g(z)}{\sin \pi z}\,dz \\
&= \frac{1}{2i}\oint_{C_N} \frac{e^{iz\phi}e^{-i\pi z}g(z)}{\sin \pi z}\,dz - \pi \operatorname{Res}\frac{e^{iz\phi}e^{-i\pi z}g(z)}{\sin \pi z} \text{ at } z_K.
\end{aligned} \tag{A6–20}$$

Here C_N is the square contour of Fig. A6–1; it is S'_N, without the detour around the pole. If $0 < \phi < 2\pi$ we can argue that as $N \to \infty$ the integral around C_N vanishes. The details of the argument are not given here. They are similar to those presented for other limiting procedures in this chapter. Passing to this limit and discarding the integral on C_N in Eq. (A6–20), we have

$$\lim_{N\to\infty} \frac{1}{2i}\oint_{S_N} \frac{e^{iz\phi}e^{-i\pi z}g(z)}{\sin \pi z}\,dz = -\pi \operatorname{Res}\frac{e^{iz\phi}e^{-i\pi z}g(z)}{\sin \pi z} \text{ at } z_K.$$

If $g(z)$ had other poles besides the one at z_K, the contour S'_N would have detours around them also. Residues at these poles would appear on the right in the preceding

---

† See, e.g., R. K. Cooper, On the Equivalence of Certain Green's Function Expressions, *American Journal of Physics*, Vol. 37, No. 10, October 1969, pp. 1032–1039.

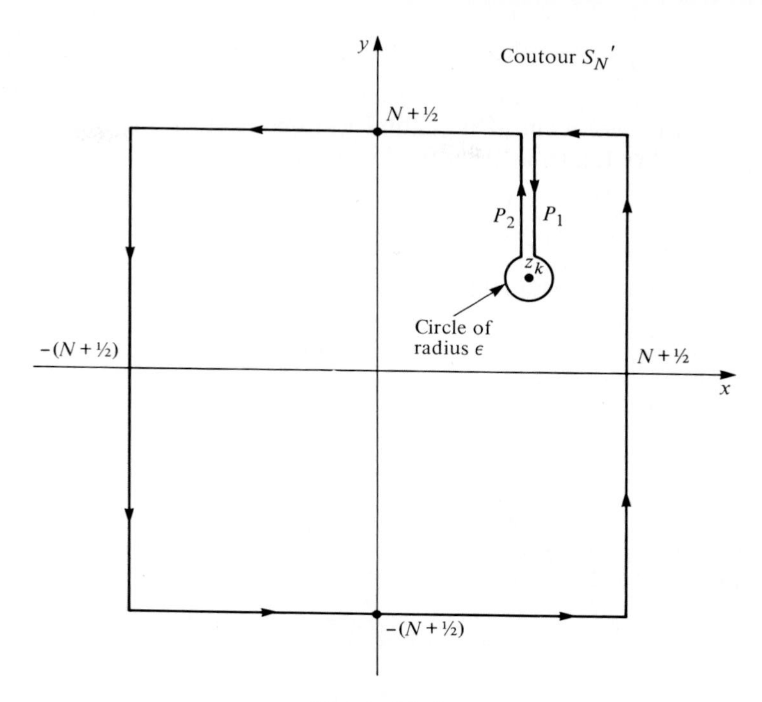

Figure A6–3

equation. Using this generalization and recognizing that the integral in the preceding equation is the Watson transformation of the complex Fourier series we have the following theorem:

THEOREM 12 Evaluation of the integral in the Watson transformation

Let $g(z)$ be analytic throughout the complex plane except for a finite number of poles, none of which is at a real value of z. Assume that there exist positive constants m, k, and R such that for $|z| \geq R$, $|g(z)| \leq m/|z|^k$. Then

$$\sum_{n=-\infty}^{+\infty} g(n)e^{in\phi} = -\pi \sum_{\text{residues}} \frac{e^{iz\phi}e^{-i\pi z}g(z)}{\sin \pi z} \text{ at all poles of } g(z), \quad \text{(A6–21)}$$

provided $0 < \phi < 2\pi$. □

EXAMPLE 2

Find a closed form expression for the sum of the complex Fourier series $\sum_{n=-\infty}^{+\infty} g(n)e^{in\phi}$, where

$$g(n) = \frac{1 - e^{-2\pi}}{2\pi i(n - i)}.$$

Solution

The preceding theorem applies. Observe that $g(z) = (1 - e^{-2\pi})/2\pi i(z - i)$ agrees with $g(n)$ when z is an integer n and that $g(z)$ has just one pole, and it is not real. Notice also that $g(z)$ satisfies Eq. (A6–19) as the following steps show:

$$|z - i| \geq |z| - 1 \qquad \text{if } |z| \geq 1 \text{ (see Eq. (1.3–20))};$$

$$|z| - 1 \geq |z| - |z|/2 \qquad \text{if } |z| \geq 2.$$

Combining the preceding inequalities,

$$|z - i| \geq \frac{1}{2}|z| \qquad \text{if } |z| \geq 2,$$

and so

$$\frac{1}{|z - i|} \leq \frac{2}{|z|} \qquad \text{if } |z| \geq 2$$

which shows that Eq. (A6–19) is satisfied.

Applying Eq. (A6–21) we have

$$\sum_{n=-\infty}^{+\infty} g(n)e^{in\phi} = -\pi \operatorname{Res} \frac{e^{iz\phi}e^{-i\pi z}(1 - e^{-2\pi})}{2\pi i(z - i)\sin \pi z} \text{ at } i$$

$$= \frac{-1}{2i} \frac{e^{-\phi}e^{\pi}(1 - e^{-2\pi})}{\sin \pi i}$$

$$= \frac{e^{-\phi}(e^{\pi} - e^{-\pi})}{-2i \sin \pi i} = e^{-\phi},$$

which is valid for $0 < \phi < 2\pi$. To check this answer we can expand $e^{-\phi}$ in a complex Fourier series using Eq. (A6–16). The given coefficients $g(n)$ are found. ◀

EXERCISES

1. a) Show that

$$\sum_{n=-\infty}^{+\infty} \frac{1}{n^2 + a^2} = \frac{\pi}{a} \coth \pi a$$

provided $n^2 + a^2 \neq 0$ for all integer n.

b) Use the above result to show that

$$\sum_{n=0}^{\infty} \frac{1}{n^2 + a^2} = \frac{1 + \pi a \coth \pi a}{2a^2}$$

and

$$\sum_{n=1}^{\infty} \frac{1}{n^2 + a^2} = \frac{\pi a \coth \pi a - 1}{2a^2}.$$

c) Assume that

$$\lim_{a\to 0}\sum_{n=1}^{\infty}\frac{1}{n^2+a^2}=\sum_{n=1}^{\infty}\lim_{a\to 0}\frac{1}{n^2+a^2}$$

and use L'Hôpital's rule together with the last result in part (b) to show that

$$\sum_{n=1}^{\infty}\frac{1}{n^2}=\frac{\pi^2}{6}.$$

2. a) Show that

$$\sum_{n=-\infty}^{+\infty}\frac{1}{n^2-i}=\frac{\pi}{\sqrt{2}}(i-1)\left[\frac{\sin\left(\frac{2\pi}{\sqrt{2}}\right)-i\sinh\left(\frac{2\pi}{\sqrt{2}}\right)}{\cosh\left(\frac{2\pi}{\sqrt{2}}\right)-\cos\left(\frac{2\pi}{\sqrt{2}}\right)}\right].$$

Hint: Eq. (3.2–14) is useful here.

b) Use the real and imaginary parts of the preceding result to show that

$$\sum_{n=0}^{\infty}\frac{n^2}{n^4+1}=\frac{\pi}{2\sqrt{2}}\left[\frac{\sinh\left(\frac{2\pi}{\sqrt{2}}\right)-\sin\left(\frac{2\pi}{\sqrt{2}}\right)}{\cosh\left(\frac{2\pi}{\sqrt{2}}\right)-\cos\left(\frac{2\pi}{\sqrt{2}}\right)}\right]$$

and

$$\sum_{n=0}^{\infty}\frac{1}{n^4+1}=\frac{\pi}{2\sqrt{2}}\left[\frac{\sinh\left(\frac{2\pi}{\sqrt{2}}\right)+\sin\left(\frac{2\pi}{\sqrt{2}}\right)}{\cosh\left(\frac{2\pi}{\sqrt{2}}\right)-\cos\left(\frac{2\pi}{\sqrt{2}}\right)}\right]+\frac{1}{2}.$$

3. a) Using Theorem 11 show that

$$\sum_{n=-\infty}^{+\infty}\frac{(-1)^n}{n^2+1}=\frac{\pi}{\sinh\pi}.$$

b) Use the preceding result to show that

$$1-\frac{1}{1^2+1}+\frac{1}{2^2+1}-\frac{1}{3^2+1}+\cdots=\frac{\pi}{2\sinh\pi}+\frac{1}{2}.$$

4. Prove Theorem 11 by following these steps:

a) Show that on the right and left sides of C_N in Fig. A6–1 we have $|1/\sin\pi z|\le 1$ and that on the top and bottom sides, $|1/\sin\pi z|\le 1/\sinh(\pi/2)\approx 0.43$. Thus $|1/\sin\pi z|\le 1$ is satisfied everywhere on C_N.

b) Let $f(z)$ be a function that is analytic except for poles, none of which is an integer and let C_N be large enough to enclose these poles. Evaluate $\oint(\pi/\sin\pi z)f(z)\,dz$ around C_N and obtain an expression similar to Eq. (A6–8).

c) What are the residues of $\pi f(z)/\sin \pi z$ at $z = n$ (an integer)?

d) Assume $f(z)$ satisfies Eq. (A6–7).

Consider $\lim_{N\to\infty}$ in the integral of part (b) and show that the integral tends to zero in the limit and thus derive Eq. (A6–13). How does Theorem 11 follow from this equation?

5. Show that for $a > 0$ we have

$$\sum_{n=0}^{\infty} \frac{(-1)^n n^2}{n^4 + a^4} = \frac{\pi^2}{4d}\left(\frac{\cos d \sinh d - \sin d \cosh d}{\sin^2 d + \sinh^2 d}\right),$$

where $d = \pi a/\sqrt{2}$.

Hint:

$$\frac{1}{\sin(x + iy)} = \frac{\sin x \cosh y - i \cos x \sinh y}{\sin^2 x + \sinh^2 y}.$$

6. Here is an alternative derivation of the result derived in Exercise 1(c).

a) Show that on the contour C_N of Fig. A6–1 we have

$$\left|\frac{1}{z^2} \cot \pi z\right| \leq \frac{\coth(\pi/2)}{(N + 1/2)^2}.$$

b) Show that

$$\oint_{C_N} \frac{\pi}{z^2} \cot \pi z \, dz = 2\pi i \operatorname{Res}\left(\frac{\pi \cot \pi z}{z^2}\right) \text{ at } z = 0$$
$$+ 4\pi i \sum_{n=1}^{N} \frac{1}{n^2}.$$

c) Let $N \to \infty$ in the preceding equation and argue that the integral on the left goes to zero. Use the result of part (a).

d) Evaluate the residue on the right in part (b) by division of series (see Section 6.3, Example 5) and obtain the result

$$\sum_{n=1}^{\infty} \frac{1}{n^2} = \frac{\pi^2}{6}.$$

7. Derive the result

$$\sum_{n=1}^{\infty} \frac{(-1)^n}{n^2} = \frac{-\pi^2}{12}$$

by following these steps: Consider $\oint \pi/(z^2 \sin \pi z)\, dz$ around the contour C_N of Fig. A6–1. Evaluate this integral using residues. Now argue that the integral around C_N vanishes as $N \to \infty$. You will need the ML inequality and you must prove that on C_N we have

$$\left|\frac{1}{z^2 \sin \pi z}\right| \leq \frac{1}{(N + 1/2)^2}.$$

Show that as $N \to \infty$ we obtain

$$0 = 2 \sum_{n=1}^{\infty} \frac{(-1)^n}{n^2} + \operatorname{Res} \frac{\pi}{z^2 \sin \pi z} \quad \text{at } z = 0.$$

Evaluate the preceding residue by series division to complete the proof.

Using Theorem 12 prove the following for $0 < \phi < 2\pi$:

8. $$\sum_{n=-\infty}^{+\infty} \frac{e^{in\phi}}{n^2+1} = \frac{\pi \cosh(\phi - \pi)}{\sinh \pi}$$

9. a) $$\sum_{n=-\infty}^{+\infty} \frac{ine^{in\phi}}{n^2+1} = \frac{\pi \sinh(\phi - \pi)}{\sinh \pi}$$

b) Use the preceding result to show that

$$\frac{1}{2} - \frac{3}{10} + \frac{5}{26} - \frac{7}{50} + \cdots = \frac{\pi \sinh \pi/2}{2 \sinh \pi}.$$

CHAPTER 7

Laplace Transforms and Stability of Systems

7.1 INTRODUCTION TO AND INVERSION OF LAPLACE TRANSFORMS

In this chapter we presuppose that the reader has some familiarity with the method of Laplace transforms in the solution of differential equations but is unfamiliar with the way in which complex variable theory can be of help when Laplace transforms are used. We will show here how residue calculus can be used in the "inversion" of Laplace transforms. We will also use our knowledge of analytic functions to determine whether the behavior of a physical system, analyzed with Laplace transforms, is stable or unstable.

We begin by briefly reviewing and listing the basic properties of Laplace transforms. Let $f(t)$ be a real or complex valued function of the real variable t. Let $s = \sigma + i\omega$ be a complex variable. Then the Laplace transform of $f(t)$, designated $F(s)$, is defined as follows:

DEFINITION Laplace transform

$$F(s) = \int_0^\infty f(t)e^{-st}\,dt. \quad \square \tag{7.1–1}$$

In general, we use lowercase letters to mean functions of t, for example, $f(t)$ and $g(t)$, and uppercase letters to denote the corresponding Laplace transforms, in this case, $F(s)$ and $G(s)$.

In anticipation of later work we define the integral in Eq. (7.1–1) as follows:

$$F(s) = \int_0^\infty f(t)e^{-st}\,dt = \lim_{\varepsilon\to 0+} \int_\varepsilon^\infty f(t)e^{-st}\,dt.$$

The lower limit is thus $0+$; that is, $t = 0$ is approached *from the right through positive values.* For ordinary functions $f(t)$ that are continuous at $t = 0$ or have jump discontinuities here (see Section 6.10), it makes no difference whether we use lower limit 0 or $0+$. However, if $f(t)$ does have a severe singularity at $t = 0$ the choice of lower limit does become significant. In the present section the lower limit is chosen so as to exclude such singular points; the function $F(s)$ is defined entirely in terms of $f(t)$ for $t > 0$. Equation (7.1–1) then is the classical definition of the Laplace transform.

For some engineering purposes it is important to define $F(s)$ so as to depend not only on the behavior of $f(t)$ for $t > 0$ but also on the behavior of $f(t)$ in a small interval around $t = 0$. This matter is treated in Section 7.5, where we deal with singular functions of t that are not functions in the usual sense. Then we will modify our definition of the Laplace transform so that the lower limit of integration is $0-$. Most of the results of Sections 7.1 and 7.2 will be applicable in Section 7.5.

The operation on $f(t)$ described by Eq. (7.1–1) is also written $F(s) = \mathscr{L}f(t)$. The function of t whose Laplace transform is $F(s)$ is written $\mathscr{L}^{-1}F(s)$. Thus $f(t) = \mathscr{L}^{-1}F(s)$. We say that *$f(t)$ is the inverse transform of $F(s)$.* Just as we have an integral, Eq. (7.1–1), defining the operator $\mathscr{L}$ we will soon regard $\mathscr{L}^{-1}$ as an operator defined by an integral.

Recall that

$$\mathscr{L}e^{-bt} = \frac{1}{s+b}, \qquad \text{if } \operatorname{Re} s > -\operatorname{Re} b, \tag{7.1–2}$$

which is derived from

$$\mathscr{L}e^{-bt} = \int_0^\infty e^{-st}e^{-bt}\,dt = \int_0^\infty e^{-(s+b)t}\,dt = \left.\frac{e^{-(s+b)t}}{-(s+b)}\right|_0^\infty$$

$$= \lim_{t\to\infty}\left[\frac{e^{-(s+b)t}}{-(s+b)}\right] + \frac{1}{(s+b)}. \tag{7.1–3}$$

Taking $s = \sigma + i\omega$, $b = \alpha + i\beta$, we obtain

$$\frac{e^{-(s+b)t}}{s+b} = \frac{e^{-(\sigma+\alpha)t}e^{-i(\beta+\omega)t}}{s+b}.$$

For $\sigma + \alpha > 0$, the preceding expression $\to 0$ as $t \to \infty$. Putting this limit in Eq. (7.1–3) establishes Eq. (7.1–2). The condition $\sigma + \alpha > 0$ is equivalent to $\operatorname{Re} s > -\operatorname{Re} b$. The inverse of Eq. (7.1–2) is

$$\mathscr{L}^{-1}\frac{1}{s+b} = e^{-bt}, \qquad t > 0. \tag{7.1–4}$$

If necessary, the reader should consult a table to again become familiar with some of the common transforms and their inverses.

Both of the operations $\mathscr{L}$ and $\mathscr{L}^{-1}$ satisfy the *linearity property*. Thus

$$\mathscr{L}[c_1 f_1(t) + c_2 f_2(t)] = c_1 \mathscr{L} f_1(t) + c_2 \mathscr{L} f_2(t) = c_1 F_1(s) + c_2 F_2(s), \tag{7.1–5}$$

where c_1 and c_2 are constants, and

$$\mathscr{L}^{-1}[c_1 F_1(s) + c_2 F_2(s)] = c_1 f_1(t) + c_2 f_2(t). \tag{7.1–6}$$

If a function $f(t)$ is piecewise continuous[†] over every finite interval on the line $t \geq 0$ and if there exist real constants k, p, and T such that

$$|f(t)| < ke^{pt}, \qquad \text{for } t \geq T \tag{7.1–7}$$

then $f(t)$ will have a Laplace transform $F(s)$ for all s satisfying $\operatorname{Re} s > p$. This transform not only exists in the half plane $\operatorname{Re} s > p$, it is an *analytic function* in this half plane.[‡] Functions satisfying Eq. (7.1–7) for some choice of k, p, and T are said to be of *order* e^{pt}.

The preceding conditions are sufficient to guarantee both the existence of the Laplace transform and its analyticity for $\operatorname{Re} s > p$. The requirement of piecewise continuity is actually overly conservative. There are functions of order e^{pt}, with integrable singularities, which possess transforms, analytic in a half plane. For example, it is proved later in this section that $\mathscr{L} 1/\sqrt{t} = \sqrt{\pi}/s^{1/2}$ for a certain branch of $s^{1/2}$. We require that $\operatorname{Re} s \geq 0$. The function $1/\sqrt{t}$ is of exponential order (one can take $p = 0$, $k = 1$, $T = 1$); however, this function is not piecewise continuous on the line $t \geq 0$ owing to its singularity at $t = 0$.

In any transformation procedure one needs to consider uniqueness. According to Eq. (7.1–1), $f(t)$ has only one Laplace transform $F(s)$. It can be shown that if $f(t)$ and $g(t)$ have the same transform $F(s)$, then for $t \geq 0$, we can almost say that $f(t) = g(t)$ over any finite nonzero interval in the variable t. We say "almost" because $f(t)$ and $g(t)$ can differ at a finite number of isolated points in each interval. No statement can be made concerning the relationship between $f(t)$ and $g(t)$ for negative t. Note that if $f(t)$ and $g(t)$ are both *continuous* for $t \geq 0$ and have the same transform $F(s)$, then $f(t) = g(t)$ for $t \geq 0$.

The usefulness of Laplace transforms relates to the ease with which we may obtain $\mathscr{L}\, df/dt$ in terms of $F(s) = \mathscr{L} f(t)$. Taking $\mathscr{L}\, df/dt = \int_0^\infty df/dt\, e^{-st}\, dt$, we integrate by parts and obtain

$$\int_0^\infty e^{-st} \frac{df}{dt}\, dt = e^{-st} f(t)\Big|_0^\infty + \int_0^\infty s f(t) e^{-st}\, dt.$$

If $f(t)$ satisfies Eq. (7.1–7), then, provided $\operatorname{Re} s > p$, $e^{-st} f(t)$ will vanish as $t \to \infty$. Also $e^{-st} f(t)$ equals $f(0)$ at $t = 0$. The integral on the right in the preceding equation is

---

[†] Piecewise continuity is discussed in Section 6.10.

[‡] See R. V. Churchill, *Operational Mathematics*, 3rd ed. (New York: McGraw-Hill, 1972), p. 186.

by definition $sF(s)$. Thus

$$\mathscr{L}\frac{df}{dt} = sF(s) - f(0). \tag{7.1–8}$$

If $f(t)$ has a jump discontinuity at $t = 0$, then df/dt will not exist at $t = 0$. However, since the Laplace transform of df/dt involves an integration only through positive values of t, we can still use Eq. (7.1–8) if we replace $f(0)$ on the right side by $f(0+)$.

The derivation of the preceding equation is valid if $f(t)$ is of order e^{pt}, $f(t)$ is continuous for $t > 0$, and $f'(t)$ is piecewise continuous in every finite interval along the line $t > 0$.

Knowing the Laplace transform of df/dt we can now find the transform of d^2f/dt^2 in a similar way. It is given by

$$\mathscr{L}\frac{d^2f}{dt^2} = s^2F(s) - sf(0) - f'(0),$$

and in general,

$$\mathscr{L}f^{(n)}(t) = s^nF(s) - s^{n-1}f(0) - s^{n-2}f'(0) - s^{n-3}f''(0) - \cdots - f^{(n-1)}(0), \tag{7.1–9}$$

provided $f(t)$ and its first, second, ..., and $(n - 1)$ derivatives are of order e^{pt}, $f(t)$ and these same derivatives are continuous for $t > 0$, and $f^{(n)}(t)$ is piecewise continuous in every finite interval on the line $t > 0$. If $f(t)$ or any of its first $n - 1$ derivatives fail to be continuous at $t = 0$, it is understood that we must use the right hand limit $0+$ in the preceding equation.

The Laplace transform of the integral of $f(t)$ is easily stated in terms of the Laplace transform of $f(t)$. If the Laplace transform of $f(t)$ exists for $\text{Re}(s) > p \geq 0$, then

$$\mathscr{L}\int_0^t f(x)\,dx = \frac{1}{s}\mathscr{L}f(t) = \frac{F(s)}{s}. \tag{7.1–10}$$

is valid for $\text{Re}\, s > p$. A formal proof is presented in Exercise 21.

Laplace transforms are of use in solving linear differential equations with constant coefficients and prescribed initial conditions. Such equations are converted to algebraic equations involving the Laplace transform of the unknown function. The following example should serve as a reminder of the method. We solve

$$\frac{df}{dt} + 2f(t) = e^{-3t}, \qquad \text{for } t \geq 0 \tag{7.1–11}$$

with the initial condition $f(0) = 4$.

From Eq. (7.1–8) we have $\mathscr{L}\,df/dt = sF(s) - 4$, and from Eq. (7.1–2), $\mathscr{L}e^{-3t} = 1/(s + 3)$. Employing the linearity property in Eq. (7.1–5) we transform both sides of Eq. (7.1–11) and obtain

$$sF(s) - 4 + 2F(s) = \frac{1}{s + 3}.$$

We solve the preceding equation and obtain

$$F(s) = \frac{1}{(s+3)(s+2)} + \frac{4}{s+2}.$$

To obtain $f(t) = \mathscr{L}^{-1}F(s)$, we could consult a table of transforms and their inverses and find that $f(t) = 5e^{-2t} - e^{-3t}$. We easily verify that this satisfies the differential equation and its initial condition.

The preceding illustrates one potentially difficult step for the Laplace transform user—performing an inverse transformation to convert $F(s)$ to the actual solution $f(t)$. Often we are lucky enough to find $F(s)$ in a table; we then read off the corresponding $f(t)$. If $F(s)$ is not listed, we must, if possible, rearrange our expression into a sum of simpler terms that do appear in our table. The reader is perhaps familiar with a set of rules for finding $\mathscr{L}^{-1}F(s)$ when $F(s) = P(s)/Q(s)$ and $P(s)$ and $Q(s)$ are polynomials in s. These rules, called the *Heaviside expansion formulas*, are based implicitly on the fact that rational expressions like $P(s)/Q(s)$ can be written as a sum of partial fractions, each of whose inverse transform is readily found.† The technique that we introduce here for finding $f(t)$ is rooted directly in complex variable theory. It is more succinct than the Heaviside method, does not involve the memorization of a set of rules, and is not limited to rational expressions.

Typically, $F(s)$ defined by Eq. (7.1–1) exists when s is confined to a half plane $\text{Re } s > p$; we observed earlier that $F(s)$ is analytic in the same half plane. For example (see Eq. 7.1–2) $\mathscr{L}e^{-2t} = 1/(s+2)$ exists and is analytic for $\text{Re } s > -2$. The analytic properties of $F(s)$ are important as they enable us to use the tools of complex variable theory.

For the moment, instead of dealing with $F(s)$, let us use $F(z)$ which is $F(s)$ with s replaced by z, that is,

$$F(z) = \int_0^\infty f(t)e^{-zt}\,dt. \tag{7.1–12}$$

Suppose that $F(z)$ is analytic in the z-plane everywhere along the line $x = a$ and to the right of this line. We also make an assumption about $F(z)$ tending to 0 as $|z| \to \infty$ along any path in the half plane $\text{Re } z \geq a$. More precisely, there exist positive numbers m, k, and R_0 so that, for $|z| > R_0$ and $\text{Re } z \geq a$ we have

$$|F(z)| \leq \frac{m}{|z|^k}. \tag{7.1–13}$$

Now let us apply the Cauchy integral formula to $F(z)$ and use the closed semicircular contour C shown in Fig. 7.1–1. The radius of the arc is b. For simplicity, we take $a \geq 0$, although an easy modification makes the discussion valid for $a < 0$ as well. We take s as some arbitrary point within C and take C_1 as the curved portion of C.

---

† See T. B. A. Senior, *Mathematical Methods in Electrical Engineering* (New York: Cambridge University Press, 1986) p. 43.

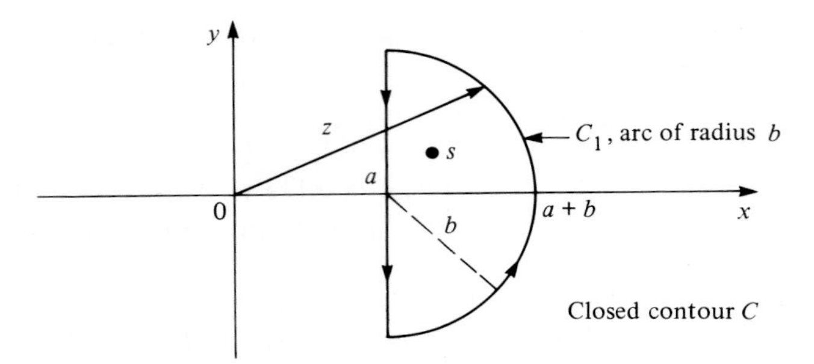

Figure 7.1–1

Integrating in the direction of the arrows we have

$$F(s) = \frac{1}{2\pi i}\oint_C \frac{F(z)}{z - s}\,dz = \frac{1}{2\pi i}\left[\int_{\substack{a+ib \\ x=a}}^{a-ib} \frac{F(z)}{z - s}\,dz + \int_{C_1} \frac{F(z)}{z - s}\,dz\right]. \tag{7.1–14}$$

Our plan is to argue that the integral over C_1 tends to zero in the limit $b \to \infty$.

Let us consider an upper bound for $|F(z)|/|(z - s)|$ on C_1. We begin with the numerator. Provided b is sufficiently large, Eq. (7.1–13) provides a bound on the numerator $|F(z)|$. As is shown in Fig. 7.1–1, on C_1 we have $|z| \geq b$ or $1/|z| \leq 1/b$. Combining this with Eq. (7.1–13), we have

$$|F(z)| \leq \frac{m}{b^k}, \qquad \text{with } z \text{ on } C_1. \tag{7.1–15}$$

Now we examine $|z - s|$ on C_1.

Some careful study of Fig. 7.1–2 reveals that on C_1 the minimum possible value of $|z - s|$ occurs when the point z lies on the line connecting points s and a. The minimum value of $|z - s|$ is indicated and is equal to $b - |s - a|$. Thus on C_1 we have

$$|z - s| \geq b - |s - a|. \tag{7.1–16}$$

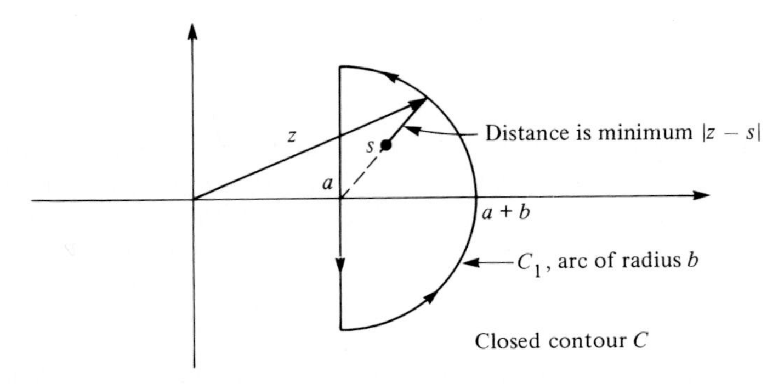

Figure 7.1–2

A triangle inequality $|s - a| \le |s| + a$ combined with Eq. (7.1–16) yields

$$|z - s| \ge b - (|s| + a). \tag{7.1–17}$$

We will assume that b is large enough so that the right side of Eq. (7.1–17) is positive. Taking the reciprocal of both sides of Eq. (7.1–17), we get

$$\frac{1}{|z - s|} \le \frac{1}{b - |s| - a}; \tag{7.1–18}$$

multiplying Eq. (7.1–18) by Eq. (7.1–15), we obtain

$$\frac{|F(z)|}{|z - s|} \le \frac{m}{b^k(b - |s| - a)}. \tag{7.1–19}$$

Now we apply the ML inequality to our integral over C_1, and notice that L, the path length, is πb. Hence

$$\left| \int_{C_1} \frac{F(z)}{z - s}\, dz \right| \le M\pi b, \tag{7.1–20}$$

where we require that $|F(z)|/|(z - s)| \le M$.

We see that M can be taken as the right side of Eq. (7.1–19). Thus Eq. (7.1–20) becomes

$$\left| \int_{C_1} \frac{F(z)}{z - s}\, dz \right| \le \frac{m\pi b}{b^k(b - |s| - a)} = \frac{m\pi}{b^k\left(1 - \dfrac{|s| + a}{b}\right)}.$$

Clearly, as $b \to \infty$, the right side of the equation $\to 0$. Therefore the integral contained on the left also goes to zero. Finally, passing to the limit $b \to \infty$ in Eq. (7.1–14) and using the result just derived for the integral on C_1, we have

$$F(s) = \frac{1}{2\pi i} \int_{a+i\infty}^{a-i\infty} \frac{F(z)}{z - s}\, dz.$$

With a reversal of limits this becomes

$$F(s) = \frac{1}{2\pi i} \int_{a-i\infty}^{a+i\infty} \frac{F(z)}{s - z}\, dz. \tag{7.1–21}$$

On the contour of integration in Eq. (7.1–21) we have $z = a + iy$, $-\infty < y < \infty$, and $dz = i\, dy$. Using y as our parameter of integration we obtain the following theorem:

THEOREM 1 Let $F(z)$ be analytic in the half plane $\operatorname{Re} z \ge a$, and in this region let $F(z)$ satisfy

$$|F(z)| \le m/|z|^k$$

whenever $|z| > R_0$. Here k, m, and R_0 are positive constants. Then if $\operatorname{Re}(s) > a$,

$$F(s) = \frac{1}{2\pi} \int_{-\infty}^{+\infty} \frac{F(a + iy)}{s - (a + iy)}\, dy. \quad \square \tag{7.1–22}$$

The preceding theorem is *not limited* to functions $F(z)$ that are Laplace transforms but is applicable to any function satisfying the conditions stated above. However, we will use the theorem to establish an integral expression that will yield $f(t)$ whenever $F(s)$ is determined.

Let $f(t)$ satisfy Eq. (7.1–7) for some p. Then its Laplace transform $F(s)$ is analytic for $\operatorname{Re} s > p$. Taking $a > p$ we will show that

$$f(t) = \frac{1}{2\pi i}\int_{a-i\infty}^{a+i\infty} F(s)e^{st}\, ds, \tag{7.1–23}$$

where the integration is performed in the complex s-plane, along a vertical line to the right of $\operatorname{Re} s = p$.

To prove Eq. (7.1–23) we take the Laplace transform of $f(t)$ as defined by this formula and show that $F(s)$ under the integral sign is recovered.

First we put $s = \sigma + i\omega$ on the right in the preceding equation, and set $\sigma = a$ on the contour of integration. Thus $ds = i\, d\omega$ and we have

$$f(t) = \frac{1}{2\pi}\int_{-\infty}^{+\infty} F(a + i\omega)e^{(a+i\omega)t}\, d\omega. \tag{7.1–24}$$

The Laplace transform of the above is

$$\begin{aligned}\mathscr{L}f(t) &= \frac{1}{2\pi}\int_0^{\infty} e^{-st}\int_{-\infty}^{+\infty} F(a + i\omega)e^{(a+i\omega)t}\, d\omega\, dt \\ &= \frac{1}{2\pi}\int_0^{\infty}\int_{-\infty}^{+\infty} e^{-(s-a)t}F(a + i\omega)e^{i\omega t}\, d\omega\, dt.\end{aligned}$$

We wish to change the order of integration in the preceding double integral. Let the following conditions be true: $\operatorname{Re} s > a$ and $\int_{-\infty}^{+\infty} |F(a + i\omega)|\, d\omega$ exists. With both these satisfied it is not hard to show that the absolute value of the integrand $e^{-(s-a)t}F(a + i\omega)e^{i\omega t}$ is integrable first from $\omega = -\infty$ to $\omega = \infty$ and then from $t = 0$ to $t = \infty$. It can then be shown that this is sufficient to guarantee the existence of this double integral and to permit us to reverse the order of integration.†

Assuming that these conditions are satisfied we have

$$\mathscr{L}f(t) = \frac{1}{2\pi}\int_{-\infty}^{+\infty} F(a + i\omega)\left[\int_0^{\infty} e^{-st}e^{(a+i\omega)t}\, dt\right] d\omega. \tag{7.1–25}$$

The inner integral is simply the Laplace transform of e^{-bt}, which we have previously evaluated (Eq. 7.1–2). Here $b = -a - i\omega$. Thus for $\operatorname{Re} s > -\operatorname{Re}(-a - i\omega) = a$ the

† The problem of justifying the reversal of the order of integration in an improper double integral is not simple. The reader is referred to the following: T. Apostol, *Mathematical Analysis*, 2nd ed. (Reading MA: Addison-Wesley, 1974), Chapter 14, and J. Pierpont, *The Theory of Functions of Real Variables* (New York: Dover 1959), pp. 479–492.

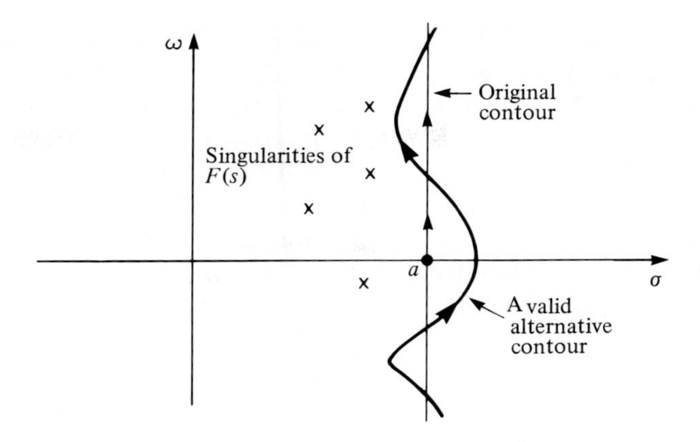

Figure 7.1–3

integral in brackets is $1/(s - a - i\omega)$. With this result we have

$$\mathscr{L}f(t) = \frac{1}{2\pi}\int_{-\infty}^{+\infty} \frac{F(a + i\omega)}{s - (a + i\omega)}\, d\omega. \tag{7.1–26}$$

Putting y in place of ω in the preceding, studying Theorem 1, and assuming its requirements to be satisfied, we see that the right side of Eq. (7.1–26) is simply the desired $F(s)$. Notice that we assumed one of the requirements of the theorem, $\operatorname{Re} s > a$, to justify a swap in the order of a double integration. Summarizing Eq. (7.1–23) and its derivation, we have

THEOREM 2 (Laplace inversion formula) Let $F(s)$ be a function analytic in the half plane $\operatorname{Re} s \geq a$ of the complex s-plane. Assume that there exist positive constants m, R_0, and k such that $|F(s)| \leq m/|s|^k$ when $|s| > R_0$ in this half plane. Then there is a function $f(t)$ whose Laplace transform is $F(s)$, and it is given by

$$f(t) = \mathscr{L}^{-1}F(s) = \frac{1}{2\pi i}\int_{a-i\infty}^{a+i\infty} F(s)e^{st}\, ds. \tag{7.1–27}$$

The integration is performed along the straight line $\operatorname{Re} s = a$ or along any other contour into which this line can legally be changed (see Fig. 7.1–3) by the principle of path independence. The preceding equation is also called the *Bromwich integral formula.* □

Comment: Our derivation of Theorem 2 presupposes the existence of $\int_{-\infty}^{+\infty} |F(a + i\omega)|\, d\omega$. A more sophisticated analysis than the one presented here shows this to be unnecessarily restrictive,[†] and we will ignore this requirement. This analysis

---

[†] See, e.g., R. V. Churchill, *Operational Mathematics*, 3rd ed. (New York: McGraw-Hill, 1972), Chapter 6.

also shows that if $F(s)$ is the Laplace transform of a function of t having a jump discontinuity at some point, say $t_0 > 0$, then the function of time produced by the right side of Eq. (7.1–27) will, when evaluated at t_0, yield the average of the right and left hand limits of $f(t)$ at t_0, i.e., $(1/2)[f(t_0+) + f(t_0-)]$. If $t_0 = 0$, the equation yields $(1/2)f(0+)$. If there is any question as to the validity of using Eq. (7.1–27) in obtaining $f(t)$ from $F(s)$, we can justify ourselves by taking the Laplace transformation of the function of t obtained from this equation and verifying that the given $F(s)$ is obtained. In certain cases the integral in Eq. (7.1–27) will be found to exist only as a Cauchy principal value. Thus the integral must be defined as

$$\lim_{b\to\infty} \frac{1}{2\pi i} \int_{a-ib}^{a+ib} F(s)e^{st}\,ds,$$

and it is this evaluation that we shall use. The definition of $F(s)$ used in Eq. (7.1–1) employs only $f(t)$ defined for $t > 0$; we thus assume $t > 0$ in applying Eq. (7.1–27). In Exercise 30 we show that the function $f(t)$ obtained from the Bromwich integral is zero for $t < 0$.

An alternate derivation of the Bromwich integral, which exploits the properties of the Fourier transform and its inverse (see Section 6.10), is developed in Exercise 29.

The function $F(s)$ is typically defined and analytic throughout some right half space of the complex s-plane, and the analytic continuation† of $F(s)$ into the remainder of this plane is often such that the Bromwich integral is evaluated by residues. For example, suppose we must find $\mathscr{L}^{-1} 1/(s+1)^2$ without a table of transforms. We have, from Eq. (7.1–27),

$$f(t) = \frac{1}{2\pi i} \int_{a-i\infty}^{a+i\infty} \frac{e^{st}}{(s+1)^2}\,ds,$$

where, because of the pole of $1/(s+1)^2$ at -1, we require $a > -1$. Let us take $a = 0$.

To evaluate our integral, we consider the contour C in Fig. 7.1–4, which consists of the straight line extending from $\omega = -R$ to $\omega = R$ and the semicircular arc C_1 on which $|s| = R$. We have

$$\frac{1}{2\pi i} \oint_C \frac{e^{st}}{(s+1)^2}\,ds = \frac{1}{2\pi i} \int_{-iR}^{iR} \frac{e^{st}}{(s+1)^2}\,ds + \frac{1}{2\pi i} \int_{C_1} \frac{e^{st}}{(s+1)^2}\,ds. \qquad (7.1\text{–}28)$$

The integral on the left, taken around the closed contour C, is readily evaluated with residues as follows:

$$\frac{2\pi i}{2\pi i} \operatorname{Res}\left[\frac{e^{st}}{(s+1)^2}, -1\right] = \lim_{s\to -1} \frac{d}{ds} e^{st} = te^{-t}. \qquad (7.1\text{–}29)$$

Passing to the limit $R \to \infty$ in Eq. (7.1–28), we see that the first integral on the right is now taken along the entire imaginary axis. One can easily show that as $R \to \infty$ the integral over the arc C_1 tends to zero whenever $t \geq 0$. The details, which involve the ML inequality, can be supplied by the reader.

---

† Analytic continuation is discussed in Section 5.7.

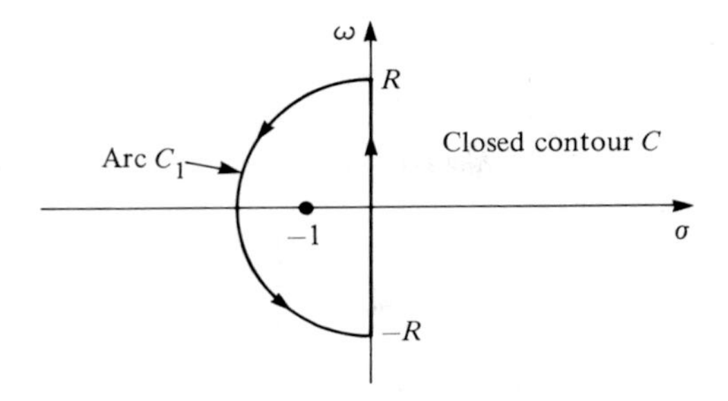

Figure 7.1–4

Thus, letting $R \to \infty$ in Eq. (7.1–28) and using Eq. (7.1–29), we have

$$te^{-t} = \frac{1}{2\pi i}\int_{-i\infty}^{i\infty} \frac{e^{st}\,ds}{(s+1)^2}.$$

On the right we have a Bromwich integral for the evaluation of $\mathscr{L}^{-1}1/(s+1)^2$. Thus $te^{-t} = \mathscr{L}^{-1}1/(s+1)^2$, which tables of transforms confirm as being correct.

The procedure just used should be generalized to permit inversion of a variety of transforms. We will therefore prove the following theorem.

THEOREM 3 Let $F(s)$ be analytic in the s-plane except for a finite number of poles that lie to the left of some vertical line Re $s = a$. Suppose there exist positive constants m, R_0, and k such that for all s lying in the half plane Re $s \le a$, and satisfying $|s| > R_0$, we have $|F(s)| \le m/|s|^k$. Then for $t > 0$,

$$\mathscr{L}^{-1}F(s) = \sum_{\text{res}} F(s)e^{st}, \qquad \text{at all poles of } F(s). \quad \square \qquad (7.1\text{–}30)$$

Theorem 3 requires that ultimately $F(s)$ falls off at least as rapidly as $m/|s|^k$ when s lies in a certain half space.

The proof proceeds as follows: Consider $(1/2\pi i)\oint_C F(s)e^{st}\,ds$ taken around the contour C shown in Fig. 7.1–5. We choose R greater than R_0 of the theorem, and a is chosen so that all poles of $F(s)$ are inside C. Recall that e^{st} is an entire function. From residue calculus we have

$$\frac{1}{2\pi i}\oint_C F(s)e^{st}\,ds = \sum_{\text{res}} F(s)e^{st}, \qquad \text{at all poles of } F(s). \qquad (7.1\text{–}31)$$

We now rewrite the integral around C in terms of integrals along the straight segment and the various arcs. Thus

$$\frac{1}{2\pi i}\oint_C F(s)e^{st}\,ds = \frac{1}{2\pi i}\int_{a-ib}^{a+ib} F(s)e^{st}\,ds + \frac{1}{2\pi i}\int_{C_1} F(s)e^{st}\,ds$$
$$+ \frac{1}{2\pi i}\int_{C_2} F(s)e^{st}\,ds + \frac{1}{2\pi i}\int_{C_3} F(s)e^{st}\,ds + \frac{1}{2\pi i}\int_{C_4} F(s)e^{st}\,ds, \qquad (7.1\text{–}32)$$

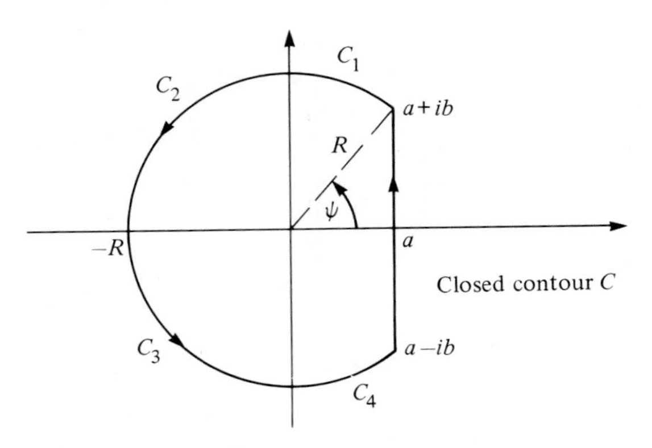

Figure 7.1–5

where, as shown in Fig. 7.1–5, C_1 extends from $a + ib$ to iR, C_2 goes from iR to $-R$, etc. Our goal is to let $R \to \infty$ and argue that the integrals taken over C_1, C_2, C_3, C_4 become zero. The first integral on the right in Eq. (7.1–32) becomes the Bromwich integral in this limit while the left-hand side of Eq. (7.1–32) is found from residues.

Let us consider I_1, the integral over C_1. We make a switch to polar variables so that $s = Re^{i\theta}$, $ds = Re^{i\theta}i\,d\theta$, and obtain

$$I_1 = \frac{1}{2\pi i}\int_{\psi}^{\pi/2} F(Re^{i\theta})e^{tRe^{i\theta}}Re^{i\theta}i\,d\theta$$

$$= \frac{1}{2\pi i}\int_{\psi}^{\pi/2} F(Re^{i\theta})e^{tR\cos\theta}e^{iRt\sin\theta}Re^{i\theta}i\,d\theta,$$

where we have put $e^{Rte^{i\theta}} = e^{Rt(\cos\theta + i\sin\theta)}$. We now use the inequality, shown in Eq. (4.2–18); it is rewritten here with different variables:

$$\left|\int_{\theta_1}^{\theta_2} u(\theta)\,d\theta\right| \le \int_{\theta_1}^{\theta_2} |u(\theta)|\,d\theta, \qquad \text{if } \theta_2 \ge \theta_1,$$

and $u(\theta)$ is any integrable function. We can thus assert that

$$|I_1| \le \left|\frac{1}{2\pi i}\right| \int_{\psi}^{\pi/2} |F(Re^{i\theta})||e^{tR\cos\theta}||e^{iRt\sin\theta}|R|e^{i\theta}||i|\,d\theta.$$

Now, $|i| = 1$ and $|e^{iRt\sin\theta}| = 1$. Further, $e^{tR\cos\theta}$ is positive, and its magnitude signs can be dropped from inside the integral. Thus

$$|I_1| \le \frac{1}{2\pi}\int_{\psi}^{\pi/2} |F(Re^{i\theta})|e^{Rt\cos\theta}R\,d\theta.$$

Since, by hypothesis, $|F(Re^{i\theta})| \le m/|s|^k = m/R^k$, then

$$|I_1| \le \frac{1}{2\pi}\int_{\psi}^{\pi/2} \frac{m}{R^k}e^{Rt\cos\theta}R\,d\theta = \frac{1}{2\pi}\frac{m}{R^{k-1}}\int_{\psi}^{\pi/2} e^{Rt\cos\theta}\,d\theta. \tag{7.1–33}$$

As θ varies from ψ to $\pi/2$, $e^{Rt\cos\theta}$ becomes progressively smaller. Thus, over the interval of integration, $e^{Rt\cos\theta} \le e^{Rt\cos\psi}$. We can substitute $e^{Rt\cos\psi}$ for $e^{Rt\cos\theta}$ in Eq. (7.1–33) and preserve the inequality there. Note that $R\cos\psi = a$ (see Fig. 7.1–5) or $\psi = \cos^{-1}(a/R)$. Rewriting the far right side of Eq. (7.1–33), we have

$$|I_1| \le \frac{1}{2\pi}\frac{m}{R^{k-1}}\int_{\cos^{-1}(a/R)}^{\pi/2} e^{at}\,d\theta = \frac{1}{2\pi}\frac{me^{at}}{R^{k-1}}\left[\frac{\pi}{2} - \cos^{-1}\frac{a}{R}\right],$$

and, because

$$\frac{\pi}{2} - \cos^{-1}\frac{a}{R} = \sin^{-1}\frac{a}{R},$$

we have

$$|I_1| \le \frac{1}{2\pi}\frac{m}{R^{k-1}}e^{at}\sin^{-1}\frac{a}{R}. \tag{7.1–34}$$

Now, we will use the inequality

$$\sin^{-1} p \le \frac{\pi}{2}p, \qquad \text{if } 0 \le p \le 1 \tag{7.1–35}$$

The validity of this is demonstrated if we sketch $\sin^{-1} p$ and $(\pi/2)p$ over $0 \le p \le 1$. Thus, with $p = a/R$, we have, by combining Eqs. (7.1–34) and (7.1–35),

$$|I_1| \le \frac{1}{2\pi}\frac{m}{R^{k-1}}e^{at}\frac{\pi}{2}\frac{a}{R} = \frac{m}{4}\frac{e^{at}}{R^k}.$$

As $R \to \infty$, the expression on the right $\to 0$. Thus I_1 must approach the same limit.

Now I_2, the integral over C_2, will be treated in a similar fashion.

$$I_2 = \frac{1}{2\pi i}\int_{\pi/2}^{\pi} F(Re^{i\theta})e^{tRe^{i\theta}}iRe^{i\theta}\,d\theta \quad \text{and} \quad |I_2| \le \frac{1}{2\pi}\int_{\pi/2}^{\pi} e^{Rt\cos\theta}|F(Re^{i\theta})|R\,d\theta.$$

Since $|F(re^{i\theta})| \le m/R^k$, we have

$$|I_2| \le \frac{1}{2\pi}\frac{m}{R^{k-1}}\int_{\pi/2}^{\pi} e^{Rt\cos\theta}\,d\theta. \tag{7.1–36}$$

A sketch of $\cos\theta$ and $1 - (2/\pi)\theta$ shows that $\cos\theta \le 1 - (2/\pi)\theta$ for $\pi/2 \le \theta \le \pi$. Thus

$$e^{Rt\cos\theta} \le e^{Rt(1-(2/\pi)\theta)}, \qquad \text{for } \frac{\pi}{2} \le \theta \le \pi. \tag{7.1–37}$$

Combining the inequalities in Eqs. (7.1–37) and (7.1–36), we get

$$|I_2| \le \frac{1}{2\pi}\frac{m}{R^{k-1}}\int_{\pi/2}^{\pi} e^{Rt(1-(2/\pi)\theta)}\,d\theta.$$

Evaluating the above, we obtain $|I_2| \le [m/(2\pi R^{k-1})][\pi/(2Rt)](1 - e^{-Rt})$. This shows that as $R \to \infty$, we have $I_2 \to 0$.

An argument much like the one just presented shows that as $R \to \infty$, the integral over arc C_3 in Eq. (7.1–32) (see Fig. 7.1–5) goes to zero. Finally, a discussion much like the one given for I_1 (the integral over C_1) can be used to show that the integral over C_4 in Eq. (7.1–32) becomes zero as $R \to \infty$.

If $R \to \infty$ in Eq. (7.1–32) with a kept constant, then b must also become infinite. Passing to this limit and using the limiting values of all integrals, we have

$$\lim_{R\to\infty} \frac{1}{2\pi i} \oint_C F(s)e^{st}\,ds = \frac{1}{2\pi i} \int_{a-i\infty}^{a+i\infty} F(s)e^{st}\,ds. \tag{7.1–38}$$

Since Eq. (7.1–31) is still valid as $R \to \infty$, it can be used to replace the integral on the left side of Eq. (7.1–38) with the result that

$$\sum_{\text{res}} F(s)e^{st} = \frac{1}{2\pi i} \int_{a-i\infty}^{a+i\infty} F(s)e^{st}\,ds, \qquad \text{at poles of } F(s).$$

The right-hand side of this equation is $\mathscr{L}^{-1}F(s)$, and so we have proved the theorem under discussion. Note that we derived the theorem by assuming $t > 0$. Equation (7.1–30) must not be applied for $t \le 0$. Indeed (see Exercise 30), if we evaluate the Bromwich integral using residues we obtain a function of t that is zero for $t < 0$.

Let $F(s) = P/Q$, where P and Q are polynomials in s whose degrees are n and l, respectively, with $l > n$. Then $|F(s)| \le c/|s|^{l-n}$ for large $|s|$, where c is a constant (see Exercise 25, Section 2.5).

The conditions of the theorem are satisfied, and $f(t)$ can be found.

EXAMPLE 1

Find

$$\mathscr{L}^{-1}\frac{1}{(s-2)(s+1)^2} = f(t).$$

Solution

With $F(s)e^{st} = e^{st}/[(s-2)(s+1)^2]$ we use Eq. (7.1–30) recognizing that this function has poles at $s = 2$ and $s = -1$. Thus

$$f(t) = \operatorname{Res}\left[\frac{e^{st}}{(s-2)(s+1)^2}, 2\right] + \operatorname{Res}\left[\frac{e^{st}}{(s-2)(s+1)^2}, -1\right].$$

The first residue is easily found to be $e^{2t}/9$ while the second, which involves a pole of second order, is

$$\lim_{s\to-1} \frac{d}{ds}\frac{(s+1)^2 e^{st}}{(s-2)(s+1)^2} = \lim_{s\to-1} \frac{(s-2)\dfrac{d}{ds}(e^{st}) - e^{st}}{(s-2)^2} = \frac{-3te^{-t} - e^{-t}}{9}.$$

Notice that the expression te^{-t} arises when we differentiate e^{st} *with respect to s*. Thus, summing residues, we obtain

$$\mathscr{L}^{-1}\frac{1}{(s-2)(s+1)^2} = \frac{e^{2t}}{9} - \frac{te^{-t}}{3} - \frac{e^{-t}}{9}.$$ ◀

A common problem in engineering is to obtain the inverse Laplace transform of a function $F(s) = (P(s)/Q(s))e^{-s\tau}$, where P and Q are polynomials in s, and τ is positive real. The technique used in the previous problem does not apply. Theorem 3 is inapplicable because $e^{-s\tau}$ is unbounded in any half plane $\text{Re}\, s \leq a$. We cannot find constants m and k such that $|F(s)| \leq m/|s|^k$. However, if the degree of Q exceeds that of P we *can* use Theorem 3 to find $\mathscr{L}^{-1}(P(s)/Q(s))$. It is shown in Exercise 15 how $\mathscr{L}^{-1}(P(s)/Q(s))e^{-s\tau}$ is now easily found from $\mathscr{L}^{-1}(P(s)/Q(s))$.

If we must find $\mathscr{L}^{-1}F(s)$ where $F(s)$ contains an infinite number of poles, Theorem 3 is again inapplicable. However, as shown in Exercise 35, we are often justified in still using Eq. (7.1–30) to obtain $f(t)$ provided that we evaluate the *infinite* summation of all the residues of $F(s)e^{st}$.

When, however, $F(s)$ is defined by means of a branch cut, then a summation of residues or a Heaviside formula will not yield $f(t)$. Instead we must return to the Bromwich integral (see Eq. 7.1–27) and deform the path of integration into some other valid contour in the s-plane along which the integration is more easily performed. An example follows.

EXAMPLE 2

Find $\mathscr{L}^{-1}(1/s^{1/2})$ by means of the Bromwich integral.

Solution

From Eq. (7.1–27) we find that

$$\mathscr{L}^{-1}\frac{1}{s^{1/2}} = \frac{1}{2\pi i}\int_{a-i\infty}^{a+i\infty} \frac{1}{s^{1/2}}\, e^{st}\, ds. \tag{7.1–39}$$

The vertical line $s = a$ must be chosen to lie to the right of all singularities of $1/s^{1/2}$. Now $1/s^{1/2}$ is a multivalued function. A single-valued branch can be established by means of a branch cut extending from the origin to infinity. A branch specified by a cut along the path $\text{Im}\, s = 0$, $\text{Re}\, s \leq 0$ will be used. When s assumes positive real values, we will take $s^{1/2} = \sqrt{s} > 0$. The path of integration can now be chosen as the vertical line $\text{Re}\, s = a > 0$ shown in Fig. 7.1–6. To evaluate Eq. (7.1–39) along this line we first consider $(1/2\pi i)\oint_C e^{st}/s^{1/2}\, ds$ taken around the closed contour C shown

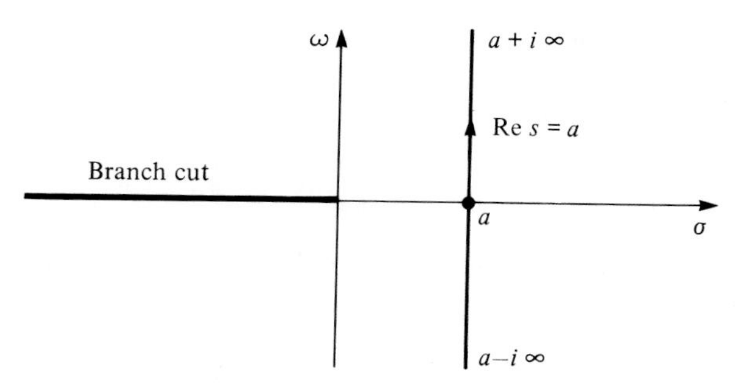

Figure 7.1–6

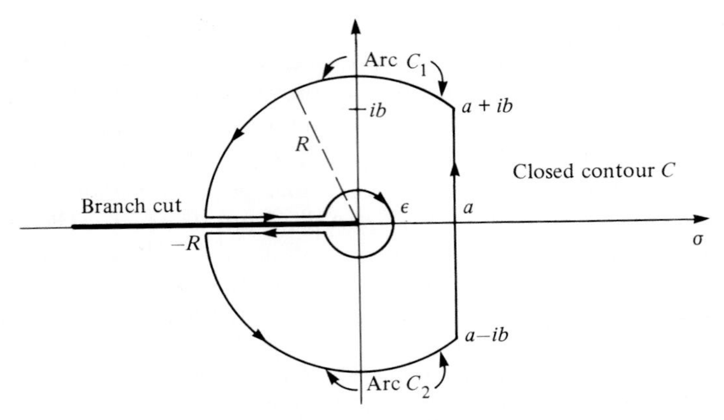

Figure 7.1–7

in Fig. 7.1–7. As $e^{st}/s^{1/2}$ is analytic on and inside C, we have $(1/2\pi i)\oint_C e^{st}/s^{1/2}\, ds = 0$. This integral is rewritten in terms of integrations taken along the various portions of C. We have

$$\frac{1}{2\pi i}\underset{\text{along Re } s=a}{\int_{a-ib}^{a+ib}} \frac{e^{st}}{s^{1/2}}\, ds + \frac{1}{2\pi i}\int_{C_1} \frac{e^{st}}{s^{1/2}}\, ds + \frac{1}{2\pi i}\underset{\text{above cut}}{\int_{-R}^{-\varepsilon}} \frac{e^{\sigma t}}{\sigma^{1/2}}\, d\sigma$$

$$+ \frac{1}{2\pi i}\oint_{|s|=\varepsilon} \frac{e^{st}}{s^{1/2}}\, ds + \frac{1}{2\pi i}\underset{\text{below cut}}{\int_{-\varepsilon}^{-R}} \frac{e^{\sigma t}}{\sigma^{1/2}}\, d\sigma + \frac{1}{2\pi i}\int_{C_2} \frac{e^{st}}{s^{1/2}}\, ds = 0, \quad (7.1\text{–}40)$$

where C_1 is the circular arc extending from $a + ib$ to $-R$, while C_2 is the circular arc extending from $-R$ to $a - ib$. For the integrals along the straight lines that are above and below the branch cuts we recognize that $s = \sigma$. Because $1/s^{1/2}$ has a branch point singularity at $s = 0$, we make a circular detour of radius ε around this point.

We now consider the limiting values of the integrals along arcs C_1 and C_2 as the radius $R \to \infty$. Referring to the derivation of Eq. (7.1–30) and to Fig. 7.1–5, we find that the discussion used there to justify setting the integrals over C_1, C_2, C_3, and C_4 to zero as $R \to \infty$ can be applied directly to the present problem. Thus, as $R \to \infty$, the integrals over C_1 and C_2 in Eq. (7.1–40) become zero.

We study now the integral in Eq. (7.1–40) that is taken around $|s| = \varepsilon$. We make our usual switch to polar coordinates $s = \varepsilon e^{i\theta}$, $ds = \varepsilon e^{i\theta} i\, d\theta$, $s^{1/2} = \sqrt{\varepsilon}\, e^{i\theta/2}$ and so

$$\frac{1}{2\pi i}\oint_{|s|=\varepsilon} \frac{e^{st}}{s^{1/2}}\, ds = \frac{1}{2\pi i}\int_{\pi}^{-\pi} \frac{e^{\varepsilon t e^{i\theta}}\varepsilon e^{i\theta} i\, d\theta}{\sqrt{\varepsilon}\, e^{i\theta/2}} = \frac{\sqrt{\varepsilon}}{2\pi}\int_{\pi}^{-\pi} e^{\varepsilon t e^{i\theta}} e^{i\theta/2}\, d\theta.$$

As $\varepsilon \to 0$, the integral on the far right is bounded and its coefficient $\sqrt{\varepsilon}/2\pi \to 0$. As $\varepsilon \to 0$, the right side of the preceding equation $\to 0$. We thus have

$$\lim_{\varepsilon \to 0} \frac{1}{2\pi i}\oint_{|s|=\varepsilon} \frac{e^{st}}{s^{1/2}}\, ds = 0.$$

Along the top edge of the branch cut in Fig. 7.1–7 $s^{1/2} = \sigma^{1/2}$ is the square root of a negative real variable. The correct value of this multivalued expression must be established. By assumption, $s^{1/2}$ is a positive real number for any s lying on the positive real axis. Here $\arg s^{1/2} = 0$. As we proceed to the top side of the branch cut, $\arg s$ increases by π (see Fig. 7.1–8). Thus $\arg s^{1/2}$ increases from 0 to $\pi/2$. Hence, when s is a negative real number and lies on the top of the cut, we have $\arg s^{1/2} = \pi/2$, which implies that $\sigma^{1/2} = i\sqrt{|\sigma|} = i\sqrt{-\sigma}$.

A similar discussion shows that along the bottom of the branch cut $\sigma^{1/2} = -i\sqrt{-\sigma}$. Note from Fig. 7.1–7 that as $R \to \infty$, with a fixed, we have $b \to \infty$. Passing to the limits $R \to \infty$, $\varepsilon \to 0$ in Eq. (7.1–40) and using the limiting values of all integrals, we obtain

$$\frac{1}{2\pi i}\int_{a-i\infty}^{a+i\infty} \frac{e^{st}}{s^{1/2}}\,ds + \frac{1}{2\pi i}\int_{-\infty}^{0} \frac{e^{\sigma t}}{i\sqrt{-\sigma}}\,d\sigma + \frac{1}{2\pi i}\int_{0}^{-\infty} \frac{e^{\sigma t}}{-i\sqrt{-\sigma}}\,d\sigma = 0. \qquad (7.1\text{–}41)$$

The second and third integrals on the left are equal and can be combined into $-(1/\pi)\int_{-\infty}^{0}(e^{\sigma t}/\sqrt{-\sigma})\,d\sigma$. Thus Eq. (7.1–41) becomes

$$\frac{1}{2\pi i}\int_{a-i\infty}^{a+i\infty} \frac{e^{st}}{s^{1/2}}\,ds = \frac{1}{\pi}\int_{-\infty}^{0} \frac{e^{\sigma t}}{\sqrt{-\sigma}}\,d\sigma.$$

The left side of this equation is the desired $\mathscr{L}^{-1}(1/s^{1/2})$. The integral on the right can be somewhat simplified with the change of variables $x = \sqrt{-\sigma}$, $x^2 = -\sigma$, $2x\,dx = -d\sigma$. Thus

$$\mathscr{L}^{-1}\frac{1}{s^{1/2}} = \frac{2}{\pi}\int_{0}^{\infty} e^{-x^2 t}\,dx.$$

The right-hand side here contains a well-known definite integral. From Exercise 25, Section 6.6 we find that $\int_0^\infty e^{-x^2 t}\,dx = (1/2)\sqrt{\pi/t}$, which means

$$\mathscr{L}^{-1}\frac{1}{s^{1/2}} = \frac{1}{\sqrt{\pi t}}$$

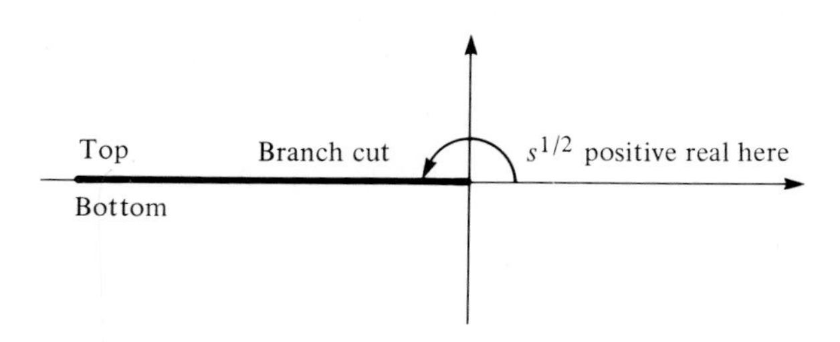

Figure 7.1–8

and also

$$\mathscr{L}\frac{1}{\sqrt{t}} = \frac{\sqrt{\pi}}{s^{1/2}}.$$

This last result can be verified when s is positive real if we use the definition of the Laplace transform, Eq. (7.1–1), make the change of variable $x^2 = t$, and again use the result of Exercise 25, Section 6.6. ◀

EXERCISES

Use residues to find the inverse Laplace transforms of the following functions for $t > 0$. Your answers should be real functions of t. Take $a \neq 0$ and $b \neq 0$ as real, $a \neq b$.

1. $\dfrac{1}{(s-1)(s+2)}$ **2.** $\dfrac{s}{(s-a)(s+b)}$ **3.** $\dfrac{1}{(s+a)^2}$

4. $\dfrac{s}{(s+a)^2(s+b)}$ **5.** $\dfrac{1}{s^2+a^2}$ **6.** $\dfrac{s}{s^2+a^2}$

7. $\dfrac{1}{(s^2+a^2)^2}$ **8.** $\dfrac{1}{(s^2+a^2)(s^2+b^2)}$ **9.** $\dfrac{1}{s^2+s+1}$

10. $\dfrac{1}{(s^2+s+1)^2}$ **11.** $\dfrac{1}{(s+1)^4}$ **12.** $\dfrac{1}{s^n}$, $n \geq 1$, integer

13. Find $\mathscr{L}^{-1}(1/(s^2 + as + b))$, where a and b are real. Consider the cases $a^2 > 4b$, $a^2 = 4b$, $a^2 < 4b$.

14. Let $F(s) = P(s)/Q(s)$, where P and Q are polynomials in s, the degree of Q exceeding that of P.

a) Suppose that $Q(s) = C(s - a_1)(s - a_2)\cdots(s - a_n)$, where $a_j \neq a_k$ if $j \neq k$. C is a constant. Thus $Q(s)$ has only first-order zeros. Show, using residues, that

$$f(t) = \mathscr{L}^{-1}F(s) = \sum_{j=1}^{n} \frac{P(a_j)}{Q'(a_j)} e^{a_j t}.$$

This is the most elementary of the Heaviside formulas and is usually derived from a partial fraction expansion of $F(s)$.

b) Use the formula derived in part (a) to solve Exercise 1 above.

15. Recall (see Section 2.2) the unit step function

$$u(t) = 0, \qquad t < 0; \qquad u(t) = 1, \qquad t \geq 0.$$

Thus if $\tau > 0$, then $f(t - \tau)u(t - \tau)$ is identical in shape to the function $f(t)u(t)$ but is displaced τ units to the right along the t-axis. Let $\mathscr{L}[f(t)u(t)] = \mathscr{L}[f(t)] = F(s)$, and show by using Eq. (7.1–1) that

$$\mathscr{L}[f(t-\tau)u(t-\tau)] = e^{-s\tau}F(s), \qquad \tau \geq 0. \tag{7.1–42a}$$

Thus conversely,

$$\mathscr{L}^{-1}[e^{-s\tau}F(s)] = u(t-\tau)f(t-\tau), \qquad \tau \geq 0. \tag{7.1–42b}$$

Use the result of Exercise 15 and Theorem 3 to find the inverse Laplace transforms of these functions:

16. $\dfrac{e^{-2s}}{s^2+1}$ **17.** $\dfrac{e^{-s}}{s}$ **18.** $\dfrac{e^{-3s}}{(s^2+1)(s^2+4)}$

19. $\dfrac{e^{-as}}{(s^2+b^2)^2}$, $a > 0$, $b > 0$

20. Let $f(t) = 0$ for $t < 1$ and $f(t) = 1$ for $t \geq 1$. This function has a jump discontinuity at $t = 1$. Use Eq. (7.1–1) to verify that $\mathscr{L}f(t) = e^{-s}/s = F(s)$. Show for this $F(s)$ that the Bromwich integral (see Eq. 7.1–27) evaluated as a Cauchy principal value has the value 1/2 when $t = 1$. Thus the inverse of $F(s)$ yields, at the jump $t = 1$, the average $(1/2)[f(1+) + f(1-)]$.

Hint: Consider $(1/2\pi i)\int_{a-ib}^{a+ib} 1/s\, ds$, $a > 0$, taken along $\text{Re}(s) = a$. Evaluate, and let $b \to \infty$ in your result.

21. a) Let $f(t) = \int_0^t g(t')\, dt'$. Recall the fundamental theorem of real integral calculus and use Eq. (7.1–8) to show that

$$\mathscr{L}g(t) = s\mathscr{L}\int_0^t g(t')\, dt',$$

from which we obtain $\mathscr{L}\int_0^t g(t')\, dt' = G(s)/s$.

b) Using the preceding result and the transform $\mathscr{L}1 = 1/s$, find $\mathscr{L}t$. Finally, extend this procedure to find $\mathscr{L}t^n$, $n \geq 0$ is an integer.

22. For the series L, C, R circuit shown in Fig. 7.1–9 the charge on the capacitor for time $t \geq 0$ is $q(t)$ coulombs. The switch is open for $t < 0$ and closed for $t \geq 0$. When the switch is closed the capacitor C contains an initial charge q_0. After the switch is closed the charge on the capacitor is

$$q(t) = q_0 + \int_0^t \imath(t')\, dt', \qquad t \geq 0,$$

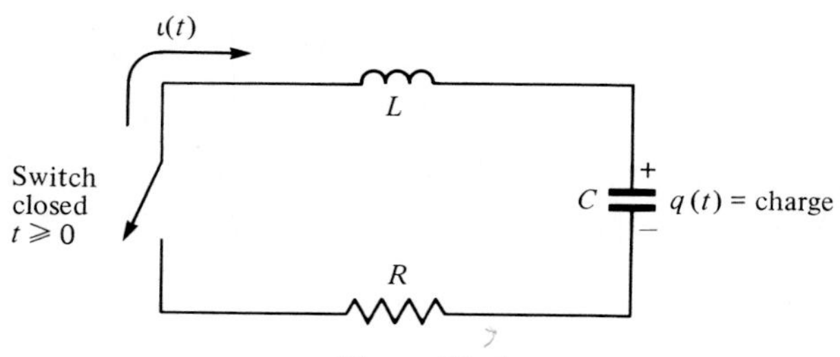

Figure 7.1–9

where $\imath(t')$ is the current in the circuit. From Kirchhoff's voltage law, when $t \geq 0$, the sum of the voltage drops around the three elements must be zero. The voltages across the inductor L, capacitor C, and the resistor R are, respectively, $L\,d\imath/dt$, $q(t)/C$, $\imath(t)R$. Thus

$$0 = L\frac{d\imath}{dt} + \frac{1}{C}\left[q_0 + \int_0^t \imath(t')\,dt'\right] + \imath(t)R.$$

From physical considerations we also require that $\imath(0) = 0$ and that $\imath(t)$ be a continuous function.

a) Show that

$$I(s) = \frac{-q_0}{C\left(Ls^2 + Rs + \frac{1}{C}\right)}.$$

Hint: Transform the preceding integrodifferential equation. Recall that $\mathscr{L}1 = 1/s$ (see Eq. 7.1–2, with $b = 0$; also use Eqs. 7.1–10 and 7.1–8).

b) Use residues to find $\imath(t)$ for $t > 0$. Consider three separate cases:

$$\text{(i) } R > 2\sqrt{\frac{L}{C}}, \qquad \text{(ii) } R < 2\sqrt{\frac{L}{C}}, \qquad \text{(iii) } R = 2\sqrt{\frac{L}{C}}.$$

Describe the qualitative differences in your results. These cases are known as *overdamped, underdamped,* and *critically damped.* How, in each case, is the location and order of the poles of $I(s)$ related to the type of damping?

23. Let $f(t)$ be a function defined for $0 < t < T$, and let $f(t) = 0$ for all $t < 0$ and for all $t > T$. The periodic extension of this function for $t > 0$ is $g(t) = \sum_{n=0}^{\infty} f(t - nT)$, as illustrated in Figs. 7.1–10(a) and (b). Show that if $f(t)$ has Laplace transform $F(s)$ then $g(t)$ has transform $G(s) = F(s)/(1 - e^{-Ts})$.

Hint: $f(t - T)$ has Laplace transform $e^{-sT}F(s)$ (see Exercise 15), $f(t - 2T)$ has transform $e^{-2sT}F(s)$, etc. Thus $\mathscr{L}g(t) = F(s)[1 + e^{-sT} + e^{-2sT} + \cdots]$. Sum the series in the brackets. What restriction applies to s for your summation to be valid?

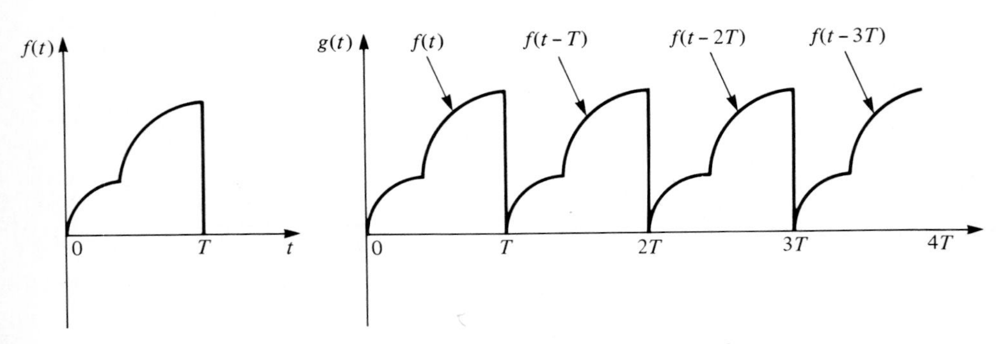

Figure 7.1–10a **Figure 7.1–10b**

Use the result of Exercise 23 to show that the Laplace transforms of the functions $g(t)$ sketched below are as follows:

24. (see Fig. 7.1–11) $G(s) = \frac{1}{s} \tanh\left(\frac{sT}{4}\right)$

25. (see Fig. 7.1–12) $G(s) = \frac{1}{s^2 T} - \frac{e^{-sT/2}}{2s \sinh\left(\frac{sT}{2}\right)}$

26. (see Fig. 7.1–13) $G(s) = \frac{\pi T}{T^2 s^2 + \pi^2} \coth\left(\frac{sT}{2}\right)$

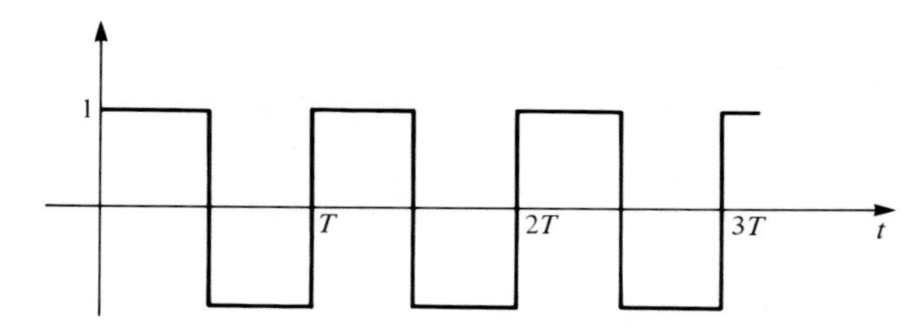

Figure 7.1–11

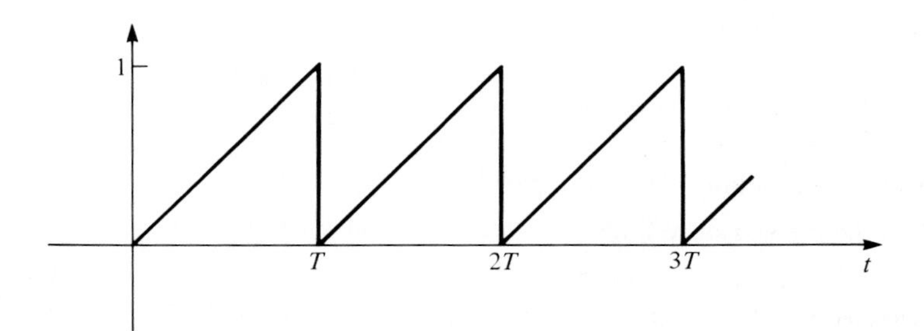

Figure 7.1–12

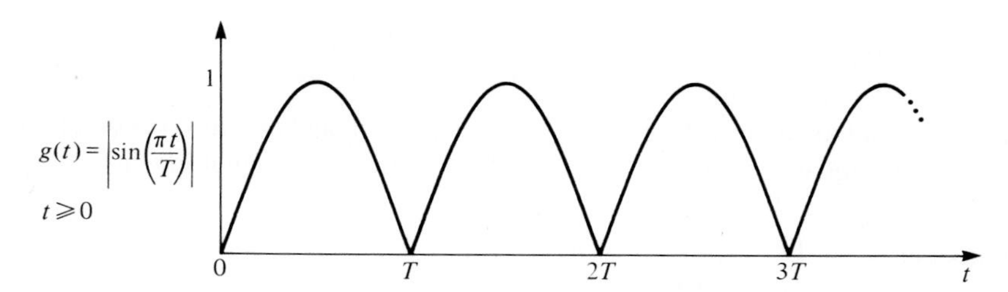

Figure 7.1–13

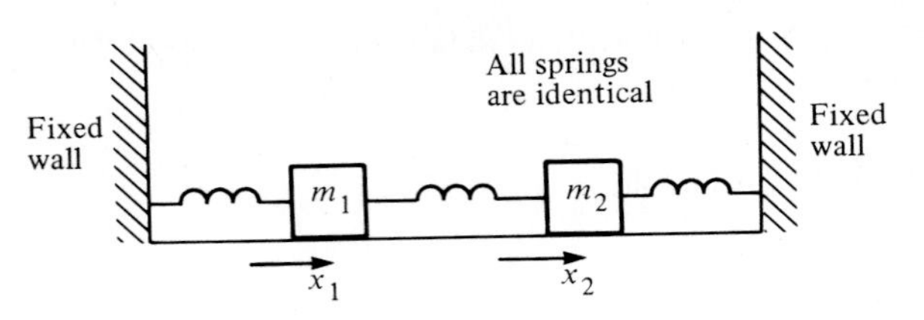

Figure 7.1–14

27. Figure 7.1–14 illustrates a mechanical problem whose solution requires coupled differential equations. A pair of masses m_1 and m_2 lie on a perfectly smooth surface and are separated by the three identical springs having elastic constant k. Mass m_1 is located by the coordinate x_1, and mass m_2 is located by the coordinate x_2. These coordinates are measured from the equilibrium configuration of the system, that is, with m_1 at $x_1 = 0$ and m_2 at $x_2 = 0$, neither mass experiences any *net* force. From Newton's second law the motion of these masses, as a function of time, is governed by the following pair of coupled differential equations:

$$m_1\frac{d^2x_1}{dt^2} = -2kx_1 + kx_2, \qquad m_2\frac{d^2x_2}{dt^2} = kx_1 - 2kx_2,$$

where $x_1(t)$ and $x_2(t)$ are continuous functions.

a) Suppose $k = 1, m_1 = 1, m_2 = 2, dx_1/dt = dx_2/dt = 0$ at $t = 0, x_1(0) = 1, x_2(0) = 0$. Take the Laplace transform of the above differential equations and obtain the simultaneous algebraic equations

$$s^2X_1(s) - s = -2X_1(s) + X_2(s),$$
$$2s^2X_2(s) = X_1(s) - 2X_2(s).$$

b) Solve the equations derived in part (a) for $X_1(s)$ and $X_2(s)$.

c) Use the method of residues to obtain $x_1(t)$ and $x_2(t)$ for $t > 0$.

28. A pair of electrical circuits are coupled by means of a transformer having mutual inductance M and self-inductances L_1 and L_2 (see Fig. 7.1–15). If these terms are unfamiliar, see any standard textbook on electric circuits. One can show that the time-varying constants $\imath_1(t)$ and $\imath_2(t)$ circulating around the left- and right-hand circuits in the directions shown satisfy the differential equations

$$v_1(t) = R_1\imath_1(t) + L_1\frac{d\imath_1}{dt} + M\frac{d\imath_2}{dt}, \qquad 0 = M\frac{d\imath_1}{dt} + R_2\imath_2 + L_2\frac{d\imath_2}{dt}.$$

a) Perform a Laplace transformation on these equations and obtain a pair of simultaneous algebraic equations for $I_1(s)$ and $I_2(s)$. Assume the initial conditions $\imath_1(0) = \imath_2(0) = 0$. Take $v_1(t) = e^{-\alpha t}$, $t \geq 0$, $\alpha > 0$.

b) Assume $L_1 = L_2 = 1$, $M = 0.5$, $R_1 = 1$, $R_2 = 1$, $\alpha = 1$. Solve the equations obtained in part (a) for $I_1(s)$ and $I_2(s)$.

c) Use residues to obtain $\imath_1(t)$ and $\imath_2(t)$.

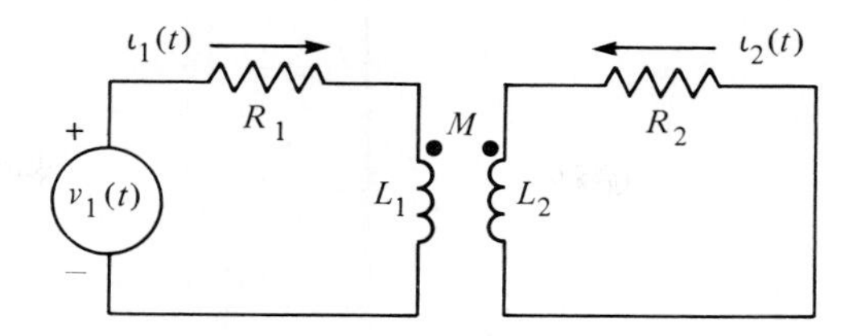

Figure 7.1–15

29. This exercise deals with connection between Fourier and Laplace transforms. In Chapter 6 we presented the Fourier transform pair

$$F(\omega) = \frac{1}{2\pi}\int_{-\infty}^{+\infty} f(t')e^{-i\omega t'}\,dt', \qquad f(t) = \int_{-\infty}^{+\infty} F(\omega)e^{i\omega t}\,d\omega,$$

which implies

$$f(t) = \int_{-\infty}^{+\infty} \frac{1}{2\pi}\left(\int_{-\infty}^{+\infty} f(t')e^{-i\omega t'}\,dt'\right)e^{i\omega t}\,d\omega. \tag{7.1–43}$$

a) Suppose in the preceding equation we take $f(t) = g(t)e^{-at}$, where $g(t) = 0$ for $t < 0$. We must, of course, also replace $f(t')$ on the right in Eq. (7.1–43) by $g(t')e^{-at'}$. Using Eq. (7.1–43), show that

$$g(t) = \frac{1}{2\pi}\int_{-\infty}^{+\infty}\left[\int_0^{\infty} g(t')e^{-(a+i\omega)t'}\,dt'\right]e^{(a+i\omega)t}\,d\omega.$$

b) Make the change of variables $s = a + i\omega$, where a is a real constant. Show that the equation derived in part (a) can be written

$$g(t) = \frac{1}{2\pi i}\int_{a-i\infty}^{a+i\infty}\left[\int_0^{\infty} g(t')e^{-st'}\,dt'\right]e^{st}\,ds.$$

c) Show that the equation derived in part (b) can be written

$$g(t) = \frac{1}{2\pi i}\int_{a-i\infty}^{a+i\infty} G(s)e^{st}\,ds,$$

where $G(s)$ is the Laplace transform of $g(t)$. Thus we have derived the Bromwich integral (see Eq. 7.1–27) for the inversion of Laplace transforms.

30. Let $F(s)$ be analytic in the s-plane in the region $\operatorname{Re} s \geq a$. Suppose there exist positive constants m, R_0, and k such that in this region $|F(s)| \leq m/|s|^k$ when $|s| \geq R_0$. Show that the Bromwich integral, Eq. (7.1–27), will yield a function $f(t)$ that is zero for $t < 0$.

Hint: The derivation is similar to that for Theorem 3. Use the contour shown in Fig. 7.1–16. Argue that the integral of $F(s)e^{st}$ on the arc goes to zero as $R \to \infty$.

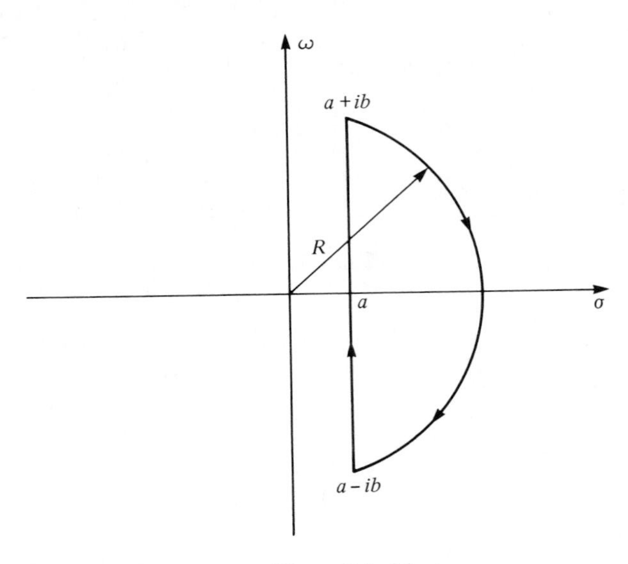

Figure 7.1–16

31. The convolution of $f(t)$ with $g(t)$, written $f(t) * g(t)$, is defined as

$$f(t) * g(t) = \int_0^t f(t - \tau)g(\tau)\, d\tau.$$

(It is easy to prove that $f(t) * g(t) = g(t) * f(t)$.) Show that

$$\mathscr{L}[f(t) * g(t)] = [\mathscr{L} f(t)][\mathscr{L} g(t)], \tag{7.1–44}$$

from which it follows that

$$\mathscr{L}^{-1}[F(s)G(s)] = f(t) * g(t). \tag{7.1–45}$$

Hint:

$$\mathscr{L}[f(t)* g(t)] = \int_0^\infty e^{-st}\left[\int_0^t f(t-\tau)g(\tau)\, d\tau\right] dt.$$

Explain why we can write the inner integral as $\int_0^\infty f(t-\tau)u(t - \tau)g(\tau)\, d\tau$, where $u(t)$ is the unit step function defined in Exercise 15. Use this expression in the preceding double integral and, assuming it is legal, reverse the order of integration; then employ Eq. (7.1–42a) in Exercise 15.

32. By starting with the contour shown in Fig. 7.1–17 and passing to the appropriate limits, show that

$$\mathscr{L}^{-1} \frac{1}{s^{1/2}(s - k)} = \frac{e^{kt}}{\sqrt{k}} - \frac{1}{\pi}\int_0^\infty \frac{e^{-ut}}{\sqrt{u}(u + k)}\, du,$$

where $k > 0$ and $s^{1/2}$ is the principal branch of this function.

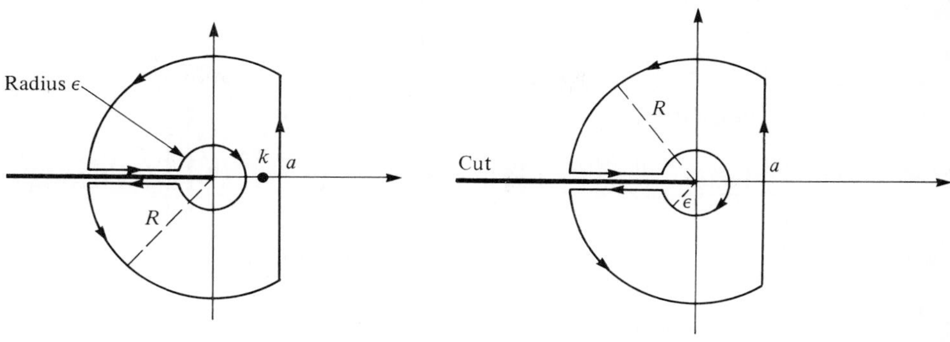

Figure 7.1–17

Figure 7.1–18

33. Show that

$$\mathscr{L}^{-1}\frac{e^{-b(s^{1/2})}}{s} = 1 - \frac{1}{\pi}\int_0^\infty \frac{e^{-xt}\sin(b\sqrt{x})\,dx}{x},$$

where $b \geq 0$, and $s^{1/2}$ is the principal branch.

Hint: Take the branch cut of s as shown in Fig. 7.1–18 using $s^{1/2} > 0$ on the positive real axis. Do an integration around the contour indicated and then allow $R \to \infty$, $\varepsilon \to 0$. Since Re $s^{1/2} \geq 0$ on the arcs of radius R, we have, for $b \geq 0$,

$$|e^{-b(s^{1/2})}| \leq 1 \quad \text{and} \quad \left|\frac{e^{-b(s^{1/2})}}{s}\right| \leq \left|\frac{1}{s}\right|.$$

Thus an argument much like that leading to Eq. (7.1–30) can be used to assert that the integrals over the two curves of radius R become zero as $R \to \infty$. Note that as $\varepsilon \to 0$, the integral around $|s| = \varepsilon$ approaches a nonzero value.

34. Show that

$$\mathscr{L}^{-1}\frac{1}{(s^2+1)^{1/2}} = \frac{1}{\pi}\int_{-1}^{+1}\frac{e^{i\omega t}\,d\omega}{\sqrt{1-\omega^2}}.$$

The integral on the right is $J_0(t)$, the Bessel function of zero order. A branch cut connecting i with $-i$ defining $(s^2+1)^{1/2}$ is shown in Fig. 7.1–19. We take $(s^2+1)^{1/2} > 0$ on the positive real axis.

Hint: Recall the principle of deformation of contours (see Chapter 4). Use this concept in order to show that $(1/2\pi i)\int F(s)e^{st}\,ds$ taken around the inner contour in Figure 7.1–19 equals this same integration performed around the closed contour C. Allow $R \to \infty$. Notice that at points such as P and Q the values assumed by $1/(s^2+1)^{1/2}$ are identical but opposite in sign.

The derivation of Eq. (7.1–30) was based on the assumption that $F(s)$ has a finite number of poles. If $F(s)$ contains an infinite number of poles, the proof given is invalid since, as R goes continuously to infinity (see Fig. 7.1–5), the contour of integration will pass through singularities of $F(s)$.

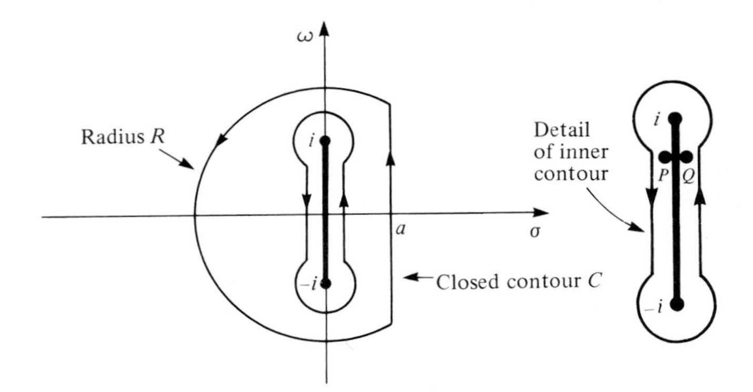

Figure 7.1–19

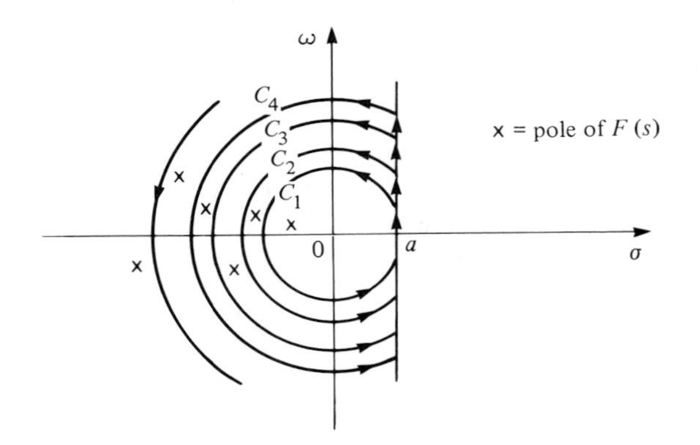

Figure 7.1–20

However, Eq. (7.1–30) is often still valid provided we use on the right the *infinite sum* of all the residues of $F(s)e^{st}$. The justification for this procedure involves replacing the contour C in Fig. 7.1–5 by an infinite sequence of expanding contours $C_1, C_2, \ldots, C_n$ (see Fig. 7.1–20) that are chosen in such a way that no contour passes through any pole of $F(s)$. If $\oint_{C_n} F(s)e^{st}\,ds$ tends to zero along the curved portion of C_n as $n \to \infty$, then one can show that the required Bromwich integral along the line $s = a$ equals the sum of all the residues of $F(s)e^{st}$. This technique is used in the following examples.

35. Show that

$$\mathscr{L}^{-1}\frac{1}{s\cosh s} = 1 + \frac{4}{\pi}\sum_{n=1}^{\infty}\frac{(-1)^n}{2n-1}\cos\frac{(2n-1)\pi t}{2}.$$

Hint: $F(s) = 1/(s\cosh s)$ has simple poles at $s = 0$ and at $s = \pm i(k\pi - \pi/2)$, $k = 1, 2, 3.\ldots$ Now consider $(1/2\pi i)\int F(s)e^{st}\,ds$ around the closed contour C_n shown in Fig. 7.1–21. Evaluate this integral by summing residues at the enclosed poles. The

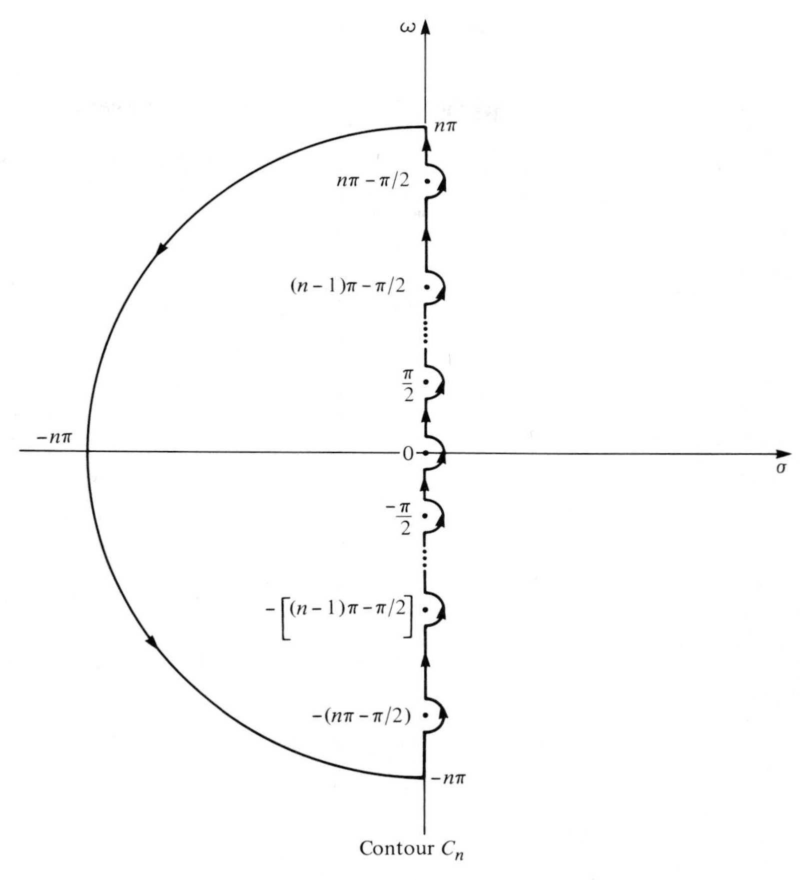

Figure 7.1–21

radius of the arc, $n\pi$, will go to infinity through discrete values as n passes to infinity through the positive integers. Argue that the integral along the curved portion of the contour goes to zero in this limit. The remaining integral, which is along the imaginary axis (with indentations), is the Bromwich integral for $f(t)$ and is thus evaluated as the sum of the residues of $F(s)e^{st}$ at all poles in the complex plane. To argue that the integral along the arc tends to zero as $n \to \infty$ it is sufficient to find a constant m such that $|1/\cosh s| \leq m$ is satisfied on the arc. If m is independent of n (for n sufficiently large) the argument used for discarding the integral along the arc in the derivation of Theorem 3 applies. To find m, recall that $|\cosh s|^2 = \cosh^2 \sigma - \sin^2 \omega$ (see Section 3.3, Exercise 17). A sketch shows that for integer n, $|\sin \omega| \leq |n\pi - \omega|$. The Maclaurin expansion of $\cosh \sigma$ reveals that $\cosh \sigma \geq 1 + \sigma^2/2$, which implies that $\cosh^2 \sigma \geq 1 + \sigma^2$. Thus $|\cosh s|^2 \geq 1 + \sigma^2 - (n\pi - \omega)^2$. Use the equation of the arc in the preceding expression to show that $|\cosh s|^2 \geq 1 + 2\omega(n\pi - \omega)$, and show that on that portion of C_n lying in the second quadrant $|\cosh s|^2 \geq 1$, so that $1/|\cosh s| \leq 1$. Note that $|\cosh s| = |\cosh \bar{s}|$. Thus the constant m is established.

36. a) Show that

$$\mathscr{L}^{-1}\frac{1}{s\sinh s} = t + \frac{2}{\pi}\sum_{n=1}^{\infty}\frac{(-1)^n}{n}\sin n\pi t.$$

Hint: The solution is similar to that used in Exercise 35. A contour like that in Fig. 7.1–21 must be used, except there are indentations at $s = 0, \pm i\pi, \pm i2\pi, \ldots$. The radius of the arc is now $n\pi + \pi/2$. Notice that $|\sinh s|^2 = \cosh^2\sigma - \cos^2\omega$, and that $\cosh^2\sigma \geq 1 + \sigma^2$, $|\cos\omega| \leq |(n\pi + \pi/2) - \omega|$.

b) Extend the preceding result to show that

$$\mathscr{L}^{-1}\frac{1}{s\sinh bs} = \frac{t}{b} + \frac{2}{\pi}\sum_{n=1}^{\infty}\frac{(-1)^n}{n}\sin\frac{n\pi t}{b},$$

where b is real.

37. Show that

$$\mathscr{L}^{-1}\frac{\cosh sx}{s\cosh s} = 1 + \sum_{n=1}^{\infty}\frac{4}{\pi}(-1)^n\frac{\cos[(2n-1)\pi x/2]}{2n-1}\cos\left[(2n-1)\frac{\pi t}{2}\right]$$

for $-1 < x < 1$.

Hint: Evaluate the Bromwich integral by using a sequence of closed rectangular contours C_n (see Fig. 7.1–22). Evaluate

$$\int_{C_n}\frac{\cosh sx}{s\cosh s}e^{st}\,ds$$

using residues. Argue that the integrals along C_{1n} and C_{2n} vanish as $n \to \infty$. (The integrals along C_{3n} and C_{4n} will vanish by similar arguments.) Note that on C_{1n} we have

$$\left|\int_{C_{1n}}\frac{\cosh sx\, e^{st}}{s\cosh s}ds\right| \leq \int_{-n\pi}^{0}\frac{|\cosh sx|e^{\sigma t}}{|s|\cosh\sigma}d\sigma,$$

where $s = \sigma + in\pi$, $|\cosh sx| \leq \cosh\sigma x$, and $\cosh\sigma \geq e^{-\sigma}/2$, $e^{\sigma t} \leq 1$, $1/|s| \leq 1/n\pi$. On C_{2n},

$$\left|\int_{C_{2n}}\frac{\cosh sx\, e^{st}}{s\cosh s}ds\right| \leq \int_{0}^{n\pi}\frac{|\cosh sx|}{|s||\cosh s|}d\omega,$$

where $s = -n\pi + i\omega$. Note that $|\cosh sx| \leq e^{n\pi x}$, $|\cosh s| \geq \sinh n\pi$.

38. Use a technique like that employed in Exercise 37 to show that

$$\mathscr{L}^{-1}\frac{\sinh sx}{s\sinh s} = x + \frac{2}{\pi}\sum_{n=1}^{\infty}\frac{(-1)^n}{n}\sin n\pi x\cos n\pi t, \qquad \text{for } -1 < x < 1.$$

Hint: Use a contour like that in Exercise 37, but the corners are now at $\pm i(n\pi + \pi/2)$ and $(n\pi + \pi/2)(-1 \pm i)$.

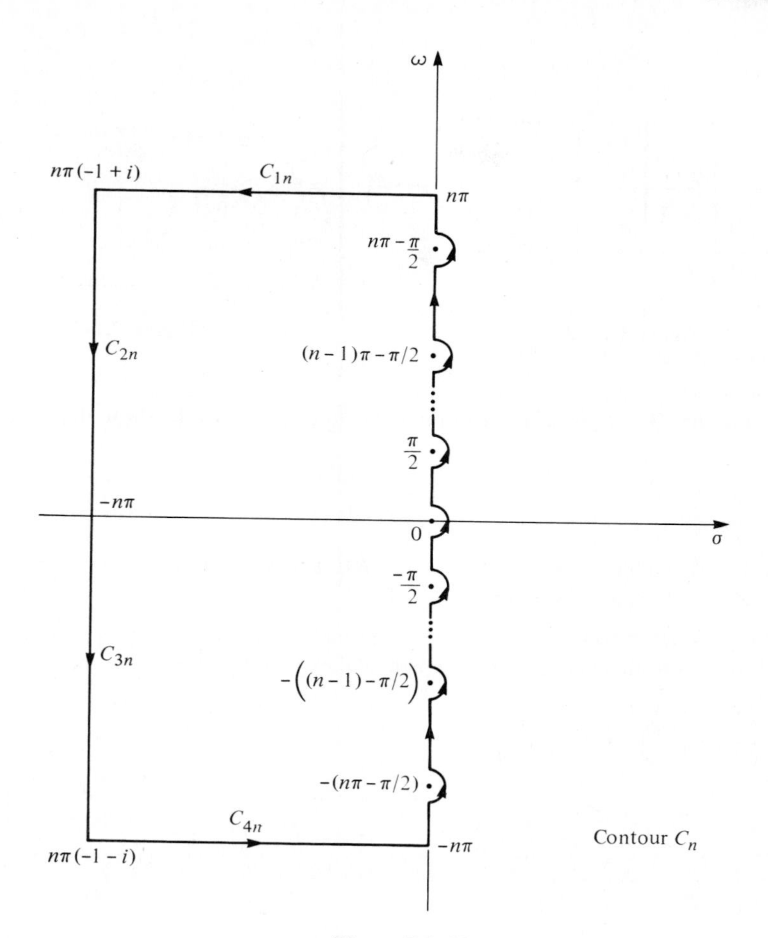

Figure 7.1–22

39. a) Show for $f(t)$, as defined in Fig. 7.1–23, that

$$F(s) = \frac{v_0}{s} \tanh\left(\frac{s\tau}{2}\right).$$

Hint: See Exercise 24.

b) For the electric circuit in Fig. 7.1–24 we can use the Kirchhoff voltage law to show that the current $\imath(t)$ satisfies

$$v(t) = L\frac{d\imath}{dt} + \imath(t)R.$$

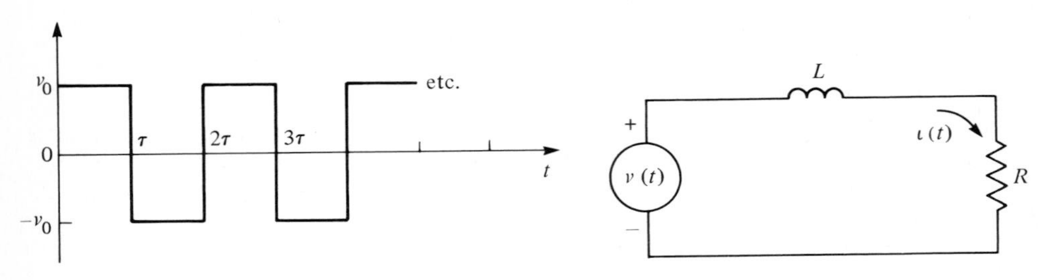

Figure 7.1–23

Figure 7.1–24

Assume that $v(t)$ is given by $f(t)$ in part (a), and that $\imath(0) = 0$. Show that

$$I(s) = \frac{v_0}{s} \frac{\tanh(s\tau/2)}{Ls + R}.$$

c) Show that $I(s)$ has simple poles at $s = -R/L$ and $s = \pm i\pi(2m - 1)/\tau$, $m = 1, 2, 3, \ldots$. Show that $I(s)$ has a finite limit as $s \to 0$.

d) Notice that $I(s)$ satisfies the conditions described in the paragraph preceding Exercise 35 with regard to its behavior on a sequence of expanding arcs. Show that

$$\imath(t) = \frac{v_0}{R} e^{-Rt/L} \tanh\left[\frac{R\tau}{2L}\right] + \frac{4\tau}{\pi} v_0 \sum_{m=1}^{\infty} \frac{\tau R \sin\left[(2m-1)\dfrac{\pi t}{\tau}\right] - (2m-1)\pi L \cos\left[(2m-1)\dfrac{\pi t}{\tau}\right]}{(2m-1)[\tau^2 R^2 + (2m-1)^2 \pi^2 L^2]}.$$

7.2 STABILITY—AN INTRODUCTION

One of the needs of the design engineer is to distinguish between two different kinds of functions of time—those that "blow up," that is, become unbounded, and those that do not. We will be a bit more precise. Let us consider a function $f(t)$ defined for $t > 0$.

DEFINITION A function $f(t)$ is *bounded* for positive t if there exists a constant M such that

$$|f(t)| < M, \qquad \text{for all } t > 0. \quad \square \tag{7.2–1}$$

We will usually just use the word "bounded" to describe an $f(t)$ satisfying Eq. (7.2–1). If no constant M can be found that remains larger than $|f(t)|$, we will say that $f(t)$ becomes *unbounded*. Such a function "blows up."

Although $f(t) = 1/(t-1)^2$ is unbounded because of a singularity at $t = 1$, our concern here is primarily with functions that fail to satisfy Eq. (7.2–1) because they grow without limit as $t \to \infty$. Thus $f(t) = e^{-t}$ is bounded because this function is less than 1, but $f(t) = e^t$ is unbounded since, for sufficiently large t, it will exceed any preassigned constant. The same is true of the functions $t \sin t$ and $e^t \cos t$, which exhibit oscillations of steadily growing size as t increases. The main subject of this and the following section is the relationship between bounded and unbounded functions and their Laplace transforms. We will also be concerned with knowing whether the response of a system (typically electrical or mechanical) is bounded or unbounded.

EXAMPLE 1

It is easy to devise an innocent-looking physical problem whose response or solution is unbounded. In Fig. 7.2–1 a mass of size m is attached to a spring whose elasticity constant is k. The mass is subjected to a harmonically varying external force $F_0 \cos \omega_0 t$ for $t \geq 0$. Let y be the displacement of the mass from that position in which the spring exerts no force. Then, if there are no frictional losses, Newton's second law asserts that

$$m \frac{d^2 y}{dt^2} + ky = F_0 \cos(\omega_0 t). \qquad (7.2\text{–}2)$$

We will assume that at $t = 0$, $dy/dt = 0$, and $y(t) = 0$. Taking the Laplace transform of Eq. (7.2–2), subject to the initial conditions, we have

$$ms^2 Y(s) + kY(s) = F_0 \frac{s}{s^2 + \omega_0^2},$$

or

$$Y(s) = \left(\frac{1}{ms^2 + k}\right)\left(\frac{F_0 s}{s^2 + \omega_0^2}\right) = \left(\frac{1}{s^2 + \dfrac{k}{m}}\right)\left(\frac{F_0 s}{m(s^2 + \omega_0^2)}\right).$$

Studying this result, we see that for $\omega_0 \neq \sqrt{k/m}$, $Y(s)$ has simple poles at $s = \pm i\omega_0$ and $s = \pm i\sqrt{k/m}$. If $\omega_0 = \sqrt{k/m}$, then $Y(s)$ has a pole of order 2 at $s = \pm i\omega_0$. In

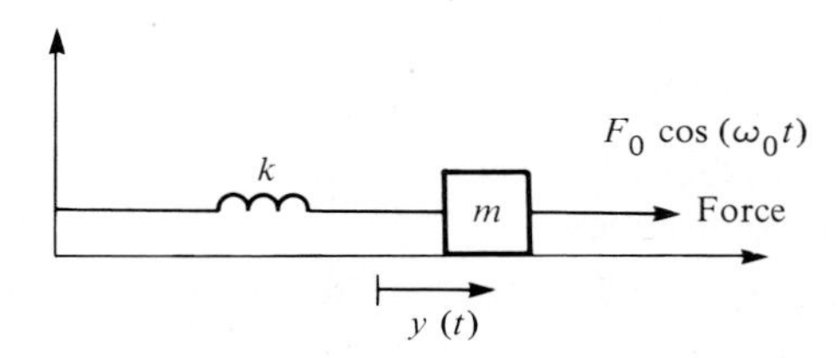

Figure 7.2–1

either case we find $y(t)$ from Eq. (7.1–30) with the result that

a) $$\omega_0 \neq \sqrt{\frac{k}{m}}, \qquad y(t) = \frac{F_0}{\left(\omega_0^2 - \frac{k}{m}\right)m}\left[-\cos(\omega_0 t) + \cos\left(\sqrt{\frac{k}{m}}\,t\right)\right],$$

b) $$\omega_0 = \sqrt{\frac{k}{m}}, \qquad y(t) = \frac{F_0}{2m\omega_0}\,t\sin(\omega_0 t).$$

Expression (a) consists of 2 cosine waves, while (b) is a sine wave whose amplitude grows with increasing t. The first result is bounded. The second is unbounded. ◀

The qualitatively different results found in cases (a) and (b) of Example 1 have something to do with the kinds of poles possessed by $Y(s)$. The response $y(t)$ of many physical systems, including the one just studied, has a Laplace transform of the form

$$Y(s) = \frac{P(s)}{Q(s)}, \tag{7.2–3}$$

where P and Q are polynomials in s having real coefficients, and the degree of Q exceeds that of P. We will assume that $Y(s)$ is an irreducible expression, that is, any identical factors of the form $(s - s_0)$ belonging to both P and Q have been divided out.

From Eq. (7.1–30) we have

$$\mathscr{L}^{-1}\frac{P(s)}{Q(s)} = f(t) = \sum_{\text{res}}\left(\frac{P(s)}{Q(s)}\,e^{st}\right), \qquad \text{at all poles.} \tag{7.2–4}$$

The poles of P/Q occur at the values of s for which $Q(s) = 0$. Those roots, designated $s_1, s_2, \ldots,$ are also called the *zeros* of $Q(s)$. We can write $Q(s)$ in the factored form

$$Q(s) = k(s - s_1)^{N_1}(s - s_2)^{N_2}\cdots(s - s_n)^{N_n}, \tag{7.2–5}$$

where k is a constant. The number N_j tells the multiplicity of the root s_j; it indicates the number of times this root is repeated. From Section 5.7 we see that N_j is also the order of the zero of $Q(s)$ at s_j.

If $Q(s) = 0$ has a root at $s = a + ib$ and if this root occurs with multiplicity N, then $e^{st}P/Q$ has a pole of order N at $s = a + ib$. The residue of $e^{st}P/Q$ at $a + ib$ contributes functions of time to $f(t)$ in Eq. (7.2–4) that can vary as

$$t^{N-1}e^{at}\cos(bt), \quad t^{N-1}e^{at}\sin(bt), \quad t^{N-2}e^{at}\cos(bt),$$
$$t^{N-2}e^{at}\sin(bt), \quad \ldots, \quad e^{at}\cos(bt), \quad e^{at}\sin(bt). \tag{7.2–6}$$

The reader can verify this fact by direct calculation or consultation with a table. Let us now consider three possibilities:

1. The root is in the right half of the s-plane. Thus $a > 0$. Each of the terms in Eq. (7.2–6) represents an unbounded function of time of either an oscillatory ($b \neq 0$) or nonoscillatory ($b = 0$) nature.

2. The root is in the left half of the s-plane, which means $a < 0$. With $a < 0$ the decay of e^{at} with increasing t causes each term in Eq. (7.2–6) to become zero as $t \to \infty$. Each term is bounded.
3. The root of $Q(s) = 0$ is on the imaginary axis. This means $a = 0$. The terms contained in Eq. (7.2–6) are now of the form

$$t^{N-1}\cos(bt),\ t^{N-1}\sin(bt),\ t^{N-2}\cos(bt),\ t^{N-2}\sin(bt), \ldots, \cos(bt), \sin(bt). \tag{7.2–7}$$

If the root is repeated, that is, if $N > 1$, then the amplitude of all these oscillatory terms except the last two, $\cos(bt)$ and $\sin(bt)$, grow without limit and are therefore unbounded. If $N = 1$, only the bounded terms $\cos(bt)$ and $\sin(t)$ are present.

A root at the origin is a special case of a root on the imaginary axis. We put $b = 0$ in Eq. (7.2–7) and again find that for a repeated root there is an unbounded contribution to $f(t)$ while for a nonrepeated root the contribution is bounded and is simply a constant.

The presence of one or more unbounded contributions to $f(t)$ on the right in Eq. (7.2–4) causes $f(t)$ to be unbounded. Our findings for possibilities 1–3 can be summarized in the following theorem.

THEOREM 4 Condition for bounded $f(t)$

Let $f(t) = \mathscr{L}^{-1}F(s) = \mathscr{L}^{-1}P(s)/Q(s)$, where P and Q are polynomials in s having no common roots, and the degree of Q exceeds that of P. Then, $f(t)$ is bounded if and only if $Q(s) = 0$ has no roots to the right of the imaginary axis and any roots on the imaginary axis are nonrepeated; in other words, if and only if poles of $F(s)$ do not occur in the right half of the s-plane and any poles on the imaginary axis are simple. □

In case (b) of the oscillating mass problem just considered (see Example 1) the unbounded result was caused by the second-order poles of $Y(s)$ lying on the imaginary axis at ω_0 and also at $-\omega_0$. In case (a) the result was bounded. The poles of $Y(s)$ were simple and lay on the imaginary axis at $\pm i\omega_0$ and $\pm i\sqrt{k/m}$.

EXAMPLE 2

Consider

$$f(t) = \mathscr{L}^{-1}\frac{s}{s^2 + \beta s + 1}, \qquad \text{where } \beta \text{ is real.}$$

Discuss the boundedness of $f(t)$ for the cases $\beta = 1$, $\beta = -1$, $\beta = 0$.

Solution

We take

$$F(s) = \frac{P(s)}{Q(s)} = \frac{s}{s^2 + \beta s + 1}.$$

The roots s_1 and s_2 of $Q(s) = 0$ are found from the quadratic formula. Thus for $s^2 + \beta s + 1 = 0$ we have

$$s_{1,2} = \frac{-\beta + (\beta^2 - 4)^{1/2}}{2}.$$

a) With $\beta = 1$,

$$s_{1,2} = \frac{-1 \pm i\sqrt{3}}{2}.$$

Here $F(s)$ has poles only in the left plane. Thus $f(t)$ is bounded.

b) With $\beta = -1$,

$$s_{1,2} = \frac{1 \pm i\sqrt{3}}{2}.$$

Now $F(s)$ has poles in the right half plane. Thus $f(t)$ is unbounded.

c) With $\beta = 0$,

$$s_{1,2} = \pm i.$$

These roots are on the imaginary axis. Now $Q(s) = (s + i)(s - i)$. The poles of $F(s)$ lie only on the imaginary axis and are simple. Thus $f(t)$ is bounded. ◀

In control theory, electronics, and often in biology and medicine one deals with systems (electrical, mechanical, or animal) subjected to an input or excitation that produces some kind of output or response. The systems analyst must answer this question: If the input $x(t)$, defined for $t > 0$, is bounded for $t > 0$, will the output $y(t)$ also be bounded?

To answer this question we will employ transforms. The Laplace transforms $Y(s)$ and $X(s)$, of $y(t)$ and $x(t)$, for the systems we will be studying are related through an expression of the form

$$Y(s) = G(s)X(s). \tag{7.2–8}$$

Here $G(s)$ is known as the transfer function of the system.† We make the following definition.

DEFINITION Transfer function

The *transfer function* of a system is that function that must be multiplied by the Laplace transform of the input to yield the Laplace transform of the output. The term *system function* is often used in lieu of transfer function. □

---

† Like $X(s)$ and $Y(s)$, $G(s)$ is the Laplace transform of a function of t, as shown in Section 7.5. In the present discussion we do not need to know this function.

To see how a relationship like Eq. (7.2–8) can arise, let us consider a system in which the variables $x(t)$ and $y(t)$ satisfy a linear differential equation with constant coefficients, that is,

$$a_n\frac{d^ny}{dt^n} + a_{n-1}\frac{d^{n-1}y}{dt^{n-1}} + \cdots + a_1\frac{dy}{dt} + a_0y = x(t). \tag{7.2–9}$$

An elementary example of such a system is the spring and oscillating mass considered earlier in this section. The input is the force $F_0 \cos \omega_0 t$ while the output is the displacement of the mass $y(t)$.

Besides Eq. (7.2–9) we are typically given initial conditions at $t = 0$ for the function $y(t)$ and its time derivatives. We assume here, as we do in the remainder of this section, that all such values are zero. This is equivalent to requiring that there be no energy stored in the system at $t = 0$.

Taking the Laplace transform of Eq. (7.2–9) in the usual way we obtain the algebraic equation

$$a_ns^nY(s) + a_{n-1}s^{n-1}Y(s) + \cdots + a_0Y(s) = X(s),$$

which yields

$$Y(s) = \frac{X(s)}{a_ns^n + a_{n-1}s^{n-1} + \cdots + a_0}. \tag{7.2–10}$$

Comparing Eq. (7.2–10) with Eq. (7.2–8) we see that, for the systems described by the differential equation (7.2–9), the transfer function is given by

$$G(s) = \frac{1}{a_ns^n + a_{n-1}s^{n-1} + \cdots + a_0}. \tag{7.2–11}$$

Any system in which the input and output are related by a linear differential equation with constant coefficients has a corresponding transfer function.

Some common systems are characterized not by differential equations but by integral equations or integrodifferential equations (see, for example, Exercise 15 of this section). Here the transfer function relating output to input is not simply the reciprocal of a polynomial expression in s, as in Eq. (7.2–11), but is the ratio of two polynomials in s. This complication also occurs in the feedback systems considered in Exercise 8, Section 7.4. To allow for such systems we will assume a transfer function of the form

$$G(s) = \frac{A(s)}{B(s)}, \tag{7.2–12}$$

where A and B are polynomials in s. The coefficients in these polynomials are invariably real numbers. Note that Eq. (7.2–11) is a special case of Eq. (7.2–12).

Now we return to our original question. If it is given that the input $x(t)$ is a bounded function, what conditions must be imposed on the transfer function $G(s)$ so that the output $y(t)$ is a bounded function? This leads naturally to the following definition.

DEFINITION Stable system

A *stable system* is one that produces a bounded output for *every* bounded input. □

A system that is not stable will, of course, be called *unstable*.

Let us study Eq. (7.2–8) for a moment: $Y(s) = G(s)X(s)$. If $x(t)$ is bounded, then $X(s)$ has no poles to the right of the imaginary axis, and any poles on the axis are simple. Now, if $G(s)$ has all its poles lying to the left of the imaginary axis, then the product $G(s)X(s) = Y(s)$ has no poles to the right of the imaginary axis. Whatever poles the product GX possesses on the imaginary axis are the poles of $X(s)$ and are simple. Thus the poles of $Y(s)$ are such that $y(t)$ is bounded. This leads to the following theorem.

THEOREM 5 Poles of stable and unstable systems

The transfer function $G(s)$ of a stable system has all its poles lying to the left of the imaginary s-axis. A system whose $G(s)$ has one or more poles on, or to the right of, the imaginary axis is unstable. □

A pole of $G(s)$ lying to the right of the imaginary s-axis results, in general, in $Y(s)$ having such a pole. Thus $y(t)$ will be unbounded. Now, comparing Theorems 4 and 5 we see that the conditions required for a bounded function are not quite the same as those required for a stable system. The transform of a bounded function can have simple poles on the imaginary axis, while the transfer function of a stable system cannot. To see why this is so, consider Eq. (7.2–8). If $G(s)$ has a simple pole on the imaginary axis and if $X(s)$ also has a simple pole at the same location, then the product $G(s)X(s)$ would have a second-order pole at this point. With $Y(s)$ now having a second-order pole on the imaginary axis the output $y(t)$ would be unbounded. For the $G(s)$ just described a bounded output is obtained for all bounded inputs except those whose transforms $X(s)$ have a simple pole coinciding with the simple pole of $G(s)$ on the imaginary axis. To describe this situation, the following definition is useful.

DEFINITION Marginal instability

An unstable system with a transfer function whose poles on the imaginary axis are simple and with no poles to the right of this axis is called *marginally unstable*. □

Marginally unstable systems are thus special kinds of unstable systems. The term "marginally unstable" is not a universal one. Some authors use the form "marginally stable" to mean the same thing.

EXAMPLE 3

The mass–spring system considered in Example 1 is marginally unstable. We had

$$Y(s) = \left(\frac{1}{ms^2 + k}\right)\left(\frac{F_0 s}{s^2 + \omega_0^2}\right).$$

The first term on the right is $G(s)$, the second, the transformed input $X(s)$. Now $G(s)$ has poles on the imaginary axis at $s = \pm i\sqrt{k/m}$. The poles of $X(s)$ are at $s = \pm i\omega_0$. When $\omega_0 = \sqrt{k/m}$, the poles of $X(s)$ are identical to those of $G(s)$, and an unbounded output varying as $t \sin(\omega_0 t)$ occurs. For $\omega_0 \neq \sqrt{k/m}$ the output is bounded and varies with both $\cos(\omega_0 t)$ and $\cos(\sqrt{k/m}\, t)$. ◀

EXAMPLE 4

The input $x(t)$ and the output $y(t)$ of a certain system are related by

$$\frac{d^3y}{dt^3} - a\frac{d^2y}{dt^2} + b^2\frac{dy}{dt} - ab^2y(t) = x(t).$$

For what real values of a and b is this a stable system?

Solution

With all initial values taken as zero we transform this equation and obtain

$$(s^3 - as^2 + b^2s - ab^2)Y(s) = X(s),$$

which we can rewrite as

$$(s-a)(s^2+b^2)Y(s) = X(s) \quad \text{or} \quad Y(s) = \frac{X(s)}{(s-a)(s^2+b^2)}.$$

We see that

$$G(s) = \frac{1}{(s-a)(s^2+b^2)}$$

and that $G(s)$ has simple poles at $s = a$ and at $s = \pm ib$. If $a > 0$, $G(s)$ has a pole to the right of the imaginary axis. The system is unstable. If $a < 0$, $G(s)$ has no poles to the right of the imaginary axis. Now, with $a < 0$, assume $b \neq 0$. $G(s)$ has simple poles on the imaginary axis at $\pm ib$. Thus the system is marginally unstable.

If $a = 0$ and $b \neq 0$, the poles of $G(s)$ are simple and lie on the imaginary axis at $s = 0$ and $\pm ib$. The system is marginally unstable.

If $b = 0$ and $a \neq 0$, $G(s)$ has a second-order pole on the imaginary axis at $s = 0$. The system is unstable.

If $b = 0$ and $a = 0$, there is a third-order pole at $s = 0$. Thus the system is unstable. ◀

EXERCISES

Which of the functions $F(s)$ given below have inverse Laplace transforms $f(t)$ that are bounded functions of t?

1. $\dfrac{1}{s(s^2+1)(s+2)}$ **2.** $\dfrac{1}{(s+1)(s^2+3s+2)}$ **3.** $\dfrac{s}{(s^2+1)(s-2)}$

4. $\dfrac{1}{(s^2+s+1)^2}$

5. $\dfrac{s+1}{s^4+s^2+1}$

6. $\dfrac{s-1}{(s^3-1)(s^2+s+1)}$

7. $\dfrac{1}{s^4+s^3+s^2+s+1}$

8. $\dfrac{s-\sqrt{2}}{s^4-s^2-2}$

Hint: See Eq. (5.2–8).

9. Let

$$F(s) = \frac{s+1}{s^2+\beta s+1}, \qquad \text{where } \beta \text{ is a real number.}$$

a) For $\beta \geq 0$ show that $f(t)$ is bounded, and for $\beta < 0$ show that $f(t)$ is unbounded.

b) For $-2 < \beta < 2$ show that $f(t)$ oscillates with t. For which values of β do the oscillations grow with t and for which values do they decay?

10. Assume that the expressions given in Exercises 1–8 are transfer functions $G(s)$. In each case is the system stable or unstable? If the system is unstable, is it marginally unstable?

For the marginally unstable systems characterized by the following transfer functions find a bounded input $x(t)$ that will produce an unbounded output $y(t)$.

11. $\dfrac{s}{s^2+9}$

12. $\dfrac{1}{s(s^2+1)}$

13. For a certain system the output $y(t)$ lags behind the input $x(t)$ by T time units and is m times the input, that is,

$$y(t) = mx(t-T), \qquad T > 0.$$

Assume $x(t) = 0$ for $t < 0$. Thus $x(t-T) = 0$ for $t < T$, and

$$y(t) = mx(t-T)u(t-T), \qquad -\infty < t < \infty,$$

where $u(t)$ is the unit step function defined in Section 2.2.

a) Show using the definition of the Laplace transform that

$$Y(s) = mX(s)e^{-sT}.$$

Hint: See Exercise 15 of the previous section.

b) Find the transfer function of this system. This is an example of a system whose transfer function is not a rational function.

c) A certain system has transfer function

$$G(s) = \frac{A(s)}{B(s)}\, e^{-sT},$$

where $T > 0$ and $A(s)/B(s)$ is a rational function in s. Show that this system can be regarded as two systems in tandem (see Fig. 7.2–2) with the output $y_1(t)$ of the first

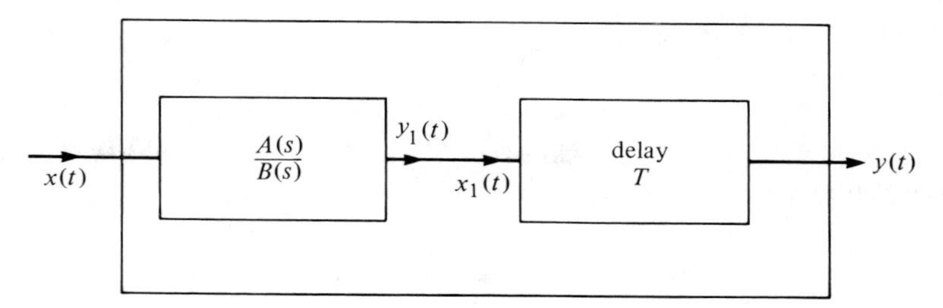

Figure 7.2–2

system fed as input $x_1(t)$ to the second system. The first system has as its transfer function the rational expression $A(s)/B(s)$. The output $y(t)$ of the second system is a delayed version of its input $x_1(t)$, i.e., $y(t) = x_1(t - T)u(t - T)$.

d) Explain why in studying the stability of the system we can ignore the factor e^{-sT}.

14. In Example 1 (see Fig. 7.2–1) assume that the mass is moving through a fluid that exerts a retarding force on the object proportional to the velocity of motion dy/dt. Thus the differential equation describing the motion is now given by

$$m\frac{d^2y}{dt^2} + ky + \alpha\frac{dy}{dt} = x(t), \qquad m, k > 0,$$

where $\alpha \geq 0$ is a constant, and $x(t)$ is the external force applied to the mass. Assume that $y = 0$ and $dy/dt = 0$ at $t = 0$.

a) Show that the transfer function relating the input $x(t)$ and the response $y(t)$ is

$$G(s) = \frac{1}{ms^2 + \alpha s + k}.$$

b) Show that for $\alpha > 0$ the system is stable and that for $\alpha = 0$ the system is marginally unstable.

15. For the R, L circuit in Fig. 7.2–3 the input is the voltage $v_i(t)$ and the output is the voltage $v_0(t)$. The current is $\imath(t)$. From Kirchhoff's voltage law we have

$$v_i(t) = v_0(t) + \imath(t)R.$$

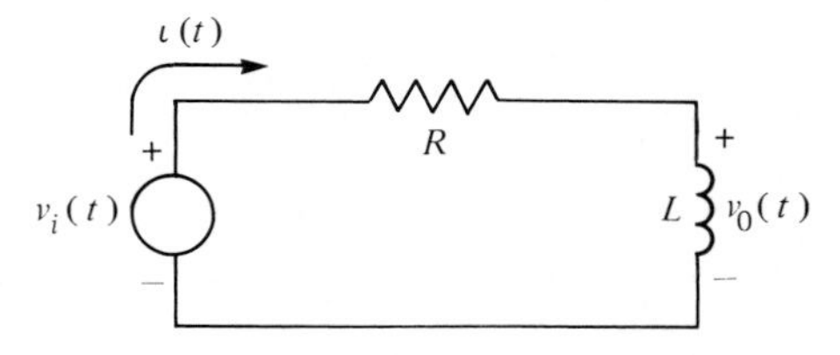

Figure 7.2–3

If $\imath(0) = 0$, then by Faraday's law,

$$\imath(t) = \frac{1}{L}\int_0^t v_0(t')\,dt'.$$

Thus v_i and v_0 are related by the integral equation

$$v_i(t) = v_0(t) + \frac{R}{L}\int_0^t v_0(t')\,dt'.$$

Show that the transfer function is

$$\frac{V_0(s)}{V_i(s)} = \frac{Ls}{Ls + R} = G(s).$$

Is the system stable? Assume R and L are positive.

16. Certain electrical devices, for example, tunnel diodes, exhibit a negative electrical resistance. A certain negative resistance diode is characterized by the equivalent circuit shown in the broken box (see Fig. 7.2–4). The indicated resistance R has a negative numerical value $-R_d$, where $R_d > 0$. The applied voltage $v(t)$ is a bounded excitation, while the supplied current $\imath_1(t)$ is the response. Writing Kirchhoff's voltage law around the two electrical meshes shown, we obtain the coupled equations

$$v(t) = L\frac{d\imath_1}{dt} + \frac{1}{C}\int_0^t [\imath_1(t) - \imath_2(t)]\,dt,$$

$$0 = \frac{1}{C}\int_0^t [\imath_2(t) - \imath_1(t)]\,dt - \imath_2 R_d.$$

We assume that L and C are both positive.

a) Assume all initial conditions are zero, and take the Laplace transform of this pair of equations. Obtain a pair of algebraic equations involving $I_1(s)$ and $I_2(s)$. Show that $G(s) = (1 - sCR_d)/(Ls(1 - sCR_d) - R_d) = I_1(s)/V(s)$.

b) Examine the poles of the transfer function $G(s)$, and show that the system is unstable.

c) Show that if $R_d > (\sqrt{L/C})/2$, the current $\imath_1$ exhibits oscillations that grow exponentially in time, and that if $R_d < (\sqrt{L/C})/2$, the current exhibits a nonoscillatory exponential growth.

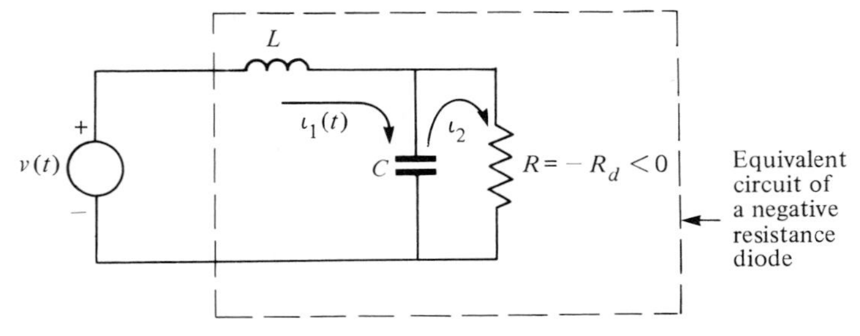

Figure 7.2–4

7.3 PRINCIPLE OF THE ARGUMENT

We have just seen that a stable system has a transfer function $G(s)$ whose poles lie entirely to the left of the imaginary axis. If $G(s)$ is a rational function, for example,

$$G(s) = \frac{A(s)}{B(s)}, \tag{7.3–1}$$

the task of determining whether $G(s)$ describes an unstable system is equivalent to seeing whether the polynomial $B(s)$ satisfies

$$B(s) = 0 \tag{7.3–2}$$

for any s on or to the right of the imaginary axis.

The exercises given in Section 7.2 were carefully selected so that the roots of Eq. (7.3–2) would be easily found. However, it is comparatively easy to construct a problem where the location of roots is not so obvious; for example, consider

$$B(s) = s^4 + s^3 - 2s + 1.$$

The problem of determining the presence of the roots of a polynomial in a half plane or, equivalently, the problem of determining the presence of poles of an expression like Eq. (7.3–1) in a half plane is important in control theory. The engineer uses a variety of mathematical tools to deal with this question. One convenient method, which uses complex variable theory, will be discussed here. The reader is referred to more specialized texts for other techniques.†

We will be using the Nyquist method. It is based on a theorem called the *principle of the argument*, which we now derive and study. An introduction to the Nyquist method is reserved for Section 7.4.

Consider a function $f(z)$ that is analytic and nonzero everywhere on a simple closed contour C. In addition we assume that $f(z)$ is analytic in the domain inside C except possibly at a finite number of pole singularities. The preceding guarantees that, as we make one complete circuit around C, the initial and final numerical values assumed by $f(z)$ are identical. As C is completely negotiated, there is no reason, however, why the initial and final values of the *argument* of $f(z)$ must be identical.

Suppose we write

$$f(z) = |f(z)|e^{i(\arg f(z))}. \tag{7.3–3}$$

We will use the notation $\Delta_C \arg f(z)$ to mean the *increase in argument* of $f(z)$ (final minus initial value) as the contour C is negotiated once in the positive direction.

Let us consider an elementary example. We take $f(z) = z$, and, as a contour C, the circle $|z| = 1$. On C we have $z = |z|e^{i\arg z} = e^{i\arg z}$. As we proceed around C once in the clockwise direction, we see from Fig. 7.3–1 that $\arg z$ progresses from 0

---

† See, for example, J. DiStefano, A. Stubberud, I. Williams, *Feedback and Control Systems*, 2nd ed., Schaum's Outline Series (New York: McGraw-Hill, 1990).

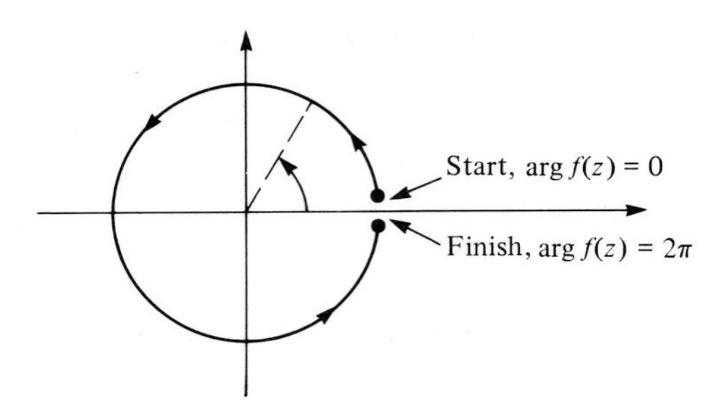

Figure 7.3–1

to 2π. Thus in this case $\Delta_C \arg f(z) = 2\pi$. Note that $f(z) = x + iy$ returns to its original numerical value after C is negotiated.

To choose another example, if $f(z) = 1/z^2$ and C is any closed contour encircling the origin, then the reader should verify that $\Delta_C \arg f(z) = -4\pi$.

These two examples illustrate something that will always be true: since our assumptions about $f(z)$ require that this function return to its starting value after C is negotiated, then $\Delta_C \arg f(z)$ must be an integer times 2π.

Now consider

$$I = \frac{1}{2\pi i} \oint_C \frac{f'(z)}{f(z)}\, dz. \tag{7.3–4}$$

We observe that

$$\frac{d}{dz} \log f(z) = \frac{f'(z)}{f(z)}$$

if we use some analytic branch of $\log(f(z))$. If we require that C not pass through any zero or pole of $f(z)$, we see that our integral I can be evaluated by a standard procedure (see Eq. 4.4–4), and thus

$$\begin{aligned}\frac{1}{2\pi i} \oint_C \frac{f'(z)}{f(z)}\, dz &= \frac{1}{2\pi i} \oint_C \frac{d}{dz} \log f(z)\, dz = \frac{1}{2\pi i} \oint_C d(\log f(z)) \\ &= \frac{1}{2\pi i} [\text{increase in } \log f(z) \text{ in going around } C] \\ &= \frac{1}{2\pi i} [\text{increase in } [\log|f(z)| + i \arg f(z)] \text{ in going around } C].\end{aligned}$$

Now $|f(z)|$ necessarily returns to its original numerical value as C is negotiated. However $\arg f(z)$ need not. Thus

$$\frac{1}{2\pi i} \oint_C \frac{f'(z)}{f(z)}\, dz = \frac{1}{2\pi} \Delta_C \arg f(z). \tag{7.3–5}$$

One can also evaluate Eq. (7.3–4) by residues. If $f(z)$ is analytic on C and at all points interior to C except at poles, then $f'(z)$ will be analytic on and interior to C except at these same poles. (Recall from Section 4.5 that the derivative of an analytic function is analytic.) The quotient $f'(z)/f(z)$ is thus analytic on C and interior to C except where $f'(z)$ has a pole or when $f(z) = 0$. Thus to evaluate Eq. (7.3–4) with residues we must determine the residue of $f'(z)/f(z)$ at all zeros and poles of $f(z)$ lying interior to C.

Suppose $f(z)$ has a zero of order n at ζ. Recall (see Section 5.7) that this means $f(z)$ has a Taylor expansion about ζ of the form

$$f(z) = a_n(z - \zeta)^n + a_{n+1}(z - \zeta)^{n+1} + \cdots, \qquad a_n \neq 0.$$

Thus factoring out $(z - \zeta)^n$, we have

$$f(z) = (z - \zeta)^n \phi(z), \tag{7.3–6}$$

where $\phi(z)$ is a function that is analytic at ζ and has the series expansion

$$\phi(z) = a_n + a_{n+1}(z - \zeta) + a_{n+2}(z - \zeta)^2 + \cdots.$$

Note that $\phi(\zeta) = a_n \neq 0$. Differentiating Eq. (7.3–6), we arrive at

$$f'(z) = n(z - \zeta)^{n-1}\phi(z) + (z - \zeta)^n \phi'(z). \tag{7.3–7}$$

Dividing Eq. (7.3–7) by Eq. (7.3–6), we obtain

$$\frac{f'(z)}{f(z)} = \frac{n}{z - \zeta} + \frac{\phi'(z)}{\phi(z)}. \tag{7.3–8}$$

The first term on the right in Eq. (7.3–8) has a simple pole at ζ in the z-plane. The residue is n. Recalling that $\phi(\zeta) \neq 0$, we see that the second term on the right has no singularity at ζ. Thus, at ζ, the residue of $f'(z)/f(z)$ is identical to the residue of $n/(z - \zeta)$ and equals n. In other words, the residue of $f'(z)/f(z)$ at ζ is equal to the order (multiplicity) of the zero of $f(z)$ at ζ.

Suppose that $f(z)$ has a pole of order p at a point α inside C. Proceeding much as before, (see Exercise 7 of this section) we find that the residue of $f'(z)/f(z)$ at α is equal to (-1) times the order p of the pole of $f(z)$ at α.

We can use the information just derived to sum the residues of $f'(z)/f(z)$ at all the singularities that this expression possesses inside C and thus evaluate the integral I in Eq. (7.3–4) (see Fig. 7.3–2). If $f(z)$ has zeros of order n_1 at ζ_1, n_2 at $\zeta_2, \ldots$ and poles of order p_1 at α_1, p_2 at $\alpha_2, \ldots$, we have

$$\frac{1}{2\pi i} \oint_C \frac{f'(z)}{f(z)}\, dz = N - P, \tag{7.3–9}$$

where

$$N = n_1 + n_2 + \cdots \tag{7.3–10}$$

is the total number of zeros of $f(z)$ inside C and

$$P = p_1 + p_2 + \cdots \tag{7.3–11}$$

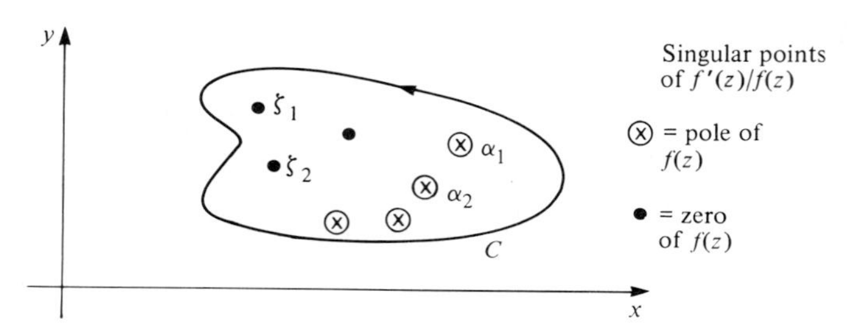

Figure 7.3–2

is the total number of poles of $f(z)$ inside C. In both Eqs. (7.3–10) and (7.3–11) zeros and poles are counted, according to their multiplicities; for example, a zero of order 2 at some point contributes the number 2 to the sum on the right in Eq. (7.3–10) and a pole of order 3 results in a contribution of 3 in Eq. (7.3–11). We will assume that $f(z)$ has a finite number of poles inside C. One can show that $f(z)$ has a finite number of zeros inside C.†

Equations (7.3–9) and (7.3–5) provide two different ways of evaluating the same integral. We dispense with the integral and utilize the right side of each equation. This provides the following theorem.

THEOREM 6 Principle of the argument

Let $f(z)$ be analytic on a simple closed contour C and analytic inside C except possibly at a finite number of poles. Also, assume $f(z)$ has no zeros on C. Then

$$\frac{1}{2\pi}\Delta_C \arg f(z) = N - P, \tag{7.3–12}$$

where N is the total number of zeros of $f(z)$ inside C, and P is the total number of poles of $f(z)$ inside C. In each case the number of poles and zeros are counted according to their multiplicities. □

The preceding theorem is called the *principle of the argument.* We should also recall that $\Delta_C \arg f(z)$ in Eq. (7.3–12) is computed when C is traversed in the positive sense. This quantity can be positive, zero, or negative depending on the relative sizes of N and P.

EXAMPLE 1

Let $f(z) = z^2 - 1$, and let C be the circle $|z - 1| = 1$. Verify the correctness of Eq. (7.3–12) in this case.

---

† It can be shown that a function that is not identically zero and is analytic in a bounded region, except at pole singularities, has a finite number of zeros in that region.

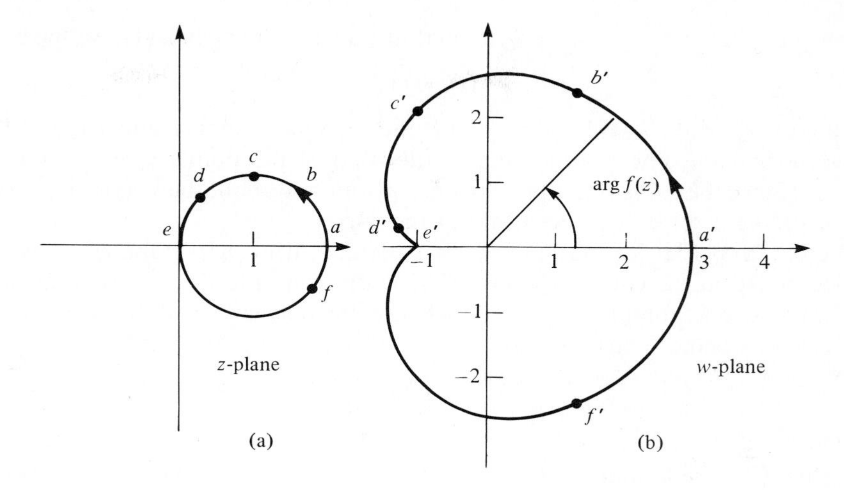

Figure 7.3–3

Solution

We will use two planes, the usual z-plane and the w-plane, the latter showing values assumed by $w = f(z)$ as z travels around C. We write $w = u + iv$.

A few points $a, b, c, \ldots$ (see Fig. 7.3–3a) lying on C are considered. We determine the corresponding image points $a', b', \ldots$ under the transformation $w = f(z)$ (see Table 2) and plot these points in the w-plane (see Fig. 7.3–3b). Using the image points we can quickly sketch the locus C' of all the values that $f(z)$ assumes on C. Notice

TABLE 2

point in z-plane	z	$z^2 - 1 = w$	point in w-plane
a	2	3	a'
b	$1 + \frac{1}{\sqrt{2}} + \frac{i}{\sqrt{2}}$	$\sqrt{2} + i(1 + \sqrt{2})$	b'
c	$1 + i$	$2i - 1$	c'
d	$1 - \frac{1}{\sqrt{2}} + \frac{i}{\sqrt{2}}$	$-\sqrt{2} + i(\sqrt{2} - 1)$	d'
e	0	-1	e'
f	$1 + \frac{1}{\sqrt{2}} - \frac{i}{\sqrt{2}}$	$\sqrt{2} - i(1 + \sqrt{2})$	f'

that because $f(z) = z^2 - 1$ is a polynomial in z with real coefficients, we have

$$f(\bar{z}) = \overline{f(z)}.$$

For example, the values assumed by $f(z)$ at the conjugate points b and f are complex conjugates of each other. Notice the relationship of the points b, b', f, and f' in Fig. 7.3–3. Since the curve C in the xy-plane is symmetric about the x-axis, the curve C' in the uv-plane must be symmetric about the u-axis.

We see from Fig. 7.3–3(a) and (b) that the argument of $f(z)$ increases by 2π as C is negotiated in the counterclockwise direction from a to b to $c\ldots$ and back to a again; that is, one complete counterclockwise encirclement of the origin has been made in the w-plane. Thus on the left in Eq. (7.3–12) we have

$$\frac{\Delta_C \arg f(z)}{2\pi} = 1.$$

To evaluate the right side of Eq. (7.3–12) we see that $f(z) = z^2 - 1$ has no pole singularities. We have $P = 0$. Since $(z^2 - 1) = (z - 1)(z + 1)$, we see (compare with Eq. 7.3–6) that $f(z)$ possesses two zeros, each of order (or multiplicity) 1. Only the zero at $z = 1$ lies within C. Thus $N - P = 1$. The correctness of Eq. (7.3–12) has been verified in this case. ◀

EXAMPLE 2

Verify Eq. (7.3–12), where $f(z) = z/(z + 1)^2$ and C is the circular contour $|z| = 20$.

Solution

A typical point on C is described by $z = 20e^{i\theta}$ (see Fig. 7.3–4). The corresponding point in the w-plane is $20e^{i\theta}/(1 + 20e^{i\theta})^2$. For our purposes we make an excellent approximation by ignoring 1 in the denominator. Thus

$$f(z) \approx \frac{20e^{i\theta}}{400e^{2i\theta}} = \frac{1}{20}e^{-i\theta}.$$

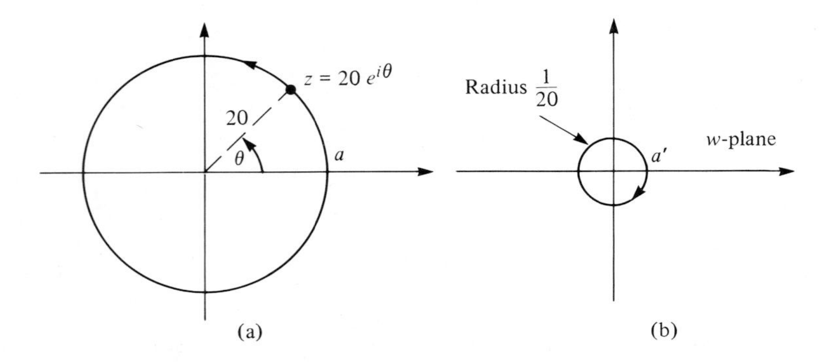

Figure 7.3–4

All the values of $f(z)$ that we will encounter on C therefore lie approximately on a circle in the w-plane of radius 1/20. As we move around C in the positive direction the angle θ increases by 2π. However, the argument of $f(z)$, which is $-\theta$, *decreases* by 2π, that is, it increases by -2π. Thus $\Delta_C \arg f(z) = -2\pi$. The left side of Eq. (7.3–12) is -1.

Now $f(z) = z/(z+1)^2$ contains a zero of multiplicity 1 at the origin of the z-plane. A vanishing denominator causes $f(z)$ to have a pole of order 2 at $z = -1$. Both the zero and the pole are inside C. Thus the right side of Eq. (7.3–12) is $1 - 2 = -1$. The formula is verified. ◀

Comment: In many texts the left side of Eq. (7.3–12) is written in a different form. If, as C is traversed, the locus of $f(z)$ makes one complete encirclement of the origin in the w-plane, then the argument of $f(z)$ increases by 2π. Every such additional encirclement results in an additional contribution of 2π to the expression $\Delta_C \arg f(z)$. Thus, on the left side of Eq. (7.3–12), $\Delta_C \arg f(z)/2\pi$ tells the net number of counterclockwise encirclements that $f(z)$ makes about the point $w = 0$.

Letting

$$E = \frac{\Delta_C \arg f(z)}{2\pi}$$

be this number, we have, from Eq. (7.3–12),

$$E = N - P. \tag{7.3–13}$$

In Example 2, $E = -1$ since the origin in Fig. 7.3–4(b) was encircled once in the *clockwise* direction, while in Example 1 we had $E = 1$.

EXAMPLE 3

Use the principle of the argument to determine how many roots $e^z - 2z = 0$ has inside the circle $|z| = 3$.

Solution

Mapping $|z| = 3$ into the w-plane by means of the transformation $w = e^z - 2z$ is a somewhat tedious (but not impossible) procedure if we follow the method employed in Example 1. For this reason, some inexpensive computer software† was used to carry out the mapping (see Fig. 7.3–5). We see that, if we move counterclockwise along $|z| = 3$ and encircle $z = 0$ once, then the corresponding image point encircles $w = 0$ twice in the positive sense. Since $w(z)$ has no poles, we have, according to Eq. (7.3–13), two solutions inside $|z| = 3$. ◀

The preceding problem illustrates the utility of the principle of the argument in determining whether an equation $f(z) = 0$ has solutions in a given region of the

---

† The software is called **f(z)** and is available for IBM and Macintosh desktop computers. It is published by Lascaux Graphics, 3771 East Guthrie Mountain Place, Tuscon, AZ 85718.

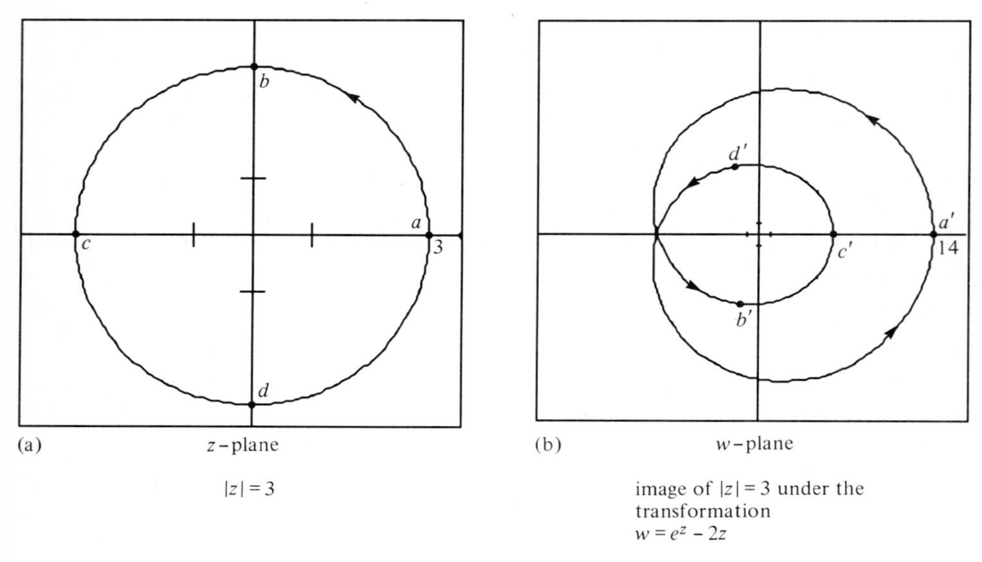

Figure 7.3–5

complex plane. This property is further developed in the following section, where we study the Nyquist method.

We proved the Fundamental Theorem of Algebra in Section 4.6 with the aid of Liouville's theorem; i.e., we showed that the equation $P(z) = 0$, where $P(z) = a_n z^n + a_{n-1} z^{n-1} + \cdots + a_0$ ($n \geq 1$, $a_n \neq 0$), has a root in the complex plane. An extension of the theorem, Exercise 18 in Section 4.6, shows that there are n roots. If we call them $z_1, z_2, \ldots, z_n$, then $P(z)$ is a constant times $(z - z_1)(z - z_2) \cdots (z - z_n)$. An alternate and simple proof of the fundamental theorem, based on the principle of the argument, is now available to us. Using this principle, we first derive Rouché's theorem (Exercise 8), from which the fundamental theorem follows (Exercise 9). Rouché's theorem, by itself, is also useful in locating the roots of both algebraic and transcendental equations (see Exercises 10–13).

EXERCISES

Let $f(z)$ be each of the following functions and take C as the circle indicated. Sketch $f(z)$ in the w-plane as z moves counterclockwise around the circle. Without using the argument principle, determine the number of zeros and poles of $f(z)$ inside C. Check your result by using the principle of the argument, Eq. (7.3–12).

1. $f(z) = z$, C is $|z - 2| = 3$
2. $f(z) = 1/z$, C is $|z - 2| = 3$
3. $f(z) = (z + 1)/z$, C is $|z| = 4$

4. $f(z) = \dfrac{1}{(z-1)^2}$, C is $|z| = 2$

5. $f(z) = \text{Log } z$, C is $|z - e| = 2$

6. $\dfrac{\sin z}{z}$, C is $|z| = \pi/2$

7. Show that if $f(z)$ has a pole of order p at α, then the residue of $f'(z)/f(z)$ at α is $-p$.

Hint: $f(z)$ can be expressed as $g(z)/(z-\alpha)^p$, where $g(\alpha) \neq 0$ and $g(z)$ is analytic at α. Why?

8. Let $f(z)$ and $g(z)$ be analytic on and everywhere inside a simple closed contour C. Suppose $|f(z)| > |g(z)|$ on C. We will prove that $f(z)$ and $(f(z) + g(z))$ have the same number of zeros inside C. This is known as *Rouché's theorem.*

a) Explain why

$$\frac{\Delta_C \arg f(z)}{2\pi} = N_f,$$

and

$$\frac{\Delta_C \arg(f(z) + g(z))}{2\pi} = N_{f+g},$$

where N_f is the number of zeros of $f(z)$ inside C, and N_{f+g} is the number of zeros of $f(z) + g(z)$ inside C.

b) Show that

$$N_{f+g} = \frac{1}{2\pi}\Delta_C \arg f(z) + \frac{1}{2\pi}\Delta_C \arg\left(1 + \frac{g(z)}{f(z)}\right).$$

Hint: $f + g = f[1 + g/f]$.

c) If $|g|/|f| < 1$ on C, explain why $\Delta_C \arg[1 + g/f] = 0$.

Hint: Let $w(z) = 1 + g/f$. As z goes along C, suppose that $w(z)$ encircles the origin of the w-plane. This implies that $w(z)$ assumes a negative real value for some z. Why does this contradict our assumption $|g|/|f| < 1$ on C?

c) Combine the results of parts (a), (b), (c) to show that $N_f = N_{f+g}$.

9. Let $h(z) = a_n z^n + a_{n-1} z^{n-1} + \cdots + a_0 z^0$ be a polynomial of degree n. We will prove that $h(z)$ has exactly n zeros (counted according to multiplicities) in the z-plane. This is a version of the *Fundamental Theorem of Algebra,* which was discussed in Section 4.6.

a) Let

$$f(z) = a_n z^n,$$
$$g(z) = a_{n-1} z^{n-1} + a_{n-2} z^{n-2} + \cdots + a_1 z + a_0 z^0.$$

Note that $h = f + g$. Consider a circle C of radius $r > 1$ centered at $z = 0$. Show that on C

$$\left|\frac{g(z)}{f(z)}\right| < \frac{|a_0| + |a_1| + \cdots + |a_{n-1}|}{|a_n| r}.$$

How does this inequality indicate that for sufficiently large r we have $|g(z)| < |f(z)|$ on C?

b) Use Rouché's theorem (see Exercise 8) to argue that, for C chosen with a radius as just described, the number of zeros of $h(z) = f(z) + g(z)$ inside C is identical to the number of zeros of $f(z)$ inside C. How many zeros (counting multiplicities) does $f(z)$ have?

10. Show that all the roots of $z^4 + z^3 + 1 = 0$ are inside $|z| = 3/2$.

Hint: Use Rouché's theorem (Exercise 8), taking $f(z) = z^4$, $g(z) = z^3 + 1$. Note that $|g(z)| \le 1 + |z|^3$.

11. Show that all roots of the equation in Exercise 10 are outside $|z| = 3/4$.

Hint: Same as above, but take $f(z) = z^3 + 1$, $g(z) = z^4$. Note that $|f(z)| \ge 1 - |z|^3$.

12. Use Rouché's theorem to show that $3z^2 - e^z = 0$ has two solutions inside $|z| = 1$.

Hint: Take $f(z) = 3z^2$.

13. Use Rouché's theorem to show that $5 \sin z - e^z = 0$ has one solution inside the square $|x| \le \pi/2$, $|y| \le \pi/2$. Explain why this root must be real.

Hint: Recall that $|\sin z| = \sqrt{\sinh^2 y + \sin^2 x}$.

7.4 THE NYQUIST STABILITY CRITERION

The Nyquist stability criterion is a special application of the principle of the argument that can often be used to ascertain whether a system characterized by a transfer function is stable. The method was first described in 1932 by Harry Nyquist (1889–1976), an engineer at Bell Telephone Laboratories, and is still widely used.

We know that if the transfer function of a system is an irreducible rational function of the form $G(s) = A(s)/B(s)$, then the system is unstable if $B(s)$ has any zeros in the right half plane (abbreviated r.h.p.) $\text{Re}\, s > 0$ or on the imaginary axis of this plane.

As in the examples of the preceding section, the Nyquist procedure involves two planes; here they are the s-plane with real and imaginary axes σ and ω and the w-plane, having axes u and v, in which values of $B(s)$ are plotted.

We first determine whether $B(s)$ has any zeros in the r.h.p. We consider $\Delta_C \arg B(s)$, where C, depicted in Fig. 7.4–1, is the closed semicircular contour of "large" radius R. The diameter of C lies on the ω-axis. R is taken large enough so that C encloses all possible zeros of $B(s)$ lying in the r.h.p. For the moment we postpone consideration of what happens if $B(s) = 0$ on the imaginary axis.

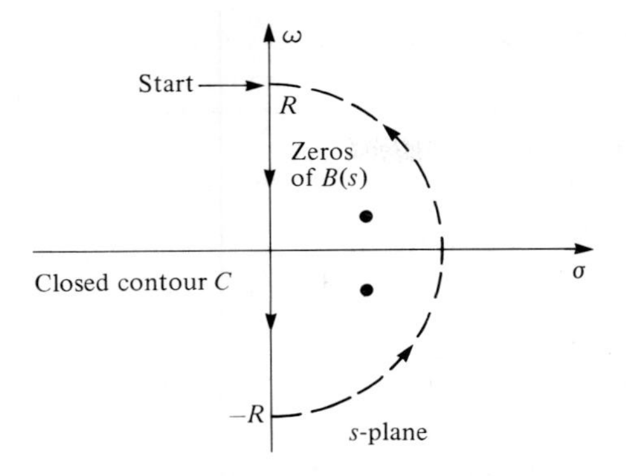

Figure 7.4–1

Now, beginning high on the ω-axis in Fig. 7.4–1 at $\omega = R$ (see "Start"), we allow ω to become less positive, shrink to zero, become increasingly negative, and stop at $\omega = -R$. While negotiating this straight line in the s-plane, we compute values of $B(s) = u(s) + iv(s)$ and plot these quantities in the w-plane (that is, the uv-plane) to trace the locus of $B(s)$. Since, typically, $B(s)$ is a polynomial with real coefficients, this locus is symmetric about the u-axis (see Example 1 of the previous section).

The next step involves our moving, in the direction of the arrow, along the dotted semicircular arc shown in Fig. 7.4–1. As s proceeds along here and returns to the point marked "start," we continue to trace the locus of $B(s)$ in the w-plane. Our task here is easy. If $B(s)$ is a polynomial of degree n,

$$B(s) = a_n s^n + a_{n-1} s^{n-1} + \cdots + a_0.$$

On the arc, $s = \text{Re}^{i\theta}$ so that

$$B(s) = a_n R^n e^{in\theta} + a_{n-1} R^{n-1} e^{i(n-1)\theta} + \cdots + a_0,$$

or

$$B(s) = a_n R^n e^{in\theta}\left[1 + \frac{a_{n-1}e^{-i\theta}}{a_n R} + \frac{a_{n-2}e^{-i2\theta}}{a_n R^2} + \cdots + \frac{a_0}{a_n R^n} e^{-in\theta}\right].$$

For arbitrarily large R the expression in the brackets can be made arbitrarily close to 1, and so

$$B(s) \approx a_n R^n e^{in\theta},$$

which means

$$|B(s)| \approx a_n R^n \quad \text{and} \quad \arg B(s) \approx n\theta.$$

Thus as s moves along the arc of radius R, $B(s)$ is closely confined in the w-plane to an arc of radius $a_n R^n$. As the argument of s changes from $-\pi/2$ to $\pi/2$ along the arc in Fig. 7.4–1, $\arg B(s)$ increases by $\approx [n\pi/2 - -n\pi/2] = n\pi$. Since we are letting $R \to \infty$, we will replace $\approx$ here by $=$.

The quantity $\Delta_C \arg B(s)$, which is the *total* increase in the argument of $B(s)$ as the closed semicircle C is traversed, is the sum of two parts: The increase in the argument of $B(s)$ as the diameter of C is negotiated, plus $n\pi$, which arises from the contribution along the curved path.

The function $B(s)$ has no singularities. Therefore Eq. (7.3–12) becomes for our contour C

$$\frac{\Delta_C \arg B(s)}{2\pi} = N, \tag{7.4–1}$$

where N is the total number of zeros of $B(s)$ in the r.h.p. Thus if $\Delta_C \arg B(s)$ is found to be nonzero, then $G(s) = A(s)/B(s)$ describes an unstable system. This determination is called the Nyquist criterion applied to polynomials, and the locus of $B(s)$ employed is called a *Nyquist diagram.* Other kinds of Nyquist criteria and diagrams are used when we deal with feedback systems. This subject is briefly treated in Exercise 8 of this section.

Regarding $\Delta_C \arg B(s)/2\pi$ in terms of encirclements, we can state the *Nyquist criterion for polynomials:*

> Suppose as s traverses the closed semicircle of Fig. 7.4–1, the locus of $w = B(s)$ makes, in total, a nonzero number of encirclements of $w = 0$ (for $R \to \infty$). Then $B(s) = 0$ has at least one root in the right half of the s-plane.

If, as s moves along the diameter of C in Fig. 7.4–1, $B(s)$ passes *through* the origin in the w-plane, then $\arg B(s)$ becomes undefined. We cannot then compute $\Delta_C \arg B(s)$. However, such an occurrence indicates that $B(s)$ has a zero on the imaginary axis of the s-plane. Since $G(s) = A(s)/B(s)$ has a pole on the imaginary axis, the system in question is unstable. To see whether it could be classified as marginally unstable, we can use a technique illustrated in Example 3.

EXAMPLE 1

Discuss the stability of the system whose transfer function is

$$G(s) = \frac{s+1}{s^3 + s^2 + 9s + 4}.$$

Solution

We must see whether $B(s) = s^3 + s^2 + 9s + 4$ has any zeros in the r.h.p. If s lies on the ω-axis, we have $s = i\omega$. Thus

$$B(s) = u + iv = -i\omega^3 - \omega^2 + 9i\omega + 4,$$

which implies

$$u = -\omega^2 + 4, \qquad u = 0, \qquad \text{when } \omega = \pm 2;$$
$$v = -\omega^3 + 9\omega, \qquad v = 0, \qquad \text{when } \omega = \pm 3 \text{ and } 0.$$

At the point "Start" in Fig. 7.4–1 ω is very large and positive. Thus u and v are large negative numbers with v (having a higher power of ω) dominating u. As ω becomes less positive, both u and v diminish in magnitude. Ultimately, when $\omega = 3$, we have $v = 0$ and u is still negative. When ω diminishes to 2, $u = 0$ while v here is positive. Finally, when $\omega = 0$, we have $v = 0$ while u is positive. The locus of $B(s)$ just described is shown in Fig. 7.4–2.

Because $B(s)$ has real coefficients, the locus generated by $B(s)$ as s moves along the negative imaginary axis is the mirror image of that just obtained for the positive axis. The result is also shown in Fig. 7.4–2.

The entire path of $B(s)$ when $s = i\omega$, $-\infty < \omega < \infty$, is shown by the solid line in Fig. 7.4–2. For $\omega \to \infty$, the argument of $B(i\omega)$ is $3\pi/2$. As ω falls to 3, arg $B(\omega)$ becomes π. At $\omega = 0$, arg $B(i\omega) = 0$, etc. When $\omega \to -\infty$, arg $B(i\omega) = -3\pi/2$. Thus the net increase of arg $B(s)$ as s moves over the imaginary axis in Fig. 7.4–1 is the final value less the initial value:

$$-\frac{3\pi}{2} - \frac{3\pi}{2} = -3\pi.$$

Now, along the large semicircular arc in Fig. 7.4–1 we have $B(s) \approx s^3$, which, as noted previously, means that the increase in argument of $B(s)$ as s moves along this arc is 3π. The *total* increase in argument of $B(s)$ as s ranges over the contour C in Fig. 7.4–1 is $\Delta_C \arg B(s) = 3\pi - 3\pi = 0$. Thus $B(s)$ has *no* roots in the r.h.p.

Since the solid curve in Fig. 7.4–2 does not pass through the origin, $B(s)$ has no zeros on the imaginary axis. The system described by $G(s)$ is stable.

The arc indicated by the broken line in Fig. 7.4–2 is the locus taken by $B(s) \approx s^3$ as s moves along the semicircular arc of Fig. 7.4–1. The net change in arg $B(s)$ over the entire path (broken and solid) in Fig. 7.4–2 is zero—a fact we have already noted. Equivalently, observe that this path makes zero net encirclements of the origin. ◀

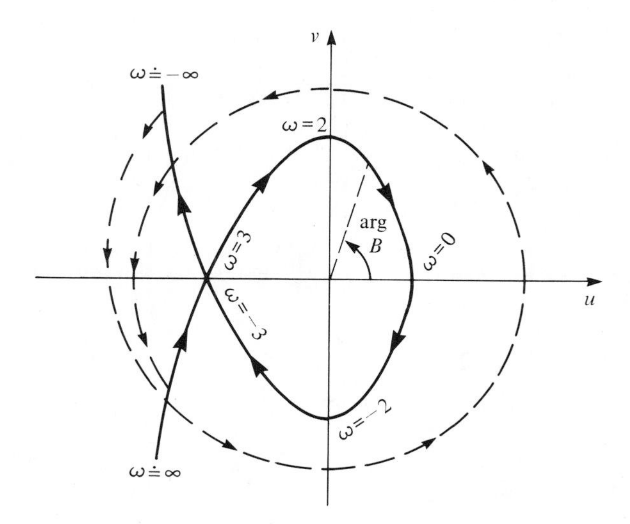

Figure 7.4–2

EXAMPLE 2

Discuss the stability of the system with transfer function

$$G(s) = \frac{s + 2}{s^3 + s^2 + 3s + 16}.$$

Solution

Letting $B(s) = s^3 + s^2 + 3s + 16$, we proceed as in Example 1. With $s = i\omega$ we have

$$B(s) = u + iv = -i\omega^3 - \omega^2 + 3i\omega + 16,$$

so that

$$u = 16 - \omega^2, \qquad u = 0, \qquad \text{when } \omega = \pm 4;$$
$$v = 3\omega - \omega^3, \qquad v = 0, \qquad \text{when } \omega = \pm\sqrt{3} \text{ and } 0.$$

A sketch in Fig. 7.4–3, indicated by the solid line, shows the locus taken by $B(s)$ as s ranges downward along the ω-axis is the s-plane. The argument of B ranges from $-\pi/2$ (for $\omega \to \infty$) to $\pi/2$ (for $\omega \to -\infty$). Thus the argument of B increases by π. Notice that this computation is unrelated to the fact that $B(s)$ executes a loop over the range $\sqrt{3} > \omega > -\sqrt{3}$. We can ignore the loop since it fails to encircle the origin of the w-plane (compare this with Example 1). As s moves along the semicircular arc in Fig. 7.4–1, the argument of $B(s)$ increases by 3π (see broken line in Fig. 7.4–3). Thus, adding contributions, $\Delta_C \arg B(s) = \pi + 3\pi = 4\pi$. According to Eq. (7.4–1), $B(s)$ has *two* zeros in the r.h.p. The system is unstable. ◀

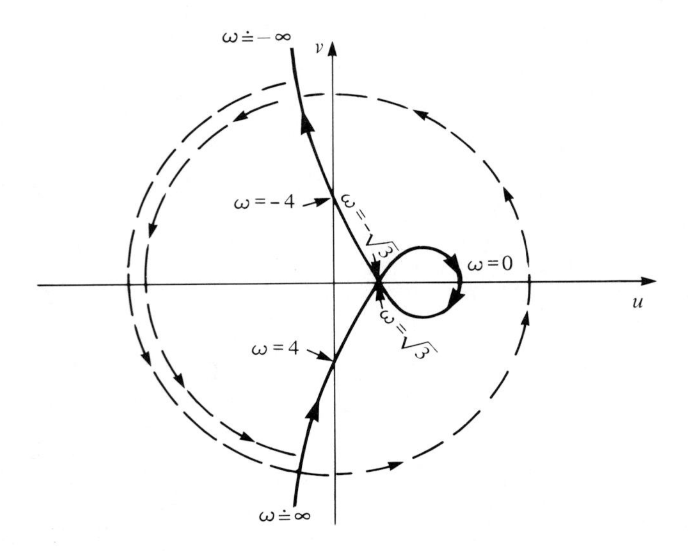

Figure 7.4–3

EXAMPLE 3

Discuss the stability of the system whose transfer function is

$$G(s) = \frac{s}{s^5 + 6s^4 + 12s^3 + 12s^2 + 11s + 6}.$$

Solution

Proceeding as before, with $B(s) = u + iv$ as the denominator of $G(s)$ and $s = i\omega$, we have on the imaginary s-axis

$$u = 6\omega^4 - 12\omega^2 + 6 = 6(\omega^2 - 1)^2,$$
$$v = \omega^5 - 12\omega^3 + 11\omega = \omega[\omega^4 - 12\omega^2 + 11] = \omega(\omega^2 - 11)(\omega^2 - 1).$$

As s moves downward along the imaginary axis, $B(s)$ traces out the path shown in Fig. 7.4–4. When $s = \pm i$ (or $\omega = \pm 1$), notice that $u = 0$ *and* $v = 0$, that is, $B(s) = 0$. The system is unstable since $B(s)$ has zeros on the imaginary axis.

Perhaps the system is marginally unstable. This would occur if both $s - i$ and $s + i$ are nonrepeated factors of $B(s)$ and, in addition, if $B(s)$ has no zeros in the r.h.p. Writing

$$B(s) = (s - i)(s + i)R(s),$$

we find $R(s)$ by long division:

$$R(s) = \frac{B(s)}{(s - i)(s + i)} = \frac{B(s)}{s^2 + 1}$$
$$= \frac{s^5 + 6s^4 + 12s^3 + 12s^2 + 11s + 6}{s^2 + 1} = s^3 + 6s^2 + 11s + 6.$$

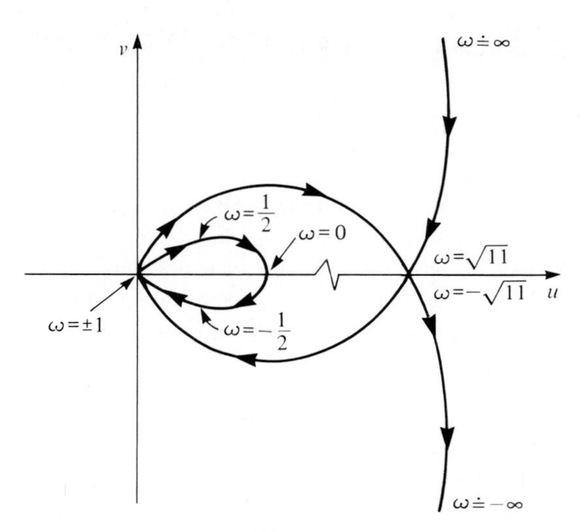

Figure 7.4–4

An application of the Nyquist method shows that $R(s)$ has no zeros on the imaginary axis or in the r.h.p. Thus $B(s)$ has zeros of order 1 on the imaginary axis and no zeros in the r.h.p. The system is marginally unstable. ◀

EXERCISES

Use the Nyquist method to investigate the stability of the systems whose transfer functions are given below. State the number of zeros of the denominator that occur in the right half plane and on the imaginary axis. State which systems are marginally unstable.

1. $\dfrac{1}{s^3 + 2s^2 + s + 4}$

2. $\dfrac{s^2 + 1}{s^4 + s^3 + 3s^2 + 2s + 1}$

3. $\dfrac{s + 3}{s^5 + 3s^4 + 5s^3 + 5s^2 + 3s + 1}$

4. $\dfrac{s^2 - 1}{s^5 + 2s^4 + 2s^3 + 3s^2 + s + 1}$

5. $\dfrac{se^{-s}}{s^5 - 2s^4 + 3s^3 + s^2 + s + 1}$

Hint: See Exercise 13, Section 7.2.

6. $\dfrac{s - 1}{s^5 + s^4 + 12s^3 + 3s^2 + 27s + 2}$

7. By making Nyquist plots, determine those values of a for which the system having transfer function $1/(s^3 + s^2 + s + a)$ is stable. Assume a is real.

8. The kinds of systems discussed in this and the previous section can be represented schematically as shown in Fig. 7.4–5. Here, $G(s)$ is the transfer function of the system, $x(t)$ and $y(t)$ the input and output, $X(s)$ and $Y(s)$ their Laplace transforms. Note that $G(s) = Y(s)/X(s)$.

A more complicated system employs the principle of *feedback*. Such systems are often used to control a physical process requiring continuous monitoring and adjustment, for example, the regulation of a furnace so as to maintain a house within a comfortable range of temperature. A block diagram of a feedback system is shown in Fig. 7.4–6. We see that an additional path, called a *feedback path* or *loop*, has been added to the original system of Fig. 7.4–5. The original system function $G(s)$ is now here called the *forward transfer*

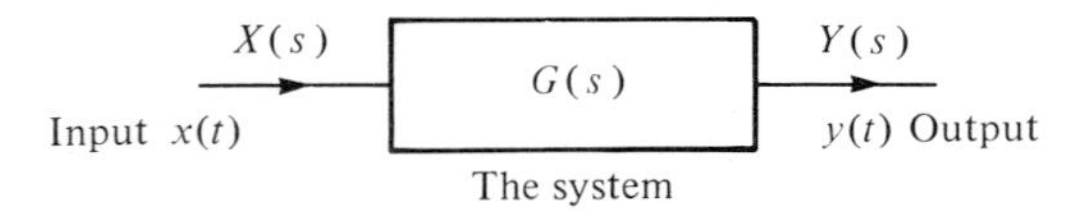

Figure 7.4–5

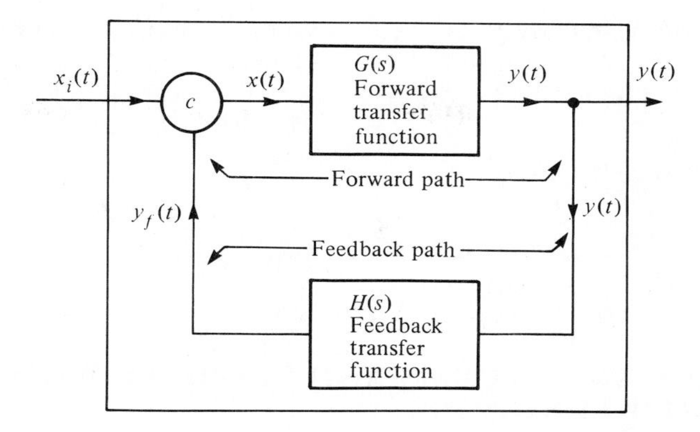

Figure 7.4–6

function. The input to the total system is $x_i(t)$ and the output is $y(t)$. The output $y(t)$ is monitored and sent down the feedback path into the system whose transfer function is $H(s)$. The output $y_f(t)$ of this subsystem is called the *feedback signal.* This feedback signal is fed into the device designated c, a comparator. The comparator provides an input signal $x(t)$ for the subsystem described by $G(s)$. Here $x(t) = x_i(t) - y_f(t)$ is the difference between the overall input signal and the feedback signal. Note that $X(s) = X_i(s) - Y_f(s)$ and $H(s) = Y_f(s)/Y(s)$ and, as before, $G(s) = Y(s)/X(s)$. The transfer function of the whole feedback system is defined as $T(s) = Y(s)/X_i(s)$.

a) Show that

$$T(s) = \frac{G(s)}{1 + G(s)H(s)}.$$

b) Typically, both $G(s)$ and $H(s)$ are ratios of polynomials in the variable s and describe stable systems. Thus the poles of $G(s)$ and $H(s)$ lie in the left half of the s-plane. A plot of the locus of $1 + G(s)H(s)$ as s negotiates the semicircle of Fig. 7.4–1 (with $R \to \infty$) can tell us whether $T(s)$ has any poles in the plane $\text{Re}\, s \geq 0$ and thus whether the feedback system is unstable. Explain why we can instead plot $w = G(s)H(s)$ as s negotiates the same semicircle; the feedback system is unstable if this locus encircles the point $w = -1$, in the positive sense, a total of one or more times or if this locus passes through $w = -1$. Otherwise the system is stable. This is a form of the Nyquist test for feedback systems.

c) Let

$$G(s)H(s) = \frac{-8}{(s+1)(s+4)(s+3)}.$$

Determine whether the feedback system is stable or unstable by investigating encirclements of -1 as described in part (b). Note that the locus of GH as s negotiates the *arc* in Fig. 7.4–1 degenerates to a point as $R \to \infty$.

d) Repeat part (c) but take

$$GH = \frac{-16}{(s+1)(s+4)(s+3)}.$$

e) Repeat part (c) but take

$$GH = \frac{-1/2}{(s+1)^3(s+3/4)^2}.$$

7.5 LAPLACE TRANSFORMS AND STABILITY WITH GENERALIZED FUNCTIONS

Generalized Functions

The word *function* has a precise meaning in mathematics. However, in the solution of problems in physical and engineering sciences this word is frequently applied to symbols that are manipulated like functions but which are not functions according to the mathematician's definition. For example, the *Dirac delta function*† (also called the *impulse function*), written $\delta(x)$, is such a symbol, one that the reader has probably already encountered. If a value is assigned to x, the value assumed by $\delta(x)$ is not known if x happens to be zero. And yet the behavior of $\delta(x)$ in an interval containing $x = 0$ is important, and we treat the symbol $\delta(x)$ as if we are dealing with a function that can be differentiated at $x = 0$ as well as integrated between limits containing $x = 0$.

The terms delta function and impulse function are in fact misnomers. Although the delta function and related symbols have been used for most of the twentieth century, it was not until around 1950 that the concept of *generalized function* was devised to deal with these symbols and to place them under the rubric of the word function.‡ We shall therefore apply this word in the present section to symbols that are not functions except in the generalized sense.

The delta function arises in physical problems where some quantity that exists with great intensity over a brief interval is to be approximated mathematically. Although the precise values assumed by this quantity are not of concern, the integral of this quantity must be known and finite for it to be represented by a delta function. For example, refer to Fig. 7.5–1(a), which depicts a bead of length L, diameter d, and mass m lying along a massless string, which we take to be the x-axis. The center of the bead is at $x = 0$. Let $\rho(x)$ be the density of mass per unit length everywhere along the string. Notice that $\rho(x) = 0$ for $|x| > L/2$ (there is no bead here). The detailed

---

† Named for the English physicist Paul A. M. Dirac (1902–1984), who popularized its use in quantum mechanics.

‡ Credit for this work belongs to the French mathematician Laurent Schwartz (1905–). For more information on a rigorous treatment of delta and related functions, see H. Bremermann, *Distributions, Complex Variables, and Fourier Transforms* (Reading, MA: Addison-Wesley, 1965), Part I.

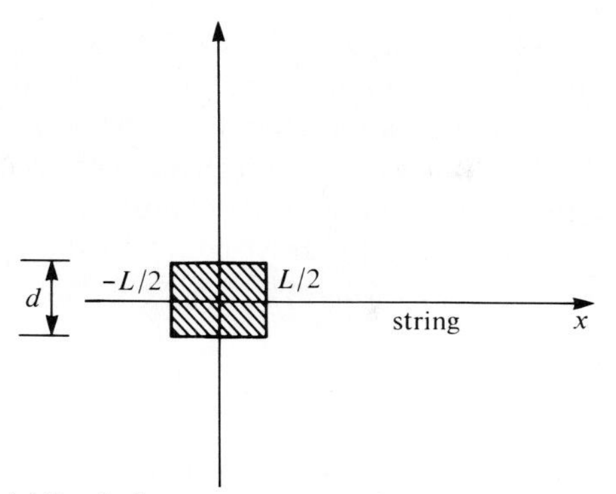

(a) Bead of mass m on a massless string

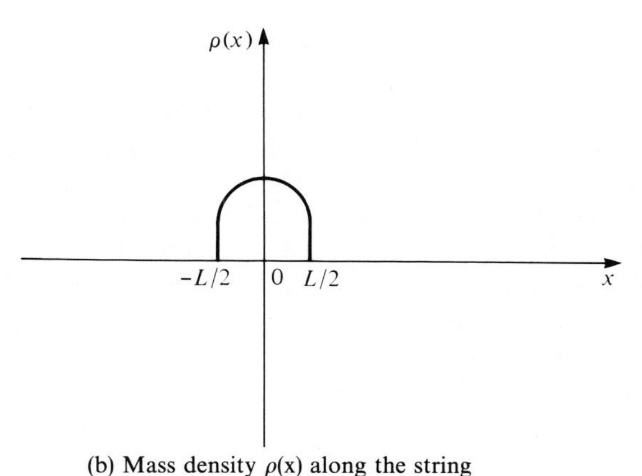

(b) Mass density ρ(x) along the string

Figure 7.5–1

behavior of $\rho(x)$ for $-L/2 < x < L/2$ is not of importance to us. A sketch of a possible distribution of $\rho(x)$ is shown in Fig. 7.5–1(b). We do know that $\int_{-\infty}^{+\infty} \rho(x)\,dx = \int_{-L/2}^{+L/2} \rho(x)\,dx = m$. Moreover the average value of $\rho(x)$ in the bead is m/L.

Now keeping the mass m and diameter d of the bead constant, we compress the bead so that $L \to 0+$. The mass density of the bead now becomes infinite. However, the bead only exists at $x = 0$, and it is here that $\rho(x) = \infty$. Otherwise, $\rho(x) = 0$. Since the mass of the bead is still m we continue to assert that

$$\int_{-\infty}^{+\infty} \rho(x)\,dx = m. \tag{7.5–1}$$

And because the bead is now of zero physical length,

$$\int_{a}^{b} \rho(x)\,dx = m, \tag{7.5–2}$$

where a and b are any two reals satisfying $a < 0 < b$. The integrals in Eqs. (7.5–1) and (7.5–2) cannot be expressed as Riemann sums because of the behavior of $\rho(x)$ at and near $x = 0$. The *theory of distributions* has been devised to describe precisely what such integrals mean. We will content ourselves with using the value of the integral and try to ignore our vagueness as to its definition.

A situation such as this one is usually described with the delta function $\delta(x)$. It possesses these properties:

$$\int_{-\infty}^{+\infty} \delta(x)\, dx = 1; \tag{7.5–3}$$

$$\delta(x) = 0, \qquad x \neq 0; \tag{7.5–4}$$

$$\int_{a}^{b} \delta(x)\, dx = 1 \qquad \text{if } a < 0 < b. \tag{7.5–5}$$

One way to visualize the delta function is as a limit of functions in the original nongeneralized sense. Each of these conventional functions is hill-shaped and encloses an area of 1 between its curve and the x-axis. The successive elements of the sequence form higher and narrower hills. An example of some elements of a possible sequence† is shown in Fig. 7.5–2, where we display $f(x) = P/[\pi(x^2 + P^2)]$ for shrinking values of the positive number P. Note that $f(0) = 1/\pi P$ and the reader should by now easily be able to show with residues that

$$\int_{-\infty}^{+\infty} \frac{P}{\pi(x^2 + P^2)}\, dx = 1,$$

i.e., the area under each curve is unity.

Integrating each function in the sequence between finite limits,

$$\int_{a}^{b} \frac{P}{\pi(P^2 + x^2)}\, dx, \qquad \text{where } a < 0 < b,$$

we obtain a result that is less than 1. However, the integral will tend to 1 as $P \to 0+$ because of the narrowing of the peak in the integrand at $x = 0$.

Multiplying the function $\delta(x)$ by a constant m has the same effect as multiplying the functions in an approximating sequence, such as Fig. 7.5–2, by the constant m; i.e., the area enclosed is no longer 1 but m. Thus

$$\int_{-\infty}^{+\infty} m\delta(x)\, dx = m \int_{-\infty}^{+\infty} \delta(x)\, dx = m, \tag{7.5–6}$$

† The reader should not think that the sequence of functions used here is the only permissible one for such a discussion. Rectangular pulses, exponentials, and almost an unlimited supply of other functions are available and used.

and it is apparent that $\rho = m\delta(x)$ describes the mass density along the string in Fig. 7.5–1(a) when $L \to 0+$.

Thus when we multiply generalized functions by constants the effect is the same as when we multiply ordinary functions by constants. Similarly, generalized functions can be added and subtracted (but not multiplied and divided) just as we do with ordinary functions.

Suppose the delta function is integrated between limits, one of which we regard as a variable. We have

$$\int_{-\infty}^{x} \delta(x')\,dx' = 0, \qquad \text{if } x < 0, \tag{7.5–7a}$$

while

$$\int_{-\infty}^{x} \delta(x')\,dx' = 1, \qquad \text{if } x > 0. \tag{7.5–7b}$$

The first result comes about because $\delta(x') = 0$, $x' < 0$, while the second is just Eq. (7.5–5) with different variables. Recalling the definition of the unit step function (Section 2.2), we have, from the preceding two equations,

$$\int_{-\infty}^{x} \delta(x')\,dx = u(x), \qquad x \neq 0. \tag{7.5–8}$$

Applying formally the fundamental theorem of integral calculus to this result, we obtain

$$\delta(x) = \frac{du}{dx}; \tag{7.5–9}$$

i.e., *the delta function is the derivative of the unit step function.*

The graph of the ordinary function $f(x - x_0)$ is a translation of the graph of the function $f(x)$ by x_0 units to the right along the x-axis. Similarly $\delta(x - x_0)$ represents a translation of $\delta(x)$. The function $\delta(x - x_0)$ becomes infinite when $x = x_0$. In addition

$$\delta(x - x_0) = 0, \qquad x \neq x_0;$$

$$\int_{-\infty}^{+\infty} \delta(x - x_0)\,dx = 1;$$

$$\int_{a}^{b} \delta(x - x_0)\,dx = 1, \qquad a < x_0 < b;$$

$$\delta(x - x_0) = \frac{du(x - x_0)}{dx}.$$

If $f(x)$ is an ordinary function (i.e., a function in the strict sense) and is continuous at $x = 0$, we assert that

$$f(x)\delta(x) = f(0)\delta(x). \tag{7.5–10}$$

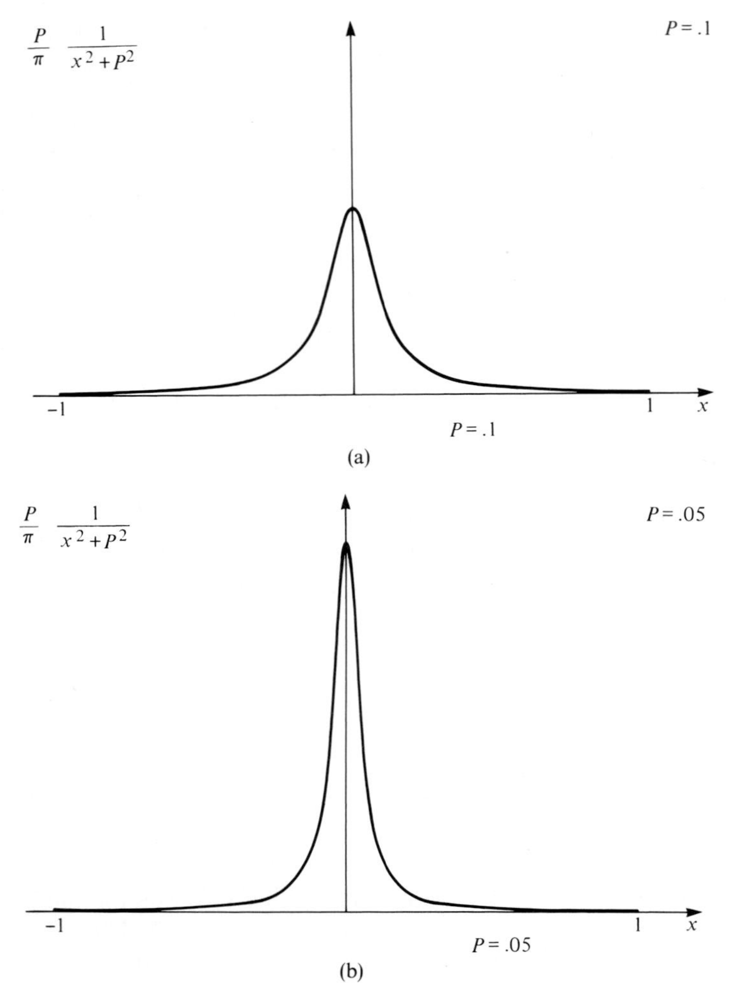

Figure 7.5–2
A sequence to approximate the delta function

The justification is this: If any member of the sequence of hill-shaped functions used in approximating $\delta(x)$ (see, e.g., Fig. 7.5–2) is multiplied by $f(x)$, the resulting curve is scaled upward by a factor of approximately $f(0)$ and the area under this new curve is now no longer unity but is approximately $f(0)$. The approximation improves as we use narrower and higher members of the sequence. In the limit as we use successively better approximations to $\delta(x)$, the area obtained under the curve is exactly $f(0)$. Moreover, the resulting function is zero for $x \neq 0$. Thus we can say

$$\int_a^b f(x)\delta(x)\,dx = f(0), \qquad a < 0 < b, \tag{7.5–11}$$

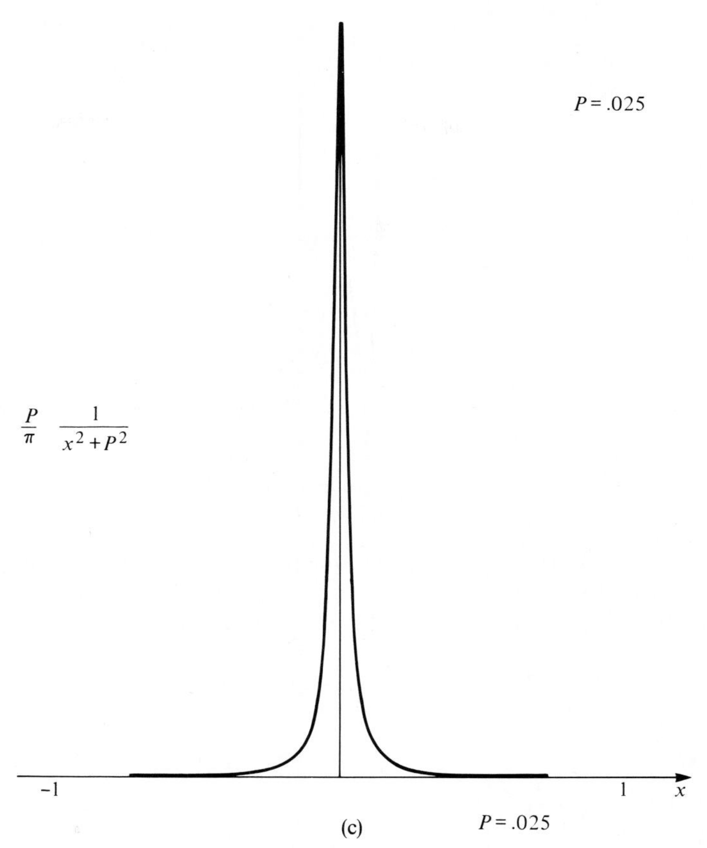

Figure 7.5–2 cont.
A sequence to approximate the delta function

and since

$$\int_a^b f(0)\delta(x)\,dx = f(0), \qquad a < 0 < b,$$

we conclude that $f(x)\delta(x) = f(0)\delta(x)$. By a similar argument we find that

$$f(x)\delta(x - x_0) = f(x_0)\delta(x - x_0) \tag{7.5–12}$$

and

$$\int_a^b f(x)\delta(x - x_0)\,dx = f(x_0), \qquad a < x_0 < b. \tag{7.5–13}$$

By taking first, second, and higher derivatives of the delta function we obtain other generalized functions. What is the meaning of $\delta'(x)$? Intuitively we should feel that it is the limit of a sequence of functions obtained by taking the derivatives of a sequence of functions like that shown in Fig. 7.5–2. One such member of the sequence

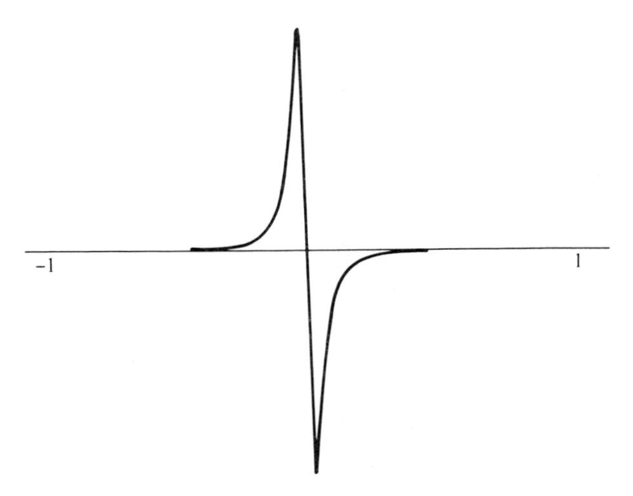

Figure 7.5–3
Derivative of curve in Fig. 7.5–2(b)

is shown in Fig. 7.5–3. In the limit of such sequences we obtain a hill of infinite height at $x = 0-$, followed by a corresponding valley of infinite depth at $x = 0+$. Now let $f(x)$ be a function of x whose first derivative is continuous at $x = 0$. Thus

$$\frac{d}{dx}[f(x)\delta(x)] = f'(x)\delta(x) + f(x)\delta'(x),$$

so that

$$\int_a^b \frac{d}{dx}[f(x)\delta(x)]\,dx = \int_a^b (f'(x)\delta(x) + f(x)\delta'(x))\,dx, \qquad a < 0 < b.$$

The left side of the preceding equation can be evaluated. Thus

$$\int_a^b \frac{d}{dx}[f(x)\delta(x)]\,dx = f(x)\delta(x)\Big|_a^b = f(b)\delta(b) - f(a)\delta(a) = 0.$$

(Recall that $\delta(x) = 0$, $x \neq 0$.) Now we expand the derivative on the above left and use the preceding result to obtain

$$\int_a^b f'(x)\delta(x)\,dx + \int_a^b f(x)\delta'(x)\,dx = 0.$$

Thus

$$-\int_a^b f'(x)\delta(x)\,dx = \int_a^b f(x)\delta'(x)\,dx.$$

The left side of the above can be found (replace $f(x)$ with $f'(x)$ in Eq. 7.5–11). We get

$$-f'(0) = \int_a^b f(x)\delta'(x)\,dx. \tag{7.5–14}$$

This result can be generalized through repetition of the procedure used. Thus

$$(-1)^n f^{(n)}(0) = \int_a^b f(x)\delta^{(n)}(x)\,dx, \qquad a < 0 < b, \tag{7.5–15}$$

provided $f^{(n)}(x)$ is continuous at $x = 0$. Here the superscript (n) refers to the nth derivative.

A result equivalent to Eq. (7.5–14) is

$$f(x)\delta'(x) = -f'(0)\delta(x),$$

which can be verified by integrating both sides of the preceding along the interval $a \le x \le b$. Similarly

$$f(x)\delta^{(n)}(x) = (-1)^n f^{(n)}(0)\delta(x).$$

The function $\delta'(x - x_0)$ is $\delta'(x)$ translated to $x = x_0$. We have

$$-f'(x_0) = \int_a^b f(x)\delta'(x - x_0)\,dx, \qquad a < x_0 < b, \tag{7.5–16}$$

and

$$(-1)^n f^{(n)}(x_0) = \int_a^b f(x)\delta^{(n)}(x - x_0)\,dx, \qquad a < x_0 < b. \tag{7.5–17}$$

To summarize, we will obtain the value of $f(x)$ at $x = x_0$ if we integrate $\delta(x - x_0)f(x)$ through any interval containing x_0, while $\delta'(x - x_0)f(x)$ integrated through the same interval yields $-f'(x_0)$, etc. Integration through other intervals yields zero.

EXAMPLE 1

Evaluate these integrals:

a) $\displaystyle\int_{-\infty}^{+\infty} \cos x\, \delta(x)\,dx$

b) $\displaystyle\int_0^{\infty} \cos x\, \delta(x - 1)\,dx$

c) $\displaystyle\int_0^{\infty} \cos x\, \delta(x + 1)\,dx$

d) $\displaystyle\int_{-\infty}^{+\infty} \cos x\, \delta(x + 1)\,dx$

e) $\displaystyle\int_0^{\infty} \cos(x + 1)\, \delta'(x - 2)\,dx$

Solution

a) $\displaystyle\int_{-\infty}^{+\infty} \cos x\, \delta(x)\, dx = \cos x\Big|_{x=0} = 1$, from Eq. (7.5–11).

b) $\displaystyle\int_{0}^{\infty} \cos x\, \delta(x-1)\, dx = \cos x\Big|_{x=1} = \cos 1$, from Eq. (7.5–13).

c) $\displaystyle\int_{0}^{\infty} \cos x\, \delta(x+1)\, dx = 0$, since $\delta(x+1) = 0$ for all x between the limits of integration.

d) $\displaystyle\int_{-\infty}^{+\infty} \cos x\, \delta(x+1)\, dx = \cos x\Big|_{x=-1} = \cos(-1)$, from Eq. (7.5–13).

e) $\displaystyle\int_{0}^{\infty} \cos(x+1)\, \delta'(x-2)\, dx = -(d/dx)[\cos(x+1)]\Big|_{x=2} = \sin(x+1)\Big|_{x=2} = \sin 3$, from Eq. (7.5–16). ◀

Laplace Transforms of Generalized Functions

Many problems in engineering and the physical sciences are solved through the approximation of some physically realizable quantity by an idealization—a delta function or one of its derivatives. For example, a brief strong current pulse, varying with time t and centered about $t = 0$, is often described by $q\delta(t)$, where q is the area (integral) of the original pulse. The charge delivered by the pulse is q. Laplace transforms are often used in the solution of problems involving these idealized physical quantities.

Because the value of $\delta(0)$ is unspecified (although the integral of $\delta(t)$ is established), we must change our original definition of the Laplace transform to the following:

$$F(s) = \int_{0-}^{\infty} f(t)e^{-st}\, dt = \lim_{\varepsilon\to 0-} \int_{\varepsilon}^{\infty} f(t)e^{-st}\, dt. \tag{7.5–18}$$

The lower limit of integration shrinks to zero through *negative* values. Thus if $f(t) = \delta(t)$, the interval where $\delta(t) \neq 0$ is included within the limits of integration. Hence by Eq. (7.5–11),

$$\mathscr{L}\delta(t) = \int_{0-}^{\infty} \delta(t)e^{-st}\, dt = e^{-st}\Big|_{t=0} = 1, \tag{7.5–19}$$

and so

$$\mathscr{L}^{-1}1 = \delta(t). \tag{7.5–20}$$

Our changing the lower limit from $0+$ (Section 7.1) to $0-$ will have no effect on our computation of Laplace transformations of functions that are continuous at $t = 0$ or have jump discontinuities at $t = 0$. The same function $F(s)$ is obtained in both cases. However,

$$\int_{0-}^{\infty} \delta(t)e^{-st}\,dt = 1 \neq \int_{0+}^{\infty} \delta(t)e^{-st}\,dt = 0;$$

i.e., the choice of limit can make a difference when we deal with generalized functions.
In this section the lower limit of integration in the Laplace transformation will be understood to be 0−, and we will drop the minus sign from next to the zero. Note that

$$\mathscr{L}\delta(t-t_0) = \int_{0}^{\infty} \delta(t-t_0)e^{-st}\,dt = e^{-st_0}, \qquad t_0 \geq 0, \tag{7.5–21}$$

and

$$\mathscr{L}^{-1}e^{-st_0} = \delta(t-t_0), \qquad t_0 \geq 0. \tag{7.5–22}$$

With the aid of Eq. (7.5–16), we have

$$\mathscr{L}\delta'(t-t_0) = -\frac{d}{dt}\, e^{-st}\bigg|_{t_0} = se^{-st_0}, \tag{7.5–23}$$

so that

$$\mathscr{L}^{-1}se^{-st_0} = \delta'(t-t_0). \tag{7.5–24}$$

In general (see Eq. 7.5–17),

$$\mathscr{L}\delta^{(n)}(t-t_0) = s^n e^{-st_0}, \qquad t_0 \geq 0. \tag{7.5–25}$$

Note the special cases:

$$\mathscr{L}\delta'(t) = s; \tag{7.5–26}$$

$$\mathscr{L}^{-1}s = \delta'(t); \tag{7.5–27}$$

$$\mathscr{L}\delta^{(2)}(t) = s^2; \tag{7.5–28}$$

$$\mathscr{L}^{-1}s^2 = \delta^{(2)}(t); \tag{7.5–29}$$

$$\mathscr{L}\delta^{(n)}(t) = s^n; \tag{7.5–30}$$

$$\mathscr{L}^{-1}s^n = \delta^{(n)}(t). \tag{7.5–31}$$

The preceding results for the transformation of generalized functions can be useful in the evaluation of inverse transformations that could not be treated by Theorem 2, as the following example demonstrates.

EXAMPLE 2

Find

$$f(t) = \mathscr{L}^{-1}\frac{s^2+s+1}{s^2+1}.$$

Solution

Theorem 2 is not applicable to $(s^2 + s + 1)/(s^2 + 1)$ because it fails to satisfy $|F(s)| \le m/|s|^k$ $(k > 0)$ throughout some half plane. Note that

$$F(s) = 1 + \frac{s}{s^2 + 1}.$$

Now $f(t) = \mathscr{L}^{-1}F(s) = \mathscr{L}^{-1}1 + \mathscr{L}^{-1}(s/(s^2 + 1))$. From Eq. (7.5–20), $\mathscr{L}^{-1}1 = \delta(t)$. We find from Theorem 3 that

$$\mathscr{L}^{-1}\frac{s}{s^2 + 1} = \sum_{\text{res}} \frac{s}{s^2 + 1} e^{st} = \cos t.$$

Thus $f(t) = \delta(t) + \cos t$. As a check, $\int_0^\infty \delta(t)e^{-st}\,dt + \int_0^\infty \cos t\, e^{-st}\,dt = 1 + s/(s^2 + 1)$, $\operatorname{Re} s > 0$, as required. ◀

Comment: More complicated rational expressions of the form $F(s) = P(s)/Q(s)$, where $P(s)$ and $Q(s)$ are any polynomials in s, can also be inverted. For the transformations performed in Section 7.1, we required that the degree of Q exceed that of P. This is no longer necessary. If the degree of Q is less than or equal to the degree of P we first perform a long division and obtain a result of this form:

$$\frac{P(s)}{Q(s)} = a_0 + a_1 s + a_2 s^2 + \cdots + a_n s^n + \frac{p(s)}{q(s)},$$

where $a_0, a_1, \ldots$, etc., are constants, and $p(s)$ and $q(s)$ are polynomials in s such that the degree of q exceeds that of p. The inverse transform of each term in the polynomial $a_0 + a_1 s + \cdots + a_n s^n$ can be found from Eq. (7.5–31), while $p(s)/q(s)$ can be inverted with Theorem 3.

Use of our present definition of the Laplace transform involves knowing the limits of functions to be transformed as they pass to zero through negative values of their argument. For a function $f(t)$ this requires our having some knowledge of its behavior for $t < 0$. It is convenient to assume in situations involving generalized functions and Laplace transforms that all functions used and sought are zero for $t < 0$.

There is a reason that we can do this. Functions that vanish for $t < 0$ are said to be *causal*. Most of the useful systems in engineering are said to be nonanticipatory or causal systems. They produce no response until there is an excitation. If we restrict ourselves to causal excitations for such systems (no excitation until $t = 0$) the output (response) will vanish for $t < 0$ and thus will also be causal.

With our assumption that functions vanish for $t < 0$ we can rewrite Eqs. (7.1–8) and (7.1–9). Taking the lower limit of integration as $t = 0-$ (instead of $t = 0+$) in the derivations of these equations, we have

$$\mathscr{L}\frac{df}{dt} = sF(s), \tag{7.5–32}$$

and in general,

$$\mathscr{L}\frac{d^n f}{dt^n} = s^n F(s). \tag{7.5–33}$$

Comparing our results with Eq. (7.5–30), we see that the preceding equations are valid for both ordinary and generalized functions. Note that Eq. (7.1–10), $\mathscr{L}\int_0^t f(x)\,dx = (1/s)\mathscr{L}f(t)$ still holds for ordinary and generalized functions. We now interpret the lower limit of integration as $0-$.

EXAMPLE 3

Use Eq. (7.5–32) to obtain the Laplace transform of the delta function $\delta(t)$ from the Laplace transform of the unit step function $u(t)$.

Solution

Let $F(s) = \mathscr{L}u(t) = \int_0^\infty e^{-st}\,dt = 1/s$. Now recall that $du/dt = \delta(t)$. Thus, from Eq. (7.5–32), $\mathscr{L}\delta(t) = s/s = 1$. This agrees with Eq. (7.5–19). ◀

EXAMPLE 4

Let $f(t) = e^{-(t-1)}u(t-1)$.

a) Find $\mathscr{L}\,df/dt$ from Eq. (7.5–32).

b) Check your result by first finding df/dt and then taking its Laplace transform.

Solution

Part (a): A sketch of $f(t)$ is shown in Fig. 7.5–4(a). Because of the jump discontinuity at $t = 1$ we expect that df/dt will contain a delta function $\delta(t-1)$. Now

$$\mathscr{L}f(t) = \int_0^\infty e^{-(t-1)}u(t-1)e^{-st}\,dt = \int_1^\infty e^{-(t-1)}e^{-st}\,dt = \frac{e^{-s}}{s+1} = F(s).$$

Applying Eq. (7.5–32), we have

$$\mathscr{L}\frac{df}{dt} = \frac{se^{-s}}{s+1} = \frac{(s+1)e^{-s}}{s+1} - \frac{e^{-s}}{s+1} = e^{-s} - \frac{e^{-s}}{s+1}.$$

Part (b): Check on solution: We have, differentiating a product in the usual way,

$$\begin{aligned}\frac{df}{dt} &= \frac{d}{dt}[e^{-(t-1)}]u(t-1) + e^{-(t-1)}\frac{d}{dt}u(t-1)\\ &= -e^{-(t-1)}u(t-1) + e^{-(t-1)}\delta(t-1).\end{aligned}$$

With the aid of Eq. (7.5–12) we can simplify the term $e^{-(t-1)}\delta(t-1)$. Thus

$$\frac{df}{dt} = -e^{-(t-1)}u(t-1) + \delta(t-1).$$

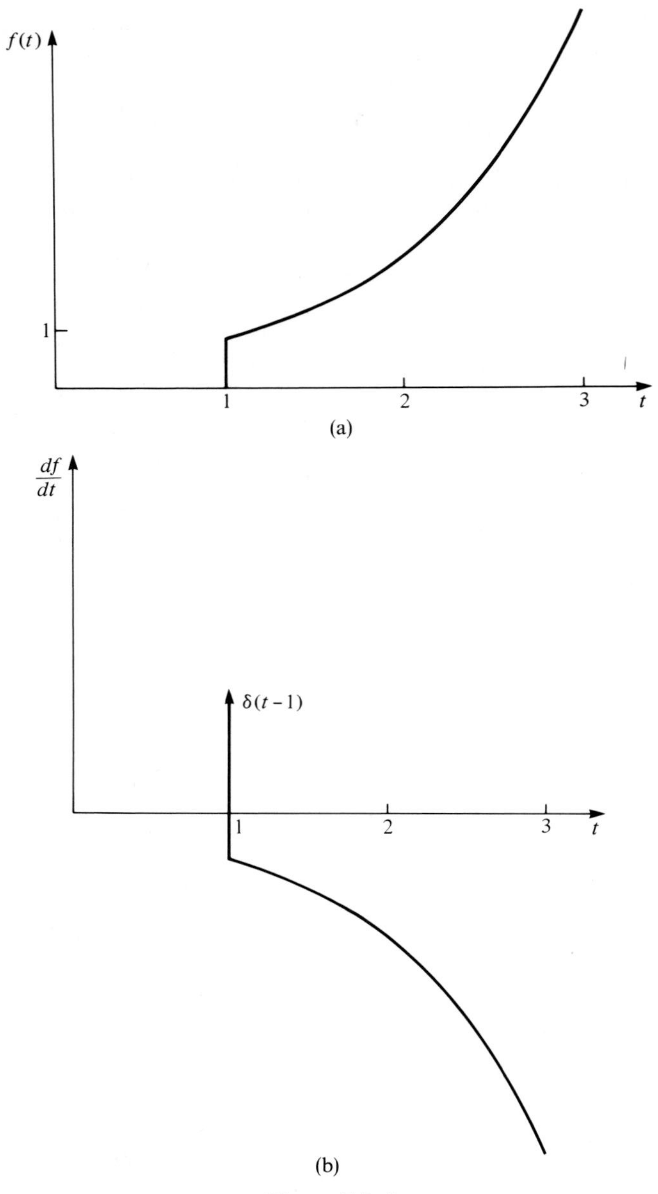

Figure 7.5–4

This function is sketched in Fig. 7.5–4(b). As expected, there is an impulse at $t = 1$. The Laplace transform of df/dt is easily obtained:

$$\mathscr{L}\frac{df}{dt} = -\mathscr{L}e^{-(t-1)}u(t-1) + \mathscr{L}\delta(t-1).$$

The first transformation on the right was actually performed in part (a), and from Eq. (7.5–21) we obtain $\mathscr{L}\delta(t-1) = e^{-s}$. Thus

$$\mathscr{L}\frac{df}{dt} = \frac{-e^{-s}}{s+1} + e^{-s},$$

which agrees with the result of part (a). ◀

The following problem illustrates the utility of both the delta function and the Laplace transformation in the solution of a physical problem.

EXAMPLE 5

A voltage pulse $v(t)$ is applied to the series resistor-capacitor circuit shown in Fig. 7.5–5. The pulse is high and narrow and is thus approximated by $\delta(t)$. The response of the circuit is $\imath(t)$. We assume that $\imath(t) = 0$ for $t < 0$. Applying Kirchhoff's voltage law to the circuit, we obtain

$$\frac{1}{C}\int_0^t \imath(t')\,dt' + \imath(t)R = \delta(t), \qquad \text{for } t \geq 0.$$

Find $\imath(t)$ for $t \geq 0$.

Solution

Using Eqs. (7.1–10) and (7.5–19), we take the Laplace transformation of both sides of the preceding equation and obtain

$$\frac{I(s)}{sC} + I(s)R = 1,$$

where $I(s) = \mathscr{L}\imath(t)$. Thus

$$I(s) = \frac{1}{\dfrac{1}{sC} + R} = \frac{sC}{1 + RsC}.$$

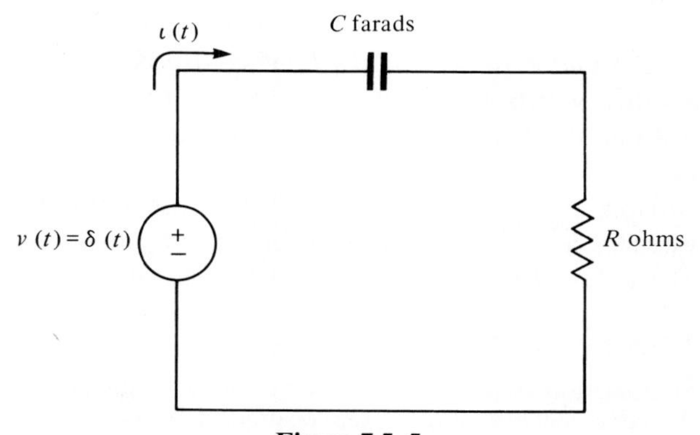

Figure 7.5–5

We note that $I(s)$ is a rational expression and the degree of the numerator and denominator are equal. We rewrite $I(s)$ as follows:

$$I(s) = \frac{sC}{1 + RsC} = \frac{1}{R} - \frac{1/R}{1 + RsC}.$$

Now

$$\imath(t) = \frac{1}{R}\mathscr{L}^{-1}1 - \frac{1}{R}\mathscr{L}^{-1}\frac{1}{1 + RsC}.$$

We know that $\mathscr{L}^{-1}1 = \delta(t)$, and we can find $\mathscr{L}^{-1}(1/(1 + RsC))$ from Theorem 3 because we are taking the inverse transform of a rational function in which the degree of the denominator exceeds that of the numerator. Thus

$$\mathscr{L}^{-1}\frac{1}{1 + RsC} = \text{Res}\left[\frac{e^{st}}{1 + RsC}, -\frac{1}{RC}\right] = \frac{e^{-t/RC}}{RC}.$$

Finally we have

$$\imath(t) = \frac{\delta(t)}{R} - \frac{e^{-t/RC}}{R^2C} \qquad \text{for } t \geq 0.$$

Since the response is zero for $t < 0$, we have

$$\imath(t) = \frac{\delta(t)}{R} - \frac{e^{-t/RC}}{R^2C}u(t),$$

valid for $-\infty < t < \infty$.

Transfer Functions with Generalized Functions

In Section 7.2 we defined $G(s)$, the transfer function of a system, as being that function which when multiplied by $X(s)$, the Laplace transform of the input, will yield $Y(s)$, the Laplace transform of the output. Thus $Y(s) = G(s)X(s)$.

Now if the input $x(t)$ is a delta function $\delta(t)$, we have $X(s) = 1$ and so $Y(s) = G(s)$. Thus *the transfer function of a system is the Laplace transform of the output when the input is* $\delta(t)$. In other words *the transfer function is the Laplace transform of the impulse response*. As an illustration, in Example 5, the transfer function of the system is $sC/(1 + RsC)$.

In our discussion of stability in Section 7.2 we observed that if the transfer function of a system is a rational expression in s, then the system will be stable if and only if all the poles of the function are to the left of the imaginary axis.[†] We

[†] Although that discussion assumed that the degree of the denominator exceeds that of the numerator, the same conclusion about the location of the poles can be extended to any rational expression with the aid of generalized functions.

now see that the same must be true of the poles of the Laplace transform of the impulse response. Referring to the treatment of bounded functions of time in Section 7.2, we realize that the locations of these poles dictate that the impulse response is a function of t that decays to zero as $t \to \infty$. In summary, for any system having transfer function $G(s)$, the impulse response is $\mathscr{L}^{-1}G(s) = g(t)$, and for a stable system, $\lim_{t\to\infty} g(t) = 0$. Example 5 illustrates a stable electrical system since the impulse response decays exponentially with time. A mechanical example of such behavior might be an ordinary bell, which when given a brief strong blow with a hammer (the impulse input) exhibits a ringing whose amplitude diminishes with time.

In the case of marginally unstable systems, $G(s)$, the Laplace transform of the impulse response, will have simple poles on the imaginary axis and no poles to the right of this axis. Then $g(t)$ will contain terms of the form $\cos bt$ and $\sin bt$ (where b is real and nonzero) corresponding to poles of $G(s)$ that lie on the imaginary axis at $s = \pm ib$. If $G(s)$ has a pole at $s = 0$, then $g(t)$ contains a constant term, independent of t. In any case, for marginally unstable systems $\lim_{t\to\infty} g(t) \neq 0$ and $\lim_{t\to\infty} |g(t)| \neq \infty$; i.e., the impulse response is bounded but does not decay to zero as t tends to infinity. A mechanical illustration of such behavior is a hypothetical ideal bell, which when struck, rings forever. An electrical example is given in Exercise 23.

EXERCISES

Evaluate the following integrals.

1. $\displaystyle\int_{-\infty}^{+\infty} \delta(x)\cos(x-1)\,dx$

2. $\displaystyle\int_{-\infty}^{+\infty} \delta(x)\sin x\,dx$

3. $\displaystyle\int_{-\infty}^{+\infty} \delta'(x)\sin x\,dx$

4. $\displaystyle\int_{-\infty}^{+\infty} \delta(x)\left[\frac{1}{x^2+1} + \tan(x+1)\right]dx$

5. $\displaystyle\int_{0}^{\infty} \delta(x+3)\cos x\,dx$

6. $\displaystyle\int_{-\infty}^{1} \delta(x+3)\cos x\,dx$

7. $\displaystyle\int_{0}^{10} \delta^{(2)}(x-1)e^{2x}\,dx$

8. $\displaystyle\int_{-3}^{+3} \delta'(x-1)[\cos(x+1)+\sin(x-1)]\,dx$

Find the Laplace transforms of the following functions $f(t)$.

9. $\delta(t) + \delta'(t-1)$

10. $\delta(t-1)\cos 3t$

11. $\delta(t-1) + u(t-2)$

12. $\delta(t-1)u(t-2)$

13. $\delta(t-2)u(t-1)$

14. $\sum_{n=0}^{4} \delta(t-n\tau),\ \tau > 0$

15. $\sum_{n=0}^{\infty} \delta(t-n\tau),\ \tau > 0$. Give your answer in closed form; i.e., sum an infinite series of terms in s. Assume $\text{Re}\, s > 0$.

With the aid of generalized functions and Theorem 3 in Section 7.1, find the inverse transforms of the following functions.

16. $\dfrac{s}{s+1}$ **17.** $\dfrac{s^2+2s+1}{s-1}$ **18.** $\dfrac{s^2+s}{s^2+s+2}$

19. $\dfrac{s^3}{s^2+s+1}$ **20.** $\dfrac{s^2e^{-2s}}{s^2+s+1}$ (See Exercise 15, Section 7.1.)

21. The differential equation satisfied by the current $\imath(t)$ in Fig. 7.5–6 is

$$L\frac{d\imath}{dt} + \imath R = v(t).$$

Here the input is the voltage $v(t)$, while the output is $\imath(t)$. If $v(t) = \delta(t)$, find $\imath(t)$ (the impulse response) by using Laplace transforms. State the transfer function of the circuit.

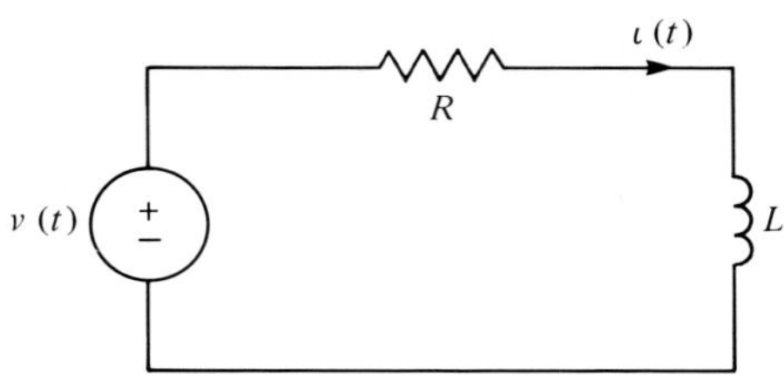

Figure 7.5–6

22. The integral equation satisfied by the voltage $v(t)$ in the circuit of Fig. 7.5–7 is

$$\frac{v(t)}{R} + \frac{1}{L}\int_0^t v(t')\,dt' = \imath(t).$$

Here the input is the current $\imath(t)$, while the output is $v(t)$. If $\imath(t) = \delta(t)$, find $v(t)$ (the impulse response) by using Laplace transforms. State the transfer function of the circuit.

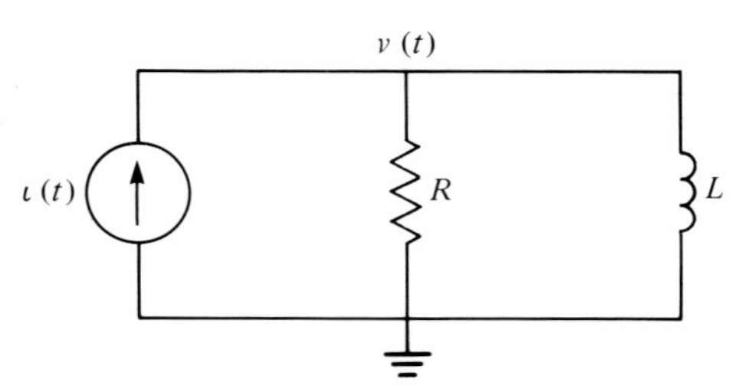

Figure 7.5–7

23. The integrodifferential equation satisfied by the current $\imath(t)$ in the circuit of Fig. 7.5–8 is

$$L\frac{d\imath}{dt} + \frac{1}{C}\int_0^t \imath(t')\,dt' = v(t), \qquad \text{where } L \text{ and } C > 0.$$

If $v(t) = \delta(t)$, find $\imath(t)$ by using Laplace transforms. Use your result to explain why the system is marginally unstable.

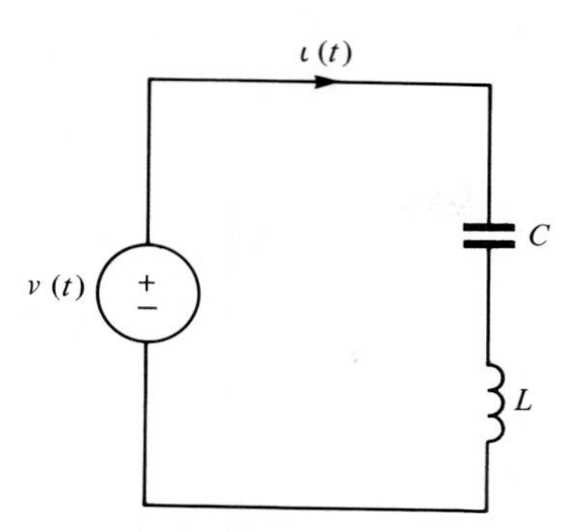

Figure 7.5–8

24. In Exercise 31 of Section 7.1, we showed how the inverse Laplace transform of the product of two functions $F(s)$ and $G(s)$ is given by

$$\mathscr{L}^{-1}[F(s)G(s)] = \int_0^t f(t-\tau)g(\tau)\,d\tau.$$

The right side of the preceding equation is the convolution of $f(t)$ with $g(t)$. Here $f(t) = \mathscr{L}^{-1}F(s)$, and $g(t) = \mathscr{L}^{-1}G(s)$.

a) Prove that if a system has impulse response $g(t)$, then when the input is $x(t)$ the output $y(t)$ is given by

$$y(t) = \int_0^t x(t-\tau)g(\tau)\,d\tau = \int_0^\infty x(t-\tau)u(t-\tau)g(\tau)\,d\tau.$$

b) Using the preceding result and the impulse response derived in Example 5, find the output of the system in that example when the input is $e^{-\alpha t}u(t)$. Here α is real, and $\alpha \neq 1/RC$.

CHAPTER 8

Conformal Mapping and Some of Its Applications

8.1 INTRODUCTION

When we first began our discussion of functions of a complex variable in Section 2.1, we learned that a functional relationship $w = f(z)$ cannot be studied by the conventional graphing procedure of elementary algebra. Instead, two planes were used, the z-plane (with axes x and y) and the w-plane (with axes u and v). We found that $w = f(z)$ sets up a correspondence between points in the z-plane and points in the w-plane. Corresponding points in the two planes are called *images* of each other.

We can also take the view that $w = f(z)$ *maps* or *transforms* points from the z-plane into points in the w-plane. Sometimes we will superimpose the w-plane on top of the z-plane so that their axes and origins coincide. We can imagine that the vector representing a point, say A, in the z-plane has been rotated, stretched (or some combination of the two) by $w = f(z)$ in order to create the vector for A' (the image of A). A typical case is shown in Fig. 8.1–1, where we see that counterclockwise rotation *and* stretching are required to obtain the vector representing A' from that for A.

If we use $w = f(z)$ to map all the points lying in a domain D_1 of the z-plane, they may form a domain D_2 in the w-plane. If this is the case, we say that D_1 is mapped *onto* D_2 by the transformation $w = f(z)$ (see Fig. 8.1–2). (Similarly, we can also speak of one region being mapped onto another.) If a curve C_1 is constructed in D_1 and all the points on this curve are mapped into the w-plane, we typically find that the image points form a curve C_2. Thus one curve is transformed into another.

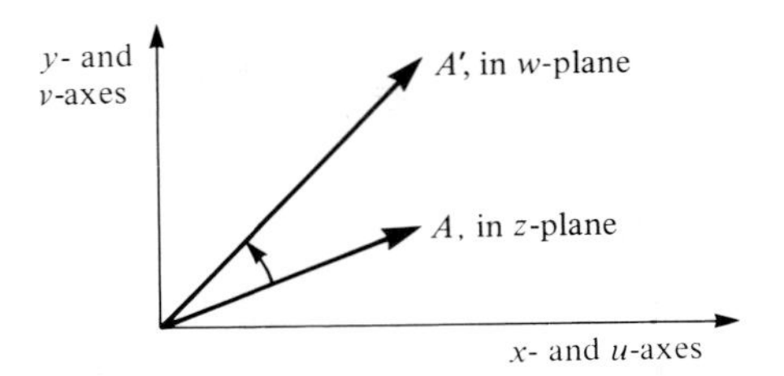

Figure 8.1–1

In Section 7.4 we had some experience in transforming large semicircular curves from the s-plane into the w-plane by way of $w = B(s)$.

In this chapter we will study, in some detail, how points, domains, and especially curves are mapped from the z-plane into the w-plane when the transformation $w = f(z)$ is an analytic function of z. Later, we will take a real function $\phi(x, y)$ and by a change of variables convert $\phi(x, y)$ to $\phi(u, v)$ defined in the w-plane. If $\phi(x, y)$ is a harmonic function in the z-plane, which implies

$$\frac{\partial^2 \phi}{\partial x^2} + \frac{\partial^2 \phi}{\partial y^2} = 0, \tag{8.1–1}$$

and if

$$w = u(x, y) + iv(x, y) = f(z),$$

which defines the change of variables is analytic, we can show that

$$\frac{\partial^2 \phi}{\partial u^2} + \frac{\partial^2 \phi}{\partial v^2} = 0, \tag{8.1–2}$$

where $\phi(u, v)$ is harmonic in the w-plane. We will find that the preservation of the harmonic property when ϕ is "transferred" from one plane to another, together with a knowledge of how contours are transformed from one plane to another by $w = f(z)$ will enable us to solve a greater variety of physical problems in electrostatics, fluid flow, etc. than those treated in Section 4.7.

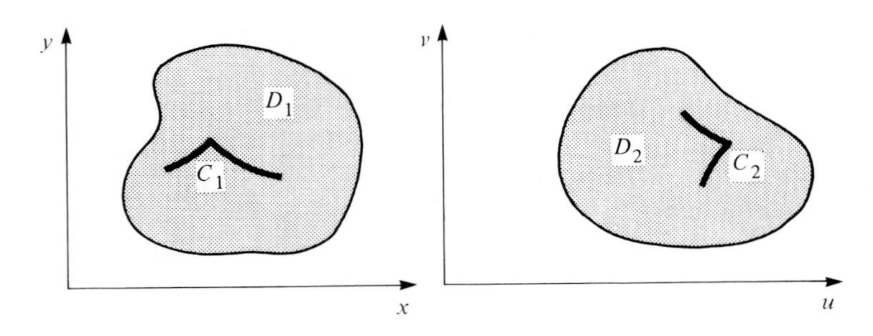

Figure 8.1–2

8.2 THE CONFORMAL PROPERTY

To see how curves can be transformed by an analytic function, let us consider the specific case

$$w = \text{Log}\, z \tag{8.2–1}$$

applied to the arc $|z| = 1$, $\pi/6 \leq \arg z \leq \pi/4$ and also to the line segment $\arg z = \pi/6$, $1 \leq |z| \leq 2$. Both the arc and the line are shown in Fig. 8.2–1(a). Each point on the arc is described by $e^{i\theta}$, where $\pi/6 \leq \theta \leq \pi/4$ and the corresponding image point is $\text{Log}\, e^{i\theta} = i\theta$. As θ advances from $\pi/6$ to $\pi/4$ along the circle, w traces out the vertical line segment $A'B'$ shown in Fig. 8.2–1(b).

Under the transformation in Eq. (8.2–1) each point on the line $\arg z = \pi/6$, $1 \leq |z| \leq 2$ has an image

$$w = \text{Log}|z| + i \arg z = \text{Log}|z| + i\,\frac{\pi}{6}.$$

As $|z|$ advances from 1 to 2, the locus of w is the horizontal line $A'D'$ shown in Fig. 8.2–1(b). The two original curves in Fig. 8.2–1(a) intersect at point A with a 90° angle (that is, their tangents, at A, have this angle of intersection). In Fig. 8.2–1(b) the image curves intersect at A' with a 90° angle. Moreover, the *sense* of the angle of intersection is preserved, that is, the tangents to the two curves at their intersection in Fig. 8.2–1(a) have an angular displacement from each other that is in the same direction as the corresponding tangents in Fig. 8.2–1(b). The preservation of both the magnitude and sense of the angle of intersection of these curves under this transformation is not an accident and will occur extensively throughout this chapter. The following definition will be useful in our discussion.

DEFINITION Conformal mapping

A mapping $w = f(z)$ that preserves the size and sense of the angle of intersection between any two curves intersecting at z_0 is said to be *conformal* at z_0. A mapping that is conformal at every point in a domain D is called conformal in D. □

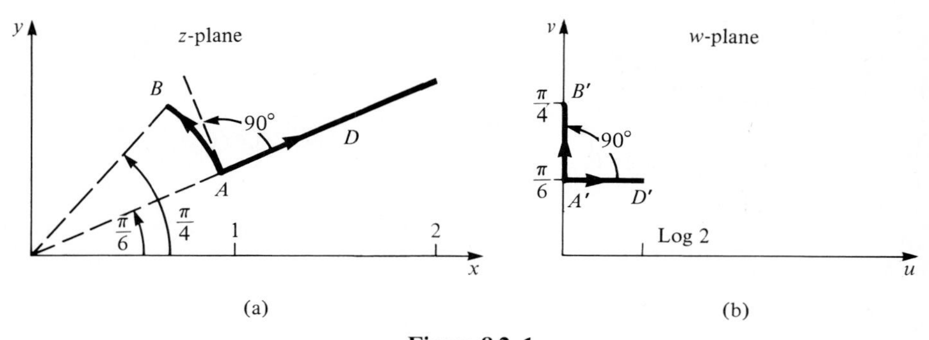

Figure 8.2–1

Occasionally one speaks of *isogonal mappings.* In this case the magnitudes of angles of intersection are preserved but not necessarily their sense.

In a moment we will be able to show why $w = \text{Log}\, z$ is conformal at the point A and also decide when functions $f(z)$ are conformal in general. The following theorem will be proved and used:

THEOREM 1 Condition for conformal mapping

Let $f(z)$ be analytic in a domain D. Then $f(z)$ is conformal at every point in D where $f'(z) \neq 0$. □

The proof requires our considering a curve C that is a smooth arc in the z-plane. The curve is generated by a parameter t, which we might think of as time. Thus

$$z(t) = x(t) + iy(t)$$

traces out the curve C as t increases (see Fig. 8.2–2a). We assume $x(t)$ and $y(t)$ to be differentiable functions of t. The curve C can be transformed into an image curve C' (see Fig. 8.2–2b) by means of the analytic function

$$w = f(z) = u(x, y) + iv(x, y).$$

The arrows on C and C' indicate the sense in which these contours are generated as t increases. At any point on C or C' we can define a directed tangent. This is a vector that is tangent to the curve and points in the direction in which the curve is being generated.

At time t_0 we are at $z(t_0) = z_0$ on C, and at the later time $t_0 + \Delta t$ we are at $z(t_0 + \Delta t) = z_0 + \Delta z$. The vector Δz connecting $z(t_0)$ with $z(t_0 + \Delta t)$ is shown in Fig. 8.2–2(a).

Now refer to Fig. 8.2–2(b). The point z_0 is mapped into the image $w_0 = f(z_0)$ on C' and $z_0 + \Delta z$ has the image point $f(z_0 + \Delta z) = w_0 + \Delta w$ on C'. If $\Delta t \to 0$, then

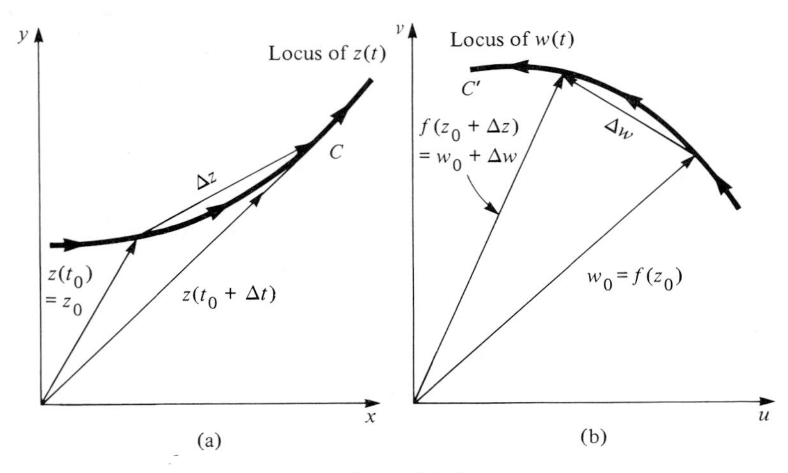

Figure 8.2–2

Δz, and consequently Δw, both shrink to zero. In Fig. 8.2–2(a) we see that, as the vector Δz shortens, its direction approaches that of the directed tangent to the curve C at z_0. Similarly, as the vector Δw shortens, its direction approaches the tangent to the image curve C' at w_0. Since Δt is real, both the vectors $\Delta z/\Delta t$ and $\Delta w/\Delta t$ have the same direction as the vectors Δz and Δw, respectively. Thus $\lim_{\Delta t \to 0} \Delta z/\Delta t = dz/dt$ and $\lim_{\Delta t \to 0} \Delta w/\Delta t = dw/dt$ are tangent to C and C' at z_0 and w_0, respectively. Note that

$$\left.\frac{dz}{dt}\right|_{z_0} = \left.\frac{dx}{dt}\right|_{z_0} + i\left.\frac{dy}{dt}\right|_{z_0}$$

and that the slope of this vector is $(dy/dx)|_{z_0}$, the slope of the curve C at z_0. Similarly dv/du_{w_0} is the slope of the curve C' at w_0.

From the chain rule for differentiation,

$$\frac{dw}{dt} = \frac{dw}{dz}\frac{dz}{dt} = f'(z)\,\frac{dz}{dt}.$$

Setting $t = t_0$ so that $z = z_0$ and $w = w_0$ in the preceding we have

$$\left.\frac{dw}{dt}\right|_{w_0} = f'(z_0)\left.\frac{dz}{dt}\right|_{z_0}.$$

Equating the arguments of each side of the above we obtain

$$\arg \left.\frac{dw}{dt}\right|_{w_0} = \arg f'(z_0) + \arg \left.\frac{dz}{dt}\right|_{z_0}. \tag{8.2–2}$$

Let

$$\phi = \arg \left.\frac{dw}{dt}\right|_{w_0}, \qquad \alpha = \arg f'(z_0), \qquad \theta = \arg \left.\frac{dz}{dt}\right|_{z_0}.$$

Thus Eq. (8.2–2) becomes

$$\phi = \alpha + \theta. \tag{8.2–3}$$

We should recall that θ and ϕ specify the directions of the tangents to the curves C and C' at z_0 and w_0, respectively (see Fig. 8.2–3). Using Eq. (8.2–3), we realize that under the mapping $w = f(z)$ the directed tangent to the curve C, at z_0, is rotated through an angle $\alpha = \arg f'(z_0)$. The rotation of the tangent is shown in Fig. 8.2–3(b).

Another smooth arc, say C_1, intersecting C at the point z_0 with angle ψ (the angle between the tangents to the curves) can be mapped by $w = f(z)$ into the image curve C'_1. The tangent to C_1 at z_0 is also rotated through the angle $f'(z_0) = \alpha$ by the mapping.

The mapping $w = f(z)$ rotates the tangents to C and C_1 by identical amounts in the same direction. Thus the image curves C' and C'_1 have the same angle of intersection ψ as do C and C_1. The sense (direction) of the intersection is also preserved, as shown in Fig. 8.2–4.

If $f'(z_0) = 0$, the preceding discussion will break down since the angle $\alpha = \arg(f'(z_0))$, through which tangents are rotated, is undefined. There is no

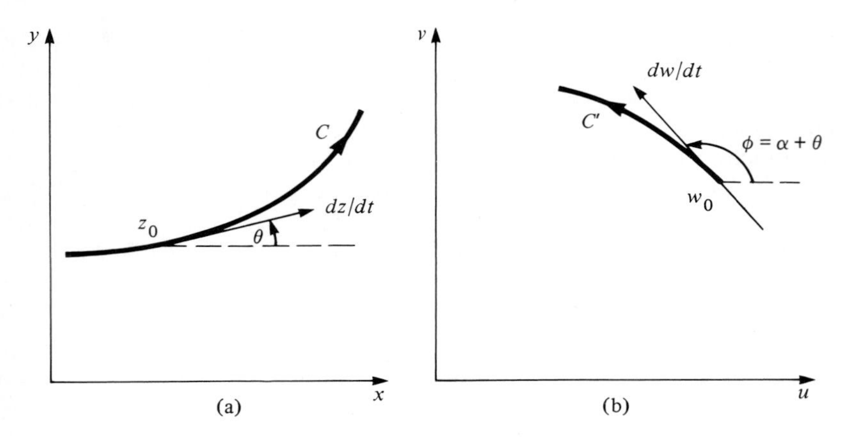

Figure 8.2–3

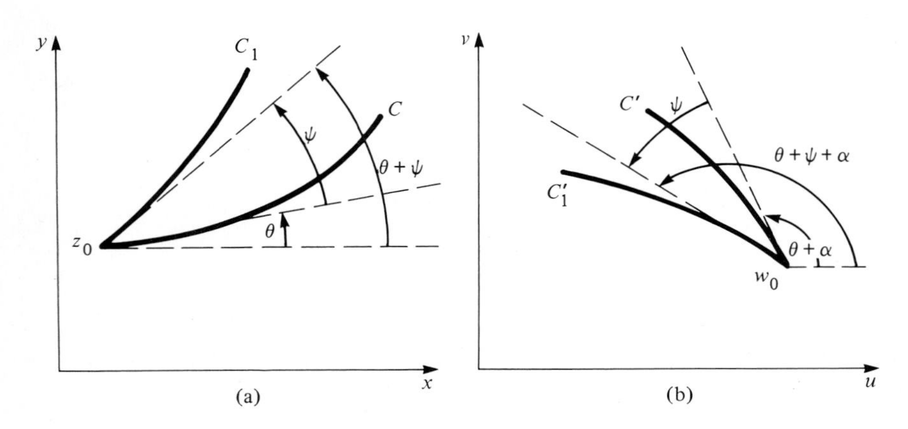

Figure 8.2–4

guarantee of a conformal mapping where $f'(z_0) = 0$. One can show that if $f'(z_0) = 0$ the mapping cannot be conformal at z_0. A value of z for which $f'(z) = 0$ is known as a *critical point* of the transformation.

EXAMPLE 1

Consider the contour C defined by $x = y$, $x > 0$ and the contour C_1 defined by $x = 1$, $y \geq 1$. Map these two curves using $w = 1/z$ and verify that their angle of intersection is preserved in size and direction.

Solution

Our transformation is

$$w = \frac{1}{z} = u + iv = \frac{1}{x + iy} = \frac{x}{x^2 + y^2} - \frac{iy}{x^2 + y^2},$$

so that

$$u = \frac{x}{x^2 + y^2}, \tag{8.2–4}$$

$$v = \frac{-y}{x^2 + y^2}. \tag{8.2–5}$$

On C, $y = x$, which, when substituted in Eqs. (8.2–4) and (8.2–5), yields

$$u = \frac{1}{2x} = -v. \tag{8.2–6}$$

Since $x > 0$, we have $u \geq 0$ and $v \leq 0$. The line defined by Eq. (8.2–6) is shown as C' in Fig. 8.2–5(b). As we move outward from the origin along C, the corresponding image point moves toward the origin on C' since, according to Eq. (8.2–6), both u and v tend to zero with increasing x.

On C_1, $x = 1$ which, when used in Eqs. (8.2–4) and (8.2–5), yields

$$u = \frac{1}{1 + y^2}, \tag{8.2–7}$$

$$v = -\frac{y}{1 + y^2}. \tag{8.2–8}$$

This implies that

$$v = -uy. \tag{8.2–9}$$

From Eq. (8.2–7) we easily obtain $y = \sqrt{1/u - 1}$ which, combined with Eq. (8.2–9), yields $v = -\sqrt{u - u^2}$. We can square both sides of this equation and make some algebraic rearrangements to show that $(u - 1/2)^2 + v^2 = (1/2)^2$.

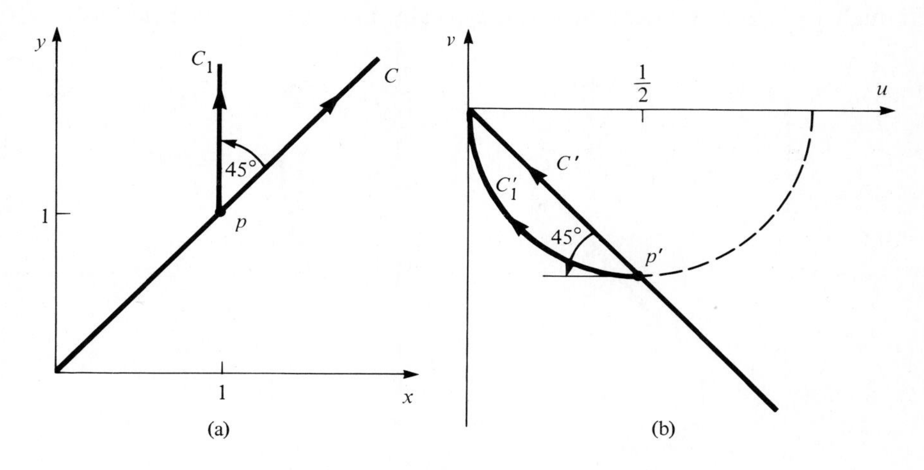

Figure 8.2–5

Thus points on C_1 have their images on a circle of radius 1/2, centered at (1/2, 0) in the w-plane. As y increases from 1 to ∞ along C_1, then, according to Eq. (8.2–7), the u-coordinate of the image point varies from 1/2 to 0 along the circle. Since v remains negative (see Eq. 8.2–8), the image of C_1 is the arc C_1' shown in Fig. 8.2–5(b).

From plane geometry we recall that the angle between a tangent and a chord of a circle is 1/2 the angle of the intercepted arc. Thus the angle of intersection between C_1' and C' in Fig. 8.2–5(b) is 45°, the same angle existing between C_1 and C. Observe in Fig. 8.2–5(a) and (b) that the sense of the angular displacement between the tangents to C and C_1 is the same as for C' and C_1'. ◀

Suppose a small line segment, not necessarily straight, connecting the points z_0 and $z_0 + \Delta z$ is mapped by means of the analytic transformation $w = f(z)$ (see Fig. 8.2–6). The image line segment connects the point $w_0 = f(z_0)$ with the point $w = f(z_0 + \Delta z)$.

Now consider

$$|f'(z_0)| = \lim_{\Delta z \to 0} \left| \frac{f(z_0 + \Delta z) - f(z_0)}{\Delta z} \right|. \tag{8.2–10}$$

Equation (8.2–10) follows from the definition of the derivative, Eq. (2.3–3), and this easily proved fact: $|\lim_{z \to z_0} g(z)| = \lim_{z \to z_0} |g(z)|$ when $\lim_{z \to z_0} g(z)$ exists. The expression $|(f(z_0 + \Delta z) - f(z_0))/\Delta z|$ is the approximate ratio of the lengths of the line segments in Fig. 8.2–6(b) and (a). Thus a small line segment starting at z_0 is magnified in length by approximately $|f'(z_0)|$ under the transformation $w = f(z)$. As the length of this segment approaches zero, the amount of magnification tends to the limit $|f'(z_0)|$.

We see that if $f'(z_0) \neq 0$, all small line segments passing through z_0 are approximately magnified under the mapping by the same nonzero factor $R = |f'(z_0)|$. A "small" figure composed of line segments and constructed near z_0 will, when mapped into the w-plane, have each of its sides approximately magnified by the same factor $|f'(z_0)|$. The shape of the new figure will conform to the shape of the old one although its size and orientation will typically have been altered. Because of the

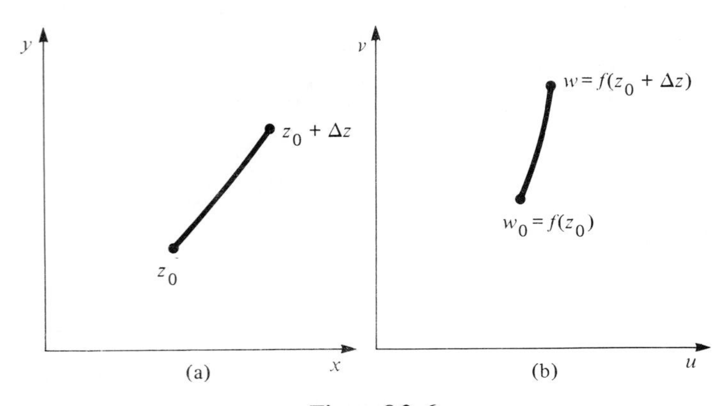

Figure 8.2–6

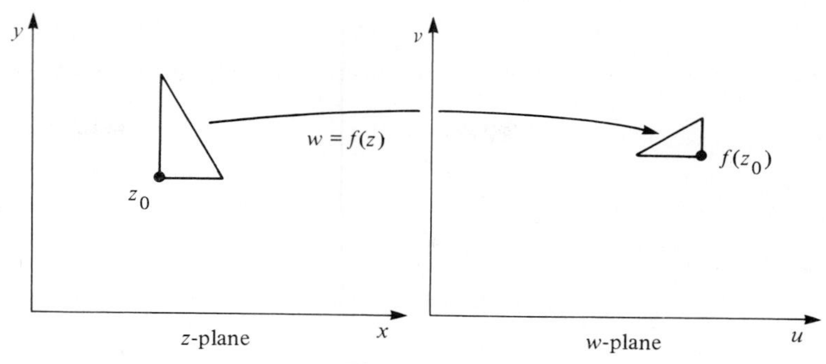

Figure 8.2–7

magnification in lengths, the image figure in the w-plane will have an area approximately $|f'(z_0)|^2$ times as large as that of the original figure. The conformal mapping of a small figure is shown in Fig. 8.2–7. The similarity in shapes and the magnification of areas need not hold if we map a "large" figure since $f'(z)$ may deviate significantly from $f'(z_0)$ over the figure.

EXAMPLE 2

Discuss the way in which $w = z^2$ maps the grid $x = x_1, x = x_2, \ldots; y = y_1, y = y_2, \ldots$ (see Fig. 8.2–8a) into the w-plane. Verify that the angles of intersection are preserved and that a small rectangle is approximately preserved in shape under the transformation.

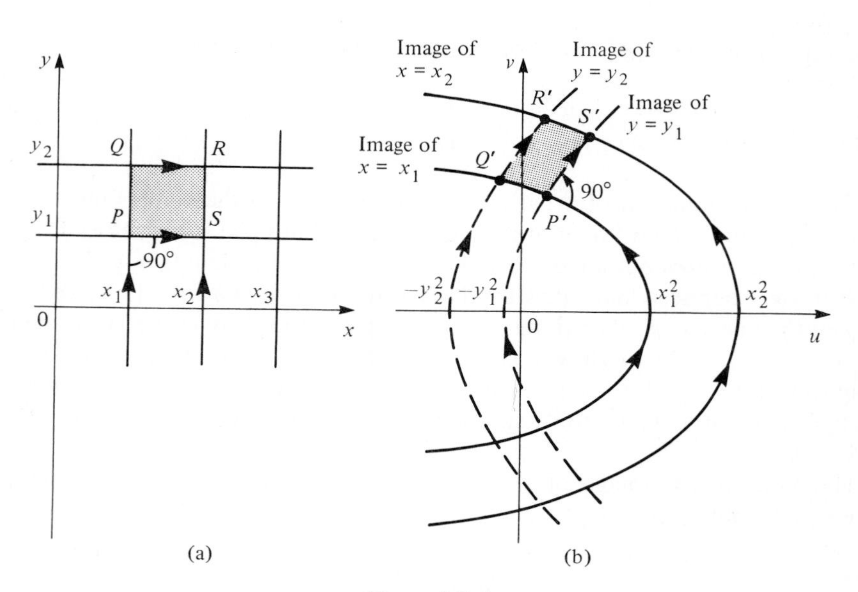

Figure 8.2–8

Solution

With $w = u + iv$, $z = x + iy$, the transformation is $u + iv = (x + iy)^2 = x^2 - y^2 + i2xy$, so that

$$u = x^2 - y^2, \tag{8.2–11}$$

$$v = 2xy. \tag{8.2–12}$$

On the line $x = x_1$, $-\infty \leq y \leq \infty$ we have

$$u = x_1^2 - y^2, \tag{8.2–13}$$

$$v = 2x_1 y. \tag{8.2–14}$$

We can use Eq. (8.12–14) to eliminate y from Eq. (8.2–13) with the result that

$$u = x_1^2 - \frac{v^2}{4x_1^2}. \tag{8.2–15}$$

As the y-coordinate of a point on $x = x_1$ increases from $-\infty$ to ∞, Eq. (8.2–14) indicates that v progresses from $-\infty$ to ∞ (if $x_1 > 0$). A parabola described by Eq. (8.2–15) is generated. This curve, which passes through $u = x_1^2$, $v = 0$, is shown by the solid line in Fig. 8.2–8. This parabola is the image of $x = x_1$. Also illustrated is the image of $x = x_2$, where $x_2 > x_1$.

Mapping a horizontal line $y = y_1$, $-\infty \leq x \leq \infty$, we have from Eqs. (8.2–11) and (8.2–12) that

$$u = x^2 - y_1^2, \tag{8.2–16}$$

$$v = 2xy_1. \tag{8.2–17}$$

Using Eq. (8.2–17) to eliminate x from Eq. (8.2–16), we have

$$u = \frac{v^2}{4y_1^2} - y_1^2. \tag{8.2–18}$$

This is also the equation of a parabola—one opening to the right. One can easily show that, as the x-coordinate of a point moving along $y = y_1$ increases from $-\infty$ to ∞, its image traces out a parabola shown by the broken line in Fig. 8.2–8(b). The direction of progress is indicated by the arrow. Also shown in Fig. 8.2–8(b) is the image of the line $y = y_2$. The point P at (x_1, y_1) is mapped by $w = z^2$ into the image $u_1 = x_1^2 - y_1^2$, $v_1 = 2x_1 y_1$ shown as P' in Fig. 8.2–8(b). P' lies at the intersection of the images of $x = x_1$ and $y = y_1$. Although these curves have two intersections, only the upper one corresponds to P' since Eq. (8.2–12) indicates that $v > 0$ when $x > 0$ and $y > 0$.

The slope of the image of $x = x_1$ is found from Eq. (8.2–15). Differentiating implicitly, we have

$$du = -\frac{2v\,dv}{4x_1^2},$$

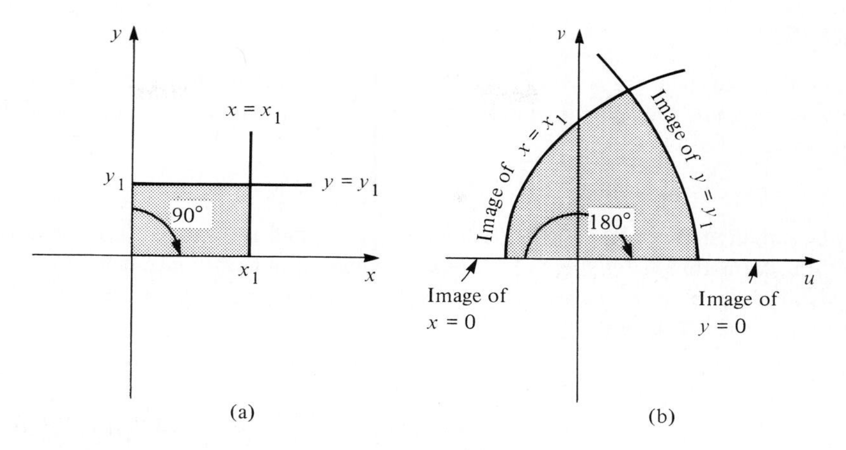

Figure 8.2–9

or

$$\frac{du}{dv} = \frac{-v}{2x_1^2}. \tag{8.2–19}$$

Similarly, from Eq. (8.2–18), the slope of the image of $y = y_1$ is

$$\frac{du}{dv} = \frac{v}{2y_1^2}. \tag{8.2–20}$$

Substituting $v_1 = 2x_1y_1$, which is valid at the point of intersection, into Eqs. (8.2–19) and (8.2–20), we find that the respective slopes are $-y_1/x_1$ and x_1/y_1. As these values are negative reciprocals of each other, we have established the orthogonality of the intersection of the two parabolas at P'. Since $x = x_1$ intersects $y = y_1$ at a right angle, the transformation has preserved the angle of intersection. Notice that the rectangular region with corners at P, Q, R, and S shown shaded in Fig. 8.2–8(a) is mapped onto the nearly rectangular region having corners at P', Q', R', and S' shown shaded in Fig. 8.2–8(b).

With $f(z) = z^2$, we have $f'(z) = 0$ at $z = 0$. Our theorem on conformal mapping no longer guarantees a conformal transformation at $z = 0$. Lines intersecting here require special attention. The vertical line $x = 0$, $-\infty < y < \infty$ is transformed (see Eqs. 8.2–11 and 8.2–12) into $u = -y^2$, $v = 0$, the negative real axis. The horizontal line $y = 0$, $-\infty < x < \infty$ is, by the same equations, mapped into $u = x^2$, $v = 0$, the positive u-axis. The lines $x = 0$ and $y = 0$, which intersect at the origin at 90°, have images in the uv-plane intersecting at 180° (see Fig. 8.2–9b). Notice that the small rectangle $0 \leq x \leq x_1$, $0 \leq y \leq y_1$ in Fig. 8.2–9(a) is mapped onto the nonrectangular shape in Fig. 8.2–9(b). The breakdown of the conformal property is again evident. ◀

EXERCISES

1. Show that the mapping $w = (\bar{z})^2$ preserves the magnitude of the angle of intersection between two line segments intersecting at any $z \neq 0$. Explain why the mapping is not conformal.

Hint: First consider $w = z^2$.

2. Two semiinfinite lines, (see Fig. 8.2–10) $y = ax$, $x \geq 0$ and $y = bx$, $x \geq 0$, are mapped by the transformation $w = u + iv = z^2$. Find the equation of each image curve in the form $v = g(u)$. If the two given lines intersect at angle α as shown, what is the angle of intersection of their images? Take $b > a \geq 0$.

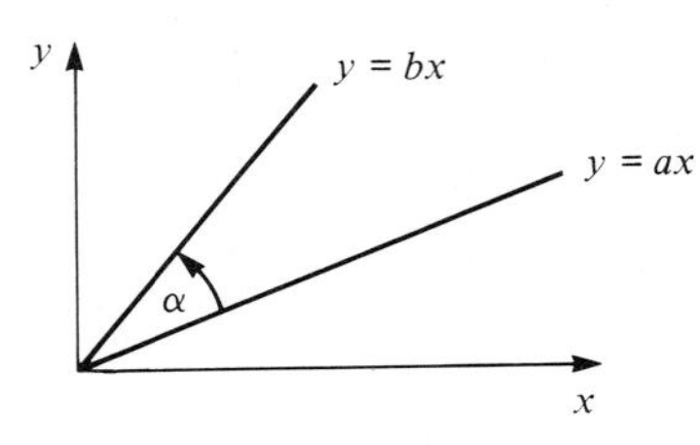

Figure 8.2–10

3. a) Consider the semiinfinite lines $y = 1 - x$, $x \geq 0$, and $y = 1 + x$, $x \geq 0$. Where in the x, y-plane do these lines intersect, and what is their angle of intersection?

b) Each semiinfinite line is mapped into the w- (or u, v-) plane by the transformation $w = z^2$. Find the equation and sketch the image in the w-plane of each line under this transformation. Give the equations in the form $v = f(u)$.

c) Using the equation of each image in the w-plane, find their point of intersection and prove, using these equations, that the angle of intersection of these image curves is the same as that found for the lines in part (a).

What are the critical points of the following transformations?

4. $w = z + 1/z$ **5.** $w = \sin z$ **6.** $w = e^{\cos z}$

7. $w = \dfrac{z - 1}{z + 1}$ **8.** $w = \text{Log}(ez) - 2z$

9. a) What is the image of the semicircular arc, $|z| = 1$, $0 \leq \arg z \leq \pi$, under the transformation $w = z + 1/z$?

Hint: Put $z = e^{i\theta}$.

b) What is the image of the line $y = 0$, $x \geq 1$ under this same transformation?

c) Do the image curves found in parts (a) and (b) have the same angle of intersection in the w-plane as do the original curves in the xy-plane? Explain.

10. Show that under the mapping $w = 1/z$ the image in the w-plane of the infinite line $\text{Im } z = 1$ is a circle. What is its center and radius?

11. Find the equation in the w-plane of the image of $x + y = 1$ under the mapping $w = 1/z$. What kind of curve is obtained?

12. Consider the straight line segment directed from (2, 2) to (2.1, 2.1) in the z-plane. The segment is mapped into the w-plane by $w = \text{Log } z$.

a) Obtain the approximate length of the image of this segment in the w-plane by using the derivative of the transformation at (2, 2).

b) Obtain the exact value of the length of this image. Use a pocket calculator to convert this to a decimal, and compare your result with part (a).

c) Use the derivative of the transformation to find the angle through which the given segment is rotated when mapped into the w-plane.

13. The square boundary of the region $1 \le x \le 1.1$, $1 \le y \le 1.1$ is transformed by means of $w = e^z$.

a) Use the derivative of the transformation at (1, 1) to obtain a numerical approximation to the area of the image of the square in the w-plane.

b) Obtain the exact value of the area of the image, and compare your result with part (a).

8.3 ONE-TO-ONE MAPPINGS AND MAPPINGS OF REGIONS

It is now necessary to study with some care the correspondence that the analytic transformation $w = f(z)$ creates between points in the z-plane and points in the w-plane. Let all the points in a region R be mapped into the w-plane so as to form an image region R'. Let z_1 be any point in R. Since $f(z)$ is single valued in R, z_1 is mapped into a unique point $w_1 = f(z_1)$. Given the point w_1, can we assert that it is the image of a unique point, that is, if $w_1 = f(z_1)$ and $w_1 = f(z_2)$, where z_1 and z_2 are points in R, does it follow that $z_1 = z_2$? The following definition is useful in dealing with this question.

DEFINITION One-to-one mapping

If the equation $f(z_1) = f(z_2)$ implies, for arbitrary points z_1 and z_2 in a region R of the z-plane, that $z_1 = z_2$, we say that the mapping of the region R provided by $w = f(z)$ is *one to one.*

When a one-to-one mapping $w = f(z)$ is used to map the points of a region R and the resulting image is a region R' in the w-plane, we say that R is mapped *one to one onto* R' by $w = f(z)$. □

A hypothetical mapping that fails to be one to one is shown in Fig. 8.3–1. Some specific, obvious cases of failure are not hard to find. For example, let R be the disc $|z| \le 2$, and let $w = z^2$. Without considering how the entire disc is mapped into the w-plane, observe that $w = 1$ is the image of both $z = -1$ and $z = +1$, that is, $1 = z^2$ implies $z = \pm 1$. Clearly $w = z^2$ cannot map the region R one to one.

The failure of $w = z^2$ to establish a one-to-one mapping for R is easily demonstrated if we solve this equation for z and obtain $z = w^{1/2}$. Given w, we see that two values of z are possible whose arguments are 180° apart. Since the given R

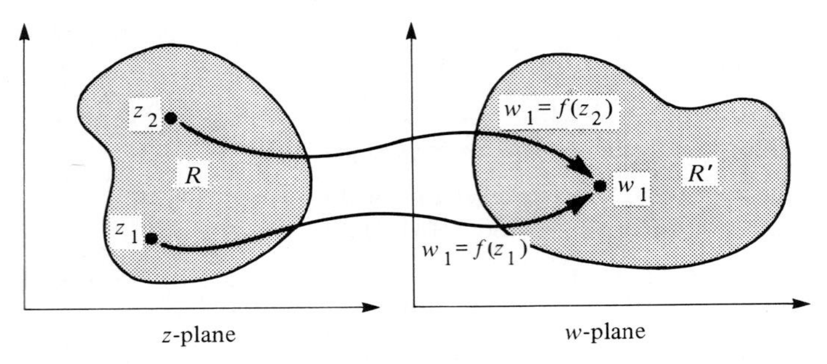

Figure 8.3–1

contains numbers whose arguments differ by 180°, a one-to-one transformation is not possible. However, by using an R in which this condition cannot occur, a one-to-one mapping can be obtained. (Example 1 will provide further discussion.)

A solution for the *inverse mapping*, $z = g(w)$, as in the previous paragraph, allows us to decide if a mapping is one to one. The analytic transformation $w = iz + 2$, for example, can be solved to yield $z = (w - 2)/i$. A point w_1 has a unique inverse point $z_1 = (w_1 - 2)/i$. Thus $w = iz + 2$ can map any region of the z-plane one to one onto a region in the w-plane.

The transformation $w = u + iv = f(z)$ can be regarded as a pair of equations $u = u(x, y)$ and $v = v(x, y)$. Thus x_0, y_0 is mapped into (u_0, v_0), where $u_0 = u(x_0, y_0)$ and $v_0 = v(x_0, y_0)$. In texts in advanced calculus it is shown that if the Jacobian of the mapping, given by the determinant†

$$\begin{vmatrix} \dfrac{\partial u}{\partial x} & \dfrac{\partial u}{\partial y} \\[2ex] \dfrac{\partial v}{\partial x} & \dfrac{\partial v}{\partial y} \end{vmatrix}$$

is not zero at (x_0, y_0), then $w = f(z)$ yields a one-to-one mapping of a neighborhood of (x_0, y_0) onto a corresponding neighborhood of (u_0, v_0).

Expanding the above determinant, we have the requirement

$$\left[\frac{\partial u}{\partial x}\frac{\partial v}{\partial y} - \frac{\partial u}{\partial y}\frac{\partial v}{\partial x}\right]_{x_0, y_0} \neq 0 \tag{8.3–1}$$

for a one-to-one mapping. Using the Cauchy–Riemann equations $\partial v/\partial y = \partial u/\partial x$ and $-\partial u/\partial y = \partial v/\partial x$, we can rewrite Eq. (8.3–1) as

$$\left[\left(\frac{\partial u}{\partial x}\right)^2 + \left(\frac{\partial v}{\partial x}\right)^2\right]_{x_0, y_0} \neq 0. \tag{8.3–2}$$

† See W. Kaplan, *Advanced Mathematics for Engineers* (Reading, MA: Addison-Wesley, 1981), Section 9.10.

From Eq. (2.3–6) we observe that the left side of Eq. (8.3–2) is $|f'(z_0)|^2$, where $z_0 = x_0 + iy_0$. Hence, the requirement for one to oneness in a neighborhood of z_0 is $f'(z_0) \neq 0$.

The preceding is summarized in Theorem 2.

THEOREM 2 One-to-one mapping

Let $f(z)$ be analytic at z_0 and $f'(z_0) \neq 0$. Then $w = f(z)$ provides a one-to-one mapping of a neighborhood of z_0. □

It can also be shown that if $f'(z) = 0$ in any point of a domain, then $f(z)$ cannot give a one-to-one mapping of that domain.

A corollary to Theorem 2 asserts that if $f'(z_0) \neq 0$, then $w = f(z)$ can be solved for an inverse $z = g(w)$ that is single valued in a neighborhood of $w_0 = f(z_0)$.

One must employ Theorem 2 with some amount of caution since it deals only with the *local* properties of the transformation $w = f(z)$. If we consider the interior of a *sufficiently small* circle centered at z_0, the theorem can guarantee a one-to-one mapping of the interior of this circle.† However, if we make the circle too large, the mapping can fail to be one to one even though $f'(z) \neq 0$ throughout the circle.

EXAMPLE 1

Discuss the possibility of obtaining a one-to-one mapping from the transformation $w = z^2$.

Solution

We make a switch to polar coordinates and take $z = re^{i\theta}$, $w = \rho e^{i\phi}$. Substituting these into the given transformation, we find that $\rho = r^2$ and $\phi = 2\theta$. We observe that the wedge-shaped region in the z-plane bounded by the rays $\theta = \alpha$, $\theta = \beta$, $r \geq 0$, (where $0 \leq \alpha < \beta$) shown in Fig. 8.3–2(a) is mapped onto the wedge bounded by the rays $\phi = 2\alpha$, $\phi = 2\beta$, $\rho \geq 0$ shown in Fig. 8.3–2(b).

The wedge bounded by the rays $\theta = \alpha + \pi$, $\theta = \beta + \pi$, $r \geq 0$ shown in Fig. 8.3–2(a) is mapped onto the wedge bounded by the rays $\phi = 2(\alpha + \pi) = 2\alpha$ and $\phi = 2(\beta + \pi) = 2\beta$, $\rho \geq 0$ shown in Fig. 8.3–2(b). Thus both wedges in Fig. 8.3–2(a) are mapped onto the identical wedge Fig. 8.3–2(b).

The inverse of our transformation is $z = w^{1/2}$. Applying this to w_1 shown in Fig. 8.3–2(b), we obtain $w_1^{1/2}$ whose values z_1 and $-z_1$ lie in the upper and lower wedges of Fig. 8.3–2(a).

Either of the wedges in Fig. 8.3–2(a) can be mapped one to one since $w^{1/2}$ has only one value in each wedge. Similarly, any domain in either wedge in Fig. 8.3–2(a) can be mapped one to one onto a domain in the w-plane. Notice that any domain

† Properties that are not limited to interiors of "sufficiently small circles" are called *global* properties of the transformation. Perhaps the most important global property of analytic functions, not proven here, is that a nonconstant analytic function must always map a domain onto a domain. For a proof see E. B. Saff and A. D. Snider, *Fundamentals of Complex Analysis* (Englewood Cliffs, N.J.: Prentice-Hall, 1976), p. 305.

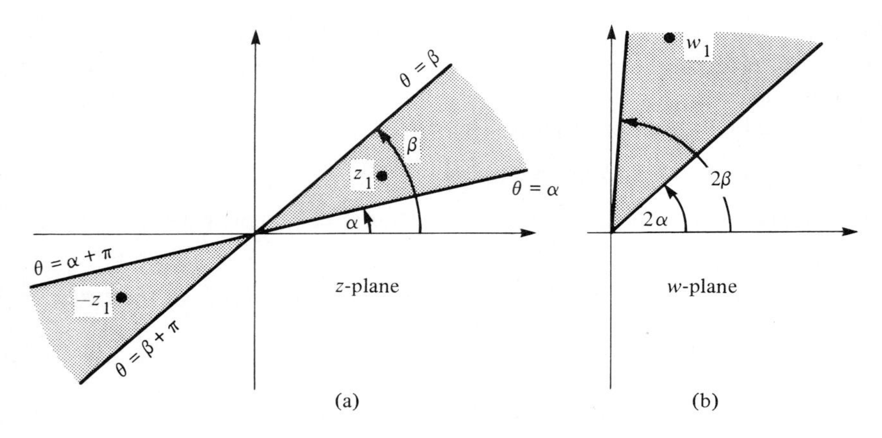

Figure 8.3–2

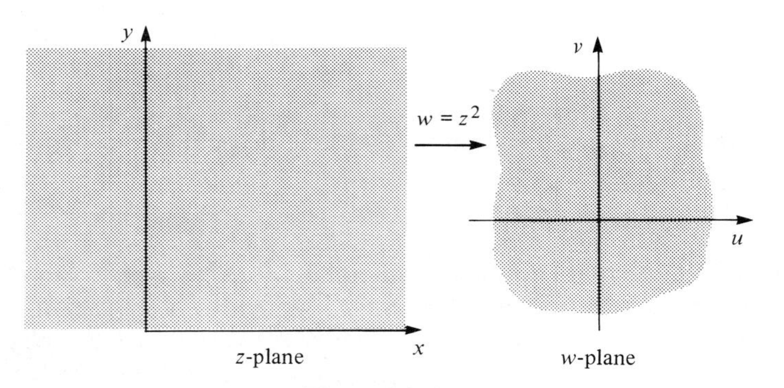

Figure 8.3–3

containing $z = 0$ must necessarily contain points from both wedges in Fig. 8.3–2(a) and cannot be used for a one-to-one mapping. However, we know that such a domain must be avoided since it contains the solution of $f'(z) = 2z = 0$.

The region $0 \leq \theta < \pi$, $r > 0$, which is the upper half of the z-plane plus the axis $y = 0$, $x \geq 0$, can be mapped one to one since it contains no two points that are negatives of each other (observe the necessity for excluding the negative real axis). The image of this region is the entire w-plane (see Fig. 8.3–3).

An alternative solution to this example, not using polar coordinates, is given in Exercise 1 of this section. ◀

EXAMPLE 2

Discuss the way in which the infinite strip $0 \leq \operatorname{Im} z \leq a$, is mapped by the transformation

$$w = u + iv = e^z = e^{x+iy}. \tag{8.3–3}$$

Take $0 \leq a < 2\pi$.

Solution

We first note the desirability of taking $0 \le a < 2\pi$. It arises from the periodic property $e^z = e^{z+2\pi i}$. By making the width of the strip (see Fig. 8.3–4a) less than 2π, we avoid having two points inside with identical real parts and imaginary parts that differ by 2π. A pair of such points are mapped into identical locations in the w-plane and a one-to-one mapping of the strip becomes impossible.

The bottom boundary of the strip, $y = 0$, $-\infty < x < \infty$, is mapped by our setting $y = 0$ in Eq. (8.3–3) to yield $e^x = u + iv$. As x ranges from $-\infty$ to ∞, the entire line $v = 0$, $0 \le u \le \infty$ is generated. This line is shown in Fig. 8.3–4(b). The points A', B', and C' are the images of A, B, and C in Fig. 8.3–4(a).

The upper boundary of the strip is mapped by our putting $y = a$ in Eq. (8.3–3) so that

$$u = e^x \cos a, \tag{8.3–4}$$

$$v = e^x \sin a. \tag{8.3–5}$$

Dividing the second equation by the first, we have $v/u = \tan a$ or

$$v = u \tan a, \tag{8.3–6}$$

which is the equation of a straight line through the origin in the uv-plane. If $\sin a$ and $\cos a$ are both positive $(0 < a < \pi/2)$, we see from Eqs. (8.3–4) and (8.3–5) that, as x ranges from $-\infty$ to ∞, only that portion of the line lying in the first quadrant of the w-plane is generated. Such a line is shown in Fig. 8.3–4(b). It is labeled with the points D', E', and F', which are the images of D, E, and F in Fig. 8.3–4(a). The slope of the line is $\tan a$, and it makes an angle a with the real axis. If a satisfied the condition $\pi/2 < a < \pi$ or $\pi < a < 3\pi/2$ or $3\pi/2 < a < 2\pi$, lines lying in respectively the 2nd or 3rd or 4th quadrant would have been obtained. The cases $\pi/2 = a$, $3\pi/2 = a$, and $\pi = a$ yield lines along the coordinate axes.

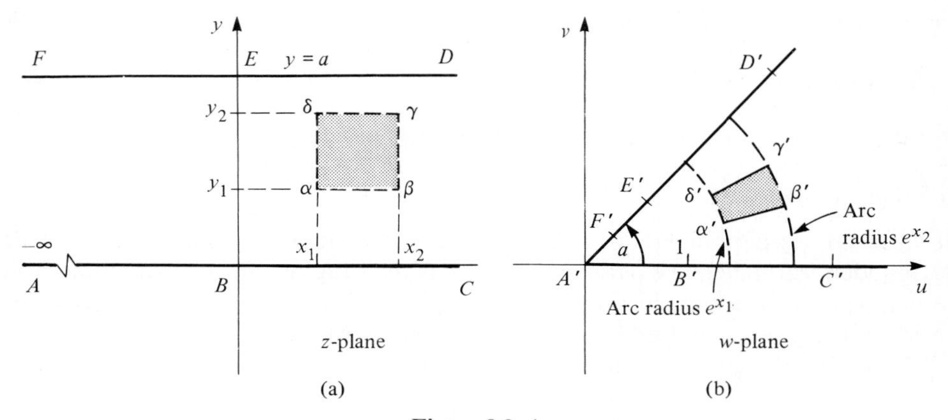

Figure 8.3–4

The strip in Fig. 8.3–4(a) is mapped onto the wedge-shaped region shown in Fig. 8.3–4(b). An important mapping occurs if the strip is chosen to have width $a = \pi$. The upper boundary passing through F, E, D in Fig. 8.3–4(a) is transformed into the negative real axis in the w-plane. The wedge shown in Fig. 8.3–4(b) evolves into the half plane $v \geq 0$, which is now the image of the strip.

The inverse transformation of Eq. (8.3–3), that is,

$$z = \log w \tag{8.3–7}$$

can be used to obtain the image in the z-plane of any point in the wedge of Fig. 8.3–4(b). Of course $\log w$ is multivalued, but there is only one value of $\log w$ that lies in the strip of Fig. 8.3–4(a). The shaded rectangular area bounded by the lines $x = x_1$, $x = x_2$, $y = y_1$, $y = y_2$ shown in Fig. 8.3–4(a) is readily mapped onto a region in the w-plane. With $x = x_1$ we have from Eq. (8.3–3) that

$$u = e^{x_1} \cos y,$$
$$v = e^{x_1} \sin y,$$

so that

$$u^2 + v^2 = e^{2x_1},$$

which is the equation of a circle of radius e^{x_1}. The line segment $x = x_1$, $0 \leq y \leq a$, is transformed into an arc lying on this circle and illustrated in Fig. 8.3–4(b). The line segment $x = x_2$, $0 \leq y \leq a$, $(x_2 > x_1)$ is transformed into an arc of larger radius, which is also shown.

The images of the lines $y = y_1$ and $y = y_2$ are readily found from Eqs. (8.3–4) and (8.3–5) if we replace a by y_1 or y_2. Rays are obtained with slopes $\tan y_1$ and $\tan y_2$, respectively. These rays (see Fig. 8.3–4b) together with the arcs of radius e^{x_1} and e^{x_2} form the boundary of a nonrectangular shape (shaded in Fig. 8.3–4b) that is the image of the rectangle shown in Fig. 8.3–4(a). Notice that the corners of the nonrectangular shape have right angles as in the original rectangle. ◀

EXAMPLE 3

Discuss the way in which $w = \sin z$ maps the strip $y \geq 0$, $-\pi/2 \leq x \leq \pi/2$.

Solution

Because $\sin z$ is periodic, that is, $\sin z = \sin(z + 2\pi)$, any two points in the z-plane having identical imaginary parts and real parts differing by 2π (or its multiple) will be mapped into identical locations in the w-plane. This situation cannot occur for points in the given strip (see Fig. 8.3–5a) because its width is π.

Rewriting the given transformation using Eq. (3.2–9), we have

$$w = (u + iv) = \sin(x + iy) = \sin x \cosh y + i \cos x \sinh y,$$

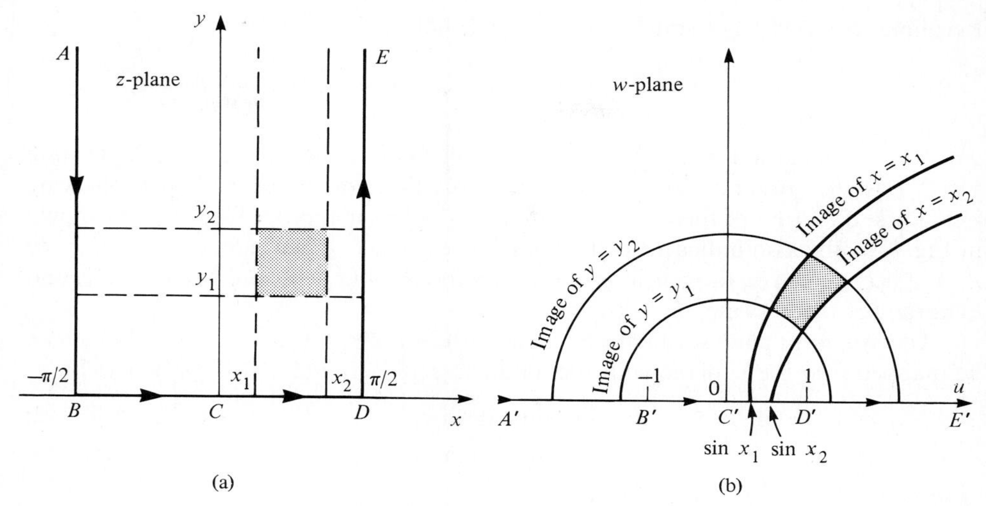

Figure 8.3–5

which means

$$u = \sin x \cosh y, \tag{8.3–8}$$

$$v = \cos x \sinh y. \tag{8.3–9}$$

The bottom boundary of the strip is $y = 0$, $-\pi/2 \leq x \leq \pi/2$. Here $u = \sin x$ and $v = 0$. As we move from $x = -\pi/2$ to $x = \pi/2$ along this bottom boundary, the image point in the w-plane advances from -1 to $+1$ along the line $v = 0$. The image of the line segment B, C, D of Fig. 8.3–5(a) is the line B', C', D' in Fig. 8.3–5(b).

Along the left boundary of the given strip $x = -\pi/2$, $y \geq 0$. From Eqs. (8.3–8) and (8.3–9) we have

$$u = \sin\left(-\frac{\pi}{2}\right)\cosh y = -\cosh y,$$

$$v = \cos\left(-\frac{\pi}{2}\right)\sinh y = 0.$$

As we move from $y = \infty$ to $y = 0$ along the left boundary, these equations indicate that the u-coordinate of the image goes from $-\infty$ to -1 along $v = 0$. The image of this boundary is thus that portion of the u-axis lying to the left of B' in Fig. 8.3–5(b). Similarly, the image of the right boundary of the strip, $x = \pi/2$, $0 \leq y \leq \infty$, is that portion of the u-axis lying to the right of D' in Fig. 8.3–5(b).

The image of the semiinfinite vertical line $x = x_1$, $0 \leq y \leq \infty$ is found from Eqs. (8.3–8) and (8.3–9). We have

$$u = \sin x_1 \cosh y, \tag{8.3–10}$$

$$v = \cos x_1 \sinh y. \tag{8.3–11}$$

Recalling that $\cosh^2 y - \sinh^2 y = 1$, we find that

$$\frac{u^2}{\sin^2 x_1} - \frac{v^2}{\cos^2 x_1} = 1,$$

which is the equation of a hyperbola. We will assume that $0 < x_1 < \pi/2$. Because $y \geq 0$, Eqs. (8.3–10) and (8.3–11) reveal that only that portion of the hyperbola lying in the first quadrant of the w-plane is obtained by this mapping. This curve is shown in Fig. 8.3–5(b); also indicated is the image of $x = x_2$, $y \geq 0$, where $x_2 > x_1$. If x_1 or x_2 had been negative, the portions of the hyperbolas obtained would be in the second quadrant of the w-plane.

The horizontal line segment $y = y_1(y_1 > 0)$, $-\pi/2 \leq x \leq \pi/2$ in Fig. 8.3–5(a) can be mapped into the w-plane with the aid of Eqs. (8.3–8) and (8.3–9), which yield

$$u = \sin x \cosh y_1, \tag{8.3–12}$$

$$v = \cos x \sinh y_1. \tag{8.3–13}$$

Since $\sin^2 x + \cos^2 x = 1$, we have

$$\frac{u^2}{\cosh^2 y_1} + \frac{v^2}{\sinh^2 y_1} = 1,$$

which describes an ellipse. Because $y_1 > 0$ and $-\pi/2 \leq x \leq \pi/2$, Eq. (8.3–13) indicates that $v \geq 0$, that is, only the upper half of the ellipse is the image of the given segment. In Fig. 8.3–5(b) we have shown elliptic arcs that are the images of the two horizontal line segments inside the strip in Fig. 8.3–5(a).

The rectangular area $x_1 \leq x \leq x_2$; $y_1 \leq y \leq y_2$, in the z-plane is mapped onto the four-sided figure bounded by two ellipses and two hyperbolas, which we see shaded in Fig. 8.3–5(b). The four corners of this figure have right angles.

It should be evident that the interior of our semiinfinite strip, in the z-plane, is mapped by $w = \sin z$ onto the upper half of the w-plane. The transformation of other strips is considered in Exercise 2 of this section.

The transformation $w = \sin z$ fails to be conformal where $d \sin z/dz = \cos z = 0$. This occurs at $z = \pm\pi/2$. The line segments AB and BC in Fig. 8.3–5(a) intersect at $z = -\pi/2$ at right angles. However, their images intersect in the w-plane at a 180° angle. The same phenomenon occurs for segments CD and DE. ◀

Suppose we needed to map a large number of vertical lines $x = x_1$, $x = x_2$, etc., and horizontal lines $y = y_1$, $y = y_2$, etc., using $w = \sin z$. Using this transformation, we could find and laboriously plot in the w-plane the image of each line, as was just done in a few cases. However, as mentioned in Section 7.3, there is some useful computer software available that can save us much work. Using the complex variables program entitled **f(z)** (available from Lascaux Graphics; see footnote, p. 461, Section 7.3), we have mapped an extensive grid lying in the space $-\pi/2 \leq x \leq \pi/2$, $0 \leq y \leq \pi/2$ into the u, v-plane. The result is illustrated in Fig. 8.3–6. The large horizontal and vertical "tic" marks are at $x = \pm 1$, $y = \pm 1$, $u = \pm 1$, $v = \pm 1$ and serve to establish the scale. Other functions are available with the software to perform different mappings.

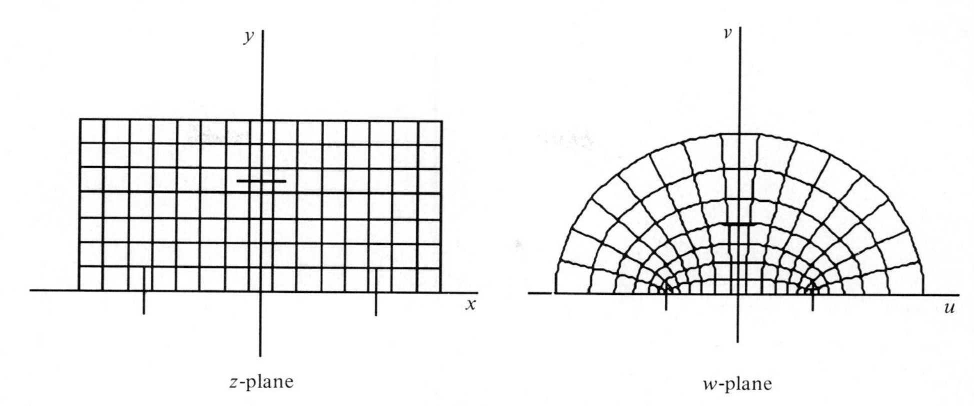

Figure 8.3–6

EXERCISES

1. a) The transformation $w = u + iv = z^2$ is applied to a certain region R in the first quadrant of the z-plane. The rectangular-shaped image region R' satisfying $u_1 \le u \le u_2$, $v_1 \le v \le v_2$ is obtained as the image of R. Here u_1 and v_1 are positive. Describe R, giving the equations of all boundaries. Does $w = z^2$ establish a one-to-one mapping of R?

b) The same transformation is applied to a region R_1 in the third quadrant of the z-plane. The region R' given in part (a) is still obtained. Describe R_1. Does $w = z^2$ establish a one-to-one mapping of R_1?

2. a) Consider the infinite strip $|\text{Re } z| \le a$, where a is a constant satisfying $0 < a < \pi/2$. Find the image of this strip, under the transformation $w = \sin z$, by mapping its boundaries.

b) Is the mapping in part (a) one to one?

c) Suppose $a = \pi/2$. Is the mapping now one to one?

How does the transformation $w = \cos z$ map the following regions? Is the mapping one to one in each case?

3. The infinite strip $a \le \text{Re } z \le b$, where $0 < a < b < \pi$

4. The infinite strip $0 \le \text{Re } z \le \pi$

5. The semiinfinite strip $0 \le \text{Re } z \le \pi$, $\text{Im } z \ge 0$

6. Consider the region consisting of an annulus with a sector removed shown in Fig. 8.3–7. The region is described by $\varepsilon \le |z| \le R$, $-\pi + \alpha \le \arg z \le \pi - \alpha$. The region is mapped with $w = \text{Log } z$.

a) Make a sketch of the image region in the w-plane showing A', B', C',... (the images of A, B, C,...). Assume $0 < \varepsilon < R$, $0 < \alpha < \pi$.

b) Is the mapping one to one? Explain.

c) What does the image region of part (a) look like in the limit as $\varepsilon \to 0+$?

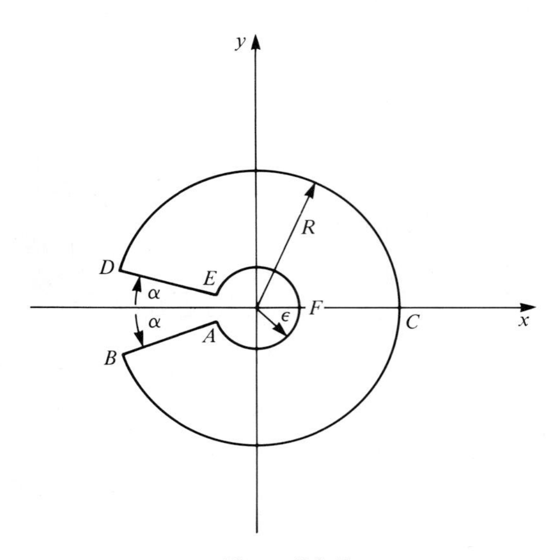

Figure 8.3–7

7. Consider the wedge-shaped region $0 \leq \arg z \leq \alpha$, $|z| < 1$. This region is to be mapped by $w = z^4$. What restriction must be placed on α to make the mapping one to one?

8. a) Refer to Example 3 of this section. Show that at their point of intersection the images of $x = x_1$ and $y = y_1$ are orthogonal. Work directly with the equation of each image.

b) In this same example what inverse transformation $z = g(w)$ will map the upper half of the w-plane onto the semiinfinite strip of Fig. 8.3–5(a)? State the branches of any logarithms and square roots in your function, and verify that point D' is mapped into D, that C' is mapped into C, and that $w = i$ has an image lying inside the strip.

9. The semiinfinite strip $0 < \operatorname{Im} z < \pi$, $\operatorname{Re} z > 0$ is mapped by means of $w = \cosh z$. Find the image of this domain.

10. a) Consider the half-disc–shaped domain $|z| < 1$, $\operatorname{Im} z > 0$. Find the image of this domain under the transformation

$$w = \left(\frac{z-1}{z+1}\right)^2.$$

Hint: Map the semicircular arc bounding the top of the disc by putting $z = e^{i\theta}$ in the above formula. The resulting expression reduces to a simple trigonometric function.

b) What inverse transformation $z = g(w)$ will map the domain found in part (a) back onto the half disc? State the appropriate branches of any square roots.

11. Following Theorem 2 there is a remark asserting that if $f'(z) = 0$ at any point in a domain, then $w = f(z)$ cannot map that domain one to one. However, in Example 1 we found that a wedge containing $z = 0$ can be mapped one to one by $f(z) = z^2$ even though $f'(0) = 0$. Is there a contradiction here? Explain.

8.4 THE BILINEAR TRANSFORMATION

The bilinear transformation defined by

$$w = \frac{az + b}{cz + d}, \qquad \text{where } a, b, c, d \text{ are complex constants,} \tag{8.4–1}$$

which is also known as the *linear fractional transformation* or the *Möbius transformation*, is especially useful in the solution of a number of physical problems, some of which are discussed in this chapter. The utility of this transformation arises from the way in which it maps straight lines and circles.

Equation (8.4–1) defines a finite value of w for all $z \neq -d/c$. One generally assumes that

$$ad \neq bc. \tag{8.4–2}$$

If we take $ad = bc$, we can readily show that Eq. (8.4–1) reduces to a constant value of w, that is, $dw/dz = 0$ for all z, and the mapping is not conformal nor especially interesting since all points in the z-plane are mapped into one point in the w-plane.

In general, from Eq. (8.4–1), we have

$$\frac{dw}{dz} = \frac{a(cz + d) - c(az + b)}{(cz + d)^2} = \frac{ad - bc}{(cz + d)^2}, \tag{8.4–3}$$

which is nonzero if Eq. (8.4–2) is satisfied.

The inverse transformation of Eq. (8.4–1) is obtained by our solving this equation for z. We have

$$z = \frac{-dw + b}{cw - a}, \tag{8.4–4}$$

which is also a bilinear transformation and defines a finite value of z for all $w \neq a/c$.

For reasons that will soon be evident, we now employ the extended w-plane and the extended z-plane (see Section 1.5), that is, planes that include the "points" $z = \infty$ and $w = \infty$.

Consider Eq. (8.4–1) for the case $c = 0$. We have

$$w = \frac{a}{d} z + \frac{b}{d}, \tag{8.4–5}$$

which defines a value of w for every finite value of z. As $|z| \to \infty$, we have $|w| \to \infty$. Thus $z = \infty$ is mapped into $w = \infty$.

Suppose however that $c \neq 0$ in Eq. (8.4–1). As $z \to -d/c$, we have $|w| \to \infty$. Thus $z = -d/c$ is mapped into $w = \infty$. If $|z| \to \infty$, we have $w \to a/c$. Thus $z = \infty$ is mapped into $w = a/c$.

Referring to the inverse transformation (see Eq. 8.4–4), if $c = 0$, we have

$$z = \frac{d}{a} w - \frac{b}{a}, \tag{8.4–6}$$

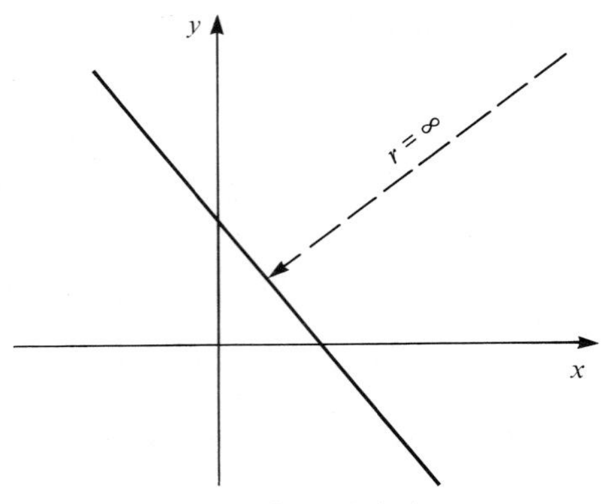

Figure 8.4–1

which also indicates that $z = \infty$ and $w = \infty$ are images (for $c = 0$). If $c \neq 0$, Eq. (8.4–4) shows that $w = \infty$ has image $z = -d/c$, whereas $w = a/c$ has image $z = \infty$. In summary, *Eq. (8.4–1) provides a one-to-one mapping of the extended z-plane onto the extended w-plane.*

Suppose now we regard infinitely long straight lines in the complex plane as being circles of infinite radius (see Fig. 8.4–1). Thus we will use the word "circle" to mean not only circles in the conventional sense but infinite straight lines as well. Circle (without the quotation marks) will mean a circle in the conventional sense. We will now prove the following theorem.

THEOREM 3 The bilinear transformation always transforms "circles" into "circles." □

Our proof of Theorem 3 begins with a restatement of Eq. (8.4–1):

$$w = \frac{a}{c} + \frac{bc - ad}{c}\frac{1}{cz + d}, \tag{8.4–7}$$

where we assume $c \neq 0$. If we put Eq. (8.4–7) over a common denominator its equivalence to Eq. (8.4–1) becomes apparent.

The transformation described by Eq. (8.4–7) can be treated as a sequence of mappings. Consider a transformation involving a mapping from the z-plane into the w_1-plane, from the w_1-plane into the w_2-plane, and so on, according to the following scheme:

$$w_1 = cz \tag{8.4–8a}$$

$$w_2 = w_1 + d = cz + d \tag{8.4–8b}$$

$$w_3 = \frac{1}{w_2} = \frac{1}{cz + d} \tag{8.4–8c}$$

$$w_4 = \frac{bc - ad}{c} w_3 = \frac{bc - ad}{c(cz + d)} \tag{8.4–8d}$$

$$w = \frac{a}{c} + w_4 = \frac{a}{c} + \frac{bc - ad}{c(cz + d)} \tag{8.4–8e}$$

Equation (8.4–8e) confirms that these five mappings are together equivalent to Eq. (8.4–7).

There are three distinctly different kinds of operations contained in Eqs. (8.4–8a–e). Let k be a complex constant. There are translations of the form

$$w = z + k, \tag{8.4–9}$$

as in Eqs. (8.4–8b) and (8.4–8e). There are rotation-magnifications of the form

$$w = kz, \tag{8.4–10}$$

as in Eqs. (8.4–8a) and (8.4–8d). And there are inversions of the form

$$w = \frac{1}{z}, \tag{8.4–11}$$

as in Eq. (8.4–8c).

If we can show that "circles" are mapped into "circles" under each of these three operations, we will have proved Theorem 3. It should be apparent that under a displacement the geometric character of any shape (circles, triangles, straight lines) is preserved since every point on whatever shape we choose is merely displaced by the complex vector k (see Fig. 8.4–2).

We can rewrite Eq. (8.4–10) as

$$w = |k|e^{i\theta_k}z, \tag{8.4–12}$$

where $\theta_k = \arg k$. Under this transformation a point from the z-plane is rotated through an angle θ_k and its distance from the origin is magnified by the factor $|k|$. The process of rotation will preserve the shape of any figure as shown in Fig. 8.4–3. We can show that under magnification "circles" are mapped into "circles." For a magnification

$$w = u + iv = |k|z = |k|(x + iy),$$

and so

$$u = |k|x, \qquad v = |k|y. \tag{8.4–13}$$

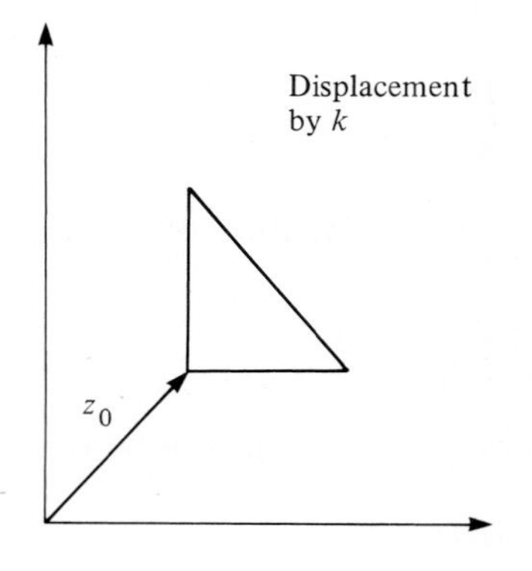

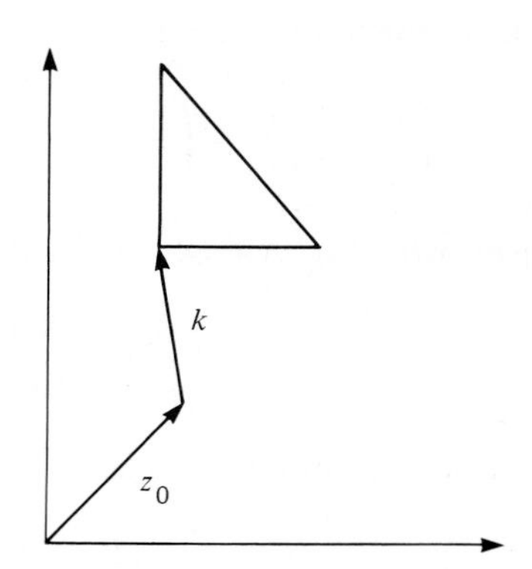

Figure 8.4–2

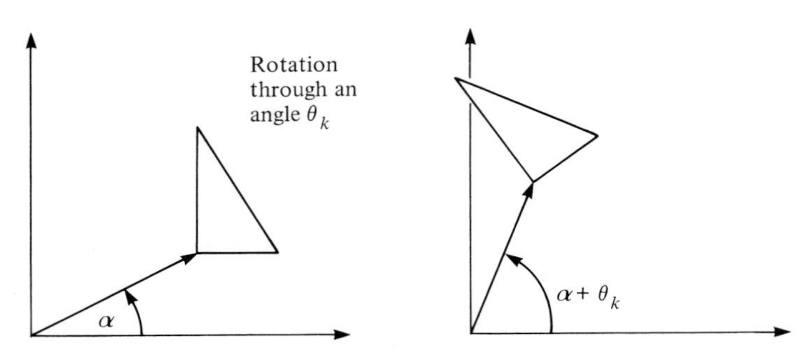

Figure 8.4–3

A circle in the xy-plane is of the form

$$(x - x_0)^2 + (y - y_0)^2 = r^2. \tag{8.4–14}$$

The center is at (x_0, y_0) and the radius is r. Solving Eq. (8.4–13) for x and y and using these values in Eq. (8.4–14), we can obtain

$$(u - |k|x_0)^2 + (v - |k|y_0)^2 = r^2|k|^2,$$

which is the equation of a circle, in the w-plane, having center at $|k|x_0$, $|k|y_0$ and radius $r|k|$. A similar argument shows that the straight line $y = mx + b$ is mapped into the straight line $v = mu + b|k|$.

To prove that Eq. (8.4–11) maps "circles" into "circles," consider the algebraic equation

$$A(x^2 + y^2) + Bx + Cy + D = 0, \tag{8.4–15}$$

where A, B, C, and D are all real numbers. If $A = 0$, this is obviously the equation of a straight line. Assuming $A \neq 0$, we divide Eq. (8.4–15) by A to obtain

$$(x^2 + y^2) + \frac{B}{A}x + \frac{C}{A}y + \frac{D}{A} = 0,$$

which, if we complete two squares, can be rewritten

$$\left(x + \frac{B}{2A}\right)^2 + \left(y + \frac{C}{2A}\right)^2 = -\frac{D}{A} + \left(\frac{B}{2A}\right)^2 + \left(\frac{C}{2A}\right)^2. \tag{8.4–16}$$

A comparison with Eq. (8.4–14) reveals this to be the equation of a circle provided

$$-\frac{D}{A} + \left(\frac{B}{2A}\right)^2 + \left(\frac{C}{2A}\right)^2 \geq 0$$

(the squared radius cannot be negative). The above condition can be rearranged as

$$B^2 + C^2 \geq 4AD. \tag{8.4–17}$$

In Eq. (8.4–15) we make the following well-known substitutions:

$$x^2 + y^2 = z\bar{z}, \qquad x = \frac{z + \bar{z}}{2}, \qquad y = \frac{1}{i}\frac{(z - \bar{z})}{2}.$$

Thus

$$Az\bar{z} + \frac{B}{2}(z + \bar{z}) + \frac{C}{2i}(z - \bar{z}) + D = 0. \tag{8.4–18}$$

With A, B, C, D real numbers Eq. (8.4–18) is the equation of a circle if $A \neq 0$, and if in addition Eq. (8.4–17) is satisfied. It is the equation of a straight line if $A = 0$.

We now replace z by $1/w$ in Eq. (8.4–18) in order to find the image of our "circle" under an inversion. The "circle" is transformed into a curve satisfying

$$A\left(\frac{1}{w\bar{w}}\right) + \frac{B}{2}\left(\frac{1}{w} + \frac{1}{\bar{w}}\right) + \frac{C}{2i}\left(\frac{1}{w} - \frac{1}{\bar{w}}\right) + D = 0.$$

Multiplying both sides by $w\bar{w}$ and rearranging terms slightly, we get

$$Dw\bar{w} + \frac{B}{2}(w + \bar{w}) - \frac{C}{2i}(w - \bar{w}) + A = 0. \tag{8.4–19}$$

Equation (8.4–19) is identical in form to Eq. (8.4–18) with D in Eq. (8.4–19) now playing the role of A, A now playing the role of D, and $-C$ taking the part of C. The meaning of B is unaltered.

With these changes in Eq. (8.4–17) it is found that this inequality remains unaltered. Thus if $D \neq 0$, Eq. (8.4–19) describes a circle as long as Eq. (8.4–18) describes one.

If $D = 0$, Eq. (8.4–19) describes a straight line, that is, a "circle." Thus we have shown that $w = 1/z$ maps "circles" into "circles."

Notice that if $D = 0$, Eq. (8.4–15) is satisfied for $z = 0$, that is, the "circle" described in the z-plane passes through the origin. With $D = 0$ Eq. (8.4–19) yields a straight line. Thus a "circle" passing through the origin of the z-plane is transformed by $w = 1/z$ into a straight line in the w-plane.

Under the assumption $c \neq 0$ we have shown that the bilinear transformation (see Eq. 8.4–1 and, equivalently, Eq. 8.4–7) can be decomposed into a sequence of transformations, each of which transforms "circles" into "circles." If $c = 0$ in Eq. (8.4–1), it is an easy matter to show that the resulting transformation

$$w = \frac{a}{d}z + \frac{b}{d},$$

which involves a rotation-magnification and displacement also has this property. Thus our proof of Theorem 3 is complete.

Because the inverse of a bilinear transformation is also a bilinear transformation (see Eq. 8.4–4), a "circle" in the z-plane is the image of a "circle" in the w-plane (and vice versa). An elementary example of this, involving the simple bilinear transformation $w = 1/z$, was studied in Section 8.2 in Example 1 as well as in Exercise 10 of that section.

A sequence of two or more bilinear transformations can be reduced to a single bilinear transformation. Thus if a bilinear transformation provides a transformation from the z-plane to the w_1-plane and another bilinear transformation provides a transformation from the w_1-plane to the w-plane, then there is a bilinear transformation that gives a mapping from the z-plane to the w-plane. This fact is proved in Exercise 29.

EXAMPLE 1

a) Find the image of the circle $|z - 1 - i| = 1$ under the transformation $w = 1/z$.

b) Find the image of the same circle under the transformation $w = (z + 2)/z$.

Solution

Part (a): The given circle is shown in Fig. 8.4–4(a). Since it does not pass through $z = 0$ the expression $w = 1/z$ is never infinite for any z lying on the circumference of the circle. Because the image of the circle does not pass through $w = \infty$, we know that this image is not a straight line in the w-plane. Thus the required image must be a circle. To find this circle we need map only three points lying on $|z - 1 - i| = 1$ into the w-plane since, as we recall from elementary geometry, the center and radius of a circle are determined by three points on its circumference.

Using $w = 1/z$, we map the points A, B, and C from Fig. 8.4–4(a) into the w-plane. These points are at $z = 1$, $z = i$, and $z = 2 + i$, respectively, and have images A', B', and C' located at $w = 1$, $w = -i$, and $w = (2 - i)/5$. Points A', B', and C', as well as the circle passing through them, are shown in Fig. 8.4–4(b). It should be apparent that this circle must be tangent to the u-axis at $w = 1$, i.e., at A'. This is because the point A is the only point on $|z - 1 - i|$ whose value is real. Under the inversion $w = 1/z$, this is the only point with a real image.

To summarize, we see from Fig. 8.4–4(b) that the required image is $|w - 1 + i| = 1$.

Part (b): The transformation $w = (z + 2)/z$ is identical to $w = 2/z + 1$. We can regard this as a sequence of three transformations: $w_1 = 1/z$, $w_2 = 2w_1$, $w = w_2 + 1$. The first transformation has already been performed and is the answer to part (a). The second, $w_2 = 2w_1$, requires merely magnifying the answer to part (a) by a factor of 2. Thus a circle of radius 2 and center $2 - 2i$ is obtained. The third transformation $w = w_2 + 1$ just involves displacing the preceding circle one unit parallel to the real axis. The center is now at $3 - 2i$. Thus our answer is $|w - 3 + 2i| = 2$, and this is shown in Fig. 8.4–4(c). The points A'', B'', and C'' are the images of A, B, and C in Fig. 8.4–4(a). ◀

EXAMPLE 2

Figure 8.4–5 shows an elementary electric circuit. V_i and V_0 are the phasor input and output potentials for the electric circuit. (See the appendix to Chapter 3 for a discussion of phasors.) The amplification factor $A = V_0/V_i$ for the circuit is given by

$$A(s) = \frac{1 + s}{2 + s}, \tag{8.4–20}$$

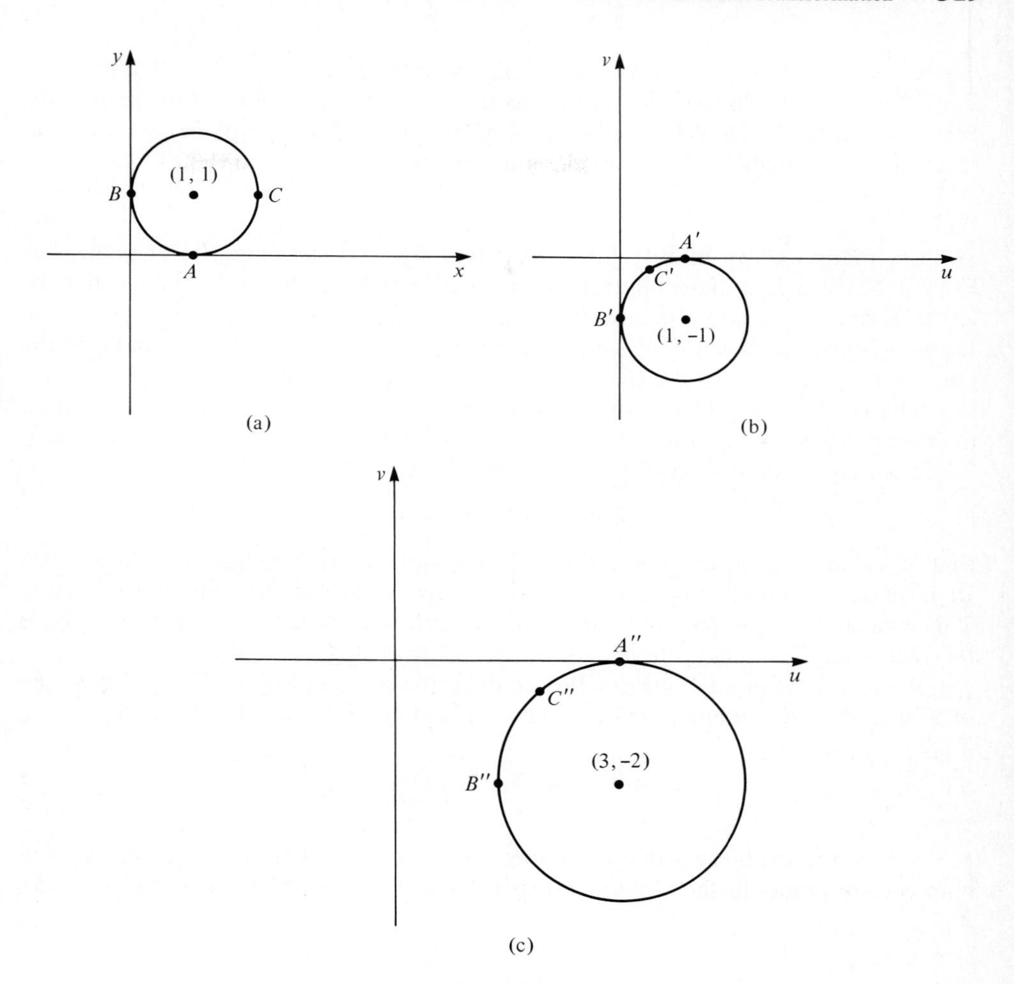

Figure 8.4–4

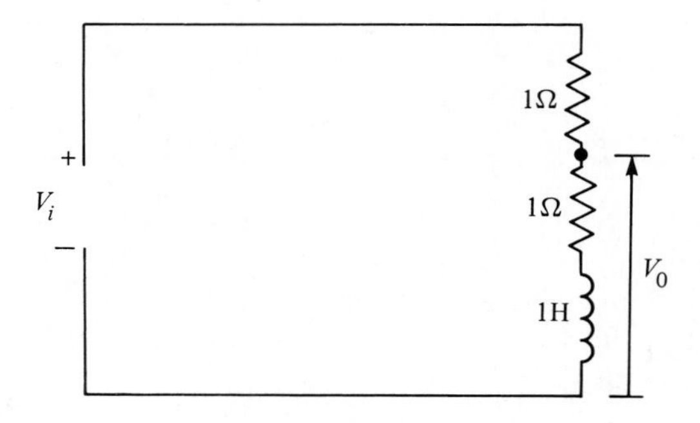

Figure 8.4–5

where $s = \sigma + i\omega$ is the complex frequency describing the potentials. If we restrict ourselves to potentials that vary in time as sine or cosine functions of fixed amplitude, then $\sigma = 0$ and $s = i\omega$. What is the locus of $A(s)$ in the complex plane as s varies in the complex frequency plane over the line $s = i\omega$, $-\infty \le \omega \le \infty$?

Solution

We are mapping the infinite line $\sigma = 0$ (see Fig. 8.4–6a) into the A-plane (see Fig. 8.4–6b) by means of the bilinear transformation in Eq. (8.4–20). The image must be either a circle or a straight line. A straight line must pass through $A = \infty$. From Eq. (8.4–20) we see that $A = \infty$ is the image of $s = -2$. But $s = -2$ does not lie on the line $s = i\omega$ $(-\infty \le \omega \le \infty)$. Hence the image we are seeking is a circle.

Observe from Eq. (8.4–20) that the image of $s = (0, 0)$ is $1/2$. As $|s| \to \infty$ along the ω-axis, the same equation shows that $A \to 1$, that is, the image of $s = \infty$ is $A = 1$.

It is easy to show from Eq. (8.4–20) that

$$A(\sigma + i\omega) = \bar{A}(\sigma - i\omega),$$

that is, values of A at conjugate points in the s-plane are conjugates of each other. In particular, a pair of conjugate points on the ω-axis have images that are conjugates. This means the circle that is the image of the entire ω-axis must be symmetric about the real A-axis.

We now have enough information to draw the circle in Fig. 8.4–6(b). The images of a few individual points in the complex A-plane are indicated; for example, if $s = i$,

$$A(i) = \frac{1+i}{2+i} \doteq \frac{\sqrt{2}}{\sqrt{5}} \underline{/18.43^\circ}. \qquad ◀$$

A common problem is that of finding a specific bilinear transformation that will map certain points in the z-plane into preassigned images in the w-plane. One also

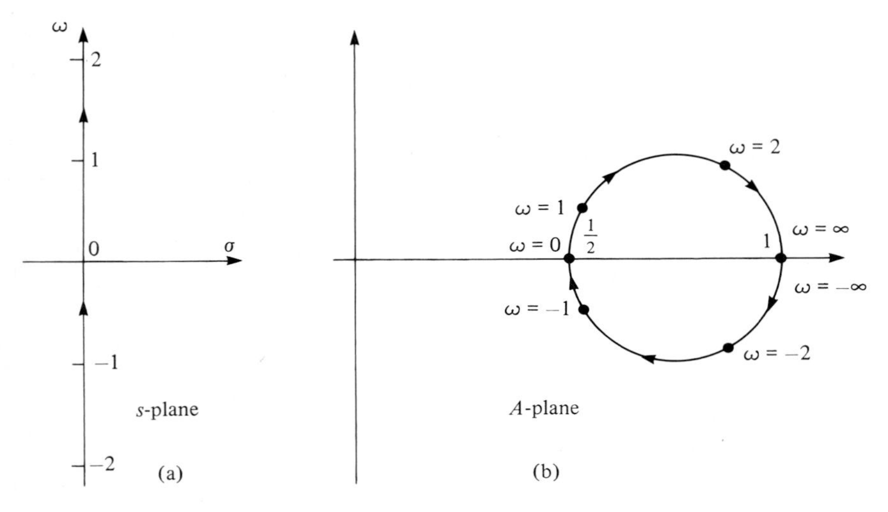

Figure 8.4–6

seeks transformations capable of mapping a given line or circle into some other specific line or circle. In these problems one must establish the constants a, b, c, d in the bilinear transformation

$$w = \frac{az + b}{cz + d}. \tag{8.4–21}$$

Let us assume that one of the coefficients, say a, is nonzero. Then we can rewrite Eq. (8.4–21) as

$$w = \frac{z + \frac{b}{a}}{\frac{c}{a} z + \frac{d}{a}} = \frac{z + c_1}{c_2 z + c_3}. \tag{8.4–22}$$

Thus only three coefficients, c_1, c_2, and c_3 need be found. Given three points z_1, z_2, and z_3, which we must map into w_1, w_2, and w_3, respectively, we replace w and z in Eq. (8.4–22) first of all by w_1 and z_1, respectively, then by w_2 and z_2, and finally by w_3 and z_3. Three simultaneous linear equations are obtained in the unknowns c_1, c_2, and c_3. Solving for these unknowns, we have determined our bilinear transformation (Eq. 8.4–22). If a solution does not exist, it is because $a = 0$ in Eq. (8.4–21). We then could obtain b, c, and d by simultaneously solving three equations obtained by substituting w_1 and z_1, w_2 and z_2, and finally w_3 and z_3 into Eq. (8.4–21) with a set equal to zero.

A more direct way of solving for a bilinear transformation involves the cross ratio.

DEFINITION Cross ratio

The *cross ratio* of four distinct complex numbers (or points) z_1, z_2, z_3, z_4 is defined by

$$(z_1, z_2, z_3, z_4) = \frac{(z_1 - z_2)(z_3 - z_4)}{(z_1 - z_4)(z_3 - z_2)}. \tag{8.4–23}$$

If any of these numbers, say z_j, is ∞, the cross ratio in Eq. (8.4–23) is redefined so that the quotient of the two terms on the right containing z_j, that is $(z_j - z_k)/(z_j - z_m)$, is taken as 1. □

The cross ratio of the four image points w_1, w_2, w_3, w_4 is obtained by replacing z_1, z_2, etc. in the preceding definition by w_1, w_2, etc. The order of the points in a cross ratio is important. The reader should verify, that, for example, $(1, 2, 3, 4) = -1/3$, whereas $(3, 1, 2, 4) = 4$.

We will now prove the following theorem.

THEOREM 4 Invariance of cross ratio

Under the bilinear transformation (Eq. 8.4–21) the cross ratio of four points is preserved, that is,

$$\frac{(w_1 - w_2)(w_3 - w_4)}{(w_1 - w_4)(w_3 - w_2)} = \frac{(z_1 - z_2)(z_3 - z_4)}{(z_1 - z_4)(z_3 - z_2)}. \quad \square \tag{8.4–24}$$

The proof of Theorem 4 is straightforward. From Eq. (8.4–21) z_i is mapped into w_i, that is

$$w_i = \frac{az_i + b}{cz_i + d}, \tag{8.4–25}$$

and similarly

$$w_j = \frac{az_j + b}{cz_j + d},$$

so that

$$w_i - w_j = \frac{az_i + b}{cz_i + d} - \frac{az_j + b}{cz_j + d} = \frac{(ad - bc)(z_i - z_j)}{(cz_i + d)(cz_j + d)}. \tag{8.4–26}$$

With $i = 1, j = 2$ in Eq. (8.4–26) we obtain $(w_1 - w_2)$ in terms of z_1 and z_2. Similarly, we can express $w_3 - w_4$ in terms of $z_3 - z_4$, etc. In this manner the entire left side of Eq. (8.4–24) can be written in terms of z_1, z_2, z_3, and z_4. After some simple algebra we obtain the right side of Eq. (8.4–24). The reader should supply the details.

If one of the points $z_1, z_2, \ldots$ is at infinity, the invariance of the cross ratio must be proved differently. If, say, $z_1 = \infty$, its image is $w_1 = a/c$ (see, for example, Eq. 8.4–25 as $z_i = z_1 \to \infty$). Thus the left side of Eq. (8.4–24) becomes

$$\frac{\left(\frac{a}{c} - w_2\right)(w_3 - w_4)}{\left(\frac{a}{c} - w_4\right)(w_3 - w_2)}.$$

If Eqs. (8.4–26) and (8.4–25) are used in this expression, the values w_2, w_3, and w_4 can be rewritten in terms of z_2, z_3, z_4. After some manipulation, the expression $(z_3 - z_4)/(z_3 - z_2)$ is obtained. This is the cross ratio (z_1, z_2, z_3, z_4) when $z_1 = \infty$.

The invariance of the cross ratio is useful when we seek the bilinear transformation capable of mapping three specific points z_1, z_2, z_3 into three specific images w_1, w_2, w_3. The point z_4 in Eq. (8.4–24) is taken as a general point z whose image is w (instead of w_4). Thus our working formula becomes

$$\frac{(w_1 - w_2)(w_3 - w)}{(w_1 - w)(w_3 - w_2)} = \frac{(z_1 - z_2)(z_3 - z)}{(z_1 - z)(z_3 - z_2)}, \tag{8.4–27}$$

which must be suitably modified if any point is at ∞. The solution of w in terms of z yields the required transformation.

EXAMPLE 3

Find the bilinear transformation that maps $z_1 = 1$, $z_2 = i$, $z_3 = 0$ into $w_1 = 0$, $w_2 = -1$, $w_3 = -i$.

Solution

We substitute these six complex numbers into the appropriate location in Eq. (8.4–27) and obtain

$$\frac{(0-(-1))(-i-w)}{(0-w)(-i+1)} = \frac{(1-i)(0-z)}{(1-z)(0-i)}.$$

With some minor algebra we get

$$w = \frac{i(z-1)}{z+1}. \tag{8.4–28}$$

This result can be checked by our letting z assume the three given values 1, i, and 0. The desired values of w are obtained. ◄

EXAMPLE 4

For the transformation found in Example 3 what is the image of the circle passing through $z_1 = 1$, $z_2 = i$, $z_3 = 0$, and what is the image of the interior of this circle?

Solution

The given circle is shown in Fig. 8.4–7(a). From elementary geometry, its center is found to be at $(1 + i)/2$, and its radius is $1/\sqrt{2}$. The circle is described by

$$\left|z - \frac{(1+i)}{2}\right| = \frac{1}{\sqrt{2}}.$$

The image of the circle under Eq. (8.4–28) must be a straight line or circle in the w-plane. The image is known to pass through $w_1 = 0$, $w_2 = -1$, $w_3 = -i$. The circle determined by these three points is shown in Fig. 8.4–7(b) (no straight line can connect

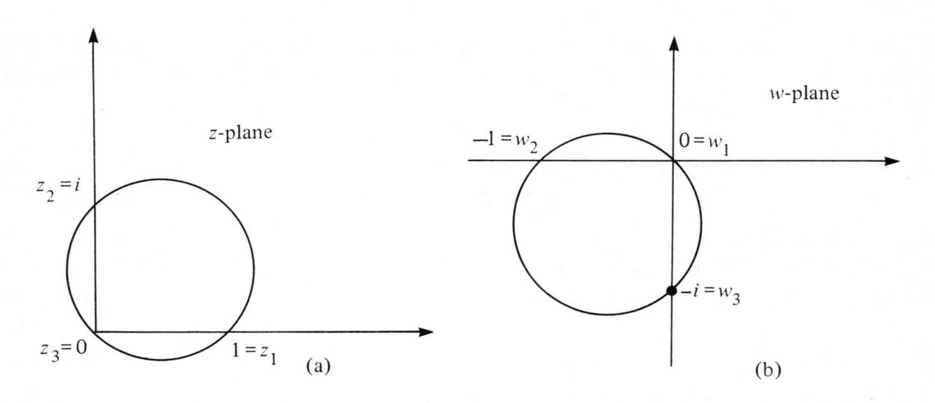

Figure 8.4–7

w_1, w_2, and w_3) and is described by

$$\left|w + \frac{(1+i)}{2}\right| = \frac{1}{\sqrt{2}}. \tag{8.4–29}$$

This disc-shaped domain

$$\left|z - \frac{(1+i)}{2}\right| < \frac{1}{\sqrt{2}},$$

which is the interior of the circle of Fig. 8.4–7(a) has, under the given transformation (Eq. 8.4–28), an image that is also a domain.† The boundary of this image is the circle in Fig. 8.4–7(b). Thus the image domain must be either the disc interior to this latter circle or the annulus exterior to it.

Let us consider some point inside the circle of Fig. 8.4–7(a), say $z = 1/2$. From Eq. (8.4–28) we discover that $w = -i/3$ is its image. This lies inside the circle of Fig. 8.4–7(b) and indicates that the domain

$$\left|z - \frac{(1+i)}{2}\right| < \frac{1}{\sqrt{2}}$$

is mapped onto the domain

$$\left|w + \frac{(1+i)}{2}\right| < \frac{1}{\sqrt{2}}.$$

◀

EXAMPLE 5

Find the bilinear transformation that maps $z_1 = 1, z_2 = i, z_3 = 0$ into $w_1 = 0, w_2 = \infty$, $w_3 = -i$.

Solution

Note that z_1, z_2, and z_3 are the same as in Example 3. We again employ Eq. (8.4–27). However, since $w_2 = \infty$, the ratio $(w_1 - w_2)/(w_3 - w_2)$ on the left must be replaced by 1. Thus

$$\frac{-i - w}{-w} = \frac{(1-i)(-z)}{(1-z)(-i)},$$

whose solution is

$$w = \frac{1-z}{i-z}. \tag{8.4–30}$$

---

† Recall that a domain is always mapped onto a domain by a nonconstant analytic function (see footnote, page 505).

Note that the circle in Fig. 8.4–7(a) passing through 1 and i and 0 is transformed into a "circle" passing through 0 and ∞ and $-i$, that is, an infinite straight line lying along the imaginary axis in the w-plane. The half plane to the left of this line is the image of the interior of the circle in Fig. 8.4–7(a), as the reader can readily verify. ◀

EXAMPLE 6

Find the transformation that will map the domain $0 < \arg z < \pi/2$ from the z-plane onto $|w| < 1$ in the w-plane (see Fig. 8.4–8).

Solution

The boundary of the given domain in the z-plane, that is, the positive x- and y-axes, must be transformed into the unit circle $|w| = 1$ by the required formula. A bilinear transformation will map an infinite straight line into a circle but cannot transform a line with a 90° bend into a circle. (Why?) Hence, our answer cannot be a bilinear transformation.

Notice however, that the transformation

$$s = z^2 \tag{8.4–31}$$

(see Example 1, Section 8.3) will map our 90° sector onto the *upper* half of the s-plane. If we can find a second transformation that will map the upper half of the s-plane onto the interior of the unit circle in the w-plane, we can combine the two mappings into the required transformation.

Borrowing a result derived in Exercise 31 of this section and changing notation slightly, we observe that

$$w = e^{i\gamma}\frac{(s - p)}{(s - \bar{p})}, \qquad \text{where } \gamma \text{ is a real number} \quad \text{and} \quad \operatorname{Im} p > 0, \tag{8.4–32}$$

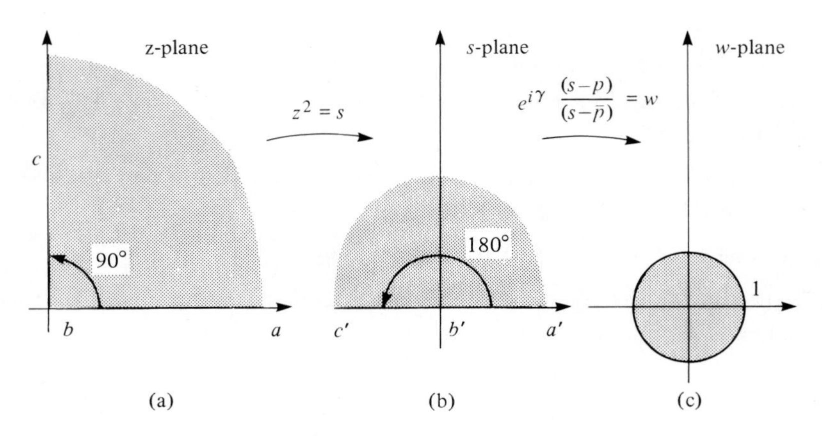

Figure 8.4–8

will transform the real axis from the s-plane into the circle $|w| = 1$ and map the domain $\text{Im } s > 0$ onto the interior of this circle.

Combining Eqs. (8.4–31) and (8.4–32), we have as our result

$$w = e^{i\gamma}\frac{(z^2 - p)}{(z^2 - \bar{p})}, \quad \text{where } \gamma \text{ is a real number} \quad \text{and} \quad \text{Im } p > 0. \tag{8.4–33}$$

A particular example of Eq. (8.4–33) is

$$w = \frac{z^2 - i}{z^2 + i}.$$

The method just used can be modified so that angular sectors of the form $0 < \arg z < \alpha$, where α is not constrained to be $\pi/2$, can be mapped onto the interior of the unit circle (see Exercise 33 of this section). ◀

EXERCISES

1. a) Derive Eq. (8.4–4) from Eq. (8.4–1).

b) Verify that Eq. (8.4–7) is equivalent to Eq. (8.4–1).

2. Suppose that the bilinear transformation (see Eq. 8.4–1) has real coefficients a, b, c, d. Show that a curve that is symmetric about the x-axis has an image under this transformation that is symmetric about the u-axis.

3. Derive Eq. (8.4–24) (the invariance of the cross ratio) by following the steps suggested in the text.

4. If a transformation $w = f(z)$ maps z_1 into w_1, where z_1 and w_1 have the same numerical value, we say that z_1 is a *fixed point* of the transformation.

a) For the bilinear transformation (Eq. 8.4–1) show that a fixed point must satisfy

$$cz^2 - (a - d)z - b = 0.$$

b) Show that unless $a = d \neq 0$ and $b = c = 0$ are simultaneously satisfied, there are at most two fixed points for this bilinear transformation.

c) Why are all points fixed points if $a = d \neq 0$ and $b = c = 0$ are simultaneously satisfied? Refer to Eq. (8.4–1).

Using the result of Exercise 4(a), find the most general form of the bilinear transformation $w(z)$ that has the following fixed points:

5. $z = -1$ and $z = 1$ **6.** $z = 1$ and $z = i$

For the transformation $w = 1/z$, what are the images of the following curves? Give the result as an equation in w or in the variables u and v, where $w = u + iv$.

7. $y = 1$ **8.** $x - y = 1$ **9.** $|z - 1 + i| = 1$

10. $|z + 1 + i| = \sqrt{2}$ **11.** $y = x$ **12.** $|z - 3 - 3i| = \sqrt{2}$

13. $|z - \sqrt{3} - i| = 1$

For the transformation $w = (z + 1)/(z - 1)$ what are the images of the following curves? Give the result as an equation in w or in the variables u and v.

14. $|z| = 1$ **15.** $|z| = 2$ **16.** $|z + 1| = 2$

Onto what domain in the w-plane do the following transformations map the domain $|z - 1| < 1$?

17. $w = \dfrac{z}{z - 1}$ **18.** $w = \dfrac{z - 1}{z}$ **19.** $w = \dfrac{z - 1}{(1 + i)z}$

Onto what domain in the w-plane do the following transformations map the domain $1 < \text{Re } z < 2$?

20. $w = \dfrac{z}{z - 1}$ **21.** $w = \dfrac{z}{2z - 3}$ **22.** $w = \dfrac{z - 1}{z - 2}$

Find the bilinear transformation that will map the points z_1, z_2, and z_3 into the corresponding image points w_1, w_2, and w_3 as described below:

23. a) $z_1 = 0$, $z_2 = i$, $z_3 = -i$; $w_1 = 1$, $w_2 = i$, $w_3 = 2 - i$.
b) What is the image of $|z| < 1$ under this transformation?

24. a) $z_1 = i$, $z_2 = -1$, $z_3 = -i$; $w_1 = 1 + i$, $w_2 = \infty$, $w_3 = 1 - i$.
b) What is the image of $|z| > 1$ under this transformation?

25. a) $z_1 = \infty$, $z_2 = 1$, $z_3 = -i$; $w_1 = 1$, $w_2 = i$, $w_3 = -i$.
b) What is the image of the domain $\text{Re}(z - 1) > \text{Im } z$ under this transformation?

26. a) $z_1 = i$, $z_2 = -i$, $z_3 = 1$; $w_1 = 1$, $w_2 = -i$, $w_3 = -1$.
b) What is the image of $|z| < 1$ under this transformation?

27. The complex impedance at the input of the circuit in Fig. 8.4–9 when driven by a sinusoidal generator of radian frequency ω is $Z(\omega) = R + i\omega L$. When ω increases from 0 to ∞, $Z(\omega)$ progresses in the complex plane from $(R, 0)$ to infinity along the semiinfinite line $\text{Re } Z = R$, $\text{Im } Z \geq 0$. The complex admittance of the circuit is defined by $Y(\omega) = 1/Z(\omega)$. Use the properties of the bilinear transformation to determine the locus of $Y(\omega)$ in the complex plane as ω goes from 0 to ∞. Sketch the locus and indicate $Y(0)$, $Y(R/L)$ and $Y(\infty)$.

Figure 8.4–9

28. a) A circle of radius $\rho > 0$ and center $(x_0, 0)$ is transformed by the inversion $w = 1/z$ into another circle. Locate the intercepts of the image circle on the real w-axis and show that this new circle has center $x_0/(x_0^2 - \rho^2)$ and radius $\rho/|x_0^2 - \rho^2|$.

b) Is the image of the center of the original circle under the transformation $w = 1/z$ identical to the center of the image circle? Explain.

c) Does the general bilinear transformation (see Eq. 8.4–1) always map the center of a circle in the z-plane into the center of the image of that circle in the w-plane? Explain.

d) Consider the special case of Eq. (8.4–1), $w = az + b$. Show that the circle $|z - z_0| = \rho$ is mapped by this transformation into a circle centered at $w_0 = az_0 + b$ with radius $|a|\rho$. Thus in this special case the original circle and its image have centers that are images of each other under the given transformation.

29. A bilinear transformation

$$w_1 = \frac{a_1 z + b_1}{c_1 z + d_1}$$

defines a mapping from the z-plane to the w_1-plane. Additionally, a second bilinear transformation

$$w = \frac{a_2 w_1 + b_2}{c_2 w_1 + d_2}$$

yields a mapping from the w_1-plane to the w-plane. Show that these two successive bilinear transformations, which together relate z and w, can be combined into a single bilinear transformation

$$w = \frac{az + b}{cz + d}.$$

What are a, b, c, d in terms of a_1, a_2, etc.?

30. For the electric circuit shown in Fig. 8.4–10 the ratio of the phasor output voltage to the phasor input voltage is given by

$$\frac{V_0}{V_i} = A(s) = \frac{1 + s}{1 + 2s},$$

where $s = \sigma + i\omega$ is the complex frequency.

a) Draw the locus in the complex plane of $A(s)$ as the complex frequency varies in the s-plane along the line $\sigma = 0, -\infty < \omega < \infty$. What is the equation of the locus? Indicate on the locus the values of A when $\omega = 0, \pm 1/2, \pm 1, \pm\infty$.

b) Suppose the complex frequency is of the form $s = -1/2 + i\omega$ (which implies a simultaneously oscillating and decaying signal). As ω varies from $-\infty$ to ∞, indicate the locus of A in the complex plane.

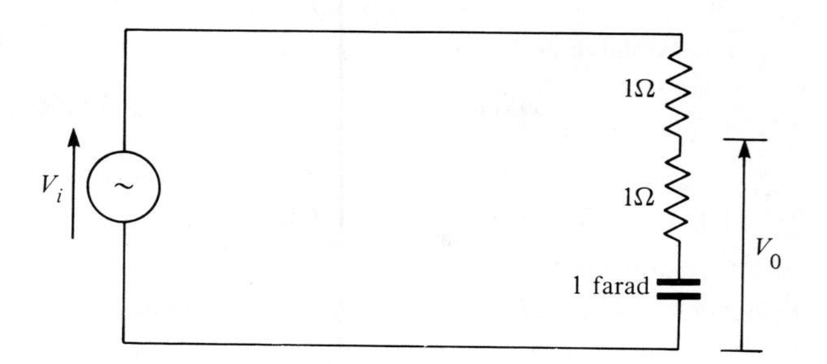

Figure 8.4–10

31. This exercise establishes the general bilinear transformation that maps the upper half of the z-plane (Im $z > 0$) onto the unit disc $|w| < 1$.

a) Put Eq. (8.4–1) in the form

$$w = \left(\frac{a}{c}\right)\frac{z + b/a}{z + d/c}.$$

Explain why the desired transformation must transform the x-axis into $|w| = 1$. Let $|z| \to \infty$ on this axis and argue that $|a|/|c| = 1$. Thus $a/c = e^{i\gamma}$, where γ is any real number.

b) From

$$w = e^{i\gamma}\,\frac{z + b/a}{z + d/c}$$

we have with $z = x$ and the use of magnitudes that

$$1 = \left|\frac{x + b/a}{x + d/c}\right|,$$

or

$$\left|x + \frac{d}{c}\right| = \left|x + \frac{b}{a}\right|.$$

Explain why this equation can be satisfied only if

$$\frac{d}{c} = \frac{b}{a},$$

or

$$\frac{d}{c} = \overline{\left(\frac{b}{a}\right)}.$$

Explain why the first choice must be discarded.

c) Taking $-p = b/a$, we have now

$$w = e^{i\gamma} \frac{(z - p)}{(z - \bar{p})}. \tag{8.4–34}$$

Note that $z = p$ has image $w = 0$. Explain why we require $\text{Im } p > 0$ in Eq. (8.4–34) so that $|w| < 1$ will be the image of $\text{Im } z > 0$.

d) Suppose in Eq. (8.4–34) we take $\text{Im } p < 0$. What is the image of $\text{Im } z > 0$ under this transformation?

32. a) Find a bilinear transformation capable of mapping the domain to the right of $x + y = 1$ onto the disc $|w| < 1$.

Hint: Transform $x + y = 1$ into the real axis, then refer to Exercise 31(c).

b) Repeat part (a), but use the domain $|w| > 1$.

33. a) Find a transformation that will map the wedge-shaped domain $0 < \arg z < \pi/6$ onto the disc $|w| < 1$.

Hint: The transformation $w_1 = z^n$ (n is a suitable integer) will map this wedge onto the upper half of the w_1-plane. Now use the result of Exercise 31.

b) Find a transformation that will map the wedge $0 < \arg z < \alpha$ onto the same disc. Take $\alpha < 2\pi$.

34. We wish to find a conformal mapping that will map the oval-shaped domain shared by the two discs (see Fig. 8.4–11a) $|z - 1| < 2$ and $|z + 1| < 2$ onto the upper half of the w-plane. We will use the following steps:

a) First find the bilinear transformation that maps the points z_1, z_2, and z_3 into ∞, 0, and 1, respectively. Why does this transform the boundaries of the oval into the pair of lines in Fig. 8.4–11(b) that intersect at an angle α? What is the numerical value of α?

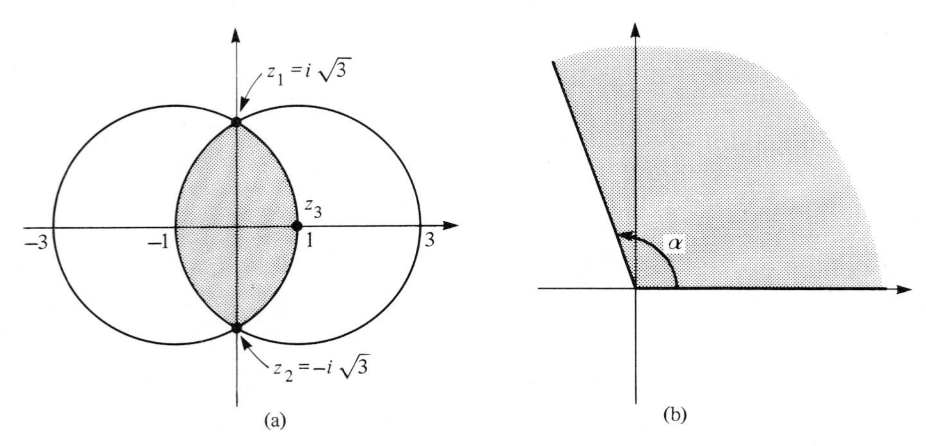

Figure 8.4–11

b) By finding the image of $z = 0$ verify that the transformation found in part (a) maps the oval-shaped domain onto the sector of angle α, shown in Fig. 8.4–11(b).

c) Obtain the solution to the exercise by mapping the sector of angle α onto the upper half-plane.

Hint: See previous problem.

35. a) Consider the domain lying between the circles $|z| = 2$ and $|z - i| = 1$. Find a bilinear transformation that will map this domain onto a strip $0 < \operatorname{Im} w < k$, where the line $\operatorname{Im} w = 0$ is the image of $|z| = 2$. The reader may choose k.

Hint: Transform the outer circle into $\operatorname{Im} w = 0$ by finding the bilinear transformation that maps $-2i$ into 0, 2 into 1, and $2i$ to ∞. Why will the inner boundary $|z - i| = 1$ be transformed into a line parallel to $\operatorname{Im} w = 0$ by this transformation? Verify that the domain bounded by the circles is mapped onto the strip. What is the value of k for your answer?

b) Use the transformation derived in part (a) and a modification of the transformation in Example 2, Section 8.3, to obtain a transformation that will map the domain bounded by the circles in part (a) onto the upper half-plane.

36. The *Smith chart* is a graphical device used in electrical engineering for the analysis of high-frequency transmission lines. It relates two complex variables: the normalized impedance $z = r + ix$ ($r \geq 0$, $-\infty < x < \infty$) and the reflection coefficient $\Gamma = a + ib$. Here a and b are real. The mapping

$$\Gamma(z) = \frac{z - 1}{z + 1} \tag{8.4–35}$$

is applied to a grid of infinite vertical and semiinfinite horizontal lines in the right half of the z-plane. The image of this grid in the Γ-plane is the Smith chart.

a) Show that the image of the region $\operatorname{Re} z \geq 0$ under the mapping in Eq. (8.4–35) is the disc $|\Gamma| \leq 1$.

b) Sketch and give the equation of the image of the following infinite vertical lines under the transformation (Eq. 8.4–35): $r = 0$, $r = 1/2$, $r = 1$, $r = 2$.

c) Sketch the image and give the equation of each of the following semiinfinite horizontal lines under the transformation (Eq. 8.4–35): $x = 0$, $r \geq 0$; $x = 1/2$, $r \geq 0$; $x = -1/2$, $r \geq 0$; $x = 2$, $r \geq 0$; $x = -2$, $r \geq 0$. The collection of images sketched in parts (b) and (c) form a primitive Smith chart.†

d) Solve Eq. (8.4–35) for $z(\Gamma)$. Show that $z(\Gamma) = 1/z(-\Gamma)$. Thus values of z corresponding to values of Γ that are diametrically opposed with respect to the origin of the Smith chart are reciprocals of each other.

---

† For a more elaborate chart, see D. Chang, *Field and Wave Electromagnetics*, 2nd ed. (Reading, MA: Addison-Wesley, 1989), p. 490.

8.5 CONFORMAL MAPPING AND BOUNDARY VALUE PROBLEMS[†]

Earlier in this book (see Section 2.6) we established the close connection that exists between harmonic functions and two-dimensional physical problems involving heat conduction, fluid flow, and electrostatics. Later (see Section 4.7) we returned to physical configurations when we investigated Dirichlet problems. We saw that when the values of a harmonic function (for example, temperature or voltage) are specified on the surface of a cylinder, the values assumed by the harmonic function inside the cylinder can be found. A similar procedure was developed to find a function that is harmonic above a plane surface when the values taken by the function on the plane are specified. What we know now are solutions of the Dirichlet problem for two simple types of boundaries.

In this section we will combine what we know about conformal mapping, harmonic functions, analytic functions, and the complex potential to solve Dirichlet problems whose boundaries are not limited to planes and cylinders. Electrostatic and heat-flow problems will be considered here. In the section after this one we will study heat and fluid-flow problems in which we seek an unknown harmonic function whose normal derivative is specified over some portion of a boundary. Although this is not a Dirichlet problem, we will again find that conformal mapping helps us to find a solution. The utility of conformal mapping in the solution of a wide variety of physical problems derives from the following theorem:

THEOREM 5 Let the analytic function $w = f(z)$ map the domain D from the z-plane onto the domain D_1 of the w-plane. Suppose $\phi_1(u, v)$ is harmonic in D_1, that is, in D_1

$$\frac{\partial^2 \phi_1}{\partial u^2} + \frac{\partial^2 \phi_1}{\partial v^2} = 0. \tag{8.5–1}$$

Then, under the change of variables

$$u(x, y) + iv(x, y) = f(z) = w, \tag{8.5–2}$$

we have that $\phi(x, y) = \phi_1(u(x, y), v(x, y))$ is harmonic in D, that is, in D

$$\frac{\partial^2 \phi}{\partial x^2} + \frac{\partial^2 \phi}{\partial y^2} = 0. \quad \square \tag{8.5–3}$$

Loosely, a solution of Laplace's equation remains a solution of Laplace's equation when transferred from one plane to another by a conformal transformation. Let us verify Theorem 5 in an elementary example. The function $\phi_1(u, v) = e^u \cos v$, which is $\operatorname{Re} e^w$, satisfies Eq. (8.5–1) (see Theorem 6, Chapter 2). Let

$$w = z^2 = x^2 - y^2 + i2xy = u + iv,$$

[†] The reader should review Sections 2.5, 2.6, and 4.7.

so that $u = x^2 - y^2$ and $v = 2xy$. Now $\phi(x, y) = e^{x^2-y^2} \cos(2xy)$ is readily found to satisfy Eq. (8.5–3), as the reader should verify.

We can easily prove Theorem 5 when the domain D_1 is simply connected. A more difficult proof, which dispenses with this requirement, is given in many texts.[†] We rely here on Theorem 7, Chapter 2, which guarantees that, with $\phi_1(u, v)$ satisfying Eq. (8.5–1) in D_1, there exists an analytic function in D_1:

$$\Phi_1(w) = \phi_1(u, v) + i\psi_1(u, v), \tag{8.5–4}$$

where $\psi_1(u, v)$ is the harmonic conjugate of $\phi_1(u, v)$. Since $w = f(z)$ is an analytic function in D, we have that $\Phi_1(f(z)) = \Phi(z)$ is an analytic function of an analytic function in D. Thus (see Theorem 5, Chapter 2) $\Phi(z)$ is analytic in D. Now

$$\Phi_1(w) = \Phi_1(f(z)) = \Phi(z) = \phi(x, y) + i\psi(x, y). \tag{8.5–5}$$

Since $f(z) = u(x, y) + iv(x, y)$, we have, by comparing Eqs. (8.5–4) and (8.5–5), that $\phi(x, y) = \phi_1(u(x, y), v(x, y))$ and $\psi(x, y) = \psi_1(u(x, y), v(x, y))$.

Because $\phi(x, y)$ is the real part of an analytic function $\Phi(z)$, it follows that $\phi(x, y)$ must be harmonic in D. A parallel argument establishes that $\psi(x, y)$, the imaginary part of $\Phi(z)$, must be harmonic in D.

To see the usefulness of Theorem 5 imagine we are given a domain D in the z-plane. We seek a function $\phi(x, y)$ (say, temperature or voltage) that is harmonic in D and that assumes certain prescribed values on the boundary of D. Suppose we can find an analytic transformation $w = u + iv = f(z)$ that maps D onto a domain D_1 in the w-plane, and D_1 has a simpler or more familiar shape than D. Assume that we can find a function $\phi_1(u, v)$ that is harmonic in D_1 and that assumes values at each boundary point of D_1 exactly equal to the value required of $\phi(x, y)$ at the image of that point on the boundary of D. Then, by Theorem 5, $\phi(x, y) = \phi_1(u(x, y), v(x, y))$ will be harmonic in D and also assume the desired values on the boundary of D. The method is illustrated schematically in Fig. 8.5–1. We have mapped D onto D_1 using $w = f(z)$. The boundaries of the domains are C and C_1; (x_0, y_0) is an arbitrary point on C with image (u_0, v_0) on C_1. The harmonic function $\phi_1(u, v)$ assumes the same value k at (u_0, v_0), as does $\phi(x, y)$ at (x_0, y_0).

Of course the point (x_0, y_0) need not be confined to the boundary of D. Let (x_0, y_0) be an interior point of D and let (u_0, v_0) be its corresponding image point, which is interior to D_1. Then the values assumed by $\phi(x, y)$ and $\phi_1(u, v)$ at (x_0, y_0) and (u_0, v_0), respectively, will be identical. Similarly the harmonic conjugates of $\phi(x, y)$ and $\phi_1(u, v)$, that is, $\psi(x, y)$ and $\psi_1(u, v)$, which are harmonic in the domains D and D_1, respectively, also assume identical values at (x_0, y_0) and (u_0, v_0).

We should note that it can be difficult to find an analytic transformation that will map a given domain onto one of some specified simpler shape. We can refer to dictionaries of conformal mappings as an aid.[‡] Often experience or trial and error

---

[†] See, e.g., R. Churchill and J. Brown, *Complex Variables and Applications*, 5th ed. (New York, McGraw-Hill, 1990), pp. 251–252.

[‡] See, for example, H. Kober, *Dictionary of Conformal Representations*, 2nd ed. (New York: Dover, 1957).

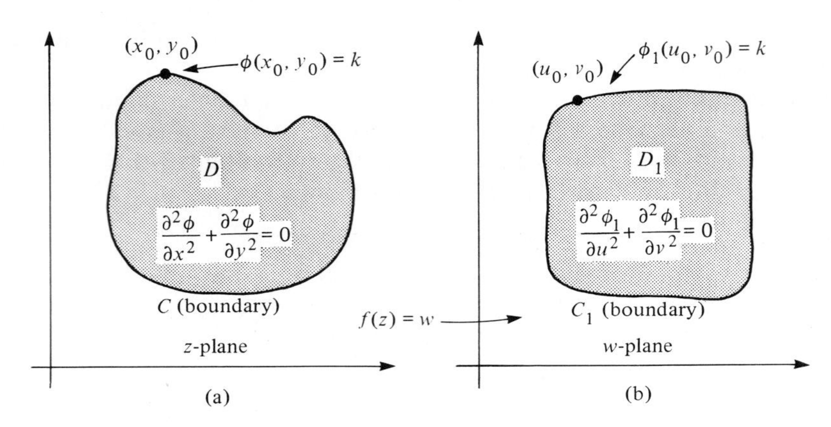

Figure 8.5–1

help. The *Riemann mapping theorem*† guarantees the existence of an analytic transformation that will map any simply connected domain (except the entire z-plane) onto the unit disc $|w| < 1$. The boundary of the domain is transformed into $|w| = 1$. The Poisson integral formula for the circle (see Section 4.7) can then be used to solve the transformed Dirichlet problem. The Riemann mapping theorem does not tell us how to obtain the required mapping, only that it exists.

EXAMPLE 1

Two cylinders are maintained at temperatures of 0° and 100°, as shown in Fig. 8.5–2(a). An infinitesimal gap separates the cylinders at the origin. Find $\phi(x, y)$, the temperature in the domain between the cylinders.

Solution

The shape of the given domain is complicated. However, because the bilinear transformation will map circles into straight lines, we can transform this domain into the more tractable infinite strip shown in Fig. 8.5–2(b). We follow the method of the previous section and find that the bilinear transformation that maps a, b, and c from Fig. 8.5–2(a) into $a' = 1$, $b' = 0$, $c' = \infty$ is

$$w = \frac{1 - z}{z}. \tag{8.5–6}$$

Under the transformation, the cylindrical boundary at 100° is transformed into the line $u = 1$, whereas the cylinder at 0° becomes the line $u = 0$.

The strip $0 < u < 1$ is the image of the region between the two circles shown in Fig. 8.5–2(a). Our problem now is the simpler one of finding $\phi_1(u, v)$, which is harmonic

---

† See R. Nevanlinna and V. Paatero, *Introduction to Complex Analysis* (Reading, MA: Addison-Wesley, 1969), Chapter 17.

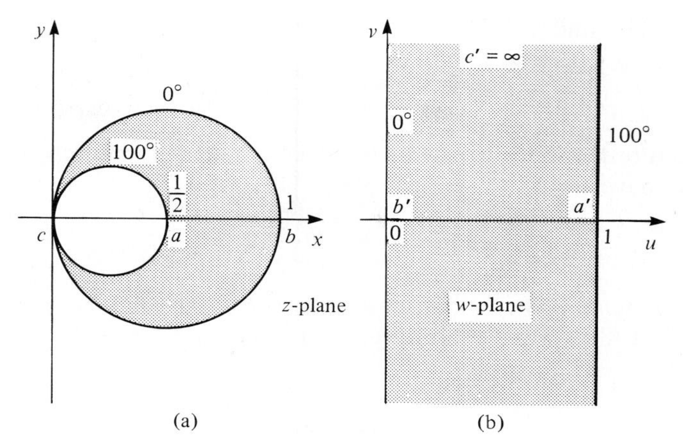

Figure 8.5–2

in the strip. We must also fulfill the boundary conditions $\phi_1(0, v) = 0$ and $\phi_1(1, v) = 100$.

The problem is now easy enough so that we might guess the result, or we can study the similar Example 1 in Section 2.6. From symmetry we expect that $\phi_1(u, v)$ is independent of v, and we notice that

$$\phi_1(u, v) = 100u \tag{8.5–7}$$

satisfies both boundary conditions and is harmonic. This is the temperature distribution in the transformed problem.

Now $\phi_1(u, v)$ is the real part of an analytic function. Using the methods of Section 2.5 or employing common sense, we see that $\phi_1(u, v) = \text{Re}(100w)$. Thus the complex temperature (see Section 2.6) in the strip is

$$\Phi_1(w) = 100w, \tag{8.5–8}$$

and the corresponding stream function is

$$\psi_1(u, v) = \text{Im}(100w) = 100v. \tag{8.5–9}$$

To obtain the temperature $\phi(x, y)$ and stream function $\psi(x, y)$ for Fig. 8.5–2(a), we must transform $\phi_1(u, v)$ and $\psi_1(u, v)$ back into the z-plane by means of Eq. (8.5–6). From Eq. (8.5–6) we have

$$w = u + iv = \frac{1}{z} - 1 = \frac{1}{x + iy} - 1 = \frac{x}{x^2 + y^2} - 1 - \frac{iy}{x^2 + y^2}, \tag{8.5–10}$$

which implies

$$u = \frac{x}{x^2 + y^2} - 1, \tag{8.5–11a}$$

$$v = \frac{-y}{x^2 + y^2}. \tag{8.5–11b}$$

From Eqs. (8.5–11a) and (8.5–7) we have

$$\phi(x, y) = 100\left(\frac{x}{x^2 + y^2} - 1\right) \tag{8.5–12}$$

for the temperature distribution between the given cylinders. Combining Eqs. (8.5–11b) and (8.5–9), we have

$$\psi(x, y) = \frac{-100y}{x^2 + y^2}. \tag{8.5–13}$$

The complex potential $\Phi(z) = \phi(x, y) + i\psi(x, y)$ can be obtained by combining Eqs. (8.5–12) and (8.5–13) or more directly through the use of Eq. (8.5–6) in Eq. (8.5–8). Thus

$$\Phi(z) = 100\,\frac{1 - z}{z}. \tag{8.5–14}$$

The singularity at $z = 0$ is typical of the behavior of complex potentials at a point where a boundary condition is discontinuous. The shape of the isotherms (surfaces of constant temperature) for Fig. 8.5–2(a) are of interest. If on some surface the temperature is T_0, the locus of this surface must be, from Eq. (8.5–12),

$$T_0 = 100\left(\frac{x}{x^2 + y^2} - 1\right). \tag{8.5–15}$$

From physical considerations we know that T_0 cannot be greater than the temperature of the hottest part of the boundary nor can it be less than the temperature of the coldest part of the boundary, that is, $0 \le T_0 \le 100°$ (see also Exercises 13 and 14, Section 4.6). We can rearrange Eq. (8.5–15) and complete a square to obtain

$$\left(x - \frac{1/2}{\left(1 + \frac{T_0}{100}\right)}\right)^2 + y^2 = \left(\frac{1/2}{1 + \frac{T_0}{100}}\right)^2.$$

Thus an isotherm of temperature T_0 is a cylinder whose axis passes through

$$y_0 = 0, \qquad x_0 = \left(\frac{1/2}{1 + \frac{T_0}{100}}\right).$$

The cross sections of a few such cylinders are shown as circles in Fig. 8.5–3.

In Exercise 1 of this section we show that the streamlines describing the heat flow are circles that intersect the isotherms at right angles. A streamline, with an arrow indicating the direction of heat flow, is shown in Fig. 8.5–3.

The complex potential (see Eq. 8.5–14) readily yields the complex heat flux density $\mathbf{q}(z)$ between the cylinders. Recall from Eq. (2.6–14) that

$$\mathbf{q}(z) = -k\overline{\left(\frac{d\Phi}{dz}\right)}, \tag{8.5–16}$$

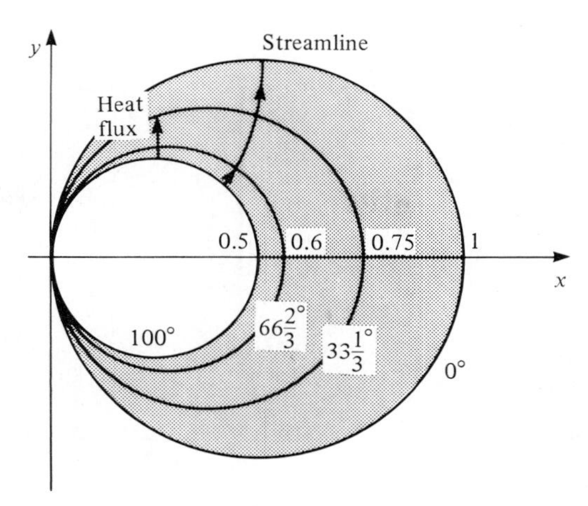

Figure 8.5–3

where

$$\mathbf{q}(z) = Q_x(x, y) + iQ_y(x, y). \tag{8.5–17}$$

We should remember that Q_x and Q_y give the components of the vector heat flow at a point in a material whose thermal conductivity is k. From Eqs. (8.5–16) and (8.5–14) we have

$$\mathbf{q}(z) = Q_x + iQ_y = 100k\left(\overline{\frac{1}{z^2}}\right) = \frac{100k}{(\bar{z})^2}.$$

Thus, for example, at $x = 1/4$, $y = 1/4$ (on top of the inner cylinder) we find

$$Q_x + iQ_y = 100k\frac{1}{\left(\frac{1}{4} - \frac{1}{4}i\right)^2} = 800ki.$$

Since $Q_x = 0$ and $Q_y = 800k$, the heat flow at (1/4, 1/4) is parallel to the y-axis, as we have indicated schematically in Fig. 8.5–3. ◀

EXAMPLE 2

Refer to Fig. 8.5–4(a). An electrically conducting strip has a cross section described by $y = 0$, $-1 < x < 1$. It is maintained at 0 volts electrostatic potential. A half cylinder shown in cross section, described by the arc $|z| = 1$, $0 < \arg z < \pi$, is maintained at 10 volts. Find the potential $\phi(x, y)$ inside the semicircular tube bounded by the two conductors.

Solution

If the boundary of the given configuration (see Fig. 8.5–4a) were transformed into the line $\operatorname{Im} w = 0$, in the w-plane, with the half disc in Fig. 8.5–4(a) mapped onto the half space $\operatorname{Im} w > 0$, we could use the Poisson integral formula for the half plane

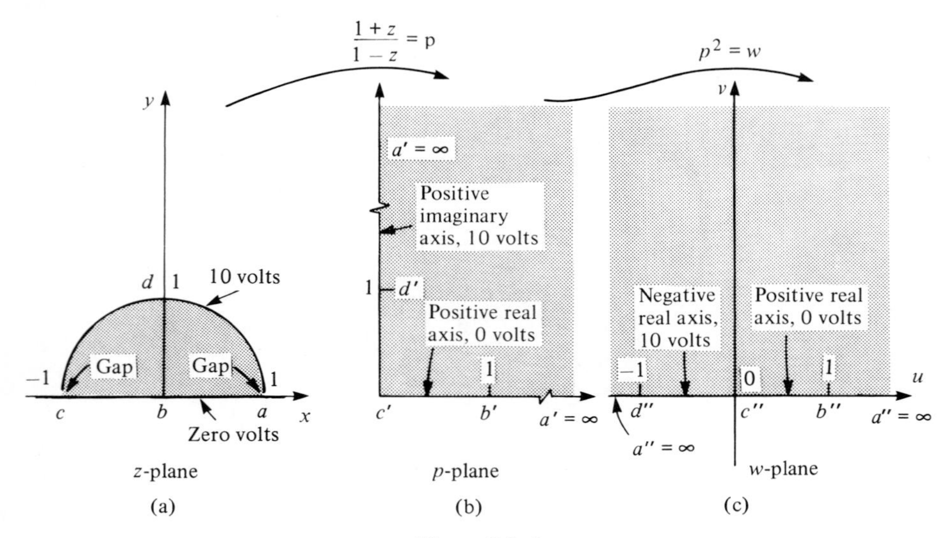

Figure 8.5–4

(see Section 4.7) to obtain $\phi_1(u, v)$ for the transformed region. A bilinear transformation by itself cannot be used to convert the semicircular boundary in Fig. 8.5–4(a) into a straight line. However, a bilinear transformation *can* map the semicircle into a pair of semiinfinite lines corresponding to the positive real and positive imaginary axes.

One readily verifies that

$$p = \frac{1 + z}{1 - z} \tag{8.5–18}$$

performs this transformation (see Fig. 8.5–4b) with the half disc being mapped onto the first quadrant of the p-plane.

Referring now to Fig. 8.4–8(a) and Fig. 8.4–8(b), we see that an additional transformation involving a square of p in Eq. (8.5–18) will map the first quadrant of Fig. 8.5–4(b) onto the upper half plane in Fig. 8.5–4(c). Combining both transformations, we find that

$$w = \left(\frac{1 + z}{1 - z}\right)^2 \tag{8.5–19}$$

maps the half disc of Fig. 8.5–4(a) onto the half space of Fig. 8.5–4(c). Corresponding boundary points and transformed boundary conditions are indicated in Fig. 8.5–4.

To find the potential in the half space of Fig. 8.5–4(c) that satisfies the transformed boundary conditions $\phi_1(u, 0) = 0, u > 0$ and $\phi_1(u, 0) = 10, u < 0$, we can use the result of Example 2, Section 4.7, which employed the Poisson integral formula for the half plane. Replacing T_0 of that example by 10 volts and x and y by u and v, we have

$$\phi_1(u, v) = \frac{10}{\pi} \tan^{-1}\frac{v}{u} = \frac{10}{\pi} \arg w, \tag{8.5–20}$$

where $0 \le \arg w \le \pi$. Notice that

$$\phi_1(u, v) = \text{Re}\left[-i\frac{10}{\pi}\,\text{Log}\, w\right],$$

which implies that the complex potential for the configuration of Fig. 8.5–4(c) is

$$\Phi_1(w) = \frac{-10i}{\pi}\,\text{Log}\, w. \tag{8.5–21}$$

To transform $\phi_1(u, v)$ of Eq. (8.5–20) into $\phi(x, y)$ for the half cylinder, we recall the identity $\arg(s^2) = 2 \arg s$. With Eq. (8.5–19) used in Eq. (8.5–20) we have

$$\phi(x, y) = \frac{10}{\pi}\arg\left(\frac{1+z}{1-z}\right)^2 = \frac{20}{\pi}\arg\frac{1+z}{1-z} = \frac{20}{\pi}\arg\frac{x+1+iy}{1-x-iy}$$

$$= \frac{20}{\pi}\arg\left(\frac{1-x^2-y^2}{(x-1)^2+y^2} + \frac{i2y}{(x-1)^2+y^2}\right),$$

and finally since $\arg s = \tan^{-1}[\text{Im}\, s/\text{Re}\, s]$, we find

$$\phi(x, y) = \frac{20}{\pi}\tan^{-1}\frac{2y}{1-x^2-y^2}. \tag{8.5–22}$$

We require that $0 \le \tan^{-1}(\cdots) \le \pi/2$ since $\phi(x, y)$ must satisfy $0 \le \phi(x, y) \le 10$. Notice, with this branch of the arctangent, that Eq. (8.5–22) satisfies the required boundary conditions, that is,

$$\lim_{(x^2+y^2)\to 1} \phi(x, y) = 10 \qquad \text{(on the curved boundary)},$$

$$\lim_{y\to 0} \phi(x, y) = 0 \qquad \text{(on the flat boundary)}.$$

One can easily show that the equipotentials are circular arcs. ◀

A common concern in electrostatics is the amount of capacitance between two conductors. If Q is the electrical charge on either conductor and ΔV is the difference in potential between one conductor and another, then the capacitance C is defined to be†

$$C = \frac{|Q|}{|\Delta V|}. \tag{8.5–23}$$

In two-dimensional problems we compute the capacitance per unit length c of a pair of conductors whose cross section is typically displayed in the complex plane. In

---

† The two conductors carry charges that are equal in magnitude and opposite in sign. Since $|\Delta V|$ is directly proportional to $|Q|$ it will be found that the ratio in Eq. (8.5–23) is independent of any assumed value for Q or of any assumed value of ΔV.

Eq. (8.5–23) we take Q_L as the charge on an amount of one conductor that is 1 unit long in a direction perpendicular to the complex plane. In the appendix to this chapter we establish Theorem 6, which is useful in capacitance calculations.

THEOREM 6 The electrical charge per unit length on a conductor that belongs to a charged two-dimensional configuration of conductors is

$$Q_L = \varepsilon \Delta \psi(z), \tag{8.5–24}$$

where ε is a constant (the permittivity of the surrounding material) and $\Delta\psi$ is the decrement (initial value minus final value) of the stream function as we proceed in the positive direction once around the boundary of the cross section of the conductor in the complex plane. □

Usually $\psi(z)$ will be a multivalued function defined by means of a branch cut. Thus $\psi(z)$ does not return to its original value when we encircle the conductor, and thus $\Delta\psi \neq 0$. The direction of encirclement is the positive one used in the contour integration, that is, the interior is on the left. Combining Eqs. (8.5–23) and (8.5–24), we have

$$c = \varepsilon \frac{|\Delta\psi|}{|\Delta V|}. \tag{8.5–25}$$

Also derived in the appendix to this chapter is this interesting fact:

THEOREM 7 The capacitance of a two-dimensional system of conductors is unaffected by a conformal transformation of its cross section. □

The usefulness of the two preceding theorems is illustrated by the following example.

EXAMPLE 3

a) The pair of coaxial electrically conducting tubes in Fig. 8.5–5(a) having radii a and b are maintained at potentials V_a and 0, respectively. Find the electrostatic potential between the tubes, and find their capacitance per unit length. This system of conductors is called a *coaxial transmission line.*
b) Use a conformal transformation and the result of part (a) to determine the capacitance of the transmission system consisting of the two conducting tubes shown in Fig. 8.5–5(b). This is called a *two-wire line.*

Solution

Part (a): We seek a function $\phi(x, y)$ harmonic in the domain $a < |z| < b$. The boundary conditions are

$$\lim_{\sqrt{x^2+y^2}\to a} \phi(x, y) = V_a, \tag{8.5–26}$$

$$\lim_{\sqrt{x^2+y^2}\to b} \phi(x, y) = 0. \tag{8.5–27}$$

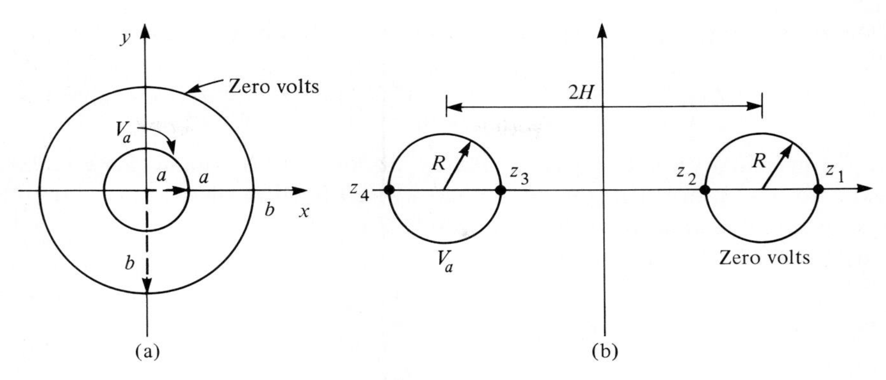

Figure 8.5–5

These requirements suggest that the equipotentials are circles concentric with the boundary. We might recall from Example 2, Section 2.5, that $(1/2)\,\text{Log}(x^2 + y^2) = \text{Log}\sqrt{x^2 + y^2}$ is harmonic and does produce circular equipotentials. However, this function fails to meet the boundary conditions. The more general harmonic function

$$\phi(x, y) = A\,\text{Log}(\sqrt{x^2 + y^2}) + B, \qquad \text{where } A \text{ and } B \text{ are real numbers,} \tag{8.5–28}$$

also yields circular equipotentials and can be made to meet the boundary conditions. From Eq. (8.5–26) we obtain

$$V_a = A\,\text{Log}\,a + B,$$

and from Eq. (8.5–27) we have

$$0 = A\,\text{Log}\,b + B.$$

Solving these equations simultaneously, we get

$$A = \frac{-V_a}{\text{Log}(b/a)}, \qquad B = \frac{V_a\,\text{Log}\,b}{\text{Log}(b/a)},$$

which, when used in Eq. (8.5–28), shows that

$$\phi(x, y) = \frac{-V_a\,\text{Log}\sqrt{x^2 + y^2}}{\text{Log}(b/a)} + \frac{V_a\,\text{Log}\,b}{\text{Log}(b/a)}. \tag{8.5–29}$$

Since $\text{Log}\sqrt{x^2 + y^2} = \text{Log}|z| = \text{Re Log}\,z$, we can rewrite Eq. (8.5–29) as

$$\phi(x, y) = \text{Re}\left[\frac{-V_a\,\text{Log}\,z}{\text{Log}(b/a)} + \frac{V_a\,\text{Log}\,b}{\text{Log}(b/a)}\right] = \text{Re}\left[\frac{V_a\,\text{Log}(b/z)}{\text{Log}(b/a)}\right].$$

The preceding equation shows that the complex potential is given by

$$\Phi(z) = V_a\,\frac{\text{Log}(b/z)}{\text{Log}(b/a)}. \tag{8.5–30}$$

The stream function, $\psi(x, y) = \text{Im}\,\Phi(x, y)$ is found from Eq. (8.5–30) to be

$$\psi(x, y) = \text{Im}\left[\frac{V_a(\text{Log}\,b - \text{Log}\,z)}{\text{Log}(b/a)}\right] = -\frac{V_a \arg z}{\text{Log}(b/a)}, \tag{8.5–31}$$

where the principal value of arg z, defined by a branch cut on the negative real axis, is used. We now proceed in the counterclockwise direction once around the inner conductor and compute the decrease in ψ (see Fig. 8.5–6).

Just below the branch cut

$$\psi = \frac{-V_a(-\pi)}{\text{Log}(b/a)} = \frac{V_a\pi}{\text{Log}(b/a)},$$

while just above the branch cut

$$\psi = \frac{V_a(-\pi)}{\text{Log}(b/a)}.$$

The decrease in ψ on this circuit is

$$\Delta\psi = \frac{2\pi V_a}{\text{Log}(b/a)}.$$

The magnitude of the potential difference between the two conductors is $|\Delta V| = V_a$ since the outer conductor is at zero potential. Thus, according to Eq. (8.5–25),

$$c = \frac{\varepsilon}{V_a}\frac{2\pi V_a}{\text{Log}(b/a)} = \frac{2\pi\varepsilon}{\text{Log}(b/a)}, \tag{8.5–32}$$

which is a useful result in electrical engineering.

Part (b): Suppose a bilinear transformation with real coefficients can be found that transforms the left-hand circle in Fig. 8.5–5(b) into a circle in the w-plane (see Fig. 8.5–7) in such a way that the points $z_4 = -H - R$ and $z_3 = -H + R$ are mapped into $w_4 = 1$ and $w_3 = -1$.

The real coefficients of the transformation ensure that the left-hand circle in Fig. 8.5–5(b) is transformed into a unit circle centered at the origin in Fig. 8.5–7 (see

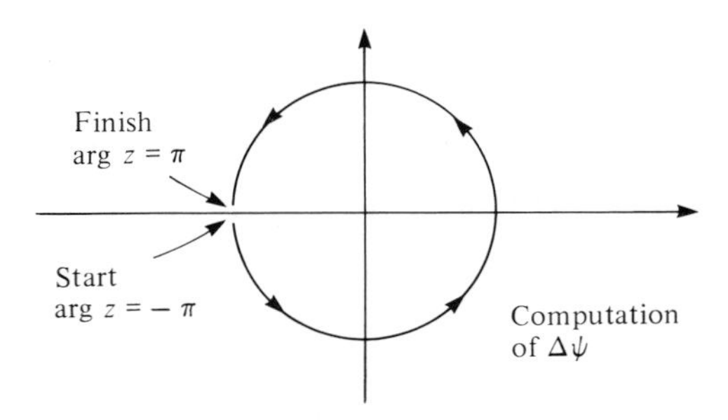

Figure 8.5–6

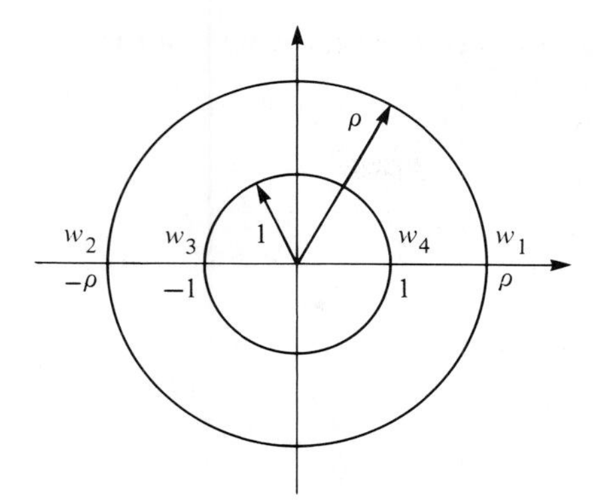

Figure 8.5–7

Exercise 2, Section 8.4). Suppose now the same formula transforms the right-hand circle in Fig. 8.5–5(b) into a circle $|w| = \rho$, with $z_1 = H + R$ having image $w_1 = \rho$, and $z_2 = H - R$ having image $w_2 = -\rho$ (see Fig. 8.5–7). We can solve for ρ by using the equality of the cross ratios (z_1, z_2, z_3, z_4) and (w_1, w_2, w_3, w_4). Thus, from Eq. (8.4–24),

$$\frac{2\rho(-2)}{(\rho - 1)(\rho - 1)} = \frac{(2R)(2R)}{(2H + 2R)(-2H + 2R)}. \tag{8.5–33}$$

Some rearrangement yields a quadratic

$$\rho^2 + \left(2 - \frac{4H^2}{R^2}\right)\rho + 1 = 0, \tag{8.5–34}$$

whose solution is

$$\rho = \frac{2H^2}{R^2} - 1 \pm \frac{2H}{R}\sqrt{\frac{H^2}{R^2} - 1}. \tag{8.5–35}$$

Since the two cylinders in Fig. 8.5–5(b) are not touching, we know that $H/R > 1$. The root containing the plus sign in Eq. (8.5–35) therefore exceeds 1. We should recall from our knowledge of quadratic equations that the product of the roots of Eq. (8.5–34) must equal 1. Thus the root containing the minus sign in Eq. (8.5–35) must lie between 0 and 1. Either root can be selected, and the same result will be obtained for the capacitance per unit length.

Let us arbitrarily choose the plus sign in Eq. (8.5–35). This corresponds to Fig. 8.5–7, where $\rho > 1$. Notice that with this choice of sign we can rewrite Eq. (8.5–35) as

$$\rho = \left(\frac{H}{R} + \sqrt{\frac{H^2}{R^2} - 1}\right)^2. \tag{8.5–36}$$

We can compute the capacitance per unit length of the coaxial system of Fig. 8.5–7 by using Eq. (8.5–32) and taking $a = 1$, $b = \rho$. Thus using Eq. (8.5–36),

$$c = \frac{2\pi\varepsilon}{\operatorname{Log}\left[\left(\frac{H}{R} + \sqrt{\frac{H^2}{R^2} - 1}\right)^2\right]},$$

or

$$c = \frac{\pi\varepsilon}{\operatorname{Log}\left(\frac{H}{R} + \sqrt{\frac{H^2}{R^2} - 1}\right)}. \tag{8.5–37}$$

By Theorem 7 this must be the capacitance of the image of Fig. 8.5–7, that is, the two-wire line of Fig. 8.5–5(b). ◀

EXERCISES

1. a) For Example 1 find the equation of the streamline along which ψ assumes a constant value β. Show that this locus, if drawn in Figure 8.5–2(a), is a circle that is centered on the y-axis and passes through the origin.

b) Use an argument based on plane geometry to show why such a circle must intersect the isotherms found in this example at right angles.

2. A heat-conducting material occupies the wedge $0 \le \arg z \le \alpha$. The boundaries are maintained at temperatures T_1 and T_2 as shown in Fig. 8.5–8.

a) Show that $w = u + iv = \operatorname{Log} z$ transforms the wedge given above into a strip parallel to the u-axis.

b) The isotherms in the strip are obviously parallel to the u-axis. Show that the temperature in this region can be described by an expression of the form

$$\phi_1(u, v) = Av + B,$$

and find A and B.

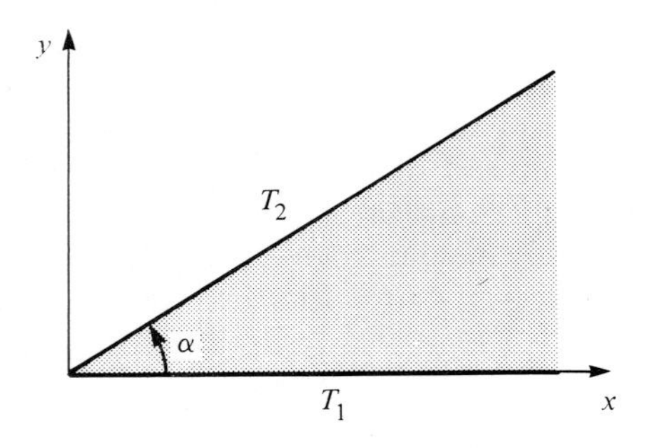

Figure 8.5–8

c) Use the result of part (b) to show that the temperature in the given wedge is

$$\phi(x, y) = \frac{T_2 - T_1}{\alpha} \tan^{-1}\left(\frac{y}{x}\right) + T_1.$$

d) Show that the complex temperature in the wedge is

$$\Phi(x, y) = -i\left[\frac{T_2 - T_1}{\alpha}\right] \operatorname{Log} z + T_1.$$

e) Describe the streamlines and isotherms in the wedge.

3. A system of electrical conductors has the cross section shown in Fig. 8.5–9. The potentials of the conductors are maintained as indicated. Determine the complex potential $\Phi(z)$ for the shaded region $\operatorname{Im} z > 0$, $|z| > 1$.

Hint: Consider $w = -1/z$, and use the result of Example 2.

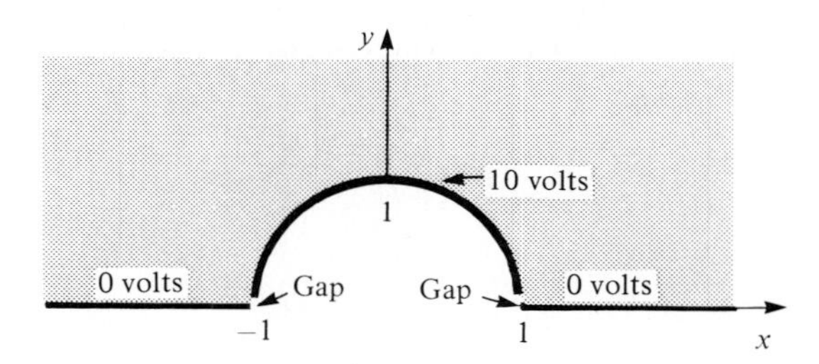

Figure 8.5–9

4. A cylinder of unit diameter is maintained at a temperature of 100°. It is tangent to a plane maintained at 0° (see Fig. 8.5–10). A material of heat conductivity k exists between the cylinder and the plane, that is for $\operatorname{Re} z > 0$, $|z - 1/2| > 1/2$.

a) Show that the temperature inside the material of conductivity k is

$$\phi(x, y) = \frac{100x}{x^2 + y^2}.$$

b) Show that the stream function is

$$\psi(x, y) = \frac{-100y}{x^2 + y^2}.$$

c) Show that the complex heat flux density vector is

$$\mathbf{q} = \frac{100k}{(x^2 + y^2)^2}(x^2 - y^2 + i2xy).$$

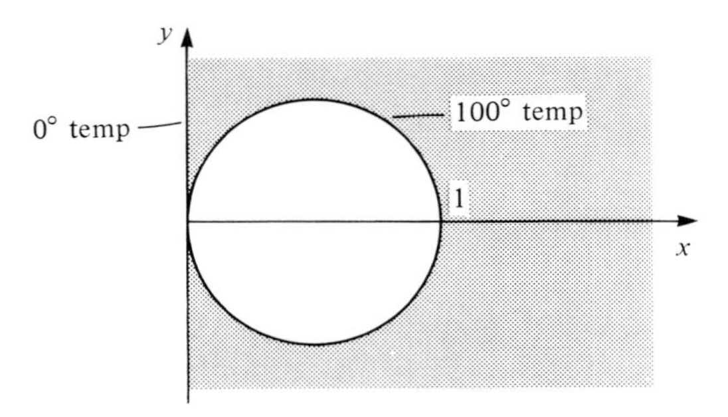

Figure 8.5–10

5. An electrically conducting cylinder of radius R has its axis a distance H from an electrically conducting plane (see Fig. 8.5–11a). A bilinear transformation will map the cross section of this configuration into the pair of concentric circles shown in the w-plane (see Fig. 8.5–11b). Image points are indicated with subscripts. Find ρ, the radius of the circle that is the image of the line $x = 0$. Assume $\rho > 1$. Use your result to show that the capacitance, per unit length, between the cylinder and the plane is

$$c = \frac{2\pi\varepsilon}{\operatorname{Log}\left(\frac{H}{R} + \sqrt{\frac{H^2}{R^2} - 1}\right)}, \qquad H > R.$$

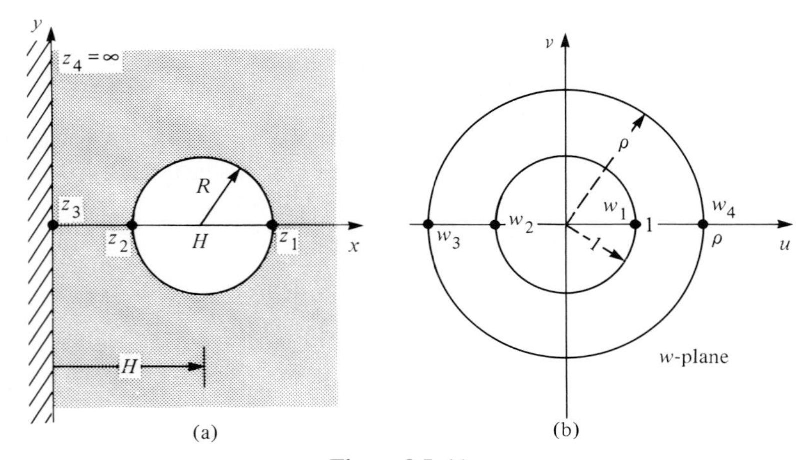

Figure 8.5–11

6. Refer to the two cylinders shown in cross section in Fig. 8.5–5(b). They are now to be interpreted as being embedded in a heat conducting material. The left-hand cylinder is maintained at temperature T_a and the right-hand cylinder is kept at a temperature of zero degrees. Let $H = 2$ and $R = 1$. Show that the temperature in the conducting material external to the cylinders is given by

$$\phi(x, y) = \frac{T_a}{2}\frac{\operatorname{Log}(\rho^2/|w|^2)}{\operatorname{Log}\rho},$$

where $\rho = 7 + 4\sqrt{3}$ and

$$|w|^2 = \frac{[x(76 + 44\sqrt{3}) + 132 + 76\sqrt{3}]^2 + y^2(76 + 44\sqrt{3})^2}{[x(20 + 12\sqrt{3}) - 36 - 20\sqrt{3}]^2 + y^2(20 + 12\sqrt{3})^2}.$$

7. a) A transmission line consists of two electrically conducting tubes with cross sections as shown in Fig. 8.5–12. Their axes are displaced a distance D. Note that $D + R_1 < R_2$. Show that the capacitance per unit length is given by

$$c = \frac{2\pi\varepsilon}{\operatorname{Log}\rho},$$

where

$$\rho = \frac{R_1^2 + R_2^2 - D^2}{2R_1R_2} + \sqrt{\left(\frac{R_1^2 + R_2^2 - D^2}{2R_1R_2}\right)^2 - 1}.$$

Express the capacitance in terms of an inverse hyperbolic function.

b) Take $D = R_1 = 1, R_2 = 3$. Let the inner conductor be at 1 volt, the outer at 0 volts. Find the electrostatic potential $\phi(x, y)$ in the domain bounded by the two circles in Fig. 8.5–12.

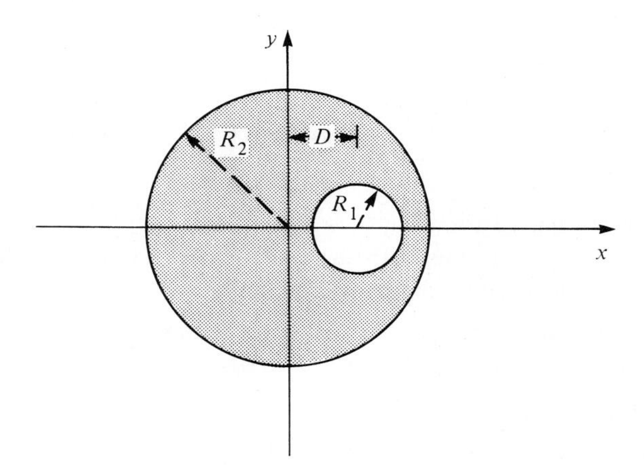

Figure 8.5–12

8. Consider the transformation $z = k \cosh w$, where $k > 0$.

a) Show that the line segment $u = \cosh^{-1}(A/k)$, $-\pi < v \le \pi$ is transformed into the ellipse $x^2/A^2 + y^2/(A^2 - k^2) = 1$ (see Fig. 8.5–13). Take $A > k$.

b) Show that this transformation takes the infinite line $u = \cosh^{-1}(A/k)$, $-\infty < v < \infty$ into the ellipse of part (a). Is the mapping one to one?

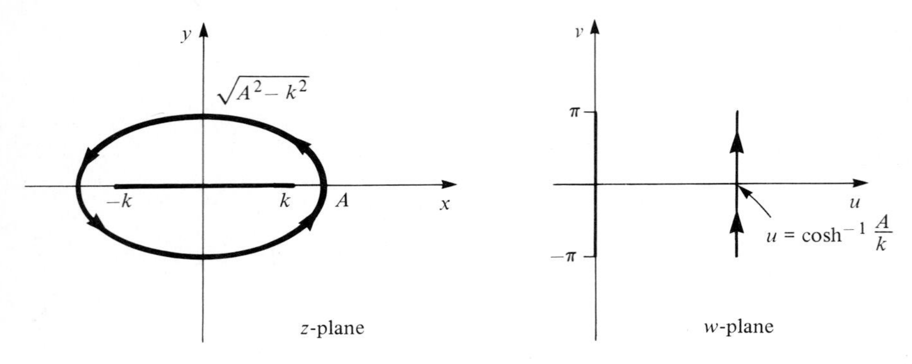

Figure 8.5–13

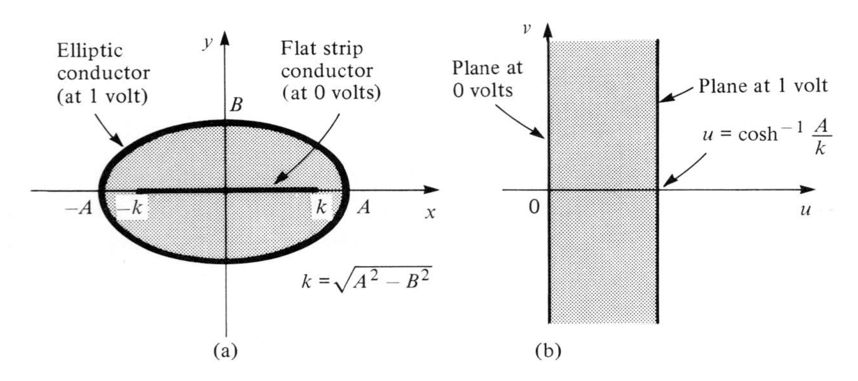

Figure 8.5–14

c) Show that the line segment $u = 0$, $-\pi < v \leq \pi$ is transformed into the line segment $y = 0$, $-k \leq x \leq k$. Is this mapping one to one? How is the infinite line $u = 0$, $-\infty < v < \infty$ mapped by the transformation?

d) Show that the capacitance per unit length of the transmission line whose cross section is shown in Fig. 8.5–14(a) is $2\pi\varepsilon/\cosh^{-1}(A/k)$.

Hint: Find the electrostatic potential $\phi(u, v)$ and the complex potential $\Phi(w)$ between the pair of infinite planes in Fig. 8.5–14(b) maintained at the voltages shown. By how much does the stream function ψ change if we encircle the inner conductor in Fig. 8.5–14(a)? Negotiate the corresponding path of Fig. 8.5–14(b).

9. a) The function

$$\phi(u, v) = A \arg w + B, \tag{8.5–38}$$

where A and B are real numbers, and $\arg w$ is the principal value, is harmonic since

$$\phi(u, v) = \operatorname{Re} \Phi(w),$$

where

$$\Phi(w) = -Ai \operatorname{Log} w + B.$$

Assume that the line $v = 0$, $u > 0$ is the cross section of an electrical conductor maintained at V_2 volts and that the line $v = 0$, $u < 0$ is similarly a conductor at V_1 volts (see Fig. 8.5–15). Find A and B in Eq. (8.5–38) so that $\phi(u, v)$ will be the electrostatic potential in the space $v > 0$.

b) Obtain $\phi(u, v)$, harmonic for $v > 0$ and satisfying these same boundary conditions along $v = 0$ by using the Poisson integral formula for the upper half plane (see Section 4.7).

c) Find A_1, A_2, and B (all real numbers) so that

$$\begin{aligned}\phi(u, v) &= A_1 \arg(w - u_1) + A_2 \arg(w - u_2) + B \\ &= \operatorname{Re}[-A_1 i \operatorname{Log}(w - u_1) - A_2 i \operatorname{Log}(w - u_2) + B] \qquad (8.5\text{–}39)\end{aligned}$$

is the solution in the space $v \geq 0$ of the electrostatic boundary value problem shown in Fig. 8.5–16, that is, $\phi(u, v)$ is harmonic for $v \geq 0$ and satisfies

$$\phi(u, 0) = V_1, \quad u < u_1; \qquad \phi(u, 0) = V_2, \quad u_1 < u < u_2;$$
$$\phi(u, 0) = V_3, \quad u > u_2.$$

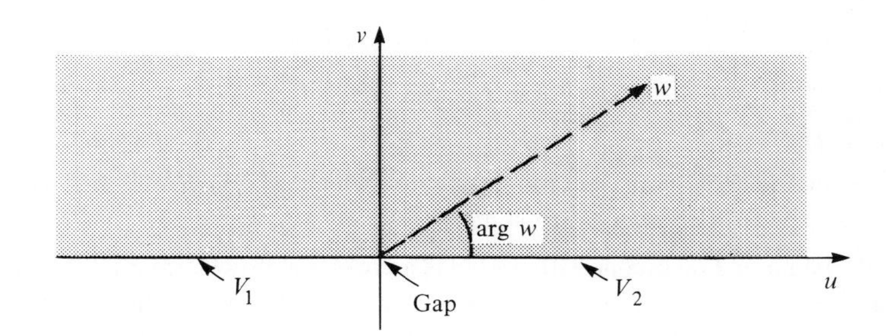

Figure 8.5–15

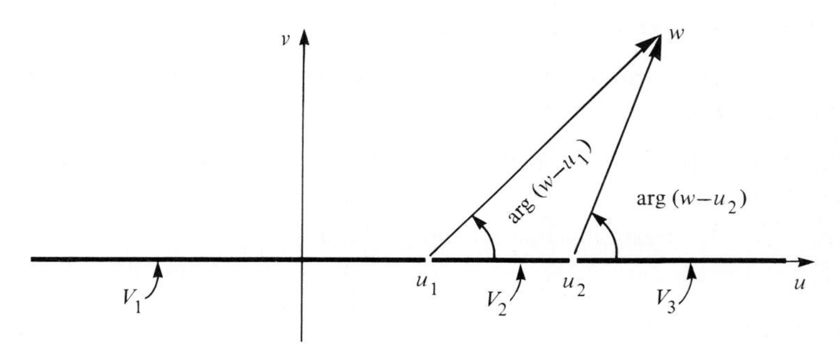

Figure 8.5–16

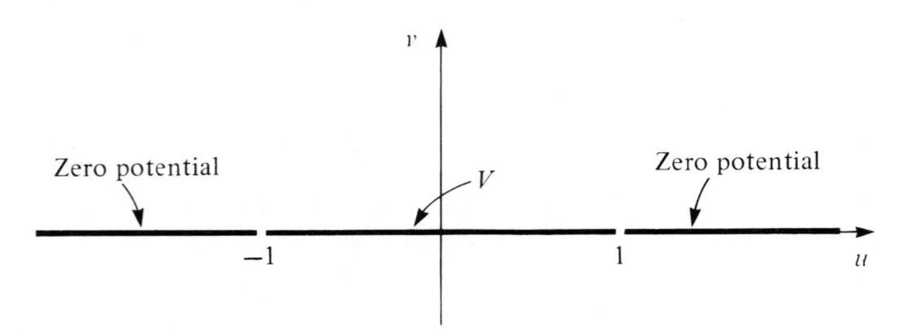

Figure 8.5–17

d) Let $u_1 = -1$ and $u_2 = 1$, $V_1 = V_3 = 0$ and $V_2 = V$ in the configuration of part (c) (see Fig. 8.5–17). Use the result of part (c) to show that $\phi(u, v)$, harmonic for $v > 0$ and meeting these boundary conditions, is given by

$$\phi(u, v) = \operatorname{Re} \Phi(w),$$

where the complex potential is

$$\Phi(w) = -i\frac{V}{\pi}\operatorname{Log}\left(\frac{w-1}{w+1}\right),$$

and

$$\phi(u, v) = \frac{V}{\pi}\tan^{-1}\frac{2v}{u^2+v^2-1}$$

for $0 \le \tan^{-1}(\cdots) \le \pi$.

e) Sketch on Fig. 8.5–17 the equipotentials for which $\phi(x, y) = V/2$ and $\phi(x, y) = V/4$. Give the equation of each equipotential.

10. a) A material having heat conductivity k has a cross section occupying the first quadrant of the z-plane. The boundaries are maintained as shown in Fig. 8.5–18. Show that inside the material the temperature is given by

$$\phi(x, y) = \frac{100}{\pi}\tan^{-1}\frac{4xy}{(x^2+y^2)^2-1},$$

where the arctangent assumes values between 0 and π.

Hint: Try to transform the given configuration into one resembling that in part (d) of Exercise 9. Use the result of that exercise.

b) Sketch the variation in temperature with distance along the line $x = y$ from $x = 0$ to $x = 2$.

c) Show that the stream function for this problem is

$$\psi = \frac{-50}{\pi}\operatorname{Log}\left[\frac{(x^2-y^2-1)^2+4x^2y^2}{(x^2-y^2+1)^2+4x^2y^2}\right].$$

d) Find $\mathbf{q}(x, y) = Q_x + iQ_y$, the complex heat flux density in the conducting material.

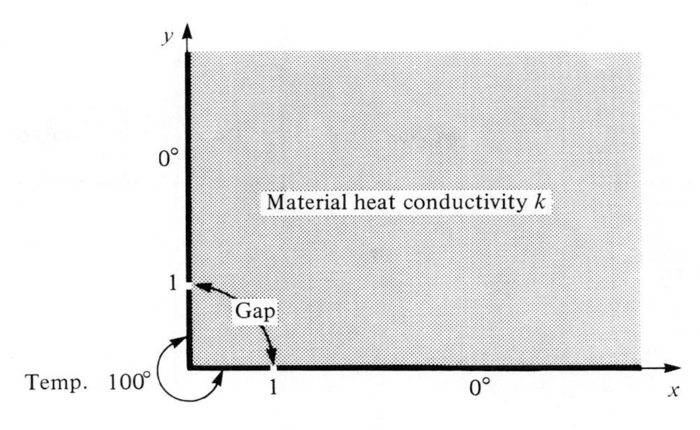

Figure 8.5–18

11. a) The boundaries of a heat conducting material are maintained at the temperatures shown in Fig. 8.5–19. Find the temperature $\phi(x, y)$, for $-a < x < a$, $y > 0$ inside the material.

Hint: Begin with a transformation like that in Example 3, Section 8.3. Also see Exercise 9 above.

b) Plot $\phi(x, a/10)$ for $-a \le x \le a$.

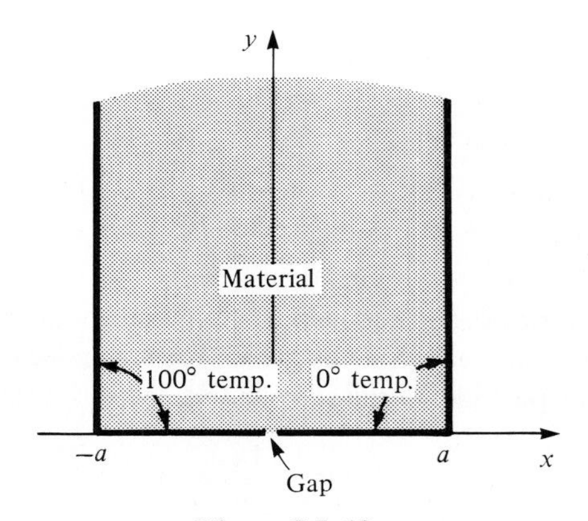

Figure 8.5–19

12. a) Show that under the transformation $w = \cos^{-1}(z/a)$ or $z = a \cos w$, that the lines $y = 0$, $-\infty < x \le -a$ and $y = 0$, $a \le x < \infty$ are mapped into the w-plane as shown in Fig. 8.5–20. Assume $a > 0$.

b) Show that the transformation maps one to one the domain consisting of the z-plane, with the points satisfying $y = 0$, $|x| \ge a$ removed, onto the strip $0 < u < \pi$ shown in Fig. 8.5–20(b). Consider the rectangular region in the strip satisfying $u_1 \le u \le u_2$, $v_1 \le v \le v_2$. What is its image in the z-plane?

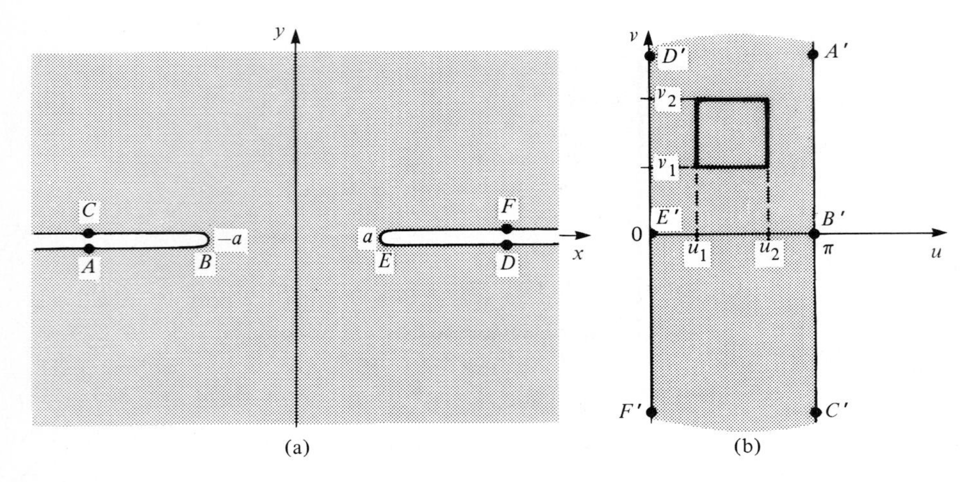

Figure 8.5–20

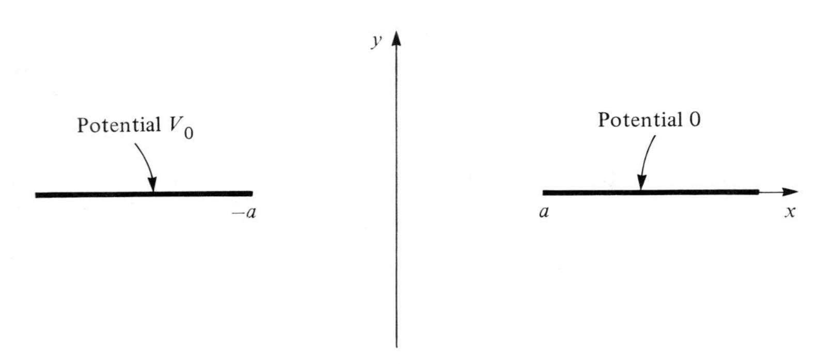

Figure 8.5–21

c) Two semiinfinite electrically conducting sheets, shown in cross section in Fig. 8.5–21, are separated by distance $2a$. The conductors are maintained at voltages V_0 and 0. Show that the complex potential in the surrounding space is given by

$$\Phi(z) = \frac{V_0}{\pi} \cos^{-1}\left(\frac{z}{a}\right).$$

d) For $a = 1$ show that the electrostatic potential is given by

$$\phi(x, y) = \frac{V_0}{\pi} \cos^{-1} \pm \sqrt{\left(\frac{x^2 + y^2 + 1}{2}\right) \pm \sqrt{\left(\frac{x^2 + y^2 + 1}{2}\right)^2 - x^2}},$$

where $0 \leq \cos^{-1}(\cdots) \leq \pi$.

Hint: Find $\phi(u)$ in the w-plane. Write u in terms of x and y by observing that $x = \cos u \cosh v$, $y = -\sin u \sinh v$ so that $(x^2/\cos^2 u) - (y^2/\sin^2 u) = 1$. Solve this for $u(x, y)$.

e) Use symmetry to argue that $\phi(0, y) = V_0/2$ for $-\infty < y < \infty$, and use this fact to establish that a minus sign precedes the inner square root in $\phi(x, y)$ given in part (d).

f) Verify that $\phi(x, y)$ in part (d) satisfies the assigned boundary conditions along the lines $x \geq 1$, $y = 0$ and $x \leq 1$, $y = 0$ provided the sign preceding the outer square root in the expression is taken as positive in quadrants 1 and 4 and negative in quadrants 2 and 3. Does this mean that $\phi(x, y)$ is discontinuous as we cross the y-axis? Explain.

g) For $a = 1$ show that the complex electric field is given by $V_0\overline{(1 - z^2)^{-1/2}}/\pi$, where $(1 - z^2)^{-1/2}$ is defined by means of branch cuts along lines corresponding to the conductors in Fig. 8.5–21.

8.6 MORE ON BOUNDARY VALUE PROBLEMS—STREAMLINES AS BOUNDARIES

In the Dirichlet problems just considered a harmonic function was obtained that assumed certain preassigned values on the boundary of a domain. In this section we will study boundary value problems that are not Dirichlet problems; a function that is harmonic in a domain will be sought, but the values assumed by this function everywhere on the boundary are not necessarily given. Instead, information regarding the derivative of the function on the boundary is supplied. We will see how this can happen in some heat-flow problems and will use conformal mapping in their solution. In the exercises we will also see how this occurs in configurations involving fluid flow.

If a heat-conducting material is surrounded by certain surfaces that provide perfect thermal insulation, then, by definition, there can be no flow of heat into or out of these surfaces. The heat flux density vector Q cannot have a component normal to the surface (see Eq. 2.6–2). We will consider only two-dimensional configurations and employ a complex temperature (see Eq. 2.6–6) of the form

$$\Phi(x, y) = \phi(x, y) + i\psi(x, y), \tag{8.6–1}$$

where $\phi(x, y)$ is the actual temperature, and $\psi(x, y)$ is the stream function. In Section 2.6 we observed that the streamlines, that is, the lines on which $\psi(x, y)$ assumes fixed values, are tangent at each point to the heat flux density vector. Fig. 8.6–1 shows the cross section of a heat-conducting material whose boundary is in part insulated.

The insulated part of the boundary must coincide with a streamline.

Otherwise, the heat flux density vector Q would have a component normal to the insulation.

We observed in Section 2.6 that the streamlines and isotherms form a mutually orthogonal set of curves. Suppose, starting at the insulated surface shown in Fig. 8.6–1, we proceed along the vector N, which is *normal* to the insulation. At the insulation we must be moving along an isotherm, and ϕ does not change. Mathematically, this is stated as

$$\frac{d\phi}{dn} = 0 \qquad \text{(at insulated surface)}, \tag{8.6–2}$$

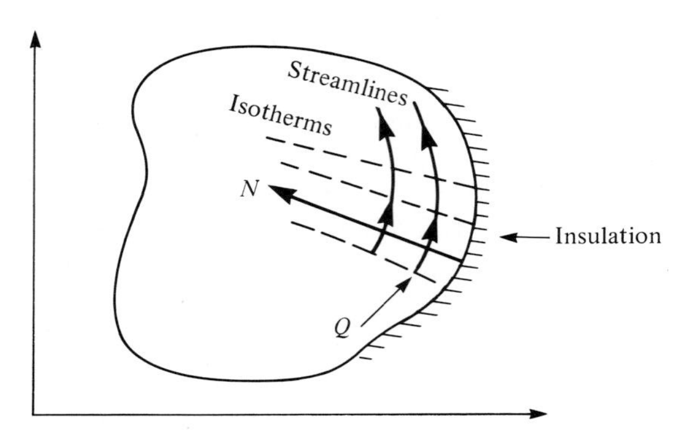

Figure 8.6–1

where n is the distance measured along the normal N. Equation 8.6–2 asserts that the "normal derivative" of the temperature vanishes at an insulated boundary.

Problems in which $d\phi/dn$ is known *a priori* everywhere on the boundary of a domain and where the harmonic function $\phi(x, y)$ is sought inside the domain are known as *Neumann problems.* In this section we will mostly consider more complicated situations in which ϕ is known on part of the boundary while $d\phi/dn$ is given on the remainder.

When we are given boundary value problems in, for example, heat conduction in which the temperature is specified on some portions of the boundary, while the remaining portions are insulated, we proceed in a manner like that used in the Dirichlet problems of the previous section. We map the cross section of the configuration from, say, the z-plane into a simpler or more familiar shape in the w-plane by means of an analytic transformation $w = f(z)$. As before, at the boundaries of the new domain in the w-plane we assign those temperatures, if known, that exist at the corresponding image points in the z-plane. There will now also be insulated boundaries in the w-plane corresponding to insulated boundaries in the z-plane.

We now seek a complex potential, an analytic function $\Phi_1(w) = \phi_1(u, v) + i\psi_1(u, v)$, such that $\phi_1(u, v)$ assumes the known assigned values at boundary points in the w-plane. We also require that $\psi_1(u, v)$ produce streamlines coinciding with the insulated boundaries in the w-plane. As before, a transformation back into the z-plane yields the analytic function $\Phi(z) = \phi(x, y) + i\psi(x, y)$, where $\phi(x, y)$ is the required temperature and $\psi(x, y)$ is the associated stream function. Now $\psi(x, y)$ will produce streamlines coinciding with the insulated boundaries and $\phi(x, y)$ will assume the prescribed values on the remaining boundaries. An example of the method follows.

EXAMPLE 1

Refer to Fig. 8.6–2(a). A heat-conducting material fills the space $y > 0$. The boundary $y = 0$, $x > 1$ is maintained at 100°, the boundary $y = 0$, $x < -1$ is at 0° while $y = 0$, $|x| < 1$ is insulated. Find the temperature distribution $\phi(x, y)$ and the complex temperature $\Phi(z)$ in the material.

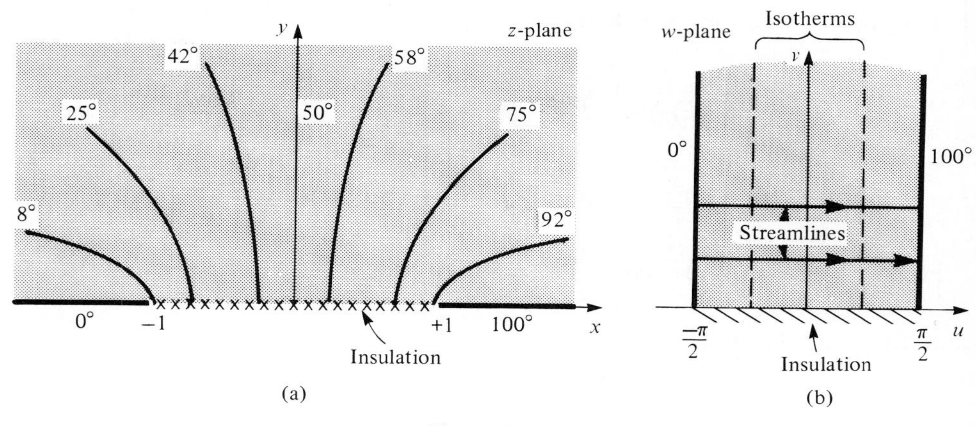

Figure 8.6–2

Solution

We seek a transformation that will take the given region into a more tractable shape. Referring to Example 3, Section 8.3 and to Fig. 8.3–5 (or to a table of transformations), we have a useful clue. Reversing the role of z and w in that example, we find that the transformation

$$z = \sin w, \tag{8.6–3}$$

or

$$w = \sin^{-1}(z)$$

maps the region shown in Fig. 8.6–2(a) onto the region in Fig. 8.6–2(b). The boundary conditions are transformed as indicated.

To solve the transformed problem we note that a temperature $\phi_1(u)$ of the form

$$\phi_1(u) = Au + B, \qquad \text{where } A \text{ and } B \text{ are real numbers} \quad \text{and} \quad -\frac{\pi}{2} \leq u \leq \frac{\pi}{2}, \tag{8.6–4}$$

will produce isotherms coincident with the boundaries along $u = -\pi/2$ and $u = \pi/2$. The associated streamlines will be parallel to the insulated boundary. In particular there will be a streamline coincident with $v = 0$, $-\pi/2 \leq u \leq \pi/2$ as required. To determine A and B, we apply the boundary conditions $\phi_1(-\pi/2) = 0$ and $\phi_1(\pi/2) = 100$ in Eq. (8.6–4). The first condition yields

$$0 = -A\frac{\pi}{2} + B,$$

and the second yields

$$100 = A\frac{\pi}{2} + B.$$

Solving these equations simultaneously, we have $A = 100/\pi$ and $B = 50$ so that Eq. (8.6–4) becomes

$$\phi_1(u) = \frac{100}{\pi} u + 50, \qquad -\frac{\pi}{2} \leq u \leq \frac{\pi}{2}. \tag{8.6–5}$$

Noticing that $\phi_1(u) = \text{Re}[(100/\pi)w + 50]$, we realize that the complex temperature is

$$\Phi_1(w) = \frac{100}{\pi} w + 50, \qquad |\text{Re } w| \leq \frac{\pi}{2}, \tag{8.6–6}$$

and the stream function is

$$\psi_1(v) = \text{Im } \Phi_1(w) = \frac{100v}{\pi}. \tag{8.6–7}$$

Since $w = \sin^{-1}(z)$, the complex temperature in the z-plane can be obtained, from a substitution in Eq. (8.6–6):

$$\Phi(z) = \frac{100}{\pi} \sin^{-1} z + 50, \tag{8.6–8}$$

where $-\pi/2 \leq \text{Re} \sin^{-1} z \leq \pi/2$.

To obtain the actual temperature $\phi(x, y)$ we have from Eq. (8.6–3)

$$z = (x + iy) = \sin w = \sin u \cosh v + i \cos u \sinh v,$$

so that

$$x = \sin u \cosh v,$$
$$y = \cos u \sinh v,$$

and since $\cosh^2 v - \sinh^2 v = 1$, we find that

$$\frac{x^2}{\sin^2 u} - \frac{y^2}{\cos^2 u} = 1.$$

We now eliminate $\cos^2 u$ from the above by employing $\cos^2 u = 1 - \sin^2 u$, which yields

$$\frac{x^2}{\sin^2 u} - \frac{y^2}{1 - \sin^2 u} = 1. \tag{8.6–9}$$

We multiply both sides of Eq. (8.6–9) by $\sin^2 u(1 - \sin^2 u)$ and obtain a quadratic equation in $\sin^2 u$ (a quartic in $\sin u$). The quadratic formula yields

$$\sin^2 u = \frac{(x^2 + y^2 + 1)}{2} \pm \sqrt{\left(\frac{x^2 + y^2 + 1}{2}\right)^2 - x^2}.$$

We take the square root of both sides of this expression and then use $u = \sin^{-1}(\sin u)$ to obtain

$$u = \sin^{-1}\left[\pm\sqrt{\frac{x^2+y^2+1}{2} \pm \sqrt{\left(\frac{(x^2+y^2+1}{2}\right)^2 - x^2}}\right]. \tag{8.6–10}$$

A substitution in Eq. (8.6–5) yields

$$\phi(x, y) = \frac{100}{\pi}\sin^{-1}\left[\pm\sqrt{\frac{x^2+y^2+1}{2} \pm \sqrt{\left(\frac{x^2+y^2+1}{2}\right)^2 - x^2}}\right] + 50. \tag{8.6–11}$$

The conditions attached to Eq. (8.6–5) here require $-\pi/2 \le \sin^{-1}(\cdots) \le \pi/2$. From physical arguments we can establish that the temperature in the heat-conducting material can be no less than 0, nor can it exceed 100. To determine the appropriate signs for the square roots note that $x = 0$, $y = 0+$ lies midway between the two conductors and by symmetry will be at a temperature of 50. This condition requires that the inner $\pm$ operator be negative. The boundary conditions $\phi(x, 0) = 0$, $x < -1$ and $\phi(x, 0) = 100$, $x > 1$ demand that the outer $\pm$ operator be positive in the first quadrant and negative in the second quadrant. Note that there is no discontinuity in temperature as we cross the positive y-axis. The isotherms in the w-plane are (from Eq. 8.6–5) those surfaces on which u is constant. According to Eq. (8.6–9) these isotherms become hyperbolas in the xy-plane. Some are sketched in Fig. 8.6–2(a) for various temperatures. ◀

EXERCISES

1. A material of heat conductivity k has a cross section that occupies the first quadrant. The boundaries are maintained at the temperatures indicated in Fig. 8.6–3.

   a) Show that the complex temperature inside the conducting material is given by

$$\Phi(z) = \frac{100}{\pi}\sin^{-1}(z^2) + 50, \qquad -\frac{\pi}{2} \le \text{Re}\sin^{-1}(\cdots) \le \frac{\pi}{2}.$$

   Hint: Map the region of this problem onto that presented in Example 1.

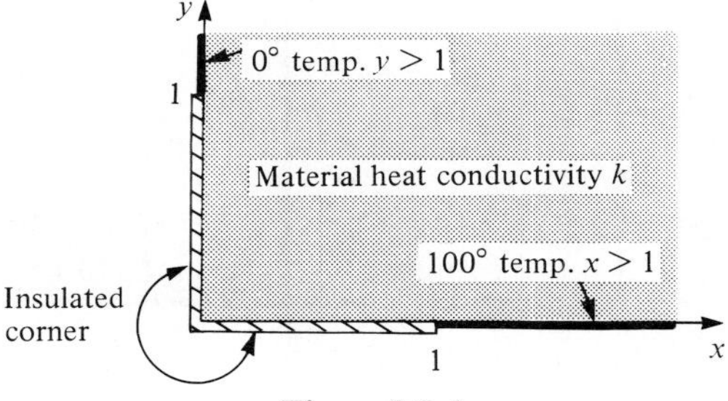

Figure 8.6–3

b) Show that the temperature inside the material is given by

$$\phi(x, y) = 50 + \frac{100}{\pi} \sin^{-1}$$

$$\left[\pm \sqrt{\frac{1 + (x^2 + y^2)^2}{2} \pm \sqrt{\frac{[1 + (x^2 + y^2)^2]^2}{4} - (x^2 - y^2)^2}} \right]$$

for $-\pi/2 < \sin^{-1}(\cdots) < \pi/2$, and the appropriate signs are used in front of each square root.

c) Plot a curve showing the variation in temperature with distance along the insulated boundary lying along the x-axis.

d) Show that the complex heat flux density is

$$\mathbf{q} = -\frac{k200}{\pi}\left(\frac{z}{(1 - z^4)^{1/2}}\right) = Q_x + iQ_y.$$

e) Let $k = 1$. By choosing the appropriate values of the square root, give the numerical values of the components of $\mathbf{q}$, that is, Q_x and Q_y, at the following locations:

$$x = 1/2,\ y = 0+; \qquad x = 2,\ y = 0+; \qquad x = 0+,\ y = 1/2;$$
$$x = 0+,\ y = 2.$$

2. a) A material of heat conductivity k has boundaries as shown in Fig. 8.6–4. Show that the complex temperature in the material is given by

$$\Phi = -\frac{200}{\pi} \sin^{-1}\left(\frac{z}{a}\right) + 100, \qquad \text{for } 0 \le \operatorname{Re} \sin^{-1}(\cdots) \le \frac{\pi}{2}.$$

Hint: Consider the mapping $z = a \sin w$ applied to the strip $0 \le \operatorname{Re} w \le \pi/2$, $\operatorname{Im} w \ge 0$.

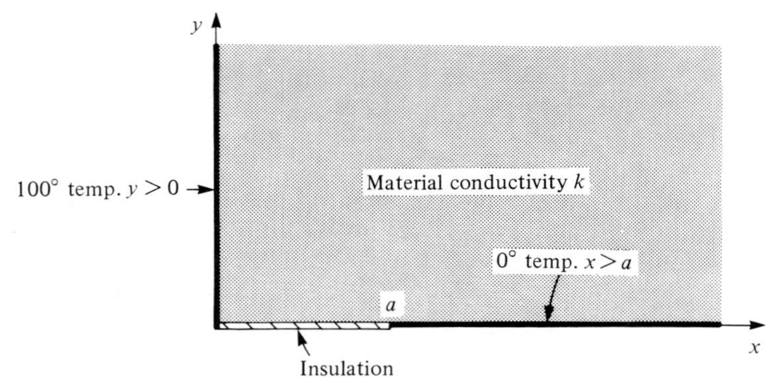

Figure 8.6–4

b) Show that the isotherm having temperature T lies on the hyperbola described by

$$\frac{x^2}{a^2 \sin^2 u} - \frac{y^2}{a^2 \cos^2 u} = 1,$$

where $u = (100 - T)\pi/200$.

c) Sketch the isotherm $T = 50$. Take $a = 1$.

3. The outside of a heat-conducting rod of unit radius is maintained at the temperatures shown in Fig. 8.6–5. One half of the boundary is insulated. Show that the complex temperature inside the rod is given by

$$\Phi(z) = 50 - \frac{100}{\pi} \sin^{-1} \frac{i(z+i)}{z-i}, \qquad -\frac{\pi}{2} \leq \text{Re} \sin^{-1}(\cdots) \leq \frac{\pi}{2}.$$

Hint: A bilinear transformation will map the configuration into that of Example 1 (see Fig. 8.6–2a).

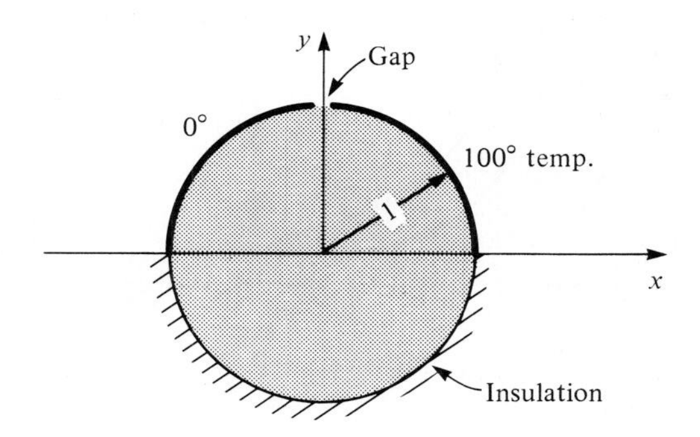

Figure 8.6–5

The following problems treat fluid flow in the presence of a rigid, impenetrable boundary.

4. When a rigid, impenetrable obstacle is placed within a moving fluid, no fluid passes through the surface of that object. At each point on the object's surface the component of the fluid velocity vector normal to the surface must vanish; otherwise, there would be penetration by the fluid. Since flow is tangential to the surface of the obstacle, its boundary must be coincident with a streamline.

The simplest type of fluid motion in the presence of a boundary is that of uniform flow parallel to and above an infinite plane (see Fig. 8.6–6). The complex potential describing the fluid flow is $\Phi = Aw$, where $w = u + iv$, and A is a real number. A is positive for flow to the right, negative for flow to the left.

a) Using Φ show that the complex fluid velocity is $A + i0$. Verify that the flow is indeed uniform and parallel to the plane, that is, parallel to the u-axis.

b) Show that the stream function for the flow is $\psi = Av$. What is the value of ψ along the boundary? Plot the loci of $\psi = 0$, $\psi = A$, $\psi = 2A$ on Fig. 8.6–6.

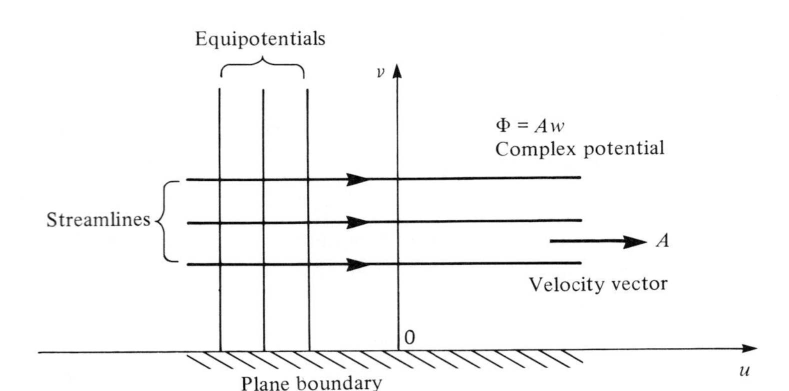

Figure 8.6–6

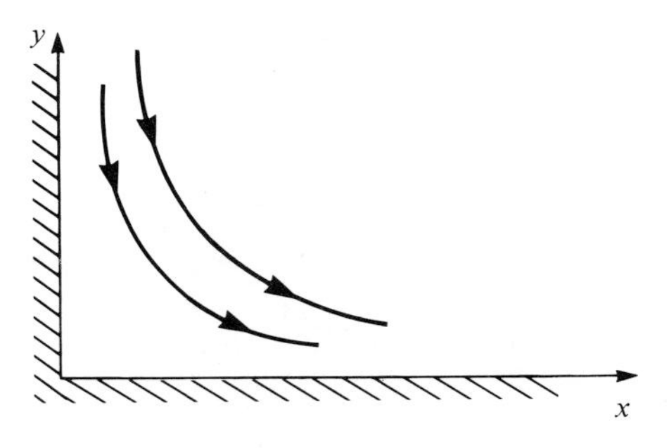

Figure 8.6–7

c) The fluid flow in the space $v \geq 0$ described in Fig. 8.6–6 is transformed into the z-plane by means of $z = w^{1/2}$, where the principal branch of the square root is used. Show that the plane boundary of Fig. 8.6–6 is mapped into the right angle boundary in Fig. 8.6–7. Show that the complex velocity potential $\Phi(w)$ of Fig. 8.6–6 is transformed into $\Phi(z) = Az^2$, which describes flow within the boundary.

d) Show that the complex velocity for flow in the corner is $2Ax - i2Ay$. Show that the speed with which the fluid moves at a point varies directly with the distance of that point from the corner. Show that fluid flow is in the negative y-direction along the wall $x = 0$, $y > 0$ and in the positive x-direction along the wall $y = 0$, $x > 0$.

e) Show that the stream function for flow into the corner is $\psi = A2xy$.

5. Fluid flows into and out of the $135°$ corner shown in Fig. 8.6–8.

a) Show that the complex potential describing the flow is of the form

$$\Phi(z) = Az^{4/3}, \qquad \text{where } A \text{ is a positive real constant.}$$

Hint: Find a transformation that will map the region $v \geq 0$ from Fig. 8.6–6 onto the region of flow in Fig. 8.6–8. Apply this same transformation to the uniform flow in Fig. 8.6–6.

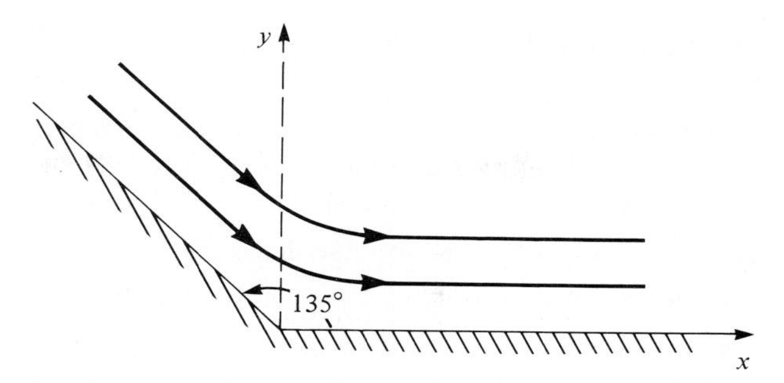

Figure 8.6–8

b) Use $z = re^{i\theta}$ to convert $\Phi(z)$ to polar coordinates, and show that the velocity potential and stream function are given, respectively, by

$$\phi(r, \theta) = Ar^{4/3} \cos \frac{4\theta}{3} \quad \text{and} \quad \psi(r, \theta) = Ar^{4/3} \sin \frac{4\theta}{3}.$$

c) Use $\psi(r, \theta)$ to sketch the streamlines $\psi = 0$ and $\psi = A$.

d) Show that the complex fluid velocity vector is $(4/3)A\sqrt[3]{r}\,\text{cis}(-\theta/3)$.

6. In this exercise we study fluid flow into a closed channel by transforming the uniform fluid flow described in Fig. 8.6–6. Use

$$z = \sin^{-1} w, \qquad -\frac{\pi}{2} \leq \text{Re} \sin^{-1} w \leq \frac{\pi}{2}.$$

a) Show that the plane boundary $v = 0$ of Fig. 8.6–6 is transformed into the closed channel shown in Fig. 8.6–9.

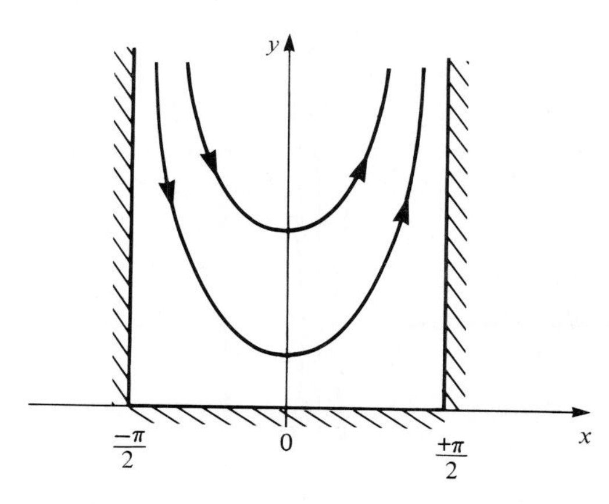

Figure 8.6–9

b) Show that the complex fluid velocity in the channel is given by $A \cos x \cosh y + iA \sin x \sinh y$.

c) Show that fluid flows in the negative y-direction along the left wall in the channel, in the positive x-direction along the end of the channel, and in the positive y-direction along the right wall of the channel. Assume $A > 0$.

d) Show that the stream function that describes flow in this channel is $\psi = A \cos x \sinh y$.

e) Plot the streamlines $\psi = 0$, $\psi = A/2$, and $\psi = A$ on Fig. 8.6–9.

7. a) A fluid flows with uniform velocity V_0 in the direction shown in Fig. 8.6–10 along a channel of width π. Show that the complex potential $\Phi = iwV_0$ describes the flow and satisfies the requirement that the walls of the channel be streamlines.

b) Use the transformation $z = \cos w$ to map this channel and its flow into the z-plane (see Exercise 12, Section 8.5). Show that the flow in the z-plane is through an aperture of width 2 within the line $y = 0$ (see Fig. 8.6–11). What is the complex potential $\Phi(z)$ describing the flow in the z-plane?

c) By using the correct branch in the velocity potential $\Phi(z)$ show that in the center of the aperture of Fig. 8.6–11 the fluid moves with velocity V_0 parallel to the positive y-axis.

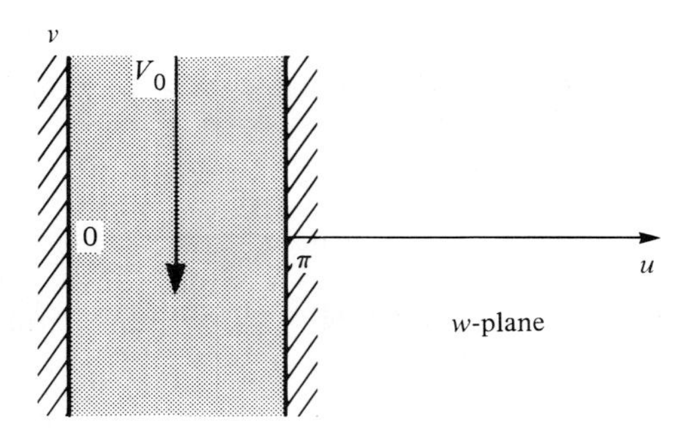

Figure 8.6–10

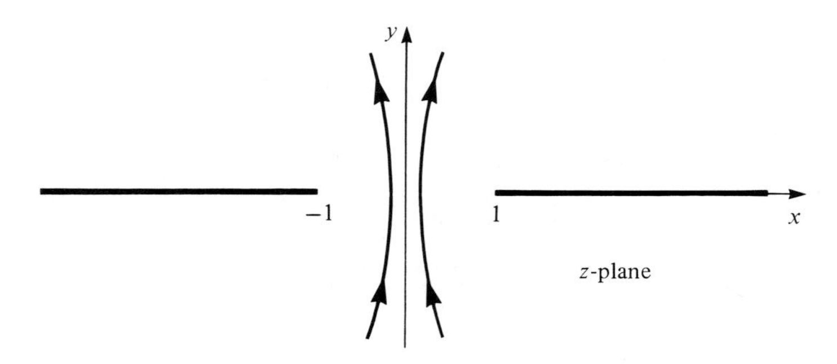

Figure 8.6–11

d) Find the equation and sketch the locus of the streamline passing through $y = 0, x = 1/2$.

e) Show that "far" from the aperture, $|z| \gg 1$, the components of velocity are given approximately by

$$V_x = \frac{V_0 \cos\theta}{r}, \qquad V_y = \frac{V_0 \sin\theta}{r}, \qquad 0 \le \theta \le \pi,$$

and

$$V_x = \frac{-V_0 \cos\theta}{r}, \qquad V_y = \frac{-V_0 \sin\theta}{r}, \qquad \pi \le \theta \le 2\pi,$$

where $z = re^{i\theta}$.

8. In this exercise we study flow around a half cylinder obstruction in a plane. Fluid flow above an infinite plane, as described in Fig. 8.6–6, is transformed by means of the formula $z = w/2 + (w^2/4 - 1)^{1/2}$. The transformation involves branch cuts extending from $w = \pm 2$ into the lower half plane. The image of $w = 0$ is $z = i$.

a) Show that the image of the axis $v = 0$ in Fig. 8.6–6 is the fluid boundary shown in Fig. 8.6–12 and that the space $v > 0$ is mapped onto the region above this boundary.

Hint: Show that the inverse of our transformation is $w = z + 1/z$. Use this to transform the boundary in Fig. 8.6–12 into $v = 0$ in Fig. 8.6–6.

b) Show that $\Phi(z) = A(z + z^{-1})$ is the complex potential describing flow in the z-plane. Assume A is real.

c) Show that the complex fluid velocity in Fig. 8.6–12 is $A(1 - 1/(\bar{z})^2)$. Why does this indicate a uniform flow of fluid to the right in Fig. 8.6–12 when we are far from the half cylinder obstruction? Assume $A > 0$.

d) Let $z = re^{i\theta}$. Show that in polar coordinates the stream function describing the flow is $\psi = A(r - 1/r)\sin\theta$. What is the value of ψ on the streamline that coincides with the fluid boundary? Sketch the streamline $\psi = A$ on Fig. 8.6–12.

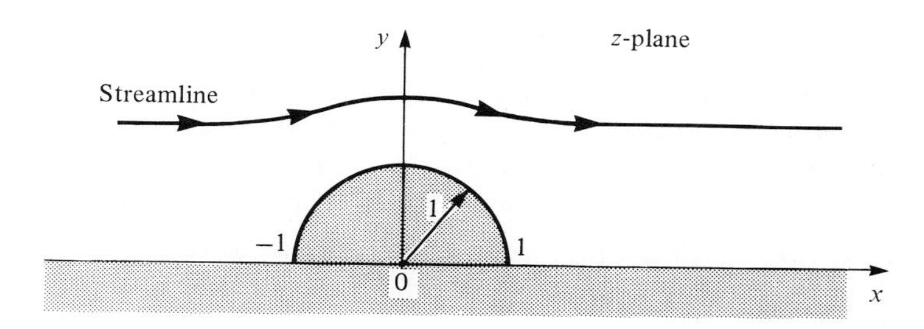

Figure 8.6–12

8.7 BOUNDARY VALUE PROBLEMS WITH SOURCES

Until now all the sources or sinks for electric flux, heat, or fluid that we have considered in our boundary value problems have either been located at infinity or embedded in the boundaries of the domain under consideration. Thus in Example 1 of Section 8.5 (see Fig. 8.5–2) heat is evolved in the inner boundary, which is maintained at 100 degrees, and moves outward where it is collected in the outer boundary, maintained at 0 degrees. There is no source or sink for heat in the domain lying between the two boundaries. Similarly as shown in Figs. 8.6–7 and 8.6–8, fluid is not generated in the domain under consideration. The flow begins at infinity and terminates at infinity. Since there are no sources or sinks present in either the thermal or fluid configurations, the net flux of heat or fluid out of any volume element whose cross section is contained in the domain under scrutiny is zero. The same situation obtains when sources of electric flux (i.e., electric charge) are maintained at infinity or in the boundaries of the domain. No net electric flux leaves any volume in the domain.

In the present section we consider what happens when a source of heat, fluid, or electric flux is placed in a domain whose boundaries are maintained in some prescribed way. We employ a particularly simple kind of source—one that is unchanging and of infinite extent in a direction perpendicular to the complex plane. (We called this the ζ direction in Fig. 2.6–1.) The source is assumed to produce its flux (heat, fluid, electricity) in a direction radially outward from itself. For a sink the flux is inward, and we will regard a sink as being simply a particular kind of source.

Our sources will have zero physical dimensions in the complex plane and can be thought of as a filament, or line, parallel to the ζ direction. For any volume containing the source, our assumption that the net outflow of flux is zero is violated. However, this assumption still holds for any volume lying outside the source but within the other boundaries of the domain. The source is represented pictorially by its cross section in the complex plane—a simple dot. (Some authors call our line sources "point" sources.)

We have two reasons for using line sources. One is that many practical sources can be well represented by their simple idealization as a line (think of a charged wire as a source of electric flux, or a slender pipe carrying hot water through the cold ground as a source of heat flux). The other is that if we know the fluid velocity, electric field, temperature, etc., produced by an idealized source in the presence of certain boundaries, then we can use this same information to obtain these same physical quantities when produced by actual, nonidealized sources, in the presence of these same boundaries.†

The complex potential associated with a line source always displays a singularity at the point marking its intersection with the complex plane. Let us study an example from heat conduction. An infinitely long "hot" filament lies perpendicular to the z-plane (see Fig. 8.7–1) and passes through $z = 0$. The environment of the filament

---

† This is the technique of the Green's function and is beyond the scope of this text. See, e.g., P. Morse and H. Feshbach, *Methods of Theoretical Physics* (New York: McGraw-Hill, 1953), pp. 791–895.

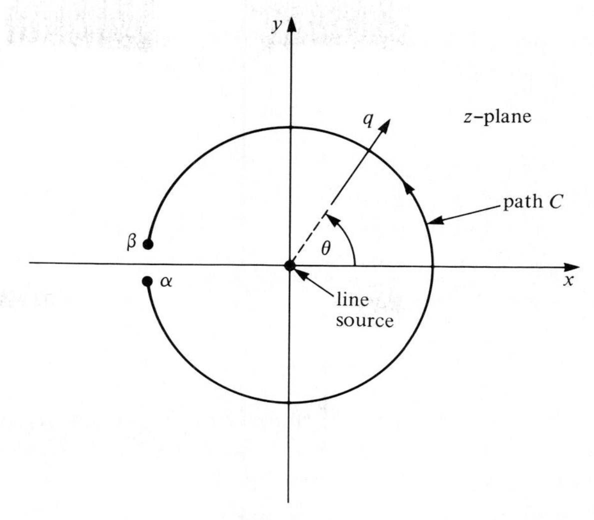

Figure 8.7–1

is an infinite uniform heat-conducting material having conductivity k. The complex temperature (or potential) created by the filament is of the form

$$\Phi(z) = A \operatorname{Log}(a/z), \tag{8.7–1}$$

where $a > 0$ and A are real constants. We choose, rather arbitrarily, to use the principal branch of the logarithm. Other branches can be used throughout this section if convenient. Now with principal values and with $\arg a = 0$, we have $\arg(a/z) = \arg a - \arg z = -\arg z$. Thus

$$\Phi(z) = \phi(x, y) + i\psi(x, y) = A \operatorname{Log}(a/|z|) - iA \arg z,$$

which means that

$$\phi(x, y) = A \operatorname{Log}(a/|z|) \tag{8.7–2a}$$

and

$$\psi(x, y) = -A \arg z, \qquad -\pi < \arg z < \pi. \tag{8.7–2b}$$

The complex heat flux density vector, introduced in Section 2.6, is readily computed from Eq. (8.7–1). We have $\mathbf{q} = -k\overline{(d\Phi/dz)} = Ak/\bar{z}$. With $z = r \operatorname{cis} \theta$ this becomes $q = Ak(\cos\theta + i \sin\theta)/r$. Thus the magnitude of the heat flux density vector varies inversely with distance r from the line source and is directed along the unit vector $\operatorname{cis} \theta$, i.e., is directed radially outward from the source as shown in Fig. 8.7–1. Its value is independent of the constant a in Eq. (8.7–1).

The quantity A in the preceding equations can be computed if we know the total heat generated per unit time by a unit length of the line source. Calling this quantity h, we surround the line source by a contour C as shown in Fig. 8.7–1. We proceed along this contour as shown, from α, which lies just below the branch cut for $\operatorname{Log} z$, to β, which lies just above the cut. As demonstrated in the appendix to this chapter, the heat flux passing outward across this contour (per unit length of source) is

$$h = k\Delta\psi, \tag{8.7–3}$$

where $\Delta\psi = \psi(\alpha) - \psi(\beta)$, i.e., the decrease in the value of the stream function as C is negotiated. This is the heat analog of Eq. (8.5–24). Now employing Eq. (8.7–2b) we have $\psi(\alpha) = -A(-\pi) = A\pi$ and $\psi(\beta) = -A\pi$, so that finally we obtain

$$h = k2\pi A \tag{8.7–4a}$$

and

$$A = h/2\pi k. \tag{8.7–4b}$$

We call $h = k2\pi A$ the *strength of the line source* of heat, since it tells the time rate of flow of heat from a unit length of the source into its surroundings. If h is negative we are dealing with a sink, and A is also negative.

We need not confine ourselves to line sources going through $z = 0$. A source passing through $z = z_0$ will have a complex potential of the form

$$\Phi(z) = A \operatorname{Log}(a/(z - z_0)), \tag{8.7–5}$$

with a corresponding heat flux density vector $Ak/\overline{(z - z_0)} = Ak \operatorname{cis}[\arg(z - z_0)]/|z - z_0|$. With $A > 0$ flow is again radially outward from the source. Equation (8.7–4b) still describes the relationship between A and h, the rate of heat flow from the source.

The preceding discussion has counterparts for electrostatic and fluid line sources. The complex potential created by a line of electrostatic charge passing through $z = 0$ is again of the form

$$\Phi(z) = A \operatorname{Log}(a/z).$$

The constant A can be computed in a manner like that used in the heat flow case. Now, however, we compute the electric flux crossing the contour C in Fig. 8.7–1 and, as discussed in the appendix, equate it to the charge enclosed. We find that

$$A = \rho/2\pi\varepsilon, \tag{8.7–6}$$

where ρ (the strength of the electric source) is the electric charge per unit length carried by the line charge and ε is the electrical permittivity of the surrounding material. Again we can displace the source away from the origin and similarly modify $\Phi(z)$.

The complex electric flux density vector created in this material is

$$\mathbf{d} = -\varepsilon\overline{(d\Phi/dz)},$$

as is explained in Section 2.6. We find that $\mathbf{d} = \rho \operatorname{cis} \theta/2\pi r$ when the line charge passes through $z = 0$.

Finally we can study a line source that passes through $z = 0$ and sends fluid radially into its surroundings. The complex velocity potential describing the fluid flow is

$$\Phi(z) = A \operatorname{Log}(z/a) = \phi(x, y) + i\psi(x, y). \tag{8.7–7}$$

Observe that this is the negative of the expression used for the heat flow and in electrostatic situations. The sign difference is explicable because of the sign difference occurring in the last line of Table 1, Section 2.6. From this table we obtain, for the complex fluid velocity, $\mathbf{v} = \overline{d\Phi/dz} = A/\bar{z} = A \operatorname{cis} \theta/r$. Thus fluid flow is radially outward from the source.

As discussed in the appendix, the outward flux of fluid through a contour C like that in Fig. 8.7–1 is

$$G = -\Delta\psi = \psi(\beta) - \psi(\alpha). \tag{8.7–8}$$

Here G is the rate of flow, with time, of a fluid of unit mass density from a unit length of the line source. This is the strength of the fluid source. Since $\psi = A \arg z$ we find that $A = G/2\pi$.

In the case of fluid mechanics, there is another type of line source, called a *vortex*. Here no fluid is evolved from the source—rather the source acts to create fluid rotation around itself much as water behaves around the axis of a whirlpool or a propeller. For a vortex, A assumes a purely imaginary value. The situation is considered in Exercise 2.

What is the value of the constant a in our three complex potentials? In no case does this constant appear in our expression for $d\Phi/dz$, and therefore its value has no influence on the complex heat flux density vector, electric flux density vector, or fluid velocity. In the case of heat conduction, we have that a line source passing through $z = 0$ creates an actual temperature $\phi(x, y) = A \operatorname{Log}(a/|z|)$. Note that on the circle $|z| = a$ the measured temperature is zero degrees. Thus the choice of a dictates how far we must move from the line source to experience a temperature of zero; this is true even when the line does not pass through $z = 0$. Similarly in electrostatics the electrostatic scalar potential created by a line of electric charge is $\phi(x, y) = A \operatorname{Log}(a/|z|)$. Hence a is the distance from the charge at which the electrostatic potential (usually called voltage) falls to zero. In fluid mechanics no particular meaning is assigned to a, and we can set its value to unity.

If a line source of electric or heat flux is placed along the axis of a rod we can choose a so as to satisfy certain simple boundary conditions on the surface of the rod. In Fig. 8.7–2 we show a "hot" filament (line source) that is coincident with the axis of a rod of radius 1. The rod is composed of material having heat conductivity k. The surface of the rod is maintained at 0 degrees. Assume the source sends out h calories per second along each meter of its length. We can create a complex potential that is associated with a source of this strength and that meets the prescribed boundary condition on the rod. We use $\Phi(z)$ as described in Eq. (8.7–1), taking $a = 1$ and $A = h/2\pi k$ (see Eq. 8.7–4a). Thus

$$\Phi(z) = \frac{h}{2\pi k} \operatorname{Log} \frac{1}{z}.$$

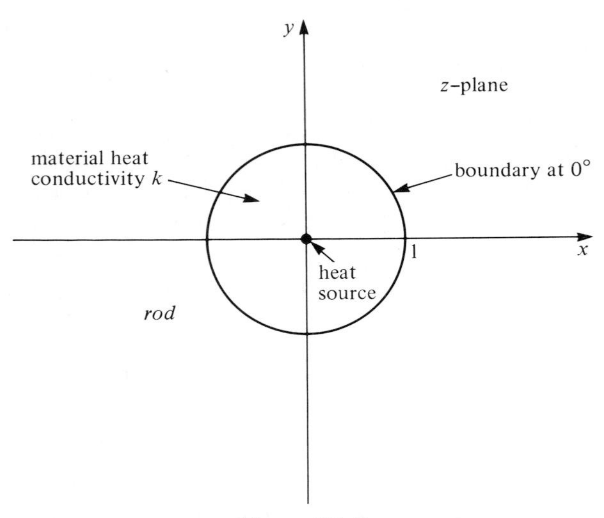

Figure 8.7–2

The actual temperature in the rod is obtained from Eq. (8.7–2a) and is found to be $\phi = (h/2\pi k)\,\text{Log}(1/|z|)$. If the surface had been maintained at some other constant temperature and/or if its radius had not been unity, we could still find the temperature inside by a suitable choice of a (see Exercise 1).

Most boundary value problems involving line sources have boundaries whose shape is more complicated than that depicted in Fig. 8.7–2. However, by making one or more conformal transformations of a simple configuration like that shown in Fig. 8.7–2, or even of the unbounded domain of Fig. 8.7–1 (see, for example, Exercise 4(c)) we can frequently acquire the solution to an interesting or practical configuration involving line sources. The technique involved is much like that used in the previous sections of this chapter in the solution of boundary value problems. As before we map the simple boundaries into the more complicated configurations. Now, however, the complex potential created by the original line source in the presence of the original boundaries must also be transformed. If in the original configuration the line source passed through $z = z_0$, then, under the transformation $w = f(z)$, we have in the new configuration a line source passing through $w_0 = f(z_0)$. If $z = g(w)$ is the inverse of the transformation $w = f(z)$, then $z_0 = g(w_0)$. Since the original complex potential created by the line source is $\Phi(z)$, the new complex potential in the w-plane is $\Phi(g(w))$. Since $\Phi(z)$ must display a singularity at z_0, the potential $\Phi(g(w))$ will display a singularity at w_0. In the new configuration, the new complex potential produces values at points on the boundary that are identical to the complex values assumed at the image points in the original, simpler configuration. Thus a boundary that was either a line of constant potential or a streamline will still have these properties under the transformation.

As we have seen, the strength of a line source is directly related to the change in value exhibited by its stream function as we negotiate a contour surrounding the source like that displayed in Fig. 8.7–1. Since the changes displayed by the stream function in the z-plane and by the transformed stream function in the w-plane (as we negotiate the image contour) will be the same, we conclude that *the strength of a line source is preserved under a conformal transformation.* The preceding assumes that the material parameters (conductivity, permittivity) used in the original configuration and in its conformal transformation are kept the same. The strength of a vortex (see Exercise 2) is also preserved under a conformal transformation.

Sometimes we are given a practical problem involving a line source in (for example) the w-plane and are fortunate enough to find an analytic transformation $z = g(w)$ that will transform this problem into a simpler problem in (for example) the z-plane. Suppose the solution in the z-plane is already known. This enables us to solve our practical problem, as we will see in the following two examples.

One further note before we give the examples: When dealing with line sources of electric, fluid, or heat evolving character we must realize that because of the laws of conservation of electric flux, fluid, and heat, there is by implication a corresponding line source of opposite strength placed at infinity. This line source acts to collect the flux generated by our original line source. This matter can be important when we do a conformal mapping involving a line source passing through a point in the finite complex plane. The mapping may succeed in bringing the line source of opposite strength at infinity into a location in the finite plane. This occurs in the following example.

EXAMPLE 1

A line source of heat generating h calories per meter of its length, per second, lies parallel to a plane maintained at a temperature of 0 degrees as shown in cross section in the w-plane (Fig. 8.7–3). The separation between source and plane is H, and there is a material of heat conductivity k filling the space $v > 0$ above the plane. Find $\Phi(w)$, the complex potential (temperature), and $\phi(u, v)$, the actual temperature, above the plane.

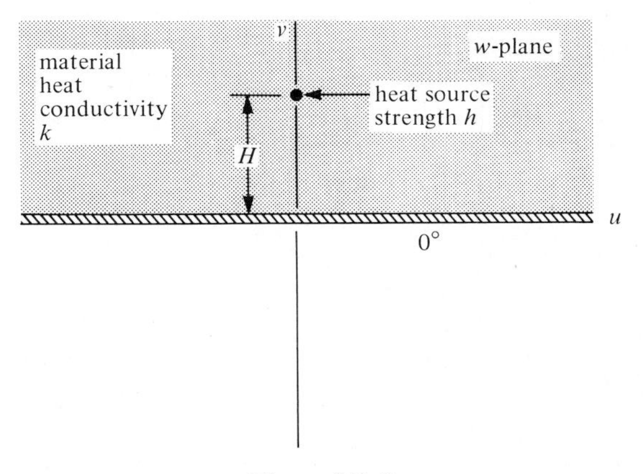

Figure 8.7–3

Solution

Exercise 32 in Section 8.4 shows how to use a bilinear transformation to map the upper half plane onto a disc. We use this transformation, Eq. (8.4–34), to solve the given problem by mapping the given configuration into one with a known solution: the configuration of Fig. 8.7–2. We must choose a transformation that will map the line source at $w = iH$ in Fig. 8.7–3 into the line source at $z = 0$ in Fig. 8.7–2.

The notation of the present problem requires that we reverse the roles of z and w in Eq. (8.4–34). Arbitrarily setting $\gamma = 0$ we obtain

$$z = \frac{w - p}{w - \bar{p}}.$$

It can be verified that another choice of γ would yield a complex potential that will differ from ours only by a constant and unimportant value in the stream function.

Since we want $w = iH$ to be mapped into $z = 0$, we choose $p = iH$. Thus our required transformation is

$$z = \frac{w - iH}{w + iH}. \tag{8.7–9}$$

From earlier discussion we know that the complex temperature inside the domain shown in Fig. 8.7–2 is

$$\Phi_1(z) = \frac{h}{2\pi k} \operatorname{Log} \frac{1}{z}.$$

Using our mapping Eq. (8.7–9) in the preceding we have, for the complex temperature in the given problem,

$$\Phi(w) = \Phi_1(z(w)) = \frac{h}{2\pi k} \operatorname{Log} \frac{w + iH}{w - iH}. \tag{8.7–10}$$

The actual temperature is

$$\phi(u, v) = \operatorname{Re} \Phi(w) = \frac{h}{2\pi k} \operatorname{Log} \frac{|w + iH|}{|w - iH|},$$

or

$$\phi(u, v) = \frac{h}{2\pi k}\left[\operatorname{Log} \frac{1}{|w - iH|} - \operatorname{Log} \frac{1}{|w + iH|}\right],$$

which is equivalent to

$$\phi(u, v) = \frac{h}{4\pi k}\left[\operatorname{Log} \frac{1}{u^2 + (v - H)^2} - \operatorname{Log} \frac{1}{u^2 + (v + H)^2}\right].$$

From the preceding we verify that on the plane $v = 0$, the temperature $\phi(u, 0)$ is indeed zero. The stream function is

$$\psi(u, v) = \operatorname{Im} \Phi(w) = \frac{h}{2\pi k}[\arg(w + iH) - \arg(w - iH)]$$

$$= \frac{h}{2\pi k}\left[\arc\tan \frac{v + H}{u} - \arc\tan \frac{v - H}{u}\right]. \quad ◀$$

Comment: Referring to Eq. (8.7–10) we notice that the complex potential for our problem can be written as:

$$\Phi(w) = \frac{h}{2\pi k} \operatorname{Log} \frac{1}{w - iH} - \frac{h}{2\pi k} \operatorname{Log} \frac{1}{w + iH}.$$

The above can be interpreted as the complex potential arising from two line sources of heat as shown in Fig. 8.7–4. The sources are at mirror image locations with respect to the line $v = 0$ and have strengths of opposite sign. It is not hard to verify, with Eq. (8.7–9), that the source at $w = iH$ is the image of the source of strength $-h$ located at $z = \infty$ in Fig. 8.7–2. The line source of heat shown in Fig. 8.7–3 creates in the space $v \geq 0$ a complex temperature that is identical to that created in identical infinite unbounded material by the original line source plus a source of equal but opposite strength (i.e., a sink) located at the mirror image of the original source.

More generally, had the original source been at $w = w_0$ (where $\operatorname{Im} w_0 > 0$) the image source would then be placed at $w = \bar{w}_0$. We would then have

$$\Phi(w) = \frac{h}{2\pi k} \operatorname{Log} \frac{1}{w - w_0} - \frac{h}{2\pi k} \operatorname{Log} \frac{1}{w + \bar{w}_0}. \quad (8.7\text{–}11)$$

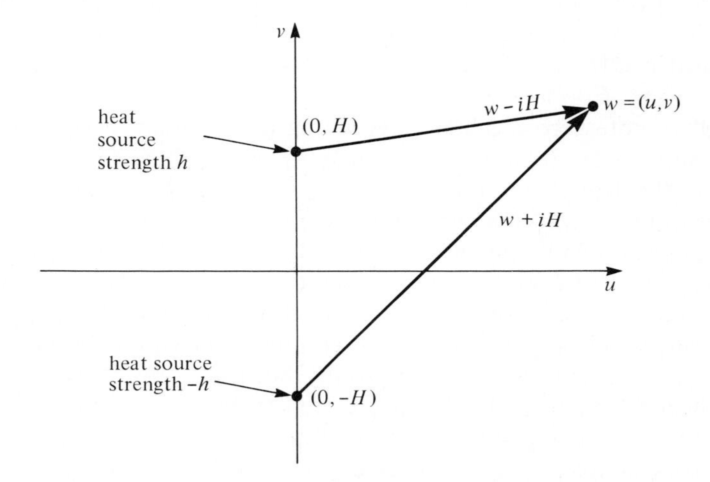

Figure 8.7–4

The preceding is an example of the *method of images*, which is used extensively in the physical sciences and is not limited to configurations with plane boundaries. As a further example of the method, we reverse the sign of the lower source in Fig. 8.7–4 and obtain a useful complex potential

$$\Phi(w) = \frac{h}{2\pi k} \operatorname{Log} \frac{1}{w - iH} + \frac{h}{2\pi k} \operatorname{Log} \frac{1}{w + iH},$$

or

$$\Phi(w) = \frac{h}{2\pi k} \operatorname{Log} \frac{1}{w^2 + H^2}. \tag{8.7–12}$$

The actual temperature and stream function are respectively

$$\phi = \frac{h}{2\pi k} \operatorname{Log} \frac{1}{|w^2 + H^2|}$$

and

$$\begin{aligned} \psi(u, v) &= \frac{h}{2\pi k} \arg \frac{1}{w^2 + H^2} \\ &= \frac{-h}{2\pi k} \arctan \frac{2uv}{u^2 - v^2 + H^2}. \end{aligned}$$

On the line $v = 0$ we have $\psi(u, 0) = (-h/2\pi k) \arctan 0$, while the temperature varies as $\phi(u, 0) = (h/2\pi k) \operatorname{Log}[1/|(u^2 + H^2)|]$. Although the temperature changes along this line, the stream function is constant. Thus the line $v = 0$ is a streamline of the complex potential, and no heat crosses the line. Hence we can conclude that $\Phi(w) = (h/2\pi k) \operatorname{Log}[1/(w^2 + H^2)]$ is the potential created by a line source of heat of strength h in the semiinfinite space $\operatorname{Im} w \geq 0$ whose boundary $\operatorname{Im} w = 0$ is *insulated.* The line source is of strength h and is located a distance H from the insulated boundary. The configuration is shown in Fig. 8.7–5. With the aid of the computer software called **f(z)**, mentioned previously, we have drawn a few streamlines on the figure for the case $h/2\pi k = 1$ and $H = 1$.

It is well to note here a subtlety pertaining to the uniqueness of our solution Eq. (8.7–12). Suppose an additional potential $\Phi_a(w) = Aw$ (where A is any real constant) were added to the expression on the right in Eq. (8.7–12). $\Phi_a(w)$ is associated with a uniform flow of heat parallel to the line $v = 0$. One of its streamlines coincides with $v = 0$. The addition of this potential to $\Phi(w)$ has no effect on the strength of the source located at $w = iH$, and the sum of the two potentials $\Phi(w) + \Phi_a(w)$ still has a streamline along $v = 0$. However, we must reject the term $\Phi_a(w)$ because it is created *entirely* by sources of heat placed at $w = \infty$ and the specification of our problem did not include any such sources. The preceding illustrates how in seeking a unique solution to a problem whose boundaries extend to infinity we must often concern ourselves with the behavior of the solution at infinity.

The above example, which involves a pair of identical line sources at mirror-image locations, can also be used to solve the problem of obtaining the complex

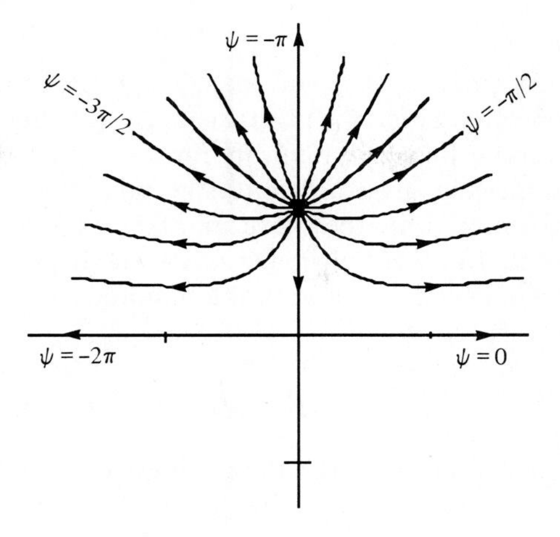

Figure 8.7–5

velocity potential caused by a line source of fluid located parallel to a rigid impenetrable barrier (see Exercise 4).

EXAMPLE 2

Shown in Fig. 8.7–6 is a filament carrying electrical charge of ρ coulombs per meter. It lies inside an electrically conducting tube of unit radius. The charge is displaced H units from the axis of the tube, which is filled with material of electrical permittivity ε. (Note that $H < 1$.) The tube is at zero electrical potential. Find the complex potential, actual potential, and complex electric flux density vector inside the tube.

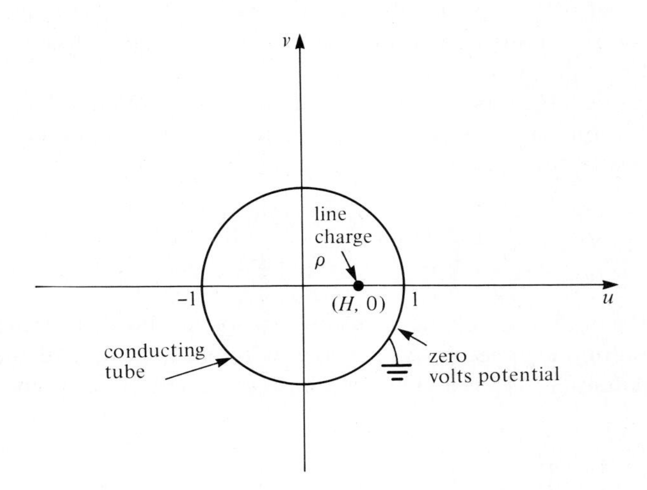

Figure 8.7–6

Solution

We seek a bilinear transformation that will map our given configuration into that of Fig. 8.7–2; i.e., the line charge is moved to the axis of a tube of unit radius and zero potential. There is more than one bilinear transformation that will accomplish this. We will assume somewhat arbitrarily that the points $w = 1$ and $w = -1$ are fixed points of the transformation, i.e., they have images $z = 1$ and $z = -1$ respectively. Furthermore we require that $w = H$ has image $z = 0$. Having established the images of three points, we use Eq. (8.4–27) to obtain the required mapping. Thus we get

$$z = \frac{H - w}{wH - 1}.$$

We know from our earlier discussion that the complex potential inside a grounded (zero potential) tube of unit radius with a line charge along its axis is $\Phi_1(z) = (\rho/2\pi\varepsilon)\,\mathrm{Log}(1/z)$. Using the transformation just found we obtain the complex potential inside the tube of the given problem:

$$\Phi(w) = \frac{\rho}{2\pi\varepsilon}\,\mathrm{Log}\,\frac{1 - wH}{w - H}. \tag{8.7–13}$$

The actual potential, or voltage, inside the tube of Fig. 8.7–6 is the real part of the preceding expression, i.e.

$$\begin{aligned}\phi(u, v) &= \frac{\rho}{2\pi\varepsilon}\,\mathrm{Log}\,\frac{|1 - wH|}{|w - H|}\\ &= \frac{\rho}{4\pi\varepsilon}\,\mathrm{Log}\,\frac{H^2[(u - 1/H)^2 + v^2]}{(u - H)^2 + v^2}.\end{aligned}$$

It is not hard to verify that on the tube, i.e., where $u^2 + v^2 = 1$ is satisfied, this voltage is zero. One can also show that inside the tube the voltage is nonnegative and lies between zero and infinity if $\rho > 0$.

The surfaces on which the voltage assumes specific constant values are of interest. To find the equations of their cross section in the u, v-plane we equate the preceding expression to the voltage of interest. Calling this V_0 we have

$$V_0 = \frac{\rho}{4\pi\varepsilon}\,\mathrm{Log}\,\frac{H^2[(u - 1/H)^2 + v^2]}{(u - H)^2 + v^2}.$$

We multiply both sides of the preceding by $4\pi\varepsilon/\rho$ and then exponentiate both sides of the resulting expression. After some manipulation we find that the surface on which the potential is V_0 is a cylinder whose cross section is circular and satisfies the equation

$$\left[u - \frac{\beta^2 H^2 - 1}{H(\beta^2 - 1)}\right]^2 + v^2 = \frac{\beta^2(H^2 - 1)^2}{H^2(\beta^2 - 1)^2}.$$

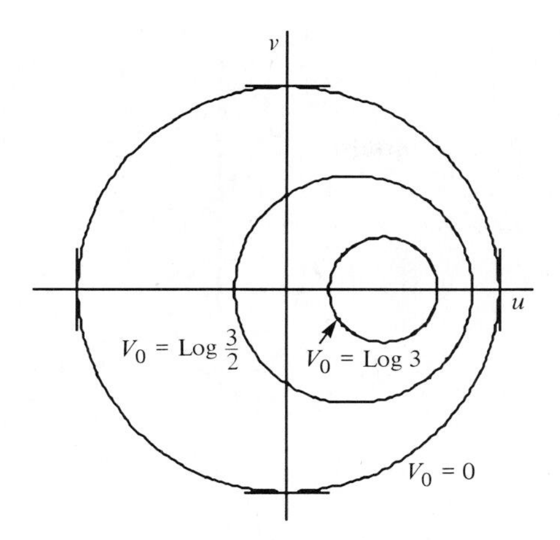

Figure 8.7–7

Here $\beta = (1/H)e^{V_0 2\pi\varepsilon/\rho}$. For $V_0 \geq 0$ the circles lie inside the tube and enclose the line charge. We have sketched a few of them in Fig. 8.7–7 by assuming that $2\pi\varepsilon/\rho = 1$ and $H = .5$. The values of V_0 are shown on the curves.

Recalling that the electric flux density vector is given by $\mathbf{d} = -\varepsilon\overline{(d\Phi/dz)}$, we use Eq. (8.7–13) and show that

$$\mathbf{d} = \frac{\rho}{2\pi}\left[\frac{1}{\overline{(w - H)}} - \frac{1}{\overline{(w - 1/H)}}\right]$$

inside the tube. ◀

EXERCISES

1. Consider the configuration of Fig. 8.7–2. The radius of the tube is now changed to b from unity, and the surface of the tube is maintained at temperature T_0 instead of zero. Show that the complex temperature in the tube is now $\Phi(z) = (h/2\pi k)\,\text{Log}(c/z)$, $|z| \leq b$, where $c = be^{T_0 2\pi k/h}$. Do this by showing that this expression yields the given flux from the heat source and satisfies the boundary condition on the surface of the tube.

2. If a fluid vortex is placed with its axis perpendicular to the complex z-plane and passing through $z = 0$ it creates the complex velocity potential $\Phi(z) = -iV_0\,\text{Log}\,z$, where V_0 is real. The strength of the vortex is defined as $2\pi V_0$ and the vortex is defined as acting at $z = 0$.

   a) Show that the actual (not complex) velocity potential for the resulting flow is $\phi(x, y) = V_0 \arg z$ and that the stream function is $\psi(x, y) = -V_0\,\text{Log}(\sqrt{x^2 + y^2})$. Describe in words the shape and location of the streamlines.

   b) Show that the complex velocity vector is $\mathbf{v} = i(V_0/r)\,\text{cis}\,\theta$ where $z = r\,\text{cis}\,\theta$. Assuming that $V_0 > 0$, explain why the fluid moves in circles counterclockwise around the vortex.

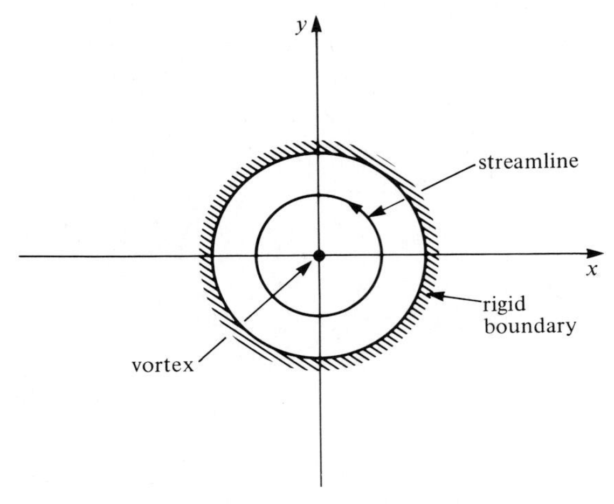

Figure 8.7–8

c) Let $\mathbf{v} = V_x + iV_y$. Show that V_x and V_y, the components of the fluid velocity vector, satisfy $\partial V_x/\partial x + \partial V_y/\partial y = 0$ if $z \neq 0$. Thus except *at* the vortex, fluid flow created by the vortex satisfies the conservation equation as described in Section 2.6.

d) The fluid vortex is placed at the center of a tube of unit radius having rigid, impenetrable boundaries as shown in Fig. 8.7–8. As described in Section 8.6, we require that the wall of the tube be a streamline. Show that this condition is met by the velocity potential $\Phi(z) = -iV_0 \operatorname{Log} z$.

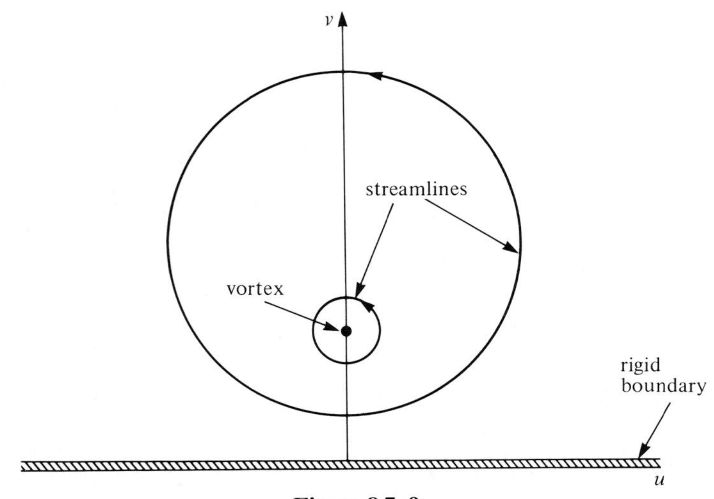

Figure 8.7–9

e) Instead of being inside the tube, the vortex is a distance H above a rigid plane as shown in Fig. 8.7–9. The u-axis must be a streamline of the resulting flow. Make a conformal mapping of the domain $|z| \le 1$ shown in Fig. 8.7–8 onto $\text{Im } w \ge 0$ and use the result to show that the complex velocity potential describing flow above the plane is given by $\Phi(z) = -iV_0[\text{Log}(w - iH) - \text{Log}(w + iH)]$ and that the complex velocity vector for the flow is

$$\frac{2V_0H[u^2 - v^2 + H^2 + 2iuv]}{u^4 + v^4 + 2u^2v^2 + H^4 + 2H^2(u^2 - v^2)}.$$

3. a) A line charge carrying ρ coulombs per meter is placed a distance H from the axis of a grounded electrically conducting tube, of unit radius, set at zero electrostatic potential. (See Fig. 8.7–10.) Assume $H > 1$. Find the complex electrostatic potential $\Phi(w)$ outside the tube.

Hint: Find a bilinear transformation that transforms the configuration inside $|z| = 1$ shown in Fig. 8.7–2 onto the region $|w| \ge 1$. A line charge is used in place of the heat source in Fig. 8.7–2.

Answer:

$$\Phi(w) = \frac{\rho}{2\pi\varepsilon} \text{Log} \frac{1 - wH}{w - H}.$$

b) Taking $\rho/2\pi\varepsilon = 1$, and $H = 2$, find the equation and sketch the equipotential along which $\phi(u, v)$, the actual potential (voltage) in Fig. 8.7–10, equals Log(5/2).

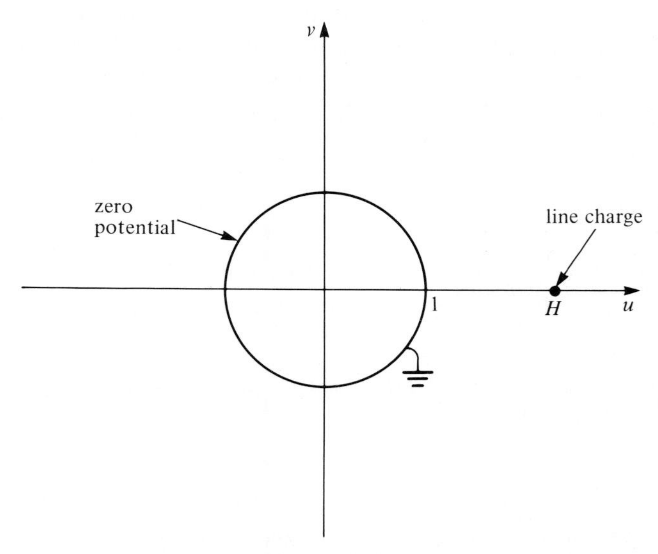

Figure 8.7–10

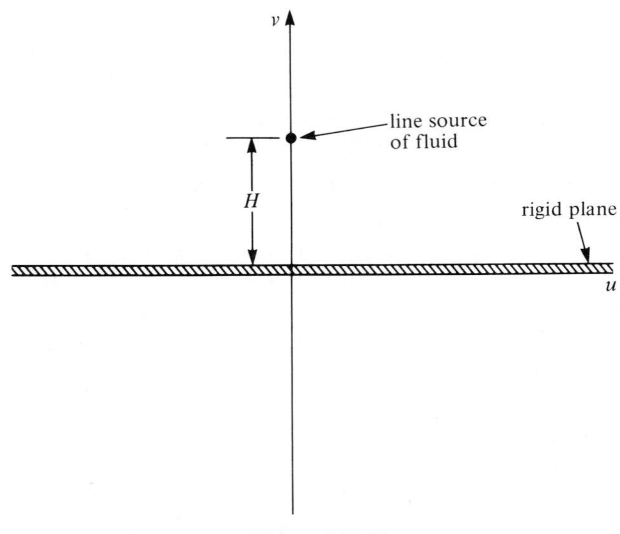

Figure 8.7–11

4. A line source of fluid of strength G lies parallel to and a distance H from a rigid impenetrable plane as shown in Fig. 8.7–11. We require that $v = 0$ be a streamline of the flow.
   a) By placing an identical line source so that it passes through the mirror-image point, $w = -iH$, show that the complex velocity potential caused by both sources, in the absence of the plane, is given by $\Phi(w) = (G/2\pi)\,\text{Log}(w^2 + H^2)$. Show that the stream function is $\psi(u, v) = (G/2\pi)\arctan[2uv/(u^2 - v^2 + H^2)]$ and that the line $v = 0$ is a streamline. Thus, a plane can be introduced along $v = 0$ without disturbing the flow produced by the original source plus its image, and we may assume that $\Phi(w)$ is the complex velocity potential created by the original line source above the plane.

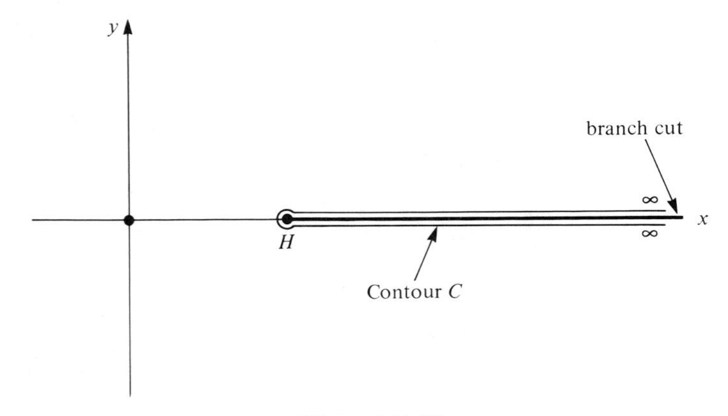

Figure 8.7–12

b) Show that the complex velocity vector created by the fluid source above the plane is

$$\mathbf{v} = \frac{G}{2\pi}\left[\frac{u + i(v - H)}{u^2 + (v - H)^2} + \frac{u + i(v + H)}{u^2 + (v + H)^2}\right].$$

c) We can solve this problem without resorting to images. Consider a line source of fluid of strength G perpendicular to the complex z-plane and passing through $z = 0$. It provides a complex potential $\Phi = (G/2\pi)\,\text{Log}(z)$. Now apply the conformal transformation $w(z) = (z - H^2)^{1/2}$ to the z-plane. Use the branch cut $x \geq H^2$, $y = 0$ and assume that $w(0) = iH$. How is the contour C shown in Fig. 8.7–12 mapped by this transformation? It is composed of points lying just above and just below the branch cut. Explain why it is transformed into a streamline. What is the transformed complex potential?

5. A line of electrostatic charge of strength ρ coulombs/meter lies in a U-shaped channel of width π as shown in Fig. 8.7–13. The channel is composed of electrically conducting material maintained at zero electrostatic potential. The charge is centered, and located a distance α from the bottom of the channel.

a) Show that the complex electrostatic potential in the channel is given by

$$\Phi(w) = \frac{\rho}{2\pi\varepsilon}\,\text{Log}\frac{\sin w + i\sinh\alpha}{\sin w - i\sinh\alpha}.$$

Hint: Consider the transformation $z = \sin w$. Use the method of images.

b) Suppose the left- and right-hand boundaries in Fig. 8.7–13 were changed from $u = \pm\pi/2$ to $u = \pm b$ $(b > 0)$. What would $\Phi(w)$ now be?

c) For the configuration of part (a) show that the actual potential (voltage) in the channel is given by:

$$\phi(u, v) = \frac{\rho}{4\pi\varepsilon}\,\text{Log}\,\frac{\sin^2 u + \sinh^2 v + \sinh^2\alpha + 2\sinh\alpha\cos u\sinh v}{\sin^2 u + \sinh^2 v + \sinh^2\alpha - 2\sinh\alpha\cos u\sinh v}.$$

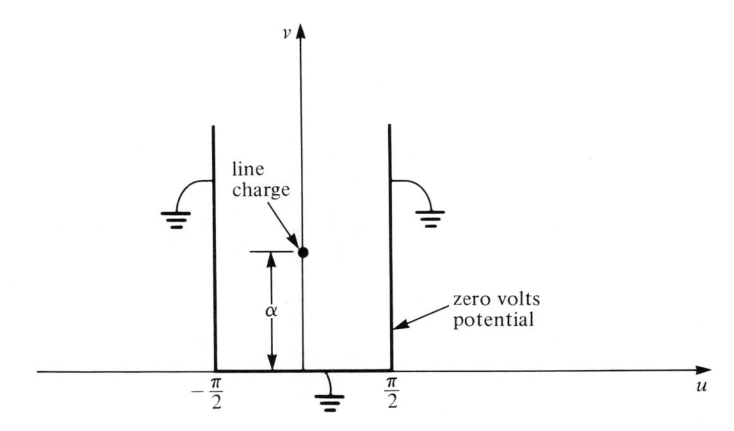

Figure 8.7–13

6. A line source of fluid of strength G lies in the middle of a channel of width π as shown in Fig. 8.7–14. The channel is open at both ends. Since the walls of the channel are rigid and impenetrable they must be streamlines of the resulting flow.

a) Show that the complex velocity potential for flow in the channel is $\Phi(w) = (G/2\pi)\,\mathrm{Log}(\cos(w))$.

Hint: Suppose the same source is perpendicular to, and passes through, the origin of the complex z-plane and that there are no boundaries. What is the complex velocity potential and what are the streamlines? Now consider the mapping $z = \cos w$ applied to the configuration of Fig. 8.7–14 and obtain $\Phi(w)$.

Note that a potential of the form $\Phi_a(w) = Aiw$ (A is any real constant) could be added to the potential $\Phi(w)$ and the boundary conditions on the walls of the channel would still be met, since the streamlines of $\Phi_a(w)$ lie parallel to the v-axis. The strength of the source would not be affected. However, $\Phi_a(w)$ is associated with a uniform flow that begins and ends at infinity in the channel and is not caused by the source placed in the channel. Thus we reject $\Phi_a(w)$.

b) Show that in the channel the complex fluid velocity vector is

$$\mathbf{v} = \frac{G}{4\pi}\left[\frac{-\sin(2u) + i\sinh(2v)}{\sinh^2 v + \cos^2 u}\right].$$

c) Show that for $v \gg 1$ the fluid velocity vector in the channel is in the direction of the positive v-axis and equals $G/2\pi$.

d) Prove that the streamlines of flow are the curves on which $(\tan u)(\tanh v) =$ real constant. Take $G/2\pi = 1$ and sketch a few streamlines in the channel, labeling them with their corresponding values of ψ.

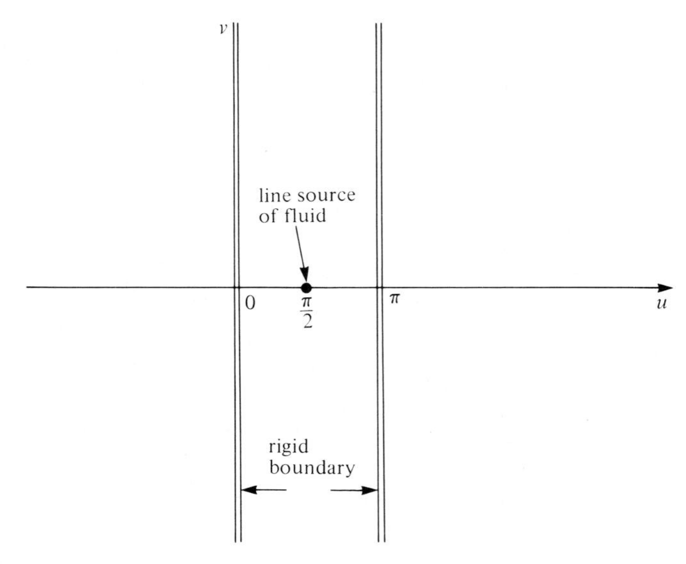

Figure 8.7–14

7. A line source of heat of strength h (calories per meter per second) is embedded in a slab of material of heat conductivity k as shown in Fig. 8.7–15. The width of the slab is π and its surfaces are maintained at a temperature of zero degrees. The line source passes through the point $w = b$, where $0 < b < \pi$.

a) Show that the complex temperature in the slab is given by

$$\Phi(w) = \frac{h}{2\pi k} \operatorname{Log} \frac{e^{iw} - e^{-ib}}{e^{iw} - e^{ib}}.$$

Hint: Consider the transformation $z = e^{iw}$.

If the right hand boundary lay along $w = \beta$ instead of $w = \pi$, what would the complex potential in the slab be? Assume $\beta > b$.

b) Inside the slab of width π, show that the actual temperature is

$$\phi(u, v) = \frac{h}{4\pi k} \operatorname{Log} \frac{\cosh(v) - \cos(u + b)}{\cosh(v) - \cos(u - b)},$$

and that if the line source is placed in the middle of the slab ($b = \pi/2$) the temperature is

$$\phi(u, v) = \frac{h}{4\pi k} \operatorname{Log} \frac{\cosh v + \sin u}{\cosh v - \sin u}.$$

c) Show that when $b = \pi/2$ the complex heat flux density vector in the slab is given by

$$\mathbf{q} = \frac{h}{2\pi}\left[\frac{-\cos u \cosh v + i \sin u \sinh v}{\cos^2 u + \sinh^2 v}\right].$$

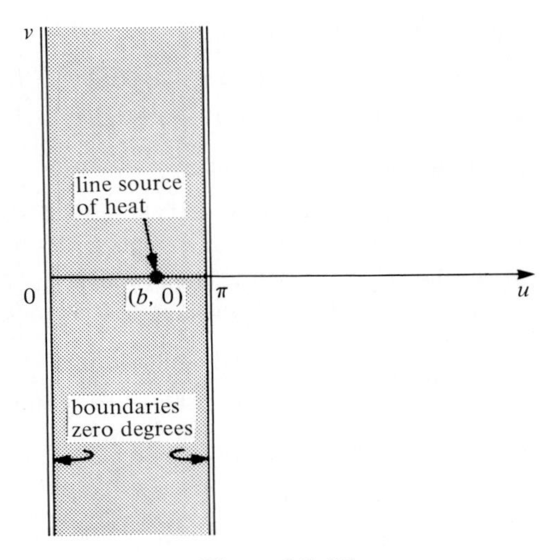

Figure 8.7–15

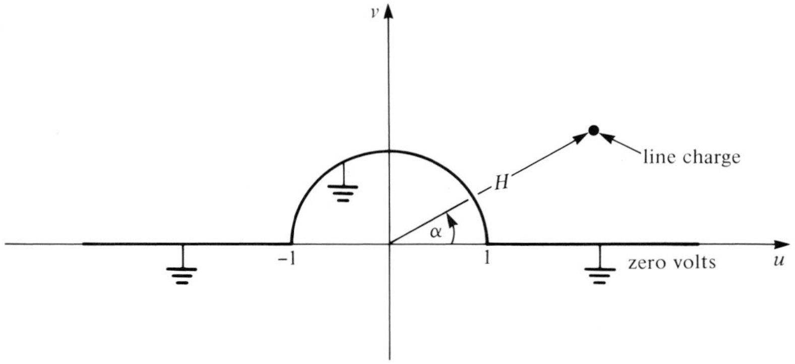

Figure 8.7–16

8. An electrostatic line charge of strength ρ is placed above a grounded (zero potential) electrically conducting plane containing a hill in the shape of a half cylinder. The arrangement is shown in cross section in Fig. 8.7–16. This might schematically represent a cloud of electrical charge above the earth and a building just before lightning moves downward from the cloud. Note that $H > 1$ is the distance of the line charge from the axis of the cylinder.

a) Show that the complex potential above the plane and external to the half cylinder is given by

$$\Phi(w) = \frac{\rho}{2\pi\varepsilon} \operatorname{Log} \frac{[w - (1/H)\operatorname{cis}(\alpha)][w - H\operatorname{cis}(-\alpha)]}{[w - (1/H)\operatorname{cis}(-\alpha)][w - H\operatorname{cis}(\alpha)]}.$$

Hint: Consider the transformation $z = w + 1/w$ applied to the given configuration. Then use the method of images applied to a plane.

b) Let $w = r \operatorname{cis} \theta$. Show that the actual potential (voltage) is

$$\phi(r, \theta) = \frac{\rho}{4\pi\varepsilon} \operatorname{Log} \frac{[r^2 + (1/H)^2 - 2r(\cos(\theta - \alpha))/H][r^2 + H^2 - 2rH\cos(\theta + \alpha)]}{[r^2 + (1/H)^2 - 2r(\cos(\theta + \alpha))/H][r^2 + H^2 - 2rH\cos(\theta - \alpha)]}$$

above the plane and outside the half cylinder.

9. a) An infinite electrically conducting plane is bent into a right angle as shown in Fig. 8.7–17. A line of electrostatic charge of strength ρ lies parallel to and a distance R from the bend, as shown in the figure. Assume that $0 < \beta < \pi/2$. Let $w = r \operatorname{cis} \theta$. Show that inside the bend $(0 \le \theta \le \pi/2)$ the complex potential is

$$\Phi(r, \theta) = \frac{\rho}{2\pi\varepsilon} \operatorname{Log} \frac{r^2 \operatorname{cis}(2\theta) - R^2 \operatorname{cis}(-2\beta)}{r^2 \operatorname{cis}(2\theta) - R^2 \operatorname{cis}(2\beta)}.$$

Hint: Consider the mapping $z = w^2$.

b) Show that the actual potential (voltage) is

$$\phi(r, \theta) = \frac{\rho}{4\pi\varepsilon} \operatorname{Log} \frac{r^4 + R^4 - 2r^2R^2\cos[2(\theta + \beta)]}{r^4 + R^4 - 2r^2R^2\cos[2(\theta - \beta)]}.$$

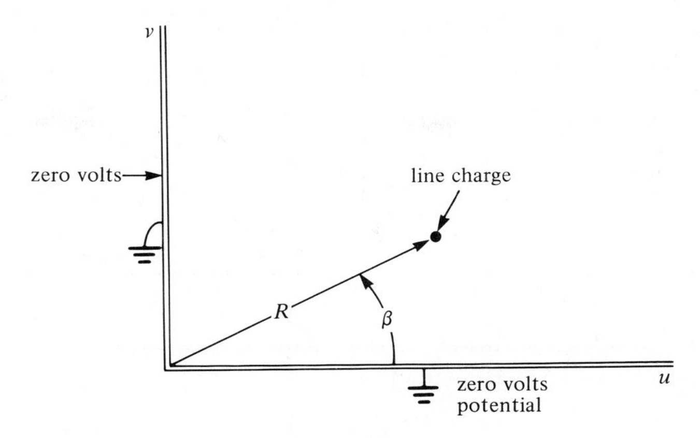

Figure 8.7–17

10. a) This problem represents a generalization of the preceding one, which you should do first. The angle of the bend is changed to α, where $0 < \alpha \leq 2\pi$, and $0 < \beta < \alpha$. Again $w = r \text{ cis } \theta$. The situation is shown in Fig. 8.7–18. Show that inside the bend $(0 \leq \theta \leq \alpha)$ the complex potential is

$$\Phi(r, \theta) = \frac{\rho}{2\pi\varepsilon} \text{Log} \frac{r^{\pi/\alpha} \text{cis}(\theta\pi/\alpha) - R^{\pi/\alpha} \text{cis}(-\beta\pi/\alpha)}{r^{\pi/\alpha} \text{cis}(\theta\pi/\alpha) - R^{\pi/\alpha} \text{cis}(\beta\pi/\alpha)},$$

where the values of $r^{\pi/\alpha}$ and $R^{\pi/\alpha}$ are taken as positive reals. Show also that the actual potential is

$$\phi(r, \theta) = \frac{\rho}{4\pi\varepsilon} \text{Log} \frac{r^{2\pi/\alpha} + R^{2\pi/\alpha} - 2r^{\pi/\alpha}R^{\pi/\alpha} \cos[(\theta + \beta)\pi/\alpha]}{r^{2\pi/\alpha} + R^{2\pi/\alpha} - 2r^{\pi/\alpha}R^{\pi/\alpha} \cos[(\theta - \beta)\pi/\alpha]},$$

where all fractional powers are evaluated as just described and $0 \leq \theta \leq \alpha$.

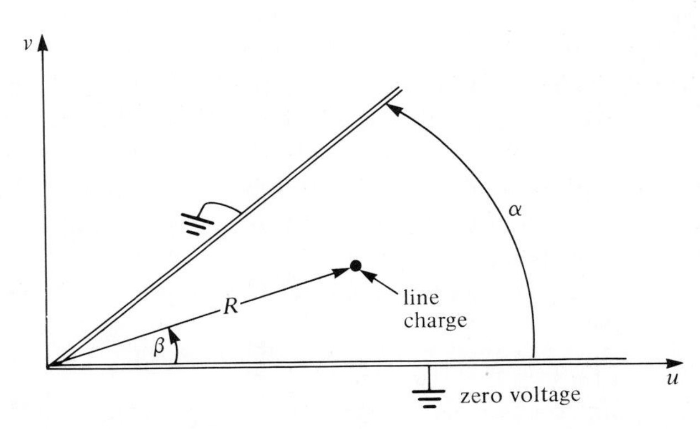

Figure 8.7–18

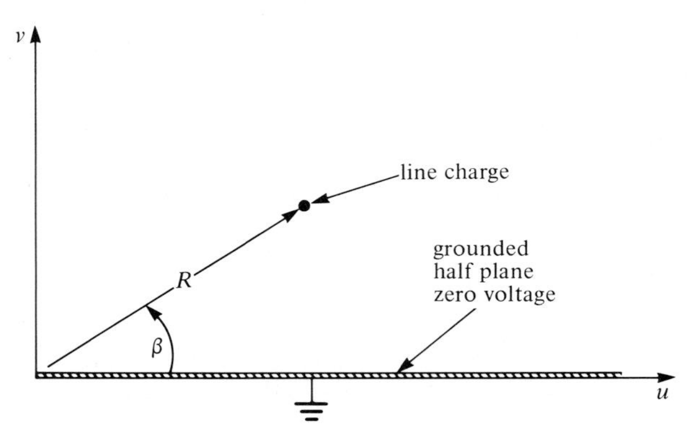

Figure 8.7–19

b) The line charge is parallel to the edge and a distance R from a grounded semiinfinite electrically conducting plane as shown in Fig. 8.7–19. Use the result derived in (a) to show that the actual electrostatic potential is

$$\phi(r, \theta) = \frac{\rho}{4\pi\varepsilon} \operatorname{Log} \frac{r + R - 2r^{1/2}R^{1/2}\cos[(\theta + \beta)/2]}{r + R - 2r^{1/2}R^{1/2}\cos[(\theta - \beta)/2]} \qquad \text{for } 0 \le \theta \le 2\pi.$$

8.8 THE SCHWARZ–CHRISTOFFEL TRANSFORMATION

Many physical problems in heat conduction, fluid mechanics, and electrostatics involve boundaries whose cross sections form a polygon. In the domain bounded by the polygon we seek a harmonic function satisfying certain boundary conditions. A one-to-one mapping $w = f(z)$ that would transform this domain from the z-plane onto the upper half of the w-plane, with the polygonal boundary transformed into the real axis, would greatly assist us in solving our problem because of the simplified shape now obtained. We discuss here something close to what is required; the Schwarz–Christoffel transformation is a formula that will transform the real axis of the w-plane (the u-axis) into a polygon in the z-plane. Once this formula is obtained (often a formidable task) an inversion can sometimes be applied that yields the desired $w = f(z)$.

A rigorous derivation of the Schwarz–Christoffel transformation will not be presented. Instead we will first convince the reader of its plausibility and then move on to some examples of its use.

To see how the formula operates, consider the simple transformation

$$z = (w - u_1)^{\alpha_1/\pi}, \qquad 0 \le \alpha_1 \le 2\pi, \tag{8.8–1}$$

where $(u_1, 0)$ is a point on the real axis of the w-plane. Equation (8.8–1) is defined by means of a branch cut originating at $w = u_1$ and going into the lower half of this plane. Equating arguments on both sides of Eq. (8.8–1), we have

$$\arg z = \frac{\alpha_1}{\pi} \arg(w - u_1). \tag{8.8–2}$$

If w is real with $w > u_1$, we take $\arg(w - u_1) = 0$, and from Eq. (8.8–2)

$$\arg z = 0. \tag{8.8–3}$$

Now if w is real with $w < u_1$, we have $\arg(w - u_1) = \pi$, and from Eq. (8.8–2)

$$\arg z = \frac{\alpha_1}{\pi} \pi = \alpha_1. \tag{8.8–4}$$

Equation (8.8–1) indicates that the points $w = u_1$ and $z = 0$ are images of each other. Refer now to Fig. 8.8–1(a) and (b). If we consider a line segment on the u-axis to the right of $w = u_1$, it must, according to Eq. (8.8–3), be transformed into a line segment in the z-plane emanating from the origin and lying along the x-axis. The line segment to the left of $w = u_1$ is, according to Eq. (8.8–4), transformed into a ray making an angle α_1 with the positive x-axis.

To summarize: $z = (w - u_1)^{\alpha_1/\pi}$ bends a straight line segment that lies on the u-axis and passes through $w = u_1$ into a pair of line segments intersecting at the origin of the z-plane with angle α_1. The more complicated transformation

$$z = c_1(w - u_1)^{\alpha_1/\pi} + c_2, \tag{8.8–5}$$

applied to the straight line of Fig. 8.8–1(a), results in the pair of line segments shown in Fig. 8.8–1(c). The angle of intersection is still α_1 but the segments no longer emanate from the origin and, in general, are rotated from their original orientation.

A transformation $z = g(w)$ that simultaneously bends several line segments on the u-axis into straight line segments in the z-plane intersecting at various angles and different locations should, in principle, transform the entire u-axis into a polygon in

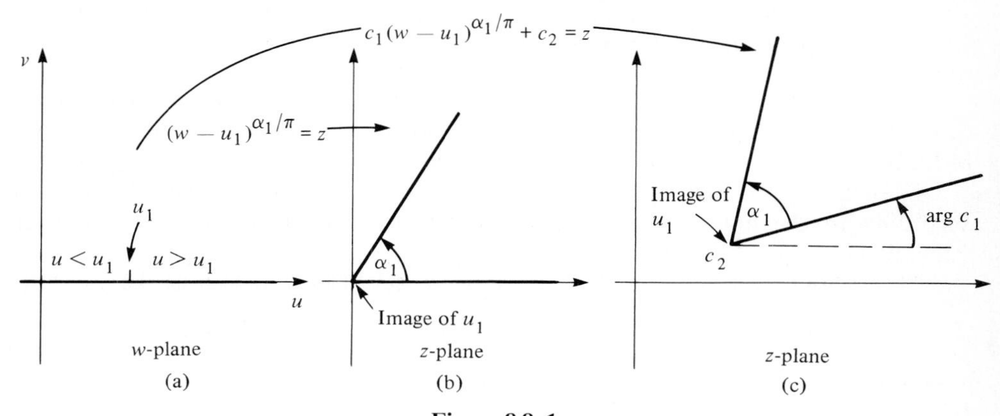

Figure 8.8–1

the z-plane. Notice from Eq. (8.8–5) that

$$\frac{dz}{dw} = c_1 \frac{\alpha_1}{\pi}(w - u_1)^{(\alpha_1/\pi)-1}.$$

This suggests our considering the following formula in order to transform the u-axis into a polygon.

$$\frac{dz}{dw} = A(w - u_1)^{(\alpha_1/\pi)-1}(w - u_2)^{(\alpha_2/\pi)-1} \ldots (w - u_n)^{(\alpha_n/\pi)-1},$$

where $(u_1, 0)$, $(u_2, 0), \ldots,$ etc. are the images in the w-plane of the vertices of the polygon and $\alpha_1, \alpha_2, \ldots, \alpha_n$ are the angles of intersection of the sides of the polygon in the z-plane. In fact, our assumption about this formula is correct and is summarized in the following theorem:

THEOREM 8 The Schwarz–Christoffel transformation

The real axis in the w-plane is transformed into a polygon in the z-plane having vertices at $z_1, z_2, \ldots, z_n$ and corresponding interior angles $\alpha_1, \alpha_2, \ldots, \alpha_n$ by the formula

$$\frac{dz}{dw} = A(w - u_1)^{(\alpha_1/\pi)-1}(w - u_2)^{(\alpha_2/\pi)-1} \ldots (w - u_n)^{(\alpha_n/\pi)-1}, \qquad (8.8\text{–}6)$$

or

$$z = A \int^{w} (\zeta - u_1)^{(\alpha_1/\pi)-1}(\zeta - u_2)^{(\alpha_2/\pi)-1} \ldots (\zeta - u_n)^{(\alpha_n/\pi)-1}\, d\zeta + B, \qquad (8.8\text{–}7)$$

where $(u_1, 0)$, $(u_2, 0), \ldots, (u_n, 0)$ are mapped into the vertices $z_1, z_2, \ldots, z_n$. If $w = \infty$ is mapped into one vertex, say z_j, then the term containing $(w - u_j)$ is absent in Eq. (8.8–6), and the term containing $(\zeta - u_j)$ is absent in Eq. (8.8–7). The size and orientation of the polygon is determined by A and B. The half plane Im $w > 0$ is mapped onto the interior of the polygon. □

A lower limit has not been specified for the integral in Eq. (8.8–7). The reader can choose this quantity arbitrarily. Note, however, that any constant this might produce can be absorbed into B. The integration is performed on the dummy variable ζ. Differentiation of both sides of Eq. (8.8–7) with respect to w yields Eq. (8.8–6) according to the fundamental theorem of integral calculus† applied to contour integrals.

To see how the transformation operates, we have from Eq. (8.8–6) that

$$dz = A(w - u_1)^{(\alpha_1/\pi)-1}(w - u_2)^{(\alpha_2/\pi)-1} \ldots (w - u_n)^{(\alpha_n/\pi)-1}\, dw.$$

---

† See Section 4.4.

Equating the arguments on both sides, we have

$$\arg dz = \arg A + \left(\frac{\alpha_1}{\pi} - 1\right)\arg(w - u_1) + \left(\frac{\alpha_2}{\pi} - 1\right)\arg(w - u_2) + \cdots$$
$$+ \left(\frac{\alpha_n}{\pi} - 1\right)\arg(w - u_n) + \arg dw. \tag{8.8–8}$$

Imagine now that the point w lies at the location marked P in Fig. 8.8–2(b). We take P to the left of $u_1, u_2, \ldots, u_n$.

As w moves through the increment $dw = du$ along the real axis toward u_1, we have $\arg dw = 0$. Since $(w - u_1), (w - u_2), \ldots, (w - u_n)$ are all negative real numbers when w is to the left of u_1, the arguments of these terms are all π in Eq. (8.8–8), and $\arg dz$ in this equation remains constant as w proceeds toward u_1. The argument of dz can only remain fixed along some locus if that locus is a straight line. Hence, as w moves toward u_1 in Fig. 8.8–2(b), the locus traced out by z, as defined in Eqs. (8.8–6) and (8.8–7), is a line segment.

When w moves through u_1, $\arg(w - u_1)$ in Eq. (8.8–8) abruptly decreases by π. However, all other arguments in this equation remain constant at their original values. According to Eq. (8.8–8), $\arg dz$ will change abruptly in value. It *decreases* by $((\alpha_1/\pi) - 1)\pi = \alpha_1 - \pi$ or *increases* by $\pi - \alpha_1$. If w, which is now to the right of u_1, moves toward u_2 along the u-axis in Fig. 8.8–2(b), $\arg dz$ remains fixed at its new value and a new line segment is traced in the z-plane. The increase in argument $\pi - \alpha_1$ just noted causes the two line segments that have been generated in the z-plane to intersect at z_1 with an angle α_1 (see Fig. 8.8–2a).

As w continues to the right along the u-axis in Fig. 8.8–2(b), we see from Eq. (8.8–8) that $\arg dz$ will abruptly increase by $\pi - \alpha_2$, and as w moves between u_2 and u_3 a new line segment is generated in the z-plane making an angle α_2 with the previous one. In this way the transformation defined by Eq. (8.8–6) or Eq. (8.8–7)

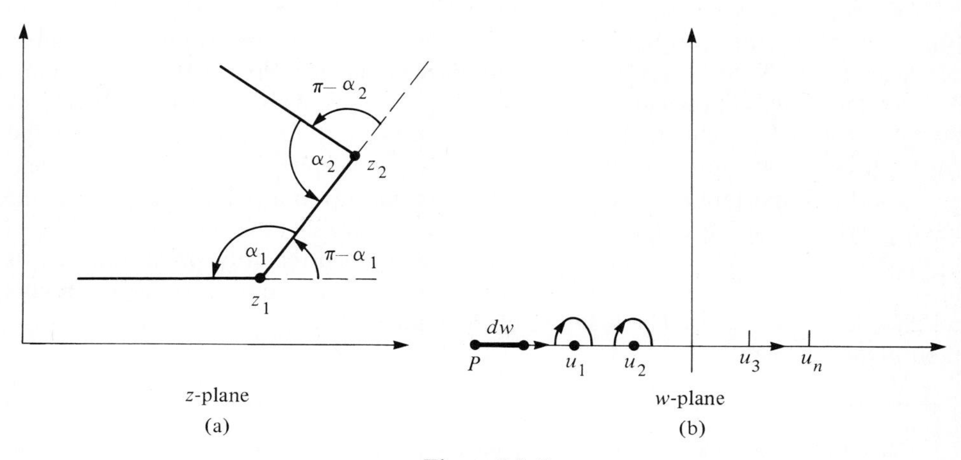

Figure 8.8–2

generates an entire polygon in the z-plane as w progresses along the whole real axis from $-\infty$ to $+\infty$ in the w-plane.

Recall from plane geometry that the sum of the exterior angles of a *closed* polygon is 2π. The exterior angle at z_1 in Fig. 8.8–2(a) is $\pi - \alpha_1$, at z_2 it is $\pi - \alpha_2$, etc. Thus

$$(\pi - \alpha_1) + (\pi - \alpha_2) + \cdots + (\pi - \alpha_n) = 2\pi.$$

If we divide both sides of this equation by π and then multiply by (-1), we obtain a relationship that the exponents in Eqs. (8.8–6) and (8.8–7) must satisfy if the u-axis is to be transformed into a closed polygon.

$$\frac{\alpha_1}{\pi} - 1 + \frac{\alpha_2}{\pi} - 1 + \cdots + \frac{\alpha_n}{\pi} - 1 = -2. \tag{8.8–9}$$

This relationship holds in Examples 1 and 2, which follow, but not in Example 3, where an open polygon is considered.

Let us see how Eqs. (8.8–6) and (8.8–7) must be modified if one vertex of a polygon, say z_n, is to have the image $u_n = \infty$. First we divide and multiply the right side of Eq. (8.8–6) by

$$(-u_n)^{(\alpha_n/\pi)-1}.$$

Thus

$$\frac{dz}{dw} = A(-u_n)^{(\alpha_n/\pi)-1}(w - u_1)^{(\alpha_1/\pi)-1}(w - u_2)^{(\alpha_2/\pi)-1}\cdots\left(\frac{w - u_n}{-u_n}\right)^{(\alpha_n/\pi)-1}.$$

As $u_n \to \infty$, the last factor on the right can be taken as 1 while the product of the first two factors $A(-u_n)^{(\alpha_n/\pi)-1}$ is maintained finite in the limit and absorbed into a new constant independent of w, which we will again choose to call A. Integration of this new equation yields Eq. (8.8–7) but with the factor $(w - u_n)^{(\alpha_n/\pi)-1}$ deleted. This procedure can be applied to any vertex: z_1, z_2, etc.

We will not prove that the Schwarz–Christoffel formula maps the domain $\text{Im } w > 0$ one to one onto the interior of the polygon. The reader is referred to more advanced texts.[†] Note, however, that for the correspondence between the two domains to exist, the images of the consecutive points $u_1, u_2, \ldots, u_n$, which are encountered as we move from left to right along the u-axis, must be $z_1, z_2, \ldots, z_n$, which are encountered in this order as we move around the polygon while keeping its interior on our left.

Another important fact, not proved here, is that given a polygon in the z-plane having vertices at specified locations $z_1, z_2, \ldots, z_n$, we find that the Schwarz–Christoffel formula can be used to create a correspondence between the polygon and the u-axis in such a way that three of the vertices will have images at any three distinct points we choose on the u-axis. The location of the images of the other $(n - 3)$ vertices are then predetermined.

[†] See, for example, R. Nevanlinna and V. Paatero, *Introduction to Complex Analysis* (Reading, MA: Addison-Wesley, 1969), Chapter 17.

EXAMPLE 1

Find the Schwarz–Christoffel transformation that will transform the real axis of the w-plane into the right isosceles triangle shown in Fig. 8.8–3(a). The vertices of the triangle have the images indicated in Fig. 8.8–3(b).

Solution

For the vertex at z_1, we have $\alpha_1 = \pi/4$ and the image point is $u_1 = -a$. For the vertex at z_2, we have $\alpha_2 = \pi/4$ and the image is $u_2 = a$. Finally, for the vertex at z_3, we have $\alpha_3 = \pi/2$ and the image is $w = \infty$. Since $w = \infty$ is mapped into z_3, a term containing w_3 and α_3 will not appear in Eq. (8.8–7). Using this formula, with the lower limit of integration chosen somewhat arbitrarily as 0, we obtain

$$z = A\int_0^w (\zeta + a)^{-3/4}(\zeta - a)^{-3/4}\,d\zeta + B$$

$$= A\int_0^w (\zeta^2 - a^2)^{-3/4}\,d\zeta + B. \tag{8.8–10}$$

This integral cannot be evaluated in terms of conventional functions. It must be found numerically for each value of w of interest. To find A and B we impose these requirements: If $w = a$ on the right side of Eq. (8.8–10), then z must equal 1, whereas if $w = -a$, the corresponding value of z is -1. Thus

$$1 = A\int_0^a (\zeta^2 - a^2)^{-3/4}\,d\zeta + B, \tag{8.8–11}$$

$$-1 = A\int_0^{-a} (\zeta^2 - a^2)^{-3/4}\,d\zeta + B. \tag{8.8–12}$$

We multiply the preceding equation by (-1) and reverse the limits of integration to obtain

$$1 = A\int_{-a}^0 (\zeta^2 - a^2)^{-3/4}\,d\zeta - B. \tag{8.8–13}$$

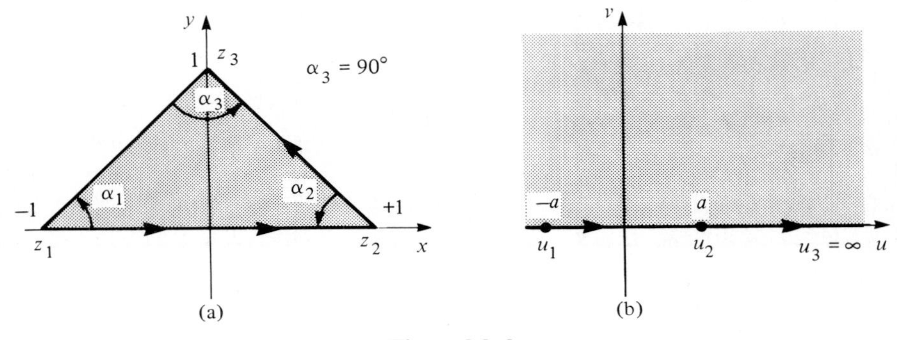

Figure 8.8–3

Adding Eqs. (8.8–11) and (8.8–13), we get

$$2 = A\int_{-a}^{+a}(\zeta^2 - a^2)^{-3/4}\,d\zeta = 2A\int_0^a(\zeta^2 - a^2)^{-3/4}\,d\zeta.$$

The last step follows from the even symmetry of the integrand. We solve the preceding equation for A to obtain

$$A = \frac{1}{\displaystyle\int_0^a(\zeta^2 - a^2)^{-3/4}\,d\zeta}, \tag{8.8–14}$$

which also must be evaluated numerically. Substituting this result in Eq. (8.8–11), we see that $B = 0$. With Eq. (8.8–14) used in Eq. (8.8–10) we have

$$z = \frac{\displaystyle\int_0^w(\zeta^2 - a^2)^{-3/4}\,d\zeta}{\displaystyle\int_0^a(\zeta^2 - a^2)^{-3/4}\,d\zeta}, \tag{8.8–15}$$

which is the required transformation. ◀

EXAMPLE 2

Find the Schwarz–Christoffel transformation that will map the half plane Im $w > 0$ onto the semiinfinite strip Im $z > 0$, $-1 <$ Re $z < 1$. The u-axis is to be mapped as indicated.

Solution

From Fig. 8.8–4(a) we see that the strip (a degenerate polygon) can be regarded as the limiting case of the triangle whose top vertex is moved to infinity. In this limit $\alpha_1 = \pi/2$, $\alpha_2 = \pi/2$, and $\alpha_3 = 0$. From Fig. 8.8–4(b) we have $u_1 = -1$, $u_2 = 1$, $u_3 = \infty$.

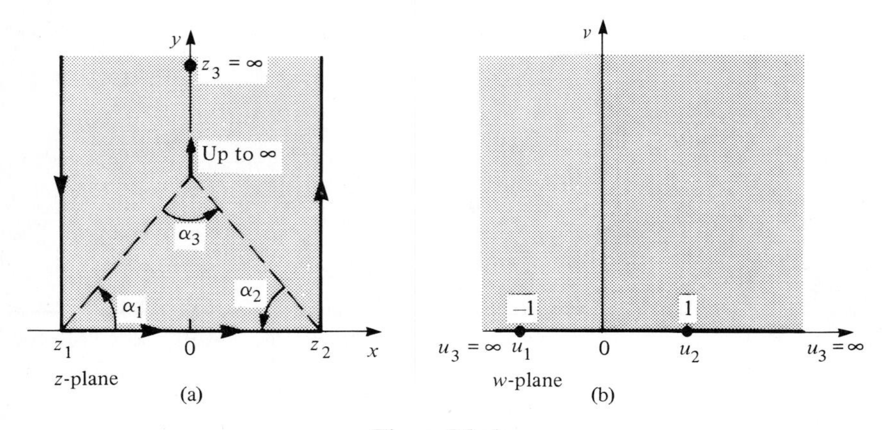

Figure 8.8–4

Substituting the preceding values for α and u in Eq. (8.8–7) (without any term involving u_3), we have

$$z = A\int^{w}(\zeta+1)^{-1/2}(\zeta-1)^{-1/2}\,d\zeta + B = A\int^{w}\frac{d\zeta}{(\zeta^2-1)^{1/2}} + B. \qquad (8.8\text{–}16)$$

To simplify the final answer we shall put $A/i = A_1$ so that

$$\frac{A}{(\zeta^2-1)^{1/2}} = \frac{A_1}{(1-\zeta^2)^{1/2}},$$

and Eq. (8.8–16) becomes

$$z = A_1\int^{w}\frac{d\zeta}{(1-\zeta^2)^{1/2}} + B.$$

The indefinite integration is readily performed (see Eq. (3.7–8)) and the constant absorbed into B. Thus

$$z = A_1 \sin^{-1} w + B. \qquad (8.8\text{–}17)$$

Since $w = 1$ is mapped into $z = 1$, the preceding implies

$$1 = A_1 \sin^{-1} 1 + B = A_1\frac{\pi}{2} + B, \qquad (8.8\text{–}18)$$

Similarly, because $w = -1$ is mapped into $z = -1$, we have from Eq. (8.8–17)

$$-1 = A_1 \sin^{-1}(-1) + B = -A_1\frac{\pi}{2} + B. \qquad (8.8\text{–}19)$$

Solving these last two equations simultaneously, we find that $B = 0$ and $A_1 = 2/\pi$. Thus Eq. (8.8–17) becomes

$$z = \frac{2}{\pi}\sin^{-1}(w). \qquad (8.8\text{–}20)$$

This same transformation (except for a change in scale) has already been studied in Example 3, Section 8.3. ◀

EXAMPLE 3

Find a transformation that maps the domain Im $w > 0$ onto the domain *outside* the semiinfinite strip shown in Fig. 8.8–5(a). Boundary points u_1 and u_2 should be mapped into z_1 and z_2 as indicated.

Solution

As we proceed from left to right along the u-axis in Fig. 8.8–5(b), the corresponding image point advances in the direction indicated by the arrow in Fig. 8.8–5(a). The "interior" of the polygon, which we regard as the domain outside the strip in Fig. 8.8–5(a), should be on our left as we negotiate the path a, b, c, d. In Eq. (8.8–7) we must include terms corresponding to $z_1 = i$ and $z_2 = 0$. The vertex of the polygon

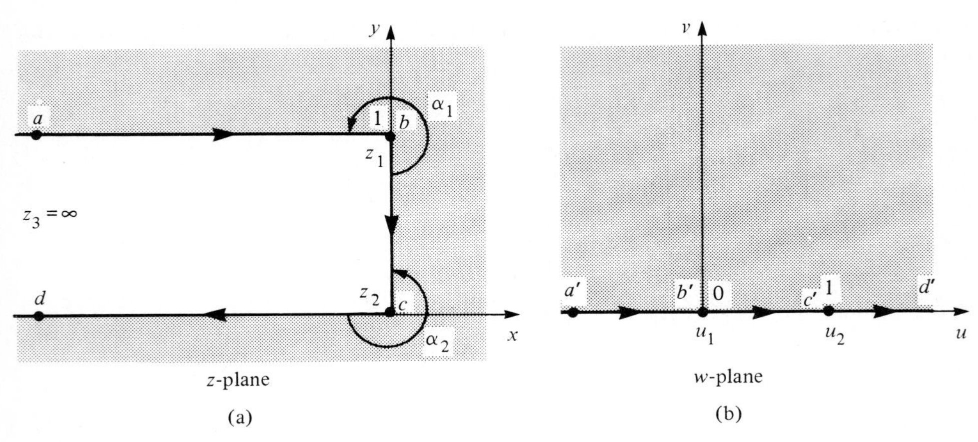

Figure 8.8–5

at $z_3 = \infty$ is mapped to $w = \infty$ and does not appear in Eq. (8.8–7). Notice that $\alpha_1 = 3\pi/2$ and $\alpha_2 = 3\pi/2$. The angles are measured along arcs passing through the *interior* of the polygon. Thus from Eq. (8.8–7) we have

$$z = A \int^{w} \zeta^{1/2}(\zeta - 1)^{1/2}\, d\zeta + B.$$

By replacing $(\zeta - 1)^{1/2}$ with $i(1 - \zeta)^{1/2}$ and absorbing i into A, we obtain

$$z = A \int^{w} \zeta^{1/2}(1 - \zeta)^{1/2}\, d\zeta + B.$$

The integral can be evaluated from tables. Thus

$$z = \frac{A}{4}\left[(2w - 1)(w(w - 1))^{1/2} - \frac{1}{2}\log\left((w(w - 1))^{1/2} + w - \frac{1}{2}\right)\right] + B. \quad (8.8\text{–}21)$$

When $w = 0$, we require $z = i$. Thus Eq. (8.8–21) yields

$$i = \frac{A}{4}\left(-\frac{1}{2}\log\left(-\frac{1}{2}\right)\right) + B.$$

Arbitrarily choosing the principal value of the logarithm we get

$$i = \frac{A}{8}\left[\text{Log } 2 - i\pi\right] + B. \quad (8.8\text{–}22)$$

When $w = 1$, we require $z = 0$, which from Eq. (8.8–21) means

$$0 = \frac{A}{4}\left[-\frac{1}{2}\log\left(\frac{1}{2}\right)\right] + B,$$

or, with the principal value

$$0 = \frac{A}{8}\text{Log } 2 + B. \quad (8.8\text{–}23)$$

Solving Eqs. (8.8–22) and (8.8–23) simultaneously, we obtain

$$A = -\frac{8}{\pi}, \qquad B = \frac{\text{Log } 2}{\pi}.$$

With these values in Eq. (8.8–21) we have

$$z = -\frac{2}{\pi}\left[(2w-1)(w(w-1))^{1/2} - \frac{1}{2}\log\left((w(w-1))^{1/2} + w - \frac{1}{2}\right)\right] + \frac{\text{Log } 2}{\pi}. \tag{8.8–24}$$

Let us verify, by appropriate choices of branches, that the point $w = 1/2$ is mapped into a point on the imaginary z-axis between 0 and 1. With $w = 1/2$ in Eq. (8.8–24) we have

$$z = -\frac{2}{\pi}\left[-\frac{1}{2}\log\left(-\frac{1}{4}\right)^{1/2}\right] + \frac{\text{Log } 2}{\pi}$$
$$= \frac{1}{\pi}\log\left(\pm\frac{i}{2}\right) + \frac{\text{Log } 2}{\pi}.$$

Using $+i$ in this expression and the principal value of the logarithm, we obtain $z = i/2$.

The reader should map several points from the u-axis into the z-plane and become convinced that the desired mapping of points in Fig. 8.8–5 can be achieved if we

a) use the principal branch of the log in Eq. (8.8–24) and

b) define $f(w) = (w(w-1))^{1/2}$ by means of branch cuts extending into the lower half of the w-plane from $w = 0$ and $w = 1$ and take $f(w) > 0$ when $w > 1$. ◀

EXERCISES

1. Is the mapping defined in Eqs. (8.8–6) and (8.8–7) conformal for $w = u_1$, $w = u_2$, etc? Explain.

2. Use the Schwarz–Christoffel formula to find the transformation that will map the sector shown in Fig. 8.8–6 onto the upper half of the w-plane. Map A into $(-1, 0)$, B into $(0, 0)$ and C to ∞.

Answer: $w = -iz^2/2$

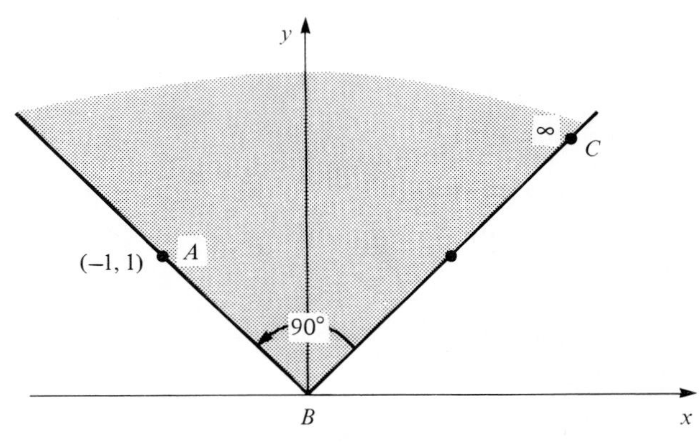

Figure 8.8–6

3. a) Use the Schwarz–Christoffel formula to find a transformation that will map the region in Fig. 8.8–7 onto the upper half of the w-plane. Map $z = 0-$ to $w = -1$, $z = i$ to $w = 0$, and $z = 0+$ to $w = 1$.

Hint: Consider the boundary in Fig. 8.8–7 as the limit of the boundary depicted in Fig. 8.8–8 as α_1, α_2, and α_3 achieve appropriate values.

Answer: $w = (z^2 + 1)^{1/2}$

b) A line charge of strength ρ passes through the point $(0, H)$, in the configuration of Fig. 8.8–7. Assume $H > 1$. The line charge is perpendicular to the plane of the paper. The boundaries shown in this figure are electrical conductors set at ground (zero) electrostatic potential. Using the result of part (a), show that the complex potential in the shaded region of the figure, which has electrical permittivity ε, is given by

$$\Phi(z) = \frac{\rho}{2\pi\varepsilon} \operatorname{Log} \frac{(z^2 + 1)^{1/2} + i\sqrt{H^2 - 1}}{(z^2 + 1)^{1/2} - i\sqrt{H^2 - 1}}.$$

c) Let $H = \sqrt{2}$ and $\rho/2\pi\varepsilon = 1$. Use the preceding result to plot the actual potential $\phi(0, y) =$ from $y = 1$ to H.

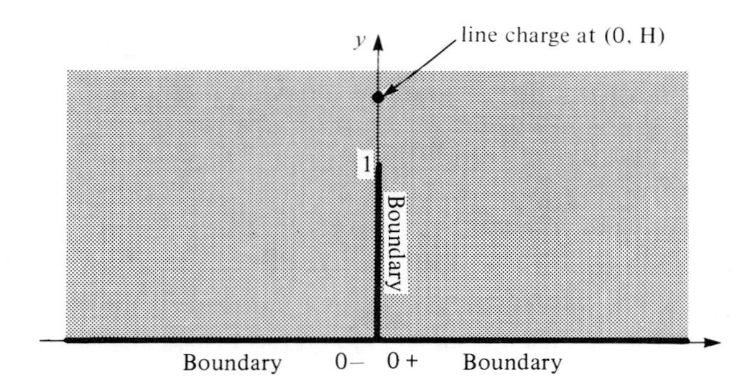

Figure 8.8–7

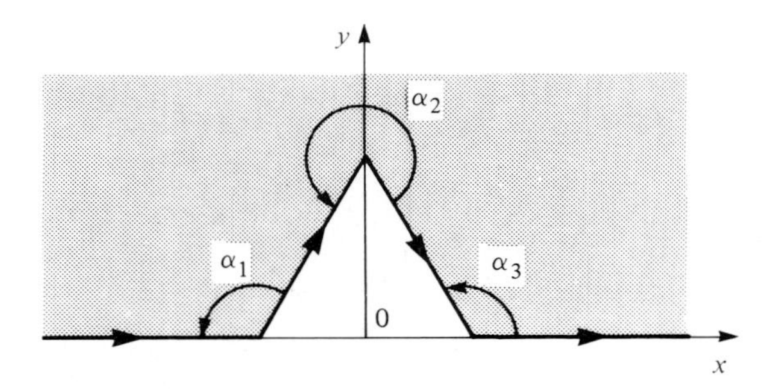

Figure 8.8–8

4. Find a transformation that maps the line $v = 0$ from the w-plane onto the open polygon in the z-plane shown in Fig. 8.8–9(a). The mapping of the boundary should be as shown in the figure.

Answer:

$$z = \frac{2}{\pi}\,[(w-1)^{1/2}w^{1/2} + \text{Log}(w^{1/2} - (w-1)^{1/2})] + i.$$

Branch points go from $w = 0$ and $w = 1$ through lower half of the w-plane to $w = \infty$; $w^{1/2} > 0$ if $w > 0$ and $(w-1)^{1/2} > 0$ if $w > 1$.

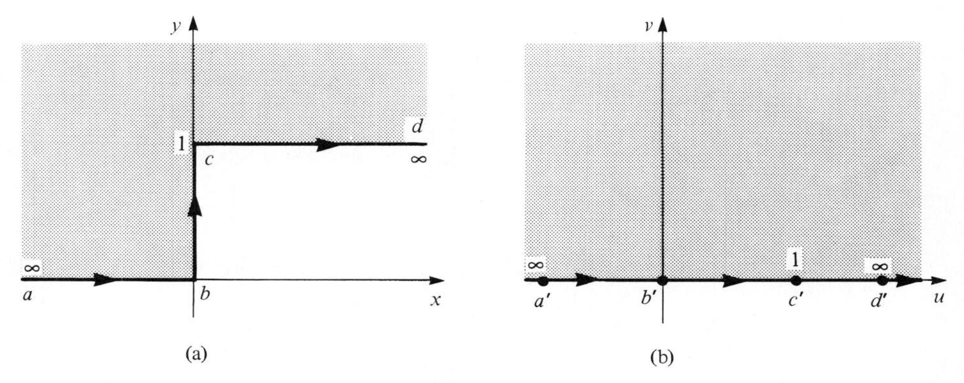

Figure 8.8–9

5. a) Find the transformation that maps the upper half of the w-plane onto the shaded region of the z-plane in Fig. 8.8–10(a). The boundary is mapped as shown in Fig. 8.8–10(a) and (b).

Hint: Consider the region in Fig. 8.8–10(a) as the limit of the region in Fig. 8.8–10(c) as $\alpha_1 \to 3\pi/2$ and $\alpha_2 \to 0$.

Answer:

$$z = \frac{2}{\pi}(1+w)^{1/2} + \frac{1}{\pi}\,\text{Log}\left(\frac{(1+w)^{1/2} - 1}{(1+w)^{1/2} + 1}\right). \tag{8.8–25}$$

b) Let the bent line a, b, c in Fig. 8.8–10(a) be the cross section of a conductor maintained at voltage V_0, and let the straight line c, d be the cross section of a conductor maintained at zero potential. Use the transformation derived in part (a) to show that if $|z| \gg 1$ and $\text{Re}\, z \geq 0$ the electrostatic potential is $\phi(x, y) \approx (2V_0/\pi) \arg z$ (principal value), and the stream function is $\psi(x, y) \approx -(2V_0/\pi)\,\text{Log}|\pi z/2|$. Sketch equipotentials and streamlines where these approximate expressions apply in the z-plane.

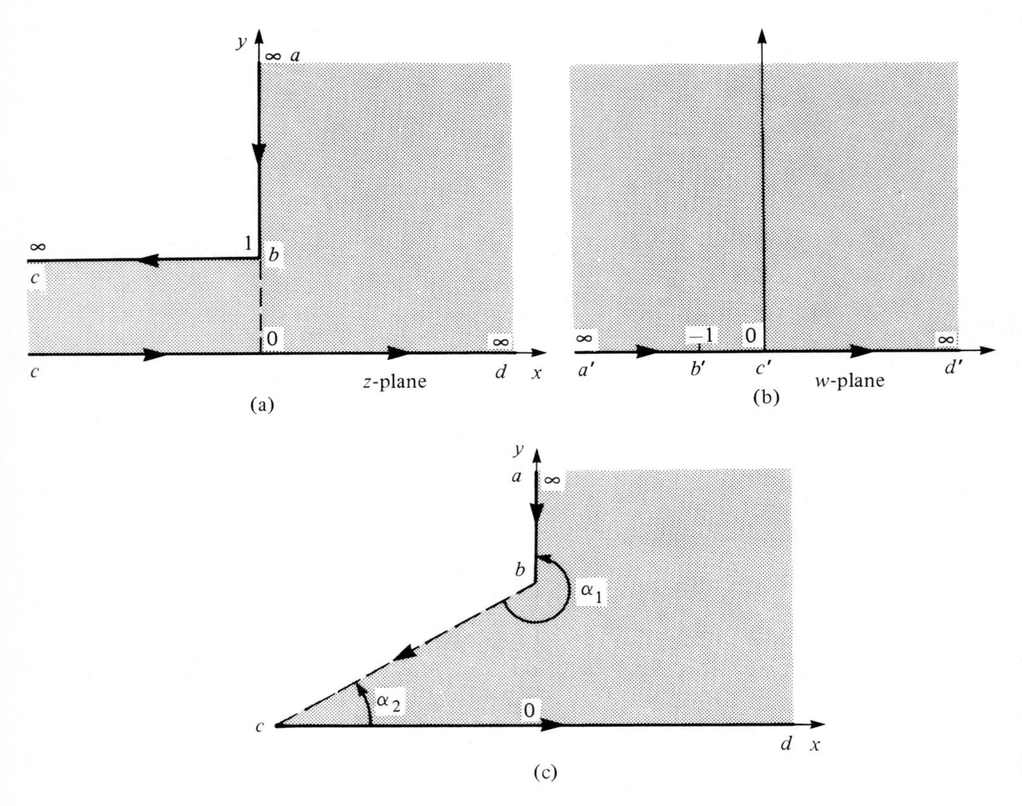

Figure 8.8–10

c) Use Eq. (8.8–25) to show that if Re $z \ll -1$ and $0 \leq \text{Im } z \leq 1$ the electrostatic potential is $\phi(x, y) \approx V_0 y$, and the stream function is $\psi \approx -V_0 x$.

Hint: For $|w| \ll 1$,

$$\frac{(1+w)^{1/2}-1}{(1+w)^{1/2}+1} \approx \frac{w}{4}$$

from a Maclaurin expansion. Sketch the streamlines and equipotentials in the z-plane, where these approximate expressions for $\phi(x, y)$ and $\psi(x, y)$ apply.

Using this sketch and the one found in part (b), guess the shape of the streamlines and equipotentials near the bend in the upper conductor.

6. a) Find the mapping that will transform the upper half of the w-plane onto the strip $0 < \text{Im } z < 1$. The boundary is to be mapped as shown in Fig. 8.8–11.

Hint: Consider an appropriate limit of the triangle formed by the broken lines in Fig. 8.8–11(a).

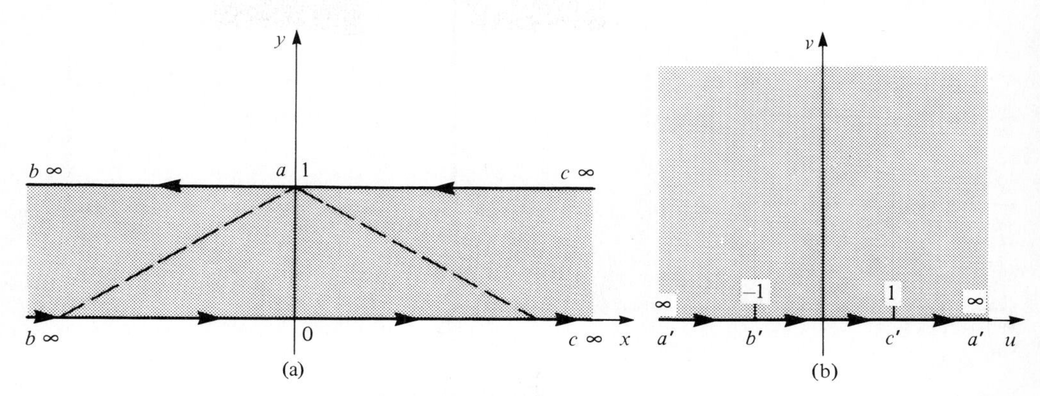

Figure 8.8–11

Answer:

$$z = \frac{-1}{\pi}\left[\mathrm{Log}(w-1) - \mathrm{Log}(w+1)\right] + i.$$

b) Assume that in the z-plane the line $y = 1$ is the cross section of a conductor maintained at an electrostatic potential of 1 volt, whereas $y = 0$ is a conductor maintained at 0 volts. One easily finds the complex potential $\Phi(z)$ in the strip $0 \le y \le 1$. Transform this result into the upper half of the w-plane using the transformation found in part (a), to find $\phi(u, v) = \mathrm{Re}\,\Phi(w)$.

Answer:

$$\Phi(w) = 1 + \frac{i}{\pi}\,\mathrm{Log}\,\frac{w-1}{w+1},$$

$$\phi(u, v) = 1 - \frac{1}{\pi}\arctan\frac{2v}{u^2+v^2-1}, \qquad 0 \le \tan^{-1}(\cdots) \le \pi.$$

c) What boundary condition does $\phi(u, v)$ satisfy on the line $v = 0$?

7. a) Find the transformation that will map the upper half of the w-plane onto the domain indicated in the z-plane in Fig. 8.8–12. The mapping of the boundaries are as shown. Take the inverse of this transformation so as to obtain w as a function of z, and describe the appropriate branch of the function obtained.

Hint: Consider Fig. 8.8–13. Let the angles shown pass to appropriate limits.

Answer: $w = k[1 - e^{\pi z}]^{1/2}$.

b) Let $k = 1$ in Fig. 8.8–12(b). Suppose the domain shown shaded in Fig. 8.8–12(a) is the cross section of a heat-conducting material. The line $y = 0$, $-\infty < x < 0$ represents a boundary maintained at 0° temperature, whereas the lines $y = \pm 1$, $-\infty < x < \infty$ are both maintained at 1°. Use the transformation found in part (a) as well as the result of Exercise 6(b) in this section or the result of Exercise 9(c) in Section 8.5 to find the complex temperature in the heated material.

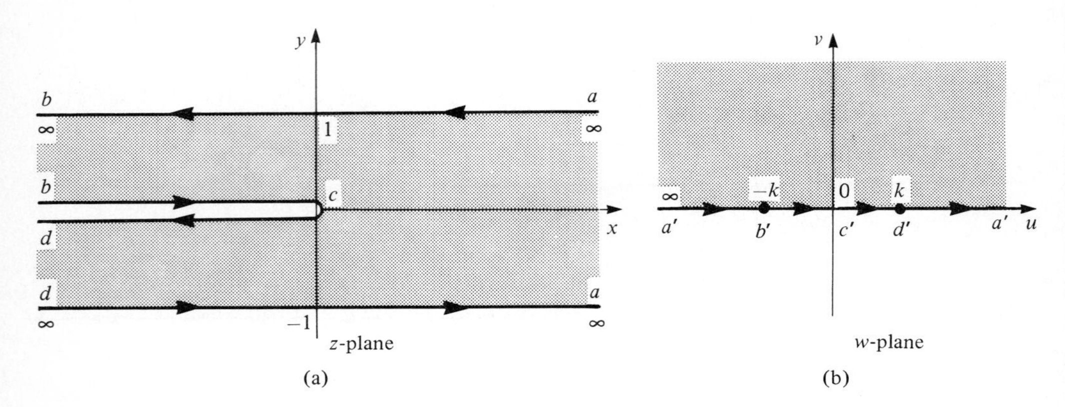

Figure 8.8–12

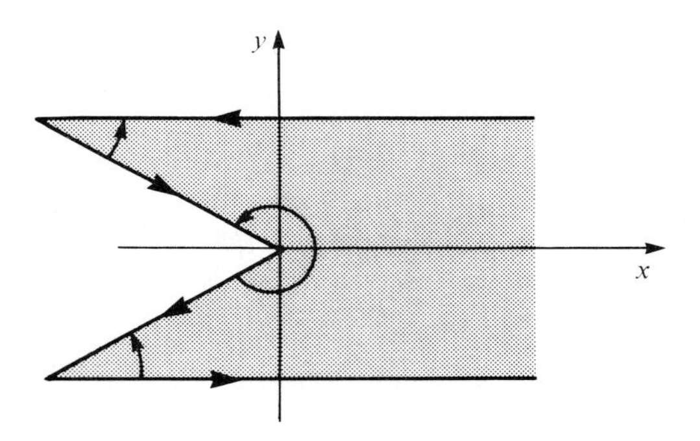

Figure 8.8–13

c) Plot the actual temperature $\phi(x, y)$ with distance along the line $x = 0$, $0 \leq y \leq 1$ and along the line $y = 0$, $0 \leq x \leq 2$.

8. a) Find the mapping that will transform the upper half of the w-plane onto the domain shown shaded in Fig. 8.8–14(a). Corresponding boundary points are shown in Fig. 8.8–10(b). Note that there is a cut in the z-plane along $y = 1$, $x < 0$ and that points a and c lie on opposite sides of this cut.

Hint: Refer to Fig. 8.8–10(c) but take the limit $\alpha_1 = 2\pi$.

Answer: $z = (1/\pi)[w + \text{Log } w + 1]$.

b) An electrical capacitor, shown in Fig. 8.8–14(b) consists of a semiinfinite plate maintained at an electrostatic potential of 1 volt and an infinite plate maintained at zero potential. Assign these same potentials to the image boundaries (in the w-plane) that arise from the transformation found in part (a). Show that the complex potential in the domain $\text{Im } w \geq 0$ is $\Phi(w) = (-i/\pi) \text{ Log } w = \phi(u, v) + i\psi(u, v)$.

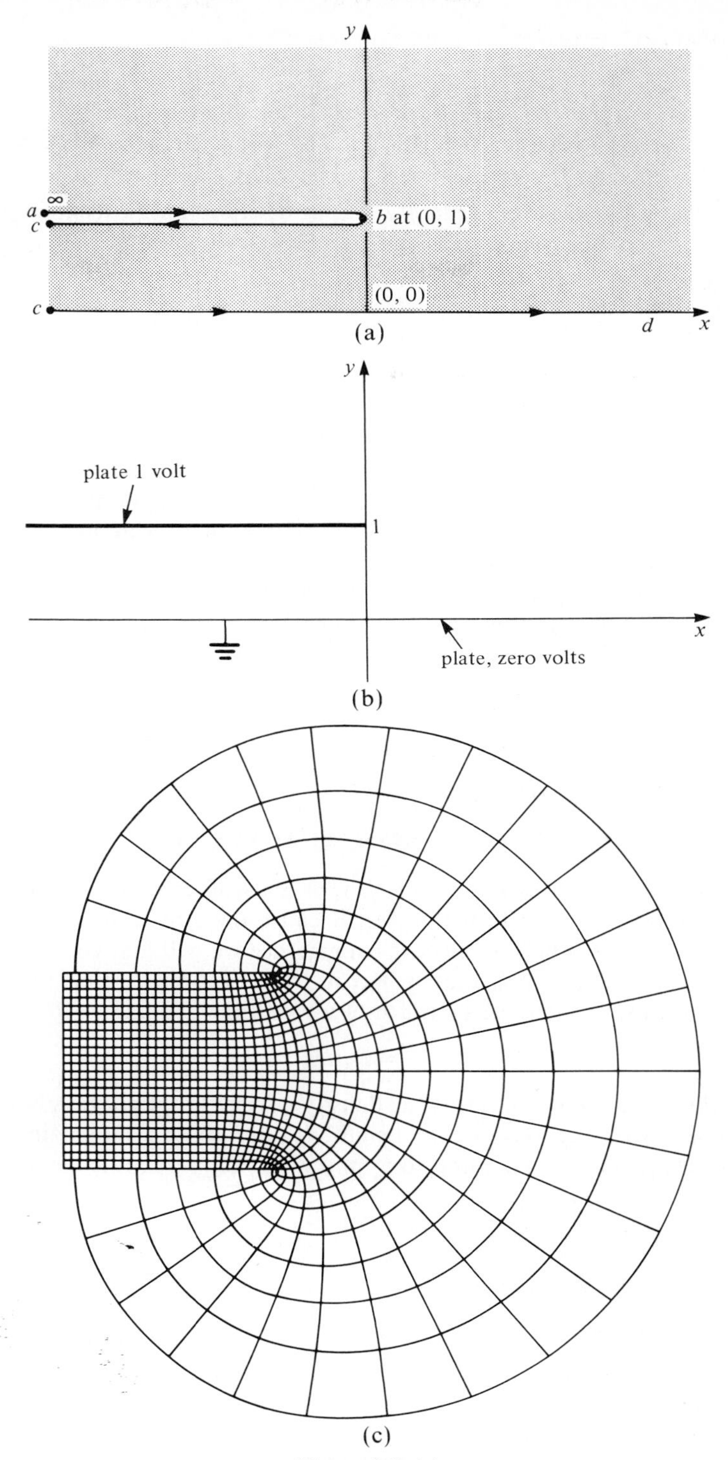

Figure 8.8–14

Show also that the value of w corresponding to given values of the electrostatic potential ϕ and stream function ψ is given by $w = e^{-\pi\psi}\,\text{cis}(\pi\phi)$. Explain why in this formula we require $0 \le \phi \le 1$ while ψ can have any real value.

c) Show that if we assign values to ϕ and ψ as described above, then the point in the capacitor where the complex potential has this value is given by

$$z = \frac{1}{\pi}\left[e^{-\pi\psi}\,\text{cis}(\pi\phi) + 1\right] - \psi + i\phi.$$

Now suppose the potential has the value $\phi = 1/3$, while the parameter ψ ranges from $-\infty$ to ∞. Plot values assumed by z on top of Fig. 8.8–14(b). This is the equipotential along which the potential is 1/3 volt. Repeat this procedure for $\phi = 2/3$.

d) Explain how the technique of this problem can be used to determine the equipotentials and streamlines of a capacitor consisting of two semiinfinite plates to which the potentials 1 and -1 are assigned. Such a capacitor, with its streamlines and equipotentials, is shown in Fig. 8.8–14(c). This plot is Plate XII taken from James Clerk Maxwell's famous book *A Treatise on Electricity and Magnetism*, 3rd edition, published in 1891 and currently available from Dover Books, New York.

APPENDIX TO CHAPTER 8

THE STREAM FUNCTION AND CAPACITANCE

We show here how the stream function $\psi(x, y)$ is useful in computations involving electrostatics, fluid flow, and heat transfer. Because the stream function has special value in the important subject of electric capacitance we will first concentrate on the applicability of $\psi(x, y)$ to electrostatics. Recall from Section 2.6 that the amount of electric flux crossing a differential surface of area dS is

$$df = D_n\,dS, \tag{A8–1}$$

where D_n is the component of electric flux density normal to the surface. Now consider surfaces whose cross sections are of length dy and dx, as shown in Fig. A8–1. Each surface is assumed to be 1 unit long in a direction perpendicular to the paper.

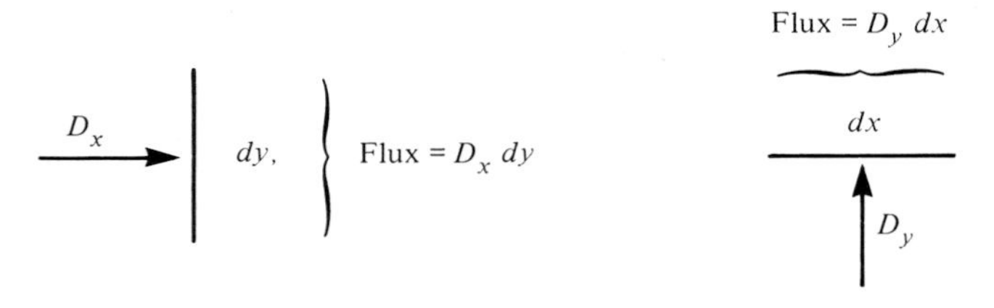

Figure A8–1

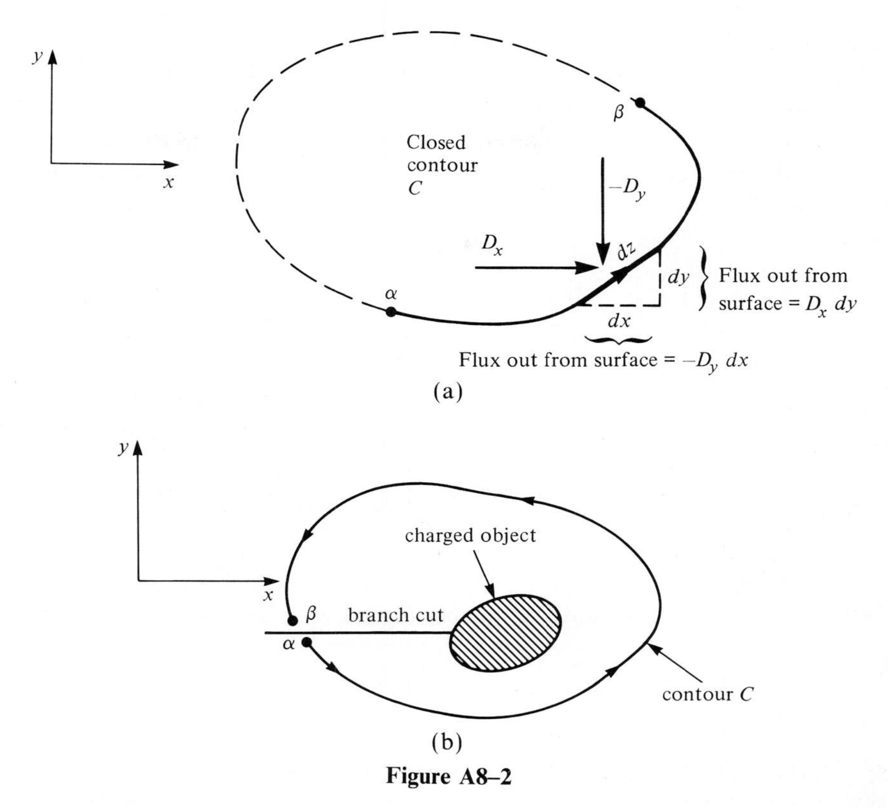

Figure A8–2

The flux crossing the vertical surface is $D_x\,dy$ while that crossing the horizontal surface is $D_y\,dx$.

Consider now the simple closed contour C shown in Fig. A8–2(a). C is the cross section of a cylinder of unit length perpendicular to the page. We seek an expression for the total electric flux emanating outward from the interior of the cylinder. Let $dz = dx + i\,dy$ be a differential distance along C, as shown. The flux df across the surface whose cross section is dz is the sum of the fluxes crossing the projections of this surface on the x- and y-axes. These projections have areas dx and dy. Thus

$$df = D_x\,dy - D_y\,dx.$$

The minus sign occurs in front of D_y because we are computing the outward flux, that is, flux passing from the inside to the outside of the cylinder. The flux passing outward along the surface whose cross section is the solid line connecting α with β in Fig. A8–2(a) is

$$f_{\alpha\beta} = \int_{\alpha}^{\beta} D_x\,dy - D_y\,dx, \tag{A8–2}$$

where we integrate in the positive sense along the solid line, that is, we keep the interior of the cylinder on our left. With the aid of Eq. (2.6–22) we rewrite Eq. (A8–2) in terms of the electrostatic potential $\phi(x, y)$. Thus

$$f_{\alpha\beta} = \int_{\alpha}^{\beta} -\varepsilon \frac{\partial\phi}{\partial x}\,dy + \varepsilon \frac{\partial\phi}{\partial y}\,dx, \tag{A8–3}$$

where ε is the permittivity of the surrounding material. Since the stream function $\psi(x, y)$ is the harmonic conjugate of $\phi(x, y)$, we have from the Cauchy–Riemann equations

$$\frac{\partial\phi}{\partial x} = \frac{\partial\psi}{\partial y},$$

$$\frac{\partial\phi}{\partial y} = -\frac{\partial\psi}{\partial x}.$$

This enables us to rewrite $f_{\alpha\beta}$ in Eq. (A8–3) in terms of ψ as follows:

$$f_{\alpha\beta} = -\varepsilon \int_{\alpha}^{\beta} \frac{\partial\psi}{\partial y}\,dy + \frac{\partial\psi}{\partial x}\,dx.$$

The integrand here is the exact differential $d\psi$, and, provided ψ is continuous along the path, the integration is immediately performed with the result that

$$f_{\alpha\beta} = -\varepsilon \int_{\alpha}^{\beta} d\psi = \varepsilon[\psi(\alpha) - \psi(\beta)]. \tag{A8–4}$$

Thus the product of ε and the decrease in ψ encountered as we move along the contour connecting α to β is the electric flux crossing the surface whose cross section is this contour (see the solid line representation in Fig. A8–2a). Let us now move completely around the closed contour C in Fig. A8–2(a) and measure the net decrease in ψ (initial value minus final value). We call this result $\Delta\psi$.

Now ψ is typically the imaginary part of a branch of a multivalued function and is defined by means of a branch cut. To use Eq. (A8–4) we require that ψ be continuous on C. Thus, as depicted in Fig. A8–2(b), we choose α to lie on one side of the cut while β lies opposite α on the other side. We integrate from α to β along C without crossing the cut. The result, $\varepsilon[\psi(\alpha) - \psi(\beta)]$, or $\varepsilon\Delta\psi$, can be nonzero.

The total electric flux leaving the closed surface whose cross section is C is $\varepsilon\Delta\psi$, which, according to Gauss's law,[†] is exactly equal to the charge on or enclosed by the surface. Thus referring to Fig. A8–2(b) we have that $\varepsilon\Delta\psi$ is exactly the charge ρ on the object lying inside C. (Because this is a two-dimensional configuration, ρ is actually the charge along a unit length of the object in a direction perpendicular to the page.) Hence

$$\rho = \varepsilon\Delta\psi. \tag{A8–5}$$

[†] See D. K. Cheng, *Fundamentals of Engineering Electromagnetics* (Reading, MA: Addison-Wesley, 1993), Section 3.4.

When two conductors are maintained at different electrical potentials, a knowledge of the stream function $\psi(x, y)$ will establish the charge on either conductor and, with the use of Eq. (8.5–23), the capacitance of the system.

Suppose we have a two-dimensional system of conductors whose capacitance per unit length we wish to determine. Let these conductors (actually their cross sections) be mapped from the xy-plane into the uv-plane by means of a conformal transformation. Then we can prove the following:

> The capacitance that now exists between the two image conductors in the uv-plane is precisely the same as existed between the two original conductors in the xy-plane.

To establish this equality, consider the pair of conductors A and B shown in the xy-plane in Fig. A8–3. They are at electrostatic potentials V_0 and zero, respectively. Under a conformal mapping these conductors become the conductors A' and B' in the uv-plane. The new conductors are assigned the potentials V_0 and zero, respectively. The complex potentials in the two planes are $\Phi(x, y) = \phi(x, y) + i\psi(x, y)$ and $\Phi_1(u, v) = \phi_1(u, v) + i\psi_1(u, v)$.

Refer now to Fig. A8–3(a), and consider the curve Γ (shown by the solid line) that goes from (x_1, y_1) to (x_2, y_2) along conductor A. The image of Γ in the w-plane is the curve Γ' that goes from (u_1, v_1) to (u_2, v_2) along the conductor A' (see Fig. A8–3b). The amount by which ψ decreases as we move along Γ is $\psi(x_1, y_1) - \psi(x_2, y_2)$, whereas the corresponding change along the image Γ' is $\psi_1(u_1, v_1) - \psi_1(u_2, v_2)$. Since the stream function $\psi(x, y)$ and its transformed version $\psi_1(u, v)$ assume identical values at corresponding image points in the two planes, the expressions for the change in ψ and ψ_1 are equal. Because (x_1, y_1) and (x_2, y_2) were chosen arbitrarily, it follows that the change in ψ occurring if we go completely around the boundary of A must equal the change in ψ_1 if we go completely around the boundary of A'. Thus the amount of charge on A and A' must be identical. The potential difference V_0 between conductors A and B is identical to the potential

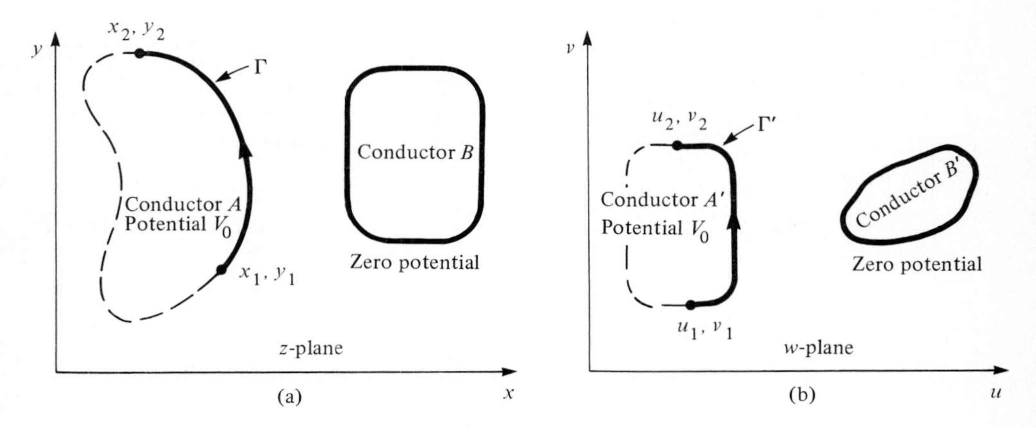

Figure A8–3

difference between conductors A' and B'. Since capacitance is the ratio of charge to potential difference, the capacitance between conductors A and B is the same as appears between conductors A' and B'.

The stream function is also of use in physical problems involving heat conduction. Reviewing Section 2.6 and especially Table 1 of that section, we see that the heat analog of Eq. (A8–4) is

$$f_{\alpha\beta} = k[\psi(\alpha) - \psi(\beta)].$$

Here ψ is the stream function associated with a complex temperature, while k is the thermal conductivity of the medium in use. Now $f_{\alpha\beta}$ is the flux of heat across a contour like that in Fig. A8–2(a) or Fig. A8–2(b). Choosing α and β as in Fig. A8–2(b), we obtain the total flux of heat across C as being $k\Delta\psi$, where $\Delta\psi$ is defined as before. If the electrically charged object in Fig. A8–2(b) is replaced by one that generates heat, we have by the law of conservation of energy that the flux of heat across C must equal the rate at which heat is being generated by the source (per unit length). Calling this rate h we have

$$h = k\Delta\psi. \tag{A8–6}$$

Finally we consider the stream function of fluid flow. Studying Table 1 in Section 2.6, we see that the Cartesian components of flux density are the real and imaginary parts of the complex velocity vector, and that the complex flux density vector is the conjugate of the derivative of the complex velocity potential, i.e., $\mathbf{v} = \overline{(d\Phi/dz)}$. Note, from the table, that the corresponding expressions for heat and electric flux, in terms of their complex potentials, contain a minus sign as well as constants k or ε. Because of this difference in sign, the equation for the flux of fluid, in terms of the fluid stream function ψ, across a contour C like that in Fig. A8–2(b) is opposite in sign from the expressions obtained in the thermal and electrostatic cases. Thus if ψ is the stream function for a complex velocity potential and α and β are chosen as in Fig. A8–2(b), we have that the total flux of fluid across C is $f_{\alpha\beta} = -\Delta\psi$, where $\Delta\psi = \psi(\alpha) - \psi(\beta)$. Again we have obtained the decrease in ψ as we proceed in the positive sense around C.

Now suppose the electrically charged object in Fig. A8–2(b) is replaced by one that radiates fluid in a direction parallel to the plane of the paper. From the law of conservation of matter, we require that the flux of fluid across C in this figure must be equal to the rate at which this source radiates fluid (per unit length). If this rate is G, we have

$$G = -\Delta\psi. \tag{A8–7}$$

Selected Answers

Chapter 1

Section 1.1

1. $x = -1/2$, rational; **3.** $x = 0$, $x = -2$, integers; **5.** $x = \pm\sqrt{2}$, reals; **7.** $x = \pm 1/2$, rational; **9.** 301/99; **11.** $a = 4$, $b = 6$; **13.** $15 + 20i$; **15.** 625; **21.** true; **23.** true; **25.** not true in general; **27.** $-i$, $2^{1537} \times i$; **29.** $x = 0$, $y = 0$ or $y = -1$; **31.** $x = 1/3$, $y = 0$.

Section 1.2

1. $\frac{1}{2} + \frac{i}{2}$; **3.** $1 + i$; **5.** $-i$; **7.** $2 + i\frac{3}{2}$; **15.** not true; **17.** not true; **19.** $\operatorname{Im} z_1 = 0$ or $\operatorname{Im} z_2 = 0$; **21.** For the identity to hold we require that $\operatorname{Re} z_2 \neq 0$ and in addition, either $\operatorname{Im} z_2 = 0$ or $\operatorname{Im} z_1 \operatorname{Re} z_2 = \operatorname{Re} z_1 \operatorname{Im} z_2$; **23.** $e = \dfrac{ac + bd}{a^2 + b^2}$, $f = \dfrac{ad - bc}{a^2 + b^2}$.

Section 1.3

1. 5; **3.** $\sqrt{2}$; **5.** 1; **7.** $-4 + i2$; **9.** $z = \sqrt{5} + 2\sqrt{5}i$; **11.** $z = \sqrt{\dfrac{-1}{2} + \dfrac{\sqrt{101}}{2}} + i\left[\dfrac{-1}{2} + \dfrac{\sqrt{101}}{2}\right]$; **13.** -3.133185 rads; **15.** $0.938 \times \pi$; **17.** zero; **19.** $-1 - i(\sqrt{3})$; **21.** $r = 2$, $\theta = -5\pi/6 + 2k\pi$; princ. value $k = 0$; **23.** $r = 32\sqrt{2}$, $\theta = -\pi/12 + 2k\pi$; princ. value $k = 0$, **25.** $r = 1/\sqrt{2}$, $\theta = -0.6435$, princ. value; **27.** $r = 1$, $\theta = -2/3\pi$, princ. value; **29.** (b) $\arg z_1 = \arg z_2 + \pi + 2k\pi$, k is an integer.

Section 1.4

1. $r = 2^9$, $\theta = \dfrac{-\pi}{2}$, $a + ib = -2^9 i$; **3.** $r\underline{/\theta} = \left(\dfrac{5}{\sqrt{2}}\right)^{12}\underline{/1.7027}$, $-5.019 \times 10^5 + i3.781 \times 10^6$; **5.** $16\underline{\Big/\dfrac{-7\pi}{15}}$, $1.6724 - i15.91$; **9.** $\dfrac{1}{\sqrt{2}} + \dfrac{i}{\sqrt{2}}$ and $\dfrac{-1}{\sqrt{2}} + \dfrac{-i}{\sqrt{2}}$; **11.** $0.86 + i0.23$, $-0.23 - i0.86$, $-0.63 + i0.63$; **13.** 1 or -1; **17.** $\pm[0.455 + i1.098]$; **19.** $w = 0.0493 - i0.2275$, $w = -1.0493 + i0.2275$; **21.** $w = 1.08 + i0.2905$, $w = -0.793 + i0.793$, $w = -0.29 - i1.084$;

$w = 1.08 - i0.2905$, $w = -0.7937 - i0.7937$, $w = -0.2905 + i1.084$; **23.** $z = 0.309 + i0.95$, $z = -0.809 + i0.587$, $z = -0.809 - i0.587$, $z = 0.309 - i0.95$; **25.** (a) not necessarily one, (b) yes; **27.** $\frac{1}{2} + \frac{i\sqrt{3}}{2}, \frac{1}{2} - \frac{i\sqrt{3}}{2}, -1$; **29.** $1.09 + i0.629$, $-i1.2599$, $-1.09 + i0.629$; **31.** 1, $\frac{-1}{2} - \frac{i\sqrt{3}}{2}, \frac{-1}{2} + \frac{i\sqrt{3}}{2}$; **33.** $2\cos\frac{\pi}{4m}$.

Section 1.5

1. $x = -2$; **3.** $y \geq x$; **5.** set of all points exterior to the circle of radius 5, center $x = -3$, $y = 4$, the circle itself is not included in the set; **7.** $z = -3 + 4i$; **9.** points on and outside circumference of circle, radius one, center $z = 1 + i0$; **11.** inequality satisfied on and inside circle $|z| = 0.567$; **13.** nearest dist. $= \sqrt{5} - 1$, furthest dist. $= \sqrt{5} + 1$; **15.** $2 < |z - 3 - i| \leq 4$; **17.** $0 < |z - 0.99i| < 10^{-2}$, deleted nehbd.; **19.** domain; **21.** the resulting set is open but not connected, not a domain; **23.** Points on $|z + 1 + i| = 3$ and points on $|z + 1 + i| = 4$ are boundary points, but do not belong to set; **25.** The point $z = 1$ is a boundary point of the set, but does not belong to the set; **27.** not closed; **29.** not closed; **31.** closed; **33.** no; **35.** (a) The unit circle $|z| = 1$ gets projected onto the equator of the sphere; (b) The points inside the unit circle get projected onto southern (lower) hemisphere; (c) The points outside the unit circle get projected onto northern hemisphere; **37.** (b) The radius of the circle on the sphere is $\frac{r}{1 + r^2}$.

Chapter 2

Section 2.1

1. $z = 0$; **3.** $z = 0$, $z = i$ and $z = -i$; **5.** $z = 0$ and $z = i$; **7.** $\frac{1}{\sqrt{5}}$; **9.** $-\frac{1}{4} - \frac{3i}{4}$; **11.** $x^2 - (y - 1)^2 + i2x(y - 1)$; **13.** $\frac{x}{x^2 + y^2} + i\left[1 - \frac{y}{x^2 + y^2}\right]$; **15.** $x^3 - 3xy^2 + i[3x^2y - y^3]$; **17.** $\frac{i}{2}[z^2 - \bar{z}^2] + iz\bar{z}$; **19.** $z\bar{z}$; **21.** $A' = i$, $B' = -1 + i$, $C' = -1$, $D' = -1 - i$, $E' = -i$; **23.** $A' = 0$, $B' = 0.347 + i\frac{\pi}{4}$, $C' = \frac{i\pi}{2}$, $D' = 0.347 + i\frac{3\pi}{4}$, $E' = i\pi$; **25.** 1; **27.** $u = x^4 + y^4 - 6x^2y^2 + 2[x^2 - y^2 + 1]$, $v = 4x^3y - 4xy^3 + 4xy$.

Section 2.2

11. $f(0) = 4$, $f(4) = 8$; **13.** (a) $M = 4$, (b) $M = 9$, (c) $M = \frac{1}{\sqrt{2} - 1}$.

Section 2.3

1. derivative does not exist at $x = \pi$, but $f(x)$ is continuous at $x = \pi$; **3.** derivative exists nowhere; **5.** $z = 0$ is only place; **7.** derivative exists for all $z \neq 0$; **9.** derivative exists for all z; **11.** derivative exists only at $z = 1$; **13.** derivative exists nowhere; **15.** $f(z)$ has a derivative everywhere except on circle $|z| = 3$; **17.** $f'(z)$ exists at $z = 0$ and also in the domain $|z| > 1$.

Section 2.4

1. $f'(z) = 4z$, entire function, $f'(1+i) = 4 + 4i$; **3.** $f'(z) = \dfrac{-4z}{(z^2+1)^3}$, $f(z)$ is analytic except if $z = \pm i$, $f'(1+i) = 0.416 + i0.288$; **5.** function is nowhere analytic; **7.** $f(z)$ is entire, $f'(z) = 3x^2 - 3y^2 + i6xy$, $f'(1+i) = i6$; **9.** entire function, $\dfrac{df}{dz} = \dfrac{2z - z^2}{e^x \cos y + ie^x \sin y}$, $\dfrac{2[\cos 1 - i \sin 1]}{e}$; **11.** analytic for all $z \neq -i$, df/dz at $1 + i = [1.31 + i1.22]10^{-2}$; **13.** $-2/3$; **15.** The function is analytic for all $z \neq 0$.

Section 2.5

1. $y = 2x^2$, these values lie on a parabola—set of points not a domain; **3.** function harmonic for all z; **5.** harmonic for $|z| > 0$; **7.** harmonic for whole complex plane; **9.** $g(x) = \dfrac{1}{2}\cos(2x)$; **13.** $v = 4xy^3 - 4x^3y - y - x + D$; **17.** $v = \dfrac{1}{2}\operatorname{Log}(y^2 + x^2) + D$; **23.** (c) intersection (0.804, 1.107), (d) 0.5, slope of $u = 1$; -2, slope of $v = 2$; **25.** (c) $v = r^2 \sin(2\theta) + \text{constant}$.

Section 2.6

1. (a) $\phi = -30x + 40y$, (b) $\psi = -30y - 40x$; **3.** (a) $E_x + iE_y = -2.39 + i1.30$, (c) $D_x = -21 \times 10^{-12}$, $D_y = 11.5 \times 10^{-12}$, (e) $\psi \approx 1.3$; **5.** (a) $\phi = x \cos \alpha + y \sin \alpha$, (b) $\psi = y \cos \alpha - x \sin \alpha$, (c) $V_x + iV_y = \cos \alpha + i \sin \alpha$, the vector makes an angle α with positive x axis.

Chapter 3

Section 3.1

1. $-13.12 - i15.2$; **3.** $-0.0325 - i(0.037)$; **5.** $1.45 + i0.79$; **7.** $1.144 + i1.279$; **9.** $1.54 + i1.317$, $0.375 - i0.32$; **13.** $z = 0 + ik\pi$, k even; **15.** $u = e^{x^2-y^2}\cos 2xy$, $v = e^{x^2-y^2}\sin 2xy$; **17.** $u = e^{e^x \cos y}\cos[e^x \sin y]$, $v = e^{e^x \cos y}\sin(e^x \sin y)$; **19.** $-11.68 - i1.714$; **21.** $\operatorname{Re} z = 0$; **23.** $\operatorname{Re} z = 0$, $\operatorname{Im} z \neq 0$, or $|z| = 1$; **25.** max $= e^2$ at $2 + i$, min $= e^0 = 1$ at i; **27.** (b) N; **29.** (a) $\phi = e^x \cos y$, $\psi = e^x \sin y$, (c) $V_x = e\cos(\pi/4)$, $V_y = -e\sin(\pi/4)$.

Section 3.2

1. $3.166 - i1.959$; **3.** $-0.243 - i1.166$; **5.** $0.507 - i4.85$; **7.** $\dfrac{1}{\sqrt{2}}$; **9.** -0.9228; **11.** $0.271 + i0.798$; **19.** solution is $z = \left(\dfrac{\pi}{4} + k\pi\right) + i0$, k integer; **21.** $z = -i\left(n\pi + \dfrac{\pi}{2}\right)$, n any integer; **23.** $z = \dfrac{\pi}{6} + k\pi + i0$, k integer.

Section 3.3

7. $-0.248 - i1.16$; **9.** $0.3028i$; **13.** $\dfrac{n\pi}{2}[1 + i]$, $n = 0, \pm 1, \pm 2, \ldots$.

Section 3.4

1. $i2k\pi$, princ. value 0; **3.** $i(\pi/3 + 2k\pi)$, princ. value $i\pi/3$; **5.** $\text{Log } 2 + i\left[\frac{2}{3}\pi + 2k\pi\right]$, princ. value is $0.693 + i\frac{2}{3}\pi$; **7.** $0.227 + i[0.986 + 2k\pi]$, princ. value with $k = 0$; **9.** $-0.841 + i(0.54 + 2k\pi)$, princ. value $-0.841 + i0.54$; **11.** $\frac{1}{\sqrt{2}} + \frac{i}{\sqrt{2}} + 2k\pi i$, $\frac{-1}{\sqrt{2}} - \frac{i}{\sqrt{2}} + i2k\pi$; **13.** $-0.1726 + i\left[\frac{\pi}{2} + 2k\pi\right]$; **15.** $2 + i + i2k\pi$; **17.** $z = -\text{Log } 2 + i2k\pi$; **19.** $i\left[\frac{1}{3}\pi + 2k\pi\right]$, $\text{Log } 2 + i2k\pi$, $i\left[\frac{-\pi}{3} + 2k\pi\right]$; **21.** $z = \text{Log}|2k\pi| + i\left[\frac{\pm\pi}{2} + 2m\pi\right]$, $k = 1, 2, \ldots$; $m = 0, \pm 1, \pm 2, \ldots$; **25.** (a) $5 \log z = \log z^5 = 5\left[\text{Log}\sqrt{2} + i\frac{\pi}{4}\right]$, (b) no, (c) yes.

Section 3.5

3. $\log 1 = 0$; **5.** $1.347 + i\frac{5\pi}{4}$; **7.** $i\frac{5\pi}{4}$; **9.** $\text{Log } 2 + i\frac{13\pi}{6}$; **11.** (a) line $y = 1$, $x \le 0$, (b) $\text{Log } 2 + i\left(\frac{-\pi}{2}\right)$; **15.** (b) $\frac{1}{z \text{ Log } z}$, (c) $y = 0$, $x \le e$.

Section 3.6

1. $e^{-2k\pi}$, p.v. with $k = 0$; **3.** $16e^{(\pi/2)+6k\pi} \text{ cis}\left[\frac{2\pi}{3} - 3 \text{ Log } 2\right]$, p.v. with $k = 0$; **5.** p.v. $= e^{\sqrt{2}\text{ Log } 2}$; **7.** $e^{-[\pi/2 + 2k\pi]\sinh 1} \text{ cis}\left[\text{Log}\frac{\pi}{2}\sinh 1\right]$, p.v. with $k = 0$; **9.** $e^{\pi - 2 + 2k\pi} \text{ cis}\left[\text{Log}\frac{1}{|\cos 2|}\right]$, p.v. with $k = 0$; **11.** $e^{[\text{Log}(\sinh 1)]^2 - \pi^2/2[1/2 + 2k\pi]} \text{ cis}[\text{Log}(\sinh 1)(\pi + 2k\pi)]$, p.v. with $k = 0$; **17.** $2.25 - i0.6039$; **19.** $e^{z \text{ Log } z}[\text{Log } z + 1]$; **21.** $0.1855 + i0.38264$; **23.** $-0.4028 - i0.1149$; **25.** $0.00464 - i0.0116$.

Section 3.7

5. $w = \text{Log}[\sqrt{2} + 1] + i\left[\frac{\pi}{2} + 2k\pi\right]$, $w = \text{Log}[\sqrt{2} - 1] + i\left[\frac{-\pi}{2} + 2k\pi\right]$; **7.** $\text{Log}[\sqrt{3} + \sqrt{2}] + i\left[\frac{\pi}{2} + 2k\pi\right]$; $\text{Log}[\sqrt{3} - \sqrt{2}] + i\left[\frac{\pi}{2} + 2k\pi\right]$; **9.** $z = \frac{-\pi}{2} - m\pi + \frac{i}{2}\text{Log}\left|\frac{1 + \pi/2 + 2k\pi}{1 - \pi/2 - 2k\pi}\right|$; **11.** $\frac{\pi}{2} + 2k\pi - i \text{ Log}(n\pi + \sqrt{n^2\pi^2 + 1})$, $n = 0, \pm 1, \pm 2, \ldots, k = 0, \pm 1, \pm 2, \ldots$; $\frac{-\pi}{2} + 2k\pi - i \text{ Log}|n\pi - \sqrt{n^2\pi^2 + 1}|$, $n = 0, \pm 1, \pm 2, \ldots, k = 0, \pm 1, \pm 2, \ldots$; **13.** (a) not true in general, (b) true in general; **15.** (b) $x = 4$, $\sinh^{-1}(x) = 2.0947$, $\text{Log } 2x = 2.0794$; **17.** $\frac{i}{2}\log\left[i \cot\left[\frac{\theta}{2} - \frac{\pi}{4}\right]\right]$; **19.** $(z^2 - 1)^{1/2}$.

Section 3.8

1. $f(z) = -3,\ f'(z) = \frac{-1}{6}$; **3.** $f(z) = 0.455 - i1.099,\ f'(z) = 0.161 + i0.388$; **5.** $f(z) = 2 - i2\sqrt{3},\ f'(z) = -0.288 - i0.1667$; **7.** $f(z) = 1.091 + i0.6299,\ f'(z) = -0.573 + i0.154$; **11.** -1.2599; **13.** $-1.2196 + i0.47177$; **15.** yes; **17.** $0.9539 + i2.33$; **19.** (c) $0.1036 - i0.25$ using princ. branch; **21.** $f(z) = \frac{\pi}{2} - i1.7627,\ f'(z) = -i0.354$; **23.** no solution; **25.** $z = -2 + i2\sqrt{3}$; **27.** $z = \pm 1, \pm 3, \pm 5, \ldots$.

Appendix A, Chapter 3

1. $3e^t \cos t$; **3.** $\frac{3e^{-t}}{\sqrt{2}}[\cos t - \sin t]$; **5.** $e^{(\sqrt{3}/2)t}\left[\cos\frac{t}{2} - \sin\left[\frac{\pi}{3} + \frac{t}{2}\right]\right]$; **7.** $2e^{3t}$; **9.** phasor $= 1$, $s = -2 + i3$; **11.** phasor $= -2ie^{-i\pi/6}$, $s = 4 + 2i$; **13.** phasor does not exist; **15.** phasor $= 2 - ie^{i\pi/4}$, $s = -1 + i$; **19.** $\frac{V_0 e^{\sigma t}}{R + L\sigma} = i(t)$; **21.** (a) $X = \frac{F_0}{-\omega^2 m + i\omega\alpha + k}$, $x(t) = \frac{F_0\cos(\omega t + \psi)}{\sqrt{(k - \omega^2 m)^2 + \alpha^2\omega^2}}$, $\psi = \tan^{-1}\frac{\alpha\omega}{\omega^2 m - k}$.

Chapter 4

Section 4.1

3. $1\frac{7}{6}$; **5.** $\frac{-1}{4}$.

Section 4.2

1. $1 + 2i$ exact; **3.** $1 + 2i$; **5.** $-i\frac{2}{3}$; **7.** (a) $e - 1$, (b) $e[(\cos 1 - 1) + i\sin 1]$, (c) $1 - e\cos 1 - ie\sin 1$; **9.** $-\pi i$; **11.** $\frac{-3}{2} + \pi i$; **13.** (a) $z = 1 + i + e^{it}$, t goes from $\frac{-\pi}{2}$ to $-\pi$, (b) $i\left[2 - \frac{\pi}{2}\right]$.

Section 4.3

1. (a) area enclosed by C, (b) $2 - 2e$, (c) $2iA$, A = enclosed area of C; **3.** no; **5.** no; **7.** no; **9.** yes; **17.** $2\pi i$; **19.** zero; **21.** πi; **23.** zero.

Section 4.4

3. $604 + i1970$; **5.** $(9 + 3i)\sin(9 + 3i) + \cos(9 + 3i) - (1 + i)(\sin(1 + i)) - \cos(1 + i)$; **7.** $\frac{1}{1 + i} - \frac{1}{(9 + 3i)}$; **9.** $\frac{-\pi}{2} - i$; **11.** $\frac{2}{3}\left[-1 - \frac{1}{\sqrt{2}} + \frac{i}{\sqrt{2}}\right]$; **13.** $-i\pi$; **17.** (a) $\frac{-1}{z - i}$ + constant, (b) $F(z) = \frac{-1}{z - i} + 1$; **19.** (b) $(i + 1)\cosh\frac{\pi}{2} - \frac{4i}{\pi}\sinh\frac{\pi}{2}$.

Section 4.5

3. $\left[\dfrac{\cos 1 - \sin 1}{26}\right]$, **5.** $\dfrac{\sinh\left[\dfrac{-1}{2} + \dfrac{i\sqrt{3}}{2}\right]}{i\sqrt{3}}$; **7.** $\dfrac{-\pi i}{2}$; **9.** $\dfrac{2\pi i[e^3(2-6i)]}{(1-2i)^2}$; **11.** $\dfrac{2\pi i 2^{20}}{20!}$; **17.** $\dfrac{\cos 1 - \cos 2}{3}$; **19.** $i\pi^2$.

Section 4.6

1. 1; **3.** 2π; **5.** $4\pi \operatorname{Log} a$; **7.** (b) average $= 1.4687006$, $e^1 \cos 1 = 1.4686939$; **9.** $|f|$ max at $z = (\sqrt{2}+1)e^{i\pi/4}$, $|f|$ max $= \sqrt{2}+1$; $|f|$ min at $z = (\sqrt{2}-1)e^{i\pi/4}$, $|f|$ min $= \sqrt{2}-1$; **11.** max value $= e^2$ at $x = 2$, $y = 1$, min value $= e^0 = 1$ at $x = 0$, $y = 1$; **15.** max at (1, 0), min at (0, 1); **17.** (e) $\dfrac{\pi}{2}\dfrac{(2n)!}{(n!)^2 2^{2n}}$.

Section 4.7

5. (c) $U(\infty) = 50$.

Chapter 5

Section 5.1

1. all $c_n = 1$, $u_n = c_n x^n$, $|x| < 1$ absolute convergence; **3.** $u_n = \dfrac{e}{n!}(x-1)^n$, abs. convergence for all x; **5.** $u_n = \dfrac{1}{n!}\sin\left(\dfrac{n\pi}{2} + \dfrac{\pi}{4}\right)\left(x - \dfrac{\pi}{4}\right)^n$, series is abs. conv. all x; **7.** cannot be used in Exercise 6 at 0 and $2e$.

Section 5.2

11. $\dfrac{1}{1-(z-1)^2}$, $|z-1| < 1$.

Section 5.3

7. $\doteq 3.21$. Agreement is 98 percent.

Section 5.4

3. $u_n = \dfrac{(-1)^n(z-1-i)^n}{(1+i)^{n+1}}$, series valid: $|z-1-i| < \sqrt{2}$; **5.** $u_n = \dfrac{(-1)^n(n+1)(z-i)^n}{(2i)^{n+2}}$, series valid: $|z-i| < 2$; **7.** $u_n(z) = \dfrac{(-1)^n(n+1)(z-1-i)^n}{i^{n+2}}$, series valid: $|z-1-i| < 1$; **9.** $u_n = \dfrac{2z^{4n+3}}{(4n+3)!}$ $n = 0, 1, 2, \ldots$, series valid all z; **13.** center 1, radius $r = \sqrt{2}\sqrt{1-1/\sqrt{2}}$; **15.** center of circle at $z = i$, radius $\sqrt{2-\sqrt{3}}$; **17.** center of expansion $z = 5$, radius of circle of conv. $= \sqrt{21}$; **19.** $-3 < x < 1$; **21.** $\dfrac{1}{2} < x < 1$; **23.** $-1 < x < 1$; **25.** $1 < x < 2$; **27.** $5(z-1) + 10(z-1)^2 + 10(z-1)^3 + 5(z-1)^4 + (z-1)^5$; **29.** (c) The radius in (b) is smaller, and equals 1.

Section 5.5

5. (b) -2.24; **9.** $c_n = (-1)^n\left[1 - \frac{1}{2} + \frac{1}{3} - \frac{1}{4} + \cdots \pm \frac{1}{n}\right]$, series valid for $|z| < 1$; **11.** $\sum_{n=0}^{\infty} c_n z^n$, $|z| < 1$, where $c_n = \left[\frac{-1}{3} + \frac{1}{3}\frac{(-1)^n}{2^n}\right]$; **13.** $\sum_{n=0}^{\infty} c_n[z - (1+i)]^n$, where $c_n = \frac{(-1)^n}{2i}\frac{(n+1)}{(i+1)^n}$, $|z - (1+i)| < \sqrt{2}$; **15.** $\sum_{n=0}^{\infty} c_n(z-2)^n$, valid for $|z - 2| < 1$, where $c_n = \frac{1}{9}(-1)^n + \frac{2}{3}(-1)^n(n+1) - \frac{1}{36}\frac{(-1)^n}{4^n}$; **17.** $d_n = \frac{-1}{6}\left(\frac{1}{2^n}\right) - \frac{1}{3}(-1)^n, c_n = \frac{d_0}{n!} + \frac{d_1}{(n-1)!} + \frac{d_2}{(n-2)!} + \cdots + \frac{d_{n-1}}{1!} + \frac{d_n}{0!}$, $\sum_{n=0}^{\infty} c_n z^n$ series valid $|z| < 1$; **19.** $\sum_{n=0}^{\infty} c_n(z-1)^n$, $|z - 1| < 1$, where $c_n = \frac{-17}{4} + \frac{1}{4}\frac{(-1)^n}{3^n}$ for $n \geq 2$, $c_0 = -1$, $c_1 = -10/3$; **21.** $z - \frac{1}{2}z^2 + \frac{5}{6}z^3 - \cdots$; **23.** (c) valid in disc $|z| < 2\pi$; **25.** (b) valid all z.

Section 5.6

1. $\sum_{n=-2}^{\infty} \frac{(-1)^n z^{2n}}{(2n+4)!}$, $|z| > 0$; **3.** $(z-1) + \sum_{n=-\infty}^{0}\left(\frac{1}{(-n+1)!} + \frac{1}{(-n)!}\right)(z-1)^n$, $|z - 1| > 0$; **5.** $\sum_{n=-\infty}^{+\infty} c_n(z-1)^n$, summary: for $n \leq 0$, $c_n = (-1)^n \sin 1$; for $n \geq 1$, n odd, $c_n = (-1)^n\left[\sin 1 - \left(1 - \frac{1}{3!} + \frac{1}{5!} - \cdots \pm \frac{1}{n!}\right)\right]$; for $n \geq 1$, n even, $c_n = (-1)^n\left[\sin 1 - \left(1 - \frac{1}{3!} + \frac{1}{5!} - \cdots \pm \frac{1}{(n-1)!}\right)\right]$; expansion valid for $|z - 1| > 1$; **7.** $\sum_{n=-\infty}^{-1} c_n(z+1)^n$, $c_n = (-1)^{n+1}$, series valid for $|z + 1| > 1$; **9.** $\sum_{n=-\infty}^{-1} c_n(z-i)^n$, $c_n = (-1)^n(2+i)^{-n-1}$, series valid for $|z - i| > \sqrt{5}$; **11.** (a) domains for Laurent series: I: $0 < |z| < 1$; II: $1 < |z| < 3$; III: $|z| > 3$, (b) for domain I, $0 < |z| < 1$, $\frac{-1}{3}z^{-1} + \sum_{n=0}^{\infty} c_n z^n$, $c_n = \frac{-1}{4} + \frac{1}{36}\frac{(-1)^n}{3^n}$; for domain II: $1 < |z| < 3$, $\sum_{n=-\infty}^{-2} c_n z^n - \frac{z^{-1}}{12} + \sum_{n=0}^{\infty} c_n z^n$, $c_n = \frac{1}{4}$ for $n \leq -2$, $c_n = \frac{1}{36}\frac{(-1)^n}{3^n}$ for $n \geq 0$; for domain III: $|z| > 3$, $\sum_{n=-\infty}^{-3} c_n z^n, c_n = \frac{1}{4} + \frac{(-1)^{n+1}}{12}3^{-n-1}$; **13.** $\sum_{n=-\infty}^{+\infty} c_n(z-1)^n, 1 < |z - 1| < 3$, $c_n = \frac{-1}{12}\frac{1}{3^n}$ for $n \geq 0$, $c_n = \frac{-1}{4}(-1)^{n-1}$ for $n \leq -1$; **15.** $\sum_{n=-\infty}^{+\infty} c_n(z-1)^n$ $(z \neq 1)$, $c_0 = 1$, $c_{-1} = (1 - i)$, all other $c_n = 0$; $|z - 1| > 0$; **17.** $f(z) = (z-1)^{-3} + 1 + 3(z-1) + 3(z-1)^2 + (z-1)^3$, $|z - 1| > 0$; **23.** (b) $\left(\frac{1}{3!}\right)^3 - \frac{2}{3!5!} + \frac{1}{7!}$.

Section 5.7

1. (b) There is no contradiction; **3.** $z = \pm i\sqrt{1 - \frac{1}{n}}$, $n = 1, 2, \ldots$, gives all zeros in domain $|z| < 1$; they are isolated; the accumulation points are at $\pm i$, do not belong to the given domain; **7.** zero of order 4 at $z = 0$, zero is of first order at $z = \pi$; **9.** zero of order 2; **11.** zero of order 8; **13.** (a) $\frac{1}{z-2}$ if $x > 2$, (b) $\frac{1}{z-2}$, which is analytic for all $z \neq 2$.

Appendix A, Chapter 5

7. Our Julia set for $c = 0$ is the disc $|z| < 1$; the boundary is $|z| = 1$, which is a smooth curve (not a fractal set).

Appendix B, Chapter 5

15. $f(nT) = \frac{1}{n!}$ for $n \geq 0$; **21.** $f(nT) = 2^{n+1}$; **23.** $f(0) = 0$, $f(n) = \frac{2^n}{3} + \frac{(-1)^{n+1}}{3}$ $n \geq 1$.

Chapter 6

Section 6.1

1. $\frac{\pi i}{2}$; **3.** 0; **5.** $\frac{\pi i}{12}$; **7.** $2\pi i$; **11.** (b) $-\pi i$.

Section 6.2

1. $\cosh\left(\frac{1}{z}\right) = \sum_{n=-\infty}^{0} c_n z^n$; $c_n = 0$, n odd; $c_n = \frac{1}{(-n)!}$, n even; residue $= 0$; **5.** $f(z_0) = 1$; **7.** $f(z_0) = \frac{i\sqrt{3}}{2}$; **11.** simple poles at $\pm i$ and ± 1; **13.** simple pole at $\left(\frac{-1}{2} + \frac{i\sqrt{3}}{2}\right)$, pole of order 2 at $\left(\frac{-1}{2} - \frac{i\sqrt{3}}{2}\right)$; **15.** simple pole at $z = -1$; pole of order 9 at $z = 0$; **17.** $z = \frac{i2k\pi}{\text{Log } 10 - 1}$, $k = 0, \pm 1, \pm 2, \ldots$ simple poles; **19.** poles of order 4 at $z = i(\pi + 2k\pi)$, k integer; **21.** poles of order 3 at $z = \pm i$; **23.** poles of order 4 at $z = ik\pi$, k any integer $\neq 0$; **25.** pole of order 4 at $z = 1$; **29.** (b) no.

Section 6.3

3. residue at $\frac{1 + i\sqrt{3}}{2}$ is $\frac{e^{1+i\sqrt{3}}}{i\sqrt{3}}$, residue at $\frac{1 - i\sqrt{3}}{2}$ is $\frac{e^{1-i\sqrt{3}}}{-i\sqrt{3}}$; **5.** pole of order 3 at i; residue $= \frac{3}{8\sqrt{2}}[-1 + i]$; **7.** residue at $z = 1$ is 1/4, residue at i is $\frac{i}{\pi^2} - \frac{1}{2\pi}$, residue at $-i$ is $\frac{-i}{\pi^2} - \frac{1}{2\pi}$; **9.** pole of order 4 at $z = 0$, residue $= 0$; **11.** simple poles, $z = \pm\sqrt{k\pi}$, $k = 1, 2, \ldots$, residue is $\frac{(-1)^k}{(\pm 2)\sqrt{k\pi}}$; simple poles, $z = \pm i\sqrt{|k|\pi}$, $k = -1, -2, \ldots$, residue is $\frac{(-1)^k}{\pm 2i\sqrt{|k|\pi}}$; pole order 2, $z = 0$, residue $= 0$; **13.** simple poles, $z = k\pi$, $k = \pm 1, \pm 2, \ldots$, residue is $(-1)^k \cos\left[\frac{1}{k\pi}\right]$; **15.** (b) $-e^2 \sin e + e \cos e$; **17.** $e - 1$; **19.** $\frac{-i \sinh 1}{4!}$; **21.** $\frac{-i}{2}$; **23.** $\frac{-1}{2}$; **25.** $\frac{\sin 1}{9!}$; **27.** $-2\pi i$; **29.** $2\pi i$; **31.** $\frac{\pi i}{e}$; **33.** $-4\pi^2 i$.

Section 6.4

1. $\dfrac{-2\pi}{\sqrt{k^2-1}}$ for $k < -1$; **13.** $\dfrac{\pi}{6}$; **15.** 0.

Section 6.5

1. exists; **3.** exists; **5.** does not exist; **7.** does not exist; **11.** $\dfrac{\pi}{\sqrt{3}}$; **13.** $\dfrac{2\pi}{\sqrt{3}}$; **19.** $\dfrac{4\pi}{5}\sin\left[\dfrac{2\pi}{5}\right]$;

27. (b) $\dfrac{\sqrt{2}\pi}{5}$.

Section 6.6

1. $\dfrac{\pi}{2}e^{-6}$; **3.** $\dfrac{-2\pi}{\sqrt{3}}e^{-\sqrt{3}/2}\sin\dfrac{1}{2}$; **5.** $\dfrac{2\pi}{9}e^{-\sqrt{3}/2}\left(\cos\dfrac{1}{2}\right)(3+2\sqrt{3})$;

7. $\dfrac{\pi}{2\sqrt{3}}e^{-\sqrt{3}/2}\left[\cos\dfrac{1}{2}+\sqrt{3}\sin\dfrac{1}{2}\right]$; **15.** $\dfrac{\pi}{\sqrt{2}}e^{-1/\sqrt{2}}\left[\cos\dfrac{1}{\sqrt{2}}+\sin\dfrac{1}{\sqrt{2}}\right]-i\pi e^{-1/\sqrt{2}}\cos\dfrac{1}{\sqrt{2}}$;

17. $\dfrac{\pi i}{2}e^{-i/\sqrt{2}}e^{-1/\sqrt{2}}(-2)(\cos 1)\dfrac{1+i}{\sqrt{2}}$; **19.** $\pi i e^{-\omega}\sinh 1$, $\omega \geq 1$; $\pi i e^{-1}\sinh\omega$, $-1 \leq \omega \leq 1$; $-\pi i e^{\omega}\sinh 1$, $\omega \leq -1$.

Section 6.7

1. (a) $2\pi i$, (b) $-2+\pi i$; **5.** $\dfrac{-\pi}{4}\sin 8$; **7.** $\dfrac{-\pi}{\left(1+\dfrac{\pi^2}{4}\right)}\left[1+e^{-1}\left(\dfrac{\pi}{2}\right)\right]$.

Section 6.9

1. pole of order 3; **3.** branch point; **5.** essential singularity; **7.** branch point; **9.** pole of order 2; **11.** essential singularity; **13.** zero; **15.** residue $=\dfrac{-3}{2}$; **17.** analytic at ∞; residue $=-1$; **19.** residue $=0$; **21.** residue $=-1$; **23.** analytic at ∞, residue $=-1$; **25.** $2\pi i$; **29.** The residue at $w=0$ is zero; **31.** $\dfrac{4\pi}{5}$.

Section 6.10

1. $\dfrac{\pi}{a}e^{-a|t|}$, all t; **3.** $f(t)=\dfrac{\pi}{a}e^{-a|t-b|}$, all t; **5.** $f(t)=\dfrac{-\pi}{a}\sin a|t|$, all t; **7.** $f(t)=\dfrac{\pi}{b}e^{-b|t|}\cosh(ab)$, $|t|\geq a$; $f(t)=\dfrac{\pi}{b}e^{-ab}\cosh(bt)$, $|t|\leq a$; **13.** $\dfrac{e^{-i\omega 5}}{2\pi i\omega}[1-e^{-i\omega}]$; **15.** (a) $F(\omega)=0$ for $|\omega|>\omega_0=\dfrac{2\pi}{T}$; and $F(\omega)=\dfrac{1}{2}$ for $|\omega|<\omega_0$.

Chapter 7

Section 7.1

1. $\dfrac{e^t}{3}+\dfrac{e^{-2t}}{-3}$; **3.** te^{-at}; **5.** $\dfrac{\sin at}{a}$; **7.** $\dfrac{1}{2a^3}\sin at-\dfrac{1}{2a^2}t\cos at$; **9.** $e^{-t/2}\dfrac{2}{\sqrt{3}}\sin\left[\dfrac{\sqrt{3}}{2}t\right]$; **11.** $\dfrac{t^3}{3!}e^{-t}$;

13. $\dfrac{e^{-at/2}\sinh\left(\sqrt{\left(\frac{a}{2}\right)^2 - b}\,t\right)}{\sqrt{\left(\frac{a}{2}\right)^2 - b}}$ if $a^2 > 4b$; $e^{-at/2}\dfrac{\sin\sqrt{b - \left(\frac{a}{2}\right)^2}\,t}{\sqrt{b - \left(\frac{a}{2}\right)^2}}$ if $a^2 < 4b$; $te^{-at/2}$ if $a^2 = 4b$.

17. $u(t-1)$; **19.** $\left[\dfrac{1}{2b^3}\sin b(t-a) - \dfrac{1}{2b^2}(t-a)\cos b(t-a)\right]u(t-a)$; **21.** (b) $\mathscr{L}t = \dfrac{1}{s^2}$, $\mathscr{L}t^n = \dfrac{n!}{s^{n+1}}$; **23.** Re $s > 0$; **27.** (b) $X_1 = \dfrac{(s)(s^2+1)}{s^4 + 3s^2 + 3/2}$, $X_2 = \dfrac{s/2}{s^4 + 3s^2 + 3/2}$,

(c) $x_2 = \dfrac{1}{2\sqrt{3}}\cos\left[\sqrt{\dfrac{3-\sqrt{3}}{2}}\,t\right] - \dfrac{1}{2\sqrt{3}}\cos\left[\sqrt{\dfrac{3+\sqrt{3}}{2}}\,t\right]$,

$x_1 = \dfrac{1}{2\sqrt{3}}\left[(-1+\sqrt{3})\cos\left[\sqrt{\dfrac{3-\sqrt{3}}{2}}\,t\right] + (1+\sqrt{3})\cos\left[\sqrt{\dfrac{3+\sqrt{3}}{2}}\,t\right]\right]$.

Section 7.2

1. bounded; **3.** unbounded; **5.** unbounded; **7.** unbounded; **9.** (b) $-2 < \beta < 0$ growing; $0 < \beta < 2$ decaying; $\beta = 0$ fixed amplitude; **11.** $(1/3)\sin(3t)$; **13.** (b) me^{-sT}.

Section 7.3

1. $N - P = 1$; **3.** $N - P = 0$; **5.** $N - P = 1$.

Section 7.4

1. 2 roots in right half plane, none on imaginary axis, system unstable; **3.** no roots in right half plane, none on imaginary axis, system stable; **5.** 4 roots in right half of s-plane, none on imaginary axis, system unstable; **7.** if $0 < a < 1$, system is stable.

Section 7.5

1. $\cos 1$; **3.** -1; **5.** 0; **7.** $4e^2$; **9.** $1 + se^{-s}$; **11.** $e^{-s} + \dfrac{e^{-2s}}{s}$; **13.** e^{-2s}; **15.** $\dfrac{1}{1-e^{-s\tau}}$;

17. $\delta'(t) + 3\delta(t) + 4e^t$; **19.** $\delta'(t) - \delta(t) + \dfrac{2}{\sqrt{3}}e^{-t/2}\sin\left[\dfrac{\sqrt{3}}{2}t\right]$; **21.** $\dfrac{e^{-(R/L)t}}{L}$; **23.** $\dfrac{1}{L}\cos\dfrac{t}{\sqrt{LC}}$.

Chapter 8

Section 8.2

3. (a) intersect at (0, 1), 90°, (b) image of $y = 1 - x$ is $v = \dfrac{1}{2}(1 - u^2)$, $u \geq -1$; image of $y = 1 + x$ is $v = \dfrac{u^2 - 1}{2}$, $u \leq -1$, (c) $u = -1$, $v = 0$, intersection is still 90°; **5.** $z = n\pi + \dfrac{\pi}{2}$, $n = 0, \pm 1, \pm 2, \ldots$; **7.** no critical points in finite plane; **9.** (a) image is $v = 0$, $-2 \leq u \leq 2$,

(b) Im $w = 0$, $2 \le \text{Re } w$ or $v = 0$, $w \ge 2$, (c) image curves intersect at 180°, but the original curves intersect at 90°; **11.** $\left(u - \frac{1}{2}\right)^2 + \left(v + \frac{1}{2}\right)^2 = \frac{1}{2}$; **13.** (a) 0.07389, (b) 0.081797.

Section 8.3

1. (a) mapping is one to one, boundaries: $u_1 = x^2 - y^2$, $u_2 = x^2 - y^2$, $v_1 = 2xy$, $v_2 = 2xy$ all in first quadrant of x, y plane, (b) mapping is one to one, equations of boundaries same as in (a) but are now in third quadrant; **3.** the given region mapped on and between appropriate branches of the hyperbolas: $\frac{u^2}{\cos^2 a} - \frac{v^2}{\sin^2 a} = 1$ and $\frac{u^2}{\cos^2 b} - \frac{v^2}{\sin^2 b} = 1$, mapping is one to one; **7.** need $\alpha < \frac{\pi}{2}$; **9.** image of given region is Im $w > 0$.

Section 8.4

5. $w = \frac{az + b}{bz + a}$; **7.** circle $\left|w + \frac{i}{2}\right| = \frac{1}{2}$; **9.** image is circle $|w - 1 - i| = 1$; **11.** image $v = -u$; **13.** image is circle $\left|w - \left(\frac{1}{\sqrt{3}} - \frac{i}{3}\right)\right| = \frac{1}{3}$; **15.** circle $\left|w - \frac{5}{3}\right| = \frac{4}{3}$; **17.** given domain is mapped onto $|w - 1| > 1$; **19.** given domain is mapped onto $v > u - 1/2$; **21.** given domain is mapped onto domain satisfying both $\left|w + \frac{1}{4}\right| > \frac{3}{4}$ and $\left|w + \frac{5}{4}\right| > \frac{3}{4}$; **23.** (a) $w = (1 + i)z + 1$, (b) image of given domain is $|w - 1| < \sqrt{2}$; **25.** (a) $w = \frac{z + i - 1}{z}$, (b) image of given domain $\text{Re}(z - 1) > \text{Im } z$ is the domain $|w| < 1$; **27.** get semicircle, $y(0) = 1/R$, $y\left(\frac{R}{L}\right) = \frac{1}{[R(1 + i)]}$, $y(\infty) = 0$; **29.** $a = a_1 a_2 + b_2 c_1$, $b = a_2 b_1 + b_2 d_1$, $c = a_1 c_2 + c_1 d_2$, and $d = b_1 c_2 + d_2 d_1$; **31.** (d) mapped onto $|w| > 1$; **33.** (a) $w = \frac{z^6 - i}{z^6 + i}$, (b) $w = \frac{z^{\pi/\alpha} - i}{z^{\pi/\alpha} + i}$; **35.** (a) $w = \frac{(-i)(z + 2i)}{z - 2i}$, $k = 1$, (b) $w = e^{-\pi i(z + 2i)/(z - 2i)}$.

Section 8.5

1. (a) $x^2 + \left(y + \frac{50}{\beta}\right)^2 = \frac{2500}{\beta^2}$; **3.** $\frac{-20i}{\pi} \text{Log}\left[\frac{z - 1}{z + 1}\right]$; **7.** (a) $c = 2\pi\varepsilon/\cosh^{-1}\left[\frac{R_1^2 + R_2^2 - D^2}{2R_1 R_2}\right]$, (b) $|w|^2 = \frac{[(10 + 4\sqrt{5})x - 3(5 + \sqrt{5})]^2 + [(10 + 4\sqrt{5})y]^2}{[(-5 + \sqrt{5})x + 3(5 + \sqrt{5})]^2 + [(-5 + \sqrt{5})y]^2}$, $\phi(w(x, y)) = \frac{\frac{1}{2}\text{Log}\frac{\rho^2}{|w|^2}}{\text{Log } \rho}$, $\rho = \frac{3 + \sqrt{5}}{2}$; **9.** (a) $B = V_2$, $A = \frac{V_1 - V_2}{\pi}$, (b) $\phi(u, v) = \frac{V_1 - V_2}{\pi} \tan^{-1}\frac{v}{u} + V_2$,

(c) $B = V_3$, $A_2 = \dfrac{V_2 - V_3}{\pi}$, $A_1 = \dfrac{V_1 - V_2}{\pi}$; **11.** (a) $\phi(x, y) = \dfrac{100}{\pi} \tan^{-1}\left[\dfrac{\tanh\left[\dfrac{\pi y}{2a}\right]}{\tan\left[\dfrac{\pi x}{2a}\right]}\right]$.

Section 8.7

3. (b) $\left(u - \dfrac{42}{9}\right)^2 + v^2 = \left(\dfrac{30}{9}\right)^2$; **5.** (b) $\Phi(w) = \dfrac{\rho}{2\pi\varepsilon} \operatorname{Log}\left[\dfrac{\sin\left(\dfrac{\pi w}{2b}\right) + i \sinh\left[\dfrac{\pi\alpha}{2\beta}\right]}{\sin\left(\dfrac{\pi w}{2b}\right) - i \sinh\left[\dfrac{\pi\alpha}{2\beta}\right]}\right]$.

Index